BOREAL FORESTS AND GLOBAL CHANGE

Boreal Forests and Global Change

Peer-reviewed manuscripts selected from the International Boreal Forest Research Association Conference, held in Saskatoon, Saskatchewan, Canada, September 25–30, 1994

Edited by

MICHAEL J. APPS and DAVID T. PRICE

Natural Resources Canada, Canadian Forest Service, Northwest Region, Northern Forestry Centre, Edmonton, Alberta, Canada

and

JOE WISNIEWSKI

Wisniewski & Associates, Inc., Falls Church, VA, USA

Reprinted from *Water, Air and Soil Pollution* 82 (1–2), 1995

Kluwer Academic Publishers

Dordrecht / Boston / London

A C.I.P. Catalogue record for this book is available from the Library of Congress

ISBN 0-7923-3665-8

Published by Kluwer Academic Publishers,
P.O. Box 17, 3300 AA Dordrecht, The Netherlands

Kluwer Academic Publishers incorporates
the publishing programmes of
D. Reidel, Martinus Nijhoff, Dr. W. Junk and MTP Press.

Sold and distributed in the U.S.A. and Canada
by Kluwer Academic Publishers,
101 Philip Drive, Norwell, MA 02061, U.S.A.

In all other countries, sold and distributed
by Kluwer Academic Publishers Group,
P.O. Box 322, 3300 AA Dordrecht, The Netherlands.

Printed on acid-free paper

Printed in the Netherlands

WATER, AIR AND SOIL POLLUTION / *Volume 82 Nos. 1/2 May 1995*

PART I
FOREST MANAGEMENT AND THE CHANGING ENVIRONMENT

PART II
NUTRIENT AND CARBON CYCLING

PART III
ADVANCED TECHNOLOGIES AND THE IMPACTS OF GLOBAL CHANGE

IBFRA: International Boreal
Forest Research Association

BOREAL FORESTS AND GLOBAL CHANGE

The members of the IBFRA '94 Conference Committee and the Editors of this volume wish to acknowledge the support provided by these, and other, sponsoring agencies:

Weyerhaeuser Canada

The Royal Society of Canada
La Société Royale du Canada

Le Programme Canadien des
Changements à l'Échelle du Globe
The Canadian Global Change Program

Prince Albert Model Forest
Association Inc

Canada

Saskatchewan

FOREWORD

STEPHEN H. SCHNEIDER

Department of Biological Sciences, Stanford University, Stanford, California 94305-5020, USA

Considerable controversy surrounds the possibility that increasing numbers of people demanding higher standards of living and using technologies to achieve those standards of living might inadvertently cause substantial changes to the world's climate. First, projections for future population range from as low as six or seven billion people in the middle of the 21st century to as high as 15 billion or more by 2100. The standards of living of this population and whether it will use carbon-based fuels such as coal, or much less polluting energy supply sources of a renewable nature, also engender debate. The Intergovernmental Panel on Climate Change (IPCC), accounting both for these uncertainties, and for those associated with the physical and biological responses to emissions of carbon dioxide and other radiatively active substances into the atmosphere, concluded that global warming ranging anywhere from as low as a degree to perhaps as high as 5 °C are plausible by the end of the 21st century. The possibility that sulphur-dioxide-produced atmospheric aerosols could both offset some of that warming on a regional basis and generate further regional climatic disruptions has been highlighted by the most recent IPCC summary. At the same time, human disruptions through nitrogen and sulphur biogeochemical cycles also affect the deposition of sulphuric and nitric acids on to the ecosystems and lakes of the Northern Hemisphere, with potential direct effects on a number of regions and also potential indirect effects of nitrate fertilization on the growth rate of forests. There is a wide range of plausible future scenarios, ranging from mild to potentially catastrophic alterations to the planetary environment. This has engendered an even more contentious debate over whether current information is adequate to slow down the rate at which humans modify the system or whether further scientific information should be gathered in advance of global-scale policy actions. Of course, whether any reaction to a range of probabilities and consequences is justified is never a scientific question *per se* but a value judgment that weighs the potential costs of abatement versus the potential costs of environmental disruption occurring unabated. Finally, most projections suggest that the largest climatic and ecological changes are likely to be in the land areas of the northern half of the Northern Hemisphere, in particular, the boreal forest zones.

Despite all the controversy surrounding anthropogenic climate change and its potential consequences, one clear connection between climate change and boreal forests is undisputed: that the roughly 5 °C global average warming that took place between about 15 000 and 5 000 years ago, when the last ice age gave way to the present inter-glacial, created a dramatic alteration to the landscape in the current boreal zone. Boreal species now present across Canada, Northern Europe and Siberia were found far to the south in what is now mixed hardwood areas and prime agricultural lands. As the ice receded, these species "chased the ice cap north", although recent analyses of the fossil pollen during the most rapid time of transition suggests that the structure of ecological communities was severely disrupted and that many species moved individualistically. The natural rate of climate change to which this ecological drama unfolded was for global average sustained rates of surface temperature change on the order of 1 °C per thousand years, whereas even the low estimates from the IPCC (Intergovernmental Panel on Climate Change, 1995). are on the order of 1 °C change per century. Thus, foresters and ecologists have long been concerned that the potential rates of anthropogenic climate change would be a factor of ten or more faster than the rates that current ecosystems had experienced as they settled into current configurations. This has led to a serious concern that not only might species have to respond by migration to climatic changes at much faster than accustomed rates, but they need to migrate through a landscape dramatically altered by human land use. The forests of the 21st century will have to contend with factories, farms and freeways in their migration paths in addition to the potential for rapid rates of climate change.

While no responsible scientist can provide a confident scenario of climatic or ecological change in the 21st century, neither can a responsible scientist deny the substantial probability of such change. Therefore, in order to estimate the potential consequences of human activities, or to suggest adaptation strategies to deal with such consequences, it is essential that the scientific community provide increasingly reliable estimates of both the magnitudes of climate change and ecological responses. Since the boreal forests are such an important component of this interaction, both being influenced dramatically by climate change and at the same time feeding back on climate change through effects on surface albedo and nutrient cycling, enhanced capacity to understand and ultimately forecast the interactions of boreal forests and climate is essential. While it is true that tropical forests have a higher degree of biological diversity and more endemic species, suggesting that these may be the venue for potential losses of bio-diversity under a variety of global change scenarios, the boreal forests seem to exhibit the potential for the largest magnitude of change and to have the greatest possible feedback on the climate. For these reasons it is important that both systems can be studied and better understood, not as competitors for research effort, but as complementary efforts aimed at understanding earth systems.

The papers in this volume are attempts both to summarize the state-of-the-art of our understanding of boreal forests and their interaction with the climate, and also to provide many suggestions for enhancing our capability in the years ahead. One of the principal problems that needs to be addressed is the fact that climatic information on a regional scale is often provided for areas many thousands of square kilometres in size, whereas ecological understanding and field experimentation take place in study areas on the order of tens to hundreds of meters. Methods to bridge the scale gap across climatology and ecology are problematic (Root and Schneider, 1993), but some papers in this volume demonstrate that techniques already exist for this purpose and that additional ones can yet be developed. Validation of models also requires adequate data sets, another issue considered in this volume.

Finally, in the quest for tools for "sustainable development," management strategies must recognize the potential implications of global climatic and other global changes on forests, while at the same time converting those global scale disturbances into the local context. In view of the large uncertainties, and also the potential for serious change, management strategies that emphasize flexibility and the capacity to respond to changing scientific and economic conditions should be a priority. While not all important uncertainties are likely to be resolved before the earth systems themselves "perform the experiment" of telling us precisely what will happen, enough is known already to outline a wide-range of potential consequences and their subjective probabilities. At a minimum, increased progress in the scientific study of boreal forests and their interactions with climate will help to place decision-making at all scales on a firmer scientific basis. That is the obligation the authors of this study offer to the society that supports our work.

References

IPCC Working Group I: 1995, *Second Scientific Assessment, 1995*. Cambridge University Press, Cambridge, UK. (In preparation).

Root, T. L. and Schneider, S. H.: 1993, Can large scale climatic models be linked with multi-scale ecological studies? *Conservation Biology*, **7**(2), 256-279

PREFACE

Boreal forests form Earth's largest terrestrial biome. They are rich in ecosystem and landscape diversity, though characterised by relatively few plant species, when compared to other forested regions. The long term viability and sustainability of boreal forests is influenced by many factors. In particular, they are subjected at intervals to large-scale natural disturbances and, more recently have come under increasing human pressure, both from the direct effects of forest operations and from potential changes in the global climate system. Boreal ecosystem development is typically a slow process; hence rapid changes in the global environment may invoke complex responses. Many industrial nations border, or lie within, boreal regions, and derive much of their economic wealth and culture from the forests. Understanding the responses of boreal forest ecosystems in these times of global change is therefore of considerable international interest, both for its policy implications and for its scientific challenges.

The International Boreal Forest Research Association (IBFRA) was formed in 1991 with the mission to

> *"...promote and co-ordinate research to increase the understanding of the role of the circumpolar boreal forest in the global environment and the effects of environmental change on that role."*

The participating countries of IBFRA (1994) are Canada, Finland, Norway, Russia, Sweden, and the United States of America. The impetus for the formation of IBFRA was the "White Sea Declaration" which was drafted at an International Field Expedition in Arkhangelsk, Russia, in 1990. Since its 1991 Founding Meeting in the Ukraine, IBFRA has met in Alaska (1992), and Norway and Sweden (1993). The three main priorities of IBFRA are (1) classification, monitoring, and inventory; (2) ecosystem function, anthropogenic impacts, and global change; and (3) forest management and biodiversity.

In 1994, the IBFRA meeting was held 25–30 September, in conjunction with its first International Science Conference on "Boreal Forests and Global Change", in Saskatoon, Saskatchewan, Canada. The objective of the conference was to convene a forum of scientists, engineers, economists and policy-makers to examine the long-term sustainability of global boreal forests, by considering the interactions among forest management practices, ecosystem carbon cycling and global change. In particular, how have human activities in the past influenced the boreal environment and how are the decisions we make today, likely to affect the boreal forests of the future? Approximately 200 individuals attended the Conference and nearly 100 presentations were given in both verbal and poster formats.

The distinction between global change and climate change is an important one, which we emphasised in the Conference and which should be apparent from the papers included here. The term *global change* is meant to embody the many environmental manifestations of human activity on planet Earth and not just the anticipated effects of greenhouse gas-induced climate change. *Global change* thus includes climate change, changes in land-use (in particular, deforestation and replacement by agriculture and intensively managed forests); the consequent impacts on human socio-economic structure and culture; the effects of industrialisation and urbanisation, in terms of pollution, land degradation and impacts on biodiversity; and even, the expanding roles of new technologies (e.g., computer modelling and satellite remote sensing) in understanding and mitigating these impacts.

It is IBFRA's expressed intent that data, research results and policy proposals arising from its activities be presented and published in highly-regarded scientific forums and publications. This Special Edition of the journal *Water, Air and Soil Pollution* contains close to 50 peer-reviewed papers from a world-wide group of experts providing scientific, economic and policy perspectives on global boreal forests at local, regional and global scales. The papers are grouped into four sections: Forest Management (including Biodiversity and Socio-economic issues), Carbon Cycling, Climate Change and Advanced Technologies. In particular, there are a number of Russian papers containing data and research results that provide information never before published in the western peer-reviewed literature. In this regard, this publication is truly a global perspective on *Boreal Forests and Global Change*.

We hope this book will provide some answers for the development of constructive policies for improving the management and conservation of global boreal forest resources in times of global change, as well as stimulating further research into boreal forest problems of international relevance.

Michael J. Apps
David T. Price
Joe Wisniewski

Edmonton, Alberta, April 1995.

ACKNOWLEDGMENTS

We are sure that we speak for all contributors to this volume when we thank Sue Nalder for her tireless effort and attention to schedules; truly, without her persistence and cheerful attention to many details, this publication would have been neither as complete nor as prompt.

We also wish to thank Dr. Eldon Ross (recently retired from the USDA Forest Service) who, as Past President of the International Boreal Forest Research Association, encouraged us to undertake this work.

The editors wish to gratefully acknowledge all contributors to this volume and the many reviewers of their manuscripts herein for the excellence of their work. We also especially thank them for staying within the very tight timelines we requested of them.

We acknowledge with grateful appreciation the support, in kind or in financial terms, of all the sponsors, namely: Natural Resources Canada, Canadian Forest Service; Green Plan Science and Technology; The Canadian Global Change Program of The Royal Society of Canada; International Science Foundation (Washington D.C. and Moscow); Weyerhaeuser Canada Ltd. (Prince Albert Division); Weyerhaeuser Company (Tacoma); United States Department of Agriculture, Forest Service (Anchorage); Canada-Saskatchewan Partnership Agreement in Forestry; Saskatchewan Environment and Resource Management; Prince Albert Model Forest Association Inc.; Parks Canada and the University of Saskatchewan.

Two of us (Apps and Price) wish to acknowledge the tolerance and encouragement of the management of the Northern Forestry Centre (Canadian Forest Service, Edmonton) for supporting this publication by making facilities and staff freely available during its preparation. Apps also wishes to thank the International Institute for Applied Systems Analysis where much of his effort on the final preparation of this volume was conducted while on study leave there.

Although there are many individuals who worked to make this volume come into being, we wish to specifically thank the following individuals for their very valuable assistance at various stages of this project: Ron Bronstein, Celina Campbell, Ian Campbell, Dave Cheyne, János Gál, Dave Halliwell, Regina Holehouse, Nadezhda Larionova, Peggy Pantell, Anatoly Shvidenko, Lynn Smith, Bob B. Stewart, Ken Winterberger, and Mary Louise Wright.

Finally we wish to thank Billy McCormac and the Editorial and Production Staff of Kluwer Academic Publishers for their encouragement and guidance.

Michael J Apps
David T Price
Joe Wisniewski

PART I

FOREST MANAGEMENT AND THE CHANGING ENVIRONMENT

WEST–EAST COOPERATION IN EUROPE FOR SUSTAINABLE BOREAL FORESTS

P. ANGELSTAM[1], P. MAJEWSKI[1], and S. BONDRUP-NIELSEN[2]

[1] *Grimsö Wildlife Research Station, Swedish University of Agricultural Sciences, S-730 91 Riddarhyttan, Sweden,*

[2] *Centre for Wildlife and Conservation Biology, Department of Biology, Acadia University, Wolfville, Nova Scotia, BOP 1X0, Canada.*

Abstract. The boreal forest is in transition. Large areas in remote regions are still in a pristine state, but extensive exploitation or intensive management for wood production dominates in Russia, Canada and the Nordic countries. The broadened view that forests are sources of other products than wood as well as of biodiversity is gradually becoming internationally accepted. To sustain the boreal forest for the future it must be realised that the forest has to be sustained as a system rather than as a number of utilities that can be considered separately. In northern Europe the situation is particularly striking with a strong contrast between the adjacent boreal forests in Scotland, Sweden and Finland on the one hand, and the north-west of Russia on the other. The two regions differ by history, type of land use and economic system but they are basically biologically similar. Combining sustainable wood production and maintenance of biodiversity and other values means setting limits to the intensity and extent of forest management. If the impact has been large, it is also necessary to build more natural features into managed systems. Developing a sustained boreal forest system by management of forests and forests reserves can be more efficient if research, education and management are well integrated. Scientists, teachers and managers must meet frequently and interact. In this way the time-lag between the advent of new knowledge and its implementation in the field can be reduced. We argue that co-operation among contrasting regions like the east and the west provides a unique opportunity for such integration. We review the problems and requirements in the west and the east, respectively, by contrasting Scotland, Sweden and the Komi Republic in Russia. The experiences are different and therefore complementary to each other. This assures long-term benefits of joint action.

Keywords. BOREAL FOREST, RUSSIA, SCANDINAVIA, CO-OPERATION, SUSTAINABLE USE, FOREST HISTORY, RESEARCH, EDUCATION, MANAGEMENT.

1. Introduction

Forestry in the boreal zone is in a transition from exploitation and sustainable wood production to a sustainable forest ecosystem. In Sweden and Finland where a high wood production based on intensive management has been achieved, environmental consideration are in focus, and the forest industry, scientists, and non-governmental organisations are trying to restore the biological values of forests (Esseen *et al.*, 1992; SOU, 1992; Wibe and Jones, 1992; Angelstam and Arnold, 1993 Tanninen *et al.*, 1994; Angelstam, 1995a, 1995b). In Scotland and in the western-most part of the boreal forest region in Europe most of the boreal forest disappeared a long time ago (Ritchie, 1920).

Forestry in the former Soviet Union is in a rapid and difficult transition between two different types of societies (Krankina and Dixon, 1992; Rosencrantz and Scott, 1992; Nilsson *et al.* 1992; Pruchkin *et al.*, 1993; Angelstam *et al.*, in press). At present, economic survival of individuals and companies dominate environmental considerations. In fact, even the functioning of the system for environmental considerations and nature protection in forests has been questioned (Grigoriev, 1994a).

Water, Air and Soil Pollution **82**: 3–11, 1995.

However, a study of the situation in the different regions, gradually developed over time, shows many similarities. Most examples may be placed on a continuum from naturally dynamic forests to intensively managed forests for wood production with intermediate stages of sheer exploitation and different types of management for timber. The pristine end of gradient continues to be subjected to intensive exploitation, now with the help of modern technique, whereas in the managed end of the gradient, efforts are made to rebuild biodiversity and to develop sustainable multiple-use forest systems with lower intensity of use for timber.

In northern Europe the situation is particularly striking with a strong contrast between the adjacent boreal forests in Scotland, Sweden, and Finland on the one hand, and the European north-west of Russia on the other. While they differ by history, type of land use, and economic system, they are biologically similar. In this paper we emphasise the importance of rapidly integrating research, education, and management in order to sustain the boreal forest ecosystem. We argue that initiation of long-term co-operation between the east and the west provides a novel approach and a unique possibility to speed up such an integration.

2. The Geography of Forests in Northern Europe

The circumpolar boreal forest is the most extensive forest type in the world and in Europe. In northern Europe, the forests from Scandinavia to the Ural mountains are similar in climate and tree species composition. They differ distinctly from the forests of Siberia in the east. In northern Europe the climate and vegetation zones are in the form of west-east belts from Scotland to the Ural mountains. South of the boreal zone, two main changes occur. First, the forest type changes to temperate deciduous forest and Norway spruce is absent. Second, the forest cover in the landscape decreases drastically. Between the northern tree-line and about 55 degrees northern latitude conifer-dominated boreal and hemiboreal forests are continuous and cover 50–70 % of the landscapes. Further south, the forest cover becomes fragmented due to human activity—with less than 30 % of forest cover in the landscape (ESA, 1992; Angelstam, 1995b).

3. The Management Gradient between East and West

3.1. The West—Scotland and Sweden

Six thousand years ago boreal forests covered most of Scotland (McVean and Ratcliffe, 1962). In the medieval ages, however, the forest resources were already diminished. Ritchie (1920) estimated that 54% of the land surface was forest in prehistoric ages compared to 5% at the beginning of the 20th century. At present, the remaining native pine-woods have been reduced to less than 1% and lie in remote areas where access and transport remained a problem (Steven and Carlisle, 1959; Bunce and Jeffers, 1977; Rackham, 1990). Scotland thus forms a western extreme in the west-east gradient

In Sweden, the first local industrial use of the trees in the forest was for early mining and started during the Medieval time. For a long time, from the 17th to the 20th century, this entailed an extensive exploitation of the southern boreal forest in Sweden (Wieslander, 1936; Arpi, 1959). Rather quickly the local shortage of forest became severe around the mines in south-central Sweden. The consumption of charcoal peaked during the end of the 19th century, and in 1885 it was estimated that 20–25% of the harvested timber volume was used in making charcoal (Arpi, 1959). Regrowth was left to nature. In the middle of the 19th century, large virgin forests still existed in parts of Sweden (Östlund, 1993; Larsson, 1989).

In the middle of the 19th century a fundamental transformation began in society, as well as in the forest landscape. Industrialisation received an impetus and agriculture was intensified. The forest acquired a trade value and became a commodity to an extent never previously anticipated. Up until about 1830–1840 the export of forest products from Sweden was modest, but then it increased exponentially. The liberal economy led to a great increase in the world trade. Between 1850 and 1900, Sweden's export of saw timber had shown a tenfold increase and harvest levels had reached nearly 20 million m^3, resulting in a growth of sawmills along the whole Bothnian coast and an exploitation of the primeval forest in the inland areas of northern Sweden (Östlund, 1993).

At the beginning of the 20th century there was a debate about the prevalent methods of harvesting–clearcutting versus cutting by size-class (Wallmo, 1897). This led to a predominant use of selection system in the whole country in the 1920s and 1930s. Although clearcutting was of limited extent during this period, it did not disappear entirely. The Swedish climate, frequently coupled with poor selection techniques, often resulted in poor regeneration after selection cuts, and there was a return to clearcutting in the 1940s. Today, broadly speaking, all harvesting of old forest is carried out as clearcutting with subsequent planting (Norway spruce and Scots pine) or natural regeneration under seed-trees (Scots pine).

In Sweden only 0.4% of the productive forest are protected in national parks and nature reserves. The corresponding figure for the mountain forests of little commercial value is 32% (Nilsson and Götmark, 1992). The development of forestry in Norway and Finland was similar to Sweden (Lehtinen, 1991; Tanninen *et al.*, 1994).

3.2. The East—northern Russia west to the Ural mountains

In the boreal and hemiboreal zones in European Russia the human population is traditionally concentrated along rivers (Tsvetkov, 1957; Kuusela, 1990). Exceptions exist where industries were developed along the northern rivers and where oil and minerals were found. The general trend, however, has been gradual settlement and subsequent exploitation of forests at increasing distances from rivers (Galasheva, 1961; Redko and Babich, 1993). During the 17th century, forests were used only near large cities. Industrial clearcuts along the tributaries of Northern Dvina started at the beginning of the 18th century. From the Komi region, timber was exported to the port at Arkhangelsk, starting in the 17th century. The export consisted only of high quality timber for building ships that were later sold to England and the Netherlands. These resources were quickly depleted and new, more remote, areas had to be surveyed. Therefore, investigations of the forests were started very early. By 1703 tsar Peter the Great had already ordered that an inventory of the forests should be made for military and ship building purposes. In 1768 the government, being interested in the development and demands of ship building, sent special expeditions to describe the forests. Interest focused on mast timber of pine and larch in north-western Russia all the way to the Urals, and the surveys covered an area of 980 000 km^2. In 1789 the forest department was founded to estimate the State Forest Fund.

The forest land was delimited as building-timber forest, fuel-wood forest, and areas with no tree growth. By 1830 surveys had reached the Ural mountains. Timber forests were divided into those suitable for ship building, saw-timber, and by age (mature and overmature). The ship building timber was dominated by Scots pine along the Vychegda river, a tributary of the Dvina, and by larch along the Pechora river. These descriptions were very short, but nevertheless important, and formed the first reliable forest descriptions.

During the second part of the 19th century in Komi the demand for timber began to increase. Still the majority of the land (58%) was untouched. Two types of methods were employed in forest harvesting. In an area that was selected for harvesting, 80–90% was subject to selective cutting and the remaining 10–20% was totally clearcut (Galasheva, 1961) .

In connection with development of the coal and oil industry in the Pechora basin, the volume of logging increased significantly. Areas of virgin taiga were thus cut and were regenerated naturally (Larin and Pautov, 1991). During the early stages of the succession, pioneer trees such as birches (*Betula pubescens*) and aspen (*Populus tremula*) dominate. From the point-of-view of industrial raw materials, this is valueless brush and classified as non-productive forest (Kuusela, 1990). On the other hand, from a wildlife conservation point of view, the resulting forest resembles the natural landscape in terms of dynamics and associated ecosystems.

From a wood production point-of-view, several regions have been over-harvested. On average, 110 % of the mean annual increment was cut in northern European Russia (Arkhangelsk, Vologda, Murmansk, Komi and Karelia) in 1988 (Barr, 1988; Nilsson *et al.*, 1992). However, under the present economic regression, the harvest has decreased considerably. From 1990 to 1992 the area harvested in Komi decreased from 165 600 ha to 117 000 ha—i.e., a decline of 29%. From 1987/88 to 1992 the harvest decreased from 27 to 16 million $m^3\ yr^{-1}$ (Pruchkin *et al.*, 1993).

4. The East–West Principle as a Time Machine

The review of the history shows that the forests vary considerably among different regions. In Scotland the natural coniferous forests have almost disappeared. In Norway, Sweden and Finland they exist, but have been severely altered, while in Russia natural forest landscapes can still be found. Thus, the gradual exploitation and management of forest resources in northern Europe has spread like a tidal wave from areas of demand to more and more distant regions. This timber frontier is now located in Siberia (Grigoriev, 1994a,b).

The spatial pattern has a corresponding temporal aspect which can be described in the following four steps (Angelstam and Arnold, 1993):

1. Pristine forest. Natural structures and processes are intact. Man does not dominate the system.
2. Early use. The most accessible pristine forests are selectively harvested for large and valuable trees.

3. Exploitation. The timber resource is locally or regionally depleted.
4. Sustainable wood production where management for the desired wood-based products is intensive and influences negatively other values.

In Table I we summarise the temporal development in five regions along the west–east gradient; Scotland, the mining area Bergslagen in south-central Sweden, northern Sweden, along the Russian rivers and in remote parts of European Russia. By travelling in space it is thus possible to travel in time. This provides a novel approach to share and exchange knowledge and experiences between the east and the west.

5. Combining Research, Education and Management for Development of Sustainable Forests.

A sustainable forest ecosystem is a prerequisite for sustainable forest use. This is a new idea and many managers in different countries are not ready to handle this wider problem. In the traditional way, the science first documents the results, teachers introduce it gradually to the education system, and then a new generation of managers is ready to improve forestry. This seems to be too slow a process for such an urgent issue as learning how to sustain a forest ecosystem. Long-term experience with one local type of management only, is one of the main obstacles to acceptance of new ways of thinking. We suggest that the development of sustainable forestry practices needs an integrated effort of scientists, teachers, and managers. Contrasting the west and the east can accelerate the progress by reducing the teaching time-lag and in the acceptance of new ideas. In retrospect, until recently wood production has dominated the forest debate (Remrod, 1991; Noss, 1993). During the last decades other values of forests have also been added (Swanson and Franklin, 1992), including recreation, watershed management (Malanson, 1993), biodiversity (Berg *et al.*, 1994; Esseen *et al.*, 1992), and the boreal forest as a carbon sink and reservoir (Alexeev, 1992). This has broadened the forest issues from the stand level to landscapes and regions, and world-wide.

Maintaining a sustainable forest ecosystem to meet future uncertainties will be difficult, but is nevertheless possible if appropriate knowledge about the functions of forest ecosystems and how to mimic natural disturbance regimes exists—and this knowledge is widely diffused and implemented. Management and ecological reserves for biological diversity will naturally reduce the short-term output compared with the harvestable volume if there are no other considerations. However, taking into account the increasing awareness of society's and ecological demands on the market, the short-term losses due to maintenance or improvement of diversity is likely to be beneficial in the long term.

5.1. Research Goals

Sustaining a forest ecosystems requires knowledge about the structure and dynamics of natural boreal forest. An important way to obtain this knowledge is to study still-intact forest stands and landscapes and to study the history of forests (Angelstam, 1995a; Östlund, 1993). Forest history data allow us to examine the extent of the structural differences between a natural and a managed forest, and the influence of different histories of management on the structure and processes in forests (Zackrisson, 1977; Östlund, 1993).

The remaining natural reference landscapes in remote parts of Europe allows studies of forest structures, processes as well as particular species of plants, animals and fungi. In the west smaller natural or semi-natural areas are found in remote parts of Finland, Sweden and Norway (Tanninen *et al.*, 1994). Areas sufficiently large to encompass naturally dynamic boreal landscape systems, however, are found only in boreal Russia—both on the Fennoscandian shield in Karelia; and on the Russian plain in the Arkhangelsk oblast and in the Komi Republic (Taskaev and Timonin, 1993). In such reference areas it is possible to study forest structures as well as the distribution and dynamics of species.

There are three research goals, *viz.*:

1. To assess quantitatively, in different geographical scales, the structural changes in regions with different history of land use (Table I). The importance of scale is evident since different activities and issues occur in several scales. Forests are managed at tree, stand, and landscape levels. Biodiversity can be assessed at point, alpha, beta, and gamma levels. Our societies are organised at the level of the individual, family, neighbourhood, municipality, county as well as country. Consequently research has to take into account consequences at all these levels.

2. To quantify how much of the structures at different scales that are required to maintain a sustainable forest ecosystem. This requires a system of book-keeping for genes, species, habitats and processes. To quantify how much is needed to maintain viable populations of species it is necessary to study the presence, population density and preferably also fitness of species that are specialising on different structural properties. Such indicator species should not be threatened by extinction or endangered since this will seriously hamper the possibilities of obtaining sufficient sample sizes. To cover the variation in the properties many different species, as well as processes, must be included by building a system of indicators and criteria (Noss, 1990; Angelstam *et al.*, 1993).

   The landscape level is particularly neglected and research often suffers from pseudo-replication in the larger scale levels We are used to gathering sufficient sample plots within a forest type and making measurements in different forest types. Achieving sufficient replication at the landscape scale is more time-consuming as well as intellectually more complex. Research must therefore be carried out both using a fine-grained approach (studying selected indicator species), and a coarse-grained approach (studying biomes and landscapes) in a gradient across different landscape types. Moreover, to generalise for application in management, it is necessary to achieve sufficient replication within the relevant systems.

3. To develop or re-orient existing approaches that could be used for rapidly implementing different kinds of results. This wide field ranges from strategic planning to practical management, and includes the development of education and training systems to spread knowledge quickly, as well as to develop management techniques that produce structures important for biodiversity.

TABLE I

Approximate timing (century AD) of periods separating the four forest history phases in Scotland (Ritchie, 1920), the Bergslagen mining district in southern Sweden (Wieslander, 1936), Norrland in northern Sweden (Bunte *et al.*, 1982; Östlund, 1993), North Russian plain (Galasheva, 1961; Redko and Babich, 1993) and close to the Ural mountains (Taskaev and Timonin, 1993).

	Scotland	Bergslagen South Sweden	Norrland North Sweden	Komi North Russia	Ural Mountain Foothills
Virgin forest (1)	5th-10th	15th	18th	19th	now
Local use (2)	10th	17th	late 19th	early 20th	-
Exploitation (3)	19th	late 19th	early 20th	now	-
Intensive Management (4)	now	now	now	-	-

5.2. Education

According to the forecasts regarding the future development of the forests, many structural properties at the landscape level (amount of very old forests, large areas with little impact) will decrease further into the future (SOU, 1992). Since species do not react to changes gradually, but instead suddenly at certain threshold values (Andren, 1994), it is important to implement existing knowledge rapidly.

The existing organisation for research and education are narrow in the sense that they are often based on a disciplinary differentiation. Solving complex problems requires new interdisciplinary solutions. The present issue demands co-operation as well as links between different disciplines and between research/education and practical management.

An efficient way to do this is to integrate everyday forest management with education and research in a system of Model Forests that represent the different forest history phases. Learning by seeing systems can thus be developed in parallel in the east and the west.

5.3. Management goals

The present challenge is to gather new knowledge and invent management techniques so that sufficient structural properties and balanced ecological processes (fire management, browsing pressure) in the boreal forest ecosystem can be maintained. In most regions, nature reserves only are not sufficient; instead natural features must be maintained or re-introduced into the managed landscape and combined with wood production. From a practical, management, and planning point of view, measures in forestry are taken at three levels:

1. the detailed level within stands.
2. the stand level.
3. the landscape level.

Our understanding of the range of problems is good at the detailed level, relatively good regarding theoretical and practical issues for the stand level, but insufficient at the landscape level. However, the real achievements in the field also vary with the scale of resolution. At present detailed considerations are carried out very carefully, while problems at the landscape levels have just started to be addressed. To re-orient existing and invent new techniques that should be used for quickly implementing different kinds of results are thus important.

6. Conclusions

Real achievement in the field will be enhanced if different components are integrated. We suggest that this process includes the following steps:

1. To make important decision-makers aware of the multiple-goal setting for forestry.
2. Inform and educate teachers of academic and school levels about the strong international trend towards combining sustainable forestry, local society demands and maintenance of biodiversity. To educate educators will help implement the new knowledge. Small groups of specialists in the different fields (ecologists, managers of industries and forests) from the east should "look into the future" by spending time in the west to learn about attempts to solve the forestry–biodiversity conflict. Similar categories of specialists from the west should "look into the past" by studying existing natural environments.
3. Comprehensive spread of information about the problems of present forestry and possibilities of alternative management.
4. Research that will quantify the amount and quality of different structural properties that are required for different species. Ideally, such studies should be conducted along a gradient from simplistic forest plantation landscapes in western Europe, through severely altered forests like the Scandinavian forests, to naturally dynamic forests in eastern Europe.
5. Gradual improvement of education programmes by input from research carried out in close co-operation with local enterprises.
6. Implementation of knowledge.
7. Independent monitoring of achievements.

The design of a future sustainable boreal forest system in the west and the east are similar. The east and the west differ considerably, however, with regard to the conditions at the starting point of its development. The trajectories of impoverishment of forest natural properties are different and the relative importance of different topics differ. The current rapid transition between different forest ecosystems and management regimes in both eastern and western Europe is a good opportunity for developing a balance between the production of the valuable timber resource and biodiversity and other values. To replace the traditional industrial way of managing forests and start development towards sustainable forests requires international co-operation.

References

Alexeev, V.: 1992. *The Boreal Forests of the World,* Proc. Conf. Ajette förlag, Jokkmokk, Sweden. pp. 6-9.

Andrén, H.: 1994, *Oikos* **71**, 355-366.

Angelstam, P.: 1995a, *In:* DeGraaf, R. (ed). *Wildlife conservation in forested landscapes.* Chapman and Hall. (In press).

Angelstam, P.: 1995b, *In:* Goldammer, and J., Furyaev, V. (eds.). *Fire in ecosystems of northern Eurasia.* (In press).

Angelstam, P., Arnold, G.: 1993, *In:* Saunders, D. A., Hobbs, R. J. and Ehrlich, P. (eds.). *Reconstruction of Fragmented Ecosystems: Global and Regional Perspectives.* Surrey Beatty and Sons, Chipping Norton, New South Wales, Australia, pp. 109-125.

Angelstam, Mikusinski, G., Travina, S.: *Department of Wildlife Ecology. Report No. X.* (In press).

Angelstam, P., Rosenberg, P., Pettersson, B.: 1993, *Skog och Forskning* **93**(1), 28-33.

Arpi, G.: 1959, *Ivar Hæggströms boktryckeri AB*, Stockholm, Sweden, 680 pp.

Barr, B. M.: 1988, *In:* J. F. Richards and R. P. Tucker (eds.) *World Deforestation in the 20th century.* Duke University Press, Durham, USA, pp. 230-261.

Berg, Å., Ehnström, B., Gustafsson, L., Hallingbäck, T, Jonsell, M, Weslien, J.: 1994, *Conservation Biology* **8**(3), 718-731.

Bunce, R.G.H. and Jeffers, J.N.R. (eds.): 1977, Institute of Terrestrial Ecology, Cambridge, UK, 120 pp.

ESA. (European Space Agency): 1992, Remote sensing map of Europe, *ESA/ESTEC*, Nordwijk, 18 pp.+map.

Esseen, P.-A., Ehnström, B., Ericsson, L. and Sjöberg, K.: 1992, *In:* Hansson, L.(ed.), *Ecological Principles of Nature Conservation.* Elsevier, London, UK, pp. 252-325.

Galasheva, V.A.: 1961, *Akademia Nauk SSSR Komi Filial.* Goslecbomizdat, Moskva, Russia, 395 pp.

Grigoriev, A.: 1994a, *Taiga News* **8**, 6.

Grigoriev, A.: 1994b, *Taiga News* **9**, 6-8.

Krankina, O.N. and Dixon, R.K.: 1992, *J. Forestry* **90**(6), 29-34.

Kuusela, K.: 1990, *Finnish National Fund for Research and Development.* Helsinki, Finland, 1 72 pp.

Larin, V.B. and Pautov, Yo. A.: 1991, *Nauka*, Leningrad, Russia, 144 pp.

Larsson, L. 0.: 1989, *Smålands Museum*, Växjö, 205 pp.

Lehtinen, A.A.: 1991, *Fennia* **169**(1), 57-169.

Malanson, G.: 1993, Cambridge University Press, Cambridge, UK, 296 pp.

McVean, D.N. and Ratcliffe, D.A.: 1962, *HMSO (Nature Conservation Monograph 1).* London, UK.

Nilsson, C. and Götmark, F.: 1992, *Conservation Biology* **6**(2), 232-242.

Nilsson, S., Sallnäs, O., Hugosson, M., and Shvidenko, A.: 1992, The Parthenon Publishing Group, Carnforth, UK, 407 pp.

Noss, R.F.: 1990, *Conservation Biology* **4**, 355-364.

Noss, R.: 1993, *In:* Aplet, G. H., Johnson, N., Olson, J. T. and Sample, V. A. *Defining Sustainable Forestry.* Island Press, Washington DC, USA, pp. 17-43.

Östlund, L.: 1993, *Dissertations in Forest Vegetation 4*, Swedish University of Agricultural Sciences, Umeå, Sweden, 132 pp.

Pruchkin, V. D., Bondarenko, V. V., Larin, V. B.: 1993, *Proc. Conf. " Natural resources and productive forces of the Komi Republic"*, 10-12 Nov 1993, 21pp.

Rackham, O.: 1990, *Trees and Woodland in the British Landscape*, J. M. Dent, London, UK, 234 pp.

Redko, G.I. and Babich, N.A.: 1993, *Arkhangelsk Severo-zapadnoe Knidznoe Izdatelstvo.* Russia, 153 pp.

Remröd, J.: 1991, *Skogsindustrierna.* Stockholm, Sweden, 102 pp.

Ritchie, J.: 1920, The Influence of Man on Animal Life in Scotland, The University Press, Cambridge,UK, 550 pp.

Rosencrantz, A. and Scott, A.: 1992, *Nature* **355**, 293-294.

Steven, H.M., Carlisle, A.: 1959, *The Native Pine Woods of Scotland*, Oliver and Boyd, Edinburgh, UK.

SOU,: 1992, *Statens offentliga utredningar* 1992, **76**, 313 pp.

Swanson. F.J. and Franklin, J. F: 1992, *Ecological Applications* **2**(3), 262-274.

Tannincn, T., Storrank. B., Haugen, I., Friis Møller, P., Löfgren, R., Thorsteinsson, I. and Rgnarsson, H.: 1994, *Nordisk Ministerråd* 1994, **7**, 109 pp.

Taskaev, A.I., Timonin, N.I.: 1993, *List of Protected Nature in the Komi Republic, Russian Academy of Sciences*, Syktyvkar, 190 pp.

Tsvetkov, M.A.: 1957, *Changes in Forest Cover in European Russia from the end of the 17th Century to 1914.* Nauka, Moscow, USSR, 312 pp.

Wallmo, U.: 1897, Rationell skogsafverkning. *Örebro.* 288 pp.

Wibe, S. and Jones, T. (eds.).: 1992, Forests: Market and Intervention Failures. Earthscan Publications Limited, London, UK, 204 pp.

Wieslander, G.: 1936, *Sveriges Skogsvårdsförbunds Tidskrift* **34**, 593-633.

Zackrisson, O.: 1977, *Oikos* **29**, 22-32.

PROBLEMS OF FOREST MANAGEMENT IN RUSSIA

G. KOROVIN

International Forest Institute, Novocheriomuskinskaia str. 69, Moscow 117418, Russia

Abstract. Radical political and economic changes in Russia have been accompanied by reforms with regards to the legislative base, the system of forest management and formulation of new forest policy. This paper presents a brief analysis of recently adopted forest legislation and the problems that exist in forest management, forest utilisation, protection and regeneration of forest resources. Lack of co-ordination of federal and regional legislation and the resolution of a number of questions that have significance with regards to development of the forest sector are discussed. The main focus is on improvements to the legislative base and the organisational structure of forest management during the transition to a market-based economy. The primary issues include: (1) determination of the forms of forest ownership and differentiation of the forest fund by forms of property; (2) differentiation of the jurisdiction of federal, regional and local authorities in the regulation of forest relationships; (3) reconciliation of the economics of forest management with methods of national regulation, and (4) determination of sources and mechanisms for forestry financing. Improvement and maintenance of forest sector information databases are considered prerequisites for the development of forest management and forest policy in Russia.

Keywords. RUSSIA, PHYTOMASS, CARBON, CONIFERS, CARBON DIOXIDE, BOREAL FOREST.

1. Introduction

The problems of forest management are pivotal issues for most countries of the world. They have special significance in Russia. At present, Russia, which owns enormous forest resources, is in transition to a market-economy in association with a major economic crisis. The former tightly controlled management, based exclusively on national ownership of forest resources are ineffective under these changing conditions. At present, a new system of forest management based on the principles of a market-economy in combination with national regulation, is being developed.

Creation of a new system of forest management, and matching it to changing conditions, is a very intricate process. This process is seminal with regards to the generation of a legal, economic and social basis for forest management, as well as to the resolution of a number of scientific issues and organisational problems.

2. Improvements to Forest Legislation

2.1. FORMATION OF THE LEGAL BASE FOR FOREST MANAGEMENT

A comprehensive package of legislation, including the Constitution of the Russian Federation, the Principles of Forest Legislation (PFL) of the Russian Federation and various forest codes and laws have been adopted over the last several years. A considerable number of the individual republic's legislative acts were formerly adopted before the federal laws and the PFL were issued. This has led to a lack of co-ordination and has introduced contradictions in the adopted documents. These contradictions concern the legal status of forests and authorities of the Russian Federation and its subjects regarding forest issues.

The PFL, adopted in 1993, reflects special features in the political, economic and social climate in Russia during this period of reform. The PFL contains a set of rules

Water, Air and Soil Pollution **82**: 13–23, 1995.

and guidelines, aimed at: (1) strengthening legal protection of the forest during the transition to a market-economy; (2) decentralisation of national forest management, and; (3) expansion of the domain of federal and local authorities with regards to forestry matters. At the same time the principle issues for forestry and forest management remain unsolved. The most urgent among these are issues connected with the legal status of the forests, such as the mechanism of forestry financing and the development of national forest regulation.

According to the Constitution of the Russian Federation and the PFL, the forest fund is under joint authority of the Russian Federation and the subjects of the Russian Federation, i.e., the territories, republics, and autonomous bodies. Maintenance and disposal of forests and forest utilisation, are to be carried out in the interests of the people that live in the corresponding territories, and of all peoples in the Russian Federation.

The monopoly of the national ownership of forests that existed earlier, is no longer clear. In addition, the important economic component of legal ownership remains indefinite, thus introducing uncertainties with regards to: (1) rights to the utilisation of the forest fund; (2) rights to receive income, and; (3) the right to sell property. The uncertainty of the legal status of the forest, and the lack of co-ordination between individual republic legislation with the PFL has enabled central, regional and local authorities to interpret property rights for forest resources as they want, and thus generate different structures of forest management that serve individual economic and political interests.

The PFL give wide authority to regional and local bodies of authority to regulate forest activities. The jurisdiction of the subjects of the Russian Federation extends to the following: (1) adoption of regional legislation; (2) joint ownership (with the Russian Federation) of the parts of the forest fund, situated in their territory; and; (3) establishment of guidelines for forest utilisation fees, setting the limits and areas for forest harvesting and arranging for the orderly sale of public forests. In addition, local authorities have jurisdiction over forest regulation, allowable cuts, fees and licensing.

The jurisdiction of the Russian Federation includes: determining general principles of forest ownership and directions of national policy, and provides national management on behalf of the state including utilisation, regeneration and protection of the forests.

2.2. ORGANISING NATIONAL MANAGEMENT OF THE FOREST

According to the PFL, national management of forest utilisation, regeneration and protection is carried out by executive authorities of both federal, regional and specially authorised forest management agencies. The system of specially authorised agencies includes both federal and regional forest management bodies and local subdivisions such as forest management districts and forest divisions.

Replacement of the state body of forestry by the system of specially authorised forestry management agencies was carried out to better delineate management and business structures. Forest management districts, which had the status of national enterprises, lost the rights to harvest and process the wood, but did receive ownership rights to the forest fund.

This reorganisation partially solved some transition problems but brought with it a set of negative consequences. The functions of forest management, ownership of forest fund, carrying out of forestry operations and creation of forest utilisation divorced from industrial purchasing of wood, are once again focused in a single body. This has intensified the problems associated with monopolies by forest management districts, which combine the functions of forest management with economic activities.

The overall system—the utilisation and the regeneration of forest resources—is artificially split. Although the forest management districts are the designated owners of the forest fund, they are not allowed to purchase wood for industrial use; this restricts their economic development. The institute of forest, though also land owners, have no harvesting rights.

In the process of improving the system of national forest management, it is necessary to define suitable management and business structures, to ensure that the balance of their interests is maintained and to define the extent of business activity by enumerating different types of forest utilisation, protection measures and requirements for regeneration of forest resources. State regulation of forest activities and the management of forest as state property must be carried out by specially authorised state agencies, that are free, as far as possible from the production activity. At present this last function can be carried out by both national and private enterprises, be direct participants in market activities and can carry out the functions of producer and deliverer of goods and services, as well as customers of the production.

2.3. Development of the Lease Relationship

For state owned forests, one of the most acceptable mechanisms for economic forest utilisation is through forest resource leases. The PFL of the Russian Federation provided institutional measures for the leasing of forest fund areas. It provides for transferral of areas of forest fund over long terms using licences and agreements as the basis for leasing. These licences and agreements determine the rights, duties and responsibilities of both lease-holders and lease-owners.

The Government of the Russian Federation adopted special regulations regarding the leasing of forest fund areas (Resolution, 1993), which specifies the order of forest leasing, the principles and the schedule of lease payments (Zaslavskaia, 1994). The federal forestry service developed recommendations on a set of methods to calculate minimal payment of forest rate and methods of payment for lease activities (Voronkov *et al.*, 1994).

In accordance with the legislation of the Russian Federation, lease-holders are the legal owners (which may include foreign parties and individuals) and have the rights to forest utilisation. The forest fund area leases can be extended for 1–50 years and may be renewed.

To successfully implement the forest leasing scheme it is necessary to ensure lease stability and to guarantee long term rights to forest utilisation. Combining current and long term interests of the nation, local authorities, forestry and forest industry in utilisation of the forest resource, while maintaining recreational use, are central problems. It is hoped that Canada and other countries operating under market-economy conditions will provide considerable assistance in the solution of these challenges to lease relationships.

2.4. Improvements to the Financing of Forestry

The forest legislation of the Russian Federation does not provide satisfactory solutions for questions concerning the sources and mechanisms of financing forestry activities. These are required to provide a sustainable, functioning industry during the transition to a market-economy. In fact, the previous regime of central financing of forestry is occurring at the expense of the national budget (Instruction, 1993). Allocated payments are not, in practise, connected with the total forest income based on payments for forest resources usage. Establishment of the rate and sizes of payments for forest utilisation (grant privilege by payments) is determined by competition of local authorities. In addition, payments for forest utilisation are being transferred to local or regional budgets and only part of this money need be applied to forest conservation and protection at the discretion of local authorities.

Attempts to solve the problem of financing of forestry activity through the creation of a state out-of-budget fund for recreation, protection and conservation did not work. Introduction of a surcharge on the order of 20% above the cost of wood harvesting, merely put most of forest uses in an extremely difficult financial situation. Subsequent reduction of the rate to 5% and a sharp reduction of the volumes of wood harvest, put an end to the role of the out-of-budget fund as a significant source of forestry financing activity.

The decoupling of budget allocations, from the total income from forest resources, has been a major disincentive for workers of forest service and has mitigated against increasing income of forestry and mobilisation of forest income. The State has practically no opportunity to influence the distribution of forest income, to reduce budget deficits at the expenses of payments for forest usage, or to balance the costs of forestry with incomes from forest usage.

The above indicated problems in financing present a serious barrier for moving Russian forestry to a market-based economy and makes it uncompetitive with other industries. The present conditions of sharp federal budget deficits and unstable economy offer few opportunities to ensure the protection of forest resources, thus resulting in a major negative influence on Russian forests.

The experience of countries operating in a market-economy and experiences with Russian Forestry before the revolution indicates the necessity of using different sources of financing, including the state budget. It seems logical to use forest income based on payments for forest production and services as the main sources of financing.

3. Problems of Forestry and Forest Management Organisation

3.1. Problems of Forest Utilisation

An important achievement of recent years in the forest sector of Russian is the legal and practical acceptance of the principle of sustainable development. No longer is uncontrolled over-harvesting tolerated—such as that which led to significant depletion of forest resources in some regions of the European-Ural part of the country, and aggravation of the quality and species content of these forests.

Discontinuance of indiscriminate over-harvesting was carried out against a background of the common reductions of production in Russian and sharp reduction of the volumes of wood purchasing (Goscomstat, 1990, 1993). During the period 1988–1993, the annual size of harvests both by stock and by area, was halved (Figure 1). Only one third of the annual allowable cut, totalling 529 million m^3 for the Russian Federation, is being used, testifying to a very low utilisation level of potential forest resources.

Despite sharp reductions in harvest volumes essential alterations in the harvesting methods during this period has not occurred. Large scale aerial photography and other data indicate that 10–25% of the standing wood is lost using existing methods of harvesting. Considerable damage to forest ecosystems is inflicted by disturbance of ecological conditions through improper use of heavy equipment. Thus, environmentally

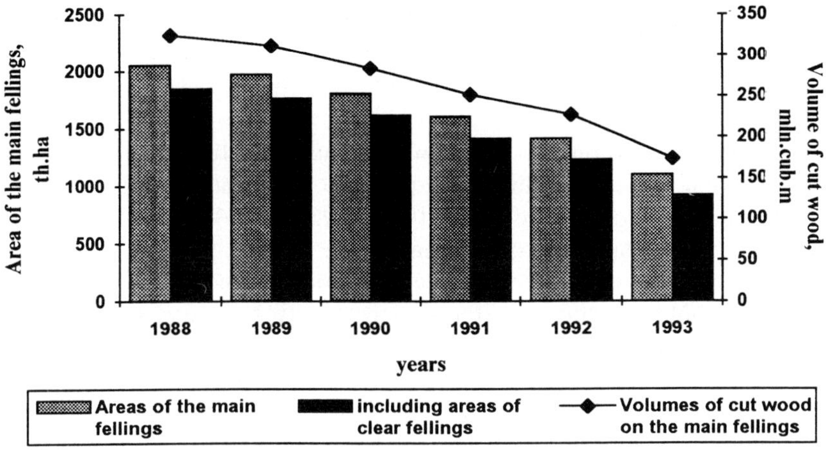

Fig. 1. Forest exploitation: annual areas and wood volumes harvested, 1988–1993.

acceptable harvest methods remains an important problem requiring improvement of both technology and equipment.

The new harvest rules adopted in 1993–1994 for most regions of the country may play an essential role in solving this problem. These rules essentially restrict permissible felling areas, and terms of contiguity adjoin, and specify concrete forestry and ecological requirements for the harvesting process.

An important problem facing the organisation of complex forest utilisation under the new conditions is a better division of forests into groups and protective categories. Within these there must be an accounting of the ecological, social and economic functions and an assessment of the priorities to be given to each. Analysis of existing distribution of forests into groups and protective categories indicates that the ecological and social role in the Asian part of the country, which accounts for 80% of the area of the Russian forest fund, are severely underestimated.

All forests within the forest fund have been classified into three groups, on the basis of their economic and ecological roles, and all forest management, utilisation, and legislation, etc., must operate within with this classification. The first group are protected forests, e.g., nature reserves, watershed areas, pre-tundra forests, and for recreational use. In this group, there are 23 "protective categories": for 13, final harvesting is prohibited, and for the remaining 10, harvesting is allowed but with very severe restrictions (usually only selective cutting is allowed). The second group are multiple-use forests with relatively strong ecological limitations on exploitation. The third group are forests used for wood production, or for other industrial exploitation, where clear-cutting is the normal harvesting method.

More than half of the above-indicated territory is situated in the permafrost zone and is represented by mountain geosystems indicating the important role of the forests, growing there. In spite of this, more than 3/4 of the area of forest fund of Siberia and Far East in relegated forests of the third group, with priority given to raw material supply, and less than 1/4 to forests of the second and first group. In the European part of the country with more favourable weather conditions, forests assigned raw material functions represents 1/3 while the forests of the first and the second group comprise 2/3 of the are of forest fund (Figure 2). Clearly the Far-East regions containing about 85% of the forest fund area, and belonging to the third group of forests with a raw materials functions, require special attention.

3.2. Problems of Reforestation

The result of the general reduction in production and inadequate financing of forestry resulted in a reduction of efforts directed to forest regeneration and increased productivity. Reforestation effort for the period 1988–1993 was reduced by 30% while production of seedlings and planting dropped nearly 50% (FSFM, 1994). The areas of thinning and sanitary cuttings dropped over the same period of time by 30% (Figure 3).

Despite the considerable reduction of reforestation activity, the harvesting rate was even lower, permitting timely reforestation of newly harvested areas. At the same time however these efforts are inadequate for timely reforestation of burned and dead stand areas to thereby reduce the areas not covered with forests; these areas total about 15% of the total forest lands of the Russian Federation.

Taking into account the importance of reforestation problems, the Government of the Russian Federation approved a state programme of reforestation for the period of 1993–1995. This program put into place a set of tasks designed to extend the volume, and raise the quality, of reforestation effort as well as to generate the necessary materials and technical base for the program.

In addition the continued creation of protected forest plantations and prevention of wind and water erosion of agricultural lands has been stipulated by the State in a complex set of programmes designed to raise soil fertility adopted in 1992. Adequate financing of the reforestation activity, stipulated in the above mentioned programmes,

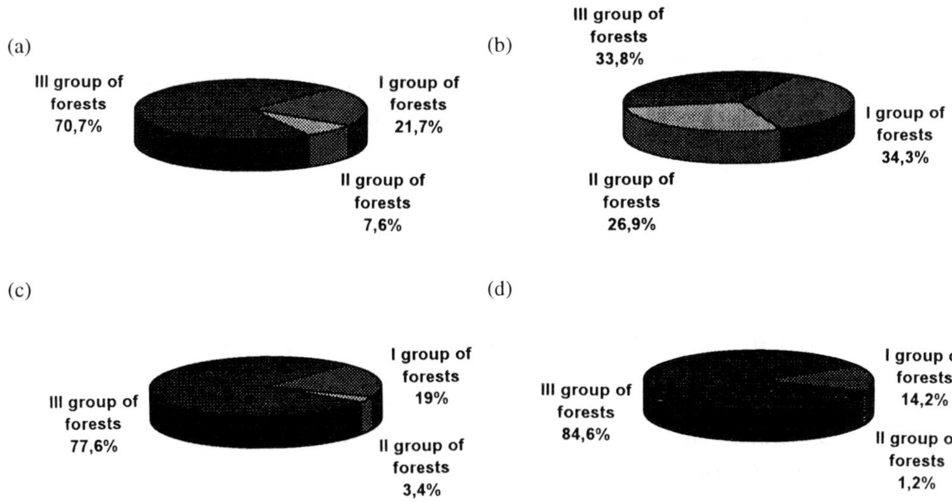

Fig. 2. Forest fund distribution by regions and forest groups. (a) Russian Federation; (b) European region; (c) Asian region; (d) Far-East region. Forest groups are based on level of economic and ecological exploitation: I - protected, II - multiple-use, III - exploited (see text for details).

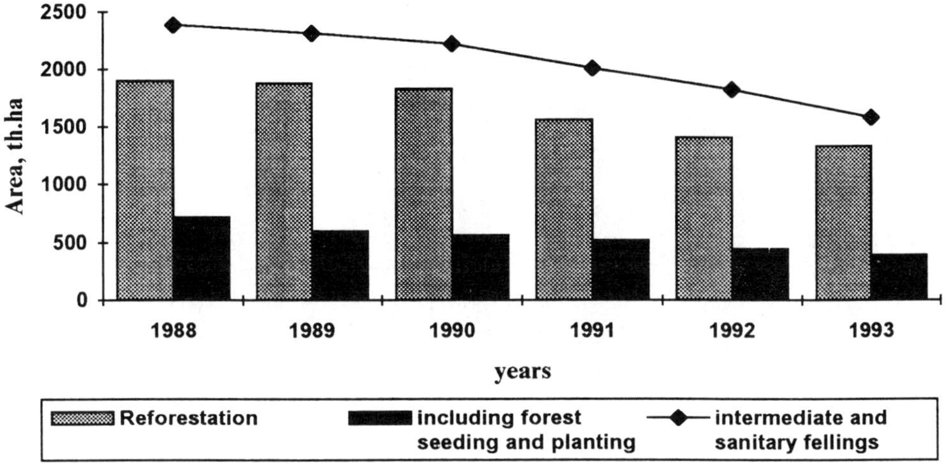

Fig. 3. Areas treated annually in forest renewal activities, 1988–1993.

unfortunately restrict the likelihood of achieving the goal of directed regulation of the forest fund structure and dynamics.

3.3. Problems of Forest Protection

The state and dynamics of Russian forests are determined significantly by the influence of accidental factors including forest fires, pests, insects, and forest disease, and other natural calamities. According to the data of the latest forest inventory the area of burned and dead stands, is approximately four times the area of harvest deforestation for the country as a whole.

The area annually damaged by fire, insect and disease outbreaks on the controlled areas of the forest fund is measured in Mha (Figure 4). The total area of forests annually disturbed by such accidental factors accounts for several hundred thousand hectares (Figure 5).

Fires play the dominant role and account for more than half of the area of dead stands. Of particular significance is the influence of fire on forest ecosystems of the Far East where the area of burned and dead stands is 7.5 times higher than the harvested area and accounts for 5.5% of the entire forest area.

The existing system of forest protection is able to control fires successfully under conditions of low and average fire danger, but has had periodic failures in conditions of high and extreme fire conditions. Raising the level of forest fire protection requires solving a complex of problems, chief of which are: strengthening of aviation and ground forest fire services; increasing manpower assignments for fire fighting crews and extending the area of fire-controlled territory. The solution of these problems is considered in the state programme of forest fire protection for the period 1993–1997. Because of inadequate funding, however, this programme is only partially carried out. Nevertheless, the measures undertaken under its auspices do permit forest fire protection to approach that achieved earlier.

The influence of pest-insects, forest disease, industrial pollution and other abiotic factors are, in general, negligible compared to the destructive influence of forest fires.

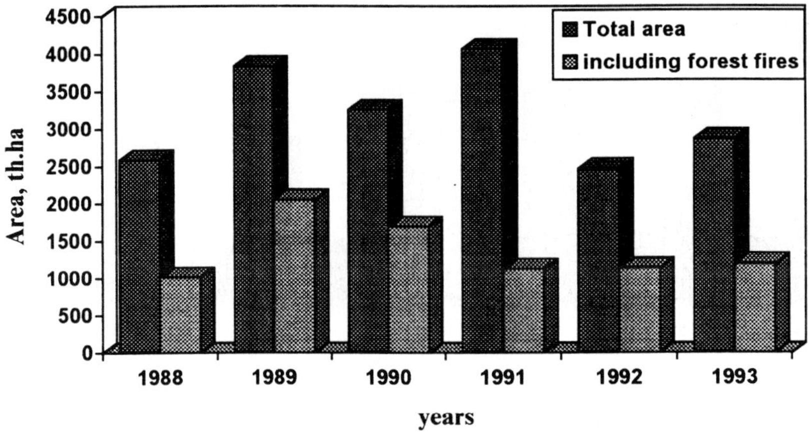

Fig. 4. Areas lost to forest fires, and foci of insect-pests and forest diseases, 1988–1993.

In some regions of the country, however, these other factors are dominant and require the creation of an effective system of forest protection.

3.4. PROBLEMS OF DEVELOPING INFORMATION DATABASES

Effective forest management requires a well-maintained information base documenting the state and dynamics of the forest fund, its ecological and resource potential, growth and yield information and ecosystem responses to the influences of anthropogenic and natural disturbances. The large size of the Russian forest fund territory and its complexity and dynamics make it necessary to accumulate, maintain and process enormous volumes of information and expert knowledge in order to manage effectively.

At the present time, inventory estimations exist for all forests of Russia, however, the degree knowledge and reliability of available data, vary substantially across the territory. The most complete and reliable data of the forest fund provides forest structure, based on a combination of ground surveys and aerial photographs. At present coverage with such data extends to about 60% of the forest fund territory, including the most economically developed regions of the country. The remaining 40% of the forest fund territory are inventoried on the basis of aero-visual assessments, and aerial and space photographs but without supporting ground surveys. Aggregation of the information obtained by these different methods, is made difficult not only by the different degree of completeness and reliability of the available data but also by the necessity of co-registering these date to a single period of time.

The creation of a national system of forest monitoring is necessary to increase the coverage and reliability of data on the forest fund and to permit efficient regulation of its state and dynamics. The enormous size of the forest fund territory and the different degrees of its economy assimilating into the market-economy make the wide use of satellite and aerospace techniques essential ingredients of the monitoring system. The potential conversion of the Russian space and military industry to such purposes may provide a successful solution for this problem.

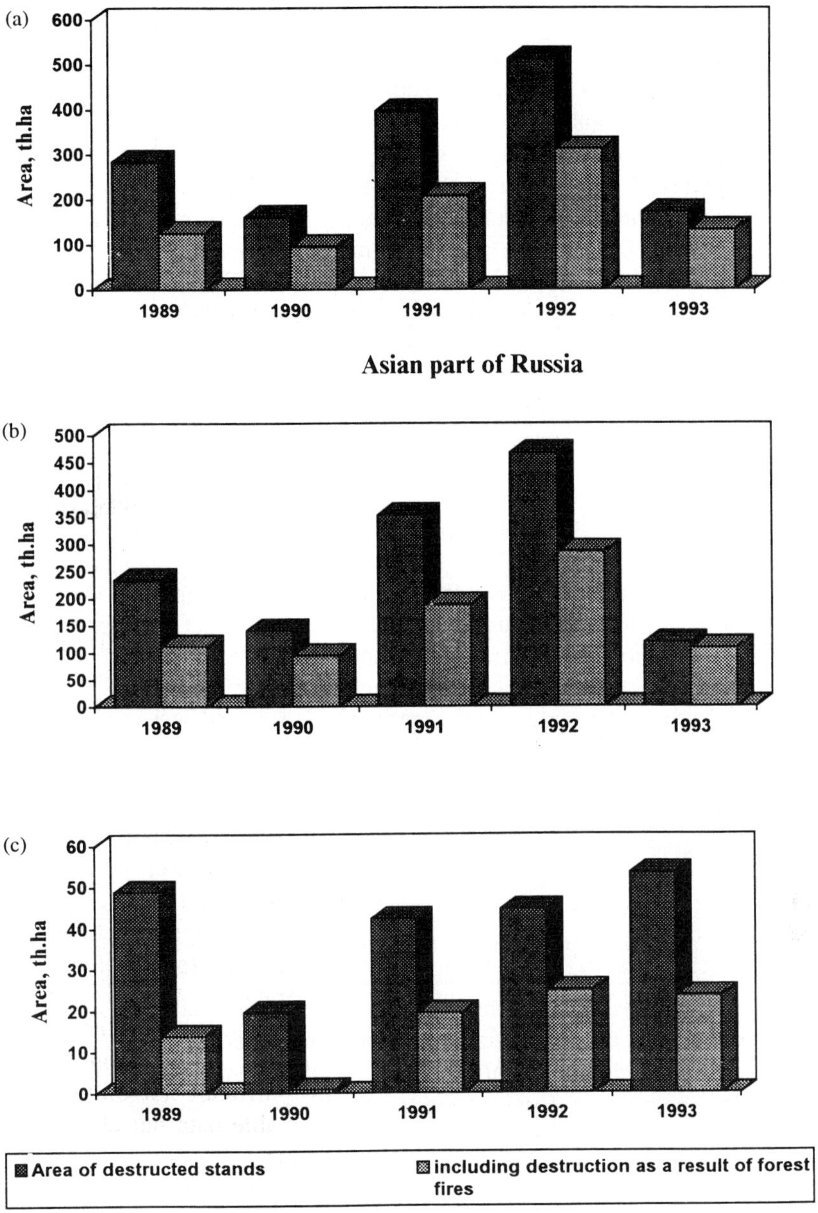

Fig,. 5. Dynamics of forest disturbances causing mortality, by region, 1989–1993. (a) Russian Federation; (b) Asian region; (c) European region.

The network of forest research institutes and organisations in Russia posses unique experimental and knowledge database on forests. The most critical problem that has faced forest science in recent years has been the failure by the forestry industry to make use of forest science expertise; there has been an absence of interest by the industry in mastering the achievements of science and technology. Generation of a new forest policy (Decree, 1994) in response to the growing requirements of society, together with development of new economic mechanisms and a new system of forest management during the transition to a market-economy, will require not only use of existing forest knowledge and databases of forests but their considerable expansion.

Underestimation of the role of science, including forest science, has been observed during the last few years, and resulted in a sharp reduction in the share of gross national product allotted to research. This has drastically affected academic, industrial and university science. The consequences have been considerable reductions in the number of forest scientists, destruction of the experimental base and a sharp reduction in scientific publications. In the medium to long-term, the most dangerous trend is the shutting down of scientific schools throughout Russia. Therefore, if scientific principles are to provide the basis for forest management, an urgent priority must be the conservation of existing data and knowledge about forests, increased usage of existing scientific potential, and the maintenance of acceptable working conditions for research groups and schools.

4. Conclusions

The transition to a market-economy demands reorganisation of the system of forest management in Russia. The new system must be a combination of market mechanisms and national regulation in areas of forest utilisation, regeneration and protection. The creation of such a system is a very complex and long process, and requires changes to legislation, economic and social structures. The issues of paramount importance, requiring immediate attention, must be the following: a clear conception of the legislative status of forests and forms of property ownership; delineation of management and business structures; and replacement of centralised budget financing by a multichannel system of financing, base on forest revenue as well as other sources of allocations.

In recent years there was a decline in wood harvesting and processing, as well as reduced scales of reforestation and forest cultivation. These changes have bad effects on social conditions but also create prerequisites for negative changes in the structure and dynamics of the forest fund. A set of urgent actions are required in order to stabilise forestry and forest purchasing enterprises, and to conserve the forest environment and its resource potential. National programmes of reforestation and protection forests foresaw the need for action but there has been insufficient financial allocation to fulfil the need.

Underestimation of the role of forest science, resulting in inadequate financial support of research has put existing scientific establishments and schools in a precarious state. Conservation and effective utilisation of accumulated scientific potential and maintenance of reasonable working conditions for forest professionals are probably the most important prerequisite for improved forest management in Russia.

Acknowledgments

The author wishes to acknowledge Russian and Canadian colleagues who assisted with the preparation of this paper, specifically A. Z. Shvidenko. He also greatly appreciates financial assistance from the USDA Forest Service and from the Canadian Forest Service (Edmonton) to attend the IBFRA '94 Boreal Forests and Global Change conference.

References

Constitution: 1993, *Constitution of the Russian Federation.* Juridical Literature Publishers, Moscow, Russia, 64 pp. [In Russian]

Decree: 1994, Basic features of Russian Federation state strategy on environmental protection (official distribution of the presidential administration), Appendix, Russian Federation Presidential Decree, No. **236**, 4 February, 1994. Moscow, Russia. [In Russian]

Federal Service of Forest Management (FSFM): 1994, Main results of activities for 1993. Russian Federation, Moscow, Russia, 26 pp. [In Russian]

Goscomstat: 1990, *Forestry of the USSR. Statistical information.* State Committee for Statistics of the USSR. Information Publishing Center, Moscow, USSR, 135 pp. [In Russian]

Goscomstat: 1993, *Forest management in the Russian Federation. Statistical information.* State Committee of the Russian Federation on Statistics. Republican Information Publishing Centre, Moscow, Russia, 135 pp. [In Russian]

Instruction: 1993, On the order of levy and procedure for using taxes in forest regeneration and protection, Federal forest issue of economical legislative and business information. *Economy and Life* (*Economica i shizn*) Special Issue, 3 March, 1993, Moscow, Russia, pp. 45-48. [In Russian]

Regulations: 1993, Forest legislation regulations in the Russian Federation. Law of the Russian Federation, No **4613-1**, signed by President 6 March, 1993. Ecos-Inform Publishers, Moscow, Russia, 64 pp. [In Russian]

Resolution: 1993, Rules on lease of forest fund lots in the Russian Federation. Resolution of the Council of Ministers, Government of the Russian Federation, 23 July, 1993, No. 712. *Legislation and Economy* (*Zakonodatelstvo i economica*) **3/4**, Moscow, Russia, pp. 122-126. [In Russian]

Voronkov, P. T., Elchev, N. M., Logazky, V. N., Maiorov, I. G., Tiashky, V. P., Vasin I. A., Dudina, E. A., Kuznetsov, I. M. and Shevelev, V. A.: 1994, Recommended method for calculating minimal rate of forest tax and lease payment. Federal forest issue of economical-legislative and business information, Issue 5, Moscow, Russia, pp. 9-70. [In Russian]

Zaslavskaja, L.A.: 1994, Forest codes and laws in the Republics of the Russian Federation. *Legislation and Economy* (*Zakonodatelstvo i economica*) **3/4**, Moscow, Russia, pp. 105-109. [In Russian]

CONSERVING THE BOREAL FOREST BY SHIFTING THE EMPHASIS OF MANAGEMENT ACTION FROM VEGETATION TO THE ATMOSPHERE

M. J. FITZSIMMONS

Parks Canada, Prince Albert National Park, P.O. Box 100, Waskesiu Lake, Saskatchewan S0J 2Y0, Canada

Abstract. Historically, the management of forest vegetation and the atmosphere have occurred in isolation. Modelling, however, suggests that boreal forests will be highly sensitive to changes in atmospheric composition, so much so that the persistence of closed forests within present boreal areas such as central Saskatchewan will depend more upon management of air than upon management of trees. This should inspire organizations responsible for the management of Canadian forests to be both advocates for, and models of, CO_2 emission reductions. But data on electricity use and fossil fuel consumption are used to show CO_2 emissions from a subset of private and public forest management institutions are not declining along a trajectory toward the international target of a 20% emission reduction relative to 1988 levels by the year 2005. Estimated emissions from the pulp and paper industry between 1989 and 1993 exceeded the linear reduction trajectory by 11.5% to 15.5%. Estimated emissions from Parks Canada for 1989 to 1991 exceeded this same trajectory by 4.4% to 17.7%. Despite commitments to ecosystem management approaches, there is little evidence that atmospheric considerations have been fully incorporated into the activities of forest management organizations.

Keywords. CARBON DIOXIDE, EMISSIONS, MANAGEMENT, BOREAL, FOREST, ATMOSPHERE, PULP, PAPER, PARKS.

1. Introduction

It was at a boreal forest lumber camp early in the 1900s, that Richard St. Barbe Baker (1889–1982) decided to qualify as a forester and dedicate his life to promoting the conservation of trees. Baker went on to receive international recognition for his decades of work protecting and restoring forests throughout the world. Fifty years ago he wrote:

> "*Wood may be regarded as merely a by-product of trees. Their greatest value is probably their beneficial effect upon life, health, climate, soil, rainfall and streams.... We live less than five minutes without air; and trees purify the air by absorbing carbon dioxide ... and throw out the pure oxygen, so necessary to life" (Baker, 1944).*

This holistic view of forests never, unfortunately, made it into the mainstream of forest management thinking during Baker's lifetime. During most of this century, forest ecosystems have been researched and managed as if timber stocks, wildlife populations, water supplies, and air quality were an assemblage of independent ecological entities.

This long-standing fragmentation of responsibility is gradually changing as scientists and decision-makers begin to unravel the ecological processes by which energy, air, water, flora, fauna, soil and other elements of ecosystems interact across various time frames and spatial scales. Today, both researchers and managers are racing to incorporate the lexicon of integrated resource management and ecosystem-based approaches into study proposals and forest plans.

Water, Air and Soil Pollution **82**: 25–34, 1995.

Integration of research and management activities for some forest ecosystem elements, timber and wildlife for example, have advanced tangibly in recent years. But there is less evidence that other system components have moved forward on the forest management agenda. The purpose of this paper is to assess the degree to which atmospheric considerations have been incorporated into the activities of organizations responsible for Canadian forests. Are forest vegetation and the atmosphere still treated as independent entities to be managed in isolation? Alternatively, have forest management institutions adopted an ecosystem approach which takes into account the relationship between trees and air recognized by Baker fifty years ago?

1.1. Atmospheric Composition and the Boreal Forest

During the last decade, our scientific understanding of the relationships between terrestrial vegetation and the atmosphere has advanced rapidly. If current trends in fossil fuel emissions and deforestation continue, the Earth's mean surface temperature is expected to rise approximately 0.3°C per decade well into the next century (Houghton *et al.,* 1990). Modellers estimate that climatic conditions conducive to boreal forests will shift poleward at rates which are an order of magnitude faster than the dispersal rates of trees (Melillo *et al.,* 1990; Solomon, 1992). As well as shifting in position, the boreal forest is predicted to constrict in area; much of the forested land in western North American boreal regions is expected to be replaced by cool temperate steppe (Bonan *et al.,* 1990; Emanuel *et al.,* 1985; Harrington, 1987; Monserud *et al.,* 1993; Rizzo and Wiken, 1992; Shugart *et al.,* 1986; Solomon, 1992).

1.2. Responses of Forest Management Institutions

There are several possible institutional responses to this new understanding of the predicted effect of global warming on boreal forests and other ecosystems. The recommendations of the Toronto Conference on the Changing Atmosphere included, among others, two important strategies: adaptation and limitation (TCCA, 1988).

Adaptation responses specific to the boreal forest include relocating forest management operations further north (Singh and Wheaton, 1991). However, adaptation was not the approach Richard St. Barbe Baker advocated to African countries confronted by a threat to forests of similar magnitude to that of global warming: desertification. Rather than working to harvest the remaining trees ahead of the advancing desert, Baker collaborated with a variety of organizations to establish a "Green Front" of planted trees in an attempt to limit desert expansion (Baker, 1970). In keeping with Baker's vision, this paper focuses on the strategy of limitation.

One approach to limitation would be to increase carbon (C) storage in forests through afforestation and the careful management of existing forest lands. However, it is unlikely that changes to carbon pools as a result of direct management of forest vegetation could curtail increases in atmospheric trace gas composition in the absence of significant reductions in fossil fuel consumption (Jarvis, 1989; Musselman and Fox, 1991). However, a second approach, steady reductions in carbon dioxide (CO_2) emissions as a result of decreases in fossil fuel consumption, could result in early stabilization of CO_2 concentrations in the atmosphere (Harvey, 1989). This limitation strategy would result in less extensive and less severe forest decline than would occur if current emission trends continued.

A reduction in CO_2 emissions to 20% below 1988 values by the year 2005 was the initial global goal recommended by the Toronto Conference on the Changing Atmosphere (TCCA, 1988). The specific objective of this paper is to compare the estimated CO_2 emissions of a subset of forest management institutions to a linear emission reduction trajectory which meets the 20% reduction by 2005 target.

Data were assembled from two institutions which, although very different, both have management responsibilities for boreal forests. The first was the Canadian Pulp and Paper Association, a private organization representing pulp and paper manufacturers across Canada. This industry has expanded its operations dramatically in western Canadian boreal regions during the 1990s. According to the environmental statement of the association, the pulp and paper industry "supports the responsible stewardship of resources, including forests, recyclable materials, fish and the aquatic habitat, wildlife, air, land and water" (CPPA, 1993).

The second organization was Parks Canada, a federal government agency which manages a system of protected areas across Canada including several national parks in the boreal forest. The objective of national parks is to protect for all time representative natural areas of Canadian significance, and manage them so that they remain unimpaired for future generations (Canadian Heritage, 1994).

2. Methods

The Canadian Pulp and Paper Association and Parks Canada both monitor their energy consumption. Available reports did not break down total energy use by ecological region. Thus, national data were generally used rather than estimates specifically for the portion of those institutions located within the boreal regions. Neither institution measures or estimates CO_2 emissions, thus it was necessary to use CO_2 emission factors to convert energy units to CO_2 mass. These emission factors were taken from published values for 1990, a year which is mid-way through the period of analysis (Table I).

Performance of these two institutions relative to emission reduction targets was evaluated by comparing estimated emissions from 1988 to 1993 to a linear reduction trajectory which started at the 1988 emissions level, and decreased annually by an amount equal to 1.7% of the initial value. This is the trajectory required to stay on track toward the target of a 20% reduction by 2005. No statistical analysis was performed on the data.

2.1. PULP AND PAPER INDUSTRY CO_2 EMISSIONS

The Canadian Pulp and Paper Association estimates energy use per Mg of pulp production through a survey of its member companies. Total energy use by the pulp and paper industry was estimated by taking the product of the CPPA's estimates of the total pulp and paper production (Mg) and the purchased energy use (GJ Mg^{-1}) by fuel type (CPPA, 1994). Total CO_2 emissions for each year were then estimated by summing the products of the energy use estimates and the 1990 CO_2 emission factors for each fuel type.

TABLE I

1990 CO_2 emission factors for common energy sources. Electricity data from Energy, Mines and Resources Canada (1991); others from Environment Canada (1992).

Fuel Type	CO_2:Energy Ratio ($Mg\ TJ^{-1}$)
Electricity	57.18[1]
Heavy Oil	74.00
Diesel Oil	70.69
Jet Fuel	70.84
Motor Gasoline	67.98
Natural Gas	49.68
Coal	88.65[2]
Propane	60.61[3]

1 Emission factor for electricity is based on 95 924 000 Mg CO_2 emitted by generation of 465 967 GWh (1677.48 PJ) energy.
2 The mid-range of values reported for Canadian bituminous (94.30 to 83.00 $Mg\ TJ^{-1}$).
3 The mid-range of values reported for liquid propane gas (59.84 to 61.38 $Mg\ TJ^{-1}$).

2.2. Parks Canada CO_2 Emissions

Parks Canada has an energy data system through which purchases of energy are tracked at each national park and historic site. These values are forwarded to Ottawa for a national compilation. Reported energy consumption estimates represent fiscal years (01 April to 31 March) rather than calendar years. Prior to 1991 the data were pro-rated to account for energy use at non-reporting locations. Total CO_2 emissions were estimated by summing the product of annual energy use for each fuel type and the 1990 CO_2 emission factors. Estimates of total energy use for the park system for 1992 and 1993 have not been reported to date.

2.3. Assumptions

The results presented below are contingent upon the following assumptions: (a) the magnitude of CO_2 emitted per unit of energy consumed for each fuel type was constant over time, and (b) the magnitude of CO_2 emitted per unit of energy consumed for each fuel type was equal to the national average. Several additional assumptions were made regarding fuel types. In the case of the pulp and paper industry, the CO_2 emission factors for diesel oil was used for energy purchases classified as "distillates" (CPPA, 1994) and the factor for propane was used those classified as "purchased steam, propane, etc." (CPPA, 1994). In the case of Parks Canada, the CO_2 emission factor for gasoline was used for transportation fuel classified as "other" (Bullard, 1990).

3. Results

3.1. Pulp and Paper Industry CO_2 Emissions

The total production of pulp and paper in Canada, and the estimated total energy purchased by fuel type for the pulp and paper industry is shown in Table II. Total pulp and paper production has varied less than 5 percent annually over the 1988 to 1993 period. Consumption of electricity, heavy oil, distillates, and coal declined and use of natural gas and "other" fuel types (purchased steam, propane, etc.) increased during the same period.

The estimated CO_2 emissions by the pulp and paper industry by fuel type for the period are shown in Table III. Total CO_2 emissions peaked in 1989 and remained above 1988 levels throughout the period of analysis. Figure 2 compares the estimated total CO_2 emissions versus the 20% reduction by 2005 trajectory. Each year, CO_2 emissions by the pulp and paper industry exceeded the linear reduction course by 11.5% to 15.5%.

TABLE II

Production and estimated energy use by the pulp and paper industry, 1988 to 1993.

Year	Pulp & paper production[1] (Tg)	Electricity (PJ)	Heavy oil (PJ)	Distillates (PJ)	Natural gas (PJ)	Coal (PJ)	Other (PJ)
1988	24.652	138.1	76.4	4.9	88.7	4.9	0.0
1989	24.690	123.5	101.2	4.9	101.2	4.9	4.9
1990	23.697	123.2	90.0	4.7	106.6	4.7	7.1
1991	24.258	128.6	87.3	4.9	114.0	4.9	4.9
1992	24.672	128.3	66.6	2.5	128.3	2.5	9.9
1993	25.535	132.8	66.4	2.6	125.1	2.6	7.7

[1] Source: CPPA 1994

3.2. Parks Canada CO_2 Emissions

Total energy use estimates by fuel type for the Canadian Parks Service for the 1988 to 1991 period are shown in Table IV. Over the period of analysis, oil and propane use declined while natural gas and diesel use increased. The total estimated CO_2 emissions derived using the CO_2 emission factors (Table I) is shown in Table V. Total CO_2 emissions peaked in 1990, and 1991 levels were slightly below those of 1988. Figure 2 compares actual CO_2 emissions to a linear reduction trajectory necessary to meet the 20% reduction by 2005 target. The CO_2 emissions by Parks Canada exceeded the linear reduction trajectory by 9.5, 17.7, and 4.4% in 1989, 1990 and 1991 respectively.

A portion of the 1990 increase in estimated emissions may be due to improved capture of energy use data for buildings within the park system (Bullard, 1991). The 1991 decline in emissions may be due in part to changes in data compilation. The 1988 to 1990 estimates were pro-rated to account for energy use by non-reporting national

TABLE III

Estimated CO_2 emissions by the pulp and paper industry, 1988 to 1993 (Tg CO_2).

Year	Electricity	Heavy Oil	Distillates	Natural Gas	Coal	Other	TOTALS
1988	7.89	5.66	0.38	4.41	0.44	0.00	18.77
1989	7.06	7.49	0.38	5.03	0.44	0.30	20.70
1990	7.05	6.66	0.36	5.30	0.42	0.43	20.22
1991	7.35	6.46	0.37	5.66	0.43	0.29	20.58
1992	7.34	4.93	0.19	6.37	0.22	0.60	19.65
1993	7.59	4.91	0.20	6.22	0.23	0.46	19.61

parks and historic sites. Over this period, the magnitude of the pro-rating was between 8.62 and 13.06% of energy use for fixed assets, which represents between 7.72% and 10.53% of total energy use. The 1991 data were not pro-rated, even though energy use data for several parks and sites was not included in the compilation.

4. Discussion

No reduction in CO_2 emissions was detected from energy use by the pulp and paper industry relative to 1988 levels. Despite the increasing use of biofuels (the emissions from which were not included in this analysis) and the decrease in purchased energy use per Mg of pulp production, the magnitude of CO_2 contributions to the atmosphere, as a result of the consumption of electrical power and fossil fuels, did not decline.

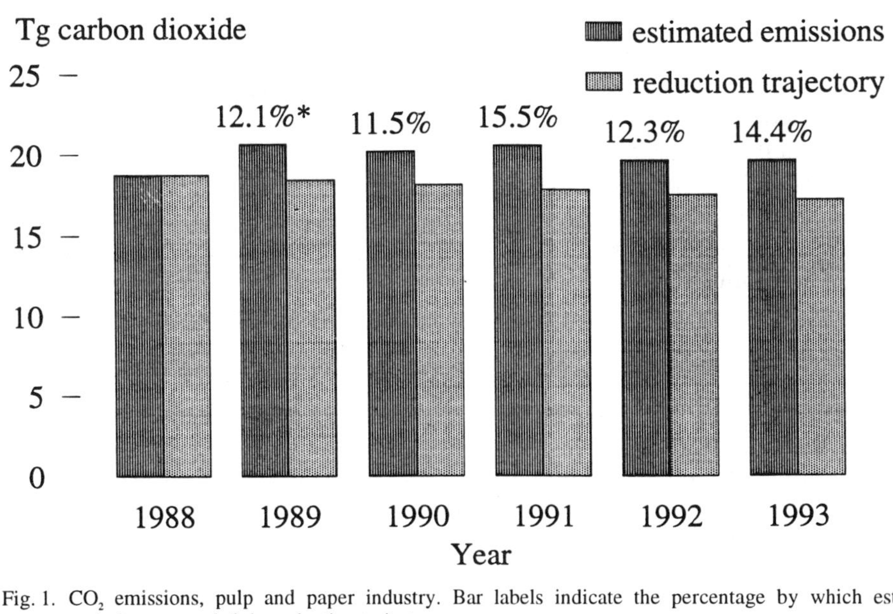

Fig. 1. CO_2 emissions, pulp and paper industry. Bar labels indicate the percentage by which estimated emissions exceeded the reduction trajectory.

TABLE IV

Estimated energy use by Parks Canada, 1988 to 1991 (TJ).

Year	Electricity	Heavy oil	Propane	Natural gas	Diesel	Gasoline	Jet fuel
1988[1]	183.03	74.64	102.61	95.39	67.37	21435	15.31
1989[1]	190.69	77.76	107.62	99.39	75.45	240.08	17.15
1990[2]	192.28	87.34	93.77	126.14	74.14	278.90	3.67
1991[3]	180.06	71.79	87.16	115.29	80.62	208.18	6.40

[1] Derived from Bullard 1990. [2]Derived from Bullard 1991. [3]Derived from Department of Environment 1993.

Schweig and Stewart (in prep.) estimated CO_2 emissions from electrical and fossil fuel energy use by the forest sector as a whole using Statistics Canada quarterly reports on energy supply and demand. These data also show that CO_2 emissions in 1989, 1992, and 1993 exceeded the 20% by 2005 linear reduction trajectory. Emissions data for 1990 and 1991 were derived using another method and were not comparable to the 1988 estimate.

No substantial reduction in CO_2 emissions was detected for Parks Canada over the 1988 to 1991 period. An apparent decline between 1990 and 1991 is due in part to the fact that the 1991 energy use estimate, unlike the that of previous years, was not pro-rated to account for non-reporting parks and sites. More recent energy consumption data is required to determine whether a decline in emissions has actually begun.

Monitoring mechanisms for energy use for fixed assets were more reliable than those for the transportation sector (Department of Environment, 1993). In some cases, transportation sector energy use was estimated from financial expenditures rather than from documented energy consumption levels. At most national parks and historic sites, energy monitoring has been ongoing throughout the period, but it has often emphasized performance assessments of individual buildings (kJ m^{-2} per degree day) and vehicles (litres per 100 km) rather than evaluation of total energy use.

Monitoring programs which measure relative rather than absolute amounts of energy use or CO_2 emissions are necessary, but may provide misleading results if analyzed in isolation. It is quite possible to reduce energy consumption per unit of pulp, per m^2 of building, or per 100 km travelled, while still increasing CO_2 emissions. Increased pulp production, larger buildings, more vehicles, and greater distances travelled can easily compensate for the decreases reported in monitoring programs for relative indicators. In the end, what is important to the atmosphere and the forest is the total magnitude of CO_2 emissions.

TABLE V

Estimated CO_2 emissions by Parks Canada, 1988 to 1991 (Gg CO_2)

Year	Electricity	Heavy oil	Propane	Natural gas	Diesel	Gasoline	Jet fuel	TOTAL
1988	10.47	5.52	6.22	4.74	4.76	14.57	1.08	47.36
1989	10.90	5.76	6.52	4.94	5.33	16.32	1.22	50.99
1990	11.00	6.46	5.68	6.27	5.24	18.96	0.26	53.87
1991	10.30	5.31	5.28	5.73	5.70	14.15	0.45	46.92

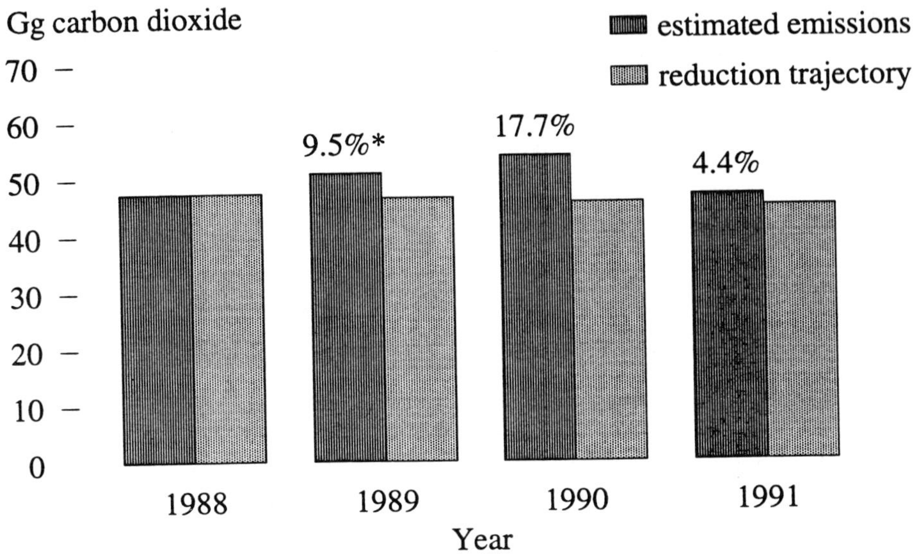

Fig. 2. CO_2 emissions, Parks Canada. Bar labels indicate the percentage by which estimated emissions exceeded the reduction trajectory.

This analysis accounted only for the direct CO_2 emissions of two forest management institutions. It did not include indirect emissions (e.g., CO_2 emitted in the production or transportation of fossil fuels prior to consumption). It also did not include other influences that these organizations could have on atmospheric composition through the regulation of emissions by other forest users (e.g., pulpwood harvesting contractors or park visitors) or the manipulation of soil and vegetation C pools. As such the analysis does not reveal the total magnitude of CO_2 emissions. Rather, the results focus initial attention on that component of total emissions which these institutions have greatest control over.

Robinson *et al.* (1993) found that, for Canadian society as a whole, emission reduction targets of the general magnitude of those recommended by the Toronto Conference on the Changing Atmosphere were both feasible and cost-effective, and that they could be achieved through energy efficiency, and to a lesser degree, fuel switching. Robinson *et al.* (1993) also reported that some institutions such as municipalities have adopted the 20% reduction by 2005 target.

5. Conclusions

If any Canadian institutions should be models of CO_2 emission reductions, it should be organizations responsible for management of boreal forests—organizations which are as threatened by global warming as the boreal forest itself. Even though the direct contribution of these institutions to changes in the composition of the atmosphere may be minimal on a global scale, a demonstration of the feasibility of emission reductions could serve as a much needed catalyst and promote similar change in other institutions and other sectors of society.

Over the three to five years for which data was available, neither of the forest management institutions studied was able to demonstrate reductions in CO_2 emissions at the rate required to reach the 20% reduction by 2005 target. If these institutions are not atypical, there is little evidence that atmospheric considerations have been fully integrated into the activities of organizations responsible for forest management in Canada. Ecosystem approaches—much touted by a variety of institutions—have not yet incorporated a crucially important component of forest systems.

At present there is a difference of several orders of magnitude between what is known about the relationship between atmospheric emissions and forests, and what is done about reducing air pollution (Woodwell, 1989). Further studies of forest management institutions are required to ascertain why CO_2 emissions are not being reduced at a more rapid rate.

Forest management institutions could reduce their CO_2 emissions by placing as much emphasis on the management of air as they place on the management of trees. However, this symbolic act would not be enough to successfully perpetuate the boreal forest over the long-term. Perhaps Canadian forest management organizations should be advocating greenhouse gas emission reductions around the world with the same intensity and determination that Richard St. Barbe Baker championed global efforts to plant trees.

Acknowledgments

I would like to acknowledge the assistance of the following individuals: Bruce Bullard, Parks Canada; Mike Frost, Canadian Pulp and Paper Association; Michel Gareau, City of Ottawa, formerly Friends of the Earth Canada; and Glenn Gustafson, Meewasin Valley Authority.

References

Baker, R. S. B.: 1944, *I Planted Trees*, Lutterworth, London, UK, p. 244.

Baker, R. S. B.: 1970, *My Life, My Trees*, Lutterworth, London, UK, p. 98.

Bonan, G. B., Shugart, H. H., and Urban, D. L.: 1990, *Climatic Change* **16**, 9-29.

Bullard, B.: 1990, *Canadian Parks Service, A&E Services, Internal Energy Report 1989/90*, Canadian Parks Service, Ottawa, Canada, 146 pp.

Bullard, B.: 1991, *Canadian Parks Service, A&E Services, Internal Energy Report 1990/90*, Canadian Parks Service, Ottawa, Canada, 198 pp.

Canadian Heritage: 1994, *Parks Canada: Guiding Principles and Operational Policies*, Supply and Services Canada, Ottawa, Canada, 125 pp.

Canadian Pulp and Paper Association (CPPA): 1993, *The Pulp and Paper Industry of Canada: Environmental Statement*, CPPA, Montreal, Canada, 1 p.

Canadian Pulp and Paper Association (CPPA): 1994, *Reference Tables*, CPPA, Montreal, Canada, 54 pp.

Department of Environment: 1993, *Annual Energy Management and Consumption Report, 1991-92: Canadian Parks Service Supplement*, Environment Canada, Ottawa, Canada, 128 pp.

Emanuel, W. R., Shugart, H. H. and Stevenson, M. P.: 1985, *Climatic Change* **7**, 29-43.

Environment Canada: 1992, *Canada's Greenhouse Gas Emissions: Estimates for 1990*, Environment Canada, Ottawa, Canada, 78 pp.

Energy, Mines and Resources Canada: 1991, *Electric Power in Canada 1990*, Supply and Services Canada, Ottawa, Canada, 162 pp.

Harvey, L. D. D.: 1989, *Climatic Change* **15**, 343-381.

Harrington, J. B.: 1987, *Can. J. For. Res.* **17**, 1313-1339.

Houghton, J. T., Jenkins, G. J. and Ephraums, J. J. (eds.): 1990, *Climate Change: The IPCC Scientific Assessment*, Cambridge University Press, Cambridge, UK, 365 pp.

Jarvis, P. G.: 1989, *Phil. Trans.* Roy. *Soc. Lond.* **B 324**, 369-392.

Melillo, J. M., Callaghan, T. V., Woodward, F. I., Salati, E. and Sinha, S. K.: 1990, *In:* Houghton J. T., Jenkins, G. J. and Ephraums, J. J. (eds.). *Climate Change: The IPCC Scientific Assessment*, Cambridge University Press, Cambridge, UK, pp. 287-310.

Monserud, R. A., Tchebakova, N. M. and Leemans, R.: 1993, *Climatic Chang*e **25**, 59-83.

Musselman, R. C. and Fox, D. G.: 1991, *J. Air Waste Manage. Assoc.* **41**, 798-807.

Robinson, J., Fraser, M., Haites, E., Harvey, D., Jaccard, M., Reinsch, A. and Torrie, R.: 1993, *Canadian options for greenhouse gas emission reduction*, Canadian Global Change Program Technical Report Series No. 93-1, Roy. Soc. Can., Ottawa, Canada, 40 pp.

Rizzo B. and Wiken E.: 1992, *Climatic Change* **21**, 37-55.

Schweig, G., and R. B. Stewart: *Inventory of Energy Use and Carbon Dioxide Emissions by the Canadian Forest Sector 1980-1993*, Natural Resources Canada, Ottawa, Canada, (Unpublished manuscript.)

Shugart, H. H., Antonovsky, M.Y., Jarvis, P. G. and Sandford, A. P.: 1986, *In:* Bolin, B., Döös, B. R., Jäger, J., and Warrick, R. A. (eds.). *The Greenhouse Effect, Climatic Change and Ecosystems*, Wiley, Chichester, UK, pp. 475-521.

Singh, T. and Wheaton, E. E: 1991, *For. Chron.* **67**, 342-348.

Solomon, A. M: 1992, *In:* Shugart, H. H., Leemans, R., and Bonan, G. B. (eds.). *A Systems Analysis of the Global Boreal Forest*, Cambridge University Press, Cambridge, UK, pp. 291-312.

Toronto Conference on the Changing Atmosphere (TCCA): 1988, *The Changing Atmosphere: Implications for Global Security*, Toronto, Canada, 12 pp.

Woodwell, G. M.: 1989, *Ecology* **70**, 14-15.

FOREST OPERATIONS AND ENVIRONMENTAL PROTECTION

A. M. FURUBERG GJEDTJERNET

Norwegian Forest Research Institute, Høgskoleveien 12, N-1432 Ås, Norway,
IUFRO P 3.08, International Union of Forestry Research Organizations

Abstract. Increased mechanisation in forestry usually means increased traffic over the ground, which also is a substrate for growth. Soil compaction, rut formation, and damage to stems and roots may all result in growth reduction. To reduce soil disturbance and tree damage, smaller machines or machines with less ground pressure must be considered. Today the traffic may cover 10–30% of the site area. This is mainly due to skidding during clear-felling and thinning. If present trends of heavy mechanisation continue, in the future there could be as many as six machine entries onto the site by the time a stand has reached an age of 50 years. Rut formation from forwarders and skidders is a problem. Harvesters seem to cause minor problems. The engine torque powering the driving wheels adds considerably to the influence of the total weight of the vehicle on the ground. High traction forces and slipping may contribute twice as much to soil disturbance as the vehicle ground pressure. When designing forest machinery for transportation of timber the payload itself is the most complicated problem. During the XIX World Congress of the International Union of Forestry Research Organisations (IUFRO) in Montreal, Canada, 1990, a Project Group on Forest Operations and Environmental Protection was established within IUFRO Division 3. From the author's point of view, concern about biodiversity is not taken into account through single-track and profit-oriented forestry in some areas and total preservation in other areas. It is the manifold of different ways of forestry that is the answer to sustainable development and conservation of biodiversity.

Keywords. SOIL COMPACTION, DAMAGE, FOREST OPERATIONS, ENVIRONMENTAL PROTECTION, PAYLOAD, DESIGN, BIODIVERSITY.

1. Introduction

During the XIX World Congress of International Union of Forestry Research Organizations (IUFRO) in Montreal, Canada, 1990, a Project Group on Forest Operations and Environmental Protection was established within IUFRO Division 3. The Project Group has three Working Parties:

1. Site impact caused by forestry operations.
2. Forest operations on sensitive sites.
3. Methods and techniques for site protection, improvement and amelioration.

The research work gives guidelines to the development of new machines: environmentally friendly forest machines. We measure site impact caused by forestry operations in practical work, and we have established research fields carrying out tests under controlled conditions. The Group deals with both basic and applied research. In addition, we incorporate active research concerning forest operations and environmental protection.

This paper presents short summaries from parts of the work of the Project Group.

2. Extent and Causes of Site Damage Due to Forestry Traffic

Dr. Iwan Wästerlund is the chairman of Working Party No. 1: Site impact caused by forestry operations. On the following pages a literature survey of Dr. Wästerlund is cited. (Wästerlund, 1988, 1989a,b).

Water, Air and Soil Pollution **82**: 35–41, 1995.

2.1. EFFECT OF SOIL AND ROOT DAMAGE

In areas with warm climate and very slow recovery (results from Oregon, USA), soil compaction after clear-felling operations could persist during the whole next rotation if no amelioration is achieved during site preparation. The reduced productivity on skid trails could reduce the total volume yield 6–16% for perhaps as much as the whole second rotation (Froelich and McNabb, 1984; Stewart *et al.*, 1988).

Root and soil damage after thinning operations could reduce the total volume production by up to 14%, and the wide openings for the strip road may give an additional loss of 6% (Wästerlund, 1990). In cold climates (e.g., Sweden) a recovery might appear within 5–10 years and the figures for growth loss after thinning is given as an average for the next 5 years after the operation.

Data concerning the effects of soil and root disturbance on forest production appear to be quite restricted to North American and Swedish conditions. Rate of recovery, effects on different sites, soil types and tree species need to be assessed much better, preferably in different climate zones as well. However, most urgent is the need for better knowledge about the relationship between the degree of soil disturbance (rut formation and compaction) and effects in the stand. A research programme was started in Norway in 1991, but it is too early to present any results.

The North American productivity loss data is represented by an average of many studies whereas the Swedish figures are more of upper limit from few experiments. Perhaps soil compaction is a more serious problem in warm climates, or could the difference in operational systems also be a factor?

Calculations show that US\$100–200 ha^{-1} could be a justifiable increase in operational costs with considerably reduced damage to roots and soil (Wästerlund, 1986; Stewart *et al.*, 1988). The amount is too small to justify skyline-yarding or aerial systems but enough to promote considerable improvement of existing systems in forest operations.

2.2. CAUSES OF DAMAGE

2.2.1. *Soil compaction*

There is a large number of papers dealing with soil compaction (Lull, 1959; Greacen and Sands, 1980; Froelich and McNabb, 1984). Very few papers deal with the whole chain of connections: machine → weight → ground pressure → traction → soil compaction. Often only a few links in the chain are well measured and the others are excluded or poorly described. There are a few exceptions, for example Dexter *et al.* (1988) and Jakobsen and Dexter (1989). However, they deal with agricultural soils without vegetation which may have different properties than the forest floor; nevertheless their papers can serve as a good basis for ideas.

2.2.2. *Machine weight and ground pressure*

A forestry machine may have a weight of 10–30 tonnes. The soil has to carry that weight at the contact areas. The total weight is shown to influence the degree of soil compaction at depth, whereas the ground pressure determines it at the surface (Danfors, 1977). On peat soil, the total weight could be the most important factor for rut formation (Wong *et al.*, 1979). Unfortunately, many of the experiments done with forestry machines contain too little variation in weight and ground pressure or contain comparisons between different actions of the machine, to be of any help for predictive models.

Rut formation after forwarder traffic has been studied both in Finland and in Sweden (Siren, 1986; Myhrman, 1990). Generally, the calculated average ground pressure of the machines, given by

$$Pressure = \frac{Load}{Tire\ radius \times Width} \qquad [1]$$

explains to a large extent the ranking between the machines in accordance with created rut depths. However, the studies are mainly comparative and contain no analyses of the deviations from the general trend in the data.

Ground pressure can be reduced by increasing the tire diameter and/or increasing the width, and to some extent by reducing the inflation pressure. Increased number of wheels, preferably in a tandem system for good traction performance (Bailey and Burt, 1981) can also be considered. Aggressive lug pattern and high inflation pressure will give high peak forces (Trabbic *et al.*, 1959). Lugs more than 20–30 mm in height are more likely to cause an increase in rolling resistance than increase the traction performance (Dwyer, 1984). Canadian experiments with high flotation tires indicate good results, especially on wet terrain, but the data are inadequate for a good evaluation (Mellgren and Heidersdorf, 1984). However, a machine with 1.5 m wide tires would require too much space for European-type thinning operations.

Wästerlund (1989a,b) has proposed a ground pressure of 30–50 kPa for an acceptable level of soil disturbance. This figure is calculated as the wheel load divided by the radius times the width of the tire (Eq. [1]). That formula underestimates the real mean ground pressure on medium to hard soils, where the pressure in reality would be 60–70 kPa. According to Dexter *et al.* (1988) the maximum pressure at the surface could be 1.6 times the mean pressure, which means about 100 kPa in this case. The suggested desirable mean ground pressure is lower than that proposed for agriculture (Taylor and Gill, 1984) but can be easily achieved with 5-tonne machines.

2.2.3. Number of trips and slippage

In a number of studies, rut formation, soil compaction, and water infiltration rate are found to be influenced by the number of trips. Often the first trip contributes more than 50% of the damage compared to the next 9 or 19 trips. This is taken as a justification for having more-or-less permanent roads on the site. This is questionable, because it implies that all machines are bad. Is it better to have completely disturbed areas at 25 m spacings than to have many slightly disturbed areas on many patches?

Wheeled machines commonly have a rather simple type of mechanical transmission without differentials. Often all wheels need to be driven and the wheels are set to travel the same distance. Curve driving or different tire size due to wear, inflation pressure or load tries to make the wheels travel different distances. This is prohibited by the "rigid" transmission and at least one of the wheels has to slip. The restraint in the transmission could be up to twice the torque needed to propel the machine (Wästerlund, 1989a,b). This transmission induced slip can occur even at low traction forces. According to Koger *et al.* (1982), slip may have a greater influence on soil compaction than tire size or inflation pressure. The usual operating value of 10–20% slip may be in the range where maximum compaction occurs (Raghavan *et al.*, 1977) (Wästerlund, 1989a,b).

3. Nature Conservation in Practical Forestry

A study of nature conservation in practical forestry was carried out by the Forest Operations Institute in Sweden. Great emphasis was placed on highlighting the costs of conservation, with the value of the residual timber being found to be the most significant cost. A rule-of-thumb for this cost is that if 5% of the volume is saved the cost will be equivalent to approximately 4% of the net conversion revenue. The significance of other cost items is such that their influence on this figure is no more than a few tenths of one percentage point.

For nature conservation to work in practice, there are four essential ingredients: Motivation, operational objectives, good planning and thorough documentation (Aldertun and Sondell, 1991).

4. Taking the Environment into Account When Designing Forest Machinery

The forwarder is clearly more problematic than the harvester as far as the environment is concerned.

When designing a machine for transporting timber, it is easy to designate the most complicated problem of the process: it is the payload itself (Gjedtjernet, 1992).

The rapid technological progress of machine construction materials, hydraulic systems, and electronic systems has given new freedom to those who design forest machines today; a lot of attention can be paid to the requirements of the environment even in the early stages of the design process (Taatila, 1994).

In Norway we are carrying out research work in using cable cranes on wet ground, ground with low carrying capacity, not only in steep terrain but also on flat, easy terrain. The limitation of use in practical forestry is the operational costs. In New Zealand, however, Terlesk (1987) found that cable crane systems were both economical and environmentally friendly. Brown and Jemison (1988) propose that further development of effective cable crane systems will be the answer for new environmentally friendly forest machines. Still, in steep terrain, the problem of erosion must be taken into account.

5. Action Research Concerning Forest Operations

In nature resource management two phrases are central: *use of nature* and *environmental protection*. What is the meaning of these phrases concerning forestry? There is dissension between different groups of forest users in a country and between countries throughout the world. The essence of the disagreement is: where is the place and position of man - *inside* or *outside* nature?

In my point of view, biodiversity is not taken into account through single-track and profit-oriented forestry in some areas and total preservation (no use by man) in other areas. It is the manifold of different ways of forestry that is the answer to sustainable development and conservation of biodiversity.

In discussions of utilisation and ecosystem perspectives of forests in the world, very seldom are the needs of local communities mentioned. When the legally binding forest convention was adopted in Rio, it was only because of India's insistence that the importance of forests for local communities came to be incorporated in the forest principles.

6. Recommendations—Soil, Tree, Machine Interactions

The IUFRO Working Party P3.08.01 "Site impact caused by forestry operations" developed the following recommendations (proposals) during the IUFRO/FAO/ECE/ILO seminar and workshop held in Feldafing, Germany, 4–8 July, 1994.

6.1. Recommendations (proposals)

6.1.1. Soil disturbance in general

1. Soil disturbance due to forest operations is a world-wide serious problem which may escalate with increased mechanisation.
2. Avoidance of soil damage is the best option and should receive the highest priority (instead of applying soil restoration as an "end-of-pipe" method).

6.1.2. Research in general

1. More insights into economic and socio-economic values and impacts of soil disturbance are needed to gain information to justify research activities in this field.
2. In soil disturbance research some areas, like physical parameters, are often measured but there exist a number of areas like soil chemistry and ecosystem changes, which we know very little about. Thus more networking and interdisciplinary research are needed, e.g., between biological and engineering scientists.
3. The influence of soil disturbance on tree growth is partially known for certain areas and tree species but a general understanding of the process and parameters is still missing.
4. Since these linkages are missing, it is difficult to advance generally proven threshold values as recommendations or rules-of-thumb.
5. It will take a long time to gather this information and meanwhile we have to rely on expert guesses.
6. Extension of present knowledge about impacts of soil disturbance towards manufacturers, forestry practice and public should be improved.
7. The final objective of the research efforts should lead to handy practical tools to assist practitioners in socially acceptable, environmentally friendly and economically feasible timber production.

6.1.3. Research methods in detail

Methods can be divided into four groups:

1. Methods which are already standardised, such as dry density measurements by means of core samples, should be kept.
2. Methods which are commonly known and used, such as cone penetration measurements, for which detailed recommendations were established during the workshop.
3. Methods which already exist, but further research is needed in order to establish standardised procedures (conductivity measurements etc.)
4. Methods to be developed due to missing apparatus or missing analytical approach (e.g., aeration)

A working group of specialists in the field of analysis should be set up in order to complete the list of parameters needed and to develop guidelines and approaches for items 3 and 4.

6.1.4. Tools and machinery in forest operations

1. The emphasis in short-term design should be adaptation and modification of existing technologies and techniques.
2. The design changes should be based on new, scientifically based, biological results.
3. In the long-term, new machine concepts will provide improvement in management practices.
4. Short-comings in training, supervision and planning have to be identified and overcome with continuous input of biological results.
5. Proper planning, assessment and audit will lead to sound implementation.
6. Awareness campaigns, training and educational programs should follow a holistic approach in order to improve the skills and public response.

6.1.5. Soil restoration

1. Soil restoration must be site-specific and sustainable.
2. Clear goals must be established for soil restoration treatments. Standards must be set (pre-treatment) to assess the success of the treatment.
3. General statements or quantification of when soil restoration is required must be developed. Known thresholds (physical, chemical, biological) should be listed and information gaps identified (these should receive priority for research). This will lead to site specific refinements.
4. Research into soil restoration must simultaneously consider physical, chemical and biological (flora and fauna) aspects.

7. Conclusions

Environmentally friendly technology and methods nearly always result in more expensive forest operations. Thus, is it necessary to improve and develop political and economical means and measures to meet future environmental challenges?

References

Aldertun, Y. and Sondell, J.: 1991, Study of Nature consideration in practical forestry—problems and possibilities. *Redogörelse* No. **1**. Skogfors, Sweden.

Bailey, A. C. and Burt, E. C.: 1981, *Trans. ASAE* **24**(5), 1103-1107.

Brown, G. W. and Jemison, G. M.: 1988, Timber harvest and environmental protection (also ind. cable yarding systems). Paper in honour of Prof. Ivar Samset, Norwegian Forest Research Institute, Ås, Norway.

Danfors, B.: 1977, Soil compaction-wheel equipment, Swedish Inst. of Agric. Eng., *Meddelande,* Uppsala, Sweden, 368 pp. [In Swedish, with English summary]

Dexter, A. R., Horn, R., Holloway, R. and Jakobsen, B.F.: 1988, *J. Terramechanics* **25**(2), 135-147.

Dwyer, M. J.: 1984, *J. Terramechanics* **21**, 19-34.

Froelich, H. A. and McNabb, D. H.: 1984, *In:* Stone E. L. (ed.), Forest soils and treatment impacts, Proc. Sixth North American Forest Soils Conference, Univ. Tenn, Knoxville, USA, June 1983, pp. 159-192

Gjedtjernet, A. M.: 1992, Environmentally friendly forest machines, Report/Stencil. Norwegian Forest Research Institute, Ås, Norway.

Greacen, E. L. and Sands, R.: 1980, *Aust. J. Soil Res.* **18**, 163-189.

Jakobsen, B. F. and Dexter, A. R.: 1989, *J. Terramechanics* **26**(2), 107-119.

Koger, J. L., Trouse, A. C. Jr., Burt, E. C. and Bailey, A. C.: 1982, *ASAE Pap.* No. **82**, 1595 pp.

Lull, H. W.: 1959, Soil compaction of forest and range lands. U.S.D.A. Misc. publ., USDA For. Serv., Washington D.C., USA.

Mellgren, P. G. and Heidersdorf, E.: 1984, The use of high flotation tires for skidding in wet and/or steep terrain, *FERIC Technical Rep.,* TR-**57**.

Myhrman, G.: 1982, Factors influencing rut formation from forestry machines, Proc. 10th Intl. Conf. ISTVS, Kobe, Japan, pp. 467-475.

Raghavan, G. S. V., McKyes, E. and Chassé, M.: 1977, *J. Agric. Eng. Res.* **22**, 79-83.

Sirén, M.: 1986, Terrängdugligheten för maskiner som lämper sig för gallring, I: Gallringsteknik i klen skog. University of Helsinki, Dept. Logging and Utilization of Forest Products, *Res. Note.* No. **47**, Helsinki, Finland, pp. 57-66. [In Swedish, with English summary]

Stewart, R., Froelich, H. and Olsen, E.: 1988, *West. J. Appl. For.* **3**(1), 20-22.

Taatila, S.: 1994, Taking the environment into account when designing forest machinery. Presented at seminar on Soil, Tree, Machine Interactions, July 4-8, Feldafing, Germany. (Unpublished manuscript).

Taylor, J. H. and Gill, W. R.: 1984, *J. Terramechanics* **21**(2), 195-213.

Terlesk, C. J.: 1987, Harvesting machines and systems evaluations: environmental considerations. SLU, Garpenberg, Sweden. *Uppsatser och resultat* No. **108**, 177-186.

Trabbic, G. W., Lask, K.V. and Buchele, W. E.: 1959, *Agric. Eng.* **40**, 678-681.

Wong, J. Y., Garber, M., Radforth, J. R. and Dowell, J. T.: 1979, *J. Terramechanics* **16**(4), 163-180.

Wästerlund, I.: 1986, *In:* Knutell, H. (ed.) *New Thoughts about Strip Roads, Damage and Techniques in Thinning Operations.* Swedish Univ. Agric. Sci., Dept. of Operational Efficiency, *Res. Note.* No. **52**, Garpenberg, Sweden, pp. 56-63. [In Swedish, with English summary]

Wästerlund, I.: 1988, *Scand. J. For. Res.* **3**, 259-272.

Wästerlund, I.: 1989a, Are present forestry machines smooth terrain-machines? SFM, Stockholm, Sweden, Meddelande Nr., **38**, pp. 21-34.

Wästerlund, I.: 1989b, *In: Impact of Mechanization of Forest Operations on the Soil,* Proc. ECE/FAO/ILO Seminar, Louvain-la-Neuve, Belgium, pp. 165-175.

Wästerlund, I.: 1990, Soil strength in forestry measured with a new kind of test rig. Proc. ISTVS 10th Intl. Conf., Kobe, Japan. 73-82 pp.

APPLICATION OF A BIOECONOMIC STRATEGIC PLANNING MODEL TO AN INDUSTRIAL FOREST IN SASKATCHEWAN

R. R. STEWART[1] and M. MARTEL[2]

[1] *Terrestrial and Aquatic Environmental Managers Ltd., Box 2290, Meadow Lake, Saskatchewan S0M 1V0*
[2] *Mistik Management Ltd., Meadow Lake, Saskatchewan S0M 1V0, Canada*

Abstract. Contemporary approaches to the issue of the allocation, conservation and management of forested land require forest managers to demonstrate the sustainability of the flow of economic, ecological and social benefits accruing from industrial use of forest resources. The integration of data sets defining these disparate resource benefits requires that they be expressed in commensurate terms such that trade-offs can be undertaken in the search for a management strategy which provides a publicly acceptable mix of benefits. To accomplish this, a new optimisation model was developed and tested for a 1.7 Mha commercial forest in west central Saskatchewan, and then used to develop 8 different forest management alternatives which were then subjected to formal evaluation procedures leading to the selection of a preferred forest management alternative.

Keywords. OPTIMISATION, FOREST MODELLING, TRADE-OFFS, INDUSTRIAL FORESTRY, LINEAR PROGRAMMING, ECONOMICS, ECOSYSTEM, BENEFIT DISTRIBUTION.

1. Introduction

A functional optimisation model for forest management activities was developed as a tool to assist Mistik Management Ltd. (Mistik) of Meadow Lake, Saskatchewan, Canada in the preparation of a Twenty Year Forest Management Plan and Environmental Impact Statement (EIS) of that plan. Project Specific Guidelines provided by Saskatchewan Environment and Resource Management (SERM, 1992) to guide Mistik in the development of the plan and EIS, emphasised a need to address long-term sustainability of the ecological, economic and social environments through the application of integrated resource management.

The requirement to address the ecological, economic and social components of the environment as part of this analysis highlighted the need for application of some form of a bioeconomic model which would permit the efficient generation of alternative forest management plans. In addition, each alternative must provide an optimum set of forest management activities to ensure that a comparison of alternatives was not obscured by uncertainty related to varying levels of management efficiency. The need for optimisation combined with the objective of this project to search for an alternative which maximises the net benefits to society from the use of forest resources led to the development of the Mistik Forest Management Model (MFMM).

This paper discusses the project requirements, the rationale for the selection and development of the MFMM, the approach to the definition of the model inputs on which net benefits analysis was based, and the analysis and the evaluation management activities to compare the alternative sets of management actions leading to the selection of a preferred forest management plan.

Water, Air and Soil Pollution **82**: 43–55, 1995.

2. Project Setting

The NorSask Forest Management License Agreement (FMLA) core on which this project is based encompasses 1.7 Mha and is situated predominantly within the Southern Boreal ecoregion. It encompasses portions of three ecodistricts, the Mixedwood, Upper Churchill, and Mixedwood-Parkland Transition zones. The Mixedwood ecodistrict is characterised by a mixture of the dominant boreal tree species (Rowe, 1972), and is the most widespread ecodistrict. The Upper Churchill ecodistrict encompasses the Northeast and northern fringes of the FMLA core. The pioneering or early successional boreal tree species such as jack pine, trembling aspen, and black spruce are the most prevalent species, occurring in both pure and mixedwood associations in the Upper Churchill (Harris *et al.*, 1983).

Planning within the FMLA will take place at both the regional and local levels. Strictly speaking, the regional planning context includes the entire FMLA. However, most of the planning takes place at a sub-regional level. Twelve sub-regions or Forest Management Units (FMUs) have been selected by aggregating parts or all of 25 Fur Conservation Areas (FCAs). FCAs are long-standing administrative districts formed in the 1940s for fur management, but which have taken on territorial significance to local peoples. To overcome some problems encountered with model dimensionality[1], some FMUs were further aggregated into larger planning units (PUs).

Approximately 23 000 people reside within the FMLA. Meadow Lake is the largest community providing services to the regional agricultural and forest-related economies and serves as the centre for regional, provincial and Indian governments. Major northern centres include Green Lake, Beauval, Buffalo Narrows, Île à-la-Crosse and La Loche, with numerous smaller villages throughout the region. Approximately 90% of the residents in these northern communities are of aboriginal descent, while approximately 100% of the population of the 10 Indian Reserves within or adjacent to the FMLA are Treaty Indian. Aboriginal residency within the agricultural settlement lands to the south of Meadow Lake varies from about 7% of the population of the rural municipalities, to about 22% of the population in the southern towns, villages and hamlets. In total, persons of aboriginal descent account for about 50% of the population of the FMLA.

Two major milling facilities rely on the FMLA landbase for fibre resources. The NorSask sawmill established in 1988 has a design capacity to mill 514 000 m^3 yr^{-1} but currently utilises approximately 300 000 m^3 of spruce and jack pine per year, while the Millar Western pulp mill which commenced operations in 1992, has a capacity for 700 000 m^3 of aspen per year. Most of the mills' requirements are secured from the NorSask FMLA crown forest landbase.

Public forest lands are likely the most complex land base on which to develop a comprehensive management plan, given that forest management planning must address (1) when, (2) where, and (3) how forest harvesting operations are to be implemented to best satisfy the varied public interests. Many members of the public have a sense of ownership of the Crown forest, and for some groups and individuals this sense of ownership is particularly deep rooted. This sense of ownership is well entrenched

[1]*dimensionality* - the size of the model matrix in terms of the number of variables, the number of elements defining each variable, and the number of permutations and combinations which are generated for analysis.

within the NorSask FMLA area given that many local residents rely either directly or indirectly on the forest for part of all of their livelihood.

The opening of the pulp mill in 1992 resulted in a significant increase in logging activity and changes in harvest technologies; for some residents, these events have been a difficult adjustment. The initiation of a road blockade protesting certain forest management practices in May 1992 along Highway #903 north of Meadow Lake, catapulted forest management operations in Saskatchewan into the national and international spotlight. Most of the concerns voiced during the blockade related to the issues of clearcut logging, use of certain harvesting technologies, and the impacts of harvesting on traditional lifestyles and future opportunities. All of these issues are underscored by a sincere desire of local groups and individuals to have more input into the planning and decision-making processes and to retain some sense of control over matters which affect individual lifestyles and livelihoods, and the structure and function of the communities in which they live.

3. Planning Requirements

The forest management plan and environmental assessment of that plan required the development of an integrated resource management approach which clearly stated the broad ecological, economic and social merits of a preferred alternative compared to other forest management alternatives which were considered less suitable.

The infinite array of potential ways in which management activities can be applied to the forest create a major challenge for the identification of candidate management alternatives from which a preferred plan can ultimately evolve. The problem is further compounded by the ecological complexity of the forest which collectively provides habitat for literally thousands of different species of plant and animal life, the levels of supply of which are inextricably linked to the forest structure[2] and the ecological processes therein. In other words, there is a jointness in the production of forest resources, the levels of which are determined by the characteristics of the forest structure. Prior to the industrial exploitation of the forest, these levels of supply were largely controlled by the natural disturbance processes of the forest, i.e., natural ageing, fire, insects and disease, blowdown, flooding, etc.

Industrial logging adds a new dimension to the disturbance regime of the forest, affording forest managers the opportunity to purposefully modify the structure of the forest through access, harvest, renewal, tending and protection activities. The successional response of the forest to the spatial and temporal distribution of both natural and planned disturbances by and large defines the suitability of the forest to support both timber and non-timber resources. For example, extensive fire or logging activities within mid- to late-successional jack pine stands on rapidly drained sandy soils would not only result in a reduction in area of this forest community, but would also dramatically diminish woodland caribou habitat and populations. The manager must be careful not to assume that logging replicates natural disturbance. The

[2]*forest structure* - characteristics of the forest in terms of species composition, age class distribution and site characteristics of a stand of trees.

regenerative response to each of fire and logging disturbance are expected to result in different forecasts of future supplies of both sawlogs and woodland caribou.

The recognition and acceptance of the simple principle that a link exists between forest structure and resource supply is fundamental to forest management planning, and obviously, models are essential to allow the forest manager to preview the predicted future effects of various forest management activities. However, for practical management, the infinite array of potential forest management alternatives must be reduced through project scoping. Three planning steps are crucial to this process:

1. Identify the resource features to be managed. The forest manager must define what key resource features are to be managed. These can be separated into forest-level[3] and site-level[4] features. The location and abundance of site features can be mapped and are not expected to change significantly over time; conversely, the supply of forest-level features are linked to the forest structure and migrate across the landscape over time. Obviously, wood supplies are a forest-level feature which are critical to mill operations. The identification of what non-timber resource features to manage is largely achieved through consultation efforts.
2. Establish the design principles on which the sets of alternative management plans are to be based. Design principles may be thought of as the guides within which each of the forest management alternatives will be developed. For example, one extreme may be to not harvest the forest while another may be to maximise the economic benefits of timber harvest; yet another may be to emphasise the maintenance of the theoretical biological carrying of the forest for woodland caribou populations.
3. Define the criteria on which the alternate sets of forest management plans will be evaluated. Examples of evaluation criteria include the levels of ecological and economic risk associated with the implementation of each of the alternative sets of management activities.

To be meaningful, the analysis and documentation on which the description of each alternative is based should meet the following principles:(1) technically sound, traceable and reproducible; (2) comprehensive in scope; (3) comparable, and (4) understandable.

An approach based on the integration of publicly held values, feelings, beliefs and prejudices using measurable objectives guided the development of this project. To achieve this, scientific methods must be applied with the intent to find a publicly acceptable forest management plan which minimises the conflicts and confrontations which have characterised forest management in the past. These considerations, combined with the need for a timely assessment, served as the criteria on which the selection of a forest resource management model was based.

---

[3]*forest-level features* - those forest resource features whose welfare largely depends on the broad forest structure. Forest-level features tend to be those which can migrate across the landscape over time. For example, moose population distribution is expected to shift across the landscape over time in concert with changes in the suitability of habitat.

4site-level features - those forest resources which tend to be fixed geographically on the landscape i.e., features like heritage resources and salt licks.

4. Model Evaluation and Selection

Two basic types of models, simulation and optimisation models, were evaluated for their suitability to address the previously discussed planning issues. With simulation models the forest manager searches for an optimum solution by sequentially adjusting management prescriptions and rejecting undesirable results. This is a laborious data laden exercise, and is plagued by the fact that the forest manager never knows when an 'optimum' solution has been achieved. Conversely, optimisation prescribes the 'when, where, how and what' management prescriptions, to satisfy the maximum or minimum value of a single objective function bounded by a set of management constraints. In this fashion, the cause-effect relationships[5] among all of the components of the forest are described mathematically. For example, the maintenance of woodland caribou habitat at a level to support 90% of the theoretical carrying capacity is a constraint on the objective function that must be achieved in defining the set of forest management activities. In optimisation, every possible combination of activities are searched, and the most efficient possible outcome is determined.

Three optimisation models built on linear programming frameworks, the New Brunswick Woodstock model, the Ontario Strategic Forest Management Model and the U.S. Forest Service FORPLAN model were considered, but various logistical and technical constraints prevented their practical application to this project which led to a decision to develop the Mistik Forest Management Model to assist in the preparation of the forest management plan and EIS.

The MFMM is an optimisation model built on the GAMS linear programming framework using matrix algebra procedures to search for all potential feasible forest structures to progressively converge on the optimum set of forest management activities to maximise the value of the objective function. The objective of the MFMM is therefore, to produce the optimum balance of net forest benefits for the least cost.

5. Definition of Alternatives

The public provided the key input to the identification of the resource features to be managed as part of the forest management plan. This resulted in the identification of (1) pulpwood and sawlog wood supply, (2) wildlife harvest supply of moose, white-tailed deer and fishers, (3) woodland caribou population supply, and (4) harvest supply of blueberries for each management unit as candidate resource features to be managed. Estimation procedures were required to determine net economic values of all resource features, as for the most part, these products are not traded openly in the marketplace. For woodland caribou, insufficient information existed on which to apply any formal economic valuation procedures and as a result, this species was excluded from the objective function. In the case of blueberries, a market does not exist beyond a maximum consumption threshold of 200 000 kg, thereby violating a critical

[5]*cause-effect relationships* - interactions between one or more components in the living and non-living environment. The concept embraces both naturally occurring interactions such as the effects of predator numbers on prey populations and the effects in which humans play a role.

assumption of linearity; as a result blueberry yields were also excluded from the net benefits analysis in the MFMM.

A number of constraints were also identified including: (1) mill capacities of 857 000 m^3 of aspen hardwood for pulp and 569 000 m^3 of softwood logs for lumber; (2) restriction of only cut and skid harvest technologies for northern PUs; (3) allocation of forest management activities by PU to achieve some pre-defined level of distribution of economic benefits to communities within the project area based on an analysis of population, employment and productive landbase data; (4) maintenance of theoretical woodland caribou populations at a maximum of 90% of the carrying capacity within all PUs currently supporting woodland caribou populations; (5) maintenance of the spatial component of ecosystem diversity at a minimum level of 50% of the area available in 1975 within each of the six PUs; (6) defining the temporal component of diversity by specifying the target age distribution based on modification of Van Wagner's (1978) description of the exponential age structure of a fire-disturbed boreal forest ecosystem (modifications included tailoring the exponential age structure to account for the inferred fire cycles and break-up ages unique to each ecosystem within each PU), and (7) imposing a two pass harvesting constraint to effectively limit the harvest area of each ecosystem class/age class combination to a maximum of 50% of what was available in the previous 20-year planning interval.

Alternative sets of forest management plans were developed by sequentially altering the management targets for benefit distribution, caribou population maintenance and ecosystem supply, and by specifying whether or not the two pass harvest constraint would be applied (Table I). 'No harvest' and 'unconstrained harvest'(excepting mill capacity) defined the two extremes among eight alternatives considered. The alternatives

TABLE I

Alternatives selected to define sets of management activities.

Resource Supply Objectives	Biodiversity maintenance	Caribou maintenance	Benefit distribution	Two pass harvesting
Management Alternatives				
1. Do nothing[1]	N/A	N/A	N/A	N/A
2. Unconstrained[2]	0%	0%	0%	No
3. Status Quo - LRSY / AAC[3]	N/A	N/A	N/A	Yes
4. Highly Constrained[4]	High	90%	75%	Yes
5. Single Pass Harvest[5]	High	90%	75%	No
6. Relaxed Benefit Distribution[6]	High	90%	25%	No
7. Relaxed Caribou Habitat[7]	High	50%	25%	No
8. Relaxed Biodiversity[8]	Moderate	50%	25%	No

1 Do nothing: all wood for mills purchased outside FMLA, no harvest.
2 Unconstrained: mill capacities the only management constraint.
3 LRSY/AAC: long run sustained yield/annual allowable cut based on Saskatchewan Forestry Branch procedures i.e., AAC= (Area at Rotation Age * Stocking at Rotation Age/Rotation Age) + (Area>Rotation Age * Stocking of Stands/ Rotation Age).
4 Highly constrained: high biodiversity violation penalty ($250/ha), 75% FMU distribution of economic benefits, two pass harvesting, 75% caribou population retention.
5 Single Pass Harvest: same as 4 minus two pass harvesting.
6 Relaxed Benefit Distribution: same as 5 except only 25% distribution of economic benefits by FMU.
7 Relaxed Caribou Habitat: same as 6 except caribou population maintenance at 50%.
8 Relaxed Biodiversity: same as 7 except biodiversity maintenance penalty reduced by 50%.

were designed such that constraints were systematically reduced from the most highly constrained' to the least constrained condition. The 'Status Quo' is the current management system based on the long run sustainable yield/annual allowable cut (LRSY/AAC) calculation used by the province, and was included to compare proposed with existing activities.

6. Data Inputs and Outputs

Both ecological and economic data are provided as output from the MFMM. The reporting of model output is governed by the level of temporal and spatial resolution which has been specified for the model analysis and the dimensionality which has been assigned to particular variables. Dimensionality is a major concern as computer hardware places practical limitations on the size of the MFMM. The spatial and temporal resolution of the output follows: (1) a total of six PUs; (2) eleven ecosystem classes aggregated from more than 350 unique productive inventory stand types; (3) eleven age classes set at intervals of 20 years, and (4) fifteen planning iterations out to 220 years based on 20-year age class intervals with the exception of the initial 20 years which provided for 1, 2–3, 4–5, 5–10 and 11–20 intervals.

For each management unit and time iteration, data on: (1) the value of the forest as defined by the objective function; (2) the area and volume of wood harvested (wood supply); (3) the area maintained in each ecosystem class/age class combination (ecosystem maintenance); (4) the area, type of treatment, and treatment cost for each hectare harvested; (5) the yield and where applicable value of all non-timber resource features, and (6) a breakdown of all other economic costs and values including employment benefits associated with application of each of the forest management activities are provided.

The key inputs required to generate the output follow.

Ecological Data - the forecasting of the supplies of future resource benefits depends largely on our ability to predict future forest structures. The following information was critical to that process: (1) initial forest structure; (2) forest successional processes and fibre yields; (3) forest structural treatment responses for each of several treatment options; (4) natural disturbance losses and responses, and (5) yields of key non-timber resource features.

Economic Data - key economic information included value and cost data for the following: (1) the derived residual value of the harvested wood at the mill gate calculated as the difference between the estimated value of the first product openly traded in the marketplace and the cost of production of those products; (2) access, harvest and haul costs; (3) silvicultural treatment costs for each treatment option; (4) mandatory crown stumpage dues and voluntary co-management fees paid to communities based on a fixed rate volume levy; (5) government administration and management costs; (6) the net value of each non-timber benefit derived using conventional economic estimation techniques where sufficient data is available, and literature sources where it was not, and (7) the public (3%) and private (5%) economic discount rates to be applied against all costs and values for calculation of the NPV (net present value) of future forest benefits.

7. Alternatives Evaluation Framework Methodology

A major objective of planning is to search for the management alternative which best satisfies societal expectations of ecological, economic and social benefits. The MFMM does not provide a comparative analysis among alternatives, and therefore a separate evaluation framework is required. The (1) total net present value, (2) economic risk, (3) ecological risk, and (4) level of community economic benefit were the four major components of this evaluation.

This assessment was achieved assigning weights to rates for each criterion thereby effecting a scoring system for comparative analysis. The weights were obtained through formal consultation methods with proactive residents of each planning unit whereby they assigned a relative importance to each of the four major decision criteria (Table II).

TABLE II

Criteria weights for aggregate FMLA assessment.

	NPV		Economic risk	Ecological risk	Benefit distribution	Total
	Public	Private				
Weight	0.12	0.15	0.13	0.38	0.22	1.00

Members of the technical planning team then assigned weights to each of the factors comprising the economic risk, ecological risk and benefit distribution criteria respectively.

Each of the key criteria are discussed below:

1. *Net Present Value of Forest Resources* - a relatively important criterion which assesses the public and private economic value of each alternative. Positive public and private NPVs were imposed as minimum conditions of acceptance as negative values would suggest the project to be uneconomical and therefore not in the public or private interests respectively.
2. *Economic Risk* - evaluated in terms of the upside risk (i.e., greater net benefit than predicted), downside risk (i.e., lower net benefit than predicted), and project viability relative to the 'Do Nothing' alternative (Table III). The upside and downside risks calculations were generated by increasing and decreasing product values and discount rates relative to the best estimate for each alternative. The viability margin provides an estimate of how much the NPV would need to decline before the 'do nothing' alternative would be preferred economically. The cumulative measure of economic risk was based on the aggregate scores for each of the risk components.
3. *Ecological Risk* - large changes in the existing forest structure increase the potential risk of unexpected negative ecological impacts. The degree of risk was estimated by consideration: (a) variance in the ecosystem class structure with particular emphasis on the loss of the white spruce ecosystem and increase in area of open forest lands; (b) landscape features with positive emphasis on the maintenance of natural landscape patterns; (c) disturbance with particular emphasis on area logged; (d) variance in the age structure with emphasis on the loss of area in late-successional forest stands, and (e) the supply of woodland caribou habitat (Table IV).

4. *Benefit Distribution* - the NPVs of each of (a) co-management fees, (b) silviculture expenditures and (c) woodlands expenditures were used as metrics of benefit distribution within each of the six PUs; scores. The individual criteria were added to estimate the overall benefit distribution rating for each alternative (Table V).

These weights were then applied against rates (direct model output) for each variable to produce absolute scores for each analysis component. To equate the absolute scores relative to each other, the absolute scores were then converted to relative scores such that the lowest score was set at 0 and the highest at 10. The sum of these scores across all planning units were then ranked creating a preference ordering for each major criterion. The final step was to conduct an aggregate analysis incorporating the relative scores all major criteria and then ranking to arrive at a preferred alternative for the FMLA. In this fashion the trade-offs among several evaluation factors are considered in the selection of a preferred management alternative.

TABLE III

Relative weighting of economic risk assessment criteria.

Economic Risk	Up Side Risk		Down Side Risk		Viability Margin		Total
	Public	Private	Public	Private	Public	Private	
Weights	0.12	0.18	0.21	0.14	0.21	0.14	1.00

TABLE IV

Relative weighting of ecological risk assessment criteria.

	Ecosystem Class Structure				Spatial Landscape Pattern	Redisturbance Potential	
	Average variance	Average range	Open forest conversion	Spruce conversion		Maximum	Average
Weights	0.06	0.045	0.075	0.16	0.16	0.0975	0.0525

	Age Class Structure				Caribou Populations		
	Average variance	Average age range	Average age	Late successional ratio	Min. Pop.	Ave. Pop.	
Weights	0.375	0.025	0.025	0.1625	0.13	0.07	1.00

TABLE V

Relative weighting of benefit distribution assessment criteria.

Benefit distribution	Co-management fees	Silvicultural expenditures	Woodland expenditures	Total
Criteria weights	0.45	0.35	0.20	1.00

8. Preliminary Results

This section provides a brief overview of the preliminary results of the Mistik evaluation of alternatives. The data is preliminary in nature and not accompanied by a detailed discussion as the formal government Environmental Impact Statement technical review was in progress at the time of preparation of this document.

The preference ranking order for the FMLA area based on the evaluation criteria discussed in the previous section resulted in the selection of Alternative #6 (Relaxed Benefit Distribution) as the set of management activities which provided the best mix of benefits. Based on individual evaluation criteria, this alternative was superior to all other alternatives only with respect to economic risk and benefit distribution (Table VI).

It is important to note that the aggregate preference as expressed across all planning units, is not necessarily mirrored by individual planning unit preferences (Table VII).

TABLE VI

Preference order ranking by evaluation criterion for the FMLA.

	Net Present Value		Economic	Ecological	Benefit	
	Private	Public	risk	risk	distribution	Overall
1. Do nothing [a]	8	8	8	1	8	8
2. Unconstrained	1	1	6	6	7	6
3. LRSY/AAC	6	5	7	8	2	7
4. Highly constrained	7	7	4	7	5	5
5. Single pass	5	6	3	2	4	2
6. Relax BDI	4	4	1	3	1	1
7. Relax caribou	3	3	2	4	3	3
8. Relax biodiversity	2	2	5			5

a See Notes to Table I for more information.

TABLE VII

Preferred alternatives by planning unit and evaluation criteria.

Criterion	PU#1	PU#2	PU#3	PU#4	PU#5	PU#6	FMLA
Economic risk	8	7	3	5	2	5	6
Ecological risk	1	1	1	1	1	1	1
Benefit distribution	6	3	3	3	2	5	6
Aggregate	6	6,7	6	5	2	5	6

Alternatives: **1**= Do Nothing / **2**= Unconstrained / **3**= LRSY/AAC / **4**= Highly Constrained / **5**= 4 + Relax Two Pass / **6**= 5 + Relaxed Benefit Distribution / **7**= 6 + Relaxed Caribou Maintenance / **8**= 7 + Relaxed Biodiversity

For example, while Alternative #6 was preferred on the basis of economic risk across all planning units, it was not the preference for any single planning unit. For benefit distribution, the overall preferred Alternative #6 was only the preferred alternative for PU #1. Conversely, the 'No Harvest' alternative resulted in the lowest ecological risk in every planning unit and obviously across all planning units. This type of analysis clearly illustrates the trade-offs that are made among the criteria in defining a preferred set of management activities for the FMLA.

Following selection of a preferred alternative for the FMLA, further analysis of the MFMM output data was conducted to illustrate some of the key relative differences among the alternatives.

8.1. Supply of Forest Products

The NPV of forest products (wood supply and wildlife harvest supply) as expected, decreases as constraints are progressively added due to a combination of higher costs and lower yields. The least preferred alternative in all cases is the "Do Nothing" alternative, is consistently followed by the 'Highly Constrained' alternative.

Respecting hardwood harvest, all active alternatives, except for the 'LRSY/AAC' alternative were capable of meeting mill requirements over 220 years (Figure 1). The failure of the 'LRSY/AAC' alternative to fulfil hardwood supply was likely related to imbalance of mid-successional age classes in some PUs, the planning level at which the AAC calculation was applied. All alternatives except the 'LRSY/AAC' and 'Highly Constrained' alternatives were able to meet softwood mill requirements until 2055, after which, the preferred alternative yielded generally lower softwood harvests than all less constrained alternatives and the LRSY/AAC alternative. In the case of the latter, the forest is being managed to maximise the yield of fibre by normalising the forest structure.

Moose and deer harvest levels both tended to be substantially higher with the preferred alternative throughout the planning horizon, except for the 'Unconstrained Alternative', which, by liquidating the softwood forest in favour of a much younger hardwood forest, improves overall habitat quality for moose and deer. Fisher harvest suffers most under the preferred and less constrained alternatives over the initial 75 years of the project, after which only the 'LRSY/AAC' alternative achieves higher harvests; the 'LRSY/AAC' alternative tends to maintain a mid-successional mixed-wood forest to which fishers are well adapted.

Woodland caribou populations are managed as a constraint in all cases except for the 'Unconstrained' and 'LRSY/AAC' alternatives, and therefore the specified population levels are achieved. For the two cases in which constraints are not applied, woodland caribou populations decline sharply in the face of declining habitat.

8.2. Supply of Forest Ecosystems and Age Structures

The maintenance of the forest ecosystem and age class structure is an important component of the biodiversity management objective. Spruce-dominated ecosystem types are the most expensive and difficult to maintain, and as a result suffer the most where biodiversity constraints are either not applied or relaxed. The 'Unconstrained' and 'LRSY/AAC' alternatives therefore exhibit the highest predicted losses of these ecosystem types; more than 50 000 and 20 000 ha yr^{-1} less are maintained respectively for these alternatives compared to the preferred alternative over the last 100 years of the planning horizon. The 'Unconstrained' alternative tends to liquidate the available softwoods throughout the planning period and the 'LRSY/AAC' alternative converts spruce-dominated ecosystem types to aspen-dominated mixedwood stands.

The area of late-successional forest stands is a surrogate metric for the age component of biodiversity maintenance. For the most part, all alternatives except the 'No Harvest' alternative maintain less area in stands at break-up age compared to the preferred alternative. These results consistently occur whenever less constraining targets are imposed on the maintenance of late successional forests. The difference is particularly striking for the 'LRSY/AAC' alternative which by 75 years has at least 20 000 ha less late successional forest; this trend is maintained throughout most of the planning horizon and exceeds 25 000 ha by 2176. This is expected as the 'LRSY/AAC' alternative encourages the removal of stands in excess of a fixed biological rotation age.

8.3. Supply of Employment Benefits

The use of mechanical harvesting technology influences the lowest levels of employment associated with the 'Unconstrained Alternative'. The 'LRSY/AAC' alternative which fixes PU harvest levels by the AAC, provide the highest benefits to those areas which may not be favoured economically by alternatives with lower benefit distribution constraints. Conversely, removal of the two pass harvesting constraint favours transferral of benefits to those management units which have large stands of homogeneous ecosystem types.

9. Conclusions

The purpose of this paper was to demonstrate the application of a new forest resources management model and associated evaluation methodology to solving the complex problem of managing a large and diverse forest landbase in the best interest of the public. Obviously, an effective forest management plan entails more than optimisation of sets of activities based on forest-level analysis of data. For example, the maintenance of ecosystem areas and age classes in the absence of a landscape management strategy will not likely effect a desired level of biodiversity management. Site management activities must be designed in concert with forest-level issues. In addition, attention to the technical issues of forest management in the absence of meaningful consultation and education processes may embellish rather than reduce the public cynicism and distrust of technological advances. However, in the absence of models like the MFMM, we can be certain that ignorance will continue to dominate the public and industrial perception of forest management.

A salient feature of this exercise has been the explicit demonstration of the nature of the trade-offs that must be made in searching for a preferred management alternative. Clearly, there is no Nirvana[6], and the quest for sustainability presents forest managers with extremely difficult choices. A recognition and acceptance of uncertainty in predictions of the future raises the spectre of risk in our pursuit of good forest management. However, both uncertainty and risk can be effectively managed once their existence is acknowledged and quantified.

The approach taken in this project was to proceed with management decisions using the best information possible and to develop a modelling tool to provide the forest manager with the opportunity to preview the possible consequences of applying sets of management activities on the supply of future resources and benefits. The development of a long-term management plan must therefore be complemented by an effective

[6] *Nirvana* - a mythical state in which the wants and needs of all members of society are fulfilled in an idealized world free of any pain or discomfort.

monitoring and research strategy designed to expand our knowledge of cause and effect relationships as the basis of constantly improving the accuracy of future forecasts of resource benefits.

The implementation of forest management is therefore a dynamic process. There is no single forest management plan that will serve societies needs indefinitely into the future, but models like the MFMM promise a process which can be continually refined and improved to further contribute to the sustainability of our communities and forests.

References

Harris, W. C., Kabzems, A., Kosowan, A., Padbury, G. and Rowe, J. S.: 1983, *Ecological Regions of Saskatchewan*, Saskatchewan Department of Natural Resources, Tech. Bull. **10**, 57 pp.

Rowe, J. W.: 1972, *Forest Regions of Canada*, For. Can., Environ. Canada, Public. No. **1300**, 171 pp.

Saskatchewan Environment and Resource Management (SERM): 1992, *Final Project Specific Guidelines for the Preparation of an Environmental Impact Statement, NorSask Forest Products Inc.* Proposed Twenty Year Forest Management Plan prepared by Mistik Management Ltd. Impact Assessment Branch, Saskatchewan Environment and Resource Management, Regina, Canada.

Van Wagner, C. E.: 1978, *Can. J. For. Res.* **8**, 220-227.

EMERGENCE OF A BIODIVERSITY CONCEPT IN SWEDISH FOREST POLICY

T. LÄMÅS[1] and C. FRIES[2]

[1]*Department of Biometry and Forest Management, Swedish University of Agricultural Sciences, S-901 87 Umeå, Sweden*

[2]*Department of Silviculture, Swedish University of Agricultural Sciences, S-901 87 Umeå, Sweden*

Abstract. Sweden's forest policy was fundamentally changed in 1993. Timber production and maintenance of biodiversity then became objectives of equal dignity. In the past, timber production had been the single primary aim of Swedish forest policy. The strategy established for maintaining biodiversity is based on a system of forest reserves and management of the matrix that mimics natural processes and builds in natural structures at the stand and, when applicable, the landscape level. One reason for not basing the strategy on reserves only is that almost all Swedish forests have been intensively managed and, as a consequence, there is a shortage of areas suitable for reserves. The new forest policy presupposes a consensus in the forestry sector of the importance of maintaining biodiversity. Training and information are as important as legislation. Maintenance of biodiversity is not an operative goal. This makes it difficult to predict the outcome of the policy and, consequently, there is a need for monitoring routines to enable changes to policy implementation. Such routines have yet to be developed.

Keywords. BIODIVERSITY, TIMBER PRODUCTION, FOREST HISTORY, FOREST POLICY, SWEDEN.

1. Introduction

Swedish forestry has been governed by production-oriented forest policies since the beginning of this century. Almost all forest land has, as a consequence, been used for production of saw-timber and pulp wood and the proportion of natural forests is only a few percentage points. This forest management has changed structures and functions in the forest ecosystems which has negatively affected a large number of organisms (Ahlén and Tjernberg, 1992; Ehnström *et al.*, 1993). Exclusive of sub-alpine forests, about 0.5% of the productive (site productivity ≥ 1 m^3 ha^{-1} yr^{-1}) Swedish forest land is protected (Statistics Sweden, 1994).

These circumstances, increasing ecological awareness, popular environmental concern, pressure from non-governmental organisations, as well as the fact that Sweden has committed itself to maintaining biodiversity, have brought about a new Swedish forest policy, aiming at more than just timber production. The policy may be illustrated by the first paragraph of the recently adopted Forestry Act which states that 'The forest is a National resource. It shall be managed in such a way as to provide a sustainable yield and at the same time maintain biodiversity. In forest management other public interests should also be considered' (Swedish Forestry Act, 1993).

To fulfil the commitment to maintain biodiversity it is necessary to modify forest management (Liljelund *et al.*, 1992). In this sense, it should be noted that the Swedish situation differs from that in other regions in the boreal biome. In Canada and Russia, there are vast forests in which natural processes operate unaffected by humans. In these countries, the natural forests play a key role in preserving biological diversity. In Sweden, on the other hand, a forest policy aiming at maintaining biodiversity must focus on the managed forests. Plant and animal diversity will be enhanced if the

Water, Air and Soil Pollution **82**: 57–66, 1995.

frequency of natural structures and processes is increased in these forests (Franklin, 1992; Liljelund *et al.*, 1992; Attiwill, 1994; Haila, 1994). However, the present array of species in Swedish forests cannot be maintained only by means of modified stand management. There are species that depend on processes and structures that cannot be preserved with the kind of disturbances that come with forestry for timber production (Ohlson, 1990; Esseen *et al.*, 1992). For those species, reserves are probably the only alternative.

In this paper we describe the emergence of a new Swedish forest policy, focusing on the developments of the last year and the means by which the goals should be attained. To facilitate understanding of the motives for the Swedish line of action, we briefly relate parts of the Swedish forest history and what large scale landscape changes forestry has caused.

2. Swedish Forest History

Human impact on Swedish forests has differed between regions. The country may be divided into three regions each with its unique forest history (Figure 1). In the most densely populated southern fifth of the country, Götaland, a large proportion of the forests has been converted to arable land. By the middle of the 19th century the forested area was probably at its lowest level (National Atlas of Sweden, 1990). Because of intense grazing and neglect of regeneration after cutting, forests were fairly open. About one hundred years ago, reforestation, mainly by planting of Norway spruce (*Picea abies* (L.) Karst.) and Scots pine (*Pinus sylvestris* L.), successively increased the forested area (Juhlin Dannfelt, 1959). Absence of grazing by domestic animals combined with systematic silviculture has, over the years, created dense forests. The major part of Götaland is hemiboreal (Ahti *et al.*, 1968) but the boreal and the temperate vegetation zones are also represented in the region. In Svealand, the region located north of Götaland and of about equal size, the iron industry required large quantities of wood and charcoal until the beginning of the present century (National Atlas of Sweden, 1990). Plots where charcoal was produced were evenly spread out in the forests and cutting for charcoaling was, during some centuries, done in a purely exploitative manner. Cutting of timber for sawing as well as grazing contributed to keeping the standing volumes in the forests at relatively low levels (Nordquist, 1959). To ensure the supply of charcoal and other essential wood products, forest management based on the clear-felling system was introduced in the middle of the 19th century. Because of this forest history, most sites carry their second or even third rotation. Svealand is a transition region, including both the hemiboreal and the boreal vegetation zones. As in the rest of boreal Sweden, Norway spruce and Scots pine are the dominant tree species in this region comprising about 85% of the standing volume (SUAS, 1993).

The northern two-thirds of Sweden, called Norrland, has always been sparsely populated. Most people live along the coast and in the large river valleys. In this region, partial cutting to a diameter limit was introduced at the beginning of the 19th century (Östlund, 1993). Developing industrialism in Europe needed large quantities of timber and a logging frontier moved northwards during that period (Bunte *et al.*, 1982).

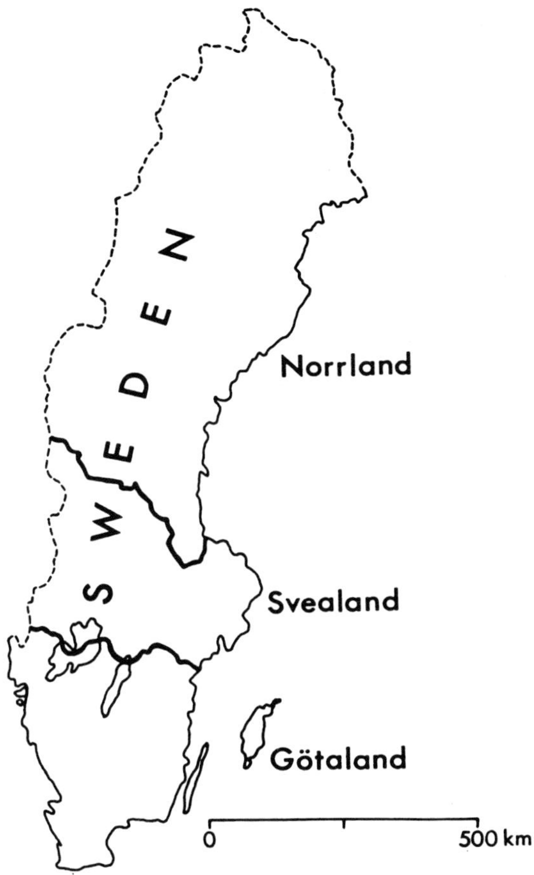

Fig. 1. Three major regions of Sweden with significantly different forest histories.

The forests were exploited for large diameter Scots pine timber which was abundant all over northern Sweden. Towards the end of the last century, Norway spruce and smaller trees also became valuable because of the developing pulp and paper industry. Until the 1940s, however, management was mainly carried out by means of selective cutting and the desired regeneration generally did not occur. The north Swedish forests therefore successively became more open and more spruce-dominated. For these supposedly well growing, but in reality low-producing forests, the expression 'green lies' was coined. During this exploitational period, standing volume in the five northernmost provinces of Sweden, which together make up Norrland, decreased by around 30-40% (Linder and Östlund, 1992).

In order to increase the future cutting level in Norrland, the state forestry launched a restoration program in which substantial resources were put into regenerating low-stocked sites (Höjer, 1954). Large clear-fellings that were thoroughly cleaned manually, mechanically, or by means of herbicides, were planted, sown or naturally regenerated. Also company and non-industrial private forestry abandoned selective cutting in favour of clear-felling. Since the clear-felling system was successively introduced on a broader scale in Norrland in the 1940s, around 40% of the productive forest land has been clear-felled and regenerated, primarily with Scots pine, with the aim of achieving pure even-aged stands. In about 1970, lodgepole pine (*Pinus contorta* Dougl. ex Loudon) was introduced on a large scale to Sweden from western Canada. In spring 1992, it was growing on around 540 000 ha predominantly in Norrland where the species covers 4% of the forest land (von Segebaden, 1992).

For Sweden as a whole, large scale impacts on the forests can be illustrated by different means. Consistent chronosequences of regeneration measures, growing stock, etc. are easily obtained from the Swedish Statistical Yearbook of Forestry which has been published since 1951 (NBF, 1951-1994). Soil scarification or prescribed burning for forest regeneration was carried out on 4.8 Mha of the 23.6 Mha of productive forest land between 1951 and 1992. During the same period, 6.1 Mha have been artificially regenerated by planting or seeding and 10.6 Mha have been pre-commercially thinned. (When evaluating these data, it should be kept in mind that the Swedish rotation period is typically one hundred years, and that some areas are pre-commercially thinned two times or more.) Also, forest drainage has influenced Swedish forests. Of today's productive wetland forests, one fifth (i.e. 1.0 Mha) have been drained (Hånell, 1990). This type of ditching has consequently influenced 4% of the total Swedish productive forest land.

This somewhat simplified and very brief Swedish forest history illustrates the fact that nearly all forest land has been managed in one way or another for at least a century. Natural or semi-natural forests comprise small islands in a matrix of managed forests (Angelstam, 1992). Today, around 3.5% of the forested area is protected—mainly subalpine forests in the north-west. In 16 out of 18 provinces in the southern third of Sweden, the protected forest area is less than 1% (Kardell and Ekstrand, 1990).

3. Ownership, Forest Inventories, and Red Lists

Some 50% percent of productive Swedish forest land is non-industrial and privately owned. The proportion is higher in the south and decreases towards the north. Approximately 42% of the forest land is owned by companies, of which about one-half belongs to a joint stock company, with the state as the largest shareholder. The remaining 8% is under other public ownership categories. The mean area of non-industrial private wood lots is about 50 ha and a minority of the forest owners depend on the forests for their living. On a national level, however, the Swedish forest industry is of great economic importance generating the largest net export value. As in Finland and Norway, the Rights of Common Access allow people to freely enter any forest land for the purpose of enjoying natural amenities, without permission from the owner of the forest (Paulsson, 1978). It is possible to hike, make a fire, pick berries and mushrooms etc., as long as the forest owner is not caused any economic damage or other significant inconveniences.

The National Forest Inventory (NFI) has been carried out since 1923 to obtain comprehensive data on Swedish forests in order to monitor and forecast the forest resources at regional or national levels. Thus, it is an important component in establishing Swedish forest policy. By means of sample plots arranged in square clusters, the whole country is now covered every year. As forestry historically has emphasised timber production, the NFI has focused on data describing volume and yield of stem wood, soil conditions, ground vegetation and forest health (Ranneby *et al.*, 1987; Söderberg, 1993). Since the first NFI, the estimated standing volume in Sweden has increased continuously from 1.8×10^9 m^3 in the 1920s to 2.8×10^9 m^3 in 1990. The estimated annual increment during 1988-1992 was about 1.0×10^8 m^3. This by far exceeds the estimated annual harvest of 0.64×10^8 m^3 during the same period (SUAS, 1993). Forecasts for long-term timber yield in the 1980s showed that it was possible to increase harvesting on a sustained yield basis, although a broader spectrum of forest products and functions were promoted (Bengtsson *et al.*, 1989).

As a tool for the implementation of the forest policy of the time, the Swedish parliament decided in 1979 that the Forest Administration (the National Board of Forestry [NBF] and 24 County Forestry Boards) should carry out an inventory on all private, non-industrial forest land, the so called General Forest Inventory (GFI). It was supposed that efficiency of forest production would increase, with a general increase in forestry activities, if detailed knowledge was available on the condition of the non-industrial privately owned forests, and on the silvicultural measures needed to keep up the production in those forests (NBF, 1981). The inventory was done by aerial photograph interpretation complemented by field surveys. Areas with conservation values were noted, although not systematically (Eckerberg, 1990). The GFI included forest condition and suggested stand treatments. They were presented on forest maps and data bases describing compartments. Costs for the GFI increased however, and it was considered that the preparation of management plans resulting from the GFI would be the forest owners' responsibility. In 1993, the parliament decided to stop the GFI. About 1.2 Mha (10%) of the non-industrial private land was therefore not covered (NBF, 1994).

For the forest ecosystem, as well as for other ecosystems occurring in Sweden, lists of threatened and care-demanding species are kept in a data base at the Swedish University of Agricultural Sciences (e.g., Ehnström *et al.*, 1993). The lists follow the IUCN categories and are established as Red Lists by the National Environmental Protection Board. A recent analysis reveals that very old, dying and dead trees, features that are changed or lacking in Swedish forests (Esseen *et al.*, 1992; Lämås and Fries, 1995), seem to be critical substrates for a large proportion of the forest species included in the Red Lists (Berg *et al.*, 1994). Most of the Red Listed forest plants are cryptogams, while insects make up the majority of animals.

Nationwide inventories of forests with high nature conservation values and of wetland forests are now underway. The inventories were started in 1991 and 1992, respectively, and will be finished in 1997 (see below).

4. A New Swedish Forest Policy

Forestry legislation has several hundred years of tradition in Sweden. The first Forestry Act (FA) was passed in 1903 and was primarily a reforestation law. Environmental considerations were included in 1975. The FA adopted in 1979 explicitly stated certain measures to be taken when clear-felling in order to meet conservation and other public interests. The primary goal of forestry was, however, still to utilise the forest's timber production capability to achieve a high and valuable yield in a sustainable manner (Swedish Forestry Act, 1979). For the first time the state-owned forests were also governed by the FA which until then had only concerned other public forests and private forests.

In 1990 the government started work on a new Swedish forest policy. A parliamentary committee was to evaluate the current forest policy and propose changes based on that evaluation and on forecasts within a broad field. In addition, the committee was instructed to include responsible stewardship in the new forest policy that would maintain biodiversity (SOU 1992:76, 1992).

An expert group analyzed different ways of maintaining biodiversity in Swedish forests. They estimated that the reserved proportion of 0.4% of the productive forest land, which are exclusively reserves in sub-alpine forests (Kardell and Ekstrand, 1990), needed to be increased to 15% if biodiversity was to be maintained (SOU 1992:76, 1992, and literature cited therein). However, because of previous management, there is a shortage of natural forests suitable for reserves, and the reserves would in any case, be isolated so that dispersal possibilities for organisms through "a hostile matrix" would be uncertain. The line of action suggested by the group, and which was accepted in 1993, was instead to reduce the proposed reserved proportion 'by more than half' and to modify management of the remaining forest land. Modifications should aim at mimicking natural processes and build in natural structures into the production forest, e.g., self-thinning, tree species mixture, and dead trees. The combination of reserves and modified forest management was, except for biological and ecological reasons, considered the most efficient solution. For example, state-owned forests that could be relatively easily reserved are mainly situated in northern Sweden. In addition, unreserved state-owned forests are managed mainly for timber production and are not suitable as reserves. Another alternative is a combination of reserves, intensively managed forests close to industries and a low intensity timber management matrix. As all forest land is accessible to people on the basis of the Rights of Common Access, intensively managed forests close to industries and, therefore also close to densely populated areas, were not found to be an appropriate alternative. Furthermore, timber production and nature conservation are not always mutually exclusive which means that from an economic viewpoint it would not be optimal to separate the two kinds of land use.

In 1993, the parliament adopted the new forest policy based on the proposal of the committee (1992/93:JoU15, 1993). Timber production and maintenance of biodiversity are now the main objectives with equal importance. As a part of the forest policy the FA was revised (Swedish Forestry Act, 1993). In the revised FA, silviculture is less regulated than in the former FA partly because of a general trend in Sweden towards less regulation and partly because fewer regulations are believed to cause more variation in the forests which in turn is believed to enhance biodiversity. Forestry is also governed by legislation other than the FA, e.g., by the Nature Conservancy Act (NCA) (NCA, 1964). This controls ditching and since January 1994 it has been possible to protect a defined set of biotopes (≤ 5 ha) of high conservation value, e.g., forests along gullies, post-fire successional forests, and natural and semi-natural old-growth forests of specific qualities, by means of the NCA (NCP, 1994).

Apart from different kinds of reserves and protected biotopes ≤ 5 ha, a complementary voluntary system was launched in January 1994, which gives the forest owner economic compensation for restrictions put on his rights to cut on sites with conservation value. Restrictions are agreed upon between the forest owner and the Forest Administration. In every case the aim of the agreement is that conservation values should be preserved, while economic sacrifice by the forest owner should be minimised. Although only tested on a small scale (about 60 cases in the whole country between January and December 1994) (G. Nordanstig, Swedish National Board of Forestry, pers. comm.), this kind of "nature conservation agreement" (Carlsson, 1994) seems to be promising from both a nature conservation and an economic point of view. A new system of subsidies to encourage the use of silvicultural treatments that enhance biodiversity is also under development. If a forest owner uses prescribed burning to regenerate a site, he may be partly compensated for extra expenses in relation to costs of the normally used mechanical soil preparation. These subsidies have replaced earlier timber production oriented subsidies.

The inventories launched in 1991 of wetland forests (Rudqvist, 1991) and in 1992 of forests with high conservation values (Nitare and Norén, 1992) are important components in the new forest policy. The latter biotopes may be called "woodland key habitats", areas with features making them possible to host Red Listed species. To some extent, the woodland key habitats, and probably also some wetland forests, will receive protection. The wetland forest inventory is done by the Forestry Administration and covers all forest land, while woodland key habitats are inventoried only on non-industrial privately owned land. On land owned by forest companies and on public land, woodland key habitats will be inventoried by the owner, at the owner's expense. As of June 1994, 7500 woodland key habitats have been mapped on 2.2 Mha out of 11.5 Mha of non-industrial privately owned forest land. It should be noted, however, that these habitats may differ in size from a fraction of a hectare to several hectares (M. Norén, Swedish National Board of Forestry, pers. comm.).

Courses and information programmes directed towards forest owners, employees in forestry, and contractors, have over a long period of time been a part of the implementation of forest policy in Sweden. From 1989 to 1993, most contractors, almost all forest workers and about one quarter of forest owners, participated in short courses ('A Richer Forest') initiated by the NBF, and focused on forest ecology, nature conservation, and silviculture (G. Nordanstig, Swedish National Board of Forestry. pers. comm.). During the last few years, many employees of forest companies and the NBF have spent 10–20 weeks at university level courses concerning these new aspects of forest management.

The Forestry Administration has since the end of the 1980s undertaken a program to assess environmental considerations after final felling. Examples of considerations are to what extent single trees, groups of trees, border zones towards streams and other open areas are left uncut (NBF, 1991; Sandström, 1991). In a comparison between one follow-up in 1989/90 and one in 1992, the proportion of acceptable clear-cuts had increased from 66% to 88% (Sandström, 1994).

5. Concluding Remarks

Timber production has affected almost all Swedish forests. Accordingly, there is very little natural forest except in the sub-alpine area. In comparison with conditions prevailing before humans affected the forests on a large scale, processes and structures on different spatial levels have been heavily altered.

In the pre-industrial era, Swedish forests had multiple-use functions. In the industrial era, on the other hand, timber production has been the dominant use of the forests. This has been encouraged by a forest policy which until the end of the 1980s had been successful; in the sense that at present there is a potential for increased cuttings and the fears of timber scarcity once prevalent in parts of Sweden have been avoided (SOU 1978:6, 1978; SOU 1992:76, 1992). The estimated future non-declining timber yield, however, depends on no severe restrictions being put on timber production or production not being affected by environmental degradation.

Increased knowledge of forest ecology, increased demand for environmentally friendly products, increased public environmental awareness, etc., has stimulated a search for strategies to maintain biodiversity. At the beginning of the 1990s the Swedish government started work on a new forest policy. The strategy that was established for maintaining biodiversity in Sweden is based on a system of forest reserves and a management of the matrix that mimics natural processes and builds in natural structures at the stand level, and, when applicable, at the landscape level. The strategy is determined by, among other things, the fact that almost all forests have been intensively managed, hence there is a shortage of natural forests suitable for reserves (Harris, 1984; Wilcove, 1989).

There are some reasons why all forests, and not only the state forests, should be subjected to modified management. Firstly, the state forests are mainly located in the northern interior parts of the country. If forestry was modified only on public land, the maintenance of biodiversity in the south and along the east coast would rely very much on reserves, of which there are very few in these parts of the country. Secondly, the state forests are to a large extent in fact privately owned in the form of a joint stock company. As the general public and institutions own as much as 49% of the shares, it could be argued that the state forest and private forests, in this meaning, should be treated similarly. Finally, the diversity among forest owners and the few restrictions on timber production practices, lead to diversity in goals of forestry and varying intensity in timber production among forest owners. Some people assume that this also promotes biodiversity. This, however, has not been established.

The new Swedish forest policy presupposes a consensus in the forestry sector of the importance of maintaining biodiversity. This is expressed in the way the policy is implemented, e.g., by courses for forest owners and forest workers instead of strict regulations. Concerning actual stand management, there are no environmental considerations explicitly stated in the FA regarding pre-commercial and commercial thinning operations. If, in the future, natural structures are to be achieved in old stands, tending of young and middle-aged stands must be done in a way that makes it possible for these structures to develop. According to today's legislation, necessary modification of the traditional management, during this phase, will be based on the land-owner's own initiative.

It has been recognised that planning at a landscape level is valuable in forest management where preservation of biodiversity is given priority (Hansson and Angelstam, 1991; Franklin, 1992; Oliver, 1992). This was acknowledged in the preparatory work for the forest policy, but not stated in written legislation. However, forest companies are now beginning to implement landscape plans on their holdings (e.g., Stora Skog, 1993). In regions with predominantly privately owned lands, it is assumed that co-ordinated achievements to preserve e.g., woodland key habitats extending across estate borders, should be done on a voluntary basis with guidance from the County Forestry Board (SUO 1992:76, 1992). So far, these kinds of agreements have not influenced nature conservation to any larger extent.

The goal to maintain biodiversity as stated in the Swedish forest policy is not operative meaning that it cannot be expressed in terms of, e.g., size of populations or areas of specific habitats. This makes it difficult to predict the outcome of the policy and, consequently, there is a need for monitoring routines to enable changes to policy implementation. Such routines have yet to be developed.

Acknowledgments

This study was funded by the Faculty of Forestry (research programme "Managing Forests for Industrial Products and Environmental Quality") at the Swedish University of Agricultural Sciences.

References

1992/93:JoU15: 1993, *En ny skogspolitik*. Jordbruksutskottets betänkande, Stockholm, Sweden. 93 pp. [In Swedish.]

Ahlén, I. and Tjernberg, M.: 1992, *Artfakta. Sveriges hotade och sällsynta ryggradsdjur 1992*, Databanken för hotade arter, Swedish University of Agricultural Sciences. Uppsala, Sweden, 346 pp. [In Swedish.]

Ahti, T., Hämet-Ahti, L. and Jalas, J.: 1968, *Ann. Bot. Fenn.* **5**, 169-211.

Angelstam, P.: 1992, *In:* Hansson, L. (ed.), *Ecological Principles of Nature Conservation, Applied in Temperate and Boreal Environments*, Elsevier, London, UK, pp. 9-70.

Attiwill, P.M.: 1994, *Forest Ecol. Manage.* **63**, 247-300.

Bengtsson, G., Holmlund, J., Lundström, A. and Sandewall, M.: 1989, *Long-Term Forecasts of Timber Yield in Sweden, AVB 85*, Swedish University of Agricultural Sciences, Dept of Forest Survey. Report 44, 329 pp. [In Swedish, with English summary.]

Berg, Å., Ehnström, B., Gustafsson, L., Hallingbäck, T., Jonsell, M. and Weslien, J.: 1994, *Conserv. Biol.* **6**, 718-731.

Bunte, R., Gaunitz, S. and Borgegård, L-E.: 1982, *Vindeln. En norrländsk kommuns ekonomiska utveckling 1800-1980*. Lund, Sweden, 410 pp. [In Swedish.]

Carlsson, M.: 1994, *Skog & Forskning* **4/94**, 24-30.

Eckerberg, K.: 1990, *Environmental Protection in Swedish Forestry*, Avebury, Aldershot, UK, and Brookfield, USA, 179 pp.

Ehnström, B., Gärdenfors, U. and Lindelöf, Å.: 1993, *Swedish Red List of Invertebrates 1993*, Databanken för hotade arter, Swedish University of Agricultural Sciences. Uppsala, Sweden, 69 pp. [In Swedish, with English abstract.]

Esseen, P-A., Ehnström, B., Ericson, L. and Sjöberg, K.: 1992, *In:* Hansson, L. (ed.), *Ecological Principles of Nature Conservation, Applied in Temperate and Boreal Environments*, Elsevier, London, UK, pp. 252-325.

Franklin, J. F.: 1992, *In:* Naiman, R. J. (ed.), *Watershed Management: Balancing Sustainability and Environmental change*, Springer-Verlag. New York, USA, pp. 25-72.

Haila, Y.: 1994, *Ann. Zool. Fenn.* **31**, 203-217.

Hånell, B.: 1990, *Torvtäckta marker, dikning och sumpskogar i Sverige*. Swedish University of Agricultural Sciences, Uppsala. Skogsfakta: Inventering och ekonomi No. **22**, 6 pp. [In Swedish.]

Hansson, L. and Angelstam, P.: 1991, *Landscape Ecol.* **5**, 191-201.

Harris, L. D.: 1984, *The Fragmented Forest*, University of Chicago Press. Chicago, USA, 211 pp.

Höjer, E. W.: 1954, 'Skogsodlingsgränsen på kronoparkerna i Norrland' in *Norrlands Skogsvårdsförbunds Tidskrift*. Stockholm, Sweden, pp. 233-253. [In Swedish.]

Juhlin Dannfelt, M.: 1959, 'Skogarna och deras vård i södra Sverige'. *In:* Arpi, G. (ed.), *Sveriges skogar under 100 år, II*. Stockholm, Sweden, pp. 263-315. [In Swedish.]

Kardell, L. and Ekstrand, A.: 1990, *Skyddad skog i Sverige. 1. Areal och virkesförråd inom nationalparker, naturreservat och domänreservat*, Swedish University of Agricultural Sciences, Dept of Environmental Forestry. Report **48**, 74 pp. [In Swedish.]

Lämås, T. and Fries, C.: 1995, *In:* Köhl, M., Bachmann, P., Brassel, P., Preto, G. (eds.), The Monte Verità Conference on *Forest Survey Designs. 'Simplicity versus Efficiency' and the Assessment of Non-timber Resources*, Swiss Federal Institute of Technology (ETH) Sect. of Forest Inventory and Planning. Zürich, Switzerland, pp. 296-311.

Liljelund, L.-E., Pettersson, B. and Zackrisson, O.: 1992, *Svensk Botanisk Tidskrift* **86**, 227-232. [In Swedish.]
Linder, P. and Östlund, L.: 1992, *Svensk Botanisk Tidskrift* **86**, 199-215. [In Swedish, with English summary.]
National Atlas of Sweden 1990, *The Forests*, Bokförlaget Bra Böcker, Höganäs, Sweden, 144 pp.
NBF (National Board of Forestry): 1951-1994, *Statistical Yearbook of Forestry 1951, ..., 1994,* Swedish National Board of Forestry, Stockholm and Jönköping, Sweden. [In Swedish, with English summary.]
NBF (National Board of Forestry): 1981, *General Forest Inventory Program*, Swedish National Board of Forestry, Jönköping, Sweden, 28 pp.
NBF (National Board of Forestry): 1991, *Tagen hänsyn vid slutavverkningar 1989-91,* Swedish National Board of Forestry. Meddelande 1991:**4**, 42 pp. Jönköping, Sweden. [In Swedish.]
NCA (Nature Conservancy Act): 1964, *Naturvårdslag,* SFS 1964:822. Stockholm, Sweden, 10 pp. [In Swedish.]
NCP (Nature Conservancy Prescriptions): 1994, *Förordning om förändring i naturvårdsförordningen. (SFS 1976:484)*, SFS 1994:1161. Stockholm, Sweden, 3 pp. [In Swedish.]
Nitare, J. and Norén, M.: 1992, *Svensk Botanisk Tidskrift* **86**, 219-226. [In Swedish, with English summary.]
Nordquist, M.: 1959, 'Skogarna och deras vård i mellersta Sverige'. *In:* G. Arpi (ed.), *Sveriges skogar under 100 år, II.* Stockholm, Sweden, pp. 316-366. [In Swedish.]
Ohlson, M.: 1990, *Dikning av näringsrik sumpskog.* Swedish University of Agricultural Sciences, Uppsala. Skogsfakta: Flora, fauna, miljö No. **14**, 4 pp. [In Swedish.]
Oliver, C. D.: 1992, *J. Forest.* **90**(9), 20-25.
Östlund, L.: 1993, *Exploitation and Structural Changes in the North Swedish Boreal Forest 1800-1992*, Swedish University of Agricultural Sciences, Dept. Forest Vegetation Ecology. Dissertation in Forest Vegetation Ecology No. **5**, 133 pp.
Paulsson, V.: 1978, *J. Roy. Swed. Acad. Agric. For.* **117**, 95-102. [In Swedish, with English summary.]
Ranneby, B., Cruse, T., Hägglund, B., Jonasson, H. and Swärd, J.: 1987, *Studia Forestalia Suecica* No. **177**. Swedish University of Agricultural Sciences, Uppsala, 29 pp.
Rudqvist, L.: 1991, *Skogseko* **3/1991**, 14-15. Jönköping, Sweden. [In Swedish.]
Sandström, E.: 1991, *Participation in the Design of a System to Assess Environmental Considerations in Forestry. A Case Study of the Greenery Project,* Swedish National Board of Forestry. Report 1991:**6**. Jönköping, Sweden, 15 pp.
Sandström, E.: 1994, *Skogseko Personal* **4/1994**, 20-21. Jönköping, Sweden. [In Swedish.]
von Segebaden, G.: 1992, *In:* Lindgren, D.(ed.), *Pinus Contorta: From Untamed Forest to Domestic Crop,* Dept. Forest Genetics and Plant Physiology, Swedish University of Agricultural Sciences, Report **11**, Umeå, Sweden, pp. 8-23.
Söderberg, U.: 1993, *In: Proceedings of Ilvessalo Symposium on National Forest Inventories, August 17-21, 1992, Finland*, Finnish Forest Research Institute. Research Papers **444**, pp. 144-149.
SOU 1978:6: 1978, *Skog för framtid. Betänkande av 1973 års skogsutredning*, Liber, Stockholm, Sweden, 194 pp. [In Swedish.]
SOU 1992:76: 1992, *Skogspolitiken inför 2000-talet*, Allmänna förlaget. Stockholm, 343 pp. [In Swedish.]
Statistics Sweden: 1994. *Protected nature*, Statistical Reports Na **41** SM 9401, Örebro, Sweden. 29 pp.
Stora Skog: 1993, *Ecological Landscape Planning. The Grangärde Area as a Model.* Falun, Sweden, 8 pp.
SUAS (Swedish University of Agricultural Sciences): 1993. *Skogsdata 1993*, Swedish University of Agricultural Sciences, Dept. Forest Survey, Umeå, Sweden, 74 pp. [In Swedish.]
Swedish Forestry Act: 1979, *Skogsvårdslag*, SFS 1979:429, Stockholm, Sweden, 5 pp. [In Swedish.]
Swedish Forestry Act: 1993, *Lag om ändring i skogsvårdslagen (SFS 1979:429)*, SFS 1993:553, Stockholm, Sweden, 9 pp. [In Swedish.]
Wilcove, D. S.: 1989, *Trend Ecol. Evolut.* **4**, 385-388.

CRITERIA AND INDICATORS OF SUSTAINABLE FOREST MANAGEMENT IN CANADA

L. F. RILEY

Criteria and Indicators Secretariat, Canadian Forest Service, 351 St. Joseph Boulevard, Hull, Québec K1A 1G5, Canada

Abstract. Sustainable forest management is an international as well as a Canadian domestic concern. Sustainability is characterised by a framework of criteria and associated indicators which identify its principal attributes and the parameters which can be measured to assess the extent to which sustainability is being achieved. Canada is developing a set of criteria and indicators of sustainable forest management for domestic purposes, and is participating in a number of similar international initiatives which will establish international norms for sustainability. Harmonisation of these various initiatives will ensure international acceptance and understanding amongst global regions of similar and dissimilar forest zones. Implementation, a follow-up goal, will ensure that the principal objectives of sustainable forest management, as espoused by the National Forest Strategy for Canada, are achieved.

Keywords. CRITERIA, INDICATORS, SUSTAINABLE, FOREST MANAGEMENT, ENVIRONMENTAL, BOREAL, TEMPERATE, ASSESSMENT, HARMONISATION.

1. Introduction

The concept of sustainable resource development is not a new one. By way of example, foresters in different parts of the world have been striving to achieve sustainability of the forest resource for a century and more. However, as a social maxim, the notion has only become part of the public lexicon since the publication of the Brundtland Report (WCED, 1987).

Nowhere has the sustainability message of the Brundtland Report been more vigorously taken up than in the global forest community. The implications for the environment, society and the economy of failing to come to grips with the sustainable management of the global forest estate were elaborated and debated at the Rio Summit, the United Nations Conference on Environment and Development (UNCED), in June 1992. In addition to action on the integrally related topics of biodiversity, climate change and desertification, UNCED placed a clear demand upon the forest nations of the world to take action on sustainable forest development through the Forest Principles (UNCED, 1992b) and Chapter 11 of Agenda 21 (UNCED, 1992a). Although not legally binding, tropical, temperate and boreal forest nations have heeded the call and set in place a number of international initiatives to advance the conservation and sustainable management of the global forest estate.

Canada has been a leader in these undertakings. Not only was it a principal proponent of sustainable forestry at UNCED, Canada had, in fact, pre-dated UNCED in its efforts to move quickly and purposefully towards the adoption of sustainable forest development as a national policy. Through a series of public fora, a broad cross-section of stakeholders and interested parties committed to a National Forest Strategy (CCFM, 1992) through the Canada Forest Accord in March, 1992. This was the impetus that has enabled Canada to take a strong lead role in the international debate and to initiate an action plan for the development of criteria and indicators of sustainable forest management for domestic forest policy.

Water, Air and Soil Pollution **82**: 67–70, 1995.

2. Criteria and Indicators

The goal of the National Forest Strategy provides essential purpose for sustainable forest management as it is viewed in the Canadian context.

> *"... to maintain and enhance the long-term health of our forest ecosystems, for the benefit of all living things both nationally and globally, while providing environmental, economic, social and cultural opportunities for the benefit of present and future generations."*

In February, 1994, the Canadian Council of Forest Ministers (CCFM) established the Canadian Criteria and Indicators Working Group and gave it the mandate to develop and recommend a framework of scientifically sound criteria and indicators of sustainable forest management in Canada. Once adopted, the framework would establish a working definition of sustainable forest management for Canada and would serve as a benchmark for the evolution of forest management policy in Canada. Also, it would guide decisions on international co-operation and assistance, facilitate international trade in forest products, provide the basis for response to "green consumerism", and be the signposts for assessment of achievement in implementing sustainable forest management.

The Working Group is comprised of a Steering Committee to provide direction, review progress and finalise the criteria and indicators framework for submission to CCFM; a Science Panel to develop scientifically sound criteria and indicators and to develop recommendations on data availability and reliability; and a Technical Committee to review recommended options for implementability and cost implications as well as to provide reliable information on data sources, monitoring and assessment capabilities across the country. Working Group membership is composed of representatives of a wide range of interest groups including federal, provincial and territorial governments, industry, non-government organisations, Aboriginal peoples, and the academic community.

Work was begun in March, 1994 supported by a Secretariat housed at the Canadian Forest Service office in Hull, Québec for co-ordination and communications functions. Meetings of the three groups have been held independently and jointly during the intervening period. A package of six criteria and associated indicators is now well advanced and preparations are being made for an interim reporting to CCFM Ministers at their meeting in October, 1994. Whereas a listing of indicators under each criterion would take inordinate space in this overview, the criteria, the primary categorisation in the process, are listed below. There are four environmental criteria and two socio-economic criteria.

- Conservation of biological diversity
- Maintenance and enhancement of forest ecosystem condition and productivity
- Conservation of soil and water resources
- Forest ecosystem contributions to global ecological cycles
- Multiple benefits to society
- Accepting society's responsibility for sustainable development

The aim is to finalise these criteria, and their associated indicators, for submission to CCFM by early January, 1995. Upon approval by CCFM, anticipated in the early part of 1995, they will be the underpinning for an implementation planning phase which will necessarily precede application by forest land management agencies across Canada. They will become also the basis for a Canadian intervention at the United Nations Commission on Sustainable Development (UNCSD) meeting in New York in mid-April, 1995.

3. Global Perspectives

On the international side, Canada continues to be an active proponent of criteria and indicators of sustainable forest management in several spheres. Following UNCED representatives of 42 countries, plus a number of international agencies and non-government organisations concerned with sustainable forest management, met in Montreal in September, 1993 to begin a process of criteria and indicator determination. The discussions of this Seminar of Experts on Sustainable Development of Boreal and Temperate Forests, held under the auspices of the Conference on Security and Co-operation in Europe, were guided in large measure by the Canadian delegation which had prepared in advance its view of essential elements. These eventually became the cornerstones of the proceedings emanating from the meeting. In turn, the Montreal proceedings, published in 1994, have become the underpinning for dialogue on criteria and indicators of sustainable forest management in several international initiatives.

As one group which took up the product of the Montreal meeting, the European Community (EC) is actively pursuing the completion of a pan-European set of criteria and indicators under direction of the 1993 Ministerial Conference on the Protection of Forests in Europe. This initiative has become known as the "Helsinki Process" by virtue of the venue for the first Ministerial conference on the topic. The EC announced criteria and associated quantitative indicators in June, 1994 and is in the process of developing greater definition as well as appropriate qualitative indicators. Canada has been a participating observer at the Helsinki Process sessions.

Because the Helsinki Process was exclusively European, non-European countries having boreal and temperate forest interests determined a need to undertake the development of criteria and indicators for conservation and sustainable management of these forest types outside Europe. This initiative, which took the name "Montreal Process" because it sprang from the 1993 Montreal meeting, began work in June, 1994 and is now well into the production of criteria and indicators for the non-European boreal and temperate region. As a full member of the Working Group formed for the purpose, Canada is playing a lead role, along with the United States, in this undertaking.

In other regions, Canada has been active in an undertaking by the International Tropical Timber Organisation to develop criteria and indicators of sustainable forest management for tropical forests. It is also the co-sponsor, with Malaysia, of an undertaking aimed at healing rifts which developed at UNCED between north and south, basically between developed and developing nations, with respect to forest management and the equitable application of international guidelines in that regard. Criteria and indicators of sustainable forest management at a global scale is one component of these discussions.

Although these international initiatives have their own unique attributes, there are strong similarities among them. Criteria are frequently identical, as might be expected where scientific principles are concerned. In all cases, there are environmental, social and economic criteria. Indicators, while demonstrating greater variation, tend to cover off the same basic concerns and rarely differ in substance except when it comes to essential regional differences.

Because Canada has been so active in each of the principal regional undertakings, it has not been difficult to ensure that these and the Canadian criteria and indicators are compatible. In fact, compatibility will be the key to harmonisation of the several global undertakings. Harmonisation, while not attempting to make the elements of each initiative the same, will ensure that, despite differences, the adherents of one group will be able to accept the precepts of the other groups and that a form of international understanding, short of international convention, will exist. It is expected that the harmonisation process will begin in earnest when UNCSD meets in New York next year.

In the meantime, the Canadian initiative will push ahead since, no matter what is undertaken internationally, Canada must have its own framework of criteria and indicators for the implementation of sustainable forest management on a cross-country, jurisdiction-by-jurisdiction basis. There is a ground-swell of interest and concern in Canada with respect to sustainable forest management and there must be common understanding of the concept if it is to have full domestic application.

The Canadian framework, as with the international frameworks, must be viewed as a living process. As our knowledge of forest ecosystems and factors promoting social and economic dimensions and enhancement improves, the criteria and indicators will evolve. Our collective efforts to monitor and report on trends and changes in indicators will identify where changes that are occurring may impact our ability to sustain the key values Canadians hold for their forests.

4. Summary

Canada is taking key steps to ensure that sustainable forest management becomes a reality in this country. The many stakeholders in the forest are bound together in the common purpose of implementing the National Forest Strategy actions, of which the development of criteria and indicators is key, that will lead us towards a forest estate that is productive and healthy for current and future generations.

References

CCFM (Canadian Council of Forest Ministers).: 1992, *Sustainable Forests: a Canadian Commitment*, Hull, Canada, 51 pp.

CSCE (Conference on Security and Cooperation in Europe).: 1994, Seminar of experts on sustainable development of boreal and temperate forests technical report—Annex 1. Canadian Forest Service. Hull, Québec, Canada, 145 pp.

UNCED (United Nations Conference on Environment and Development).: 1992a, Agenda 21, Chapter 11. Combatting Deforestation. United Nations, New York, USA.

UNCED (United Nations Conference on Environment and Development).: 1992b, Non-legally binding authoritative statement of principles for a global consensus on the management, conservation and sustainable development of all types of forests, United Nations, New York, USA.

WCED (World Commission on Environment and Development).: 1987, *Our Common Future* (The Brundtland Report), Oxford University Press, Oxford, UK, 383 pp.

FORESTRY AND THE BOREAL FOREST: MAINTAINING INHERENT LANDSCAPE PATTERNS

S. BONDRUP-NIELSEN

Centre for Wildlife and Conservation Biology, Department of Biology, Acadia University, Wolfville, Nova Scotia B0P 1X0 Canada

Abstract. The forestry industry in Canada has become sensitive to its image through pressure from environmentalists and the European markets and along with Natural Resource Departments across the country, is working hard to change its image by developing environmentally sound management. The dilemma is how to continue resource extraction from the forest yet ensure that wildlife populations do not become endangered. An important ecosystem process in the boreal forest is forest fire and the forestry industry has traditionally argued that clear cutting simulates forest fire by creating large areas devoid of vegetation that can then grow back either through natural regeneration or by planting. What is neglected in present forestry is the landscape pattern created through cutting which is often large continuous expanses of recently clear-cut land. In ecology we often talk about process and pattern. I will argue that we should emulate the pattern across the landscape. Wildlife (including all life) have evolved in landscapes with inherent patterns. If we can maintain the inherent pattern in a landscape I argue that we can maintain the wildlife. Here I outline a method for describing landscape patterns and maintaining these patterns through forest management.

Keywords. BOREAL FOREST, HARVEST DISTRIBUTION, MAINTAINING LANDSCAPE PATTERN, COMPUTER SIMULATION, WILDLIFE.

1. Introduction

Forestry has been under pressure from environmentalists, naturalists, and the general public mainly because of clear cutting practices and spraying of biocides. The attack on forestry practices probably culminated in the airing of 'Voices in the Forest' in the David Suzuki series 'The Nature of Things' in Canada. Forestry industries have become sensitive to their image and along with Natural Resource Departments across Canada and around the world, are working hard to become more "environmentally friendly" and to educate the public about forestry practices. More and more we hear terms like Ecological Forest Management, Integrated Forest/Wildlife Management, Integrated Resource Management and Managing for Biodiversity (e.g. Thompson and Welsh, 1993).

A nagging question in forestry is how to continue resource extraction from the forest yet ensure that wildlife populations do not become endangered. To add to this challenge, the public and legal perception of wildlife is changing. Once limited only to game species, the term wildlife in Canada, for example, is now defined as all life, from bacteria to plants and animals, including many species yet undescribed (Anon, 1990)! The *Wildlife Policy of Canada* states in principle that no human activity, including forestry, should cause a decline or increase in any wildlife species. In Sweden a new forestry act was passed in early 1994 and it recognises and requires that ecological functions of forest ecosystems be maintained through forestry practices (Lämås and Fries, 1995, this volume).

If we could know the habitat requirements of all species and somehow set population density goals for these, we could in theory develop forest management

Water, Air and Soil Pollution **82**: 71–76, 1995.

plans that would ensure their protection. This would require that forest management plans include population levels and specific habitat requirements for all species as specific objectives. Clearly, even if we had all the information needed, forest management would become inextricably tied in a Gordian knot! So, how do we manage forests such that they can continue to produce wood yet maintain their valuable ecological functions?

In this article I address this question. I present what I term the pattern game. This game illustrates how spatial patterns can be maintained and how difficult it is to change them. I argue that if we maintain the natural forest landscape pattern representative of a region we will take care of many of the ecological values of that region. I present a method for maintaining forest landscape patterns which brings me back to the pattern game. Finally I outline some of the problems with this approach

2. The Pattern Game

Assume we have 81 squares arranged in a 9×9 matrix. Each square has a number ranging from 1 to 27 such that 3 squares have the same numbers (i.e. three 1's, three 2's etc.) One way to describe the pattern of squares with similar numbers is by the average nearest neighbour distance between the squares labelled 1, the squares labelled 2 etc. The average nearest neighbour distance among all sets of similarly numbered squares is one measure of the pattern they create. If we introduce time, then with each time step all squares increase by a value of one such that the squares with number 1 will become 2 etc. Three squares numbered 26 or greater will be chosen and given a value of 1. Some squares may therefore increase in value above 27. Squares with a value of 30 must be selected and receive a value of 1. The choice of the three squares is critical for this may maintain or alter the pattern. If we want to maintain the present pattern we merely choose the combination of squares which when cut least disrupts the pattern; this will of course be the squares with a value of 27. However, we may wish to change the pattern. To do this we choose the combination of three squares with average nearest neighbour distance that represents the new pattern with each time step and observe how long it takes to alter the total pattern. The results of a series of runs maintaining a pattern followed by a series of runs where the pattern is changing are presented in Figure 1. Patterns can be maintained infinitely but if one wants to change the pattern it takes a long time to do so. What is the lesson for forestry?

3. Forestry and Forest Ecosystems

What exactly do we want to achieve? It is essential we ensure that ecological functions and processes in forests are not degraded through forest management. We must recognise that a forest is more than a collection of trees associated in different stand types and age classes with accompanying yield curves. Forests are ecosystems composed of dynamic and diverse communities of plants and animals - wildlife. Complex interplays among climate, soil type, soil organisms and the plants and animals in forest ecosystems have evolved over many thousands of years (Hunter 1990). All players in a forest ecosystem have important roles. Some players may be

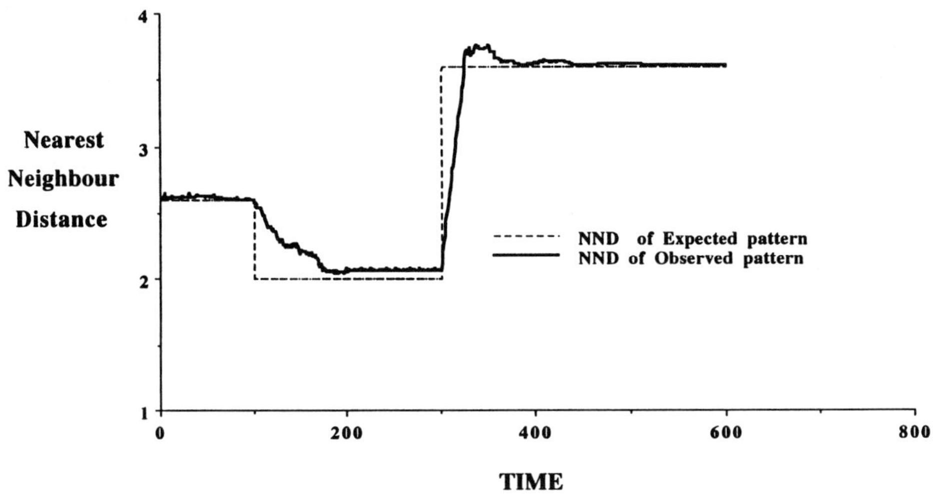

Fig. 1. Results of a run of the pattern game for 600 time periods. Stippled line represents the pattern represented by nearest neighbour distance we wish to achieve, solid line represent the actual pattern.

more essential than others, and an ecosystem may be able to compensate for the loss of a few players. However, who will take the responsibility for consequences of disrupting a system we are only beginning to understand?

How can we practice forestry with minimal disruption to forest ecosystem function? Clear cutting must be recognised as the only economically feasible approach to harvesting trees for pulp and paper production. If other forest industries were to develop where value was added to the wood harvested, such as furniture making, post-and-beam house construction or what ever, other harvesting methods than clear-cutting could become economical. Then, we would not be faced with the problems associated with clear cutting but other problems might surface such as high-grading and/or loss of dead wood in the forest. However, if we are to remain in pulp and paper production, clear cutting appears inevitable.

The mosaic of different forest stands varying in age and size in a landscape were traditionally a function of geology, soil types and disturbance regimes such as spruce bud worm outbreaks, wind storms and fires. Landscapes are thus dynamic, however, over large regions a characteristic pattern should be evident. Wildlife in a region has evolved in characteristic landscapes, however, through forest management we are altering the landscape patterns. Forest management concerns itself with patterns in the landscape. However, this relates to blocking and cost efficient ways of accessing the wood. The landscape pattern being created by present forestry activity will persist far into the future. This should be evident from the pattern game. Forest stands cut today will be ready for harvest in 60–80 years depending on rotation age, thus perpetuating the cutting pattern.

Forest management should try to mimic as much as possible, "natural" landscape patterns including tree species mixes found in an area. This has been suggested by

Rowe (1992) and Hunter (1990) calls this approach managing for a diversity of diversity. No method has been suggested as to how to achieve this. A landscape approach to forest management should be adopted as the fundamental management philosophy where the clear cutting pattern mimics the natural landscape pattern. This approach will in general be good for wildlife as it ensures a diversity of stand sizes and juxtapositions. By managing for a diverse mosaic of stand types (habitat types) across landscapes we will in fact be managing for all wildlife. Monitoring systems for a variety of wildlife will have to be in place to determine how well we are doing but fundamentally if habitats are available wildlife will survive.

How do we maintain naturally diverse forest landscape mosaics? This, I think, is not as difficult as it may seem given present computer and software technology. Geographic Information Systems (GIS) allows analysis of computer maps with relational spatial data and linear programming (LP) is used to predict forest harvest.

At the level of the landscape we wish to maintain "natural" forest stand mosaics. In order to do this we need to know what the landscape is like—we need to be able to quantify it. The fundamental unit is the forest stand. These units are often too small for practical management considerations. Classification of forests generally is hierarchical and we need to conglomerate stand types into manageable sets of forest types. Classification of forest types is often arbitrary and ultimately must be chosen such that they represent important ecological features in a region. A given region can be characterised by, among other things, the variety of forest types, size distribution of patches of forest types and spatial arrangement of the patches of the different forest types. Spatial arrangements of patches belonging to a particular forest type can be quantified by nearest neighbour distances. Each forest type will make up a particular proportion of an area, will have a particular distribution of nearest neighbour distances and sizes of patches. There are many other landscape parameters that could also be calculated (Baker and Cai, 1992). It is relatively easy to calculate these landscape parameters using GIS. GIS computer software, when properly programmed, can identify the different stand types and calculate the various parameters.

In forest management, linear programming is used to determine how much volume and which types of stands should be cut. Although forest management is very explicit about where harvesting should take place, the consideration of the spatial arrangement is strictly economic. However, if we wish to maintain the natural landscape pattern for a region the harvest distribution must be such that it disrupts this pattern the least. Here we can learn from the pattern game. Blocks which potentially can be cut must be identified. Then, it becomes a matter of running computer simulations to determine the combination of blocks which, when cut, least disrupts the "natural" landscape pattern characteristic of that region. In other words, the traditional blocking which is carried out to determine the harvesting distribution must consider the impact of the distribution on the landscape. This method of forest management might be called Ecological Forest Landscape Management.

One aspect of a natural forest landscape that will inescapably differ under forest management is age distribution. Under sustained forest management as it is presently envisioned a rectangular distribution of age classes is necessary such that a constant amount of resource can be extracted on a yearly basis. However, Ecological Forest Landscape Management can maintain stand types, sizes and spatial arrangements.

Silviculture techniques could be employed to maintain particular tree species mixes in given areas.

Good forestry has generally been equated with re-planting. In fact, clear cutting is often justified on the basis that a better and potentially more productive forest will result from planting selected kinds and strains of conifers. However, planting and the various treatments associated with planting may not be ecologically sound. This does not mean that we have to stop research into tree improvement, cloning etc. for there is certainly a need for this. What we need to evaluate is the potential for natural regeneration in association with some planting of desired species. Ultimately our interest should be to restore fully functioning forest ecosystems with an approximate mix of species of both flora and fauna which were there before we cut the area.

There is much controversy over old growth forests and rotation age in forestry. This needs thorough debate and I will not get into it here. However, one option would be to have different rotation ages within each forest management area. For example, 10 to 15% of the area could be in a 120–150 year rotation age, 10 to 15% in a 100–120 year rotation, another 10 to 15% in a 80–100 year rotation, and the remainder in a 60–80 year rotation. Some formula could be struck that would satisfy the forest industry plus the need to have old age stands within forest management areas.

There are a number of problems with this Ecological Forest Landscape Management proposal. There may not be road access to a particular stand or block of stands that should be cut. Ownership pattern may not allow implementation of the plan. However, the method is flexible in that one can always choose the combination of possible blocks slated for cutting that least disrupts the natural pattern of the landscape. Crown land might serve as a buffer to harvesting on private land that does not follow the guide-lines.

Ecological Forest Landscape Management recognises that if we take care of forest communities across a landscape we will be taking care of most wildlife species in an area. It further recognises that forests are dynamic and that if you do not create some large clear cuts there will be no large even-aged young stands or mature stands in the future which all adds to biodiversity. This approach also recognises that the majority of biodiversity will be outside parks and other protected areas where it needs to be protected. Monitoring of selected wildlife species will be critical to the Ecological Forest Landscape Management plan and should be carried out to determine how successful we are. If densities of particular wildlife species change drastically, special management action must be taken and can easily be superimposed on the Ecological Forest Landscape Management plan. Target wildlife species needing special action may be best protected by creating a preserve.

4. Conclusions

Major impacts by humans includes fragmentation of habitat and altering landscape patterns. Once a landscape pattern has been altered it may take a long time to restore traditional landscape patterns particularly when the processes affecting landscape patterns are slow by human time scale such as the growing of a forest. The pattern game has shown this convincingly. Harvest distribution has traditionally largely been determined based on economic considerations. However, harvest distribution should consider the landscape pattern created and how this affects the regional landscape pattern we wish for an area. This can be done by computer simulation determining the harvest distribution that least disrupts the chosen landscape pattern by comparing the resulting distribution of landscape parameters with the distribution characteristic of the region. This approach I call Ecological Forest Landscape Management and it is pivotal to conservation because it represents recognition that we are a part of nature and not apart from it.

Acknowledgments

I thank Drs. Tom Herman and Ian Methven, plus two anonymous reviewers for commenting on earlier versions of this manuscript.

References

Anon: 1990, *A Wildlife Policy for Canada.* Environment Canada, Canadian Wildlife Service Cat. No. CW66-59/1990E, 29 pp.

Baker, W. L. and Cai, Y. 1992, *Landscape Ecology* **7**, 291-302.

Lämås T. and Fries, C.: 1995, *Water, Air, Soil Pollut.* **82**, 69-74.

Hunter, M.: 1990, *Forests, Forestry and Wildlife.* Prentice Hall, Englewood Cliffs, USA, 370 pp.

Rowe, J. S.: 1992, *For. Chron.* **68**, 222-224.

Thompson, I. A. and Welsh, D. A.: 1993, *For. Chron.* **69**, 32-39.

FOREST HEALTH MONITORING IN CANADA: HOW HEALTHY IS THE BOREAL FOREST?

J. P. HALL

Canadian Forest Service, 351 St. Joseph Boulevard, Hull, Quebec K1A 1G5, Canada

Abstract. The Canadian boreal forest covers 299.2 Mha which is two-thirds of Canada's forest and runs in a continuous belt from Newfoundland north and west to the Yukon. The major species are spruce, pines, balsam fir, white birch and trembling aspen often occurring in extensive monocultures. Wildfire is the driving successional force in the boreal forest and has remained so despite fire suppression activities and extensive harvesting. Insects and diseases also cause extensive damage. In order to ensure the sustainability of forests, it is necessary for the forest manager to know the condition of the health of these forests. The CFS established in 1984 the Acid Rain National Early Warning System in order to monitor the health of the forests. National results show that mortality is generally in the normal range of 1-3% and is caused by known stresses; insects, diseases and abiotic damage. No signs of pollution damage have been yet been detected in boreal forests by the system. An early warning system to detect and monitor conditions remains an essential part of our commitment to the sustainability of Canada's forests.

Keywords. BOREAL FORESTS, FOREST HEALTH, AIR POLLUTION, DETECTION.

1. Introduction

Canada's forests are a valuable economic and sociological resource whose sustainability is essential to our well-being. To ensure sustainability forest managers need to know if forests are healthy. During the 1970s and 1980s concerns began to evolve concerning the risks to forests from air pollution, most specifically acid rain. Part of this concern arose from damage observed in European forests which was attributed to pollution. In Canada, it was the opinion of air pollution scientists and other experts on forest productivity that the continued deposition of sulfates and nitrates at ambient levels would have deleterious long term consequences on forest productivity (Fraser, 1989). As a result of these concerns and perceptions about forest health, the Canadian Forest Service (CFS) established the Acid Rain National Early Warning System (ARNEWS) in 1984 (Hall and Addison, 1991).

It was primarily the hardwood/mixedwood forests in southern Canada exposed to potentially damaging levels of pollution which prompted the establishment of the network. These forests were exposed to >20 kg ha^{-1} yr^{-1} of sulfate deposition; levels known to cause damage to aquatic ecosystems. Boreal forests are far from the sources of pollution and receive less than the southern forests, but they are also situated on the more acid-sensitive soils of the Canadian Shield. In these instances a lower amount of pollution could have significant effects on these poorly buffered soils.

Monitoring biological responses of forests to pollutant deposition in the field is difficult since biological systems respond to a complex of environmental factors rather than to a single factor such as air pollution (Bormann, 1985; Manion, 1981). Other stresses affect forests with such large impacts that subtle responses, such as may be caused by pollutants, are lost in the "noise" of the data. Thus by the time that pollution-caused damage is obvious, extensive damage to the ecosystem may already have occurred. The approach taken by the CFS was to assess the state of the health of

Water, Air and Soil Pollution **82**: 77–85, 1995.

the forest using a common set of measurements taken on permanent sample plots. The plots were established by the CFS - Forest Insect and Disease Survey (FIDS). FIDS has provided an annual national overview of forest conditions in the survey of the 220 Mha of productive forest since 1936.

The objective of ARNEWS is to detect damage to forest trees and soils caused by acid rain by identifying the damage sustained by Canadian forests that are not attributable to natural causes or management practices and to monitor over the long-term, vegetation and soils to detect changes attributable to acid deposition and other pollutants in representative forest ecosystems.

The ARNEWS was designed to detect damage caused by acid rain so that action could be taken to forestall anticipated damage. The ARNEWS consists of 150 permanent sample plots throughout the forested ecosystems of Canada. There are 59 plots in the Boreal Shield Ecozone, and 11 plots in the Boreal Plains Ecozone. The 70 plots are located in semimature, natural forests where the number of trees would normally be expected to diminish by 1–2% annually from competition as the stand matures. Mortality in excess of this also often occurs, caused by defoliating insects, bark beetles, root or stem diseases or weather extremes.

2. Methods

Several types of measurements are made in the plots:

The slope, aspect and forest cover type are described for each plot. Every five years, crown structure, crown density and chemical composition of foliage and soils are measured. Tree condition, defoliation and the presence of insects and diseases are measured annually. Insects and diseases are surveyed at various times of the year to detect the major forest pests.

Insects and diseases are identified and their levels of damage are determined. The trees in the plots are assessed visually; sampling and detailed measurements requiring hand-held branch samples are made on off-plot trees to ensure that trees in the plot are not disturbed. Branches are collected from the ground using pole pruners. Specimens collected for analysis in the laboratory are shipped in bags.

Conifer tree crowns, except tamarack, are classified according to the amount of defoliation in the crown (Table I). Defoliation is defined as foliage missing, for whatever reason, from the normal foliage complement of the tree species. This takes into account the natural loss of needles as a twig matures.

Hardwood tree crowns, as well as tamarack, are assessed based upon the visible parts of the outer crown; the quantity and quality of foliage in the outer crown, and the percentage of the outer crown that contains bare twigs and dead branches (Table I).

The condition of the foliage is one of the major contributing factors to tree health, both in the short and long term. It also influences the assessment of tree condition, and it is the foliage which exhibits the visual symptoms caused by various air pollutants, the detection of which is one of the basic objectives of the ARNEWS (Magasi, 1988, D'Eon *et al.*, 1994).

Possible symptoms of air pollution are compared with known symptoms of emission toxicity on vegetation (Malhotra and Blauel, 1980). The symptoms of air pollution damage are not highly specific and frequently resemble damage from other causes.

The strategy of the ARNEWS program is to detect signs of damage to forests by determining the cause(s) of the observed change. This is the critical part of the monitoring, and here the experience of FIDS field technicians trained to recognise symptoms of damage such as those caused by climatic conditions, nutrient deficiencies, air pollution, and the effects of insects and diseases is crucial. The ability of FIDS to call on a cadre of forest health specialists further enhances the reliability of the system.

The ARNEWS network is not enough in itself to provide a reliable assessment of the state of health of the boreal forest with 70 plots. The ARNEWS and its analysis are supported by other FIDS surveys which assess the health of the forest thus enhancing the credibility of the results.

Recent classification into ecozones by the State of the Environment Reporting of Environment Canada has provided a method of reporting based on the physical and biological characteristics forests (Anonymous, 1993). Consequently, results are described for two Ecozones, the Boreal Shield and the Boreal Plains Ecozones.

The evaluation of tree health is based on mortality rates, the condition of tree crowns and the presence of symptoms of damage from air pollution for each Ecozone.

3. Description of Boreal Ecozones

The **Boreal Shield Ecozone** extends from northern Saskatchewan east to Newfoundland, passing north of Lake Winnipeg the Great Lakes and the St. Lawrence River. The zone is dominated by broadly rolling uplands and lowlands. Its surface is characterised by precambrian granitic bedrock outcrops interspersed with ridged to hummocky, discontinuous veneers and blankets of glacial moraine, fluvio glacial and colluvium. Soils range from Humo-Ferric Podzols in the south to Brunisols in the north. The zone generally supports closed stands of conifers, largely white spruce, black spruce, balsam fir, and tamarack. Towards the south, there is a wider distribution of broadleaf trees; white birch, trembling aspen and balsam poplar; and conifers, white pine, red pine, and jack pine. Throughout the contrasting areas of exposed bedrock, this mosaic of soils and non-soils tends to be covered with a range of communities dominated by lichens and mosses, shrubs, forbs and grasses. Small to medium-sized lakes are common. Climate in the Ecozone ranges from humid, cold cryoboreal in the northwestern half to prehumid, moderately cool boreal climate in the east. The continental climate is modified by maritime conditions in its eastern coastal regions and by cold arctic air masses.

There are 59 plots in this ecozone. The health of the following species are described; eastern white pine, jack pine, tamarack, black spruce, white spruce, balsam fir, trembling aspen, white birch, and red maple.

The **Boreal Plains Ecozone** extends as a wide band from the Peace River country of British Columbia in the northwest to the southeastern corner of Manitoba.

This nearly level to gently rolling plain consists largely of hummocky to kettled glacial moraine and lacustrine deposits. It is Underlain by Cretaceous shales, and soils are largely Luvisols which grade southward into Black Chernozems and northward into Brunisols and Organics.

The main conifer species are white spruce, black spruce, jack pine, and tamarack. Broadleaf trees, particularly white birch, trembling aspen and balsam poplar, are most

TABLE I

Crown classification for conifers (a) and hardwoods (b).

(a) Conifers (except tamarack) are classified according to the amount of defoliation in the crown. Defoliation is defined as foliage missing for whatever reason from the normal foliage complement of the tree species. Consider bare and dead tops as defoliated. The health of conifer crowns is assessed as the percentage of foliage missing, for whatever reason, from the 'normal' foliage complement of the tree. This takes into account the natural loss of needles as a twig matures.

Code	Assessed condition description
1	No defoliation.
2	Only current foliage defoliated, total defoliation <25%.
3	Current and/or older foliage defoliated but total <25%.
4	25-50% total defoliation.
5	51-75% total defoliation.
6	76-90% total defoliation
7	More than 90% total defoliation.
8	Tree died since last assessment.
9	Dead tree.

(b) The hardwood classification system is based upon the visible parts of the outer crown; the quantity and quality of foliage in the outer crown, the percentage of the outer crown that contains bare twigs, and dead branches. Tamaracks are assessed as if they were hardwoods. The condition of hardwood tree crowns is assessed according to the following:

Code	Assessed condition description
10	Full complement of foliage, tree without crown damage.
20	Foliage thin, off colour, no dead branches or bare twigs.
30	No dead branches and bare twigs in <5% of crown.
35	No dead branches and bare twigs in more than 6% of crown.
40	Dead branches and bare twigs in <15% of the crown.
45	Dead branches and bare twigs in 16-25% of the crown.
50	Dead branches and bare twigs in 26-37% of the crown.
55	Dead branches and bare twigs in 38-50% of the crown.
60	Dead branches and bare twigs in 51-75% of the crown.
65	Dead branches and bare twigs in > 76% of the crown.
70	More than 50% of the crown dead, only small adventitious branches present, usually at the base of the crown or stem.
08	Tree died since last assessment.
09	Dead tree.

numerous in the transitional section leading to the prairie grasslands. Black spruce and tamarack increase in dominance along the more northerly sections of the ecozone. The subhumid, moderately cold cryoboreal climate is typified by cold winters and moderately warm summers. This ecozone is strongly influenced by continental climatic conditions.

There are 11 plots in the Boreal Plains Ecozone. Results are described for jack pine, lodgepole pine, black spruce, white spruce and trembling aspen.

4. Results and Discussion

4.1. BOREAL SHIELD ECOZONE - TREE MORTALITY

The annual mortality rates for species in this ecozone have ranged from 0% to 2.5% with most less than 1%. Mortality was caused by natural factors; suppression, damage from the combined effects of defoliating insects, diseases and from storm damage. The highest mortality rates occurred in trembling aspen (2.5%) and balsam fir (1.8%), and was caused by damage from outbreaks of the large aspen tortrix (*Choristoneura conflictans* Wlk.) and the eastern spruce budworm (*Choristoneura fumiferana* Clem.) respectively. Defoliation from these insects, coupled with subsequent attack by other insects, caused higher rates of mortality than normal.

4.2. BOREAL SHIELD ECOZONE - CROWN CONDITION

Eastern white pine (*Pinus strobus* L.)

Tree Condition is variable throughout the ecozone. Crowns are in poor condition in some areas and have been deteriorating rapidly in recent years. Most trees have been in class 4 since 1989. In 1992 66.7% of trees were in Class 4, but in 1993 58.8% were in class 5 and 29.4% in class 4. The missing foliage is a result of various defoliating insects. White pine blister rust (Cronartium ribicola Fisch.) has persisted in some areas for several years generally affecting younger trees.

Jack pine (*Pinus banksiana* Lamb.)

Tree condition is generally good with 63.8% in classes 1 and 2. This is slightly down over the past five years, 79.5% were in these classes in 1992. Tree condition was best in Manitoba and Ontario, crown damage was higher in Quebec and Alberta. The most pre-valent and damaging pests were western gall rust (*Endocronartium harknessii* [J.P. Moore] Y. Hiratsuka) and red squirrels (*Tamiasciurus hudsonicus*). Trees were also affected by the Olethreutid moth (*Rhyacionia* sp.) with the larvae of this species damaging many trees.

In some areas in Ontario tree condition has improved since 1989 with 90% of dominant/codominant trees having no apparent damage. In Quebec, damage from defoliation by jack pine budworm (*Choristoneura pinus pinus* Free.) occurred frequently. In the Manitoba portion of the Boreal Shield Ecozone about 97% of the trees were in the healthiest classes.

Tamarack (*Larix laricina* (Du Roi) K. Koch)

Most trees are healthy and have been damaged only at trace levels. Most damage has been caused by the larch sawfly (*Pristiphora erichsonii* Htg.).

Black spruce (*Picea mariana* (Mill.) B.S.P.)

Tree condition was generally good in 1993 with 45.8% of trees in classes 1 and 2 and 33.5% in class 3. However, this is down from previous years when about 70% were in classes 1 and 2. Tree condition was best in Newfoundland with over 80% in the healthiest classes (1-3). Crown damage was more prevalent in Quebec where over half the sampled trees were damaged by defoliating insects and rust damage on needles. In some areas crown condition actually improved coincident with a reduction in spruce budworm activity. Evidence of past defoliation by spruce budworm was common in Ontario plots.

White spruce (*Picea glauca* (Moench) Voss.)

Tree condition overall was good with 74.8% in classes 1 and 2 which is less than 25% defoliation. Tree condition was best in Newfoundland where trees had recovered from past infestations of the spruce budworm. Here, the percentage of trees with no defoliation was 86%. Higher levels of damage occurred elsewhere and defoliation by the spruce budworm was present at low levels.

Balsam fir (*Abies balsamea* (L.) Mill.)

Overall tree condition in balsam fir was good with 51.7% in classes 1 and 2 and 36.8% in class 3. This is a slight deterioration from previous years; in 1991 and 1993 the corresponding percentages were 73.2% and 76.2%. This decline in crown condition was caused by a variety of insect pests.

Trees in Newfoundland have been defoliated in two successive years by balsam gall midge (*Paradiplosis tumifex* Gagne) and the hemlock looper, (*Lambdina fiscellaria fiscellaria* Guen.. Infestation by the balsam woolly adelgid, (*Adelges piceae* Ratz.) increased in 1993 also influencing tree condition. A needle rust (*Melampsorella caryophyllacearum* [Schroet.] Hedwiga), infected 24% of trees in one plot.

Some intermediate trees showed some decline symptoms and were in classes 4 or 5. Light damage caused by winter drying affected 11% of balsam fir in one plot in Newfoundland, this was the only incidence of abiotic damage on balsam fir in 1993.

In some areas, particularly in Ontario, tree condition improved slightly as a result of reduction in the eastern spruce budworm infestation. Balsam fir had been subjected to moderate to severe levels of budworm defoliation since 1988 resulting in the deterioration in crown condition, but the budworm defoliation is reduced in 1993 compared to previous years.

Trembling aspen (*Populus tremuloides* Michx.)

Overall tree condition was fair as might be expected in a species attractive to many insects and diseases. About a third of the trees were in classes 1 and 2 (35.5%) with an additional 56.0% in classes 3 and 4.

Most trees had some foliar damage and about a third had trace foliar damage caused by diseases. Damage has been caused by the large aspen tortrix, the forest tent caterpillar (*Malacosoma disstria* Hbn.), and leaf beetles (*Chrysomela* sp.) for several years. Decay was caused by false tinder conk (*Phellinus tremulae* [Bond.] Bond. & Boriss.) and poplar Peniophora (*Peniophora polygonia* (Pers.:Fr.) Boud.). Stem cankers were also common.

White birch (*Betula papyrifera* Marsh.)

Overall tree condition of white birch indicates considerable crown damage with about two-thirds of the trees (62.0%) in condition classes 3 and 4. This damage was caused by various defoliating insects, by the birch casebearer, (*Coleophora serratella* L.) and stress from competition resulting in numbers of dead branches.

Red maple (*Acer rubrum* L.)

Tree condition improved in 1993. Overall tree condition had deteriorated steadily from 1989 caused by drought-induced decline, physical damage and associated decay. 1993 is the first year when no further deterioration occurred. Insect defoliation levels have been at trace to light levels since 1989. Leaf diseases, such as tar spot have also occurred at trace levels. Neither stress is a major factor in causing the decline.

4.3. Boreal Shield Ecozone - Air Pollution Damage

No symptoms of damage by air pollution were observed on any sampled trees in the Boreal Shield Ecozone. Tree health was affected however by a variety of insects, diseases and occasionally by abiotic factors.

4.4. Boreal Plains Ecozone - Tree Mortality

Mortality of species described for this ecozone has been 2% or less annually since measurements began in 1985. Mortality was caused primarily by natural thinning, with some damage by insects and diseases. Defoliating insects caused little damage in this ecozone.

4.5. Boreal Plains Ecozone - Crown Condition

Jack pine (*Pinus banksiana* Lamb.)

Tree condition has improved from 1992 with 85% of the jack pine in the healthiest classes. In addition to Armillaria root rot, two other pests damaged the trees: lodgepole pine dwarf mistletoe (*Arceuthobium americanum* Nutt. ex. Engelm.) and western gall rust. Both lodgepole pine dwarf mistletoe and western gall rust are causing minor damage to jack pine. In previous years the jack pine budworm had damaged trees but the incidence of this pest is declining.

Lodgepole pine (*Pinus contorta var. latifolia* Engelm.)

Some decline in tree condition occurred in 1993. Trees in the healthier classes (1-3) comprised 76.5% in 1993 compared to 94.5% in 1992. In 1993, 19.7% of assessed trees were in class 4 with 25-50% defoliation.

Light to moderate foliar discoloration was caused by a needle cast, *Lophodermella concolor* (Dearn) Darker, and an associated hyperparasite, *Hendersonia pinicola* (Wehmeyer). This is part of a widespread occurrence of pine needle cast which, in 1993, occurred over more than 193 000 ha in western Canada.

Black spruce (*Picea mariana* (Mill.) B.S.P.)

Trees are generally healthy with 92.3% of assessed trees in the healthiest class. The only pest found on black spruce was large-spored spruce-Labrador tea rust

(*Chrysomyxa ledicola* Lagh.). The rust was causing minor damage to black spruce in Manitoba in Duck Mountain Provincial Park.

White spruce (*Picea glauca* (Moench) Voss)

Trees were generally healthy in 1993 with 93% of the trees in the healthiest classes. Tree condition improved from 1992 because of a lower incidence of insects and diseases.

In northern Saskatchewan, trees were damaged by an outbreak of spruce budworm which had infested many trees at light to moderate levels of defoliation. Other pests damaging trees included the spruce gall midge (*Mayetiola piceae* [Felt]), the pale spruce gall adelgid (*Adelges strobilobius* [Kltb.]), and yellow witches' broom (*Chrysomyxa arctostaphyli* Diet.).

Trembling aspen (*Populus tremuloides* Michx.)

Tree health declined in 1993 from 1992, primarily because of defoliation by the large aspen tortrix which caused moderate to severe defoliation throughout large parts of the Ecozone. Other pests observed of significance were false tinder conk and poplar Peniophora. Overall, 49.6% of the trees were in classes 1 and 2 and 34.6% were in Class 3 in 1993. In 1992, 89.7% were in class 1, indicating a marked deterioration in one year. Tree condition has fluctuated widely over the past five years coincident with insect infestations and local frost occurrences which also vary widely from year to year.

4.6. Boreal Plains Ecozone - Air Pollution Damage.

The trees were affected by a variety of known insects and diseases but no symptoms of pollution damage were observed.

5. Conclusions

Results to date show that tree mortality of boreal species has generally been at normal levels, within the 1-3% range common in unmanaged natural forests. Slightly higher than normal mortality rates occurred occasionally. The eastern spruce budworm was responsible for the added mortality in balsam fir, and the large aspen tortrix and several decay fungi caused higher mortality in trembling aspen. Opportunistic diseases attacking the weakened trees also contributed to the increased mortality. All mortality was attributed to identifiable and known stresses; insects, diseases, and weather extremes.

No signs of pollution damage from the long range transport of pollutants have yet been observed in any of the 70 plots in the boreal ecozones (Hall, 1993). Thus by these standards we can say that there is no evidence that the boreal forests are not healthy, although they are impacted by a variety of insects, diseases and climatic events. Forests are adapted to these stresses and can be expected to recover. Experience of insect and disease monitoring in these forests shows that this is so.

An early warning system to detect and monitor conditions remains essential to the assessed of tree health, and ensuring forest sustainability.

Acknowledgments

The author wishes to acknowledge the members of the Forest Insect and Disease Survey for supplying the data for this paper, and Environment Canada for supplying the Ecozone Descriptions.

References

Anonymous: 1993, *Ecological Science Centres News*, Vol. **1**, No.1.

Bormann, F. H.: 1985, *Bioscience* **35**(7), 434-441.

D'Eon, S. P., Magasi, L. P., Lachance, D., and DesRochers, P.: 1994, *ARNEWS Manual on Plot Establishment and Monitoring*. Can. For. Serv. Inf. Rep. P-X-117, Canadian Forest Service, Petawawa Natl. For. Inst., Chalk River, Canada, 10 pp.+App.

Fraser, G. A.: 1989, *Acid Rain Control: Potential Commercial Forestry Benefits to Canada*, For. Can. Inf. Rep. E-X-42, Forestry Canada, Ottawa, Canada, 31 pp.

Hall, J. P.: 1993, *Acid Rain National Early Warning System (ARNEWS) Annual Report 1992*. For. Can. Inf. Rep. ST-X-7, Forestry Canada, Ottawa, Canada. 50 pp.

Hall, J. P. and Addison, P. A.: 1991, *Response to Air Pollution: ARNEWS assesses the health of Canada's forests*. For. Can. Inf. Rep. DPC-X-34, Forestry Canada, Ottawa, Canada, 13 pp.

Magasi, L. P.: 1988, *Acid Rain National Early Warning System: Manual on Plot Establishment and Monitoring*, Can. For. Serv Inf. Rep. DPC-X-25, Canadian Forestry Service, Ottawa, Canada, 17 pp.+App.

Malhotra, S. S., Blauel, R. A.: 1980, *Diagnosis of Air Pollutant and Natural Stress Symptoms on Forest Vegetation in Western Canada*, Can. For. Serv. Inf. Rep. NOR-X-228, Canadian Forestry Service, North. For. Res. Centre, Edmonton, Canada,. 84 pp.

Manion, P. D.: 1981, *Tree Disease Concepts*. Prentice-Hall, Englewood Cliffs, USA.

BIOPRODUCTIVITY OF SPRUCE STANDS IN NORTHERN EUROPEAN RUSSIA

G. A. CHIBISOV

Institute of Forestry and Forest Chemistry, Arkhangelsk, Russia

Abstract. The bioproductivity of forest ecosystems in terms of forest succession and biodiversity has been studied in northern European Russia near Arkhangelsk since the 1930s. The relationship between forest productivity and forest treatments (thinning and other silvicultural practices) is still not well understood. In this paper, results are reported for spruce–birch stands of natural origin that have undergone thinnings of various intensities over the past 30 years. Changes in stand aboveground and root biomass were recorded in a series of re-measurements and for a series of different thinning intensities and are reported here. A number of biotic and abiotic factors play important roles in determining the productivity and the distribution in biomass components. These factors are changed by thinning treatments. Thus the productivity and structure of foliage as well as other aboveground and belowground biomass components were found to be dependent on the density and spatial distribution of the trees in the stand; production of merchantable stem-wood was also influenced by thinning treatments. The optimum density needed for maximum stand production, however, need not always be best from the economic perspective.

Keywords. BIOPRODUCTIVITY, DENSITY, BIOMETRY, BIOMASS, PHYTOCOENOSES, SPRUCE, THINNING, TAIGA, BIRCH–SPRUCE, PHYTOMASS, CARBON

1. Introduction

Bioproductivity of forest ecosystems and its relation to succession and biodiversity in northern European Russia is of both theoretical and practical importance. Many aspects of forest bioproductivity are used not only as biophysiological principles in studying phytocoenoses, i.e., in the study of how plant communities 'work' in the production of biological resources, but also as methodological basis for determining the effectiveness of different forestry measures (Utkin, 1975). According to Protopopov (1975) it is necessary to extend the definition of forest types with their biomass, which changes with density, age, etc. To find out the effects of environmental factors on forests, it is necessary to study the biometric characteristics, which in combination with mensurational parameters can give a complete description of forest phytocoenoses. The most important biometric characteristic is the stand biomass including stem, roots, branches and foliage.

1.1. Research History

In the Northern Region, bioproductivity of forests has been investigated by the Forestry Institute of Karelia (Morozova, 1973; Kazimirov *et al.*, 1978); by the Komi Branch (Bobkova *et al.*, 1977) and by the Arkhangelsk Forest Institute. In the 1930s Molchanov (1971) investigated this topic in the Arkhangelsk area.

From analysis of previous studies of forest bioproductivity, one can state that the least studied area is the relationship between bioproductivity and different forest management treatments (e.g., thinnings). Most studies deal with the development of foliage mass and assimilation surfaces and their relation to photosynthesis and growth (Georgievsky, 1957; Savina and Shuravleva, 1978). Some data on the effect of

Water, Air and Soil Pollution **82**: 87–96, 1995.

thinnings on phytomass are presented in the papers of Davydov (1976), Plokhov (1971), Vatkovsky (1976), Smirnov (1974), Sennov (1977), Molchanov (1978) and Astrologova and Nakvasina (1983). According to Atrokhin and Ievin (1985), the main objective of thinnings is to increase the quantity of the illuminated leaves on the trees, and to form a forest canopy with optimal productivity and optimal radiation usage.

When analysing pure spruce stands with 5-year thinning intervals Smoljanikov (1969) came to the conclusion that biomass undergoes a considerable change following thinning. Root biomass increases proportionally to thinning intensity (from weak to strong), while total aboveground biomass decreases with intensity. The increase in production of stem biomass is less than the increase in production of auxiliary biomass, i.e., roots, branches and foliage. Savina and Shuravleva (1978) found that seven years after thinning of a 14-year old Scots pine stand, the biomass of the needles exceeded the needle biomass of the control plots; this increase was found to be proportional to thinning intensity.

2. Materials and Methods

The investigations were carried out in the middle zone of the European taiga in natural spruce–birch stands, and in stands which had been thinned two or three times with different intensities during the past 30 years. The dynamics of biomass and its distribution was investigated on the basis of two or three repeat measurements.

As a typical process in the northern taiga, spruce–birch stands are established after large-scale clearcutting or forest fires. The species composition of these forests undergoes a substantial change both spatially and territorially. Natural regeneration of pure spruce stands spans a considerable time period. The main importance of biodiversity is the effect of biocoenoses on the soil and biological accumulation. At the same time, the high proportion of birch in the stand has a considerable negative physical and biophysical effect on the growth and productivity of the spruce. For this reason, determining the bioproductivity of the spruce as a function of age and species composition, is of both theoretical and practical importance. The changes in bioproductivity over time and site factors are given in Table I.

It is useful to compare the productivity of spruce stands, originating from spruce–birch stands, to that of pure, even-aged spruce stands, using yield Tables. Unfortunately such comparisons cannot be made because the origin of the pure spruce stands is unknown.

Enhancing the productivity of spruce stands using forest management techniques (primarily thinning) is an important objective both for resource management and for investigating the stability of forest ecosystems and the cause-effect relationships among thinnings, site and growth.

The biological productivity of spruce stands was investigated based on methods developed and discussed by many authors under the organisation of investigations in the frame of IGBP (Posdniakov *et al.*, 1968; Protopopov, 1975; Utkin, 1975; Semechkina, 1978; and others).

Different thinning methods (intensity, frequency) were examined. The time of last thinning was between 4 and 30 years before measurement. Measurements were taken on 112 sample plots, measurement time period varied between two to four inter-thinning intervals (5–30 years). Sample plot size depended on age, species composition and density and varied from 0.02 ha to 0.5 ha. The basic stand parameters and biomass components were estimated with 2 to 4% accuracy assuming 0.8 to 0.9 confidential probability.

To measure phytomass components (stem, bark, branches, needles, wood below ground both in green and oven-dry state) sample trees were cut, about 3 to 14 trees under each account. The total number of spruce sample trees was 470 and the number of birch trees was 65. The selection of sample trees was based on stem maps and the size distribution of breast height diameter, height and crown diameter.

TABLE I

Bioproductivity (Mg ha^{-1}) of spruce stands in birch–spruce phytocoenoses

Age	Canopy layer	Composition*	Stem over bark	Bark	Crown	Needles	Live crown	Branches	Above-ground
A. Wood sorrel–spruce forests									
30	I	100 b							
	II	77 s	8.63	1.84	10.40	5.41	7.36	4.99	19.00
40	I	100 b							
	II	79 s	11.70	2.04	10.30	5.12	6.91	5.08	21.90
50	I	100 b							
	II	82 s	15.10	2.40	11.20	5.43	7.30	5.77	26.30
60	I	100 b							
	II	85 s	21.20	2.99	13.00	6.11	8.18	6.89	34.20
70	I	98 b 2 s	2.29	0.23	1.03	0.44	0.60	0.59	3.32
	II	87 s	28.40	3.64	14.90	6.86	9.17	8.04	43.30
80	I	97 b 3 s	3.22	0.31	1.38	0.58	0.79	0.80	4.60
	II	89 s	35.80	4.27	17.20	7.77	10.40	9.43	53.00
90	I	96 b 4 s	3.86	0.36	1.63	0.68	0.93	0.95	5.49
	II	91 s	43.60	11.30	19.90	8.39	11.80	11.50	63.50
100	I	95 b 5 s	4.61	0.93	2.01	0.83	1.14	1.18	6.62
	II	93 s	49.8	5.44	22.50	9.89	13.30	12.60	72.30
B. Grass–spruce forests									
30	I	100 b							
	II	77 s	6.35	1.37	8.0	4.18	5.71	3.82	14.4
40	I	100 b							
	II	80 s	8.64	1.59	7.42	3.71	4.99	3.71	16.1
50	I	100 b							
	II	83 s	12.2	1.90	8.69	4.20	5.64	4.49	20.9
60	I	100 b							
	II	86 s	17.1	2.39	10.4	4.87	6.53	5.53	27.5
70	I	92 b 8 s	7.16	0.74	3.60	1.53	2.11	2.07	10.8
	II	88 s	22.7	2.94	12.5	5.76	7.73	6.74	35.2
80	I	90 b 10 s	9.44	0.95	4.77	2.01	2.78	2.76	14.2
	II	90 s	28.3	3.51	15.3	6.94	9.35	8.36	43.6
90	I	88 b 12 s	11.6	1.16	6.00	25.2	3.50	3.48	17.6
		91 s	32.4	3.90	17.6	7.88	10.7	9.72	50.0
100	I	86 b 12 s	13.9	1.37	7.30	3.04	4.24	4.26	21.2
	II	93 s	35.4	4.20	20.00	8.88	12.1	11.1	55.4

* b = birch, s = spruce, numbers are % area coverage

TABLE I (continued)

Bioproductivity (Mg ha^{-1}) of spruce stands in birch–spruce phytocoenoses.

Age	Canopy layer	Composition*	Stem over bark	Bark	Crown	Needles	Live crown	Branches	Above-ground
C. Fresh blueberry–spruce forests									
30	I	100 b							
	II	74 s	7.32	1.54	7.94	4.15	5.58	3.79	15.3
40	I	100 b							
	II	80 s	10.3	1.89	9.19	4.64	6.23	4.55	19.5
50	I	100 b							
	II	83 s	14.2	2.31	10.4	5.11	6.84	5.30	24.6
60	I	100 b							
	II	84 s	19.0	2.82	12.2	5.83	7.79	6.37	31.2
70	I	95 b 5 s	4.1	0.45	2.25	0.97	1.34	12.8	6.35
	II	86 s	25.5	34.7	14.6	6.81	9.10	7.79	40.1
80	I	91 b 9 s	7.17	0.76	3.91	1.67	2.31	2.24	11.1
	II	87 s	30.3	3.88	15.9	7.31	9.77	8.59	46.2
90	I	88 b 12 s	11.0	1.14	5.83	2.48	3.43	3.35	16.8
	II	88 s	35.5	4.32	17.5	7.93	10.6	9.57	53.0
100	I	85 b 15 s	15.0	1.53	7.81	3.29	4.56	4.52	22.8
	II	89 s	40.8	4.74	18.9	8.46	11.3	10.4	59.7
D. Wet blueberry–spruce forests									
30	I	100 b							
	II	70 s	5.79	1.34	9.73	5.07	7.09	4.66	15.5
40	I	100 b							
	II	74 s	7.77	1.48	8.15	4.11	5.60	4.04	15.9
50	I	100 b							
	II	78 s	11.2	1.88	9.47	4.64	6.28	4.83	20.7
60	I	100 b							
	II	80 s	15.7	2.36	11.1	5.29	7.14	5.81	26.8
70	I	82 b 18 s	12.6	1.42	7.85	3.41	4.74	4.44	20.5
	II	82 s	20.0	2.74	12.2	5.37	7.64	6.83	32.2
80	I	80 b 20 s	15.5	1.71	9.39	4.04	5.62	5.35	24.9
	II	83 s	25.1	3.25	14.3	6.58	8.86	7.72	39.4
90	I	78 b22 s	18.7	2.01	11.0	4.68	6.52	6.32	29.7
	II	85 s	29.8	3.68	16.0	7.27	9.80	8.73	45.8
100	I	75 b25 s	21.6	2.66	12.4	5.27	7.34	7.13	34.0
	II	86 s	33.5	4.01	17.7	7.95	10.8	9.75	51.2

* b = birch, s = spruce, numbers are percentage area coverage

Root systems were investigated by the methods of vertical and horizontal excavation on 49 sample plots in stands with ages up to 70 years (120 trees in total). The classification of roots was the following: coarse root (diameter (D) > 2 cm), semi-coarse roots ($2 \geq D > 0.2$) and fine roots ($D \leq 2$). To define the root density the method of monoliths was used. Monoliths of $30 \times 30 \times 40$ cm (the last figure is depth) and $50 \times 50 \times 40$ cm were taken out in 15 cm layers, and the roots were washed out and classified. There were 3 or 5 monoliths selected in each sample plot randomly distributed over the sample plot area.

Correlation and regression analyses were used to calculate relations between phytomass components, biometric, and mensurational parameters of the stands. The systematic errors of the equations were non-significant at 5% significance level, and the standard errors did not exceed 3–5% for different components in the zones of calculation (confidence level: 0.95).

3. Results

The change of spruce biomass with density is presented in Table II. Stem biomass increased with increasing density (stem ha^{-1}).The mass of needles and branches, and total aboveground biomass increased in both ends of the density range relative to those at intermediate densities. Root biomass was at the minimum at the maximum density, however, inversely related to stand density and caused total tree mass to decrease with increasing density. Spruce biomass increased 3.5 to 4 times that of the control plots.

TABLE II

Biomass of spruce stands (absolute dry weight, Mg ha^{-1}).

Stem density (stem ha^{-1})	Stem with bark	Needles	Crown including branches	Total Above Ground	Root system	Total tree mass
1000	40.4	12.4	32.0	72.4	15.5	87.9
2000	42.2	11.8	27.0	69.2	17.0	86.2
3000	43.0	11.9	26.5	69.5	16.8	86.3
4000	43.3	12.1	26.8	70.1	15.5	85.6
5000	43.5	12.4	27.0	70.5	13.9	84.4
6000	43.7	12.7	27.9	71.6	12.3	83.9

Needle biomass increased with increasing density, although branch biomass and total crown biomass both decreased. The percentage biomass proportions are presented in Table III.

Biomass composition is influenced by the rank of the trees in the canopy in each stand. For suppressed trees, stem biomass contributed only 40% of total aboveground biomass, with a consequently greater proportion in crown biomass. The crown biomass of dominant trees, however, increased with thinning intensity. The accumulation of these different biomass components can be analysed from two aspects. First, there is the general biological aspect, including soil, climate and vegetation conditions, carbon cycling and the prognosis for potential productivity of forest soils. Second, there is the economic aspect, i.e., merchantable stand productivity. For a specific production goal, tree size is just as important consideration as total aboveground biomass. For the production of large material, the size of the average tree is clearly important. Considering the biomass of the average tree, it is evident that a greater number of larger trees can be grown in stands with lower densities (Table IV).

An index of foliage assimilation rates may be inferred by relating the current annual stem volume increment to the volume and weight of the needles (Table V). Maximum volume increment per Mg of dry needles evidently occurs in stands of about 2000 stem ha^{-1}.

TABLE III

Biomass components of spruce stands (% of total tree mass).

Stem density (stem ha^{-1})	Stem	Branches	Needles	Crown	Part above ground	Roots
1000	46.0	22.3	14.1	36.4	82.4	17.6
2000	49.0	17.6	13.7	31.3	80.3	19.7
3000	49.8	16.9	13.8	30.7	80.5	19.5
4000	50.6	17.2	14.1	31.3	81.9	18.1
5000	51.4	17.2	14.6	31.8	83.2	16.8
6000	52.1	18.1	15.1	33.2	85.3	14.7

TABLE IV

Average tree component biomass (kg, absolute dry weight).

Stem density (stem ha^{-1})	Stem with bark	Branches	Needles	Crown	Part above ground	Root system	Total
1000	40.4	19.6	12.4	32.0	72.4	15.5	87.9
2000	21.1	7.6	5.9	13.5	34.6	8.5	43.1
3000	14.3	4.9	4.0	8.9	23.2	5.6	28.8
4000	10.8	3.7	3.0	6.7	17.5	3.9	21.4
5000	8.7	3.0	2.5	5.5	14.2	2.8	17.0
6000	7.3	2.5	2.1	4.6	11.9	2.0	13.9

TABLE V

Productivity of the assimilation equipment in spruce stands.

Stand density (stem ha^{-1})	Current annual increment I ($m^3\ ha^{-1}\ yr^{-1}$)	Needle mass R_{needle} ($Mg\ ha^{-1}$)	I/R_{needle}
1000	10.1	12.5	0.814
2000	11.2	11.9	0.907
3000	10.8	11.9	0.907
4000	9.7	12.1	0.802
5000	8.4	12.4	0.677
6000	7.1	12.7	0.559

4. Discussion

Table VI presents some relationships among biometric data for the bioproductivity of spruce stands. The specific weight of the canopy and the relative weights of crown and roots are determined using the method suggested by Protopopov (1975). Specific crown mass (U_{crown}) is the ratio of the absolute dry biomass of the crown (R_{crown}) to the volume occupied by the canopy:

TABLE VI

Biometric characteristics of the stands.

Density stem ha^{-1}	R_{crown}/R_{stem}	Specific crown mass U_{crown}	Relative crown mass (Mg m^{-2})	Relative root mass (Mg m^{-2})	Canopy productivity (U_{crown}/R_{stem})	Stability (R_{above}/R_{root})
Control	0.66	18.2	0.97	0.50	1.54	4.86
1.0	0.79	32.5	2.00	0.98	0.80	4.67
2.0	0.64	33.3	1.40	0.89	0.79	4.07
3.0	0.62	37.5	1.25	0.80	0.87	4.14
4.0	0.62	41.0	1.21	0.70	0.95	4.52
5.0	0.62	45.0	1.81	0.60	1.03	5.07
6.0	0.64	18.0	1.17	0.53	1.09	5.82

$$U_{crown} = \frac{R_{crown}}{l} \times 10^{-4} \qquad [1]$$

where l is the average crown length.

The relative masses of the crown and root, i.e., the relation of crown or root weight to stand basal area respectively provides an indication of the occupancy of the air and soil by phytomass.

The ratio between aboveground biomass (R_{above}) and the mass of the root system (R_{root}) is an important measure in determining the stability of forest ecosystems. As can be seen from Table VI, maximum productivity and stability occur in 55 year old stands at a density of 2000 stem ha^{-1}.

Based on the analyses of sample trees of different ages, equations were derived to relate biomass to diameter, height, number of stems, basal area and the proportion of the species in the stand, as follows:

$$M_{stem} = \frac{13.63 DBH \times H^{1.05} \times C}{(H - 1.16)^{-0.48}} \qquad [2]$$

$$M_{bark} = \frac{4.88 DBH \times H^{0.61} \times C}{(H - 1.16)^{-0.51}} \qquad [3]$$

$$M_{crown} = \frac{8.45 DBH \times H^{0.86} \times C}{(H - 1.16)^{0.60}} \qquad [4]$$

$$M_{needles} = \frac{5.40 DBH \times H^{0.74} \times C}{(H - 1.16)^{0.57}} \qquad [5]$$

where M is wet biomass, the subscripts indicate biomass type (stem biomass includes bark), DBH is breast height diameter, H is height and C is the percentage proportion of the species in the stand. The equations can be applied for the following intervals of the independent variables: DBH: 4 to 40 cm, H: 4 to 26 m. The absolute dry contents in the fresh weights are: bark 43.6%; crown 50.0%; needles 46.1%.

Based on the information obtained from the repeated measurement of the sample trees, managed stand bioproductivity tables were compiled for the spruce stands (Table VII). With systematic thinning treatments applied up to rotation age it is possible to increase the bioproductivity of spruce stands relative to the untreated stands by 2.2 to 1.6 (in forest types wood sorrel, grass, blueberry fresh and blueberry wet respectively).

Bioproductivity of the phytocoenoses is determined by a number of abiotic and biotic factors. After thinnings, stability and dynamic balance of the phytocoenoses is regained by the reaction of the individual trees both in time and space. The intensity of this response is determined by the initial stage (species composition, age, density, growth conditions) and the level of impact, i.e., thinning regime (intensity, frequency, method). For each age and growing condition, there is an optimal structure of maximum or most desirable bioproductivity.

Besides the substantial difference between the thinned stands and the control stands, there is a consistent increase in all elements of the biomass with decreasing density. Hence stability of the forest phytocoenoses is provided for a given density level for all ages.

Given their responses to the treatments, the different biomass components can be sorted into the following descending order: needles, branches, stems. The order according to the rate of mass accumulation would be: branches, needles, stems. The composition of biomass depends mainly on age, intensity and the timing of thinnings. With increasing age, the proportion of woody tissue increases, while the masses of crown and foliage decrease. The needle production for trees of different canopy position varies considerably; this variation is reduced by greater thinning intensity. At a given stand density, the biomass of the fraction depends on age. The fraction composition has a good correlation with the diameter–height ratio.

Productivity and structure of the foliage depends on the density and the spatial distribution of trees. At the optimal density, maximum bioproductivity occurs, an observation confirmed in other regions of the country (Kairiukshtis and Iodvalkis, 1976; Kuzmichev, 1977; Buzikhin and Pshenichnikova, 1980; Tiabera, 1982; Usoltsev, 1985).

5. Concluding Remarks

There are several possible meanings for the term ‘optimal density’. The optimal density needed for maximum stand production may not always be the best from the economic perspective.

Biological productivity can be regulated with thinnings under middle taiga conditions. Based on continuous observations of phyto-climate over the past 25 years it can be concluded that the productivity of spruce stands will increase with anticipated climate warming.

The main directions of research into spruce bioproductivity research can be summarised as follows:

- biophysical principles of phytocoenoses and ecosystems, interactions with human influences;
- phytomass, mensurational and biometric properties;
- succession, biological and economic valuation, organic balance, forest management objectives;
- biodiversity and ecosystem stability;

- dynamics of organic matter production, potential productivity of forest soils;
- macroclimate, carbon storage.

General bioproductivity combined with commonly observed biometric parameters can be used as the basis of a universal means for monitoring the state of forest and the potential effect of global climate change.

TABLE VII

Bioproductivity (Mg ha^{-1}) of thinned spruce stands (absolute dry weight).

Age (yr)	Density (stem ha^{-1}).	Stem over bark	Bark	Total crown	Needles	Live crown	Branches	Total aboveground
A. Wood sorrel–spruce forests								
30	10 960	25.9	4.0	14.5	7.13	9.33	7.37	40.4
40	4 330	40.5	5.38	21.9	10.2	13.6	11.7	62.4
50	2 243	54.8	6.57	29.3	13.1	17.8	16.2	84.1
60	1 463	67.8	7.57	35.5	15.5	21.2	20.0	103.0
70	1 009	80.8	7.50	41.9	17.9	24.7	24.0	123.0
80	745	93.1	9.34	47.9	20.1	27.9	27.8	141.0
90	594	104.0	10.0	52.9	21.9	30.6	31.0	157.0
100	492	113.0	10.6	57.5	23.6	33.0	33.9	171.0
B. Grass–spruce forests								
30	13 083	24.0	3.80	13.4	6.67	8.68	6.73	37.4
40	5 018	37.5	5.12	20.5	9.64	12.8	10.9	58.0
50	2 615	51.2	6.28	27.4	12.4	16.7	15.0	78.6
60	1 632	63.3	7.23	33.3	14.7	20.0	18.6	96.6
70	1 171	74.8	8.08	39.2	16.9	23.2	22.3	114.0
80	869	86.5	8.89	44.7	19.0	26.2	25.7	131.0
90	692	86.0	9.53	49.5	20.7	28.7	28.8	146.0
100	577	105.0	10.1	53.6	22.2	30.9	31.4	159.0
C. Fresh blueberry–spruce forests								
30	15 909	22.1	3.60	12.3	6.19	8.03	6.11	34.4
40	6 226	34.4	4.83	18.7	8.89	11.8	9.81	53.1
50	3 244	45.9	5.85	24.9	11.4	15.3	13.5	70.8
60	2 124	56.9	6.74	30.1	13.4	18.2	16.7	87.0
70	1 463	67.8	7.57	35.5	15.5	21.2	20.0	103.0
80	1 080	78.1	8.31	40.6	17.4	24.0	23.2	119.0
90	882	87.0	8.93	44.9	19.0	26.3	25.9	132.0
100	715	94.9	9.45	48.8	20.5	28.4	28.3	144.0
D. Wet blueberry–spruce forests								
30	17 720	20.3	3.40	11.7	5.91	7.65	5.79	32.0
40	7 267	32.0	4.60	17.5	8.38	11.0	9.12	49.5
50	3 775	43.4	5.63	23.3	10.8	14.4	12.5	66.7
60	2 418	53.0	6.43	28.4	12.8	17.3	15.6	81.4
70	1 692	63.3	7.23	33.3	14.7	20.0	18.6	96.6
80	1 235	73.1	7.95	38.3	16.6	22.7	21.7	111.0
90	988	81.0	7.04	42.2	18.0	24.9	24.2	123.0
100	824	88.7	9.04	45.8	19.3	26.8	26.5	135.0

Acknowledgments

The author wishes to acknowledge Canadian and Russian colleagues who assisted with the preparation of this paper, specifically Drs J. Gál and A. Z. Shvidenko. He also greatly appreciates financial assistance from the USDA-Forest Service and the Canadian Forest Service (Edmonton) to attend the IBFRA '94 Boreal Forests and Global Change Conference.

References

Astrologova, L. E. and Nakvasina, E. N.: 1983, Change of phytomass in the *Pinus*-Deciduous young stand due to thinning. *Forest Science (Lesovedenie)*, No. **5**, pp. 64-66 [In Russian]

Atrokhin, V. G. and Ievin, I. K.: 1985, Thinning. *Agroproizdat*. Moscow, USSR, 254 pp. [In Russian]

Bobkova, K. S., Verkholanzeva, L. A. and Galenko, E. P.: 1977, Primary productivity of forests of green mosses forest group in northern taiga. In: *Ecological Investigations of Nature Resources of the North of the Unblacksoil Zone*. Syktivkar, USSR, pp. 60-69. [In Russian]

Buzikhin, A. I. and Pshenichnikova, L. S.: 1980, *Formation of Pinus-deciduous young stands. Nauka*. Novosibirsk, USSR, 175 pp. [In Russian]

Georgievsky, N. P.: 1957, Thinning. *Goslesbumizda*. Moscow, USSR, 119 pp. [In Russian]

Davydov, A. V.: 1976, Thinning. *Forest Industry*. Moscow, USSR, 180 pp. [In Russian]

Kairjukstis, L. A. and Iodvalkis, A. I.: 1976, Criteria of optimal density for modelling of maximum productive biogeocenoses. In: *Biophysical and System Investigations in Forest Biogeocoenology*. Petrosavodsk, USSR, p. 89 [In Russian]

Kazimirov, N. I., Kabanov, V. V., Presnukhin, J. V.: 1978, Growth of pine stands in Northern Karelia. *In: Formation and Productivity of Pine Stands in Karelia Republic and Murmansk Oblast*. Petrozavodsk, USSR, pp. 102-107. [In Russian]

Kuzmichev, V. V.: 1977, *Regularities of Stand Growth*. Nauka. Novosibirsk, USSR, 160 pp. [In Russian]

Molchanov, A. A.: 1971, *Productivity of Green Masses in Forests of Different Zones, Nauka*. Moscow, USSR, 274 pp. [In Russian]

Molchanov, A. A.: 1978, *Impact of Anthropogenic Factors on Forests. Nauka*. Moscow, USSR, 139 pp. [In Russian]

Morozova, R. M.: 1973, Biological turnover of nitrogen and mineral elements in *Pineta vaccinosum* ecosystems. *In: Reports of State Natural Reserves "Kivach"*, Issue 2, Petrosavodsk, USSR, pp. 209-228. [In Russian]

Plokhov, A. A.: 1971, Interaction of pine and birch by root systems in Middle Taiga Subzone. *In: Investigations in Forest Management and Wood Processing*. Moscow, USSR, part 1, pp. 41-45. [In Russian]

Posdniakov, L. K., Protopopov, V. V. and Gorbatenko, V. M.: 1969, *Biological Productivity of Forests in Middle Siberia and Jakutia*. Book Publishing, Krasnojarsk, USSR, 155 pp. [In Russian]

Protopopov, V. V.: 1975, *Environmental Role of Dark Coniferous Forests*. Nauka, Novosibirsk, USSR, 326 pp. [In Russian]

Savina, A. V. and Shuravleva, M. V.: 1978, *Physiological Background of Thinning*. Forest Industry, Moscow, USSR, 103 pp. [In Russian]

Semechkina, M. G.: 1978, *Structure of Phytomass of Pine Stands*. Nauka, Novosibirsk, USSR, 165 pp. [In Russian]

Sennov, S. N.: 1977, *Thinning*. Forest Industry, Moscow, USSR, 160 pp. [In Russian]

Smirnov, N. T.: 1974, Development of assimilation organs of *Pinus sylvestris* under thinning. *In: Forests and Tree Species of Northern Kasakhstan*. Leningrad, USSR, pp. 87-92. [In Russian]

Tiabera, A. P.: 1982. Modelling productivity of pine stands of different density. *Forest Industry (Lesnaja promishlennost)*, No. **5**, pp. 59-62. [In Russian]

Vatkovsky, O. S.: 1976, *Analysis of Generation of Primary Productivity*. Nauka, Moscow, USSR, 116 pp. [In Russian]

Usoltsev, V. A.: 1985, *Modelling of Structure and Dynamics of Phytomass of Stands*. Krasnojarsk State University, Krasnojarsk, USSR, 192 pp. [In Russian]

Utkin, A. I.: 1975, Biological Productivity of Forests (Methods of Investigations and Results), Series *"Results of Science and Technology: Forest Science and Forestry"*. VINITI (All-Russian Institute of Science and Technology), Moscow, USSR, Vol. **1**, pp. 9-189. [In Russian]

LONG-TERM EXPERIMENTS IN SELECTIVELY CUT NORWAY SPRUCE (*Picea abies*) FOREST

K. ANDREASSEN

Norwegian Forest Research Institute, Høgskoleveien 12, N-1432 Ås, Norway

Abstract. Selective cutting is one of many harvesting methods alternative to clear-cutting. Both the influence on different ecological systems and the effect on forest productivity of this harvesting method are important. In this study, the productivity in 16 long-term experimental plots in selectively cut Norway spruce (*Picea abies*) forest is analysed.

Keywords. SELECTIVE CUTTING, UNEVEN-AGED FOREST, STAND DEVELOPMENT, YIELD, *Picea abies*.

1. Introduction

Selective cutting is one possible harvesting method that will probably reduce nutrient leaching of forest areas compared to clear-cutting. This cutting method also maintains most of the tree biomass of the forest compared with clear-cutting. In a selectively cut forest, large chemical or physical changes are greatly reduced because of the partial cutting. The influence of selective cutting on different ecological systems is comprehensive, and the effect on the growth and productivity of the trees is considered to be very important. In addition, the regeneration costs are larger and the wood quality is poorer in fast growing even-aged plantations. Little is known about selective cutting in Norway other than the results from Böhmer (1957). Today the interest in different types of partial cutting is increasing.

Selective cutting provides an uneven-aged and multi-storied forest, which is characterised by trees in all diameter classes (Trepp, 1974; Schütz, 1989). In selective cutting, scattered trees or small groups are harvested at a time, and the method is further characterised by combining the cutting, initiation and tending of regeneration in one single operation. Some investigations of productivity of selectively cut forests indicate yields 5–15% lower (m^3 ha^{-1} yr^{-1}) than in even-aged forests (Böhmer, 1957; Mitscherlich, 1963; Lundqvist, 1989).

The objective of this investigation was to study stand development in uneven-aged Norway spruce (*Picea abies* (L.) Karst.) forests treated by selective cutting. A second objective was to compare the yield in an uneven-aged Norway spruce forest with the yield capacity of an even-aged forest. A more detailed version of the main results of the experiment is reported in a Norwegian journal (Andreassen, 1994).

2. Sites and Methods

The sites consist of 16 experimental plots established from 1921 to 1939 (Table I). The plots were treated several times by selective cutting during the experimental period (Table I). The ratio between mean diameter of removed trees and diameter before cutting, and cutting grades are reported in Table II. Initial volume (after first removals), volume at last inventory, total volume yield, and total volume increment are

Water, Air and Soil Pollution **82**: 97–105, 1995.

TABLE I

Location and inventories of the experiments.

Plot	Area (ha)	Latitude	Longitude	First inventory	Last inventory	Number of inventories	No. of cuts
36	0.4	64.4	11.4	1 921	1 990	13	6
56	1.0	59.2	11.3	1 922	1966	10	7
61	1.0	65.5	12.7	1 922	1 990	12	9
64	0.6	64.2	12.4	1 922	1 966	9	6
90	1.0	59.5	11.3	1 923	1 991	12	11
107	0.7	59.2	9.6	1 924	1 967	11	10
129	0.6	61.7	11.8	1 924	1 969	10	7
141	1.0	60.8	10.2	1 925	1 966	10	6
145	0.7	64.3	11.5	1 925	1 990	13	4
178	0.6	65.8	12.4	1 927	1 990	10	4
329	1.2	61.7	10.6	1 935	1 989	7	2
394	0.2	61.1	11.7	1 937	1 964	6	2
395	0.3	61.1	11.7	1 937	1 991	7	2
453	0.5	64.6	12.4	1 939	1 989	9	4
2 311	0.5	59.9	9.9	1 929	1 991	10	7
2 312	0.5	59.9	9.9	1 929	1 957	7	5
Mean	0.7	62.1	11.3	1 929	1 980	9.8	5.8

listed in Table III. According to Tables I, II and III, the cutting programmes are different both in frequency and intensity. The harvested trees were selected subjectively, which is important in selective cutting.

The observed mean annual increment (MAI) is calculated by dividing the increment (Table III) by the corresponding period (Table I) of the experiment. The observed increment is compared with the yield capacity (YC) of even-aged Norway spruce stands (the control). Yield tables and yield capacity for even-aged Norway spruce stands is estimated by Braastad (1975) and Tveite and Braastad (1981).

If the yield capacity is to be estimated, it is necessary to first estimate the site index (SI). In this study, SI was estimated in each of 16 plots as the mean of four methods: dominant height and corresponding age were used in both uneven-aged Norway spruce plots and in a neighbouring even-aged Norway spruce forest (Tveite, 1977). In addition the SI was estimated according to vegetation type, and according to vegetation type and age of dominant trees (Nilsen and Larsson, 1992). The method of SI estimation is further analysed and discussed by Andreassen (1994).

3. Results

The standing volume and volume of removed trees (vertical lines) during the experimental period are shown for plots 36, 56, 90 and 141 in Figure 1. In Figure 2, stand development (diameter distribution at different periods) is shown for the same four plots. In Figure 2 the number of remaining trees (after cutting) is plotted and interpolated for each fifth year.

Stand development in even-aged and uneven-aged forests is compared in Figure 3. Plot 62 is not described in Table I and Table II despite being even-aged; however, it is located just 10 m from plot 61, and the structure and volume before first cutting were about the same as in this plot.

A regression model of the form MAI = aYC + b(YC)^2 was established where a and b are estimated parameters and the intercept is zero. Figure 4 shows the observed mean annual increment (MAI) compared to estimated yield capacity (YC) of even-aged forests on similar sites using this regression model.

The yield of uneven-aged forest was on average 15–20% lower than the estimated yield capacity. The observed increment (MAI) was analysed at different sites given the following regression relationships between MAI and yield capacity (YC) and between SI and regeneration condition (RC):

TABLE II

Mean values of cutting method /cutting grade.
D1: Mean diameter before cutting.
D2: Mean diameter of removed trees.
V1: Volume before cutting.
V2: Volume of removed trees.

Plot	D2/D1	V2/V1
36	1.24	
56	1.12	0.16
61	1.22	0.19
64	1.08	0.30
90	1.08	0.19
107	1.25	0.25
129	1.14	0.15
141	1.03	0.14
145	0.98	0.29
178	1.08	0.21
329	1.02	0.14
394	0.97	0.25
395	1.13	0.32
453	1.38	0.27
2311	1.06	0.30
2312	1.14	0.32
Mean	1.13	0.22

$$MAI = 1.035YC - 0.033(YC)^2 \qquad [1]$$

with $r^2 = 0.46$ and RMSE = 1.5 (m^3 ha^{-1} yr^{-1}).

$$MAI = -1.18 + 0.370H_{40} + 0.942RC \qquad [2]$$

with $r^2 = 0.51$ and RMSE = 1.5 (m^3 ha^{-1} yr^{-1}). Site index at age 40, H_{40} (top height at 40 years), of 11 m, 17 m and 23 m correspond to YCs of 3.5, 7.5 and 12.0 m^3 ha^{-1} yr^{-1} respectively. The regeneration condition (RC) was set to 0 for unfavourable regeneration conditions, or 1 for favourable regeneration conditions.

4. Discussion

According to Table II, all plots except 145 and 394 have a ratio of D1:D2 greater than 1.0, although the ratios of these latter two plots are very close to 1.0 (0.96 and 0.99). The cuttings are on average 22% of the volume before cutting (Table II). These two results show the plots were treated with selective cutting (thinning from above) during the experimental period. In most of the plots, the volume has increased during the experimental period (Table III).

TABLE III

Standing volume, harvested volumes and increment (m^3 ha^{-1}).

Plot	After first cut	After last cut	Total cut	Total increment
36	176	196	277	297
56	113	149	230	266
61	127	152	303	328
64	182	55	273	145
90	151	182	456	488
107	152	93	268	209
129	164	171	211	218
141	117	194	200	277
145	118	150	178	209
178	66	163	133	230
329	37	119	44	126
394	45	177	55	187
395	20	408	50	438
453	121	200	177	256
2311	215	358	454	597
2312	205	53	322	170
Mean	126	176	227	278

The annual increment was on average 3.7% per year of the standing volume (m^3 ha^{-1}) (Tables I and III). Barth (1929) and Böhmer (1957) found increment rates of about 3% per year. Lundqvist (1989) indicates increment rates between 2 and 4% per year. The increment rate depends strongly, however, on the site index.

The plots show different diameter distributions during the experimental period (Figures 2 and 3). The diameter distributions in plots 56, 61 and 141 are most clearly J-shaped. Other plots have approximately this distribution, but often there are too few trees in the lowest diameter classes. A condition to maintain the structure in a selectively cut forest is to have enough trees in the lower diameter class which can substitute for trees in the upper classes when trees are cut or die. Without an inverse J-shaped distribution, the structure cannot be stable. Several scientists emphasise the importance of an inverse J-shaped diameter distribution in selectively cut forests (Liocourt, 1898; Böhmer, 1957; Indermühle, 1978; Lundqvist, 1989). To maintain the structure, the cuttings must be done regularly and in a way that promotes the development of an "inverse J-shaped" diameter distribution. Probably the time between cuttings should be 10–20 years under Nordic conditions, and most often at high SIs. Schütz (1989) recommends about 8 years between the cuttings at SIs of 8–10 m^3 ha^{-1} yr^{-1}.

In plot 36 and particularly in plot 90, the density was too high in the beginning of the experimental period to maintain the structure (see Figure 2). Too few medium sized trees were cut, and this was one of the reasons for the lack of regeneration in the middle of the experimental period. This resulted in stability problems for the diameter distribution. Today, the diameter distribution corresponds better to the theoretical "inverse J-shaped" graph. In plot 141, the structure is more successful (Figure 2), but even here the diameter distribution varies during the experimental period. Such small variations in the diameter distribution must be considered normal.

Stand development of even-aged and uneven-aged forest is compared in Figure 3. In an uneven-aged forest, the structure is, or should be constant as discussed above. In an even-aged forest the diameter distribution changes over time (Figure 3).

The results of this study also emphasise the lower volume production in selectively cut forests (means of 15–20% less per year) (Figure 4 and Table IV). The regression model (dotted line in Figure 4) definitely lies below the slope MAI = YC (solid line), which shows lower yield in the uneven-aged forest compared to the even-aged forest. This also corresponds with other investigations in other countries (Mitscherlich, 1963; Lundqvist, 1989). Conversely, Böhmer (1957) indicated almost no discrepancy between observed increment and yield capacity. The yield capacity is a range rather than an exact value, however, and there was a lot of variation in the yield. The yield capacity also reflects "good" forestry treatments, e.g., high density (high number of stem ha^{-1}) and regular thinning. Often in practical even-aged forestry, it is difficult to achieve a volume production as high as yield capacity (Lundqvist, 1989), because the growth model is often based on the assumptions of dense stands and regular thinning at appropriate intervals.

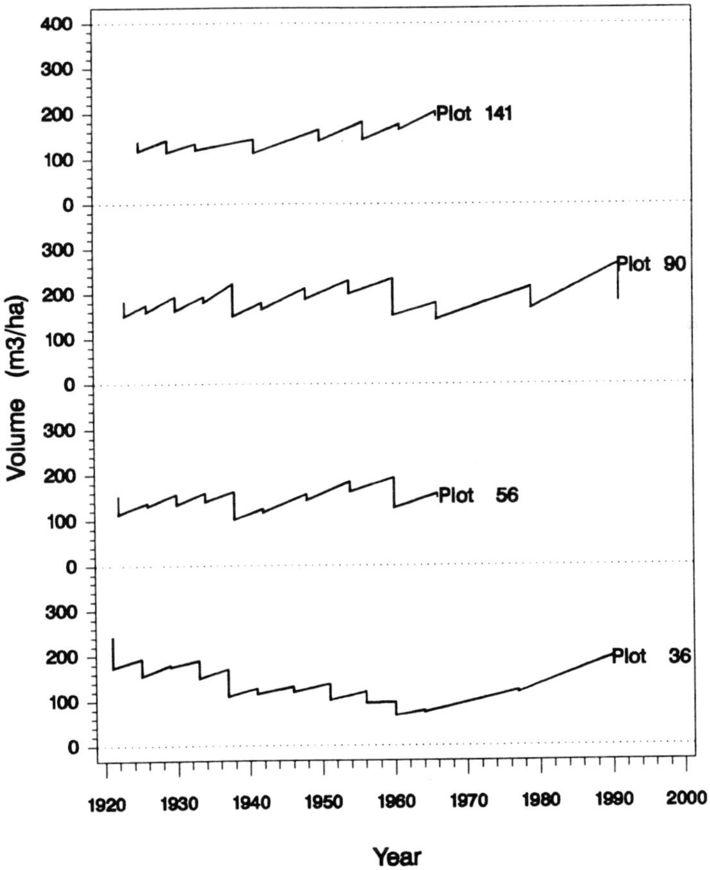

Figure 1. Standing volume (m^3 ha^{-1}) for four forest plots managed under selective cutting regimes.

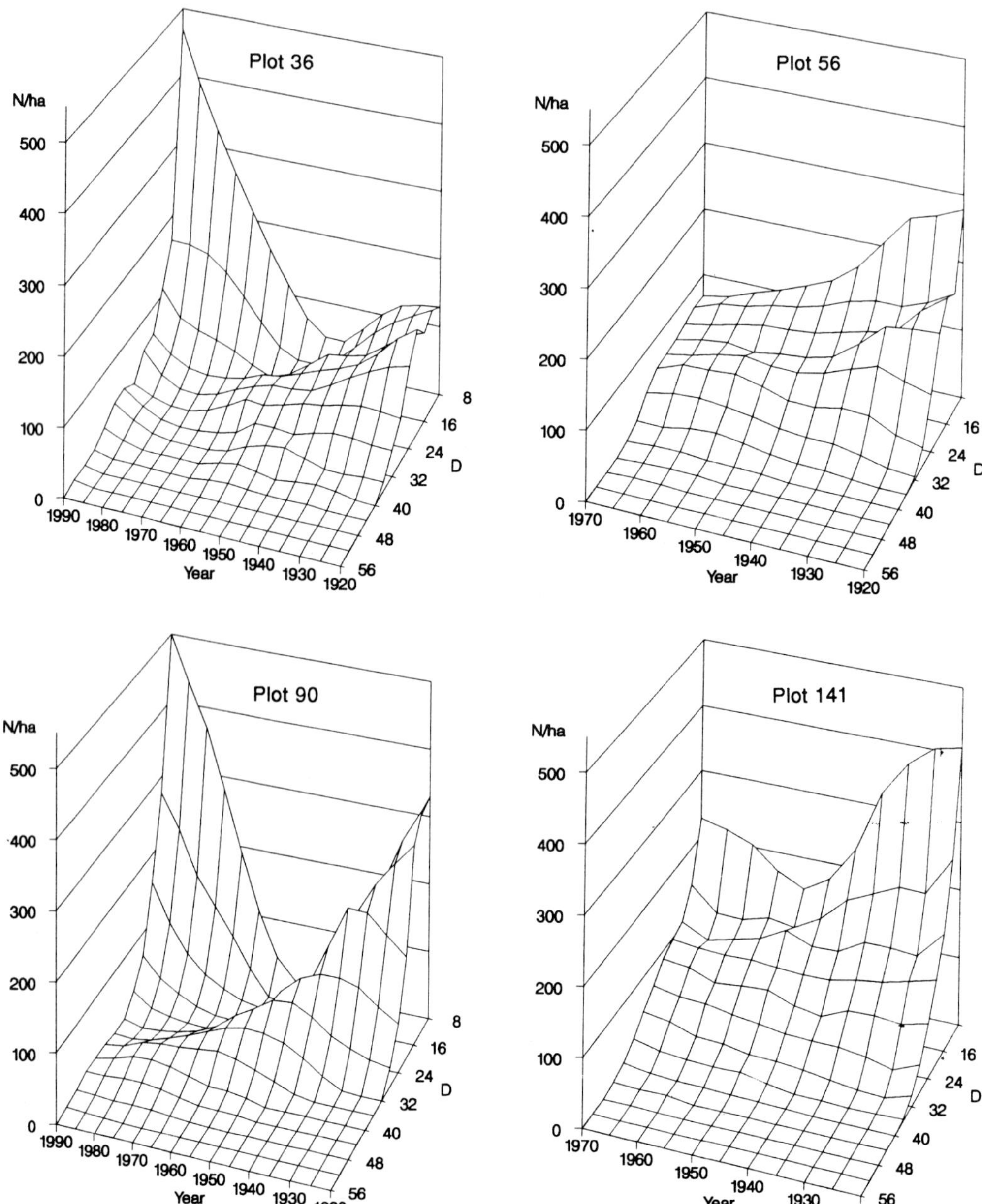

Figure 2. Stand development (number of tree stem ha^{-1} in 4 cm diameter classes).

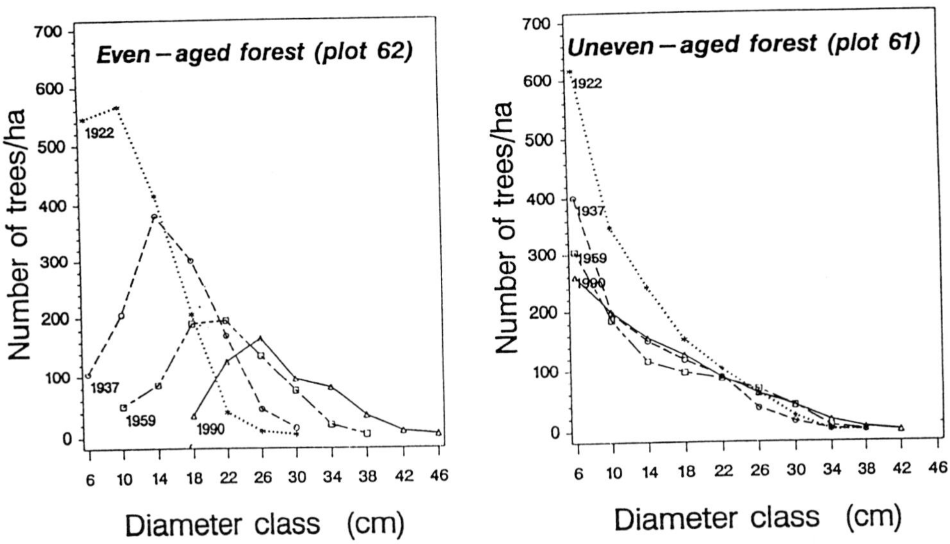

Figure 3. Comparison of stand development in even-aged and uneven-aged forest.

The observed increment was studied according to different SIs and different regeneration conditions. By separating the plots into two groups, favourable and unfavourable regeneration conditions corresponding to Skoklefald (1983), it was possible to analyse the increment in different sites. The deviation between observed increment and estimated yield capacity increases with SI and more unfavourable regeneration conditions. In sites with medium SI and favourable regeneration conditions the loss was 10–20% of estimated yield capacity. In sites with unfavourable regeneration conditions and/or high SI, the loss was 20–40%. This indicates that selective cutting is most successful on sites where natural regeneration is easy to achieve.

After clear-cutting, the forest floor will be exposed to rain, wind and large temperature changes. In selective cutting, the trees and structure of the forest will be maintained. In this way the forest floor is not so strongly exposed and nutrient leaching can be expected to be reduced compared to that following clear-cutting. By maintaining the forest as a continuously covered area, the biomass balance (both flora and fauna) will not be as disturbed. The amount of carbon bound in the humus layer will probably be higher with this kind of harvesting method because of reduced humus destruction. In this way the total variation will probably be reduced. However, in some parts of the world, forest disturbances (e.g., forest fire, windthrow) in large areas are a natural means of regeneration.

5. Conclusions

Selective cutting in Norway spruce (*Picea abies*) forests gives multi-storied stands and is most successful on sites favouring good natural regeneration. To maintain stand structure, stem removals must be done regularly. It is important to harvest in such a way as to promote the development of an "inverse J-shaped" diameter distribution. The stand must be opened up sufficiently to reduce competition for light and root space and spruce allow regeneration to be initiated. In this study, volume production in uneven-aged forests was on average 15–20% lower than the estimated production of even-aged forest on similar sites. Losses were lower in vegetation types where regeneration conditions were good, and on sites with medium and low site indices. If conditions are "good", the loss of volume production is small (10–20%), and if the conditions are "difficult" the loss is greater (20–40%).

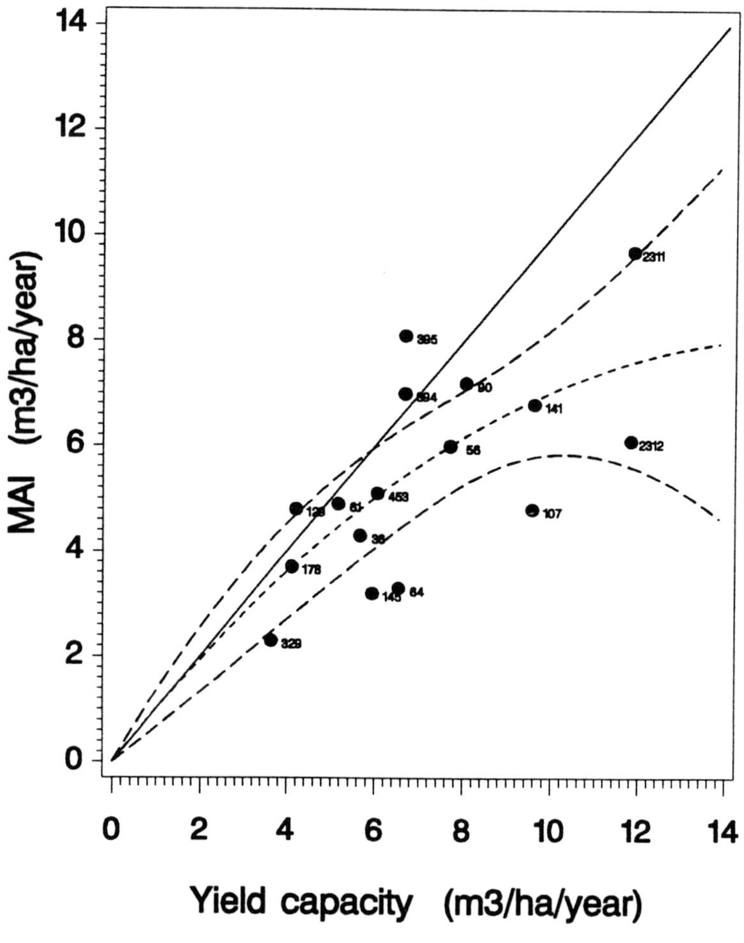

Fig. 4. Comparison of observed increment (MAI) in uneven-aged forest with yield capacity (YC) of even-aged forest. The solid line shows the slope MAI = YC, and the dashed middle line shows the regression model MAI = aYC + b(YC)^2. The other two dashed lines represent the 95% confidence limits. The regression model has no intercept, and a and b are estimated parameters (see text).

References

Andreassen, K.: 1994, Utvikling og produksjon i bledningsskog. (Development and yield in selection forest.) *Medd. Nor. inst. skogforsk.* **47**(5), 1-37.

Barth, A.: 1929, *Acta Forestalia Fennica* **34**(15), 1-33.

Braastad, H.: 1975, Produksjonstabeller og tilvekstmodeller for gran. (Yield tables and growth models for *Picea abies*) Medd. Nor. inst. skogforsk. **31**(9), 356-537.

Böhmer, J. G.: 1957, *Bledningsskog II. Tidsskr. Skogbr.* **65**, 203-247.

Indermüehle, M.-P.: 1978, Struktur-, Alters- und Zuwachs- untersuchungen in einem Fichten- Plenterwald der subalpinen Stufe. (Diss.ETHZ) In: Beih. nr. 60, Z. *Schweiz.Forstwer*, pp. 1-98.

Liocourt, F.: de. 1898, De l'aménagement des sapinières. *Bull. soc. forest. Franche-Comté et Belfort* **4**, 396-409.

Lundqvist, L.: 1989, *Blädning i granskog. Strukturförändringar, volymtillväkst, inväksning och föryngring på försöksytor skötta med stamvis blädning.* Institutionen för skogsskötsel, Sveriges lantbruksuniversitet. Avhandling, pp. 1-100.

Mitscherlich, G.: 1963, Untersuchungen in Schlag- und Plenterwältern. *Allg. Forst- u. J.-Ztg.* **134**(1), 1-12.

Nilsen, P. and Larsson, J.: 1992, Bonitering av skog ved hjelp av vegetasjonstype og egenskaper ved voksestedet. Rapp. *Skogforsk* **22**/92, 1-43.

Schütz, J.-Ph.: 1989, *Der Plenterbetrieb.* Fachbereich Waldbau ETH, Zürich, Switzerland, pp. 1-54.

Skoklefald, S.: 1983, Naturlig foryngelse på ulike skogtyper. *Aktuelt fra statens fagtjeneste for landbruket* **3**, 12-18.

Trepp, W.: 1974, Der Plenterwald. Mitt. HESPA, *Luzern.*, **24**(66), 1-65.

Tveite, B.: 1977, Bonitetskurver for gran. (Site index curves for Norway spruce [Picea abies (L.) Karst]) *Medd. Nor. inst. skogforsk.* **33**(1), 1-84.

Tveite, B. and Braastad, H.: 1981, Bonitering for gran, furu, og bjørk. (Site-index for spruce, pine and birch). *Norsk Skogbr.* **27**(4), 17-22.

REFORESTATION TRIALS IN THE KHABAROVSK TERRITORY, RUSSIA

R. LOWERY[1] and E. ZABUBENIN[2]

[1]*International Reforestation, Weyerhaeuser Company, WWC 2F2, Tacoma, WA 98477, USA*
[2]*Khabarovsk Forest Service, 71 Frunze St., 680620 Khabarovsk, Russia.*

Abstract. Enormous areas of primary forest in Russia have been and continue to be destroyed by fire and harvesting. Almost all of these areas have been left to reforest naturally. Little reforestation by planting has been attempted and these efforts have met with limited success. Consequently vast areas are without forest cover and are unlikely to have a positive carbon balance. Weyerhaeuser Company initiated co-operative forest planting trials in the Khabarovsk Territory in 1989. Native seeds from this area were brought to the U.S. and used to produce high quality containerised seedlings. These seedlings were then used in research trials and small operational plantings in 1991 and 1992 near Vanino and Khabarovsk. In both 1993 and 1994, seedlings were provided for large-scale operational plantings in the Khabarovsk Territory. Seedling survival has generally been greater than 90% and growth has been excellent over a wide range of site conditions. The operational planting results demonstrate that contrary to local opinion, reforestation practices proven elsewhere can be used to successfully reforest sites that have been burned or harvested in this region. Key to the operational planting success were high quality seedlings, good logistical support and a well trained, organised and positively motivated work force.

Keywords. RUSSIA, KHABAROVSK, REFORESTATION, RESEARCH, OPERATIONAL, PLANTING, CONTAINERISED SEEDLINGS.

1. Introduction

The existing natural forest cover in the Khabarovsk territory is well developed except in the vicinity of the larger cities and along transportation routes where repeated fires and harvesting have destroyed millions of hectares of forest (Barr and Braden, 1988; Cardellichio *et al.*, 1990). Throughout Russia in the 1980s, some 3 Mha of forests were damaged or destroyed annually by wildfire (Gusewelle, 1992).

Reforestation by planting in Russia was very limited prior to 1950 and the natural reforestation that took place following fires or harvesting saw birch and aspen replace the more valuable conifers (Krankina and Dixon, 1992). During the 1950s, some 300 000 ha were regenerated annually in southern Russia; and from 1970-1987 approximately 1 Mha were established (Burdin, 1991). However, annual harvest levels of 3.5 to 4 Mha resulted in a 138 Mha non-forested backlog by 1983 (Barr and Braden, 1988). Forest cover losses due to fire, insects and disease added considerable additional area to this figure. This forest destruction also has likely contributed to the build-up of greenhouse gases in the atmosphere.

While large areas in Russia are reported to have been regenerated in the 1980s using direct seeding (Burdin, 1991), more efficacious efforts with seedling planting have not been used extensively as was done in boreal Alaska (Zasada, 1990; Zasada *et al.*, 1977) and adjacent Heilongjiang Province, China (Jingwen and Hongzhen, 1989). Seedling availability is one factor limiting replanting in Russia. Only 7 million seedlings, of some 60 different species, were being produced annually in nurseries across Russia in the 1980s (Burdin, 1991). The success of these replanting efforts and earlier reforestation efforts also

Water, Air and Soil Pollution **82**: 107–113, 1995.

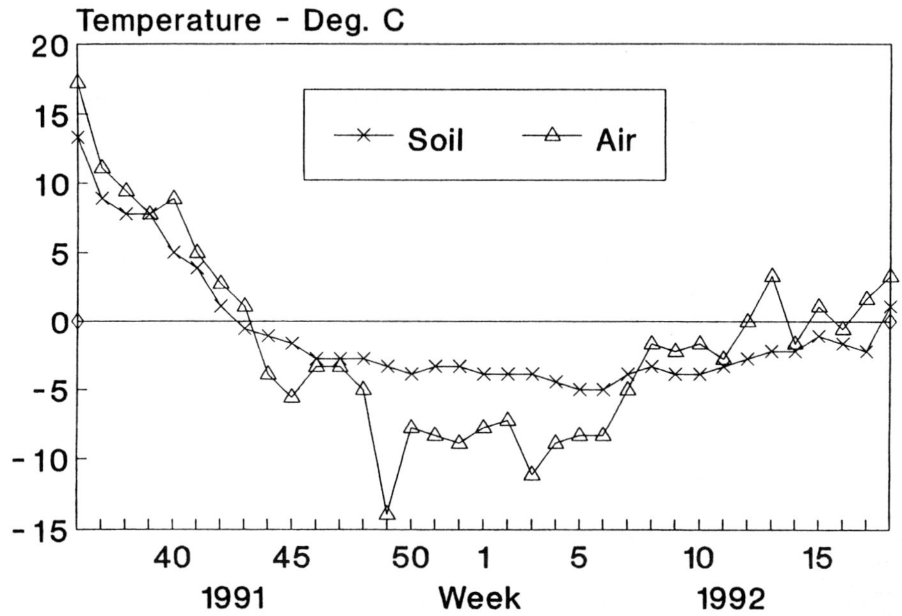

Fig. 1. Weekly mean soil temperature at -10 cm on the Gatka research site and air temperature at +3 m in Vanino during the 1991–1992 winter season.

is not well documented, but few regenerated sites have developed into healthy, properly-stocked stands. During the past five years, one-third of the trees in state plantations reportedly have perished due to several factors such as low quality nursery stock, inadequate or incorrect site preparation, inappropriate planting techniques, and lack of competition control (Krankina and Dixon, 1992).

Winters in the Khabarovsk Territory are long and cold (Figure 1). However, local conjecture notwithstanding, there were no readily apparent biological reasons why proven reforestation practices combined with high quality nursery-grown seedlings of local origin would not lead to successful reforestation. In December 1989, the Khabarovsk Territory Governor, the Khabarovsk Territory Forest Production Conglomerate, the Far East Forestry Research Institute, a Vanino forest products organisation (Koppensky LPK), and Weyerhaeuser Company representatives, agreed to begin co-operative reforestation research experiments near the port city of Vanino. Vanino is located on the western shore of the Tatar Strait in Khabarovsk Territory at 49°01'N, 140°15'E.

2. Objectives

These trials were undertaken to demonstrate that high quality seedlings of economically important forest tree species of the Khabarovsk Territory could be produced in tree seedling nurseries, then used to successfully reforest burned and harvested areas in the Territory. Additionally we wanted to build our reforestation knowledge base and refine reforestation technology for the Russian Far East.

3. Methods

Since a local forest nursery was not available near Vanino, seeds of the three principal coniferous species: Yeddo spruce (*Picea ajanensis* Fisher), Korean pine (*Pinus koraiensis* Sieb. and Zucc.) and Dahurian larch (*Larix dahurica* Turcz.) were obtained from Khabarovsk Territory officials and returned to the U.S. for seedling production. White spruce (*Picea glauca* Moench) and lodgepole pine (*Pinus contorta* var. *latifolia* (Engelm.)) were also included in these experimental plantings. The seedlings were grown at the Weyerhaeuser Company, Rochester Forest Regeneration Center greenhouses near Rochester, Washington.

3.1. Research Plantings

For the first planting trial, approximately 9800 seedlings were shipped from freezer storage to Vanino by air freight in early May, 1991. Following training in seedling handling and planting procedures, Koppensky foresters planted the first research trial of 3600 seedlings. The trial site is located 2 km east of the village of Gatka (48°54'N, 140°09'E; elevation 100 m) (Figure 2). The research trial site had burned in human-caused, stand-replacing fires some 15 years earlier and again in 1990. Soils on the site are stony, silty, loams developing from dark basalt. The climate of the area is humid-continental, with a cold dry winter and a warm, wet five-month summer. A supplemental experimental planting was done by Far East Forestry Research Institute personnel at Khektsir, near the territorial capitol, Khabarovsk (48°19'N, 135°03'E; elevation 170 m).

A second research planting was done at the Gatka site in September, 1991, and a third was established in May 1992 adjacent to the 1991 plantings. Each of the Gatka research plots contain 300 trees planted at a 2-m spacing and cover 5.1 ha in total. Seedlings remaining after installation of the research trials were operationally planted by Koppensky foresters on a variety of harvested and burned sites in the area. A total of 112 00 seedlings were planted in 1991 and 1992 (Table I).

3.2. Large-scale Operational Plantings - 1993 and 1994

In late April 1993, some 796 000 seedlings were donated and shipped to the Khabarovsk Territorial Forest Service for operational plantings in the Territory. In June 1994, an additional 444 000 seedlings were sent for planting (Table I). These seedlings were also from Russia-collected seed and grown at the Weyerhaeuser greenhouses. They were placed in freezer storage in late-December 1992, and ocean-shipped in refrigerated containers to Vanino. Planting started in early June and continued until mid-July. In 1993, seedlings were supplied to Koppensky, and the Gurskii-, Sovetskii-, Tumninskii-, Vysokogornyi-, and Sakhalin-*leskhozes*. Seedlings were planted on a

TABLE I

Weyerhaeuser-grown seedlings shipped to the Khabarovsk Territory for planting from 1991 through 1994.

Year	1991		1992		1993	1994	
Season	spring	fall	spring	fall	spring	spring	Total
Number	9778	7560	87 54	7360	795 765	443 621	1 351 238

wide variety of site conditions from near sea level to 1100 m elevation, and on sites ranging from fresh mineral surfaces to sites burned some 10 years ago and covered with a dense grass and shrub layer. In 1994, all seedlings were supplied to Gurskii-*leskhoz*. A Russian-language seedling planting training video and detailed written reforestation operations plans were prepared and served to guide the planting operations.

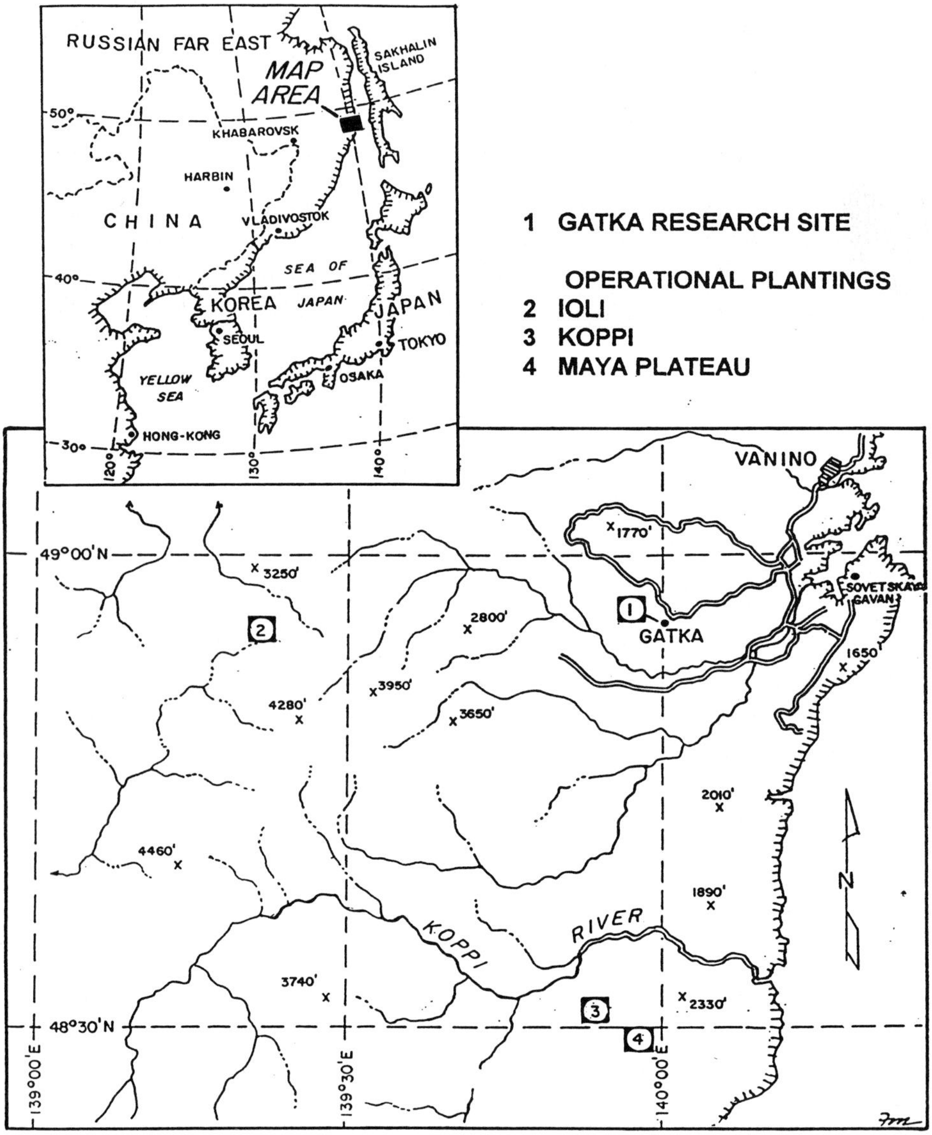

Fig. 2. Region of Khabarovsk Territory where Weyerhaeuser reforestation trials were installed in 1991 and 1992. Inset: location of Gatka research site and operational plantings near Vanino and Sovetskaya Gavan.

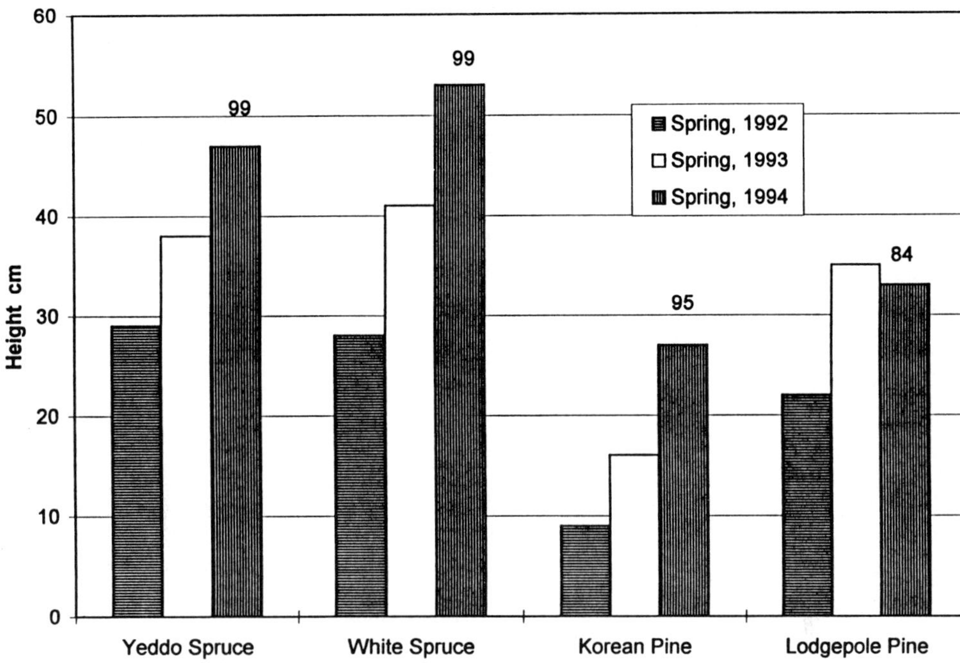

Fig. 3. Initial, first-year and second-year seedling heights of the spring 1991 Gatka research plantings. Numbers above bars are spring 1994 survival percentages for each species.

4. Results and Discussion

Survival to age three in the plantings of local species generally has been 90% or better. The exception is the 1992-planted Korean pine where survival dropped from 100 to 42% during the 1993 summer drought. Summers are typically warm and humid with frequent rainfall. Survival of the small Dahurian larch planted in the fall of 1991 is also low and now stands at 78%. The other Korean pine and Dahurian larch plantings show excellent survival.

The original lodgepole pine seed source was not frost-hardy and has suffered considerable winter damage and mortality (Table II). It was replaced with a frost-hardy source in June, 1994. The only other seedling damaging agent noted to date in the research plantings is rabbit clipping. The incidence is very low though, having occurred on only 2 of the 2670 permanent sample trees. Growth of the research plantings has also been very good (Figure 3).

Survival of the 1993 operational plantings was consistently high across all the reporting areas (Table III), in spite of 1993 having been the driest summer in memory according to local foresters. Inadequate root development prior to the 1993 drought, and retarded on-site hardening, were the probable causes of desiccation or frost damage noted in spring 1994 on some white spruce planted at a high elevation Koppensky site and at the Gurskii-*leskhoz*. Temperatures at the latter site may drop to -45 °C before significant snow accumulation occurs. The local species were not visibly affected, nor were earlier white spruce plantings

TABLE II

Annual, post winter-season seedling survival by species at the Gatka research site. Each plot contains 300 trees, 60 are permanent sample trees.

Plant Date/Species	No. of Plots	Spring Survival - % 1992	1993	1994
May, 1991				
Picea ajanensis	6	100	100	99
Picea glauca	3	100	99	99
Pinus koraiensis	1	100	97	95
Pinus contorta	2	100	98	84*
September, 1991				
Picea ajanensis	3	100	100	99
Pinus koraiensis	3	100	100	98
Larix amurensis	2	96	96	91
Larix dahurica	2	81	81	78
May, 1992				
Picea ajanensis	2		100	100
Picea glauca	5		99	99
Pinus koraiensis	3		99	42
Pinus contorta	4		100	70*
Larix dahurica	5		93	82
Picea glauca P+1	2		100	100
Pinus contorta P+1	2		100	53*
Larix dahurica P+1	2		100	100
September, 1992				
Picea ajanensis	2		100	100

* Replaced with a frost-hardy source in June, 1994.

The research plantings will be measured periodically and if fire protection is successful, the plots will eventually be converted to growth and yield plots. The operational plantings will be monitored into the future. Emphasis is now on improving reforestation capabilities within the Khabarovsk Territorial Forest Service with technical assistance and personnel training.

Increased forest fire losses above historical levels and increased forest harvesting since the 1950s have left large areas in the Khabarovsk Territory without forest cover. These changes have likely contributed significantly to greenhouse gases. The situation has been exacerbated by the lack of timely reforestation on most of these areas. The plantings reported here have been successful over a wide variety of site conditions and demonstrate that reforestation by planting is technically feasible in the Khabarovsk Territory. If approached and carried out as an integrated set of steps, as in these demonstration plantings, most deforested areas could be brought back into forest cover and become net accumulators of carbon much more quickly than if natural revegetation processes are allowed to continue.

TABLE III

First growing season survival of Weyerhaeuser-grown seedlings operationally planted by various Russian forestry organisations in 1993. The four organisations reporting survival results received 95% of the 796 000 seedlings provided.

Organisation	No. received (1000s)	Survival - percent
Koppensky	430	98
Sovetskiileskhoz	100	89
Gurskiileskhoz	200	99
Tumninskiileskhoz	35	N.A.
Vysokogornyileskhoz	25	97
Sakhalinleskhoz	6	N.A.
Total	796	Average 97

5. Conclusions

The conjecture that reforestation by planting will not work in the Khabarovsk Territory has been refuted. High quality, genetically adapted forest tree seedlings can be used to successfully reforest recently harvested and burned areas in the Khabarovsk Territory. Large-scale reforestation projects also can be successful here once the capacity for high quality seedling production is developed. Also of key importance, though will be a properly trained work force, organised for the task, positively motivated to do quality work and given good logistical support.

References

Barr, B. M. and Braden, K. E.: 1988, *The Disappearing Russian Forest. A Dilemma in Soviet Resource Management.* Rowman and Littlefield, Hutchinson, London, 252 pp.

Burdin, N. A.: 1991, *Unaslyva* **165**, 43-50.

Cardellichio, P. A., Binkley, C. S. and Zausev, U. K.: 1990, *J. Forestry* **88**, 12-36.

DaKo, C. and Ning, Z. *et al.*: 1989, *In:* Gasbarro, A. and Lawson, T. (eds.), *Forest Regeneration at Northern Latitudes Close to Timber Line.* USDA For. Serv. PNW Res. Sta. Gen. Tech. Rep. PNW-GTR-247, pp. 61-65.

Gusewelle, C. W.: 1992, *American Forests* (May/June), 17-20 pp.

Jingwen, L. and Hongzhen, Z.: 1989, *In:* Gasbarro, A. and Lawson, T. (eds.), *Forest Regeneration at Northern Latitudes Close to Timber Line.* USDA For. Serv. PNW Res. Sta. Gen. Tech. Rep. PNW-GTR-247, pp. 66-67.

Krankina, O. N. and Dixon, R. K.: 1992, *J. Forestry* **90**, 29-34.

Revel, J., Lavender, D. P. and Charleson, L.: 1990, *FRDA Rep. 145*, Forestry Canada and British Columbia Ministry of Forests, Victoria, Canada, 14 pp.

Zasada, J.: 1990, *Forest Industry Lecture,* Fac. Agric. and For., Univ. Alberta, Edmonton, Canada **25**, 44 pp.

Zasada, J. C., Van Cleve, K., Werner, R. A., McQueen, J. A. and Nyland, E.: 1977, *In: Proc. Symp. North American Forest Lands at Latitudes North of 60 Degrees.* U. of Alaska, Fairbanks, USA, pp. 137-195.

ARTIFICIAL REGENERATION OF SPRUCE ON COLD, WET SOIL: 10 YEARS ALONG

C. HAWKINS, T. LETCHFORD and M. KRASOWSKI

Red Rock Research Station, RR #7, RMD #6, Prince George, British Columbia V2N 2J5, Canada

Abstract. Ten year results are presented from two reforestation sites in the Bowron drainage of British Columbia. Spruce (*Picea glauca, P. engelmannii*) of the same seedlot and stocktype, was planted at the same time on the same site-types. Site 1 (S1) utilized different sizes of inverted mineral mounds and Site 2 (S2) used mounding coupled with herbicide (glyphosate) site preparation or brushing treatments. Ten year survival on mounds was 100% at S2 and greater than 90% at S1. Control treatment survival was 34 and 66% at S1 and S2, respectively. Mounding increased 10 year stem volume (SV) by more than 300% over control at S1. Mounding or herbicide site preparation approximately doubled the SV over control while mounding plus herbicide site preparation resulted in a three fold increase of SV at S2. *Pissoides* leader weevil incidence was at endemic levels at both sites but it is expected to increase if global warming scenarios hold. This would reduce future site productivity. The most costly treatment to establish at S2, mounding plus herbicide site preparation, was the most cost effective at year 10. These data suggest that spending more at establishment can result in greater survival, growth and return on investment.

Keywords. ARTIFICIAL REGENERATION, MOUNDS, GLYPHOSATE, *Picea, Pissoides.*

1. Introduction

In the late 1970s, spruce bark beetle populations went from endemic to epidemic levels in the Bowron River watershed in central British Columbia. A massive clearcut logging operation was done to eliminate this infestation. By the time salvage logging was completed in 1987, approximately 50 000 ha had been disturbed by logging and wildfire (area lies between 53°N and 54°N, and between 121°W and 122°W) (Anonymous, 1991). Planting started in the early 1980s and most openings were replanted by 1992. However, early regeneration attempts on productive sites on north and east aspects above 1000 m were often not successful. Biological constraints to regeneration on these sites were thought to be: (1) low soil temperatures; (2) excess soil moisture; (3) poor soil aeration; (4) low light due to dense herbaceous vegetation cover; and (5) vegetation and snow press (Draper *et al.*, 1985, Spittlehouse, 1994). Operational constraints included: (1) excessive soil moisture; (2) a large volume of residual debris; (3) short and inconsistent prescribed fire windows; and (4) an inappropriate location for large scale chemical vegetation control (Draper *et al.*, 1985).

Earlier work has shown soil warming treatments improve the growth of white spruce (Dobbs and McMinn, 1977; McMinn, 1982) and soil aeration is crucial for good growth (Zinken *et al.*, 1974). Clearcutting results in soil temperature increases. However it was postulated that soil temperature at 10 cm would still be too cold to allow good root growth following logging (Draper *et al.*, 1988). Partial shading by competing vegetation may initially promote spruce survival (Eis, 1967) but in the longer term, it reduces spruce growth (Eis, 1981). The effects of herbaceous competition after logging far exceeded the effects of partial shading and warmer soil temperature (Draper *et al.*, 1988). Therefore, research was to be directed at temperature-vegetation and light-vegetation relationships.

Water, Air and Soil Pollution **82**: 115–124, 1995.

A series of cooperative regeneration research trials were established in the early 1980s on the west side of the Bowron River Valley. Approaches included: (1) inverted mineral capped mounds (mechanical site preparation); (2) herbicide site preparation; (3) herbicide brushing and weeding treatments; (4) prescribed fire site preparation; (5) planting stocktype selection (size); and (6) timing of planting, all with appropriate controls. Some trials were established on similar biogeoclimatic units (Meidinger and Pojar, 1991). This allowed informal comparisons among treatments and sites. The cooperating agencies were the British Columbia Forest Service (BCFS) and the Canadian Forest Service (CFS). All sites are extant and 15 year measurements are planned for 1998.

This paper reports on 10 year field results (1993) for: (1) a mechanical site preparation trial; and (2) a mechanical site preparation and herbicide site preparation and brushing trial. Treatment cost effectiveness was calculated.

2. Materials and Methods

2.1. Sites

The two sites are approximately 80 km southeast of Prince George on the west side of the Bowron River. The climate is continental and similar to that of Prince George which has mean January and July temperatures of -12 and 15 °C respectively (Environment Canada, 1981). Both sites are in the SBSvk subzone but are transitional to the higher elevation, less productive ESSF zone (Meidinger and Pojar, 1991). The dominance of lady fern following logging indicated the potential productivity of both sites (C. DeLong, 1993, pers. comm.). Site one (S1) is at about 1100 m elevation on a northeast aspect with a 10–15% slope. The soils are gleyed grey luvisols, silt loam in texture, with 20% coarse fragments, subhygric moisture and permesotrophic nutrient regimes (Draper *et al.,* 1985). The site was logged in the winter of 1981–1982 and unsuccessfully treated with a prescribed burn in the fall of 1982. Site two (S2) is 2 km from S1, at about 1160 m elevation with a 20% slope on a north aspect. The soils are medium textured originating from morainal till with a compacted layer at 0.5 m, and have subhygric moisture and permesotrophic nutrient regimes (Herring and Draper, 1984). The area was winter logged in 1980–1981 and lightly burned in the fall of 1981. Site two was operationally planted in the spring of 1982 but success was limited.

2.2. Site preparation and planting

The study at S1 is a randomized complete block design with five treatments replicated in 18 blocks. A treatment unit was a 15-seedling row (Letchford, 1994a). Treatments were: (1) *control*: no site preparation; (2) *organic mat*: LFH layer (40 × 60 cm plan area) inverted by a Bräcke patch scarifier in August 1983; (3) *6 cm mound*: organic mat created by Bräcke scarifier with 6 cm deep mineral soil cap (7.2 L) added by hand in August 1983; (4) *12 cm mound*: organic mat with 12 cm cap (14.4 L); and (5) *24 cm mound*: organic mat with 24 cm cap (28.8 L). Mineral caps were constructed by hand to ensure uniformity within and among treatments. Organic mat thickness (May 1984) averaged 9 cm (range 6–15 cm). Nine of the 18 blocks were retained as permanent sample plots.

The study at S2 is a split plot factorial replicated five times in 0.16 ha blocks (Letchford, 1994b). A treatment unit was a 10-seedling row. The main plot factor is mounding at two levels (first digit in 2-digit treatment code) *0*: unmounded; and *1*: inverted organic mat with 24 cm mineral soil cap constructed by hand in the summer of 1983 by inverting 40 × 60 cm plan area patches of LFH and capping with about 30 L of mineral soil. These mounds were similar to the 24 cm mounds at S1. Herbicide is the subplot factor at four levels (second digit in the 2-digit treatment code, for example: (0,1) is unmounded and herbicide site preparation in August 1983): *0*: no herbicide; *1*: site preparation with glyphosate (2.25 kg active ingredient per ha) in August 1983 (applied using two backpack sprayers attached to a 5 m boom); *2*: brushing and weeding glyphosate spray (as in *1*) applied in August 1985; and *3*: glyphosate site preparation (*1*) plus a glyphosate brushing and weeding when needed. Herbicide treatment *3* was never needed. This doubled the number of seedlings in herbicide treatment *1* to 200.

Both sites were planted with bareroot transplant, 2+1 white spruce (*Picea glauca* (Moench) Voss.) seedlings of BCFS registered seedlot 4097 grown at a nursery near Prince George. Site 1 was planted 30 May to 1 June 1, 1984, just after S2 was planted.

2.3. Measurements and Statistical Models

Seedlings at both sites were measured for height (HT) and ground-line stem diameter (GSD) in years 0, 1, 2, 3, 5, 7 and 10 (September 1993). Height was measured to the nearest 0.5 cm and GSD to the nearest 0.1 mm in the first 5 years, and to the nearest 5 cm and 1 mm respectively thereafter. Tree condition was scored using subjective codes (Letchford, 1994a,b). Stem volume (SV) was calculated as a cone.

Data analyses for both sites were done using SAS software (SAS Institute Inc., 1988). Analyses of variance were done on individual tree data using the general linear models (GLM) procedure (Table I). Some S1 blocks had no control treatment survivors, therefore data from only five blocks were analyzed. Block × treatment mean squares comprised almost 100% of the error term. Neither block, nor block × treatment were tested. Means were separated with the Tukey option as is recommended for unequal sample sizes (SAS Institute Inc., 1988). Large sample size differences at S2 required treatment, means to be compared by least significant difference (LSD) calculated as: $t\sigma((1/n_1) + (1/n_2))^{1/2}$ where t is TINV function (SAS Institute Inc., 1988) with $p = 0.975$ and 16 degrees of freedom, σ is the square root of the mean square error, and n_1 and n_2 are sample sizes of the two means being compared. Survival and condition codes were not analyzed but rather frequencies are presented and discussed.

TABLE I

Analysis of variance models used for testing sources of interest at S1 and S2. df signifies degrees of freedom.

Source	df	error term
Site 1		
Block	4	not tested
Treatment	4	B × T
B × T	16	not tested
Tree (B T)	225	
Total	249	
Site 2		
Block	4	not tested
Mounding	1	B × M
B × M	4	B × M × H
Herbicide	2	B × M × H
M × H	2	B × M × H
B × M × H	16	Tr(B M H)
Tree (B M H)	332	
Total	361	

TABLE II

Treatment means of HT (height), GSD (ground-line stem diameter), and SV (stem volume) at 10 years for Site 1. Mean separations were done using Tukey's Studentized Range Test. Means followed by the same letter are not significantly different, $\alpha = 0.05$.

Treatment	n	HT (m)	GSD (mm)	SV (L)
24 cm cap	44	2.276a	57a	2.333a
Organic mat	40	2.248a	53a	2.233a
6 cm cap	72	2.140a	53a	1.944a
12 cm cap	72	2.137a	53a	1.892a
Control	22	1.463b	31b	0.591a

2.4. Effective Cost of Treatments

No site preparation costs were included for control treatments. Herbicide application costs were estimated at \$154 per ha based on aerial application (C. Farnden, 1992, pers. comm.). Inverted organic mats and mineral capped mounds smaller than 10 cm can be constructed with a Bräcke mounder for \$250 per ha. A larger machine is required for mineral capped mounds greater than 10 cm and the cost is estimated[1] at \$300 per ha. Planting costs are set at \$0.30 and \$0.42 per tree for mounded and unmounded sites, respectively. Herbicide pretreatment has no effect on planting costs. All other silviculture costs are assumed equal for both sites and are not considered.

Ten year treatment volume (Trt Vol) per ha was estimated as: [mean Trt SV (m^3)] × (1400 stems planted per ha) × (Trt survival). Treatment effective cost was estimated in 1993 Canadian dollars as the cost per 1 m^3 of wood produced 10 years after plantation establishment by the following: (Trt Vol per ha)/(Site preparation + planting cost per ha).

3. Results

3.1. Site 1

There were significant treatment differences at year 10 for HT and GSD but not for SV (Table II). The control treatment had smaller trees compared to the other four treatments which were similar to one another.

Survival and condition codes are presented by treatment in Table III. The control treatment had much poorer survival (34%) than the other treatments which were similar and averaged 91%. Survival has declined in all treatments since the last assessment in 1990. A majority of trees classified as sickly in the previous assessment were dead by the present assessment. Proportionally there were more healthy trees on the non-control treatments. Almost all non-control trees were affected with *Adelges* while the leader weevil, *Pissodes*, was not often evident. Severe stem sweeps (sweep greater than the GSD), occurred on about two-thirds of the control trees and one-third of the trees in the other treatments.

[1]Estimates are based on discussions with local contractors and silviculturists. A. Todd of Integrated Silviculture Services, Prince George, and Lorne Bedford of B.C. Ministry of Forests, Victoria, are thanked for their input.

Establishment costs and effective cost estimates are expressed in 1993 dollars and relative terms by treatment in Table IV.

3.2. Site 2

Analysis of variance for tree HT, GSD and SV all showed significant mounding and herbicide effects and non-significant interaction effects (Table V). Mounding improved tree growth across all levels of herbicide treatment (Table VI). Herbicide site preparation improved growth more than herbicide brushing and weeding across the two levels of mounding.

Survival after 10 years was 100% on mounds and 81% in unmounded areas (Table III). Survival was best, 93%, when herbicide was used as a site preparation (0,1) while it was poorest with treatments (0,0) and (0,2) at 66 and 72% respectively. Survival decreased over time for the unmounded treatments. On average, 92% of the trees were scored as healthy and *Adelges* infestation and stem sweeps were similar to S1 (Table III).

TABLE III

Summary of survival and tree condition code percentages, relative to the number of live trees by treatment, and across treatments, at year 10 for Sites 1 and 2.

	Site 1 treatments					
	Control	Organic mat	6 cm cap	12 cm cap	24 cm cap	All
Total no. of trees	65	72	125	125	72	459
No. of live trees	22	58	118	121	67	386
Survival (%)	34	81	94	97	93	84
Condition code						
Healthy	59	76	76	69	78	73
Sickly	41	16	8	8	13	12
Adelges	77	93	95	93	99	94
Pissodes	-	3	6	7	3	5
Major sweep	68	28	32	33	39	35

	Site 2 treatments						
	(0,0)	(0,1)	(0,2)	(1,0)	(1,1)	(1,2)	All
Total no. of trees	50	100	50	50	100	50	400
No. of live trees	33	93	36	50	100	50	362
Survival (%)	66	93	72	100	100	100	91
Condition code							
Healthy	91	89	92	90	92	100	92
Sickly	3	2	-	-	-	-	1
Adelges	79	97	100	100	100	100	97
Pissodes	-	5	1	-	4	2	3
Major sweep	36	35	36	50	29	46	37

TABLE IV

Site preparation and planting costs per seedling in 1993 Canadian dollars by treatment. Total cost is the cost of establishing 1400 stems per ha in 1993 dollars. Relative costs are relative to the control treatment set at 100. Effective cost of each treatment is based on 10 seasons growth.

Treatment	Site prep. ($)	Planting ($)	Total cost ($ ha^{-1})	Relative total cost	Effective cost ($ m^{-3})	Relative effective cost
Site 1						
Control (0)	0.000	0.420	588	100	2091	100.0
Organic mat (1)	0.179	0.300	658	112	265	12.7
6 cm mound (2)	0.179	0.300	658	112	262	12.5
12 cm mound (3)	0.214	0.300	720	122	280	13.4
24 cm mound (4)	0.214	0.300	720	122	237	11.3
Site 2						
Control (0,0)	0.000	0.420	588	100	735	100.0
Herbicide prep. (0,1)	0.110	0.420	742	126	278	37.8
Herbicide B&W (0,2)	0.110	0.420	742	126	485	66.0
24 cm mound (1,0)	0.214	0.300	720	122	326	44.4
Mound and herb. prep. (1,1)	0.324	0.300	874	149	234	31.8
Mound and B&W (1,2)	0.324	0.300	874	149	293	39.9

4. Discussion

Mounding on wet, rich sites provides several benefits to planted seedlings: improved soil aeration; control of competing vegetation; increased light availability; and warmer soil temperatures (Spittlehouse, 1994).

4.1. SITE 1

Survival was affected significantly by site preparation. The control treatment had poorer survival than the four mounding treatments which were not different from each other (Table III). Survival continues to decrease with each assessment. More than 10% of trees across treatments were rated as sickly at year 10 (Table III). Retrospective analysis of mortality on this site indicates most dead trees were scored as sickly prior to their demise (Letchford, 1994a). Therefore, further mortality is expected, particularly in the control treatment where the major limiting factors have not been overcome.

Early results from this project suggested HT and GSD increased with increased mineral capping (Draper *et al.,* 1985). However, by year five significant differences were observed only between the control and the four mounding treatments (Hawkins and Letchford, 1993). The relationship still holds at year 10 (Table II). Control tree HT at year 10 is similar to HT for the four mounding treatments at year 7 (Hawkins and Letchford, 1993). The effectiveness of the organic mat treatment suggests that breaking up the continuity of the cold, wet, organic mat has significant benefits. The largest environmental differences in the early years were between the control and the organic mat treatments. The control and organic mat treatments respectively had 50 and 90% of the degree days (5 °C base) by the end of August relative to the 24 cm cap treatment in the first growing season (Draper *et al.*, 1985). Photosynthetically active radiation at the seedling tip in the second growing season was 49, 64 and 78% of the mean daily background total respectively for the control, organic mat and 24 cm cap treatments (Draper *et al.*, 1988). Vegetation control was marginally better on the mineral capped mounds than on the organic mat (Draper *et al.*, 1985).

Three-quarters of the trees were rated as healthy after 10 seasons (Table III). Control trees had a higher incidence of poor attributes, although *Adelges* was less common in the control than other treatments. *Adelges* infestation averaged 95% on the mounding treatments, about 20% greater than on the control. Interestingly, *Adelges* may prefer healthy rather than sickly trees. It is not expected to affect the vigour of the trees unless the infestation persists over a prolonged period (Ives and Wong, 1988).

The leader weevil, *Pissodes*, is present on about 5% of the trees (Table III). No attacks were observed on control trees. Perhaps this reflects their small size, less visible leader and lower temperatures under increased vegetation cover (Spittlehouse *et al.,* 1994). The endemic level of *Pissodes* in this ecosystem is thought to be about 5% and poses no long term problem (R. Alfaro, 1994, pers. comm.). However, weevil incidence may increase in this ecosystem if macro-climate is warming, as McMullen (1976) suggested summer heat units regulate weevil populations. Increased weevil attacks would decrease site productivity as a response to global warming.

Major stem sweep (Table III), a deformity and wood quality issue, affects about one-third of the trees with smaller control trees, having about twice the incidence. Previous observations on-site suggest stem sweep is caused by snow creep in mid winter and early spring. However, vegetation press could be a causal agent particularly on the smaller trees. Most of the sweeps are within 45 cm of the ground. Major sweep, if persistent, could reduce the quality of crop trees, thereby reducing the yield in the control treatment. However as trees increase in GSD, they become less susceptible to snow creep and may outgrow the deformation.

The control treatment is 7.5 to 9 times more costly relative to the mounding treatments at year 10 (Table IV). This reflects both the poor survival and the decreased growth of control trees. The difference is less than 19% between the poorest and the best mounding treatments (12 and 24 cm cap). Raw planting on this site may be of economic interest at establishment, but clearly it is not a viable option.

TABLE V

ANOVA table for variables HT (height), GSD (ground-line stem diameter) and SV (stem volume) at year 1 on Site 2 (see text). df is degrees of freedom.

		HT (cm)		GSD (mm)		SV (cm^3)	
Source	df	F-value	Prob. > F	F-value	Prob. > F	F-value	Prob. > F
Block	4	2.82	0.170	3.33	0.135	3.21	0.143
Mounding	1	9.63	0.036	20.44	0.011	11.74	0.027
B × M	4	1.87	0.164	1.56	0.232	1.49	0.252
Herbicide	2	14.11	0.000	24.42	0.000	19.50	0.000
M × H	2	1.05	0.372	2.07	0.159	0.03	0.971
B × M × H	16	1.26	0.223	1.19	0.271	1.17	0.289
Tree (B M H)	332						

4.2. Site 2

Two years after site preparation with glyphosate, vegetation cover was reduced by 40% and vegetation height by 50 cm (Draper and Herring, 1988). After four seasons, herbicide site preparation was more effective at promoting seedling growth than brushing and weeding with glyphosate (Draper and Herring, 1988). The effect of mounding was not clear at four years; the best overall treatment was the combination of mounding and herbicide site preparation. (Draper and Herring, 1988). After seven growing seasons the same trends were evident (Hawkins and Letchford, 1993). Relative soil warming by treatment was least in the control treatment (100), best in the mound and herbicide site preparation (175), with herbicide alone (120) and mounding alone (160) being intermediate (Draper and Herring, 1988). The growth results do not follow the early soil warming trend, suggesting soil temperature was not the dominant limiting factor.

At year 10, the gains due to mounding and/or herbicide site preparation are even more evident (Tables V and VI). In the absence of mounding, herbicide site preparation is better than a later brushing and weeding. Mounding improves the seedling response described for herbicide treatments. Relative to no treatment, mounding alone (1,0), or herbicide brushing and weeding (0,2), resulted in similar growth improvements. Wood and von Althen (1993) observed similar results for white spruce in response to the timing of vegetation control in the boreal forest in Ontario. In another mounding and herbicide trial with jack pine in the boreal forest, the positive effects of herbicide site preparation and the combination of herbicide and mound site preparation were not apparent until year 5 (Sutton and Weldon, 1993).

This plantation was generally in good health (Table III), a marked improvement over S1. Major stem sweeps were observed in about one-third of the trees as was the case at S1. The incidence of sweep was lowest on herbicide-treated mounds suggesting that vegetation press compounds the stem sweep associated with mounding. The incidence of leader weevil, while lower than S1 is considered endemic for this ecosystem (R. Alfaro, 1994, pers. comm.) and carries the same caveats described above.

TABLE VI

HT (height), GSD (ground-line stem diameter) and SV (stem volume) means by mainplot, subplot and treatment factors on S2 at year 10. Means followed by the same letter are not significantly different at $\alpha = 0.05$.

Treatment	n	HT (m)	GSD (mm)	SV (L)
Mounded (1)	200	2.36 a	56 a	2.26 a
Not mounded (0)	162	2.10 b	49 b	1.69 b
Herbicide site prep. (1)	193	2.38 a	57 a	2.37 a
Herbicide B&W (2)	86	2.24 a	52 b	1.87 b
No herbicide (0)	83	1.95 b	49 c	1.29 c
Control (0,0)	33	1.66 a	35 a	0.864 a
Herbicide prep. (0,1)	93	2.27 b	54 b	2.047 b
Herbicide B&W (0,2)	36	2.08 b	48 ab	1.516 ab
24 cm mound (1,0)	50	2.14 b	50 b	1.578 b
Mound and herb prep (1,1)	100	2.48 c	60 c	2.669 c
Mound and B&W (1,2)	50	2.35 bc	55 bc	2.127 b

The control treatment (0,0) at S2 was the least cost effective, being three times that of the best treatment (Table IV). Treatment (1,1), though the most expensive to install, was the most cost effective because of the significant increase in growth. The two-year delay in applying herbicide (0,2), increased the effective cost relative to the pre-plant application (0,1) almost two fold. Clearly, some form of site preparation and/or post-planting vegetation control results in better crop performance and is more cost effective. This view of spending money wisely at establishment is gaining acceptance in the boreal forest (Bauer, 1994).

On S2, mounding did improve tree growth but not to the extent of herbicide site preparation even though mounding was more effective at soil warming (Draper *et al.*, 1988). When herbicide site preparation or brushing and weeding were coupled with mounding, growth was further improved. This indicates that neither herbicide nor mounding alone fully overcame the site's limiting factors: cold, wet soils and vegetation competition. Site preparation with herbicide ameliorated more of the limiting factors than did mounding on S2. The relief from above and below-ground competition provided by the herbicide was more beneficial than the warmer soil temperatures and spot vegetation control effected by mounding. This counters the suggestion of Draper *et al.* (1988) that soil warming was more important than controlling competing vegetation. The results also allude to the complex dynamics of regeneration processes on these productive sites.

4.3. Sites 1 and 2

Site preparation improved growth on both sites. Site 2 had the largest saplings while S1 had the greatest growth on the mounds that were common to both sites. Control seedlings on S2 had better survival and growth than did their counterparts on S1. This resulted in marked differences in cost effectiveness of the control treatments between the two sites, with S1 being about three times more costly than S2. These observations suggest there are significant environmental differences between the two sites even though they are classified as the same biogeoclimatic site association (SBSvk/05). Furthermore, treatment responses suggest that while limiting factors may be similar on the sites, they are not the same and their impacts may differ. This is reinforced by the observations of Cleary and Kelpas (1981) that each forest site is a unique series of limiting factors: the biological constraints to survival and growth.

5. Conclusions

The above discussion leads to four conclusions. Firstly, the best or most cost effective treatment is not necessarily the same on sites of the same biogeoclimatic site association. There is no universal best treatment. Secondly, no attempts were made to identify thresholds for limiting factors, soil temperature being the exception. The other thresholds for these sites are not known. Regardless, successful regeneration is possible. Thirdly, leader weevil attacks will likely increase if global warming scenarios hold and site economic productivity will decline. Fourthly, spending money on site preparation at establishment is the most cost effective option on both sites. The inexpensive option resulted in unsatisfactory survival and growth.

Acknowledgments

This project was funded by the B.C. Ministry of Forests. Project initiators and leaders, working plan authors, and motivators were D. A. Draper and L. J. Herring of the British Columbia Ministry of Forests and R. G. McMinn of the Canadian Forest Service. We thank K. Thomas and an anonymous reviewer for constructive criticism.

References

Anonymous: 1991, Tour Guide. B. C. For. Serv., Prince George Forest District. Prince George, Canada.

Bauer, J.: 1994, *Can. Forest Indust.* (June), 20-24.

Cleary, B. D. and Kelpas, B. R.: 1981, *Special Publ. No. 1.* Oregon State Univ., For. Sci. Res. Lab., Corvallis, USA, 31 pp.

Draper, D. A. and Herring, L. J.: 1988, EP 966 and EP 972.01, File Rep., Res. Br., B. C. For. Serv., Victoria, Canada. (Unpublished manuscript.)

Draper, D. A., Binder, W. D., Fahlman, R. and Spittlehouse, D. L.: 1985, *In: Interior Spruce Seedling Performance - State of the Art.* North. Silv. Committee, B. C. For. Serv., Victoria, Canada, pp. F1-F20.

Draper, D. A., Spittlehouse D. L., Binder, W. D. and Letchford, T.: 1988, *In*: Landis, T. D. (ed.), *Proc. Combined Meeting Western Forest Nursery Associations.* USDA For. Serv. Gen. Tech. Rep. RM-167. Ft. Collins, USA, pp. 212-218.

Dobbs, R. C. and McMinn, R. G.: 1977, *In:* Black, T.A. *et al.* (eds.), *Energy, Water and the Physical Environment.* Soil Science Workshop Report. B.C. Min. Agric., Victoria, Canada, pp. 66-73.

STRUCTURE AND BIOMASS OF LARCH STANDS REGENERATING NATURALLY AFTER CLEARCUT LOGGING

I. DANILIN

Institute of Forest, Russian Academy of Sciences, Siberian Branch, Akademgorodok, Krasnoyarsk 660036, Russia

Abstract. Variations in the succession following cutting of a herbaceous *Larix sibirica* Ledeb. phytocoenosis along the southern boundary of boreal forests in southern Siberia and in Eastern Hentey, Mongolia, were studied. Morphometric methods were used to determine the dimensional hierarchies of coenopopulation individuals. Structure and productivity of the aboveground components including standing wood, herbaceous cover and litter were studied. The maximum aboveground phytomass was measured as 212.3 Mg ha^{-1} (oven dry mass). The highest total aboveground biomass productivity rate of a *Larix sibirica* phytocoenosis located at its southern limit exceeds 7 Mg ha^{-1} per year. The maximum annual phytomass increment was found to be 4.4 Mg ha^{-1} for the overstorey trees, 2.1 Mg ha^{-1}, for the herbaceous layer and 0.7 Mg ha^{-1} for forest litter.

Keywords. STRUCTURE, BIOMASS, CLEARCUTTING, SOUTHERN SIBERIA, MONGOLIA, *Larix sibirica*.

1. Introduction

The southern boundary of the boreal forest in Asia lies in southern Siberia and Mongolia. Sub-boreal, subarid and strongly continental climates are the major limiting factors for the boreal forest coenoses in this area. Here, forest ecosystems develop on seasonal frost soils and are surrounded by vast steppe areas and uplands.

Developing criteria for forest ecosystem stability is necessary for appropriate forest resource management. It is therefore important to obtain complex biometric characteristics of forest stands of different ages and thoroughly study their structure and biological productivity. This information would be vital for understanding the geographic variations in forest succession processes in disturbed vegetation communities located along the southern boundary of boreal forests in Eurasia. Information on structural and functional mechanisms in forest biogeocoenoses and their anthropogenic dynamics are also valuable for management purposes (Danilin and Tsogt, 1992).

2. Materials and Methods

The study was conducted in the eastern Baykal region (Zabaykal'skiy region) in the Eastern Hentey forest province, where larch stands occur in a range of altitudinal belts, and in the upper stream area of the Onon river. The sites are located approximately at 49°10'N, 110°0'E. All these areas are dominated by larch (*Larix sibirica* Ledeb.) stands with herbaceous ground layers, growing usually in the lower portions of mountain slopes. They comprise more than 36 000 ha (42%) of the total forest area in the Eastern Hentey (Savin *et al.*, 1988). These are highly productive stands, commercially valuable, and accessible by vehicles; over the last 30 years these stands were subjected to intensive cutting. Common cuts and pseudo-clear-cuts are practised here. The harvested wood is usually dragged by tractors leading to considerable disturbance of

Water, Air and Soil Pollution **82**: 125–131, 1995.

TABLE I

Description of sample sites for 100% larch stands. DBH, height, crown width and length are average values for the site.

Plot #	Stand age (yr)	DBH (cm)	Height (m)	Crown width (m)	Crown length (m)	Site class	Basal area (m^2 ha^{-1})	Volume live/dead (m^3 ha^{-1})	Density (tree ha^{-1}) live	Density (tree ha^{-1}) leave	Density (tree ha^{-1}) dead	Slope Aspect	Slope Angle
1	16	1.6	3.8	0.9	1.6	III	11.3	39/0.2	54 500	200	1500	NE	16°
2	28	5.9	7.9	1.6	4.0	III	14.6	74/0.1	5200	300	200	NE	13°
3	30	7.5	9.5	1.1	2.7	II	41.4	212/13.3	9100	1100	9600	W	13°
4	37	18.4	15.4	3.0	6.9	I	33.9	305/0.9	1200	800	900	E	5°
5	70	18.9	18.1	2.7	6.1	III	43.7	397/19.0	1500	30	600	E	21°

ground cover and soils. After cutting, logging residues are brought into piles evenly distributed over the cut area and left untreated. In unlogged stands, understorey trees number from 2400 to 5000 young larches ha^{-1}, their quality and distribution depending on their age, basal area, and the area occupied by the parent stand. In logged areas surviving understorey averages, as a rule, from 400 to 600 individuals per hectare.

Both parent larch stands and cut areas suffer from frequent fires that enhance natural regeneration. In small clear cuts (up to 5 ha) natural regeneration is abundant. Larch seedlings germinate annually and their resistance to environmental stresses is relatively high; in the first 5–6 years after cutting the number of seedlings exceeds 50 000 ha^{-1}, providing favourable conditions for regeneration of the cut area. By the time the young generation canopy closes (in about 15 years), the herbaceous cover and litter layer characteristic of the previous stands are regenerated completely.

Five sample plots adequately representing the stands under study were established (Table I). Understorey was absent at all plots. The plots were characterised by taiga-peat, seasonal frost floor, and loamy soils and were situated at 1550 m asl on gradual-to-steep slopes of various aspects.

All trees at the sample plots were measured and mapped, to establish their size-dependent position in the phytocoenosis and to the determine structure of aboveground biomass components, including trees, herbaceous cover, and litter. Stem analysis trees were selected from each diameter class, divided into fractions to be weighed (weight errors were ±1 g and ±100 g for fraction weights up to 10 kg and above 10 kg, respectively). Samples for determination of moisture contents were taken from each fraction and oven-dried. The values were analytically approximated (adjusted), summed by diameter class and recalculated on a 1 ha unit area basis. The approximation accuracy was not less than 0.97 ($r^2 > 0.97$). To evaluate herbaceous phytomass, fifty 20×25 cm mini-plots were established on each plot. Herbaceous vegetation was grouped by species (nomenclature according to Grubov, 1982). Litter was divided into fractions (Tables IV and V). Roots and soil were removed. Herbs and litter samples were oven-dried and weighed, with an error of ±0.01 g (Rodin *et al.*, 1968). The sampling was performed in the second half of July, which is towards the end of the period of active vegetation growth in the experimental area.

3. Results and Discussion

Biomass accumulation rates in larch stands depend partly on the size of trees and stand density. Increase in all biomass categories is in direct proportion to the increase in biometric parameter values of trunks and crowns. Correlations on these sites are strong: 0.8–0.9 for trunk diameters, 0.6–0.7 for crown lengths, and 0.6–0.8 for both tree heights and crown diameters. Tree age has less effect on phytomass; here, the correlation varies from 0.2 to 0.4. The correlation between biomass parts is much stronger than between biomass and linear tree dimensions ($r^2 > 0.95$). A decrease in stand density results in a significant increase in living branch and needle biomass. However, the standing dead branch biomass increases much more slowly. Trunk wood and bark biomass remain unchanged (Table II). Increase in average stand age leads to some changes in correlation between phytomass fractions. In 16-year old stands the crown fraction exceeds that of trunk wood. In 30- to 40-year old larch stands, crown fraction mass is less than that of trunks. In 70-year old stands the trunk wood fraction accounts for 70–80% of total tree phytomass. The living branch fraction (wood and bark) exceeds the needle phytomass for all stand ages. For the lower diameter classes, some increase in dead branch fraction is recorded. An increase in tree size and age causes an increase in trunk wood and a decrease in bark biomass.

Data in Table II indicate that the total stand aboveground phytomass, trunk wood mass, and bark mass consistently increase with stand age. This is not the tendency with the living and dead branches, and the needle mass remains at about the same level or varies insignificantly. The highest tree mortality was recorded at sample plot 3. Here, due to high stand density, trees of the lowest diameter class show very high mortality. High tree mortality at sample plot 5 is attributed to the effects of a surface fire that occurred there in 1987.

As the average stand age increases, stand structure changes. Data in Table II show that in 16-year old larch stands, crown phytomass accounts for 40% and trunk phytomass (with bark) for 51% of total stand phytomass, with the living branch fraction equal to that of trunk wood (without bark) (35%). In 30-year old stands, crowns comprise 23% and trunks 77% of total phytomass. In 70-year old stands the trunk phytomass increases up to 89%, while crown phytomass is reduced to minimal values (11%). For needles, the range of phytomass reduction with stand ageing was recorded to be 1–13% and, for the living branch phytomass, 6–35%. The maximum proportion of dead branches, was observed at sample plot 3, which may be due to very high canopy closure, inevitably leading to a high mortality rate of branches in the middle and lower portions of the stand canopy. Bark fraction values did not differ significantly: 16, 17, and 15% for 16-, 30-, and 70-year old stands, respectively.

Variations of productivity processes are best reflected in annual phytomass increment, rather than in total stock value (Utkin, 1975). Annual phytomass increase is the most important indicator of phytomass vitality level. It is derived by dividing the aboveground stand phytomass value by the average tree age in the stand. For sample plots 1–5, results were as follows: 2.1; 2.0; 4.0; 4.4; 2.7 Mg ha^{-1} yr^{-1} and 2.4; 2.6; 7.0; 8.2; 5.7 m^3 ha^{-1} yr^{-1} (oven-dry matter) respectively. Maximum estimates of annual phytomass production were acquired at sample plots 3 and 4, where the stand biological potential was realised with productivity levels of 8.2 and 7.0 m^3 ha^{-1} yr^{-1} and 4.4 and 4.0 Mg ha^{-1} yr^{-1} (oven dry matter) respectively.

Comparison with other studies shows that the stands in this study are characterised by higher values than, for example, forests in the Western Sayan, where a 40-year old *Larix sibirica* stand of productivity class IV stocked at 98 m^3 ha^{-1} accumulated 73 Mg ha^{-1} of phytomass, while in a 70-year old stand of productivity class III (*L. sibirica* with herbaceous ground cover) the phytomass value was 164 Mg ha^{-1} (Pozdniakov *et al.*, 1969). As reported by these authors, in Yakutia the value (oven dry matter) of

TABLE II

Aboveground biomass of *Larix sibirica* Ledeb. stands

Plot #	Stand age (yr)	Wood stock (m^3 ha^{-1})	Aboveground Biomass (Mg ha^{-1} oven-dry mass) Needle foliage	Live branches	Dead branches	Stem wood	Stem bark	Cones	Total
1	16	38.5[a]	4.472	11.816	0.346	11.774	5.447		33.855
		(0.2)[b]			*(0.012)*	*(0.030)*	*(0.012)*		*(0.054)*
2	28	73.7	5.205	14.311	2.994	23.667	8.377	0.643	55.197
		(0.1)			*(0.005)*	*(0.018)*	*(0.011)*		*(0.034)*
3	30	211.9	3.231	10.186	12.575	71.251	20.689	-	117.932
		(13.3)			*(2.477)*	*(3.843)*	*(1.238)*		*(7.558)*
4	37	304.8	3.628	16.543	9.446	112.367	21.738	-	163.722
		(0.9)			*(0.032)*	*(0.258)*	*(0.058)*		*(0.348)*
5	70	396.6	2.723	10.337	7.714	139.885	28.433	-	189.092
		(19.4)			*(0.234)*	*(5.218)*	*(0.873)*		*(6.325)*

a Normal font is for live trees.
b Second line (italics) is for dead trees.

aboveground phytomass for a young (14-year old) *Larix gmelinii* Rupr. stand mixed with *Betula platyphylla* Sucacz. was 13 Mg ha^{-1}, whereas for a 50-year old *L. gmelinii* stand of productivity class IV (*Vaccinium vitis-idaea* L. forest type) with a total phytomass of 110 m^3 ha^{-1}, this value was 56 Mg ha^{-1}. Krasikov (1987) noted that in southern Evenkia a 70-year old *L. sibirica* stand of productivity class III had an aboveground phytomass value of 98 Mg ha^{-1}. Molchanov (1971) reported the following phytomass data obtained for larch stands in the Arkhangelsk Region: 28 Mg ha^{-1} for 20-year old stands of productivity class I; 96 Mg ha^{-1} for 40-year old stands of productivity class II; and 148 Mg ha^{-1} for 60-year old stands of productivity class III. Comparison of these data with tree growth table values for larch stands in various regions of Siberia (Falaleev *et al.*, 1975) also indicates that in the age interval 16–70 the productivity rate in the stands of interest is much higher. This suggests that stands of *L. sibirica* at the southern limit of its distribution are characterised by higher growth rates and, over the same period of time, accumulate more biomass than larch stands of similar age, forest type, and productivity in other regions of Siberia.

The rate of accumulation of organic matter is considerably influenced by the herbaceous cover. This study determined species in the herbaceous layer using Grubov's (1982) classification. The greatest herbaceous layer phytomass was recorded at sample plot 4 (Table III). Here nutrient and moisture conditions, as well as the solar radiation regime, were most favourable for field layer growth development. The lowest herbaceous phytomass was measured at sample plot 3. The herb layer was poorly developed due to the high stand density and consequent almost completely closed canopy. According to Table III it can be concluded that herbaceous phytomass composition is to some extent dependent on stand density. The greater the gaps between adjacent trees, the greater the herbaceous phytomass. Herbaceous cover reaches its maximum in glades, i.e., in canopy openings, and conversely is sparsely developed under clumps of trees.

In the study area the most abundant herbaceous species are: *Elymus sibiricus* L., *Geranium pratense* L., *Vicia unijuga* A. Br., *Carex amgunensis* Fr. Schmidt, *Carex lanceolata* Boott, *Galium verum* L., *Chamaenerion angustifolium* (L.) Scop., *Chrysanthemum zawadskii* Herb., *Artemisia laciniata* Willd., *Poa sibirica* Roshev., and *Potentilla tanacetifolia* Willd. ex Schlecht. These species are also found to have the highest phytomass.

Dead organic matter accumulated during the life cycle of one forest generation can exceed the living phytomass by several times (Molchanov, 1971), with its role in the stand being as important and versatile (Sukachev and Dylis, 1966). The amounts of organic matter accumulated in forest litter increases with the stand age. The highest amount is found at sample plot 4. Sample plot 5 (a 70-year old stand) had reduced accumulated litter. That can be interpreted as a result of a surface fire that occurred here in 1987 and consumed a considerable portion of litter, primarily in the upper litter layer. Dead needles comprise most of the organic matter in forest litter, followed by dead branches, cones, and bark scales (Table IV). The amount of dead branches increases with the stand age, while the needle fraction is somewhat reduced. This could result from the fall of dead branches as trees grow and from the reduction in dead needle contribution to forest litter. The thickness of the litter layer appeared to be approximately the same at all the sample plots. Close to the stem the litter layer is 5–7 cm thick and consists mainly of needles. The total mass increase of all forest litter averaged 600 kg ha^{-1} yr^{-1} (oven dry matter).

4. Conclusions

The aboveground phytomass structure in *Larix sibirica* Ledeb. stands is closely related to their stand structure, age, density, and productivity. By the age of 70, larch stands accumulate up to 212 Mg ha^{-1} of organic mass. With increasing age the standing dead organic matter decreases consistently, in contrast with the living phytomass. The major contributor to forest biomass accumulation is the tree upper story which accounts for 75–92% of the total organic mass in the phytocoenosis. Dead phytomass, with values gradually decreasing from 24 to 7% with increasing stand age, consists mainly of the litter layer. Herbaceous phytomass is constant and does not exceed 1%. The highest productivity rate of the total aboveground portion of a *Larix sibirica* phytocoenosis at the southern limit of its distribution exceeds 7 Mg ha^{-1} yr^{-1} biomass. The maximum annual phytomass increment was found to be 4.4 Mg ha^{-1} for the aboveground part of the tree, 2.1 Mg ha^{-1} for herbaceous cover and 0.7 Mg ha^{-1} for forest litter.

TABLE III

Herb biomass of the sample plots (oven-dry mass, kg ha^{-1}).

Species	Sample Plot Number				
	1	2	3	4	5
Adenophora lamarckii Fisch.	-	15	-	35	-
Allium lineare L.	-	1	-	-	-
Anemone crinita Juz.	-	-	-	-	-
Artemisia laciniata Willd.	-	16	-	-	165
A. sericea Web. *ex* Stechm.	3	-	-	15	-
A. tanacetifolia L.	-	-	-	-	-
Astragalus membranaceus (Fisch) Bunge	-	16	-	-	-
Calamagrostis obtusata Trin.	61	-	-	-	-
Campanula glomerata L.	-	7	-	-	-
C. rotundifolia L.	-	11	1	8	-
Carex amgunensis Fr. Schmidt.	99	2	1	146	219
C. caespitosa L.	-	-	-	-	-
C. lanceolata Boot.	-	25	-	310	-
Chamaenerion angustifolium (L.) Scop.	38	1	-	-	131
Chryzanthemum zawadskii Herb.	16	14	2	55	121
Dasiphora fruticosa (L.) Rydl.	-	9	-	4	-
Draba sibirica (Pall.) Thell.	-	-	-	-	9
Elymus sibiricus L.	-	3	-	161	1000
Festuca ovina L.	1	-	-	-	33
Fragaria orientalis Losinsk.	61	-	-	6	-
Galium boreale L.	-	-	-	-	-
Galium verum L.	-	2	-	225	-
Geranium pratense L.	-	-	-	515	-
Geranium pseudosibiricum I. Meyer.	10	7	-	-	-
Heteropappus altaicus (Willd). Novopokr.	-	2	-	-	-
Lathyrus humilis Ser. Spreng.	43	2	-	-	-
Myosotis suaveolens Waldst. et Kit	-	5	-	21	-
Myosotis sylvatica (Ehrh.) Hoffm.	-	-	-	-	-
Poa sibirica Roshev.	-	18	-	7	146
Polygonum alopecuroides Tucz. *ex* Meissn.	-	-	-	-	-
Potentilla tanacetifolia Willd. *ex* Schelcht	-	118	-	-	-
Pulsatilla turczaninovii Kryl. *et* Serg.	-	-	-	48	-
Pyrola incarnata (DC) Freyn.	-	-1	-	-	-
Rosa acicularis Lindl.	48	-	-	-	-
Sanguisorba officinalis L.	2	38	-	57	21
Saussurea elongata DC	-	-	-	-	-
S. parviflora (Poir.) DC	-	-	-	-	-
Sedum aizoon L.	-	-	-	-	-
Sedum purpureum (L.) Schult.	-	-	-	18	-
Serratula centauroides L.	-	35	-	38	-
Stellera chamaejasme L.	-	4	-	-	-
Taraxacum officinale Wigg.	-	7	-	-	-
Thalictrum minus L.	-	-	1	-	60
T. simplex L.	-	37	-	45	11
Trisetum sibiricum Rupr.	-	-	-	-	-
Valeriana officinalis L.	-	-	-	5	-
Vicia baicalensis (Turcz.) B. Fetsch.	-	-	-	-	-
V. cracca L.	-	-	-	-	-
V. unijuga A. Br	-	-	-	410	-
Viola uniflora L.	-	1	-	-	-
Total	383	419	5	2127	1916

TABLE IV

Structure of litter in *Larix sibirica* stands. (Depth is to middle of litter layer.)

Plot No.	Stand age (yr)	Density (stem ha^{-1})	Biomass oven-dry matter (Mg ha^{-1})					Depth (cm)
			Needle	Twigs	Grass	Cones and bark	Total	
1	16	56 200	8.923 [a]	0.953	0.712	0.360	10.948	2
			(81.5) [b]	*(8.7)*	*(6.5)*	*(3.3)*	*(100)*	
2	28	5 700	11.566	1.071	1.071	0.330	14.038	2
			(82.4)	*(7.6)*	*(7.6)*	*(2.4)*	*(100)*	
3	30	19 800	15.754	2.486	-	0.005	18.245	2
			(86.34)	*(13.63)*	-	*(0.03)*	*(100)*	
4	37	2 900	17.262	3.653	1.974	1.346	24.235	2
			(71.2)	*(15.1)*	*(8.1)*	*(5.6)*	*(100)*	
5	70	2 130	8.354	5.344	1.171	0.108	14.977	2
			(55.8)	*(35.7)*	*(7.8)*	*(0.7)*	*(100)*	

a First line (normal font) is mass in Mg ha^{-1}
b Values in italics on second line are percentage of total litter mass

References

Danilin, I. M. and Tsogt, Z.: 1992, *In: Ecology and Nature Utilization in Mongolia,* Proc. Intl. Conf., Inst. Botany, Mongolian Academy of Sciences, Ulan-Bator, Mongolia, pp. 82-83. [In Russian]

Falaleev, E. N., Bezzabotnov, E. L., Danilin, M. A., Semechkin, I. V. and Sokolov, E. K.: 1975, *Growth Processes of the Main Forest-Forming Tree Species of Siberia, Part II,* Siberian Technological Institute, Krasnoyarsk, USSR, pp. 13-56. [In Russian]

Grubov, V. I.: 1982, *Classification of Mongolian Vascular Plants (With Atlas).* Nauka, Leningrad, USSR, 443 pp. [In Russian and Mongolian]

Krasikov, I. I.: 1987, *In: Larch and its Composite Processing.* Siberian Technological Institute, Krasnoyarsk, USSR, pp. 23-27. [In Russian]

Molchanov, A. A.: 1971, *In: Biomass Productivity in Forests of Different Zones.* Nauka, Moscow, USSR, 275 pp. [In Russian]

Pozdniakov, L. K., Protopopov, V. V. and Gorbatenko, V. N.: 1969, *In: Biological Productivity of the Central-Siberian and Jakutian Forests.* Institute of Forest and Wood, Siberian Branch, USSR Academy of Sciences, Krasnoyarsk, USSR, 149 pp. [In Russian]

Rodin, L. E., Remezov, N. P. and Bazilevich, N. I.: 1968, Russia, *In: Methods for Investigation of Dynamic and Biological Cycles in Phytocoenoses.* Nauka, Leningrad, USSR, 144 pp. [In Russian]

Savin, E. N., Miljutin, L., Krasnoshekov, J. N., Korotkov, I. A., Suncov, A. V., Dugarjav, C., Dashzeveg, C., Tsogo, Z., Dorjsuren, C., Jam'jansuren, S. and Gombosuren, N.: 1988, *Forests of the Mongolian People's Republic. Larch Stands of the Eastern Hentey.* Nauka, Moscow, USSR, 176 pp. [In Russian]

Sukachev, V. N. and Dylis, N. V.: 1966, *Program and Methods of Biogeocoenosical Investigations.* Nauka, Moscow, USSR, 334 pp. [In Russian]

Utkin, A. I.: 1975, VINITI *Forestry and Silviculture* **1**, 9-189. pp. 9-189. [In Russian]

DESICCATION OF WHITE SPRUCE SEEDLINGS PLANTED IN THE SOUTHERN BOREAL FOREST OF BRITISH COLUMBIA

M. J. KRASOWSKI[1], T. LETCHFORD[1], A. CAPUTA[1], and W. A. BERGERUD[2]

[1]*British Columbia Ministry of Forests (BCMoF) Research Branch, Red Rock Research Station, Rural Route #7, Rural Mail Delivery #6, Prince George, British Columbia V2N 2J5, Canada*
[2]*BCMoF Research Branch, 31 Bastion Square, Victoria British Columbia V8W 3E7, Canada.*

Abstract. Three consecutive years of seedling injury evaluation data from two research sites established in the southern part of British Columbia boreal forest are presented. In 2 out of 3 years the injury was substantial. Planted white spruce is injured during its first overwintering while naturally established seedlings are not. A brief summary of 1992/93 post-winter assessments of 1-year old operationally established plantations is also presented. Most of the injury is sustained in late March and April. Microclimatic records indicate that desiccation is the main cause of observed injury. Prompt dehydration and severe injury resulted from exposing seedlings to controlled environment conditions similar but less harsh than those recorded in late March and April on monitored forest sites. Positioning of young roots of planted seedlings deep in the soil is viewed as a factor contributing to the injury.

Keywords. FREEZE-DESSICATION, WHITE SPRUCE, OVERWINTER INJURY, ARTIFICIAL REGENERATION, BOREAL FOREST.

1. Introduction

Injuries to white spruce seedlings occur frequently in the southeastern part of the British Columbia boreal forest, approximately between 55°N and 57°N (Krasowski *et al.*, 1993). In this area, winters are long, with extreme temperatures down to -50 °C and frequent high winds. Warm Chinooks (called Foehn winds in Europe) occur several times in winter, reducing or eliminating snowpack (Krasowski *et al.*, 1993).

Possible reasons for the observed injuries are: direct freezing damage due to extremely low and rapidly changing temperature, seedling dehardening by warm temperatures occurring during winter (caused by Chinook winds), photoinjury due to high solar radiation during freezing air temperatures, and seedling desiccation due to exposure of shoots and foliage to high vapour pressure deficit while soil and roots are frozen.

Correct identification of the causes of low-temperature related injuries is difficult. Symptoms often appear long after the damage was done and the suspected cause might be a secondary effect of another injury (Tranquillini, 1982). We are investigating overwinter injury to newly planted white spruce seedlings to determine precisely what stress causes it and when it takes place. In this publication, we provide an overall assessment of the injury problem for the area and determine its cause and timing of occurrence.

Water, Air and Soil Pollution **82**: 133–146, 1995.

2. Materials and Methods

2.1. Extent and Timing of Injury

Three-year data are presented for two unsheltered research sites established near Dawson Creek. The sites were manually brushed and planting spots screefed. Each year, container-grown seedlings of local white spruce seedlots were planted in spring (late May) and in summer (early August). Seedlings were numbered and tagged. They were assessed before their first winter to account for earlier damage, then again in May or June. Seedling condition was classed as: (1) good (≤ 30% dead foliage); (2) moderate (> 30% to < 60%); (3) poor (≥ 60% to < 100%); and (4) dead. Each year, remnants of the previous year's planting were removed, then new microsites were chosen for planting. Unlike in the first two years, there was no destructive sampling from the sites in 1992/93, so many more seedlings could be assessed.

One of the affected forest districts (Fort St. John) contracted out an assessment of all their sites planted in the spring of 1992 (little summer planting took place in this district in 1992). The contractor performed the survey in the summer of 1993, recording the incidence of deaths and injury attributed to winter-related causes. A brief summary is presented.

To determine the timing of the injury, seedlings were randomly sampled throughout two winters from the research sites (two per row of planting time treatment). In 1990/91, 10 spring- and 10 summer-planted seedlings were sampled on each collection date. In 1991/92, 24 spring-planted and 48 summer-planted seedlings were sampled each time. Sampling intervals depended on weather changes. Snow, if present, was carefully removed. Seedlings were cut at the ground level, packed into plastic bags, and brought to the research station in a cooler. After overnight thawing at +4 °C in the dark, the severed ends were freshly cut under water. Seedlings in water-filled jars were placed under growth-promoting conditions. Visual assessment was done 2 weeks later.

2.2. Climate Monitoring

Observations were similar on both sites, therefore, data from one site only are presented. Microsite and weather variables were monitored with Campbell Scientific Inc. (CSI) model 21X data loggers. Air temperature at 2, 10, and 40 cm heights, and soil temperature at 5, 10, and 50 cm depths, were monitored with thermocouples at three microsites. Averaged data are presented. Relative humidity and air temperature at 1.5 m were measured with a CSI model HMP 35CF probe. Photosynthetically active radiation (PAR) was measured at 1.5 m height with a Li-Cor model 190 SB quantum sensor. Soil thermocouples were scanned every 5 minutes, while all other sensors were scanned every 30 seconds. Shown data are condensed to daily maximum and minimum for temperature and to daily total average for PAR. Vapour pressure deficit (VPD) was calculated from air temperature and relative humidity. An average VPD between 12:00 and 14:00 hours is presented. Snow depth was averaged from several measurements per site on each site visit. Between the visits, snow depth was inferred from air temperatures at different heights and soil temperature data.

2.3. Growth-Chamber Experiments

These were done to determine if conditions similar to those recorded in March/April on the research sites result in seedling injury. In 1992, two seedlots (SL) planted on the research sites were also planted at Red Rock Research Station (about 300 km to the south). The seedlots were from similar longitudes (about 121°W) and latitudes (close to 56°N), but SL 8779 was from about 1000 m and SL 8782 from about 800 m elevation. Spring-planted stock was 2+0 while summer-planted stock was 1+0. Seedlot 8782 was a larger stock than SL 8779. The former was nursery-grown in PSB 415B containers (Beaver Plastic Ltd., Edmonton, Alberta), with root cavity 18 cm deep and 105 ml volume. The 8779 was grown in PSB 313B containers, 15 cm deep and 65 ml volume. Therefore, possible stock type effect could be concealed by SL effects.

Individually numbered seedlings were planted in spring and in summer into 25 cm long segments of 10 cm diameter PVC pipe filled with field soil. The tubes were then buried in soil outdoors in 96 groups of four. Each group contained only one seedling of each seedlot / planting time combination. In January, February, and April 1993, 24 random groups were removed from the ground and thawed overnight at +5 °C in the dark. Next, water was added to each tube until soil saturation and bottom ends of pipe segments were closed and sealed. Each tube was placed into a plastic bag sealed around the seedling stem.

Two growth chambers (G) were set at +10/+4 °C (day/night), 300 µmol m^{-2} s^{-1} of incandescent plus fluorescent light, and VPD of 0.55/0.20 kPa (day/night). The lights were on for 8, 10, and 12 h in January, February, and April trials, respectively. In each growth chamber, six random tubes per SL/planting time (P) were placed into a controlled temperature bath; another six tubes were outside the bath. Bath temperature was decreased from +4 °C to -2 °C over 3–4 h and kept at the latter temperature constantly. Consequently, two treatments (T) were created: (1) frozen roots, and (2) control.

Two seedlings per SL/P were randomly sampled from each treatment in each growth chamber after 7, 14, and 21 day treatment durations (D) from the start. They were placed at recovery conditions of 25/20 °C, 400 µmol m^{-2} s^{-1}, with ample watering. Visual evaluation of injury (%) was done after two weeks. Seedling mean water content (fresh weight / dry weight) was calculated from weights of four pairs of leaves sampled from the upper 2 cm of each seedling being taken into recovery conditions, except at the beginning when two random seedlings per T/SL/P/G were sampled to match the factorial design.

SAS (SAS Institute Inc., Cary, North Carolina) software was utilised for data analyses. The experimental design was a factorial, with a seedling nested in all other factors. Seedlot and growth chamber were considered random factors.

3. Results

3.1. Extent of Injury

Table I shows the percentage of seedlings per condition class in 1991, 1992, and 1993 post-winter assessments. Data are pooled across planting dates because there were no consistent planting-time related trends. Although the immediate mortality was low, injury was massive after the 1991/92 and 1992/93 winters. Few seedlings were free from injury and many had dead foliage on parts of the stem. There was almost no direct spring frost-related injury in any year. Spring frosts kill swelling buds and young elongating shoots, and these symptoms are easy to identify.

TABLE I

Summary of post-winter seedling condition assessments made on seedlings planted in 1990, 1991 and 1992 on two research sites. Data are pooled across planting times on each site. SLK = Stewart Lake site; BM = Bear Mountain site.

Planting/ assessment year	Site	Number of seedlings assessed	Seedling condition* (% of seedlings assessed)			
			Good	Moder	Poor	Dead
90/91	SLK	265	76	13	8	3
90/91	BM	295	68	16	8	8
91/92	SLK	454	36	28	27	9
91/92	BM	347	25	38	25	12
92/93	SLK	949	24	46	26	4
92/93	BM	954	21	49	28	2

* Condition codes relate to % foliage damage as follows: Good: ≤ 30%; Moderate: > 30% to < 60%; Poor: ≥ 60% to < 100%; and Dead: 100%.

The poor state of 1-year old white spruce plantations following the 1992/93 winter was widely noticed throughout the Peace River area. The contracted survey of spring-planted sites in the Fort St. John Forest District showed an average mortality of 24% on sites planted without mechanical site preparation. Average mortality was 12% on plowed sites but only 6% on mounds. The latter had 66% uninjured seedlings, while plowed sites had 38%. Sites on south-facing slopes had the greatest average mortality (21%), while those with northern aspects had the least (12%). Site size was apparently of little relevance: average mortality on sites larger than 10 ha was 19%, and was 15% on smaller ones, while 46% and 49% of seedlings, respectively, were uninjured. Worst mortality (72%) was on a large, plowed site. One-year old (1+0) planting stock had an average mortality of 16%, compared to 21% for 2+0 seedlings. Worst mortality (72%) occurred in 2+0 stock, while the worst for 1+0 was 48%. Best condition was 98% uninjured seedlings for 1+0, and 87% for 2+0 planting stock.

3.2. Timing of Injury

Fall assessment of the research sites made in 1990 showed minimal injury occurring before November. Little injury was found in samples collected between November and the spring of 1991, but the spring injury assessment also showed low injury levels.

Few 1991 planted seedlings showed any injury in the fall of 1991 except for the pale-green appearance of some summer-planted seedlings resulting from post-planting drought. Most of these later died, contributing to higher mean injuries in November and December collections in which a few of these seedlings happened to be sampled

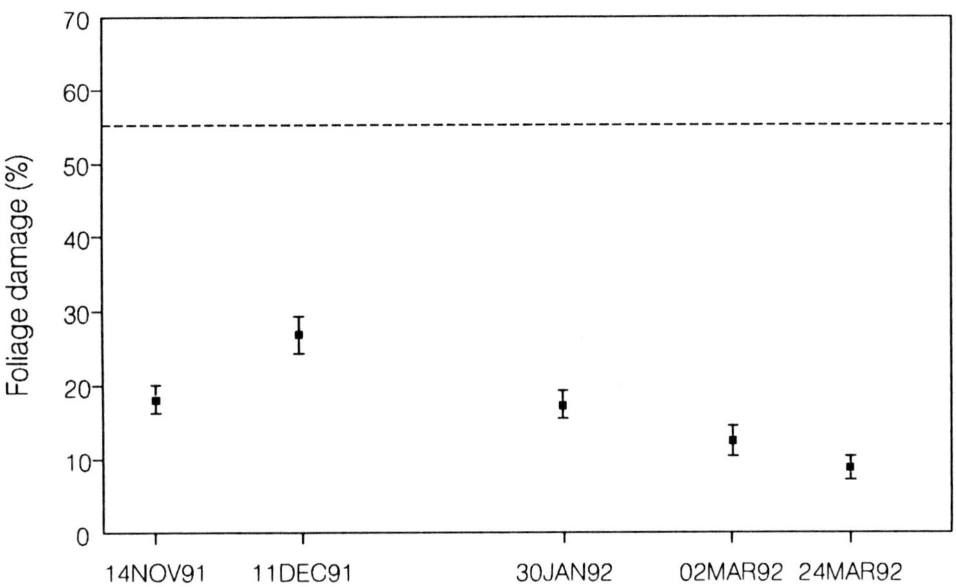

Fig. 1. Mean foliage injury (black points) and standard errors (vertical bars) in samples collected from two research sites during the 1991/92 winter. Dashed line indicates the average foliage injury across these sites calculated from seedling assessment made in June 1992.

(Figure 1). Apart from these deaths, foliage injury in all collections was very low compared to that found in June 1992 (Figure 1). This indicates that the bulk of injuries found in June took place after mid-March.

3.3. Climate Data

Figure 2 shows monthly mean air temperatures at 1.5 m over three winters on the research sites compared with the 1951–1980 averages for the Dawson Creek airport. The winters of 1990/91 and 1991/92 were warmer than average. In 1990/91, snow cover was deep and long lasting. In 1991/92 and 1992/93, snow accumulation was high in mid-winter but the snow was gone by late winter (Figure 2). This probably explains the difference in seedling injury between 1990/91 and the 2 later years.

In presentation of detailed data, we will concentrate on 1992/93. The winter started cold in late October (Figures 3a–c, 4a) and the soil froze at 5 cm in November (Figure 5a), then progressively deeper (Figure 5b), freezing at 50 cm by late December (Figure 5c). Air temperature (Figures 3a, 4a) decreased to about -40 °C in early January but it was milder near the ground (Figures 3b,c) due to snow cover (Figure 2). Two Chinook events with temperatures well above 0 °C occurred in late January and in February (Figures 3a–c, 4a) removing the snow, and soil temperature at 5 and 10 cm temporarily rose to about 0° C (Figures 5a, 5b). Then the soil depleted of snow cover froze again (Figures 5a, 5b). Diurnal air temperature fluctuation from early March until late April

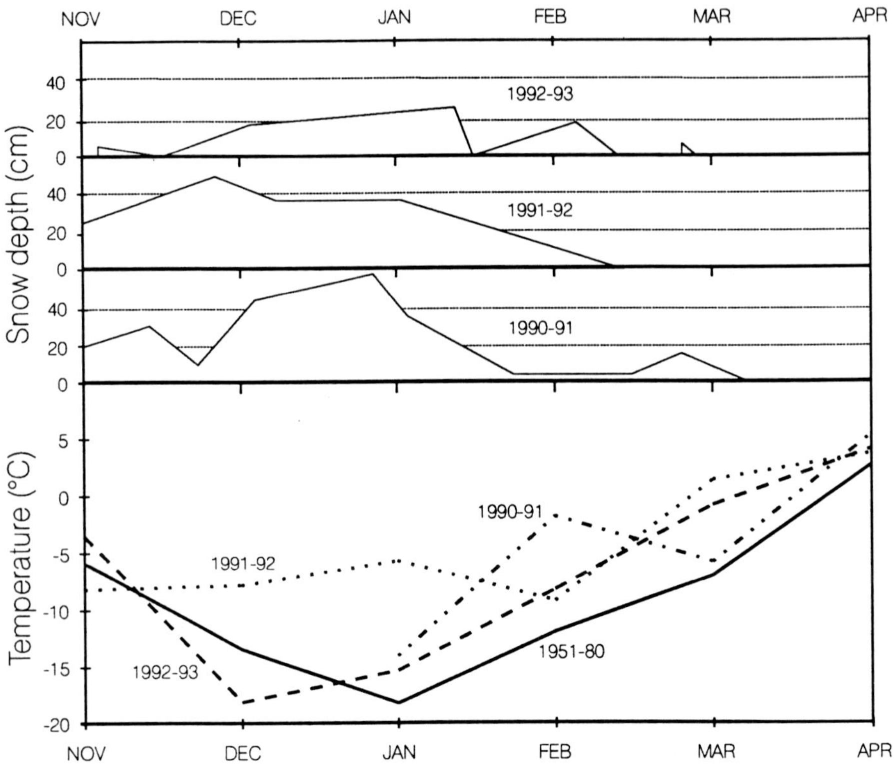

Fig. 2. Snow depth averages and monthly mean temperatures at the Bear Mountain site for the three study periods compared with 1951–1980 climate normals at the Dawson Creek airport, (Atmospheric Environment Service, 1983).

(Figures 3a–c, 4a) delayed soil thawing. The soil remained frozen at 5 cm until about April 10 (Figure 5a) and until even later at 10 and 50 cm depths (Figures 5b, 5c).

Photosynthetically active radiation (Figure 4b), and VPD (Figure 4c), increased with increasing day length and air temperature towards spring, although the January Chinook caused a peak in VPD that was not matched until early April. Springtime VPD values were in and above the range in which desiccation injury was induced to seedlings with frozen roots in growth-chamber experiments.

3.4. Growth-Chamber Experiments

Data collection in the April experiment was terminated after 1 week due to a power failure. Since results of all three experiments were similar, the February experiment details are shown as an example. Mean seedling injury after 1, 2, and 3 weeks of experiment duration are shown in Figure 6a. Analysis of variance (Table II) identifies treatment as the most significant source of variance for the injury; planting time is the next major one. Figure 6a well illustrates the results of the analysis shown in Table II.

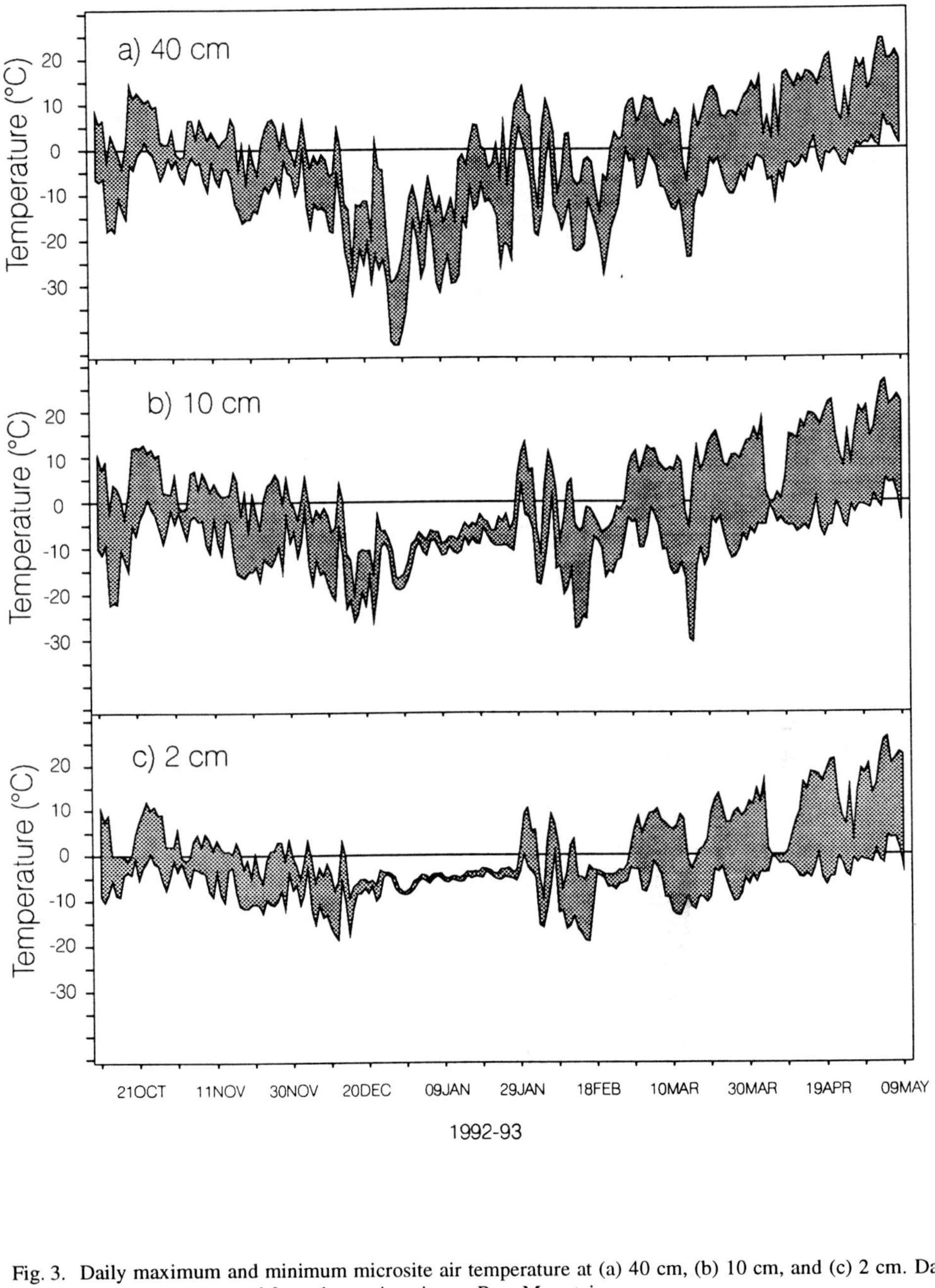

Fig. 3. Daily maximum and minimum microsite air temperature at (a) 40 cm, (b) 10 cm, and (c) 2 cm. Daily values were averaged from three microsites at Bear Mountain.

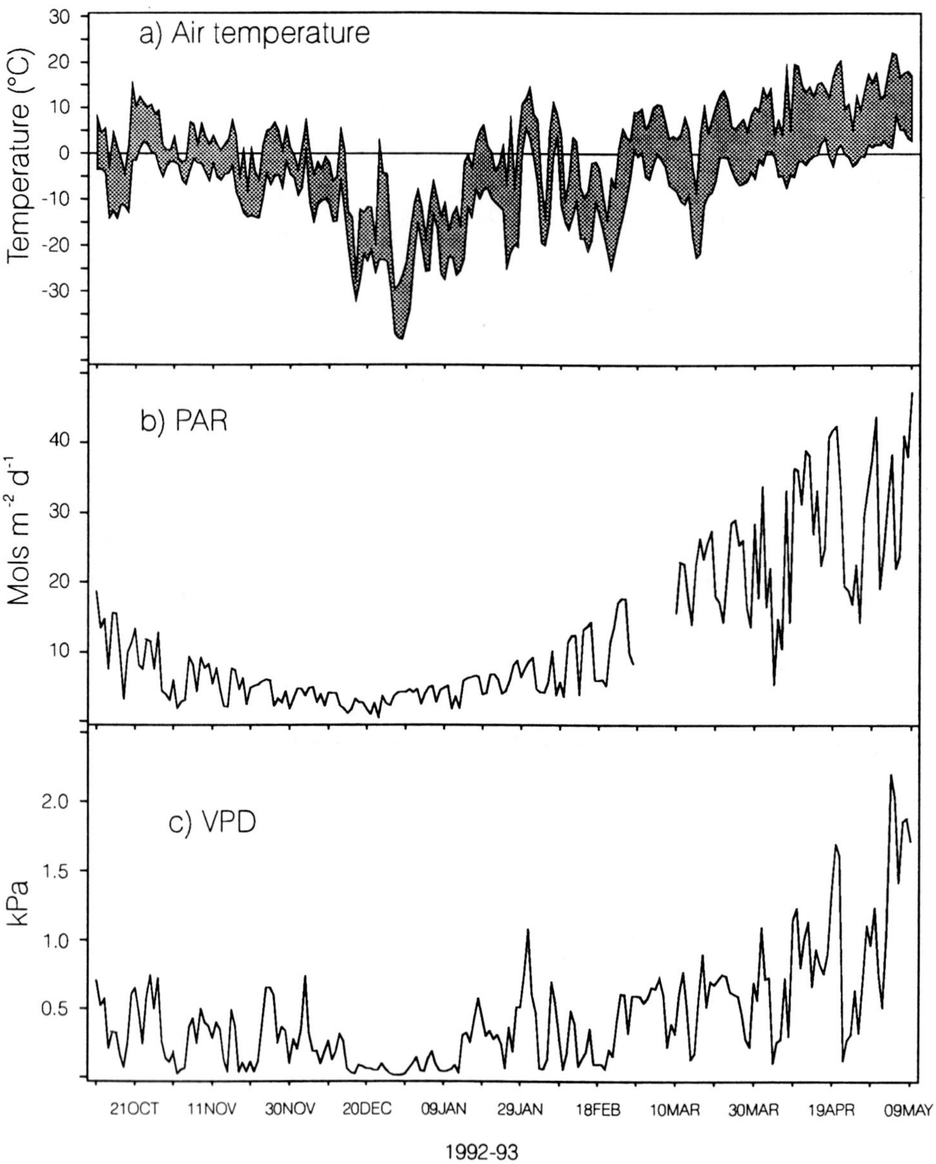

Fig. 4. Daily weather variables recorded at the Bear Mountain site at 1.5 m. (a) Maximum and minimum air temperature. (b) Total photosynthetically active radiation (PAR). (c) Average vapour pressure deficit (VPD) between 12:00 and 14:00 hours.

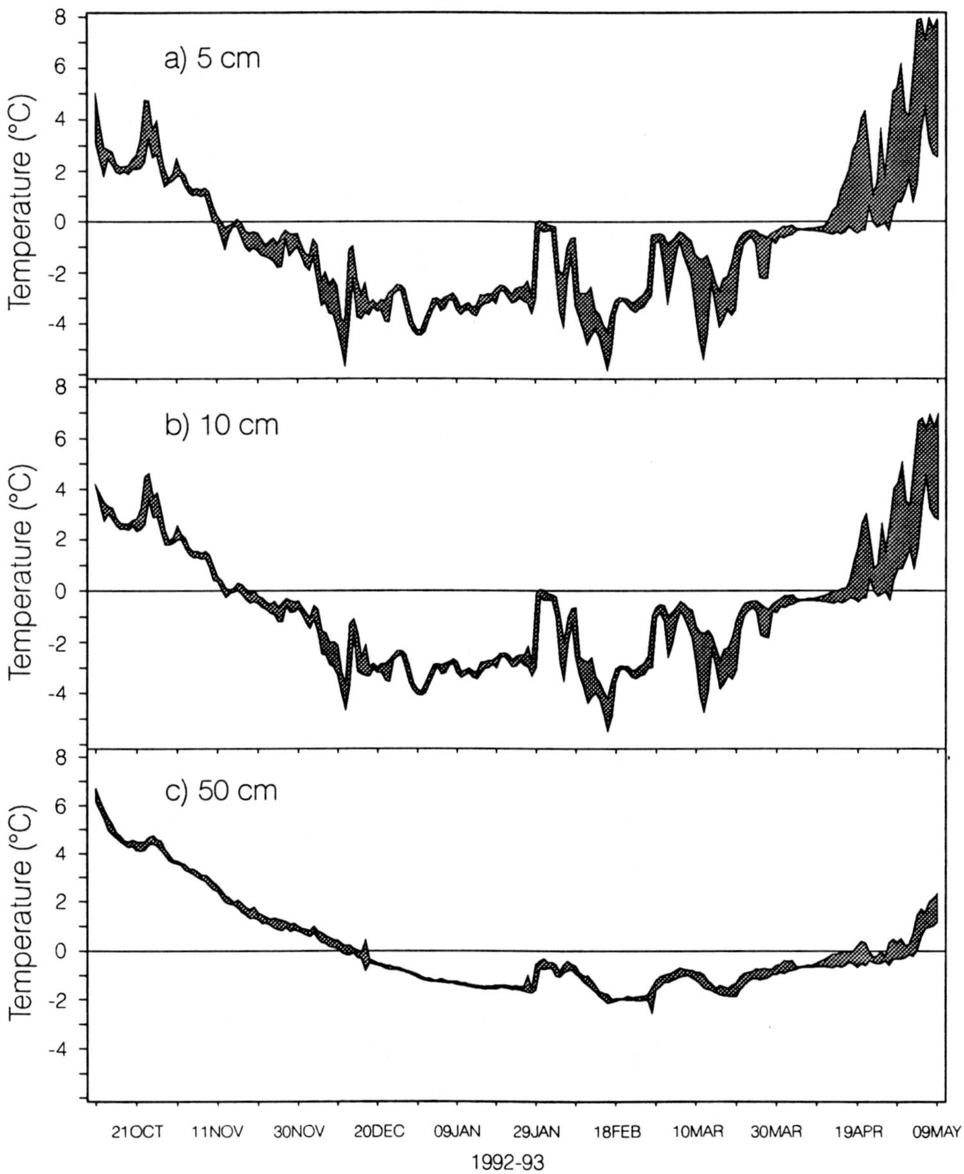

Fig. 5. Daily maximum and minimum microsite soil temperature at (a) 5 cm, (b) 10 cm and (c) 50 cm. Daily values were averaged from three microsites.

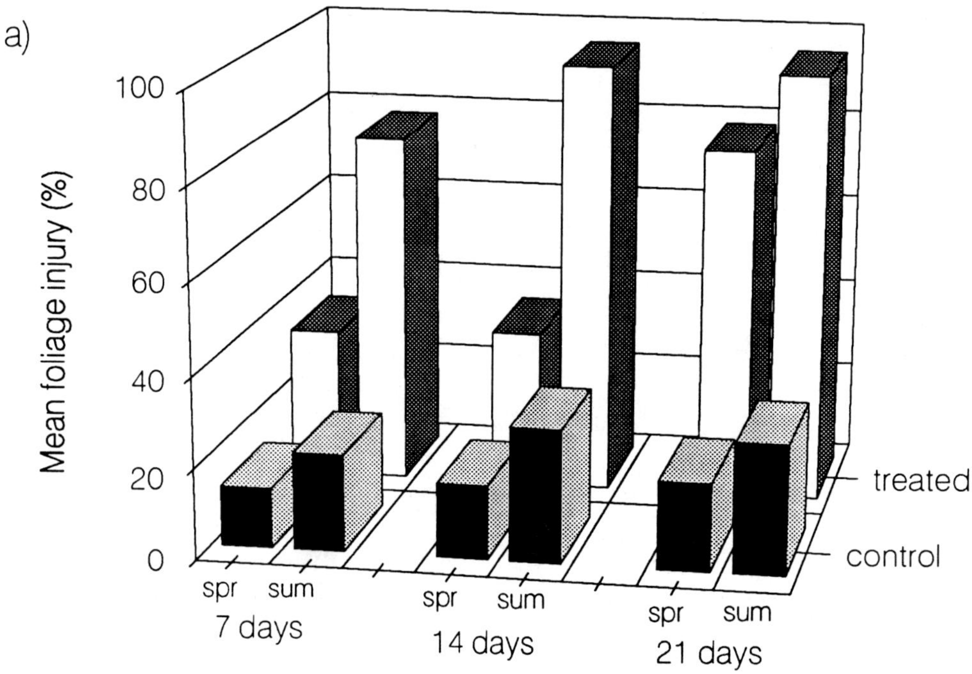

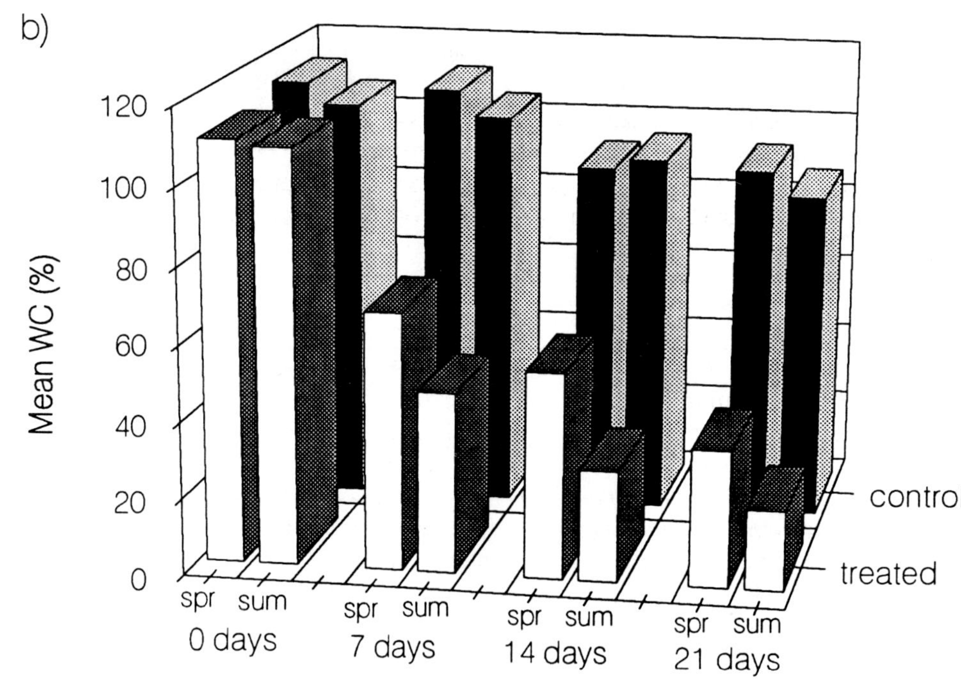

Fig. 6. (a) Mean foliage injury, and (b) mean water content of treated and control seedlings by planting time, after 7, 14, and 21 days of treatment in a growth chamber.

TABLE II

Analysis of variance on percentage foliage injury sustained in February of the growth-chamber desiccation experiments. df is degrees of freedom. SS is sums of squares (type III).

Source[a]	df	SS	F-value	P-value	Term used for error
G	1	376	1.7	0.20	b
SL	1	1 473	6.6	0.013	b
T	1	55 200	244.4	0.0016	c
P	1	15 708	69.5	0.0072	c
D	2	5 821	12.9	0.049	c
T×P	1	5 704	25.3	0.024	c
T×D	2	2 698	6.0	0.11	c
P×D	2	2 277	5.0	0.13	c
T×P×D	2	1 691	3.7	0.18	c
SL×T×P×D	11	2 723	0.8	0.63	SL×G×T×P×D
G×T×P×D	11	3 119	0.9	0.54	SL×G×T×P×D
SL×G×T×P×D	12	3 662	1.4	0.22	b
b	48	10 730			

a T = treatment; G = growth chamber; SL = seedlot; P = planting time; and D = duration.
b Sampling error.
c Pseudo F-tests output by SAS were used here. The Error term used is:
MS(SL×T×P×D) + MS(G×T×P×D) – MS(SL×G×T×P×D) = 225.88 with degrees of freedom (df) = 2.47.

Foliage water content declined much faster in the frozen root treatment than in the controls (Figure 6b). The same sources of variance as in the injury analysis were significant in water content analysis and in the same ranking order. However, in the latter analysis, T×P×D effects were significant (Table III). Figure 6b clearly explains the results of the analysis shown in Table III except the significant SL effect. This SL effect was caused by greater and faster decline of water content in SL 8782 than in SL 8779.

4. Discussion

We have shown that injury to planted white spruce seedlings in the southern boreal forest of British Columbia can be a significant problem in some years. Most of the injury is apparently sustained in late March and in April when local conditions are typical freeze-desiccation conditions (Sakai, 1970; Tranquillini, 1982; Christersson and Von Fircks, 1988). These are certainly not conditions promoting photoinjury, which requires subfreezing temperatures coupled with intense light (Öquist, 1983; Powles, 1984). The timing of the injury does not support the notion that it is caused directly by low temperature. Our unpublished data suggest that frost-hardiness of these northern seedlots is adequate until March. Although warm Chinook periods may decrease it, the decrease is not sufficient to explain the observed injuries. Finally, in the controlled-environment experiments, we showed that it took 1–3 weeks to cause heavy seedling damage or even death by imposing desiccating conditions similar (although not as

TABLE III

Analysis of variance on foliage water content (percent of dry weight) in February run of growth-chamber desiccation experiments. df is degrees of freedom. SS is sums of squares (type III).

Source[a]	df	SS	F-value	P-value	Error
G	1	39.0	0.4	0.53	Tree
SL	1	2 665.6	27.7	0.0001	Tree
T	1	65 997.2	547.0	0.0001	c
P	1	3 685.9	30.5	0.0005	c
D	3	28 217.0	228.7	0.0001	b
T×P	1	1 247.1	10.3	0.012	b
T×D	3	11 985.2	97.1	0.0001	b
P×D	3	279.9	2.3	0.17	b
T×P×D	3	561.2	4.5	0.046	b
SL×T×P×D	15	137.6	1.23	NS	
G×T×P×D	15	99.7	0.90	NS	
SL×G×T×P×D	16	111.5	1.16	0.31	
Tree(SL×G×T×P×D)	128	12 317.6			

a T = treatment; G = growth chamber; SL = seedlot; P = planting time; and D = duration.

b Pseudo F-test using the denominator of 0.9091 MS(SL×T×P×D) + 0.9091 MS (G×T×P×D) – 0.89089 MS(SL×G×T×P×D) + 0.0707MS(error) = 123.3689 with 6.90 df.

c Pseudo F-test using the denominator of 0.8182 + MS(SL×T×P×D) + 0.8182 MS(G×T×P×D) – 0.8 MS(G×SL×T×P×D) + 0.1636 MS(error) = 120.655 with 8.14 df (degrees of freedom).

harsh) to those recorded on the research sites. Sakai (1970) used similar experiments to demonstrate that desiccation was the cause of injury and mortality to white spruce planted in Japan. He concluded that desiccation was the most important factor limiting successful reforestation in Japan.

We have not found any specific recommendations for dealing with freeze-desiccation problems in reforestation. The situation we described is not a typical example of post-planting stress that can be resolved by better handling and preconditioning of planting stock (Rietveld, 1989). It appears that some types of mechanical site preparation (such as mounding) could lower the risk of desiccation; however, this type of treatment has limited applicability. The lower incidence of injury on mounds indicates that the solution to the problem is to be found below the ground.

Appropriate root growth after planting has been postulated as one of the cardinal prerequisites of successful seedling establishment (Sands, 1984; Rietveld, 1989). Revel *et al.* (1990) recommended that the summer planting program in the Prince George Forest Region (including the area of interest to this study) be completed in early August at the latest, so that there is enough time for substantial root growth before winter. In the course of this study, we observed numerous new roots in summer-planted and, in particular, in spring-planted seedlings, but most of these roots grew from the bottom of the root plug. The number of new roots that a seedling puts out after planting has been often implicated in determining its survival and performance (see Rietveld, 1989 for review), but positioning of these roots in the soil has rarely been considered. Container-grown seedlings used in British Columbia during the last decade tend to grow most new roots from the bottom of the root plug and very few from its sides (McMinn, 1978; von der Gönna, 1989). This is a very undesirable characteristic of planting stock destined for areas with freeze-desiccation problems. Krasowski *et al.* (1993) reported that naturally regenerated seedlings are not significantly affected by this injury. They tend to grow many roots in the shallow organic/mineral soil interface, which is one of the main features distinguishing naturals from planted seedlings (Eis, 1970; Halter and Chanway, 1993). Consequently, naturally regenerated seedlings may take advantage of superficial soil thawing during warm spring days and take up some water. Krasowski *et al.* (unpubl.) found that severely dehydrated seedlings can restore reasonable foliage water content after just 1 day with soil temperatures barely above zero. While seedling height and previous-year growth rate (in spring-planted seedlings) are related to the extent of desiccation injury (Krasowski *et al.*, unpubl.), we believe that the key to the successful resolution of the problem is in altering the below-ground rather than the aboveground planting stock attributes.

It is difficult to predict what impact global warming may have on the severity of desiccation injury in this geographic area. The impact would depend not only on the temperature change but also on changes in precipitation. The desiccation problem and heavy soils in the area may restrict potential influx of species from the south.

5. Conclusions

Serious injury to planted white spruce in the southern boreal forest of British Columbia apparently results from seedling desiccation caused by local climate (deep freezing of fine texture soils) and by rooting characteristics of planted seedlings. Remedy to the problem should be sought by employing shallow-rooting planting stock, and by using mechanical site preparation that promotes early soil warming.

Acknowledgments

This study was funded from the Forest Resource Development Agreement (FRDA II), project FR02 and FR05 funds granted to M.J. Krasowski. Thanks to Bonnie Hooge and Peter Ott for technical help and to Karen Dale for secretarial assistance.

References

Atmospheric Environment Service: 1983, *Canadian Climate Normals, 1951-1980, Temperature and Precipitation, British Columbia*. Environment Canada, Downsview, Canada, 268 pp.

Christersson, L. and Von Fircks, H.: 1988; *Silva Fennica* **22**, 195-201.

Eis, S.: 1970, Dep. Fish., Can. For. Serv. Publ. **#1276**, 10 pp.

Halter, M. R. and Chanway, C. P.: 1993, *Ann. For. Sci.* **50**, 71-77.

Krasowski, M. J., Herring L. J. and Letchford, T.: 1993, *Winter Freezing Injury and Frost Acclimation in Planted Coniferous Seedlings*, FRDA Report **#206,** ISSN 0835-0752; 36 pp.

Krasowski, M. J., Letchford, T., Caputa, A. and Bergerud, W.: 1995, *In: New Forests*, Spec. Issue of Proc. Intern. IUFRO Symp. on Seedling Quality "Making the Grade", Sault Ste. Marie, Ont., Canada, 11-15 September, 1994. (unpublished manuscript).

McMinn, R. G.: 1978, *In:* Van Erden, E. and Kinghorn, J. M. (eds.), Proc. Root Form Plant. Trees Symp., Victoria, British Columbia., Min. For. Rep.#8, pp. 379-412.

Öquist, G.: 1983, *Plant, Cell, Environ.* **6**, 281-300.

Powles, S. B.: 1984, *Ann. Rev. Plant Physiol.* **35,** 15-44.

Revel, J., Lavender, D. P. and Charleson, L.: 1990, *Summer Planting of White Spruce and Lodgepole Pine Seedlings*, FRDA Report #**145**, ISSN 0835-0752, 14 pp.

Rietveld, W. J.: 1989, *N.J.A.F.* **6**, 99-107.

Sakai, A.: 1970, *Ecology* **51**, 657-664.

Sands, R.: 1984, *Aust. For. Res.* **14**, 67-72.

Tranquillini, W.: 1982, *In:* Lange, O. O., Osmond, P. S., and Zeigler, H. (eds.), *Encyclopedia of Plant Physiology, New Series,* Vol. **12**B. Springer-Verlag, Berlin-Heidelberg, Federal Republic of Germany, pp. 379-400.

von der Gönna, M.: 1989, First Year Performance and Root Egress of White Spruce and Lodgepole Pine Seedlings in Mechanically Prepared and Untreated Planting Spots in North Central British Columbia, M.Sc. thesis, Faculty of Forestry. University of British Columbia, Vancouver, Canada, 130 pp.

JACK PINE (*Pinus banksiana*) SEEDLING EMERGENCE IS AFFECTED BY ORGANIC HORIZON REMOVAL, ASHES, SOIL, WATER AND SHADE

D. G. HERR and L. C. DUCHESNE[1]

Petawawa National Forestry Institute (PNFI), P.O. Box 2000, Chalk River, Ontario KOJ 1JO, Canada,
[1]*Corresponding author*

Abstract. The effects of organic horizon removal, ashes, soil water, and shade on jack pine *(Pinus banksiana* Lamb) seedling emergence were investigated. For this purpose soil monoliths were taken to the laboratory and received prescribed burning, leading to 100%, 75%, 50%, 25%, and 0% organic horizon removal. One half of each monolith contained ashes generated from burning whereas the other half was kept ash-free. Each half of every monolith was sown with jack pine seeds and the monoliths were then watered under four watering schedules (100%, 75%, 50%, and 25% of the regional average daily June rainfall) or shaded under four shading levels (100%, 75%, 50%, and 25% photosynthetically active radiation). Seedling emergence was most successful under high watering schedules, increased depth of burn, high shading, and without ashes. Ash had an inhibitory effect on seedling emergence.

Keywords. *Pinus banksiana*, FIRE, SEEDS, REGENERATION, SOIL WATER POTENTIAL, LIGHT.

1. Introduction

Jack pine (*Pinus banksiana* Lamb.) is the most widely distributed pine species in Canada (Rudolph and Laidly, 1990) and has significant economic importance to the forest industry (Foster and Morrison, 1976; Anonymous, 1980; Rudolph and Laidly, 1990). The silvicultural implications of the role of fire in jack pine ecosystems are well known (Eyre and LeBarron, 1944; Cayford *et al.*, 1967; McRae, 1979) and have been investigated in terms of fire behaviour (Weber *et al.*, 1987), seedbed preparation (Fowells, 1965; Chrosciewicz, 1974, 1990), and regeneration (LeBarron, 1944; Thomas and Wein, 1985; Weber *et al.*, 1987; Chrosciewicz, 1988).

Little of the literature describing the effects of fire on jack pine regeneration provides quantitative detail concerning the relationships among organic horizon consumption, depth of burn, and post-fire regeneration. Chrosciewicz (1974) demonstrated that jack pine seedling abundance and height correlate negatively to the post-fire organic horizon depth. Thomas and Wein (1985) have proposed a model relating the effect of shade and fire severity (depth of burn) on post-fire conifer seedling establishment. In their model, jack pine seedling establishment is most successful under high fire severity and low shade levels. Other studies concerning the effect of fire on jack pine regeneration involve far less quantitative detail. Existing data (LeBarron, 1944; Fraser and Farrar, 1953; Chrosciewicz, 1974; Thomas and Wein, 1985) need to be complemented by studies that address the relationships among factors such as jack pine seedling emergence, ash content, organic horizon removal, shade, and soil water. The objectives of our study were to investigate the effects of organic horizon thickness, ash, shading, and soil water, on jack pine seedling emergence under greenhouse conditions. Previous attempts to investigate these factors in the field failed because of difficulties caused by variable weather conditions, depth of burn, and seed predation (unpublished data).

Water, Air and Soil Pollution **82**: 147–154, 1995.

2. Materials and Methods

Soil monoliths were collected from the forest floor of the Frontier Lake experimental site (46°00'N, 77°33'W) at the Petawawa National Forestry Institute (PNFI) within the Middle Ottawa Section (L.4c) of the Great Lakes–St. Lawrence forest region (Rowe, 1972). Site characteristics and overstory composition are described in Herr *et al.* (1994).

Portions of the forest floor, measuring 18 cm × 30 cm, were outlined with 30 cm serrated knives to produce soil monoliths. A wooden board measuring 18 cm × 30 cm and 2 cm deep was placed on the litter surface while the monolith was outlined to avoid disturbing the surface litter. Each monolith was measured to a minimum depth of 25 cm, including the organic horizon, the Ae horizon, and a variable portion of the B horizon. We use the term *organic horizon* to describe the L (litter), F (fermentation), and H (humification) horizons (Spurr, 1980). Monoliths were allowed to air dry at room temperature for several weeks to ensure uniform soil moisture levels. A total of 215 monoliths were collected for this investigation: 200 were used to evaluate the effect of shading and soil water (100 monoliths each) on seedling emergence. The remaining 15 were used as unseeded controls to determine the presence and level of the jack pine soil seed bank.

The organic horizon thickness of each monolith was measured (mean thickness 5.0 ±1.44 cm std. dev.) and 100%, 75%, 50%, or 25% of the organic horizon was removed with a 30 cm serrated knife. Controls (0% removal treatment) consisted of monoliths in which the organic horizon remained undisturbed. Soil monoliths were placed in Rootrainer boxes (Spencer–Lemaire Industries Ltd., Edmonton, Alberta) lined with nursery shade cloth. The portions of organic horizon removed were combined with their respective treatments (i.e., all 25% organic horizon removal portions were combined) and burned *ex situ* using a propane torch in a fume hood until no further combustion was attainable. The total mass of ashes for each set of removal treatments (i.e., 25%, 50%, 75%, and 100%) was divided by the number of monoliths contributing to that particular treatment, and then half of that mass was spread as an uniform layer onto half of the area of each monolith. This created two treatments for each monolith: an organic horizon removal with ash present, and an organic horizon removal with ash absent. The two treatments were kept separate with plastic dividers to prevent the ashes from spilling over.

Each half (ash present and ash absent) of each of the 200 monoliths used in the shade and watering schedule experiments were seeded with 100 unstratified, viable jack pine seeds (PNFI seedlot number 75 30 200, seed year 1975, viability 99.9%, retested in 1991). The monoliths were then placed in a greenhouse at ambient air temperature at monolith surface (no shading) of 23.8 ±4.3 °C standard deviation, with maximum and minimum temperatures of 34.0 °C and 16.3 °C, respectively (Figure 1). Fifteen untreated monoliths (0% organic horizon removal) were placed under a 25% PAR level, and watered with 145 mL deionized water per day for 4 weeks to determine the presence and level of the jack pine soil seedbank.

Watering of the monoliths was conducted daily using the June precipitation records over a 30-year period from the Bransted weather station at PNFI (G. Péch, pers. comm.), which is located approximately 15 km east of the monolith collection site. June is the month of the growing season with the highest rainfall at PNFI (Weber *et al.*, 1987). Therefore, this value was used as the best case scenario for seed emergence. The average daily June precipitation of 2.7 mm per day was adjusted for the entire (18 cm × 30 cm) surface of the monolith, yielding a watering schedule of 145 mL

Fig. 1. Soils monoliths on a greenhouse bench.

deionized water per monolith per day. Moreover, one hundred monoliths were used (20 of the 0% removal treatment, and 20 of each of the 25%, 50%, 75%, and 100% removal treatments) to determine the effect of watering schedule on jack pine seedling emergence. Five monoliths from each removal treatment were randomly assigned to one of the four different watering schedules. The four watering schedules used in this portion of the study included 25%, 50%, 75% and 100% of the average daily June rainfall from the Bransted weather station. The monoliths were then placed on greenhouse benches (Figure 1) in a completely randomized design, and watered daily according to their prescribed watering schedule.

One hundred monoliths were used (20 of each of the 0%, 25%, 50%, 75%, and 100% removal treatments) to determine the effect of shading on jack pine seedling emergence by establishing four separate shading levels (25 monoliths per shade level) performed simultaneously, each in a randomised complete block design. Repetition of the shade factor was not possible due to greenhouse space limitations; therefore statistical comparisons could not be made among the four different shading levels. The shading experiments were conducted during July, 1993. The four shade levels were: 25%, 50%, 75%, and 100% of PAR in greenhouse conditions. Shading was accomplished by suspending wooden lattices 60 cm above the monoliths. Measurement of PAR at solar noon above and below the lattice was accomplished with a Li-Cor LI-190SB Quantum sensor (Li-Cor, Lincoln, NE, USA); the number of slats in the lattice was then adjusted to obtain the desired PAR level. Monoliths were watered daily using the June average daily rainfall as before. Seedlings were considered to have emerged when the seed coat was lifted above the substrate by the hypocotyl. Percent emergence was thus measured by counting the number of seed coats lifted off the surface of the monolith. Total percent jack pine seedling emergence was determined at the end of June, 1993, four weeks after the beginning of the experiment.

For statistical analysis of percent emergence, the data were arcsine-transformed, tested for homoscedasticity, and analyzed using the PROC GLM procedure of SAS (SAS Institute Inc., 1985). To test for significant differences among treatments, the Tukey's HSD multiple comparison test was used ($\alpha = 0.01$).

3. Results

No jack pine seedlings emerged from the 15 untreated monoliths. Therefore, no corrections to the final emergence data were necessary to account for the presence of a natural soil seedbank.

Jack pine seedling emergence was not observed under either the 25% or the 50% watering schedules (data not shown). The effects of watering schedule, organic horizon removal, ash, and the watering schedule × organic horizon removal interaction on jack pine seedling emergence were significant (Table I). The interaction effects of organic horizon removal × ash, watering schedule × ash, and watering schedule × organic horizon removal × ash were not significant. The effects of watering schedule, organic horizon removal, and watering schedule × organic horizon removal accounted for the majority of the total variance (Table I). Under the 75% watering schedule, the effects of organic horizon removal, ash, and organic horizon removal × ash on jack pine seedling emergence were not significant (Table I). Under the 100% watering schedule, the effects of organic horizon removal and ash on jack pine seedling emergence were significant, but the organic horizon removal × ash effect was not significant (Table I). The effect of removal accounted for most of the total variance (Table I).

Under the 75% watering schedule, jack pine emergence was variable and no differences were found at any organic horizon removal level with or without ash (Figure 1). Emergence was variable for the 75% watering schedule (Figure 2). Under the 100% watering schedule,

TABLE I

Effects of watering schedule (Water), organic horizon removal (Removal), and ash (Ash) on jack pine seedling emergence for the watering schedule experiment.

Level of ANOVA	Source of variation	% of variance accounted for	F	Significance[a]
Entire watering schedule experiment	Water	18.0	32.01	***
	Removal	12.4	5.54	**
	Ash	3.6	6.55	**
	Water × Removal	15.3	6.83	***
75% watering schedule level	Removal	10.1	1.21	NS
	Ash	3.2	1.53	NS
	Removal × Ash	2.7	0.33	NS
100% watering schedule level	Removal	38	7.83	***
	Ash	6.1	5.02	*
	Removal × Ash	6.3	1.29	NS

a NS - not significant; *** significant at $P < 0.001$; ** significant at $P < 0.01$; * significant at $P < 0.05$.

emergence was higher on the without ash treatments (Figure 2). Generally, germination increased with watering with ash (Pearson R: 0.437, $P < 0.001$) or without ash (Pearson R: 0.382, $P < 0.001$).

The effects of organic horizon removal, ash, and organic horizon removal × ash on jack pine seedling emergence under the 25% PAR level were significant (Table II). The effects of organic horizon removal and ash accounted for the majority of the total variance at the 25% PAR level (Table II). Jack pine seedling emergence was greatest under the 50%, 75%, and 100% organic horizon removal in the absence of ash under the 25% PAR level (Figure 2). The presence of ash reduced jack pine seedling emergence for all organic horizon removal levels compared to jack pine seedling emergence in the absence of ash (Figure 2). Seedling emergence increased as percent organic horizon removal increased (Figure 2).

The effects of organic horizon removal, ash, and organic horizon removal × ash on jack pine seedling emergence under the 50% PAR level were significant (Table II). The effects of organic horizon removal and ash accounted for the majority of the total variance (Table II). Jack pine seedling emergence was greater in the absence of ash under the 50% PAR level (Figure 2). Jack pine seedling emergence generally increased as increasing portions of the organic horizon were removed for the 50% PAR level (Figure 2).

The effects of organic horizon removal, ash, and organic horizon removal × ash on jack pine seedling emergence under the 75% PAR level were significant (Table II). The effect of ash accounted for the majority of the total variance (Table II). Seedling emergence was lower in the presence of ash than in the absence of ash under the 75% PAR level (Figure 2). Seedling emergence increased as percent organic horizon removal increased under the 75% PAR level (Figure 2).

The effect of ash on seedling emergence under the 100% PAR level was significant, while the effects of organic horizon removal and ash × organic horizon removal on seedling emergence were not significant (Table II). Ash accounted for the majority of the total variance under the 100% PAR level (Table II). No seedlings emerged in the presence of ash under the 100% PAR level at 25%, 50%, and 75% organic horizon removal. Seedling emergence increased as increasing portions of the organic horizon were removed; however, no significant differences in seedling emergence under the 100% PAR level were found for the various organic horizon removal treatments (Figure 2).

4. Discussion

The absence of jack pine seedlings in the soil seedbank of the unseeded control monoliths was expected. The majority of cones on older (> 10 cm diameter at breast height) jack pine trees are serotinous, and do not open unless exposed to the heat of a forest fire (Gauthier *et al.*, 1993). Because the forest stand at the Frontier Lake monolith collection site had not been exposed to fire for several decades, it is unlikely that any significant jack pine soil seedbank exists. This conclusion is strengthened by the observation that jack pine seedling recruitment is virtually non-existent at the Frontier Lake experimental site (data not shown).

In this investigation, seedling emergence improved with the removal of increasing portions of the organic horizon, particularly when shade treatments are considered. The relationship is less evident under full light or at reduced watering schedules.

Although exposed mineral soil is generally considered the most suitable seedbed for regenerating many conifer species, total organic horizon reduction by fire or scarification may not yield maximum recruitment and/or growth. The forest floor acts as a nutrient

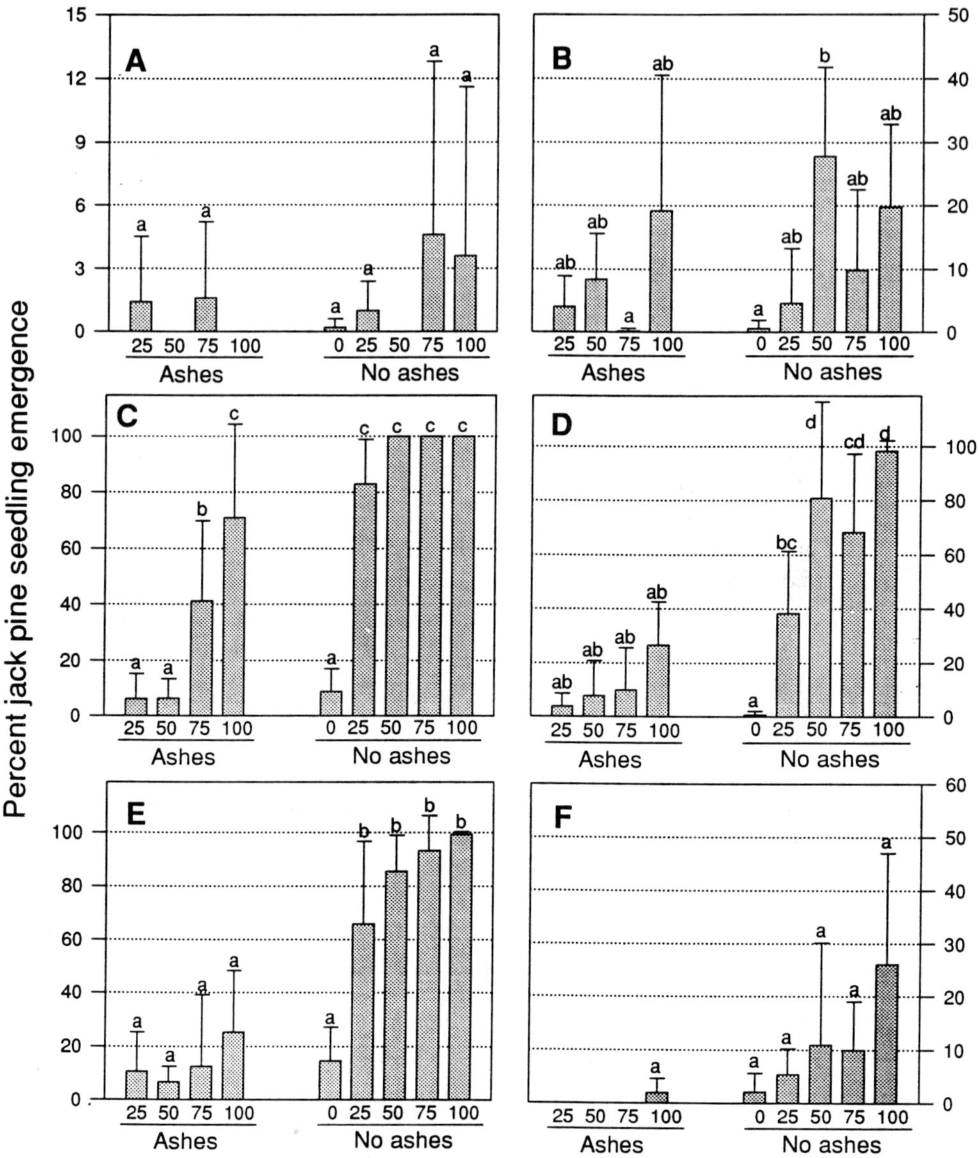

Fig. 2. Mean jack pine emergence ±1 SD (n = 5) affected by organic horizon removal and ash under the (a) 75% watering schedule, (b) 100% watering schedule, (c) 25% PAR treatment, (d) 50 PAR treatment, (e) 75% PAR treatment, and (f) 100% PAR treatment. Means followed by a different letter are significantly different (Tukey's HSD $\alpha = 0.01$).

TABLE II

Effects of organic horizon removal (Removal), Ash, and Removal × Ash on jack pine seedling emergence for the shading.

Shade experiment	Source of variation	% of total variance accounted for	F	Significance[a]
	Removal	48.8	35.80	***
25% PAR	Ash	53.2	156.14	***
	Removal × Ash	6.4	6.30	**
	Removal	34.0	20.32	***
50% PAR	Ash	35.9	85.63	***
	Removal × Ash	13.3	7.97	***
	Removal	14.1	7.50	***
75% PAR	Ash	52.0	111.48	***
	Removal × Ash	14.7	7.80	***
	Removal	10.9	2.20	NS
100% PAR	Ash	33.9	27.31	***
	Removal × ash	5.5	1.11	NS

a NS - not significant; *** significant at $P < 0.001$; ** significant at $P < 0.01$; * significant at $P < 0.05$

storage site and it has been suggested that total removal of the organic horizon may be an inappropriate management strategy (Chrosciewicz, 1974). Weber *et al.* (1985) suggested that removal of the organic horizon can have negative impacts on future site productivity. Therefore, it is critical to understand how depth of burn affects jack pine emergence and subsequent natural regeneration.

Our results show that ash reduces the emergence of jack pine seedlings. Two explanations can be advanced for this. First, the increase in soil pH caused by ash may be responsible. Although considerable research has addressed the effect of fire on soil pH in many fire driven ecosystems, little is known about the effect of pH on jack pine seedling emergence (Wright and Bailey, 1982). Second, ashes dry faster, absorb less water, and absorb more heat than unburned organic material (Ahlgren and Ahlgren, 1960; Chandler *et al.*, 1983). These factors expose seeds on the ash treatment to lower water potentials with possible detrimental effects on seedling emergence.

An analysis of soil water potential in our experiment shows that a factor or factors other than soil water potential affect jack pine seedling emergence (data not shown). An hypothesis for reduced emergence on the control monoliths is that chemical leachate(s) from the L, F, and/or H horizons have an allelopathic-like effect on seedling emergence. Alternatively, it is possible that the litter layer acts as a barrier to seed movement to more favourable germination and seedling emergence conditions on the F horizon. Further investigations are needed to determine the effect of the litter layer on seedling emergence of jack pine.

Based on the hypothesis that climate change may lead to reduced summer precipitation in parts of eastern and central Canada (Flannigan and Van Wagner, 1991), the results presented here have implications for future jack pine distribution and abundance in relation to potential climate change. We have shown that a slight reduction in the watering schedule severely limits jack pine seedling emergence. The relationship between jack pine seedling emergence on soil monoliths under different watering schedules in greenhouse conditions and jack pine seedling emergence under field conditions needs to be investigated. However, it is important to note that our observations are derived from only one soil type in a greenhouse situation and that other soil types need to be studied. Nevertheless, the models presented in this investigation are the first to present a comprehensive picture of the relationships among ash, soil water, shading, and depth of burn on jack pine germination. Future work should focus on the long term effects of these factors on seedling survival and growth. Such information could be used by forest managers interested in optimising stand regeneration after disturbance.

Acknowledgments

This work was supported by research grants from Parks Canada and Canada's Green Plan climate change initiative. Thanks are addressed to L. Clark, G. Jones, J. Ahearn, and E. Audet for laboratory and field assistance. Constructive criticisms by staff members of the Fire Ecology and Behaviour project, and editorial review by A. Yapa at the Petawawa National Forestry Institute are appreciated.

References

Ahlgren, I. F. and Ahlgren, C. E.: 1960, *Bot. Rev.* **26**: 483-533.
Anonymous: 1980, Statistics. Ont. Min. Nat. Res., Toronto, Canada, 29 pp.
Cayford, J. H., Chrosciewicz, Z. and Sims, H. P.: 1967, A review of silvicultural research in jack pine. Canada. Dept. For. Rural Develop., Forestry Branch, Departmental Publication No. **1173,** 255 pp.
Chandler, C., Cheney, P., Thomas, P., Trabaud, L. and Williams, D.: 1983, *Fire in Forestry, Volume 1, Forest Fire Behavior and Effects*. Wiley, New York, USA, 450 pp.
Chrosciewicz, Z.: 1974, *Can. J. For. Res.* **4**, 455-457.
Chrosciewicz, Z.: 1988, *Forest. Chron.* **64**, 315-319.
Chrosciewicz, Z.: 1990, *Forest. Chron.* **66**, 579-584.
Eyre, F. H. and LeBarron, R. K.: 1944, Management of jack pine stands in the Lake States. USDA For. Serv. Technical Bulletin No. **863**. U.S. Forest Service, Lake States Forest Experiment Station, pp. 2550-2554.
Flannigan, M. D. and Van Wagner, C. E.: 1991, *Can. J. For. Res.* **21**, 66-72.
Foster, N. W. and Morrison, I. K.: 1976, *Ecology* **57**, 110-120.
Fowells, H. A.: 1965, *Silvics of Forest Trees of the United States.* USDA Agric. Handbook No. **271**.
Fraser, J. W. and Farrar, J. L.: 1953, Effect of watering, shading, seed-bed medium, and depth of sowing on jack pine germination. Silvic. Leaflet No. **90**., Div. For. Res., For. Branch, Dept. Resources. Develop., Ottawa, Canada, 4 pp.
Gadd, N. R.: 1962, Surficial geology - Chalk River, descriptive notes. Marginal notes to map 1132A, Canadian Geological Survey Committee, Ottawa, Canada, 8 pp.
Gauthier, S., Bergeron, Y. and Simon, J. P.: 1993, *Can. J. For. Res.* **23**, 394-401.
Herr, D., Duchesne, L. C., Tellier, R. and McAlpine, R. S.: 1994, *Intl. J. Wildland Fire* **4**, 95-102.
LeBarron, R. K.: 1944, *J. Agric. Res.* **68**, 97-117.
McRae, D. J.: 1979, Prescribed burning on jack pine logging slash: a review. Dept. Environ., Can. Forest Serv., Great Lakes For. Res. Centre Inf. Rep. **O-X-289**, Sault St. Marie, Canada, 57 pp.
Rowe, J. S.: 1972, Forest regions of Canada. Can. For. Serv. Publ. No. **1300**.
Rudolph, T. D. and Laidly, P. R.: 1990, *Pinus banksiana* Lamb. Jack pine. *In*: Burns R. M. and Honkala B. H. (eds.) *Silvics of North America. Volume 1, Conifers.* USDA For. Serv. Agric. Handbook 654, Washington D.C.
SAS Institute Inc.: 1985, *SAS User's guide: Statistics, Version 5.* SAS Institute Inc. Cary, USA, 1028 pp.
Spurr, S. H.: 1980, *Forest Ecology*. Wiley, Toronto, Canada, 687 pp.
Thomas, P. A. and Wein, R. W.: 1985, *Can. J. Forest Res.* **15**, 148-155.
Van Wagner, C. E.: 1966, Three experimental fires in jack pine slash. Dept. For., Govt. Can. Publ. No. **1146**.
Weber, M. G.: 1988, Fire and Ecosystem Dynamics in Eastern Canadian *Pinus banksiana* Forests. *In*: Verboeven J. T. A., Heil G. W., and Werger M. J. A. (eds.) *Vegetation Structure in Relation to Carbon and Nutrient Economy.* S.P.B. Academic Publishing, The Hague, The Netherlands, pp. 93-105.
Weber, M. G., Hummel, M. and Van Wagner, C. E.: 1987, *Forest Chron.* **63**, 340-346.
Weber, M. G., Methven, I. R. and Van Wagner, C. E.: 1985, *Can. J. Forest Res.* **15**, 313-318.
Wright, H. A. and Bailey, A. W.: 1982, *Fire Ecology. J.*, Wiley, New York, USA, 501 pp.

MODELLING FOREST REGENERATION PROCESSES IN CLEAR-CUT AND BURNED AREAS IN THE ANGARA REGION

V. SOKOLOV, S. FARBER and I. DANILIN

Institute of Forest, Russian Academy of Sciences, Siberian Branch, Akademgorodok, Krasnoyarsk 660036, Russia.

Abstract. A method to select measures for enhancing natural forest-regeneration process in areas under extensive management is discussed. Using available information on geomorphological site characteristics and stand parameters, indices are determined using a specific criterion that describes the forest regeneration process. Using these indices as model inputs, clearcuts and burns can be classified; within each stratum the natural forest regeneration process is similar. Analysing these strata, one can select a regime of forest treatment that is appropriate for a given stratum and, consequently, for the clearcuts (or burns) it includes.

Keywords. CLEAR-CUTS, BURNS, FOREST REGENERATION.

1. Introduction

In practical forest management, various forest regeneration regimes are classified according to groups of sites having similar environmental conditions. Soil-typological classifications can be used for intensively managed areas, schemes of forest types, however, are more appropriate for heavily forested taiga regions.

Analysis of forest regeneration in clearcuts and burns, conducted for Central Angara Region in the forest resource management regions of the Ustilimsk Forest Industrial Complex (UIFIC), indicated that significant differences exist only between forest types contrasting in their drainage features (Farber and Sokolov, 1991). This supposes that the use of forest-type classification to determine forest-regeneration regimes is correct only for sites of sphagnum- and lichen-forest types.

2. Experimental Material

In determining forest regeneration regimes, we must first identify the groups of clearcuts and burns that are characterised by similar, natural, forest-regeneration processes. It is common knowledge that classification can be based on any parameters or descriptors which are then used as classification system inputs. These may be chosen using best-available information sources, namely: space and aerial images, forest plantation maps, topographic maps, and site inventories. Most important are the following parameters: slope angle and aspect, stand phytomass production-class, and species composition of over- and under-story. The parameter that has greatest influence on forest regeneration process is given higher priority in the classification. We suggest that the selection of parameters should be improved by taking into account the quantitative criterion (K):

Water, Air and Soil Pollution **82**: 155–160, 1995.

$$K = \sum_{i=1}^{n} \frac{a_{1i} - a_{2i}}{a_{1j} + a_{2j}} \qquad [1]$$

where a_{1i} and a_{2i} are numerical values for one of the forest-regeneration descriptors (n in number) for extreme treatments of the input parameter of interest. For example, to evaluate *slope aspect* as a classification system input parameter we might consider natural regeneration success (or any other descriptor characterising forest regeneration process) on south facing (a_{1i}) and north-facing (a_{2i}) slopes.

Using the data in Table I, quantitative criteria (K) are calculated for each parameter whose usefulness as a classification system input is under investigation (Table II). The ranges of significance for the analysed parameters depends on the forest regeneration processes specific to the different clearcuts and burns. The reasons for the range in significance are differences in soil drainage, compaction, peat content, and degree of decomposition.

On the basis of the calculated K values, the following parameters were chosen as the classification system inputs: forest phytomass production-class and presence of pine in a pre-fire stand understory (for burned pine stands); availability of pine in the tree-stand and; regrowth of a pre-cut stand for clearcuts. Note that these parameters are not directly related to each other. Forest regeneration processes are very dependent on specific site conditions (bedrock type, soil differences, drainage peculiarities, topography, precipitation regime etc.) and on more general features, such as geographic location and climate. The selected inputs should be considered as indicators of particular site conditions.

Basically, the system content of clearcuts (burns) considers all natural forest regeneration. However, for the objective of this study—investigation of distribution of forest regeneration regimes according to the classification model strata—it is sufficient to examine only the general trends in variations of the main statistics (Table III).

3. Results and Discussion

Analysis of the model characteristics allows a correct correlation of each strata with the forest-regeneration regime.

3.1. Burns

The total number of pine, birch and aspen in the understory is 6025; 3778; 4492 and 4771 for I, II, III and IV strata, respectively. An increase in the amount of regenerating stems (self-sown crop) suggests more favourable site conditions. Thus, it is clear that conditions in I and IV strata are relatively better than in II and III strata.

A considerable fraction (53%) of non-regenerated burns is characteristic of the I stratum. This is due to relatively rich soils which provide conditions for enhanced herb development that, in turn, leads to intensive sod development that hampers the regeneration process. On the other hand, in the I stratum the number of burned sites invaded by pine after a fire is greater than the number invaded by hardwood species. This suggests that in these sites, soil conditions are generally favourable for the development of a pine understory. Once the degree of understory development in the I stratum is sufficient, these data indicate that such burn sites should be left untreated and to allow natural regeneration to proceed. In this case, however, about a half of the burn sites would fail to regenerate satisfactorily. Thus, for burns on sites previously occupied by pine stands of III and higher productivity class with pine understory, management intervention to enhance the natural regeneration process is recommended.

TABLE I

Average values for forest regeneration process descriptors according to analysed input parameters.

	Slope				Presence of Pine					
	Aspect		Angle		In Stand		In previous Understory		Productivity Class	
Natural forest regeneration descriptor	N	S	0–2	> 2	7–10	< 7	Yes	No	III	IV
Pine burns										
Average number of self-sown pine and larches per ha	909	1662	2400	1237	1600	1239	2700	824	927	2309
Pine and larch regeneration probability	0.14	0.41	0.50	0.28	0.52	0.47	0.48	0.24	0.16	0.58
Species composition[a] of naturally regenerating forest (average), %	70 B 20 A 10 P	20 P 10 L 50 B 20 A	30 P 10 L 50 B 10 A	20 P 10 L 50 B 20 A	20 P 10 L 10 S 50 B 10 A	20 P 10 L 50 B 20 A	30 P 20 L 40 B 10 A	60 B 20 A 10 P 10 L	60 B 20 A 10 P 10 L	30 P 10 L 40 B 20 A
Number of observations	37	70	18	89	59	48	35	72	68	38
Pine clearcuts										
Average number of self-sown pines and larches per ha	1035	1141	1420	979	1436	605	1301	944	964	1976
Pine and larch regeneration probability	0.25	0.35	0.50	0.25	0.44	0.15	0.54	0.18	0.28	0.56
Species composition[a] of naturally regenerating forest (average), %	30 P 10 L 10 S 30 B 20 A	30 P 10 S 30 B 30 A	40 P 10 L 30 A 20 B	20 P 10 S 40 B 30 A	30 P 10 L 10 S 30 B 20 A	20 P 10 S 40 B 30 A	40 P 10 L 30 A 20 B	20 P 10 S 40 B 30 A	20 P 10 L 10 S 30 B 30 A	50 P 10 L 30 B 10 A
Number of observations	63	130	56	137	114	79	76	117	167	25

a P-pine, L-larch, F-fir, S-spruce, B-birch, A-aspen.

In stratum II the percentage of burns that have recovered through pine regeneration is considerable. This species amounts, in this case, to 2000 self-sown individuals per hectare. This is a good reason to recommend a natural regeneration regime for burns of IV and V productivity classes, that were previously dominated by pine stands, had a pine understory beneath the canopy, and where site conditions were relatively arid.

In stratum III, the number of non-regenerated burns roughly equals the number of burns that have become overgrown with hardwood species. Burns overgrown with pine account for only 12% of the total area burned. Moreover, as the number of self-sown pines here does not exceed 400 individuals per hectare, artificial regeneration will be the only appropriate solution for burns of the III productivity class that, previous to fire, were covered by stands without pine understory below the canopy.

TABLE II

Selection of input parameters to form the forest regeneration model strata.

	Slope		Presence of pine		Productivity Class
K-input	Aspect	Angle	Previous stand	Previous understory	
Pine burns					
K	29.3	32.0	12.7	53.2	42.7
K	49.1	28.2	5.0	33.3	56.8
K	50.0	14.3	0.0	42.9	33.3
Total	128.4	74.5	17.7	129.4	132.8
Pine clearcuts					
K	4.9	18.4	40.7	15.9	34.4
K	16.7	33.3	49.1	50.0	33.3
K	14.3	42.9	33.3	42.9	33.3
Total	35.9	94.6	123.1	108.8	101.0

In stratum IV, the fraction of burns overgrown with pine and the fraction of burns overgrown with hardwood species are approximately equal (48% and 38%, respectively). Non-regenerated burns make up only 14% of the area under consideration. In comparison with stratum III, the number of self-sown trees increases–in general–up to 1200 individuals per hectare for pine, and up to 3600 for hardwood species. These figures suppose that the burns included in stratum IV not to be recovered through natural regeneration; they, however, indicate that the site condition potential could be realised more effectively—given an appropriate method of enhancing regeneration artificially. Thus, for burns of IV and V productivity classes previously dominated by pine stands with pine understory beneath the canopy, management intervention to enhance regeneration is recommended.

3.2. Clearcuts

Analysis of forest regeneration process (Table III) provides the following insights. Stratum I incorporates relatively dry sites, characterised by poor forest productivity, and mostly covered by pure pine stands. Stratum II is distinguished for the sites that provide favourable conditions for regeneration of highly productive mixed pine stands. Stratum III includes wet sites with poor soils that support relatively pure pine stands. Stratum IV incorporates clearcuts located primarily on loamy soils that are richer than those in stratum II. This is due to the fact that the variations in the model strata location characteristics, depend on bedrock type and topography. It is evident that within similar Quaternary deposits, bedrock influences both soil drainage and richness in the same way. Under these conditions, *relief* may be considered one of the most important parameters governing forest regeneration processes; more specifically *slope angle*—and, at its extreme values—*slope aspect*. In reality, however, quite a range of soils can be observed on similar sites, due to the mosaic pattern of Quaternary deposits in the area under study. These peculiarities are responsible for the relatively minor role of relief and slope parameters.

TABLE III

Forest regeneration characteristics by model strata (I, II, III, IV) for pine burns and pine clearcuts.

	I		II		III		IV	
Regeneration	X[a]	V[b](%)	X	V(%)	X	V(%)	X	V(%)
Pine burns								
Pine	1975	86	1521	67	386	69	1195	75
Larch	575	54	1236	88	146	69	78	75
Cedar							56	
Fir	200		100		100			
Spruce	75		571		64		22	
Birch	3700	52	1764	44	3053	53	2282	84
Aspen	350	33	493	68	1071	56	995	84
Species composition,[c] %	30P10L50B10A		30P20L10S30B10A		60B20A10P10L		30P50B20A	
Pine clearcuts								
Pine	1544	107	753	115	920	109	734	83
Larch	243	89	72	50	100	80	109	
Cedar	3		50		78		73	
Fir	53		62		91		164	
Spruce	43		100		449		264	
Birch	664	96	950	49	1228	73	2430	89
Aspen	755	109	1247	81	781	59	1568	91
Species Composition,[c] %	50P10L20B20A		20P10S40A30B		30P10S40B20A		50B30A10P10S	

a X-average regenerated tree number
b V-variation coefficient
c P-pine, L-larch, F-fir, S-spruce, B-birch, A-aspen.

A characteristic of stratum I is the considerable fraction of clearcuts overgrown with pine. This is, however, to be expected for light soils and relatively dry site conditions. Stratum I is also the leader in terms of the number of self-sown pines. For clearcuts characterised by favourable site conditions, where 7–10 mature pines were recorded in species composition of stands having pine understory beneath the canopy, natural regeneration regime is recommended.

Analysis of the characteristics of forest regeneration process in strata II, III, and IV reveals that site conditions in these strata are generally favourable for light coniferous forest regeneration. Hence, the recommendation is as follows: for stands that had 7–10 mature pines in their species composition, but had no pine understory beneath the canopy, as well as for mixed pine stands that had 4–6 mature pines in the overstory, it is necessary to support natural regeneration process after cutting.

In order to determine whether forest management recommendations are correct for a given clearcut (burn), it is reasonable to introduce some special criterion, such as *measure of correctness probability*. To develop this criterion, an analysis of natural forest regeneration process in pine clearcuts was conducted using the data from 88 site observations. The final decision—the forest regeneration measures to utilise in each specific case—was based on the comparison of measured data concerning the quantity and quality of economically valuable understory, with the quantity of the understory needed for normal regeneration of clearcuts. This probability criterion turned out to be 0.7, suggesting a correlation between the model input parameters and regeneration process descriptors. Unfortunately, the use of inaccurate forest inventories has reduced the probability criterion estimated value.

4. Conclusions

Practical application of the classification system for clearcuts and burns does not require knowledge of either its structural components, or its general forestry implementations. If a model to determine forest regeneration regime is available, then geographic descriptors of clearcuts (burns) and the inventory characteristics of the pre-event stand that influence the choice of input parameters are all the system user needs to know.

References

Farber, S. K. and Sokolov, V. A.: 1991, *In:* Forest Resource Management and Regeneration in the Angara Yenisei Region (collected papers), Krasnoyarsk, *Lesosibirsk* **1**, 240-247 [In Russian].

USING FUEL CHARACTERISTICS TO ESTIMATE PLANT IGNITABILITY FOR FIRE HAZARD REDUCTION

J. C. HOGENBIRK and C. L. SARRAZIN-DELAY

Department of Biology, Laurentian University, Sudbury, Ontario, P3E 2C6, Canada

Abstract. There are areas in the boreal forest where the combination of highly flammable vegetation and frequent ignition events create a high fire hazard. The resultant fires cause considerable economic and social damage. During global change, fire frequency may increase in parts of the boreal forest. We are investigating the feasibility of using less-flammable vegetation to reduce the number of people-caused fires in herbaceous communities of northern Ontario. Herbaceous species abundant in these areas were ranked according to potential ignitability. Ignitability was estimated from chemical and physical fuel characteristics measured on live and dead tissue. *Achillea millefolium, Aster macrophyllus* and *Anaphalis margaritacea* had the lowest potential ignitability. Their rank was consistent for fuel characteristics measured on live and dead leaf tissue. Many of the commercially available forbs tested (e.g., *Trifolium* spp.) were of low to intermediate ignitability. Grass species such as *Poa compressa* and *Agrostis scabra* had the highest potential ignitability. Physical characteristics were not significantly correlated with chemical characteristics ($p > 0.05$, $n = 20$). There was a significant but trivial correlation between surface area to volume ratio and volume ($r = -0.80$, $p \leq 0.05$, $n = 20$). There was a significant correlation between surface area to volume ratio and density ($r = 0.54$, $p \leq 0.05$, $n = 20$). There was a significant correlation between total ash and silica-free ash content ($r = 0.61$, $p \leq 0.05$, $n = 47$), between total ash and energy content ($r = -0.55$, $p \leq 0.05$, $n = 47$) and between silica free ash and energy content ($r = -0.28$, $p \leq 0.05$, $n = 47$). Research is continuing to evaluate the feasibility of using less-flammable plants for fire hazard reduction. Planting less-flammable vegetation in fire prone areas, or around property and fire-sensitive natural areas, may help prevent ignition or slow fire spread. This technique could be a long term, cost effective and environmentally friendly method of fire prevention and control in selected areas of the boreal forest.

Keywords. *Achillea millefolium, Anaphalis margaritacea, Aster macrophyllus, Trifolium* spp., BOREAL FOREST, FLAMMABILITY, HERBACEOUS COMMUNITIES,

1. Introduction

There are areas in the boreal forest ecosystem where the combination of graminoid-dominated vegetation and a steady supply of anthropogenic ignition agents increases fire frequency. During global change, wildfire frequency is expected to increase in parts of the boreal forest (summarized in Stocks, 1993). It is likely that the frequency of people-caused fires will increase if fire hazard and human population density increase. Economic and social damage would also be expected to increase. Current methods of fire prevention may not be effective in reducing this damage. As well, current methods, such as prescribed burning and herbicide application, often favour the return of flammable graminoid vegetation (Corns and Schraa, 1962; Blackmore and Corns, 1979). We have targeted areas along railway right-of-ways where spring fires are frequent and where the remoteness of the sites makes detection and fire control more difficult. The overall research objective was to determine the feasibility of using less-flammable vegetation to reduce anthropogenic fire frequency.

Flammability is not yet rigorously defined. Current thinking suggests that flammability is composed of ignitability, sustainability, combustibility and consumability (Anderson, 1970; Mak, 1988; Martin *et al.*, 1994). Ignitability is often

Water, Air and Soil Pollution **82**: 161–170, 1995.

measured as time to ignition. Sustainability is the ability of the fire to continue burning after the external heat source is removed (usually recorded as fully, partially or not sustained). Combustibility is measured as the rapidity with which a fuel is consumed (mass loss per unit time). Consumability is measured as the completeness of combustion (mass loss as proportion of total mass). These components are often measured in different ways under different conditions and are related in poorly defined ways to overall flammability. Plant characteristics can be related to one or several aspects of flammability. Our emphasis was on plant characteristics related to time to ignition although in our evaluation of species we considered traits related to other components of flammability.

High moisture content (Blackmarr, 1972; Stockstad, 1976; Gill *et al.*, 1978; Xanthopoulos and Wakimoto, 1993), high tissue density (Gardner and Thomson, 1991) and low surface area to volume ratio (Montgomery and Cheo, 1971) are known to increase time to ignition. Ash content, in particular silica-free ash content, influences combustion (Philpot, 1970). High silica-free ash content (up to ca. 12%) slows the rate of pyrolysis (combustibility) but may extend burning time (Philpot and Mutch, 1968). Its effect on ignitability of plant tissue is largely unknown. Ignition tests on salt-impregnated paper strips show that high salt content has a variable effect on ignitability, and may reduce flaming but increase charring (Montgomery and Cheo, 1969). Recent work by Frandsen (1987) on peat shows that increasing total ash content reduces the sustainability of smouldering combustion. High total and silica-free ash should increase time to ignition.

Low concentration of ether extractives may increase time for piloted ignition but decrease it for spontaneous ignition (Mutch, 1964). Uncontrolled differences in moisture content in Mutch's experiment make this conclusion tentative. Low energy content should also reduce ignitability. However, the concentration and energy content of low-volatile compounds may be more important than total energy content (Chandler *et al.*, 1983).

Sylvester and Wein (1981) measured fuel characteristics of arctic plant species and used Rothermel's fire spread model (Rothermel, 1972) to estimate Byram's intensity (Byram, 1959) which they used as a measure of flammability. This approach is of limited use in our research because the fire spread model does not deal with ignition and initial fire spread, nor with herbaceous plant species, which are our objectives.

In this preliminary study we use simple relationships between fuel properties and time to ignition (discussed above) to rank chosen plant species by potential ignitability. We also examined the correlations between fuel characteristics to determine the interrelationships and possible trade-offs among physical and chemical fuel characteristics.

2. Methods

We measured fuel characteristics of some herbaceous species abundant along railroad right-of-ways near Sudbury, Ontario, Canada. We also measured fuel characteristics for some commercially available species used for reclamation and highway seeding. Collections were made in June 1994. Plant material was sorted into live and dead classes and then into leaf and stem tissue classes: there was no reproductive tissue at sampling time. Samples were dried at 80 °C for 48 hours. Stems were sorted into 1 mm diameter classes.

Physical fuel characteristics, volume, density and surface area to volume ratio, were calculated from length, diameter and weight measured for 10–15 stems from each stem class. Species means were calculated as a weighted average using stem class values multiplied by the proportion of biomass in that stem class. Physical fuel characteristics were measured on stems only.

Chemical fuel characteristics, total ash, silica-free ash and energy content, were measured separately for all tissue types although dead tissue was not available for all species. Dried samples were ground in a Wiley mill to pass through a 1 mm^2 mesh. Total and silica-free ash were determined for two to eleven 1 g samples using the methods of Allen *et al.* (1986). Energy content was measured for two 1 g samples with a Parr adiabatic bomb calorimeter using standard methods (Anonymous, 1969; American Society for Testing and Materials, 1993). Energy content was not corrected for acids produced during combustion as this correction is negligible for plant material (Chapman, 1986). Energy content calculated in cal g^{-1} was converted to J g^{-1} using 1 thermochemical calorie = 4.1840 J (Chapman, 1986).

Species were ranked by the measured value of the fuel characteristic. Rankings were done separately within each fuel characteristic. Species were ranked from 1 up to 16, with a rank of 1 given to the species with the fuel characteristic value associated with longer time to ignition. The maximum rank was given to the species with the fuel characteristic value associated with shorter time to ignition. For example, the species with the lowest surface area to volume ratio was given a rank of 1. Ranks were assigned in the same way for the species with the lowest energy content. For the remaining fuel characteristics, the species with the highest value was given a rank of 1. This ranking system ensured that species with lower rank had fuel characteristic values indicative of lower potential ignitability.

There were no ties in the ranks. In this simple ranking, fuel characteristics were assumed to have an equal and independent effect on time to ignition. Analysis was conducted separately for live stem, dead stem, live leaf and dead leaf. We used Friedman's method to test whether species mean ranks were significantly different from one another and Kendall's coefficient of rank concordance to test whether a species ranking was consistent for different fuel characteristics (Sokal and Rohlf, 1981). Kendall's coefficient of rank concordance ranges from 0 to 1, with 1 indicating perfect concordance. Spearman's rank correlation coefficients and Pearson's product-moment correlation coefficients (Sokal and Rohlf, 1981) were computed to show relationships between fuel characteristics.

3. Results

Plant species mean ranks were significantly different from one another for physical and chemical characteristics ($p = 0.04$) and for chemical characteristics ($p < 0.01$) measured on live stem tissue (Table I). There was no significant difference in mean ranks for physical characteristics measured on live stem tissue ($p = 0.28$). Plant species ranks were strongly concordant for chemical characteristics ($W = 0.87$) (Table I).

Species ranks were weakly concordant for physical and chemical characteristics (W = 0.30) or physical characteristics (W = 0.40) measured on live stem tissue.

Plant species mean ranks were not significantly different for physical ($p = 0.17$), chemical ($p = 0.60$) or all fuel characteristics measured on dead stem tissue ($p = 0.78$). Species ranks within the dead stem tissue group were moderately concordant for physical characteristics (W = 0.52) and weakly concordant for chemical (W = 0.25) and all fuel characteristics (W = 0.08).

There were significant differences in species mean ranks for chemical characteristics measured on live leaf tissue ($p < 0.01$). Differences in species mean ranks for chemical characteristics measured on dead leaf tissue were marginally significant ($p = 0.10$). Plant species ranks were strongly concordant for chemical characteristics measured on live leaf tissue (W = 0.70) and moderately concordant for chemical characteristics measured on dead leaf tissue (W = 0.54) (Table I). Absolute values for fuel characteristics are in Appendix I.

There were significant or marginally significant ($p \leq 0.10$) differences in mean rank for the following tissue types: live stem tissue (physical plus chemical characteristics, and chemical characteristics), and live or dead leaf (chemical characteristics). *Achillea millefolium, Anaphalis margaritacea* and *Aster macrophyllus* had lower mean rank in these tissue types (Table I). *Trifolium pratense* and *T. repens* had intermediate rank. Graminoid species such as P*oa compressa* and *Agrostis scabra* had higher mean rank.

Physical characteristics were not significantly correlated with chemical characteristics for either absolute (measured) values or ranks based on absolute values ($p > 0.05$) (Table II). Significant correlations only occurred between physical characteristics or between chemical characteristics.

Chemical characteristics were significantly correlated ($p \leq 0.05$) with one another over the whole data set. A species rank for total ash was positively correlated with its rank for silica-free ash and its rank for energy content ($r = 0.73$ for both) (Table II). Silica-free ash rank was significantly and positively correlated with energy content rank ($r = 0.46$). Correlations among ranks based on chemical fuel characteristics were positive and usually significant for live stem and live leaf tissue. For dead stem and dead leaf tissue, correlation coefficients between chemical characteristics were significant and positive only for ranks based on total ash and energy. Species ranks based on physical fuel characteristics, with one exception, were not significantly correlated with other physical or chemical characteristics. Ranks based on live or dead stem volume were significantly and positively correlated with surface area to volume ratio ranks ($r \geq 0.84$) (Table II). Pearson product-moment correlation coefficients computed from absolute values were generally similar in magnitude (Table II). These correlation coefficients were opposite in sign when either energy content or surface area to volume ratios were involved because of the way in which species were ranked within these characteristics.

4. Discussion

Published relationships between fuel characteristics and time to ignition were used to rank plant species by estimated (potential) ignitability. *Achillea millefolium, Anaphalis margaritacea* and *Aster macrophyllus* had lower potential ignitability based on chemical characteristics measured on live stem, live leaf and dead leaf tissues. Species with intermediate ignitability included *Trifolium pratense* and *T. repens*. Grass species had chemical fuel characteristics consistent with higher ignitability. Rankings based on physical fuel characteristics, although not statistically significant, tend to support rankings based on chemical characteristics.

In general, species had similar rank across chemical characteristics within a given tissue type; species ranks in the dead tissue types were the exception. Ranks based on total ash, silica-free ash and energy content were usually significantly and strongly correlated ($r^2 \cong 0.53$); the correlation between silica-free ash and energy content was weaker ($r^2 \cong 0.21$).

TABLE I

Mean rank of herbaceous plant species. Rank based on physical and chemical fuel characteristics. Physical characteristics were mean stem volume, stem density and stem surface area to volume ratio. Chemical characteristics were total ash, silica-free ash and energy value. Low rank is indicative of low potential ignitability. See text for ranking method. The number of species (i.e., the maximum possible rank), the x^2 statistic calculated by Friedman's method, probability (p) from the χ^2 distribution and Kendall's coefficient of rank concordance (W) are given.

	Live stem			Dead stem			Leaf Chem.	
Species*	Phys. & Chem.	Phys.	Chem.	Phys. & Chem.	Phys.	Chem.	Live	Dead
Achillea millefolium L.	5	4	6	3	2	5	3	1
Agropyron repens (L.) Beauv.	10	6	13				11	
Agrostis gigantea Roth.	9	10	9	3		3	5	4
Agrostis scabra Willd.							13	9
Anaphalis margaritacea (L.) Benth and Hook.	3	5	2	5	4	5	4	7
Aster macrophyllus L.	4	6	1	4	2	5	2	
Bromus inermis Leysser	8	5	11				9	5
Calamagrostis canadensis (Michaux) Beauv.	10	11	9	4	5	4	10	6
Galium mollugo L.	9	12	5	4	5	2	7	
Lotus corniculatus L.	9	6	11				11	
Oryzopsis asperifolia Michaux							12	7
Phleum pratense L.	8	5	10	4	4	4	12	5
Poa compressa L.	11	9	12				14	8
Trifolium hybridum L.	7	10	3				12	
Trifolium pratense L.	8	7	8				7	3
Trifolium repens L.	7	9	5				4	
Number of Species	14	14	14	7	6	7	16	10
x^2	23.0	15.5	33.7	2.5	7.8	4.6	31.4	14.6
p	0.04	0.28	<0.01	0.78	0.17	0.60	<0.01	0.10
W	0.30	0.40	0.87	0.08	0.52	0.25	0.70	0.54

*Nomenclature as per Gleason and Cronquist (1991) for forbs and Dore and McNeill (1980) for grasses.

Species ranks were less similar across physical characteristics than across chemical characteristics within the live stem tissue type. Stem surface area to volume ratio and stem volume were significantly and strongly correlated ($r^2 > 0.70$). This is expected because both characteristics were derived from stem length and diameter. Stem density is weakly correlated with surface area to volume ratio ($r^2 < 0.30$).

Physical characteristics were weakly correlated with chemical characteristics ($r^2 < 0.17$). It has been suggested that density could be correlated with extractive (energy) content through differences in chemical composition (Martin *et al.*, 1994). Changes in energy and ash content could affect density but we did not find a strong, direct correlation. The lack of significant correlation between physical and chemical characteristics suggests the hypothesis that these characteristics are independent measures of potential ignitability.

Species mean rank was not significantly different for fuel characteristics measured on dead stem tissue. This could be due to a small number of species sampled. This tissue type was also unusual in that absolute values for total ash and silica-free ash were poorly correlated. Ranks based on these values were negatively correlated, in contrast to the positive and usually significant correlation for other tissue types. Interestingly, species with low potential ignitability in other tissue types had high potential ignitability based on chemical characteristics measured on dead stem tissue. These species, *Achillea millefolium, Anaphalis margaritacea* and *Aster macrophyllus*, had some of the lower total ash and higher energy values in the dead stem tissue group (see Appendix). It remains to be seen if the relative ignitability ranking of these species will change when values for all species are obtained for the dead stem tissue group.

In this preliminary analysis we assigned each fuel characteristic equal weight to determine overall potential ignitability. Differences in physiochemical properties are believed to be important for low temperature and low heat capacity ignition sources (Merzhanov and Averson, 1971). Research is underway at our laboratory to test hypotheses about the relative importance of fuel characteristics.

It should be noted that species mean physical fuel characteristics depend on the relative proportion of biomass in each stem class. These proportions would be expected to change over the growing season. Our sampling period in early June corresponds to the railway fire season so we are confident that our values are appropriate. Studies to measure the fuel characteristics of litter are underway.

Studies will measure other plant attributes related to spring fires in herbaceous vegetation. Data on spring growth rates, litter production and decomposition should help refine the list of potential species. Long-term field tests are being set up to measure the ignitability and fire spread rates of the top ranked species under natural conditions.

In summary, published relationships between ignitability and physical and chemical fuel characteristics were used to provide an estimate of ignitability. These rankings rely primarily on total ash, silica-free ash and energy content to determine ignitability. Rankings based on physical fuel characteristics, although not statistically significant, tend to support rankings based on chemical characteristics. The strategic use of less-flammable plants has the potential to be an effective method of fire hazard reduction in parts of the boreal forest where human activity provides both the fuel and ignition agents. Less-flammable plants will not reduce all people-caused fires but could be effective in areas such as right-of-ways, campgrounds or around dwellings, when used in conjunction with other fire hazard reduction methods.

TABLE II

Spearman's rank correlation coefficients and Pearson's product-moment correlation coefficients (italics) computed between fuel characteristics. Sample size (n) and critical value for p = 0.05 are given. For correlations calculated for data from all tissue types the sample size is 20 for correlations involving physical characteristics and sample size is 47 for correlations between chemical characteristics.

Plant tissue type	Fuel characteristic	Volume	Density	Area/ Volume	Total ash	Silica-free ash	n	Critical Value (p = 0.05)
ALL TYPES	Density	-0.09[a]	1.00				20	0.43
		-0.38[b]	*1.00*					
	Area/Volume	0.88	0.10	1.00				
		-0.80	*0.54*	*1.00*				
	Total ash	0.09	0.04	0.17	1.00		47	0.28
		-0.39	*0.34*	*0.28*	*1.00*			
	Silica-free ash	0.19	0.22	0.23	0.73	1.00		
		-0.34	*0.41*	*0.29*	*0.61*	*1.00*		
	Energy content	0.06	0.08	0.16	0.73	0.46		
		0.38	*-0.32*	*-0.30*	*-0.55*	*-0.28*		
LIVE STEM	Density	-0.41	1.00				14	0.51
		-0.36	*1.00*					
	Area/Volume	0.84	-0.15	1.00				
		-0.84	*0.36*	*1.00*				
	Total ash	-0.08	-0.12	0.04	1.00			
		-0.04	*0.01*	*-0.12*	*1.00*			
	Silica-free ash	*	-0.05	0.05	0.94	1.00		
		*	*0.09*	*-0.11*	*0.97*	*1.00*		
	Energy content	-0.13	-0.09	*	0.78	0.67		
		0.03	*0.04*	*0.08*	*-0.83*	*-0.77*		
DEAD STEM	Density	-0.09	1.00					
		0.36	*1.00*					
	Area/Volume	1.00	-0.09	1.00				
		-0.90	*-0.05*	*1.00*				
	Total ash	-0.43	-0.66	-0.43	1.00			
		-0.59	*-0.49*	*0.53*	*1.00*			
	Silica-free ash	-0.43	0.66	-0.43	-0.54	1.00	6	0.75
		-0.25	*0.50*	*0.61*	*0.05*	*1.00*		
	Energy Content	-0.20	-0.26	-0.20	0.75	-0.57		
		0.19	*0.08*	*-0.10*	*-0.82*	*0.47*		
LIVE LEAF	Silica-free Ash				0.69	1.00	16	0.48
					0.66	*1.00*		
	Energy Content				0.59	0.36		
					-0.69	*-0.40*		
DEAD LEAF	Silica-free Ash				0.37	1.00	10	0.60
					0.45	*1.00*		
	Energy Content				0.75	-0.19		
					-0.85	*		

* Correlation coefficient < ±0.01

[a] Spearman's rank correlation coeffecients

[b] Pearson's product moment correlation coefficients

Acknowledgments

The project has been funded by the Ontario Ministry of Natural Resources. Additional funding for personnel provided by: Ontario Ministries of Natural Resources, Northern

Development and Mines, Education and Training, and by Employment and Immigration Canada. The continuing co-operation of Canadian National and Canadian Pacific railway companies is appreciated. Growth experiments were conducted at the Ontario Forest Research Institute in Sault Ste Marie, Ontario. The advice and assistance of personnel from Ontario Ministry of Natural Resources and Laurentian University were greatly appreciated. We thank our field and laboratory assistants for their help.

References

Allen, S. E., Grimshaw, H. M. and Rowland, A. P.: 1986, *In:* Moore, P. D. and Chapman, S. B., (eds) *Methods in Plant Ecology*, Blackwell, Oxford, UK, 589 pp.

American Society for Testing and Materials: 1993, Standard Test Method for Gross Calorific Value of Coal and Coke by the Adiabatic Bomb Calorimeter, *In:* ASTM Committee E-5 on Fire Standards, *ASTM Fire Test Standards*, ASTM D 2015 -93, American Society for Testing and Materials, Philadelphia, USA, 1148 pp.

Anderson, H. E.: 1970, *Fire Technol.* **6**, 312-322.

Anonymous: 1969, *Instructions for 1241 and 1242 Adiabatic Calorimeters*, Parr Instrument Company, Moline, USA, 23 pp.

Blackmarr, W. H.: 1972, *Moisture Content Influences Ignitability of Slash Pine Litter*, U.S. Department of Agriculture Forest Service Southeastern Forest Experiment Station, Asheville, USA, 7 pp.

Blackmore, D. G. and Corns, W. G.: 1979, *For. Chron.* **55**, 102-105.

Byram, G. M.: 1959, *In:* Davis, K. P. (ed), *Forest Fire Control and Use*, McGraw-Hill, New York, USA, 584 pp.

Chandler, C., Cheney, P., Thomas, P., Trabaud, L. and Williams, D. (eds.): 1983, *Fire in Forestry: Volume I, Forest Fire Behavior and Effects.* Krieger Publishing Company, Malabar, USA. 450 pp.

Chapman, S. B.: 1986, *In:* Moore, P. D. and Chapman, S. B., (eds.) *Methods in Plant Ecology*, Blackwell, Oxford, UK, 589 pp.

Corns, W. G. and Schraa, R. J.: 1962, *Can. J. Plant Sci.* **42**, 651-659.

Dore, W. G. and McNeill, J.: 1980, *Grasses of Ontario*, Research Branch, Agriculture Canada, Hull, Canada, 566 pp.

Frandsen, W. H.: 1987, *Can. J. For. Res.* **17**, 1540-1544.

Gardner, W. D. and Thomson, C. R.: 1991, *Fire and Materials* **15**, 3-9.

Gill, A. M., Trollope, S. W. and MacArthur, D. D.: 1978, *Aust. J. For. Res.* **8**, 199-208.

Gleason, H. A. and Cronquist, A.: 1991, *Manual of Vascular Plants of Northeastern United States and Adjacent Canada*, The New York Botanical Garden, Bronx, USA, 910 pp.

Mak, E. H. T.: 1988, *For. Sci.* **34**, 523-529.

Martin, R. E., Gordon, D. A., Gutierrez, M. E., Lee, D. S., Molina, D. M., Schroeder, R. A., Sapsis, D. B., Stephens, S. L. and Chambers, M.: 1994, *In: Proceedings of the 12th Conference on Fire and Forest Meteorology*, Society of American Foresters, Bethesda, USA, 796 pp.

Merzhanov, A. G. and Averson, A. E.: 1971, *Combustion and Flame* **16**, 89-124.

Montgomery, K. R. and Cheo, P. C.: 1969, *Amer. J. Bot.* **56**, 1028-1032.

Montgomery, K. R. and Cheo, P. C.: 1971, *For. Sci.* **17**, 475-478.

Mutch, R. W.: 1964, *J. Appl. Chem.* **14**, 271-275.

Philpot, C. W.: 1970, *For. Sci.* **16**, 461-471.

Philpot, C. W. and Mutch, R. W.: 1968, *Flammability of Herbicide Treated Guava Foliage*, U.S. Department of Agriculture Forest Service, Intermountain Forest and Range Experiment Station, Ogden, USA, 9 pp.

Rothermel, R. C.: 1972, *A Mathematical Model for Predicting Fire Spread in Wildland Fuels*, , U.S. Department of Agriculture Forest Service, Intermountain Forest and Range Experiment Station, Ogden, USA, 40 pp.

Sokal, R. R. and Rohlf, F. J.: 1981, *Biometry. The Principles and Practice of Statistics in Biological Research*, Freeman, New York, USA, 859 pp.

Stocks, B. J.: 1993, *For. Chron.* **69**, 290-293.

Stockstad, D. S.: 1976, *Spontaneous and Piloted Ignition of Cheatgrass*, U.S. Department of Agriculture Forest Service, Intermountain Forest and Range Experiment Station, Ogden, USA, 12 pp.

Sylvester, T. W. and Wein, R. W.: 1981, *Can. J. Bot.* **59**, 898-907.

Xanthopoulos, G. and Wakimoto, R. H.: 1993, *Can. J. For. Res.* **23**, 253-258.

Appendix

Mean fuel characteristics for selected plant species collected near Sudbury, Ontario. Physical characteristics were calculated for each stem class and weighted by the proportion of biomass in that class to give a mean for the species. Total ash, silica-free ash and energy content are expressed on a dry weight basis.

	Physical Characteristics			Chemical Characteristics							
	Volume (cm^3)	Density ($g\ cm^{-3}$)	Area/Vol (cm^{-1})	Total ash content (%)			Silica-free ash content (%)			Energy content $J\ g^{-1}$	n=2
Species	Mean	Mean	Mean	Mean	S.E.	n	Mean	S.E.	n	Mean	S.E.
LIVE STEM											
Achillea millefolium	0.6	0.4	17	8.5	0.1	5	8.0	0.1	5	17259	27
Agropyron repens	0.2	0.4	28	3.4	0.1	4	2.2	0.1	4	18078	47
Agrostis gigantea	0.2	0.4	30	5.8	0.1	6	2.7	0.1	6	17014	10
Agrostis scabra				5.6		1	3.7		1		
Anaphalis margaritacea	0.2	0.4	25	13.0	0.2	5	12.5	0.2	5	16870	8
Aster macrophyllus	0.2	0.5	27	18.0	0.2	5	17.1	0.2	5	15271	34
Bromus inermis	0.3	0.4	25	6.1	0.1	3	4.3	0.1	3	18226	17
Calamagrostis canadensis	0.2	0.3	30	8.3	0.1	5	4.5	0.1	5	17804	32
Galium mollugo	0.1	0.3	38	9.2	0.1	5	8.7	0.1	5	17204	10
Lotus corniculatus	0.2	0.5	29	5.3	0.2	4	5.2	0.1	4	18830	15
Oryzopsis asperifolia											
Phleum pratense	0.3	0.3	22	4.2	0.1	6	2.8	0.1	6	17417	7
Poa compressa	0.1	0.5	42	3.8	0.1	6	2.7	0.1	6	17933	26
Trifolium hybridum	0.2	0.3	31	9.9	0.1	5	9.4	0.1	5	16981	1
Trifolium pratense	0.3	0.3	28	8.2	0.1	5	7.9	0.1	5	17648	22
Trifolium repens	0.1	0.5	36	9.2	0.1	5	8.3	0.2	5	17036	49
DEAD STEM											
Achillea millefolium	1.5	0.3	11	3.1	0.2	4	1.8	0.2	4	18713	9
Agropyron repens											
Agrostis gigantea				8.6		1	0.8		1	17385	50
Agrostis scabra											
Anaphalis margaritacea	0.3	0.3	24	2.6	0.2	3	2.5	0.2	3	19131	28
Aster macrophyllus	1.4	0.3	13	2.3	0.1	4	1.7	0.1	4	18522	60
Bromus inermis				10.0		1	2.9		1		
Calamagrostis canadensis	0.4	0.2	21	4.6	0.4	3	0.9	0.4	2	18412	12
Galium mollugo	0.2	0.3	30	4.4	0.3	2	3.3	0.6	2	18142	5
Lotus corniculatus											
Oryzopsis asperifolia											
Phleum pratense	0.5	0.3	17	3.8	0.1	3	1.0	0.1	3	17836	24
Poa compressa											
Trifolium hybridum											
Trifolium pratense										16268	n=1
Trifolium repens											

Appendix (Continued)

	Physical Characteristics			Chemical Characteristics							
	Volume	Density	Area/ Vol	Total ash content			Silica-free ash content			Energy content	
	(cm^3)	($g\ cm^{-3}$)	(cm^{-1})	(%)			(%)			$J\ g^{-1}$	n=2
Species	Mean	Mean	Mean	Mean	S.E.	n	Mean	S.E.	n	Mean	S.E.
LIVE LEAF											
Achillea millefolium				13.0	0.3	8	8.8	0.2	8	17484	11
Agropyron repens				8.6	0.1	5	5.1	0.1	5	19374	1
Agrostis gigantea				13.4	0.1	6	5.1	0.2	6	17753	38
Agrostis scabra				5.4	0.1	4	3.7	0.1	4	19000	17
Anaphalis margaritacea				11.7	0.1	5	11.0	0.1	5	18271	42
Aster macrophyllus				13.4	0.1	5	10.8	0.1	5	17681	52
Bromus inermis				9.8	0.1	8	5.5	0.1	8	19259	17
Calamagrostis canadensis				9.6	0.1	5	4.7	0.1	5	19203	60
Galium mollugo				9.4	0.1	5	8.1	0.1	5	18622	39
Lotus corniculatus				8.6	0.1	11	8.0	0.1	11	19433	14
Oryzopsis asperifolia				6.5	0.3	6	3.0	0.1	6	18075	7
Phleum pratense				8.5	0.2	6	4.5	0.3	6	19059	1
Poa compressa				6.7	0.1	3	4.2	0.1	3	19396	35
Trifolium hybridum				8.4	0.1	5	7.6	0.1	5	19377	2
Trifolium pratense				9.2	0.1	5	8.5	0.1	5	18956	17
Trifolium repens				11.7	0.1	5	10.7	0.1	5	18207	27
DEAD LEAF											
Achillea millefolium				19.6	0.3	5	6.5	0.3	4	15165	2
Agropyron repens										19286	16
Agrostis gigantea				12.7		1	2.4		1	17092	20
Agrostis scabra				5.2	0.1	4	1.7	0.1	4	19200	25
Anaphalis margaritacea				7.2	0.3	4	5.7	1.1	3	19590	12
Aster macrophyllus										16349	45
Bromus inermis				10.1	0.2	3	3.1	0.6	3	18498	43
Calamagrostis canadensis				10.8	0.1	4	1.7	0.2	4	17660	29
Galium mollugo										17694	21
Lotus corniculatus											
Oryzopsis asperifolia				10.0		1	0.9		1	17452	66
Phleum pratense				14.2		1	1.4		1	17043	39
Poa compressa				5.7		1	2.0		1	18787	41
Trifolium hybridum											
Trifolium pratense				14.3		1	9.7		1	18568	99
Trifolium repens											

SIMULATION OF MIXEDWOOD MANAGEMENT OF ASPEN AND WHITE SPRUCE IN NORTHEASTERN BRITISH COLUMBIA

J. R. WANG[1], P. COMEAU[2] and J. P. KIMMINS[1]

[1]*Faculty of Forestry, University of British Columbia, 270-2357 Main Mall, Vancouver, British Columbia V6T 1Z4, Canada*

[2]*British Columbia Ministry of Forests, Research Branch, 31 Bastion Square, Victoria, British Columbia V8W 3E7, Canada*

Abstract. FORECAST, an ecosystem simulation model, was calibrated for aspen (*Populus tremuloides* Michx) and white spruce (*Picea glauca* (Moench) Voss) stands using data collected in the Boreal White and Black Spruce biogeoclimatic zone in northeastern British Columbia and published data. Simulations were undertaken to examine the effects of initial density of aspen on yield of white spruce in an aspen and spruce mixedwood stand, and to compare the predicted stemwood biomass yields of aspen, white spruce and mixedwood stands. Results of the simulations suggest that mixedwood management regimes on the same medium quality site should have higher stemwood yield compared to pure white spruce stand. Simulated stemwood biomass yield of pure aspen stands over 240 years on medium site varied from 682.5 Mg ha^{-1} to 239.1 Mg ha^{-1} for different rotation lengths (30 to 120 years). Repeated rotations of monoculture white spruce produced much less stemwood biomass, simulated yields over 240 years ranging from 877.3 Mg ha^{-1} to 248.4 Mg ha^{-1} for rotation lengths of 60 to 240 years. Simulated aspen and white spruce mixedwood stands produced higher stemwood biomass yields than the pure white spruce stands, but less than the pure aspen stands; from 217.4 Mg ha^{-1} to 292.8 Mg ha^{-1} over 240 years. Variations in initial densities of aspen did not affect spruce stemwood biomass yield over the simulation period. This model shows potential for comparing the relative effects of different management strategies on harvestable volume and variety of other ecosystem variables. A calibrated version of the model should be useful as both a management simulator and a research tool. However, shortcomings in the representation of the canopy architecture of mixed species stands suggested the need to develop an individual tree version of this ecosystem management model for application to mixed species stands.

Keywords. BOREAL FOREST, MIXEDWOOD MANAGEMENT, ASPEN, WHITE SPRUCE, NORTHEASTERN BRITISH COLUMBIA, SIMULATION.

1. Introduction

The boreal mixedwood is a productive and important component of the forest landscape in northeastern British Columbia and there is growing interest in managing this part of the Boreal White and Black Spruce biogeoclimatic zone (BWBS) (Peterson *et al.*, 1989). The two major tree species found in British Columbia's boreal forest are aspen (*Populus tremuloides* Michx.) and white spruce (*Picea glauca* (Moench) Voss.). After mixedwood stands are disturbed by clearcutting or fire, aspen normally regenerates prolifically from root suckers. If there is an adequate seed source, white spruce regenerates naturally under the aspen canopy as long as the aspen stand is not too dense. On some sites, white spruce invades concurrently with aspen and other hardwood species (Peterson and Peterson, 1992). In contrast, monoculture plantations of white spruce are often difficult to establish on clearcut mixedwood sites because of frost damage and grass competition (Lieffers *et al.*, 1993). The proportion of white spruce and aspen in natural mixedwood stands varies from site to site. One can find: (1) pure generally even aged stands of white spruce; (2) pure even aged stands of aspen; (3) even aged aspen overstory with a variety of white spruce age classes in the

Water, Air and Soil Pollution **82**: 171–178, 1995.

understory; and (4) aspen and white spruce mixed in the canopy. While we are waiting the many decades needed to obtain empirical data from field trials of the results of alternative silvicultural systems, it is necessary to use stand-level simulation modeling. In order to predict the probable outcomes of different stand management strategies, FORECAST[1], an ecosystem simulation model, is capable of examining certain consequences of managing aspen and white spruce mixed stands, because it simulates nutrient cycling, competition for light and nutrients, and interactions between trees, shrubs, and herbs.

When comparing strategies for the management of aspen and white spruce mixed stands for timber production, the following factors must be considered: (1) cost of aspen control; (2) yield of aspen; (3) effects of aspen on regeneration, growth and yield of white spruce, and 4) the cost of control of herbs and shrubs, or their negative effects on white spruce seedling growth. The consequences of timber production, strategies on wildlife and other non-timber values should also be considered. Such comparisons require the use of appropriate ecosystem-level models which are capable of simulating the effects of critical processes such as nutrient cycling and resource competition on yield, the interaction between trees and minor vegetation, and the implication of different silvicultural systems for wildlife habitat values and ecosystem sustainability.

The objectives of this study were: (1) to prepare an initial calibration of FORECAST for aspen, white spruce and mixed aspen/white spruce ecosystems; (2) to complete a series of runs of the model to evaluate the predicted consequences of several alternative management strategies.

2. Model Calibration

FORECAST requires information on biomass accumulation in different biomass components for each species to be simulated, on the turnover of detrital materials, and on nutrient cycling for each site class simulated. A sample data set for FORCYTE-11 is presented in Kimmins *et al.* (1990). A similar data set for aspen and white spruce was assembled based on field research (Wang *et al.*, 1994) in the BWBS biogeoclimatic zone of northern British Columbia and published literature on aspen and white spruce mixed stands. The study area was located in Dawson Creek Forest District. Description of the climate, soils and vegetation in the area can be found in Lord and Green (1986).

Published papers and unpublished reports dealing with the ecology and management of aspen and white spruce in pure and mixed stands were located by computer search and through contact with researchers in B.C. The following variables were assembled: age, height, diameter at breast height or basal diameter, density (stem ha^{-1}), green crown length, stemwood volume, biomass (stem, branch, foliage, coarse roots, small and fine roots), net primary production, mortality, litterfall, decomposition rates for different components of tree litter, nutrient concentrations, light attenuation, stand

[1] FORECAST is a hybrid-simulation, ecosystem management model developed from an earlier model FORCYTE-11 (Kimmins et al., 1990; and Kimmins, 1993) developed for the Canadian Forest Service as part of the ENFOR program. The model is still under development, so a user's manual and documentation have not yet been developed. Consequently, reference is made to the user's manual and documentation for FORCYTE-11.

structure (stand table), stand origin (coppice or seed), and rates of symbiotic nitrogen fixation by shrubs. Most of the data for the calibration data set for aspen had already been collected (Wang *et al.*, 1994). A similar data set for white spruce was collected from published literature and from growth and yield information provided by the Ministry of Forests, British Columbia. Published biomass regression equations (Gholz *et al.*, 1979) were applied to stand table data to estimate the tree component biomass of white spruce. Estimates of root biomass for white spruce were based on Kimmins and Hawkes (1978) and Safford and Bell (1972). Decomposition data for spruce and aspen were from Alban and Pastor (1993), Prescott *et al.* (1993), Rustad (1994) and Berg and Tamm (1994).

After model calibration and prior to the ecosystem simulations, a file describing initial state values for forest floor organic matter components and site nutrient capital was prepared as described by Kimmins (1993). The initial starting state was the ecosystem condition predicted by the model after three periods of 240 years during each of which aspen and white spruce regenerated together and then grew until disturbance killed the stand at age 240 years leaving all the biomass and nutrients on the site. Aspen and white spruce regenerated naturally after disturbance. Site quality was set at medium site (SI = 9.2 for white spruce and SI = 21.5 for aspen). At the end of the setup runs, forest floor and soil organic matter totalled 230.9 Mg ha^{-1} and site N capital 1635 kg ha^{-1}, respectively.

Following calibration preparation of the initial state file, model behavior was evaluated and simulations of selected management scenarios were conducted. The ecological and yield consequences of these alternative approaches to management of aspen and white spruce mixed stands were evaluated. As a result of gaming with the calibrated FORECAST model, shortcomings and inadequacies of the model were identified and recommendations are made concerning modification to the model needed to increase its usefulness for management gaming with mixedwood stands in the BWBS zone.

3. Simulations

Four sets of simulations were completed to examine: (1) stemwood biomass yield of monoculture aspen stands over different rotation lengths; (2) stemwood biomass yield of monoculture white spruce stands over different rotation lengths; (3) white spruce stemwood biomass yield of mixedwood of aspen and spruce over a 120-year rotation with different initial aspen densities; and (4) effect of climatic change on competitive abilities of aspen and white spruce in a mixedwood stand. All the simulations were run for a site of medium nutritional quality.

3.1. Monoculture Aspen and White Spruce Stands: Effects of Rotation Length on Stemwood Biomass Yield

For the pure aspen stand simulations, 50 000 seedlings ha^{-1} of aspen were naturally regenerated at year 1 and harvested at different ages to represent rotation lengths. Figure 1 shows the predicted stemwood biomass yield over the 240-year simulation period. The 30-year rotation produced the highest biomass (682.5 Mg ha^{-1}). As rotation length increased, stemwood decreased from 560.2 Mg ha^{-1} in the 40-year rotation to

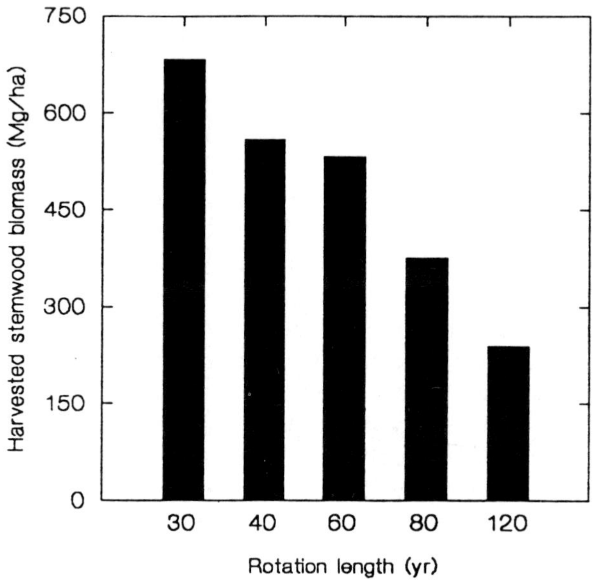

Fig. 1. Effects of different rotation lengths on cumulative stemwood biomass yield of pure aspen stands over 240 years. Aspen naturally regenerated at year 1 at a density of 50 000 seedlings ha^{-1} and was harvested at 30, 40, 60, 80 and 120 years, respectively.

239.1 Mg ha^{-1} in the 80-year rotation. The results indicated that the pure aspen stands started to decline after 60 years.

For pure white spruce stands, 1200 seedlings were planted at year 1. Figure 2 presents the harvested stemwood biomass yield of different rotation lengths over 240 years. There was an increase in stemwood biomass yield from 87.3 Mg ha^{-1} in the 60-year rotations to 248.4 Mg ha^{-1} in the 120-year rotations, but the harvested stemwood biomass yield decreased to 163.5 Mg ha^{-1} in the 240-year rotation stand. This agrees with the fact reported by Viereck *et al.* (1983) that a 250-year old white spruce stand in Alaska started to break up.

3.2. Yield of Mixedwood Management

One of the objectives of this simulation exercise was to examine the potential effect of different initial densities of aspen on stemwood biomass yield of white spruce and aspen. Initial aspen density was simulated from 500 to 50 000 stems ha^{-1}. White spruce was planted in year 1 at a density of 1200 stems ha^{-1} and harvested at 120 years. Aspen naturally regenerated at year 1 at different densities and was harvested at year 60. At the time of the harvest, 85% of stemwood and stembark was removed from the site. Branches, leaves, and roots were left on the site. Results of the simulations (Figure 3) suggested that maximum cumulative stemwood biomass yield of white spruce (98.7 Mg ha^{-1}) and aspen (79.7 Mg ha^{-1}) was obtained when initial aspen density was 20 000 seedlings ha^{-1}. In general, as initial aspen density increased, stemwood biomass yield of

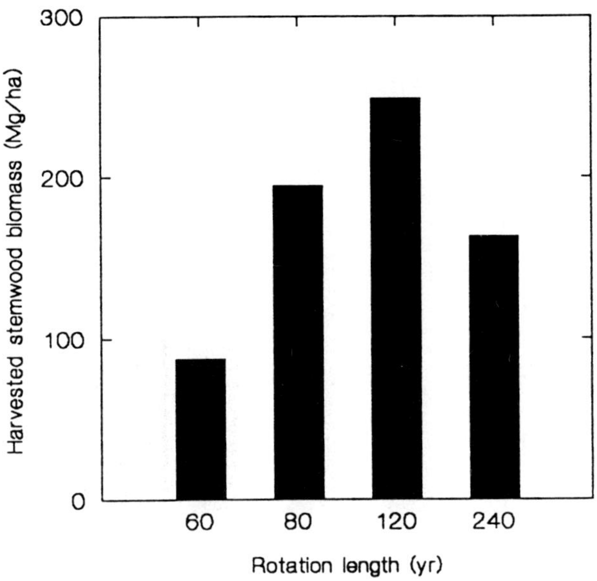

Fig. 2. Effects of different rotation lengths on cumulative stemwood biomass yield of pure white spruce stands over 240 years. White spruce was planted at year 1 at a density of 1200 seedlings ha^{-1} and was harvested at 60, 80, 120 and 240 years, respectively.

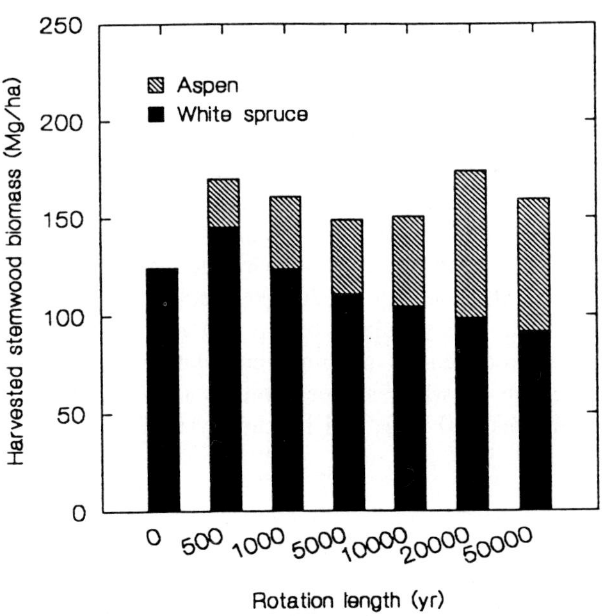

Fig. 3. Simulated effects of initial densities of aspen on cumulative stemwood biomass yield of white spruce and aspen mixedwood on a medium site over 120 years. White spruce was planted at year 1 at a density of 1200 seedlings ha^{-1} and harvested at 120 years. Aspen naturally regenerated at year 1 at different densities and was harvested at 60 years.

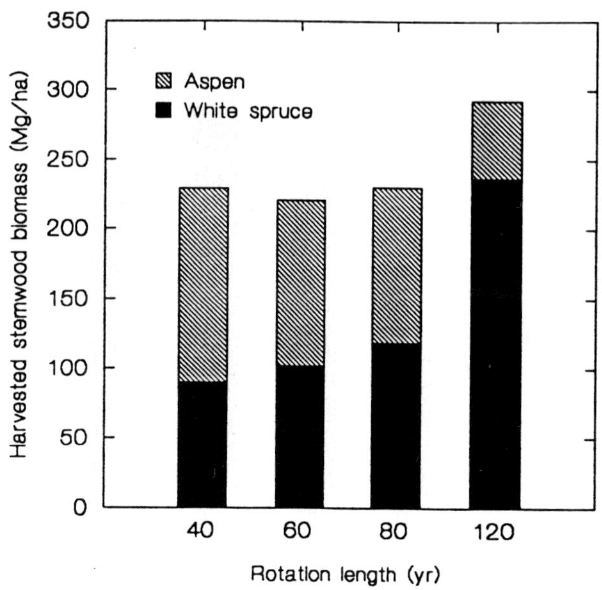

Fig. 4. Simulated effects of rotation lengths on cumulative stemwood biomass yield of white spruce and aspen mixedwood on a medium site over 240 years. White spruce was planted at year 1 at a density of 1200 seedlings ha^{-1}, and harvested at 120 years. Aspen naturally regenerated at year 1 at a density of 20 000 seedlings ha^{-1}, and was harvested at 40, 60, 80 and 120 years, respectively.

white spruce decreased. There was little increase in cumulative aspen stemwood biomass yield because most of the increased number of aspen seedlings and suckers died out very quickly. DeLong (1991) reported that after 2–5 years aspen density tends to decline fairly rapidly. Most aspen clumps thin down to one main stem within ten years.

Simulated effects of different rotation lengths on stemwood biomass yields of white spruce and aspen mixedwood are shown in Figure 4. White spruce was planted at year 1 at a density of 1200 seedlings ha^{-1} and harvested at 120 years. Aspen naturally regenerated at year 1 at a density of 20 000 seedlings ha^{-1} and was harvested at different ages. The model was simulated over a 240-year period. It was concluded that the 120-year rotation produced the maximum cumulative stemwood biomass yield and had a higher proportion of white spruce biomass than the other rotations. This is because when aspen was 120 years old its canopy broke up while growth of white spruce increased.

4. Discussion

This model shows potential for exploring the consequences of a variety of management strategies. The simulated data were considered to be consistent with the observed data if the range of simulated data overlapped with the range of the observed data. Compared with the published stemwood biomass data (11.3–172.2 Mg ha^{-1}) of white spruce at different ages (45–190 years) (Yarie and Van Cleve, 1983; Van Cleve *et al.*,

1983), the simulated stemwood biomass data varied from 21.8 Mg ha^{-1} at age 60 years to 124.2 Mg ha^{-1} at age 240 years (Figure 2). For pure aspen stands, observed stemwood biomass varied from 88 Mg ha^{-1} at age 30 to 250 Mg ha^{-1} at 90 years old on a good site (Peterson and Peterson, 1992). The simulated data were from 85.3 Mg ha^{-1} at 30 years old to 133.4 Mg ha^{-1} at 120 years old (Figure 1). However, while results of these simulations appear consistent with published information and observations on pure and mixed stands of aspen and white spruce, further verification of the model behavior and validation of the model predictions are required. Results of FORECAST simulations appear to be sensitive to input biomass data used in model calibration. The biomass data of young natural white spruce stands will be needed for further testing of the model. By comparing simulated stemwood biomass yields of monocultures of aspen (Figure 1) and white spruce (Figure 2) with that of mixed stands (Figure 4), it appeared that mixedwood stands of white spruce and aspen had higher stemwood biomass yield than pure stands of aspen or white spruce.

TABLE I

Simulated effects of altered winter photosynthesis on stemwood biomass yield of white spruce and aspen. Aspen was harvested at 60 years and spruce at 120 years.

	Biomass (Mg ha^{-1})	
Percent of winter photosynthesis	Spruce	Aspen
2% for aspen, 15% for white spruce	91.4	124.1
2% for aspen, 20% for white spruce	93.7	116.7
2% for aspen, 25% for white spruce	95.5	110.9

In FORECAST, competition between plants is a function of the relative heights of their canopies, their photosynthetic light saturation curves, their nutrient uptake demands and their soil occupancy by fine roots (Kimmins, 1993). Although there is no simulation of how climate change might alter competitive relationship among species, it is possible to explore one aspect of the possible effects of climatic change on competitive relationships between deciduous and coniferous species: the effects of warmer winter on winter photosynthesis of conifers, and the consequences for competition with deciduous species. In the model input file, one can alter the percentage of total annual photosynthesis by evergreen species that was conducted in the period when deciduous competitors are leafless. Runs were made using three different winter photosynthesis percentage values (Table I). The results showed that as the percentage of winter photosynthesis by evergreen white spruce was increased, its stemwood biomass also increased. It might be inferred from this that if the winter climate in boreal forest becomes warmer (i.e., warmer temperature and shorter winter) white spruce will become more competitive relative to aspen because of increased winter photosynthesis.

Acknowledgments

Funding for this project was provided by the British Columbia Ministry of Forests and the Canada-British Columbia Partnership Agreement on Forest Resource Development (FRDA II, 1991–1994).

References

Alban, D. H. and Pastor, J.: 1993, *Can. J. For. Res.* **23**, 1744-1749.

Berg, B. and Tamm, C. O.: 1994, *Scand. J. For. Res.* **9**, 99-105.

DeLong, C.: 1991, *Dynamics of Boreal Mixedwood Ecosystems.* FRDA Rep. **164**, British Columbia Min. For. Res. Br., Victoria, Canada, 137 pp.

Gholz, H. L., Grier, C., Campbell, A. G. and Brown, A. T.: 1979, *Equations and their Use for Estimating Biomass and Leaf Area of Plants in the Pacific Northwest.* For. Res. Laboratory, Oregon State University, Covallis. USA, 37 pp.

Lieffers, V. J., Macdonald, S. E. and Hogg, E. H.: 1993, *Can. J. For. Res.* **23**, 2070-2077.

Kimmins, J. P. and Hawkes, B. C.: 1978, *Can. J. For. Res.* **8**, 265-279.

Kimmins, J. P., Scoullar, K. A. and Apps, M. J.: 1990, *FORCYTE-11 User's Manual for the Benchmark Version. Project Report.* ENFOR Project P-370 For. Can., Northern For. Cent., Edmonton, Canada, 294 pp.

Kimmins, J. P.: 1993, *Scientific Foundations for the Simulation of Ecosystem Function and Management in FORCYTE-11.* Northwest Region, Inf. Rep. NOR-X-328, 88 pp.

Lord, T. M. and Green, A. J.: 1986, *Soils of the Fort St. John-Dawson Creek Area, British Columbia.* Agriculture Canada, Report No. 42, Victoria, Canada, 130 pp

Peterson, E. B. and Peterson, N. M.: 1992, *Ecology, Management, and Use of Aspen and Balsam Poplar in the Prairie Provinces.* Special Report, Northwest Region, Forestry Canada, Edmonton, Canada, 252 pp.

Peterson, E. B., Kabzems, R. D. and Peterson, N. M.: 1989, *Hardwood Management Problems in Northeastern British Columbia: An Information Review.* FRDA Rep. 066 For. Can. and Min. For. Victoria, Canada, 77 pp.

Prescott, C. E., Taylor, B. R., Parsons, W. F. J., Durall, D. M. and Parkinson, D.: 1993, *Can. J. For. Res.* **23**, 1576-1586.

Rustad, L. E.: 1994, *Ecology* **75**, 867-879.

Safford, L. O. and Bell, S.: 1972, *Can. J. For. Res.* **2**, 169-172.

Van Cleve, K., Oliver, L., Schlentner, R., Viereck, L. A. and Dyrness, C. T.: 1983, *Can. J. For. Res.* **13**, 747-766.

Viereck, L. A., Dyrness, C. T., Van Cleve, K., and Foote, M. J.: 1983, *Can. J. For. Res.* **13**, 703-720.

Wang, J. R., Zhong, A. L., Comeau, P., Tsze, M. and Kimmins, J. P.: 1994, *Aboveground Biomass and Nutrient Accumulation in an Age Sequence of Aspen (Populus Tremuloides) Stands in the Boreal White and Black Spruce Zone, British Columbia.* Contract Report. Ministry of Forests, Canada.

Yarie, J. and Van Cleve, K.: 1983, *Can. J. For. Res.* **13**, 767-772.

APPLICATION OF SCIENCE TO ENVIRONMENTAL IMPACT ASSESSMENT IN BOREAL FOREST MANAGEMENT: THE SASKATCHEWAN EXAMPLE

H. T. EPP

Saskatchewan Environment and Resource Management 3085 Albert Street Regina, Saskatchewan S4S 0B1, Canada

Abstract. A major objective of environmental impact assessment (EIA) is prediction. Two major goals in science are: (1) explanation of observed phenomena, and (2) prediction. The EIA process makes use of the methods of academic science to provide the explanatory framework for the required data gathering which is needed for the scientifically valid prediction of effects. Follow-up monitoring after a development proceeds is required to test the predictions for their accuracy, important in both scientific and decision-making aspects of EIA. This testing is needed not only for future EIAs in the area, but also for ongoing project management. In boreal forest management, where EIA prediction is ecosystem-oriented rather than site-specific only, the EIA process needs to be tied to long-term, ecosystem-based forest management plans. This must be done via ongoing tests of predictions, and then by feeding the test-derived information back into a dynamic decision-making process. In Saskatchewan this unique and holistic application of the EIA process to long-term forest ecosystem management has been initiated by applying EIA to the twenty-year management plan required by the holder of any forest management agreement (FMA). This is an example of harnessing scientific predictive and testing capacity of the EIA process to set the stage for ecologically sustainable, integrated forest resource management. This also is an example of local action to address a global problem by ensuring the continued integrity of ecological processes in a portion of Canada's boreal forest.

Keywords. ENVIRONMENTAL IMPACT ASSESSMENT, BOREAL FOREST, MANAGEMENT, SUSTAINABILITY, SCIENCE.

1. Introduction

The central goal of science is to increase and make more precise human knowledge; the means to this end is explanation of observed phenomena (Caws, 1969; Romesburg, 1991). What sets scientific explanation apart from other forms of explanation is its restriction to observed phenomena and its rigid adherence to formal logic.

What constitutes knowledge has been the subject of endless philosophical debate over the years. Many argue that science is only one way to knowledge, and that it may be derived traditionally via trial-and-error, or via intuitive or spiritual insight, depending on the viewpoint (Knudtson and Suzuki, 1992). In any case, even inductively derived generalizations are subject to culture-bound precepts, and must be formulated carefully to minimize bias. This paper is limited to application of scientific testing of information pertinent to environmental impact assessments.

The full sequence of scientific explanation begins with induction, observing patterns in technically accurate data sets. The next step is deductive, drawing a logical conclusion from regularities observed in the information base. This conclusion is an hypothesis, an untested explanation. Testing the hypothesis via experimentation or rigidly controlled, repetitive observations hones the explanation to being as near to a true statement, or a natural law, as is possible given available information and technology. The best testing procedure is error elimination, that is, to deduce all possible alternative hypotheses, and then attempt to falsify them all in turn. When all competing hypotheses have been eliminated save one, then that one must be the

Water, Air and Soil Pollution **82**: 179–188, 1995.

nearest to the true explanation possible at that time (Popper, 1959; Haila, 1982). When new information becomes available, the testing or verification process may begin again. Care must be taken to ensure that all hypotheses are testable, however, as untestable hypotheses remain speculation only.

Currently, Popper's (1959) thesis that predictions/hypotheses falsification leads to superior science is under some assault, because "persuasive deductions" from known facts may lead to equally valid science (Brush, 1989). Nevertheless, due to the fact that effects prediction is central to EIA, predictions falsification is the most useful test of resource management plans which aim to avoid deleterious effects on real, living ecosystems under anthropogenic stress.

Environmental impact assessment (EIA) provides major input into many development decision-making processes, entrenched in legislation in many jurisdictions around the world. EIA makes use of reductionist science to provide the building blocks for a more holistic synthesis for sound information decision-making. Unfortunately, the new information generated during the EIA process tends to be inductively derived, so that the predictions often are untested or, at best, partially verified, hypotheses, depending upon the amount of relevant information previously available.

Inherently, EIA is predictive, predicting effects of proposed developments. Two scientific weaknesses are built into the EIA process, however. One, predictions of effects cannot be tested before the decisions have to be made. The reason for this problem is that the environmental effects cannot occur before the project begins, and the decision has to be made before that. Two, after making predictions an EIA proposes mitigation procedures to minimize or eliminate the predicted effects, yet this is done prior to knowing what will be the effects. The resulting difficulty is that undesirable effects have to occur to establish any cause-effect relationship, yet the purpose of the exercise is to avoid the effects. Hence, EIA predictions or hypotheses can be tested only via post-EIA follow-up monitoring. But the EIA process rarely extends beyond the decision, with the result that EIA synthesis and decisions tend to be based on incomplete science, untested predictions. These are inherent weaknesses universal to EIA processes, clearly limiting application of science to EIA, numerous attempts to the contrary notwithstanding.

Having identified the need for improving the scientific input into the EIA process, the objectives of this paper are to: (1) prepare a design to scientifically test EIA predictions, (2) illustrate the design by applying it to test cases in Saskatchewan's boreal forest, and (3) address the two inherent scientific weaknesses in EIA to improve the process.

2. Testing Environmental Assessment Predictions

Curtis (1994), has identified EIA predictions as either qualitative or quantitative, with the latter being by far the more common. Unfortunately, qualitative predictions tend to be the more difficult to test, and many are untestable. Consequently, a more scientific EIA process which relies on testing predictions as hypotheses will require increased care in formulating predictions and in ensuring technically accurate data.

Effects predictions can be tested at one of two places during the EIA process: (1) by post-EIA monitoring and then testing the original EIA predictions against the pre-EIA baseline (Beanlands and Duinker, 1983; Kennedy and Ross, 1992); and (2) by treating project or plan site alternatives as major hypotheses, and then subjecting each to rigid falsification attempts to select the most environmentally sound option(Epp and Curtis, 1994). The former approach has the weakness of limiting use of alternative hypotheses, hence not leaving room for hypothesis falsification. The latter approach has the strength of a high level of scientific procedural rigour during the EIA rather than post-EIA; this strengthens the scientific nature of the decision-making process, but it still does not encompass falsification of more specific alternative ecological effects predictions. Obviously, the ideal approach incorporates both procedures. The test cases discussed in this paper explore the application of exactly this ideal approach.

The act of falsification follows an ordered sequence of events: (1) several competing generalizations are induced from preliminary data focused on an overall goal; (2) these generalizations are expressed as alternative preliminary explanations, untested hypotheses; (3) specific events trajectories or test predictions, which follow logically from the hypotheses, are deduced from the generalizations; (4) these predictions then are tested via experimentation or replication in actual relevant events, with the test results determining the truth or falsity of the predictions; (5) testing the predictions helps to establish causal or statistical, probabilistic relationships with the observed effects; (6) the results of these tests then are applied to each alternate hypothesis to determine if it is untrue. Verified predictions establish proximal causalities which, in turn, assist in confirming hypotheses for broader cause–effect relationships. This method of scientific testing, termed hypothetico-deductive (h-d), is the most rigorous means of establishing explanations as near to the truth as possible given the present level of knowledge (Haila, 1982; Romesburg, 1991; Epp and Curtis, 1994).

Recognizing the fact that total objectivity in deriving generalizations is unattainable, it is important to minimize this potential weakness by preparing competing hypotheses. Even then the potential for bias exists, but this can be dealt with by acknowledging that only an approximation of the truth or reality of an event is obtainable given that the event and the observer never are totally separable.

Application of this h-d procedure to EIA requires that the actual pre-development assessment and post-development monitoring be combined into one information generating and evaluation process. The first three events in the above-noted sequence normally would occur during the EIA itself, and the latter three would apply to post-EIA monitoring. This combination is relevant especially to EIAs of resource or ecosystem management plans where development is continuous, and is less applicable to site-specific projects isolated in time and space. That is why this procedure lends itself well to ecosystem and sustainability oriented boreal forest management.

3. Integrated EIA and Forest Management Planning

Forest management, in response to societal demands, recently has undergone a paradigm shift from emphasis on maximum timber productivity and sustained yield to an emphasis on (1) maintaining the forest and its site-specific diversity as an interconnected whole, the holistic view; (2) sustainable ecosystem management; and

(3) increased public involvement in decision-making (World Commission on Environment and Development, 1987; Gillis, 1990; Hammond, 1991; Rowe, 1992; Cortner and Moote, 1994). "*Acceptance of the ecosystem approach establishes common ground for those concerned with forestry, wildlife, water and recreation, thus encouraging partnerships in sustainability*" (Rowe, 1992). On the EIA front, the need to improve the scientific validity of EIA has increased with its rising public profile (cf. Beanlands and Duinker, 1983; Malik and Bartlett, 1993).

In Saskatchewan, boreal forest management provides an ideal case for testing EIA predictions and, thereby, improving the scientific input into forest management and the specific EIA process itself. Long-term (twenty-year) forest resource management plans are required under any forest management agreement (FMA) between a forest harvesting company and the provincial government, and each plan must include an EIA. Under this long-term area plan and EIA, much shorter term (five-year and annual) site-specific management plans are required; here is where the recommendations based on predictions are implemented, providing ready made testing sites. In the far north of the province, well north of the traditional "commercial" forest, uranium mining affects the boreal forest ecosystem. Each new mine requires a major EIA for consideration for approval. A new initiative begun by Saskatchewan Environment and Resource Management in consultation with local communities and in co-operation with industry, is post-EIA monitoring of traditionally valued ecosystem components and indicator species to test EIA and cumulative effects predictions. A third post-prediction testing initiative involves testing the predicted effects of spraying *Bacillus thuringiensis* var. *kurstaki (B.t.k.)* on commercial forest stands to provide biological control of spruce budworm infestations. The effects on non-target Lepidoptera and on their neo-tropical migrant passerine bird predators are not known; the ongoing tests are designed to provide such information.

The purpose of these boreal forest EIAs and post-EIA studies is to indicate how the proponents will meet the overall, future state-of-the forest goal, namely, *that the existing forest landscape, with its species, age-stand, land-water mosaic, and natural ecological processes remain extant in perpetuity*. Meeting this goal should help preserve all habitat types and ecological processes, further ensuring biodiversity for continued ecological and economic stability within an FMA area and mining environs. The mechanism whereby Saskatchewan's EIA process is applied to FMAs and their twenty-year forest management plans is described elsewhere (Epp, 1993). The emphasis here is on developing a means of testing the EIA options and predictions to enhance their scientific validity.

Clearly, the detailed information provided in a twenty-year management plan sets the stage for testing both site-specific cutting alternatives, and plan alternatives like harvesting methods and site options, to provide a scientifically sound approach to meet the overall EIA goal of maintaining ecological processes. Equally clearly, this EIA lends itself to *ongoing testing*, in five-year and annual cutting plan reviews, after completion of the EIA process. For example, each five-year plan should contain cutting area options, all subject to error elimination or falsification procedures, basic hypothesis testing.

The overall goal or end of each combined twenty-year forest management plan/EIA and the other EIAs is to maintain ecological and economic stability, a goal based on the assumption or premise that fundamental ecological process changes are undesirable environmentally. Ecological processes are landform and climate dependent. Broader than landform is the concept of landscape, now recognized as an ecosystem encompassing the human element, composed of the "geosystem, biosystem and anthroposystem" (Leser and Rodd, 1991). Hence, the general *means to the end of ecological process maintenance is to maintain landscape integrity* (Epp, 1992). The purpose of the EIA is to provide the information needed to implement the means, thereby contributing to meeting the overall end.

In order to activate the means, then, scientifically sound, verified, information is required. The EIA provides a baseline from which may be formulated generalizations or hypotheses for testing, an ongoing process.

4. Testing Procedures

Any scientific hypothesis testing procedure requires a clear statement of the overall goal and assumptions or premises under which the hypotheses are generated from preliminary data, and recognition of the alternative hypotheses. An objective test requires a clear understanding of the assumptions underlying the goals, and their explicit recognition and enunciation (Baskerville, 1994). In the case of Saskatchewan's boreal forest FMAs, the information generated by the EIAs and subsequent annual cutting plan reviews will provide an opportunity to test fundamental forest management options as alternative or competing hypotheses. A goal and assumptions for such a project described below, form the basic premise. Following this, examples of fundamental forest management options are presented as alternate or competing hypotheses for potential long term-testing. Hypothetical examples of more site-specific ecological alternatives are presented as hypotheses or possibilities for more immediate testing. Test predictions, then, are deduced, ready to be subjected to falsification procedures. In addition, an example is provided of an ongoing study of the effects of anti-spruce budworm spraying over various infested forest stands within an FMA.

4.1. General Goal

Maintain ecological stability within the FMA area, which is desirable from the economic, traditional use, and wilderness perspectives.

Assumptions: (i) Ecological and economic stability in the FMA area are dependent upon long-term maintenance of ecological processes. (ii) Maintenance of extant ecological processes is dependent upon maintenance of the existing ecological mosaic or patch dynamics, that is, landscape integrity. (iii) The landscape mosaic is controlled largely by fires caused by random lightning strikes.

Hypothesis: Clearcutting based on timber stand selection, as a harvesting method, maintains these ecological processes best while permitting economic use of the forest. Hence, clearcutting promotes optimal long-term ecological and economic stability.

Test Prediction: Clearcutting emulates natural burns in size, shape, and age of tree stands affected, so forest regeneration will proceed via successional stands similar to what occurs naturally. Application of this forest management method on a sustainable basis results in interconnected habitats sufficient to maintain the existing forest mosaic, its ecological processes, its biodiversity, and its economic potential.

Alternative Hypothesis: Modified patch cutting based on ecologically defined patches, as a harvesting method, maintains these ecological processes best while permitting economic use of the forest. Hence, modified patch cutting promotes optimal long-term ecological and economic stability.

Alternative Test Prediction: Clearcutting is not done randomly, so the resulting new landscape will not emulate one where random lightning-caused fires occur. Modified, ecological patch based cutting, based on existing fire-caused stands, more nearly emulates natural burns in size, shape, and age of tree stands affected, so forest regeneration will proceed via successional stands similar to those occurring naturally. When trees are harvested sustainably in this manner, interconnected habitats sufficient to maintain all ecological processes, biodiversity, and economic potential will be retained.

4.2. Specific Goal (Hypothetical Example)

Return aspen stands in Saskatchewan's FMA areas to the closest possible approximation of pre-harvesting stand variability and relationships with topography and other forest stands.

Assumption: The best method of ensuring natural aspen regeneration following clearcut harvesting is to allow suckering from the roots.

Hypothesis: The most efficient way to ensure aspen regeneration in Saskatchewan's FMA areas to an approximation of the original stand following clearcut harvesting is to (i) thin the stand if regrowth is very thick, and (ii) remove competition, such as shrubs.

Test Prediction: Following clearcutting, aspen will regenerate from the roots in stands so thick that return to natural stand conditions will be very slow, taking many years due to intense competition for space and light.

Alternative Hypothesis: The most efficient way to ensure aspen regeneration in Saskatchewan's FMA areas to an approximation of the original stand following clearcut harvesting is to leave it strictly alone, allowing regeneration to proceed naturally.

Alternative Test Prediction: Following clearcutting, aspen will regenerate quickly to an approximation of the original stand.

4.3. Specific Goal (Example of a Study in Progress)

Provide biological control of spruce budworm infestations in an FMA by spraying *Bacillus thuringiensis* var. *kurstaki (Btk)*, which is lethal to Lepidoptera, over the most heavily infested sites. The environmental assessment review addresses the effects of the spray on migrant bird predators of the Lepidoptera.

Assumption: *Bacillus thuringiensis* var. *kurstaki* specifically kills Lepidoptera, including the target spruce budworm larvae and non-target Lepidoptera.

Hypothesis: The killing of non-target Lepidoptera, a side-effect of spraying spruce budworm with *Btk,* will significantly lower local populations of insectivorous migrant neo-tropical passerine birds within the area affected by the spray.

Test Prediction: Bird counts after spraying will indicate significantly fewer individuals per unit area affected compared to counts done prior to the initiation of spraying.

Alternative Hypothesis: The killing of non-target Lepidoptera, a side-effect of spraying spruce budworm with *Btk,* will not significantly lower populations of insectivorous migrant neo-tropical passerine birds within the area affected by the spray.

Alternative Test Prediction: Bird counts after spraying will indicate no significant difference in numbers of individuals per unit area compared to counts done prior to initiation of spraying.

Hypothesis testing proceeds via attempting to show each test prediction to be false. Falsifying one prediction leads to the logical deduction that its alternate must be true, or at least as near to the truth as possible until data are sufficient to deduce new predictions from the hypotheses. The procedure outlined here is provided as an example of how improved scientific testing during and after the EIA process can add useful information for ecologically sustainable boreal forest management. The Saskatchewan forest management process, which incorporates EIA directly into planning, lends itself well to the application of such rigorous scientific testing. This should enhance the validity of both the EIA and planning processes and, consequently, overall management. The result should be a dynamic plan that is *adaptive* in Holling's (1978) sense of the word, that is, it incorporates prediction, testing, application of new knowledge on an ongoing basis, post-EIA monitoring for a continuing information base, and plan modification incorporating the new information.

5. Conclusions

Science, as logical explanation, can help EIA by establishing and explaining causal relationships between environmental disturbance and ecological effects. Using boreal forest management as an example, this paper provides a means for applying science to the EIA process as a vehicle to effect the new environmental management paradigm, namely, environmentally and economically sustainable ecosystem management along with increased public involvement in decision-making. Sound, sustainable ecosystem management requires scientific predictions testing, which this procedure provides, and applying the EIA process to long-term forest management plan formulation ensures that the public involvement inherent in the EIA process provides public input to the plan.

Clearly, mitigation should be an ongoing process following any EIA, making the case for continued post-EIA testing. Using a Saskatchewan boreal forest test case, the EIA follow-up testing procedures outlined here indicate how it is possible to establish the difference between the pre-development baseline environment and the post-development environment caused by the development and by mitigation, providing a means for validating or rejecting mitigation methods. Hence, this paper meets the first two goals of this discussion, namely, to devise a plan or design to test scientifically EIA predictions in Saskatchewan's boreal forest, and to illustrate the design by applying it to test cases.

The testing procedure outlined here helps establish what is the difference between mitigated and non-mitigated development effects by verifying predictions of changes to the baseline environment, leading to establishing optimal mitigation measures. The development and potential development cases used as examples in this paper involve forest harvesting and management activities, including twenty-year forest management plans and spruce budworm spraying. While improving the scientific validity of the EIA process itself via testing fundamental forest management options, this procedure also provides for testing more specific predictions during post-decision or post-EIA follow-up research. In addition, this procedure establishes a scientific base for improving predictions by future EIAs within the general area or within similar environments elsewhere. This meets the third goal of this paper, namely, that it presents a means to improve the EIA process by limiting the two inherent weaknesses in EIA, the need to make decisions before effects predictions can be tested satisfactorily, and the need to propose mitigation methods before being certain about the nature and extent of the effects.

Clearly, the hypothetico-deductive (h-d) scientific procedure, the most rigorous known way to ensure accuracy of information, can and should be applied to EIA. Using boreal forest management in Saskatchewan as its example, this paper has identified places in the EIA process where h-d science can be applied, inherent weaknesses in EIA notwithstanding, including the testing of narrow site-specific predictions and the testing of broad project options. Application of h-d science to EIA does not make the EIA process itself a science, however, as the purpose of EIA remains providing the soundest possible information for decision-making. What h-d science does for EIA is make the decision-making process more *scientific*, enhancing its overall validity and, hence, the soundness of decisions. The value of this lies in the long-term, so that future EIA decisions within a defined area like Saskatchewan's boreal forest will be based on increasingly more precise and valid cause-effect predictions, thus addressing the inherent conundrum in the EIA process as well as improving forest management efficiency.

One of the difficulties of applying hypothesis testing to EIA is that fundamentally EIA is an holistic exercise involving long-term, cumulative, and even synergistic changes which are not easily addressed by the reductionist h-d analytical method. The intent of this paper is not to introduce reductionist methodology into EIA at the expense of the holistic view. Rather, it is necessary at this time, when EIA is maturing rapidly as a decision-making tool, to recognize the need for soundly based information that forms the conceptual building blocks for addressing the holistic concerns of major EIAs. This conclusion is in line with Odum's (1988) theoretical advances which regard ecological systems as non-equilibrium complex, self-organizing systems, with higher levels of organization requiring holistic synthesis from reductionist derived, verified information.

This paper provides a new theoretical perspective on applying science to EIA. Testing of this new approach is ongoing at this time as part of several environmental assessments in Saskatchewan's boreal forest.

Acknowledgments

This paper has been reviewed by Larry Stanley, Forestry Branch, Saskatchewan Environment and Resource Management, and by Barry Peel, Mistik Management Ltd. of Meadow Lake, Saskatchewan, and has benefited substantially from their comments. In addition, I wish to thank Stan Rowe and an anonymous referee for their constructive comments which have further benefited the arguments presented in this paper. The opinions expressed in this paper are my own, however, and do not necessarily reflect those of the Government of Saskatchewan, of Mistik Management, or of any of the reviewers.

References

Baskerville, G. L.: 1994, Who has the best forest management? Faculty of Forestry, Univ. British Columbia, Vancouver, Canada. (Unpublished manuscript.)

Beanlands, G. E. and Duinker, P. N.: 1983, *An Ecological Framework for Environmental Impact Assessment in Canada.* Environment Canada, Hull, Quebec, 132 pp.

Brush, S. G.: 1989, *Science* **246**, 1124-1129.

Caws, P.: 1969, *Science* **166**, 1375-1380.

Cortner, H. J. and Moote, M. A.: 1994, *Environmental Management* **18**, 167-173.

Curtis, F.: 1994, A conceptual framework for environmental effects prediction and verification. Saskatchewan Environment and Resource Management, Regina, Canada. (Unpublished manuscript.)

Dickman, M.: 1991, *Environmental Impact Assessment Review* **11**, 171-180.

Epp, H. T.: 1992, *In:* Ingram G. B. and Moss, M. R. (eds.) *Landscape Approaches and Wildlife Ecosystem Management.* Proc. Second Symp. Can. Soc. Landscape Ecology and Management, Univ. British Columbia, Vancouver, Polyscience Publications Inc., Morin Heights, Canada, pp. 19-26.

Epp, H. T.: 1993, *In:* Marczyk, J. S. and Johnson, D. B. (eds.) *Sustainable Landscapes.* Proc. Third Symp. Can. Soc. Landscape Ecology and Management, Univ. Alberta, Edmonton, Polyscience Publications Inc., Morin Heights, Canada, pp. 69-81.

Epp, H. T. and Curtis, F. A.: 1994, Role of deductive science in environmental impact assessment. Saskatchewan Environment and Resource Management, Regina, Canada. (Unpublished manuscript.)

Gillis, A. M.: 1990, *Bioscience* **40**, 558-562.

Haila, Y.: 1982, *Annales Zoologici Fennici* **19**, 255-263.

Hammond, H.: 1991, *Seeing the forest among the trees: the case for wholistic forest use.* Polestar Press Ltd., Vancouver, Canada, 309 pp.

Harris, W. C., Kabzems, A., Kosowan, A. L., Padbury, G. A. and Rowe, J. S.: 1983, *Ecological Regions of Saskatchewan*. Tech. Bull. 10. Forestry Division, Saskatchewan Parks and Renewable Resources, Regina, Canada, 57 pp.

Holling, C. S. (ed.): 1978, *Adaptive Environmental Assessment and Management.* Wiley, New York, USA, 404 pp.

Kennedy, A. J. and Ross, W. A.: 1992. *Environmental Management* **16**, 475-484.

Knudtson, P. and Suzuki, D.: 1992, *Wisdom of the Elders*. Stoddart, Toronto, Canada, 232 pp.

Leser, H. and Rodd. H.: 1991, *In:* Esser G. and Overdieck, D. (eds.). *Modern ecology: Basic and Applied Concepts*. Elsevier, The Netherlands, pp. 831-844.

Malik, M., and Bartlett, R. V.: 1993, *The Environmental Professional* **15**, 34-45.

Popper, K. R.: 1959. *The logic of scientific discovery*. Hutchinson, London, UK.

Romesburg, H. C.: 1991. *J. Wildlife Manage.* **43**, 293-313.

Rowe, J. S.: 1992. *Forestry Chron.* **68**, 22-224.

World Commission on Environment and Development: 1987. *Our Common Future*. Oxford University Press, Oxford, UK, 400 pp.

PART II

NUTRIENT AND CARBON CYCLING

NITROGEN MINERALIZATION IN BOREAL FOREST STANDS OF ISLE ROYALE, NORTHERN MICHIGAN

R. STOTTLEMYER[1], B. TRAVIS[2], and D. TOCZYDLOWSKI[2]

[1]*National Biological Service, 240 W. Prospect, Ft. Collins, CO 80526,*
[2]*Dept. Biological Sciences, Michigan Technological University, Houghton, MI 49931*

Abstract. The correlation of soil temperature and moisture with inorganic N concentrations and net mineralization beneath major species types in mature boreal and northern hardwood forests was examined over a two year period. Soils beneath species types where the canopy was dominated by *Betula papyrifera, Picea glauca, Alnus rugosa* or, in northern hardwoods, *Acer saccharum* were studied. Net NO_3^- mineralization varied by species type and net total inorganic nitrogen (N) mineralization varied by month and the interaction of species type with month. Soil NO_3^- concentration and NO_3^- mineralization were correlated for spruce, and inversely correlated for alder and maple. Soil NH_4^+ concentration and NH_4^+ mineralization were inversely correlated for alder and maple. In laboratory temperature and moisture treatments of birch, spruce and maple soils, NH_4^+ and total inorganic N-mineralization increased with temperature. The response to moisture was most evident for NO_3^- mineralization in maple soils.

Keywords. NITROGEN MINERALIZATION, BOREAL FOREST, NORTHERN MICHIGAN, NITROGEN CYCLE, CLIMATE CHANGE.

1. Introduction

Linked forest productivity/soil process models suggest that the southern boreal ecosystem fringe could be the most sensitive of eastern North American forests to climate change (Pastor and Post, 1988; Solomon, 1988). The rate of climate change in this region may be more significant than its magnitude (Solomon and Bartlein, 1992). Global warming, and perhaps change in moisture, could increase above-ground primary production or below-ground heterotrophic respiration (Shaver *et al.*, 1992). Which process might increase the more rapidly or greater in magnitude could determine whether the ecosystem is a carbon (C) source or sink, and is a key question in assessing forest ecosystem response to climate change. Change in ecosystem C is also dependent on nutrient status, especially available inorganic N (NO_3^-, NH_4^+) (Burke, 1989; Shaver *et al.*, 1992; Yin, 1992). Forest soil nitrogen (N) availability often limits productivity and therefore C-cycling. There is evidence also that available N pools in forest soils are related to regional temperature gradients (Yin, 1992). To assess the effect of climate shift on the C cycle requires knowledge of how climate change might effect other element cycles, especially limiting nutrients such as N.

Conifers appear more responsive than hardwoods to the addition of N in northern hardwood ecosystems (Aber *et al.*, 1993; Gundersen, 1991). Due to its ecotonal nature, the Upper Great Lakes region is characterised by large differences among forest ecosystems in potential N-mineralization and nitrification rates (Zak *et al.*, 1986). Species variation in response to climate change in the region appears more important than site variation, and groups of species can be separated based upon growth response to climate over time (Graumlich, 1993). Recent evidence in this region exists showing a forest species growth response to the decline in precipitation and increase in temperature of the past decade (Graumlich, 1993; Jones *et al.*, 1993). The complexity

Water, Air and Soil Pollution **82**: 191–202, 1995.

of species interaction with climate change in an ecotonal region, where atmospheric N inputs are also high (NADP, 1991), points out the need for additional study of major forest species responses especially in the southern boreal fringe.

The objective of this two-year study, initiated in 1992, was to relate change in soil inorganic N concentration and net mineralization with soil temperature and moisture beneath dominant species types in mature boreal forest and, for comparison, northern hardwood forest located in Isle Royale National Park, MI.

2. Materials and Methods

The Wallace Lake watershed (48°03'N, 88°38'W) is located in Isle Royale National Park, north-western Lake Superior. The bedrock is metamorphosed ophitic flood basalts covered by scattered alkaline glacial till originally derived from limestone bedrock immediately south of James Bay. The soils are 3000-5000 years old, and classified as sandy to coarse loamy, mixed, frigid Alfic Haplorthods with Histic Humaquepts (Humic Gleysol) beneath alder. They are predominantly of the Michigamme-Arcadian-Rock Outcrop Association (Shetron and Stottlemyer, 1991).

Isle Royale lies within the vegetational transition zone between the boreal forest found on the north shore of Lake Superior and the northern deciduous forest which dominates the lake's south and eastern shores. The 176 ha Wallace Lake watershed is dominated by the boreal forest formation. The overstory is a mixture of white birch (*Betula papyrifera* Marsh.), quaking aspen (*Populus tremuloides* Michx.), white spruce (*Picea glauca* (Moench) A. Voss) and balsam fir (*Abies balsamea* (L.) Mill.). Scattered stands of white cedar (*Thuja occidentalis* (L.)) and alder (*Alnus rugosa* (Du Roi) Spreng. [*A. incana* (L.) Moench]) also exist within the watershed. In total, the forest has 10 canopy species with an aboveground biomass of 108 Mg ha^{-1} (Rutkowski and Stottlemyer, 1993). There has been no known direct human disturbance of the watershed except for a small portion of the drainage burned in 1936.

The study focused on forests where birch or spruce, the dominant hardwood and conifer in Isle Royale boreal forests (Rutkowski and Stottlemyer, 1993), were prevalent in the canopy. Birch prevails on summit shoulders (Natl. Coop. Soil Surv., Agric. Handbook #18, 1994) where soils are thinner, better drained and drier. Spruce prevails on foot or toe slopes where soils are cooler and drainage poorer. Canopy interception of snow by spruce can result in frozen soils (Stottlemyer, 1982). The vegetation age was between 50 and 70 years. Mature spruce stands are very rare apparently due to blowdown (Toczydlowski *et al.*, 1993) assisted by periodic outbreaks of spruce bark beetle (*Dendroctonus rufipennis*). For comparison, within the Wallace Lake watershed we also studied replicated plots dominated by alder, and in the SW end of the park replicated plots where the overstory was dominated by mature (> 150 yr) sugar maple (*Acer saccharum* Marsh.). Alder prevails in lowlands between tongues of exposed bedrock or areas of very shallow soil. Seasonally, alder soils are saturated, cool, but soils do not freeze in winter. Sugar maple is found on a deep, well-drained spodosol with fragipan > 50 cm deep with finer surface texture (silt loam cap) than beneath birch-spruce (Shetron and Stottlemyer, 1991).

Within each major species type, birch, spruce, alder, and maple, three 0.1 ha (20 × 50 m) plots were randomly located along transects for a total of 12 plots. Plots were about 100 m apart. On each plot a partial Li-Cor meteorological station (Model 1000, Li-Cor Corp., Lincoln, NE), consisting of instrument shelter, datalogger, five temperature sensors (thermistors at 5 and 10 cm depth beneath the organic layer, in the forest litter, and at 10 cm and 1.3 m above the forest floor) was installed in the last week of May, and operated until the third week of October each year. A subset (one plot/species type) of dataloggers was left in place for over-winter monitoring in all species types. Daily mean, maximum, and minimum air temperature and daily mean soil temperature were logged throughout the study. Mean temperatures for field incubation periods were determined from daily means. For soil temperatures, the mean temperature at 5 cm and 10 cm depths, and a composite of both was used when assessing field incubation response to temperature.

2.1. Field Incubations

Monthly total inorganic N-mineralization and NO_3^- mineralization in each species type were estimated using the buried polyethylene bag technique (Eno, 1960). Within each 0.1 ha plot, three randomly placed 25 m^2 subplots were established. About the middle of each month from May to October, a site for soil sampling and incubation was randomly selected within each subplot. After pulling back the surface organic layer, the upper 10 cm of mineral soil, consisting of A and upper E horizons, was collected with a 5 cm diameter soil corer. Paired cores were collected in each subplot. One sample from each pair, representing a non-incubated sample, was placed in a Whirl Pac, brought to the field laboratory, and stored at 2 °C. The other was placed in a 0.025 mm thick polyethylene bag and returned to the same hole for incubation. The original surface organic matter was then replaced. After about 30 days, the field incubated sample was collected and returned to the field laboratory for analyses. Due to limited winter access, for over-winter sampling three sets of incubation samples were placed in each subplot in late October. A bag was then removed from each subplot in mid January, mid February, and late April or early May when the monthly incubations resumed.

Following collection, all soil samples were refrigerated for no more than 3 to 4 days at 2 °C. This allowed time for completion of sample collection. Soil moisture was determined by oven drying (105 °C for 24 to 30 h) a subsample with percent moisture equal to the g H_2O/soil dry weight. Bulk density was calculated from total oven dry weight/total soil volume. A larger subsample was then separated in a 2 mm sieve. The > 2 mm fraction was weighed, and discarded. The < 2 mm material was then weighed and divided into two samples. One subsample was extracted for NO_3^- and NH_4^+ using 2M KCl, the other frozen for total N determination. Analytical results were corrected for the > 2 mm gravel content.

2.2. Laboratory Incubations

Two laboratory incubations were conducted following the basic design in Binkley *et al.* (1994). In mid July 1992, three samples from each subplot beneath birch, spruce, and alder were collected for laboratory incubations at 10, 15 and 20 °C (9 samples from each species type at each temperature). The incubations were conducted continuously from 17 July to 3 September, in Percivial growth chambers operating at 75% relative humidity with 16 h of light daily, located at Michigan Technological University. Mid–July samples were picked because the influence of snowmelt on

N-release was gone, optimal field soil moisture conditions existed, and the samples were representative of soil conditions at the midpoint of the growing season before leaf abscission (late August). In this set of incubations, we did not determine possible response to moisture change. There was a 5% moisture loss during the 10 °C and 15 °C incubations, and a 7% loss during the 20 °C incubations.

In mid-August 1993, the above incubations were repeated using three samples from each subplot beneath birch, spruce, and maple, or 27 samples for each species type. Within each of the three temperatures, three moisture treatments on replicated samples for each species type were applied: low moisture which was field moisture; medium moisture which was 15% of total sample weight; and high moisture which was 30% of total sample weight. The "high" moisture content was at field capacity. Samples were incubated continuously for 38 days. The incubator's environment was maintained as in 1992. Following incubations, samples were kept in a dark refrigerator during processing.

2.3. Laboratory Analyses of Soil Samples

Soil sub-samples each weighing 5 g were extracted with 50 ml of 2M KCl, and NO_3^- (cadmium reduction) and NH_4^+ (indophenol) determined on a Lachat autoanalyzer (Lachat Instruments, Milwaukee, WI). The net total inorganic N mineralized for each incubation period was estimated from the difference between initial and final NO_3^- and NH_4^+ content, and net NO_3^- mineralization was estimated by the difference between initial and final NO_3^- content (Binkley *et al.*, 1994). Total N was determined using Kjeldahl digestions (Page *et al.*, 1982). Quality assurance on the Lachat included running control standards every 10 samples, ten duplicates in each block of approximately 40 samples, and 10% reruns. In each total N-digestion block set, six of the 40 samples were duplicates.

2.4. Statistical Analyses

For statistical analyses, in field incubations the three subplot soil samples from each plot were averaged, giving an n of three (three plots) for each species type each sample date. Variances among species types were equal (Bartlett's test for homogeneity of variance). To test for the need to apply repeated measures analyses, each plot was checked for correlation with time (months) for all variables measured. Except for total soil N in two species types, correlation was not significant with time. Therefore, repeated measures analysis was not used. For monthly analyses among species types, results from each 1992 incubation period (month) were combined with those from the same month in 1993. We used Systat Multivariate General Linear Hypothesis (MGLH) ANOVA routines. To examine the effects of vegetation, month, and plots within vegetation, we used the split plot design (analysis of variance for profile data; Morrison, 1967).

For laboratory experiments, both years were analysed separately. Results from the three subplot samples were averaged giving an n of three (three plots) for each species type and for each treatment. Split plot analysis was used. Temperature/moisture chambers were unreplicated, and residual error from the three factor analysis was used in all tests.

3. Results

3.1. Field Temperature, Moisture and Inorganic N Concentrations

Mean soil moisture was highest ($p < 0.001$, $r^2 = 0.44$, $n = 71$) beneath alder (Table I, Figure 1). Soil moisture beneath alder was higher ($p < 0.05$, $r^2 = 0.70$, $n = 12$) all months except October, and near-saturation following snowmelt. Soil bulk density was lowest in alder ($p < 0.001$, $r^2 = 0.23$, $n = 105$). There was no overall significant difference in soil temperature by species type, but in May (Figure 1) soil temperatures beneath spruce were lower ($p < 0.001$, $r^2 = 0.90$, $n = 30$). Total N concentration differed by species type ($p < 0.001$, $r^2 = 0.31$, $n = 62$) with values higher in alder and maple ($p < 0.01$) than spruce and birch.

Soil NO_3^- concentrations were highest beneath alder ($p < 0.05$, $r^2 = 0.70$, $n = 71$) (Table I, Figure 2). Concentrations beneath alder exceeded ($p < 0.05$) those under birch. Concentrations beneath alder were highest ($p < 0.05$, $r^2 = 0.60$, $n = 18$) in June. There was no significant species type with season (month) interaction. Soil NH_4^+ concentration varied by month ($p < 0.001$, $r^2 = 0.85$, $n = 71$), and the interaction of species type and month ($p < 0.001$). Concentrations were higher beneath maple ($p < 0.05$) than birch or alder, and higher ($p < 0.05$) from mid-June to mid-August. Total inorganic N concentration also varied by month ($p < 0.001$, $r^2 = 0.84$, $n = 71$), and the interaction of species type with month ($p < 0.001$). Concentrations where higher beneath maple ($p < 0.05$) than birch or alder, and higher ($p < 0.01$) from mid-June to mid-August.

3.2. Field Incubations

Monthly net NO_3^- mineralization varied by species ($p < 0.05$, $r^2 = 0.56$, $n = 91$) (Figure 2). Net NO_3^- mineralization beneath alder exceeded that of maple ($p < 0.05$) and birch ($p < 0.01$). There was no significant monthly trend or interaction between species type by month. Net NH_4^+ mineralization varied by month ($p < 0.01$, $r^2 = 0.64$, $n = 91$) with May-June ($p < 0.001$) and June-July ($p < 0.01$) amounts exceeding those of July-August. NH_4^+ mineralization also varied by the interaction of species type and month ($p < 0.01$, $r^2 = 0.64$, $n = 91$). Net total inorganic N-mineralization differed ($p < 0.01$, $r^2 = 0.64$, $n = 91$) by month with May-July levels above ($p < 0.01$) those of July–August. The interaction of species type with month was also significant ($p < 0.01$).

There was an inverse correlation between monthly net NH_4^+ mineralization ($p < 0.05$), total inorganic N mineralized ($p < 0.05$), and temperature beneath alder. Monthly soil NO_3^- concentrations and net NO_3^- mineralization were correlated for spruce ($p < 0.05$, $r^2 = 0.25$, $n = 18$), and inversely correlated for alder ($p < 0.001$, $r^2 = 0.81$, $n = 18$) and maple ($p < 0.001$, $r^2 = 0.73$, $n = 18$). Monthly soil NH_4^+ concentrations and NH_4^+ mineralization were inversely correlated for alder ($p < 0.01$, $r^2 = 0.37$, $n = 18$) and maple ($p < 0.01$, $r^2 = 0.41$, $n = 18$). Total monthly soil inorganic N concentration and monthly total N-mineralization were inversely correlated for birch ($p < 0.05$, $r^2 = 0.25$, $n = 18$), alder ($p < 0.01$, $r^2 = 0.48$, $n = 18$), and maple ($p < 0.01$, $r^2 = 0.45$, $n = 18$).

The annual net N mineralized, averaged for all species types, from this study was 1.4 g N m^{-2}. For the two dominant boreal species, birch and spruce, the rate was 1.1 g N m^{-2}.

TABLE I

Mean field soil characteristics by species.

Species type	Percent[1] moisture	Bulk density (g cm^{-3})	NO_3^- (mg N m^{-2})	NH_4^+ (mg N m^{-2})	Inorganic N (NO_3^- + NH_4^+) (mg N m^{-2})	Total N (mg N m^{-2})	Temp. (°C)
By plot							
Birch	23 (20)	0.6 (0.2)	7.5 (13)	152 (138)	160 (145)	144 (89)	
Birch	24 (26)	0.6 (0.2)	4.3 (5)	152 (105)	156 (106)	117 (61)	
Birch	29 (12)	0.6 (0.2)	4.0 (7)	113 (108)	117 (108)	110 (57)	
Spruce	24 (20)	0.7 (0.3)	14.1 (18)	478 (326)	492 (328)	194 (114)	
Spruce	27 (16)	0.6 (0.2)	4.9 (6)	140 (86)	145 (87)	137 (85)	
Spruce	22 (29)	0.7 (0.2)	6.5 (7)	130 (115)	136 (119)	135 (71)	
Alder	42 (16)	0.6 (0.2)	64.0 (54)	186 (120)	250 (161)	388 (386)	
Alder	52 (11)	0.4 (0.2)	38.7 (26)	123 (136)	161 (156)	367 (253)	
Alder	46 (10)	0.5 (0.1)	67.9 (73)	141 (69)	209 (133)	233 (153)	
Maple	31 (4)	0.8 (0.1)	37.5 (118)	325 (351)	362 (347)	333 (29)	
Maple	31 (3)	0.7 (0.1)	1.1 (2)	664 (1094)	665 (1094)	418 (140)	
Maple	36 (5)	0.7 (0.2)	12.6 (26)	309 (244)	321 (240)	426 (70)	
			By vegetation[2]				
Birch	25 (16)[b]	0.6 (0.2)[b]	5.2 (9)[a]	139 (118)[c]	144 (121)[b]	124 (68)[b]	8.8 (5)[a]
Spruce	25 (22)[b]	0.7 (0.2)[b]	8.6 (12)[a]	255 (264)[b]	264 (268)[b]	155 (100)[b]	5.7 (5)[a]
Alder	47 (13)[a]	0.5 (0.2)[c]	56.5 (55)[a]	150 (115)[c]	207 (153)[b]	329 (198)[a]	6.8 (4)[a]
Maple	33 (5)[b]	0.7 (0.1)[a]	17.2 (70)[a]	437 (685)[a]	454 (681)[a]	392 (89)[a]	8.8 (5)[a]

1 Percent moisture is by weight determined gravimetrically. Standard deviation is in parentheses.

2 Where letters differ there are significant differences ($p < 0.05$) among means.

3.3. Laboratory Incubations

In the 1992 laboratory incubations, where the only treatment was temperature (Figure 3), among species types there was an increase ($p < 0.05$) in net inorganic N-mineralization with temperature increases from 10 to 15 °C and 10 to 20 °C. There was a significant difference in response by species type ($p < 0.001$). This was principally due to the increase in net inorganic N-mineralization for birch at the 15 °C ($p < 0.001$) and 20 °C ($p < 0.05$) incubations, spruce at 15 °C ($p < 0.01$), and alder at 15 °C ($p < 0.05$). There were no significant interactions between species type and temperature.

The 1993 laboratory incubations, where both temperature and moisture were manipulated, showed a significant increase among birch, spruce and maple in net NH_4^+ ($p < 0.01$, $r^2 = 0.51$, $n = 81$) and total inorganic N-mineralization with temperature ($p < 0.01$, $r^2 = 0.32$, $n = 81$), and a possible increase in net NO_3^--mineralization ($p < 0.05$, $r^2 = 0.17$, $n = 81$) with moisture. Most of the gain in total inorganic N-mineralization occurred at the mid-temperature range (10–15 °C) and mid-moisture level.

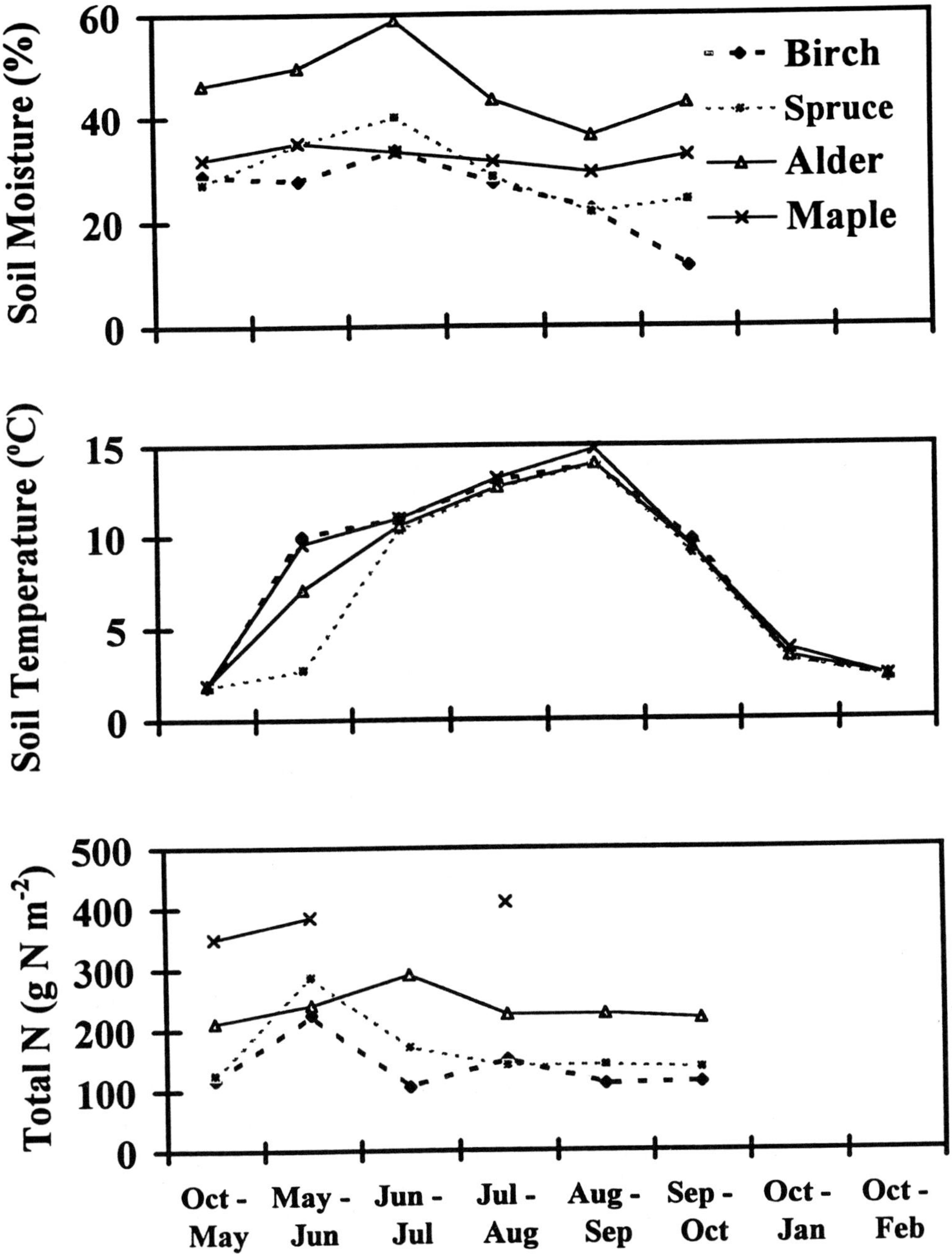

Fig. 1. Mean soil moisture, temperature at 5 cm depth, and total N concentration, for field incubation periods and species type. Values here and in Figures 2 and 3 are averages for the same time periods (months) from 1992 and 1993.

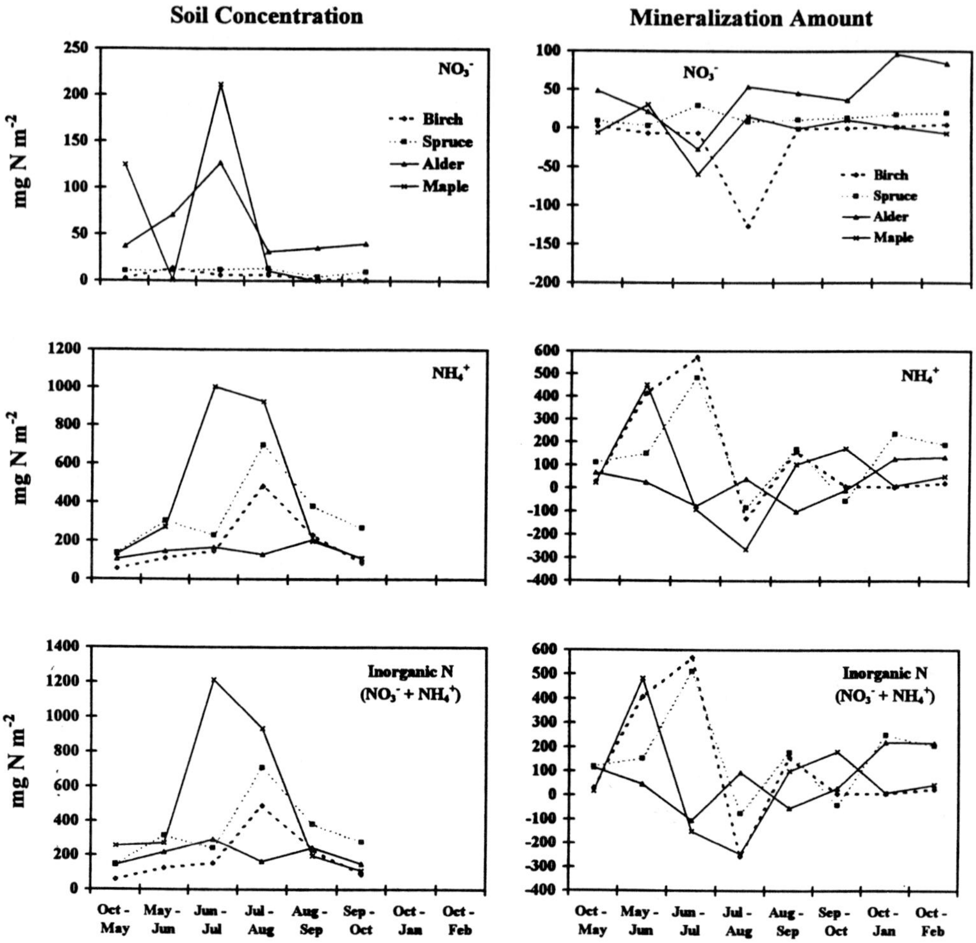

Fig. 2. Mean initial soil concentrations and monthly net amount mineralized for NO_3^-, NH_4^+, and total mineral N in soil beneath birch, spruce, alder, and maple, 1992-1993.

By species type, net NO_3^- mineralization increased for maple with temperature ($p < 0.01$, $r^2 = 0.67$, $n = 27$), moisture ($p < 0.01$), and the interaction of temperature and moisture ($p < 0.05$). The response was most prevalent at mid-temperature and all moisture levels. Soil net NH_4^+ mineralization increased in birch ($p < 0.01$, $r^2 = 0.64$, $n = 27$) and spruce ($p < 0.05$, $r^2 = 0.45$, $n = 27$) with temperature. Total net soil inorganic N-mineralization increased in birch ($p < 0.001$, $r^2 = 0.64$, $n = 27$) and maple ($p < 0.01$, $r^2 = 0.53$, $n = 27$) with temperature, and in spruce with moisture ($p < 0.05$, $r^2 = 0.47$, $n = 27$).

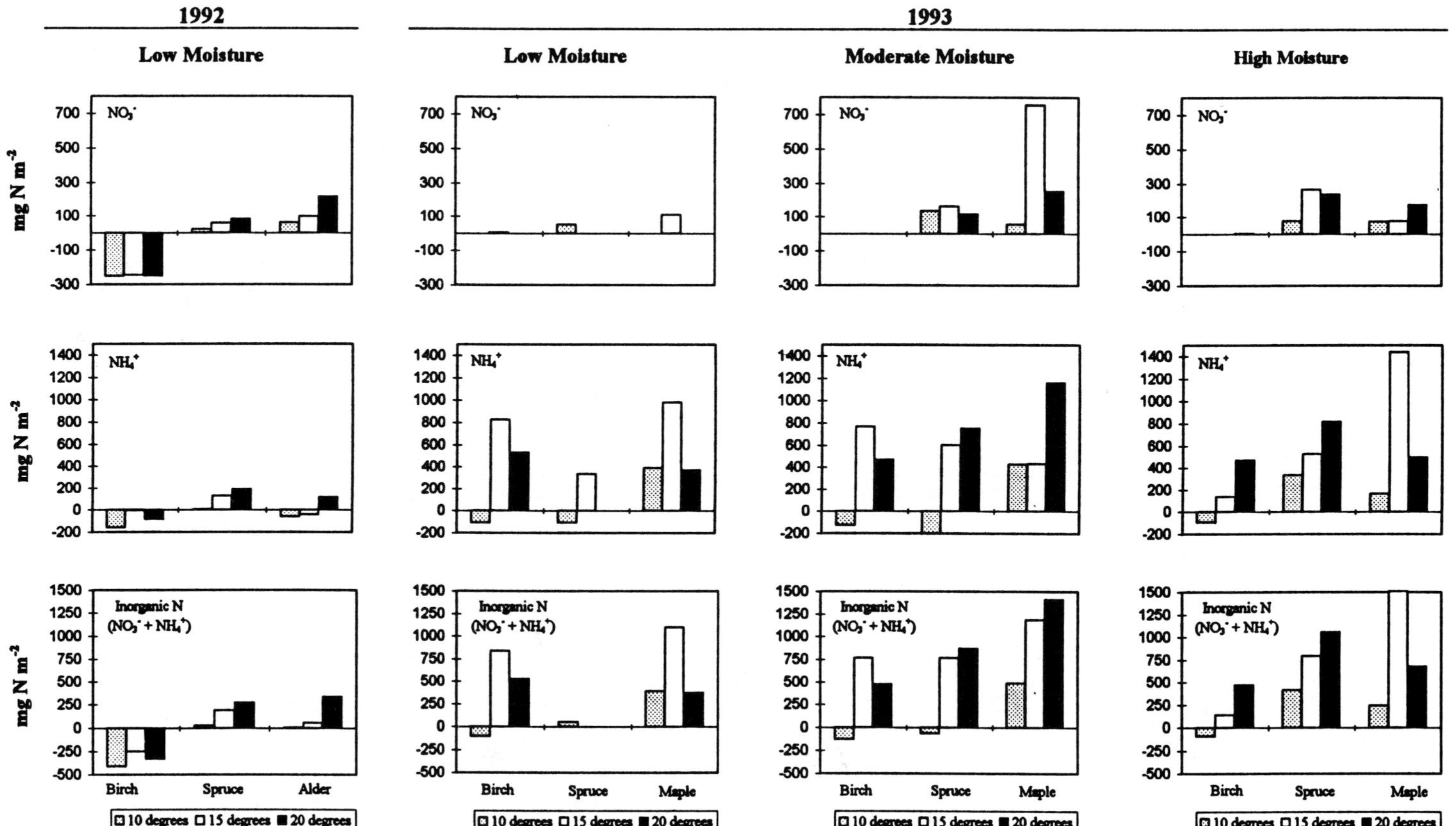

Fig. 3. Laboratory net NO_3^-, NH_4^+, and total inorganic N-mineralization amounts at low (field) moisture with temperature treatments, 1992, and low, medium, and high moisture and temperature treatments, 1993. Both incubation periods were 38 days.

4. Discussion

4.1. Soil moisture, temperature and inorganic N

The higher surface soil moisture content observed from mid-May through June reflected snowmelt inputs which occur up through mid-May (Stottlemyer and Toczydlowski, 1991), increased seasonal rain input, and low evapotranspiration due to the absence of leaves which just begin to appear in the understory by mid-May. Beneath alder, soils were saturated to within 5 cm of the surface into June. Site conditions and low soil bulk density likely contribute to the high soil moisture beneath alder. Characteristic of alder stands on Isle Royale, these plots were located on low elevation, relatively flat sites once flooded by beaver dams. The general soil moisture increase on plots in fall reflected increased precipitation, reduced evapotranspiration, and lowering temperatures. The most uniform moisture content was among maple plots while birch was the most variable. Birch is characterised by high spatial variation in distribution of understory biomass (Rutkowski and Stottlemyer, 1993), and its canopy is patchy which can result in high variation in radiation input.

The temperature increases in May reflected the absence of shading beneath hardwoods. The reduced rate of temperature increase beneath alder likely reflected both the high soil moisture and low site elevation which permits pooling of cool air.

The seasonal trend in inorganic N concentration was similar to other studies in unmanipulated forests (Nadelhoffer *et al.*, 1984; Pastor *et al.*, 1984). NH_4^+ concentrations were larger and more variable than NO_3^- with the highest levels of total inorganic N in the spring and early summer and lowest in mid-summer.

4.2. Field incubations

Gross mineralization rates were not measured in this study. While studies in some ecosystems have shown a relationship between gross and net mineralization rates (Burke *et al.*, 1989), recent studies in forest soils do not find much correlation (Davidson *et al.*, 1992; Hart *et al.*, 1994). The absence of gross mineralization data could limit interpretations, especially with regard to processes, from data sets confined to net mineralization rates. However, compared with soil microbial immobilisation and mineralization, and plant uptake, the quantitative effect of other processes on soil N-availability is generally small (Yin, 1992). So net N-mineralization data do provide a useful index of N-supply since they primarily represent the cumulative difference between N-immobilisation and mineralization.

The inverse correlation of monthly soil inorganic N concentrations and mineralization beneath alder and maple indicate the potential importance of microbial immobilisation especially later in summer and over-winter. Except for alder, high mineralization appeared to be followed by about a month with increased soil inorganic N concentrations (Figure 2).

The low mean daily soil temperatures during winter apparently inhibited microbial immobilisation more than mineralization. This was most evident beneath alder. With warming soil temperatures beneath alder, there were declines in monthly net inorganic N-mineralization during May, June, and early July. This suggests soil microbial immobilisation exceeded mineralization with warming temperatures, and that readily mineralized organic N became limiting. Monthly net N-mineralization also declined beneath maple after mid May and beneath birch and spruce after mid June. This was concurrent with warming soils and declines in soil moisture. The decline in net inorganic N-mineralization with moisture for maple and especially spruce was consistent with the 1993 laboratory incubations. The spring release of snowpack inorganic N solutes (Stottlemyer and Hanson, 1989), which can be up to 150 mg N m^{-2}, would have added to the available soil inorganic N. Much of this is NO_3^-, and soil NO_3^- concentrations did increase slightly by mid-May except under maple.

Over the relatively short one-month incubation periods during and following the growing season, June—September, net NO_3^- mineralization beneath spruce and alder remained small (spruce <10 mg N m^{-2}, alder <50 mg N m^{-2}, Figure 2). Other studies suggest that NO_3^- accumulations beneath conifers often do not occur at all (Aber *et al.*, 1989). Beneath alder, N fixation may have accounted for the high soil NO_3^- levels relative to other species types. If one estimates annual alder N-fixation using the approach of Binkley (1986), based upon alder foliage biomass in this watershed (Rutkowski and Stottlemyer, 1993) fixation would be approximately 1.5 g N m^{-2}. This is at the lower end of the range in fixation rates for this species.

Annual rates of N-mineralization for northern temperate forests range from 2–12 g N m^{-2} (Nadelhoffer *et al.*, 1983; Pastor *et al*, 1984, 1993). The annual net N-mineralization, composited for all species, from this study was at the low end of this range as might be expected since most species were boreal. The composite for the two dominant boreal species, birch and spruce, was below that found on low productivity sites in another study on Isle Royale (Pastor *et al.*, 1993), and more similar to that observed in inland Alaskan taiga sites (Hart and Gunther, 1989; Stottlemyer, 1992).

4.3. LABORATORY INCUBATIONS

The laboratory incubations showed similar, relative trends with temperature. Net mineralization for all species types increased with temperature especially in the range 10 to 15 °C. In laboratory studies, Nadelhoffer *et al.* (1991) found a similar magnitude of response in net N-mineralization in arctic soils with a temperature change from 9 to 15 °C. Such results suggest that, for these northern soils, when temperatures approach 10 °C, a threshold is reached above which mineralization rates increase more rapidly than microbial immobilisation.

In the paired field and laboratory incubations during mid-summer 1992 and 1993, the observed higher laboratory net mineralization rates relative to the field rates appears common to such studies. All methods used to assess N-mineralization are but indices (Binkley and Hart, 1989). Laboratory conditions represent the ideal where moisture and temperature levels can be kept constant likely resulting in mineralization values greater than actually occur in the field. Also net mineralization change over one incubation period does not show the change in gross mineralization rates which over time could affect net mineralization rates as gains in soil microbial biomass come into equilibrium (Hart *et al.*, 1994).

Acknowledgments

Support for this study was provided by the U.S. Dept. of Interior Global Change Program. Watershed monitoring costs were provided by the National Biological Service Watershed Research Program. The authors also thank the staff of Isle Royale National Park for providing logistical support, and Rudy King, Biometrician, USDA Forest Service Rocky Mountain Forest and Range Experiment Station, for assistance on statistical analyses.

References

Aber, J. D., Nadelhoffer, K. J., Steudler, P., and Melillo, J. M.: 1989, *Bioscience* **39**, 378-386.
Aber, J. D., MaGill, A., Boone, R., Melillo, J. M., Steudler, P., and Bowden, R.: 1993, *Ecol. Appl.* **3**(1), 156-166.
Binkley, D. and Hart, S. C.: 1989, *Adv. Soil Sci.* **10**, 57-112.
Binkley, D., Stottlemyer, R., Suarez, F., and Cortina, J.: 1994, *Ecoscience* **1**(1), 64-70.
Burke, I. C., Reiners, W. A., and Schimel, D. S.: 1989, *Biogeochemistry* **7**, 11-31.
Davidson, E. A., Hart, S. C., and Firestone, M. K.: 1992, *Ecology* **73**, 1148-1156.
Eno, F.: 1960, *Soil Sci. Soc. Amer. Proc.* **24**, 277-279.
Graumlich, L. J.: 1993, *Can. J. For. Res.* **23**, 133-143.
Gundersen, P.: 1991, *For. Ecol. Manage.* **44**, 15-28.
Hart, S. C. and Gunther, A. J.: 1989, *Oecologia* **80**, 284-288.
Hart, S. C., Nason, G. E., Myrold, D. D., and Perry, D. A.: 1994, *Ecology* **75**(4), 880-891.
Jones, E. A., Reed, D. D., Mroz, G. D., Liechty, H. O., and Cattelino, P. J.: 1993, *Can. J. For. Res.* **23**, 229-233.
Morrison, D. F.: 1967, *Multivariate Statistical Methods*, McGraw-Hill Book Co., p. 186-197.
Nadelhoffer, K. J., Aber, J. D., and Melillo, J. M.: 1983, *Can. J. For. Res.* **13**, 12-21.
Nadelhoffer, K. J., Giblin, A. E., Shaver, G. R., and Laundre, J. A.: 1991, *Ecology* **72**(1), 242-253.
Nadelhoffer, K. J., Aber, J. D., and Melillo, J. M.: 1984, *Plant Soil* **80**, 321-335.
NADP: 1991, *NADP NTN Annual Data Summary*, National Atmospheric Deposition Program, Ft. Collins, CO, 475 pp.
Page, A. L., Miller, R. H., and Keeney, D. R.: 1982, *Methods of Soils Analyses, Part 2*, Amer. Soc. Agron., p. 607-616.
Pastor, J., Aber, J. D., and McClaugherty, C. A.: 1984, *Ecology* **65**, 256-268.
Pastor, J., Dewey, B., Naiman, R. J., McInnes, P. F., and Cohen, Y.: 1993, *Ecology* **74**(2), 467-480.
Pastor, J. and Post, W. M.: 1988, *Nature* (London) **344**, 55-58.
Rutkowski, D. and Stottlemyer, R.: 1993, *Amer. Midl. Nat.* **130**, 13-30.
Shaver, G.R., Billings, W. D., Chapin III, F. S., Giblin, A. E., Nadelhoffer, K. J., Oechel, W. C., and Rastetter, E. B.: 1992, *Bioscience* **42**(6), 433-441.
Shetron, S. G. and Stottlemyer, R.: 1991, *Isle Royale National Park soil survey and map*, School of Forestry, Michigan Technological Univ., Houghton, MI 49931, 365 pp.
Solomon, A. M.: 1988, *Oecologia* **68,** 567-579.
Solomon, A. M., and Bartlein, P. J.: 1992, *Can. J. For. Res.* **22,** 1727-1738.
Stottlemyer, R.: 1982, *In*: Herrmann, R. and Johnson, A. I. (eds.), *Acid Rain: A Water Resource Issue for the 80's*, American Water Resources Association, Bethesda, USA, pp. 79-83.
Stottlemyer, R.: 1992, *Arc. Alp. Res.* **24**(4), 291-303.
Stottlemyer, R. and Hanson, Jr., D.: 1989, *Soil Sci. Soc. Am. J.* **53**(1), 270-274.
Stottlemyer, R. and Toczydlowski, D.: 1991, *Biogeochemistry* **13**, 177-197.
Toczydlowski, D., Stottlemyer, R., and Dlutkowski, S.: 1993, *Bull. Ecol. Soc. Amer.* **74**(2), 457.
Yin, X.: 1992, *Can. J. For. Res.* **22**, 707-712.
Zak, D. R., Pregitzer, K. S., and Host, G. E.: 1986, *Can. J. For. Res.* **16**, 1258-1263.

THE BOREAL FOREST TRANSECT CASE STUDY: GLOBAL CHANGE EFFECTS ON ECOSYSTEM PROCESSES AND CARBON DYNAMICS IN BOREAL CANADA

D. T. PRICE and M. J. APPS

Canadian Forest Service, Northern Forestry Centre, 5320 - 122 Street, Edmonton, Alberta T6H 3S5, Canada

Abstract. The Boreal Forest Transect Case Study (BFTCS) is a multi-disciplinary ecological study organised around a 1000 km transect located in central Canada. The transect is oriented along an ecoclimatic gradient in a region likely to undergo significant environmental change within the next few decades, and crosses the climate-sensitive boreal forest biome, including the transitions north and south into tundra and grassland respectively. Originally conceived as an extension to the BOReal Ecosystem Atmosphere Study (BOREAS), the 10-year BFTCS project projects the intensive canopy-scale measurements and modelling advances obtained from BOREAS to a wider range of sites with a longer-term perspective. In addition to considering ecophysiological processes with time-frames of the order of one year or shorter, BFTCS addresses the effects of larger scale, longer term processes including vegetation succession and ecosystem disturbances. The BFTCS currently provides practical linkages among ecosystem monitoring, field experiments and regional scale modelling. It will ultimately provide a knowledge-base of key processes and their environmental sensitivities, and assessments of possible climate feedbacks, which can be used to assess the possible consequences of global change both regionally and globally.

Keywords. BOREAL, FOREST, TRANSECT, ECOSYSTEM, MODEL, CARBON, BOREAS, CANADA.

1. Introduction

Boreal, tundra and subarctic (collectively termed "high-latitude") terrestrial biomes, cover approximately 25% of the Earth's land surface, and are estimated to contain 800–900 Pg carbon (C), or about one third of the global terrestrial ecosystem total (Gorham, 1991; Apps *et al.*, 1993; Woodwell and Houghton, 1993; Chapin and Matthews, 1993). These biomes consist of many ecosystem types, including "warm–dry" and "cool–wet" boreal forests, wetlands and grasslands, with many ecotonal intergrades, typically forming a very heterogeneous landscape. The forest component of this landscape is typically of fire-origin. Hence, the interaction of topography and fire history leads to an intimately mixed mosaic of stands of varying ages and species composition, often interspersed with patches of other vegetation communities. Many of the component ecosystems are expected to be very sensitive to changes in climate, with observable responses likely to occur at all scales of organisation i.e., in stand-level processes (energy and gas exchange) (e.g., Bonan *et al.*, 1990; Oechel *et al.*, 1993), in community-level dynamics (patch-level seeding, competition and succession) (e.g., Zoltai and Vitt, 1990; Price and Apps, 1995, unpubl. ms.) and in the frequency and intensity of disturbances (particularly fire and insect infestations) occurring at the landscape and even regional scales (e.g., Clark, 1990; Kurz *et al.*, 1992, 1995b). The expected sensitivities suggest the general hypothesis that should climate warming occur, as is commonly predicted by general circulation models (GCMs), then the combination of effects at these different scales would cause the structure (i.e., species composition and spatial distribution) of existing northern high-latitude vegetation to

Water, Air and Soil Pollution **82**: 203–214, 1995.

change. This would inevitably lead to significant effects on the functional characteristics of these ecosystems, which could be of considerable socio-economic impact both regionally and globally. For example, Bonan *et al.* (1992) simulated the complete removal of the boreal forest and concluded that effects on surface albedo and roughness might lead to significant cooling of the global climate.

While such a negative feedback may appear possible in the longer term, there is also much concern about the possible short-term (next 50–100 yr) impacts of global change on the apparent role of northern terrestrial biomes as an important contributor to the so-called "missing carbon sink", which may be retarding the accumulation of anthropogenic CO_2 in the global atmosphere (e.g., D'Arrigo *et al.*, 1987; Tans *et al.*, 1990, 1993; Apps *et al.*, 1993). Recent and ongoing work is drawing attention to the possibility that anticipated global changes (i.e., climate warming and land-use conversions, including deforestation) may result in accelerated releases of terrestrial stored C (Woodwell and Mackenzie, 1995; Oechel *et al.*, 1993), causing these biomes to become net sources of CO_2, and hence producing a feed forward to the greenhouse warming effect (Woodwell, 1989; Woodwell and Houghton, 1993; Gorham, 1991; Townsend *et al.*, 1992).

Early analyses of the sensitivity of Canadian high-latitude vegetation zones by Zoltai *et al.* (1991), Rizzo and Wiken (1992), and others, have indicated that significant shifts in vegetation boundaries and productivity would occur under GCM-derived equilibrium climate projections, possibly causing new communities to establish in the present-day boreal region, resembling those currently established under warmer climates further south. Such interpretations are generally based on the arbitrary assumptions that anthropogenic CO_2 emissions will cease at twice the pre-industrial atmospheric concentration, causing the greenhouse warming effect to stabilise, and that vegetation and soils will ultimately develop into new "steady-state" ecosystems determined by a few key climatic factors (e.g., Rizzo and Wiken, 1992; Prentice *et al.*, 1992; Smith *et al.*, 1992). As the paleo-ecological record implies, some existing vegetation communities may disappear altogether, to be replaced by new, previously unrecorded ecosystems (Solomon and Bartlein, 1992; Campbell and McAndrews, 1993).

In reality however, if global warming does occur, it is evident that plausible changes in vegetation structure may create both positive and negative feedbacks. Some feedbacks are likely to result in steady-state endpoints (e.g., changes in average albedo) while others will be more immediate and transient in nature (e.g., CO_2 release from drier organic soils and methane releases from melting permafrost). From the human socio-economic viewpoint, understanding the transient behaviour during the warming period, i.e., the next 50–100 years, will be of far greater significance than the ultimate steady-state condition (which may take several hundred years or longer to attain). Therefore it is clearly important, both scientifically and in the interests of resource management, to better understand the short-term dynamic responses of the boreal vegetation to anticipated global changes.

Shuttleworth (1991) outlined a conceptual model planet (Modellion) in which large-scale simulation experiments could be carried out on Earth's physical climate system. He argued that such model worlds are necessary because of the obvious lack of proper controls on large-scale physical experiments, and because the major impacts of global change need to be known long before they actually occur. Extension of this philosophy to a large regional transect (albeit of an area much smaller than that of the entire planet) can provide an effective means of aligning models with reality. Regional transects allow some physical experimentation (notably field manipulations, as well as long-term ecosystem monitoring over large areas) which can be combined with predictive modelling at high spatial resolutions, to validate predictions of future spatial and temporal dynamics.

2. Transect Description

The Boreal Forest Transect Case Study (BFTCS) is an extensive, long-term study based on a contiguous, mid-continental transect crossing an important climate-sensitive northern terrestrial biome. The major purpose is to support/co-ordinate integrated studies of boreal forest processes, as a means of obtaining information about the current structure and function of the Canadian boreal biomes and, in particular, their probable responses to global change

Covering approximately 1000 by 100 km, the transect is located in a region where the relief is generally flat with only 700 m elevational change and no major geophysical obstacles along its entire length (Figure 1). It is oriented approximately southwest–northeast along a pronounced ecoclimatic gradient, perpendicular to the current vegetation boundaries of the southern boreal forest with the aspen parkland of central Saskatchewan, and of the northern boreal forest with the low subarctic tundra (or "subarctic woodland") vegetation of northern Manitoba (Ecoregions Working Group, 1989).

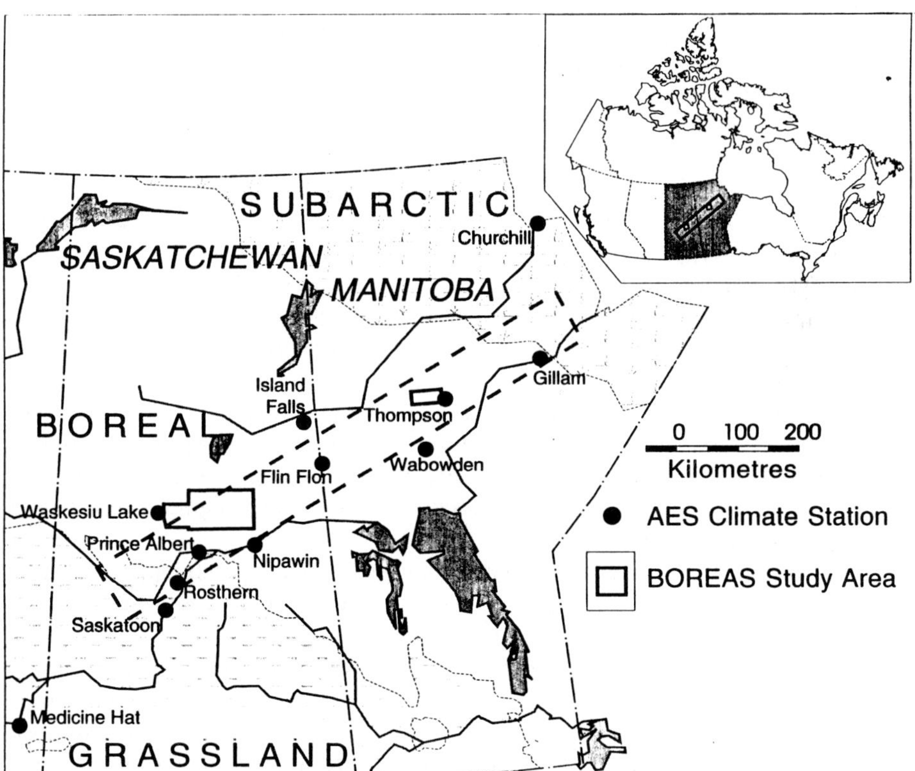

Fig. 1. Location of the Boreal Forest Transect Case Study (BFTCS) in Canada, showing major ecological zonations, positions of local long-term weather recording stations, and the BOREAS study areas.

TABLE I

Climate stations in BFTCS area, January and July mean temperature, annual precipitation and estimated annual potential evapotranspiration (PET): AES, 1951–1980 normals, and GISS 2×CO_2 climate scenario (Hansen *et al.*, 1988). PET was estimated using Jensen *et al.* (1990) following Hogg (1994).

	1951–1980				2×CO_2			
Climate station	January temp. (°C)	July temp. (°C)	Annual precip. (mm)	Annual PET (mm)	January temp. (°C)	July temp. (°C)	Annual precip. (mm)	Annual PET (mm)
Churchill	-27.5	11.8	403	130	-15.7	17.0	520	235
Gillam	-27.1	15.0	450	238	-16.5	18.9	540	342
Thompson	-26.6	15.7	544	295	-17.8	19.7	652	431
Wabowden	-24.5	16.9	464	308	-16.4	20.8	547	446
Flin Flon	-22.4	18.2	474	365	-14.9	21.8	563	492
Waskesiu Lake	-20.7	16.3	464	385	-13.3	19.8	549	522
Nipawin	-21.2	18.4	403	462	-14.6	21.6	471	614
Prince Albert	-21.5	17.4	398	444	-14.7	20.7	463	589
Rosthern	-20.1	18.3	373	500	-13.4	21.3	430	657
Saskatoon	-19.2	18.5	348	520	-12.9	21.6	396	688
Medicine Hat	-12.7	19.9	347	686	- 6.5	22.4	390	910
Means	-22.1	16.9	424	394	-14.2	20.5	502	539

At the north-eastern end, conditions are relatively cold and wet, while in the south-west, the climate is relatively warm and dry (Table I). Although the gradient in precipitation is not particularly steep, estimates of annual potential evapotranspiration (PET), show a steady increase from Churchill in north-eastern Manitoba, to Medicine Hat in southern Alberta. The gradient in PET is a major factor determining the current boreal forest zonal distribution in the southern ecotone; growing season soil moisture deficits probably play a critical role in determining forest structure and productivity (Price *et al.*, 1993; Hogg, 1994). In the north, low winter temperatures are a more important physiological limitation to growth and species survival.

Soil types vary from predominantly chernozemic in the prairie grassland areas, through mixtures of luvisols, humo-ferric podzols and fibrisols (peats), overlying glacial deposits, to cryic fibrisols (organic soils with discontinuous permafrost) overlying Canadian Shield bedrock in northern Manitoba (Clayton *et al.*, 1977). These differences in soil type result in a general gradient of decreasing site fertility moving from south to north. The interaction of climatic and nutritional gradients hence results in maximum site productivities generally occurring toward the middle of the transect area in the mid-boreal zone.

The transect area also exhibits a gradient in land-use pressure, ranging from dryland agriculture in the prairie grassland and commercial forest management in the southern boreal forest, to low-impact traditional uses such as trapping, hunting and firewood collection in the more sparsely populated north. Large uncontained forest fires of natural origin are a major disturbance factor across the entire western boreal forest (e.g., see Flannigan and Van Wagner, 1991; Sirois, 1992; Bergeron and Flannigan, 1995), but the level of investment in protection and suppression tends to be higher in the more productive and accessible areas of the south, with the result that greater losses from fires tend to occur in the more remote northerly areas.

The general projections for future climates derived from GCM simulations indicate changes in mean annual temperature and precipitation within the next 50–100 years. Table I shows that the greatest warming is anticipated during winter, at the northern end of the transect. Since the greatest seasonal temperature variation is in the north, this warming effect will therefore greatly reduce the north–south summer temperature differential. For example, the GISS model (Hansen *et al.*, 1988) predicts a change in this differential from 8 °C currently to only 3.5 °C under $2\times CO_2$. The GCMs are less consistent in their projections of future precipitation. It appears however, that the greatest increases in annual rainfall will also occur at the northern end of the transect, whereas the greatest increases in annual PET will be at the south, leading to significantly warmer and drier conditions there (Table I; see also Hogg and Hurdle, 1995, this volume).

3. Experimental Design

3.1. Rationale

The apparent climate sensitivity of high-latitude ecosystems combined with the GCM predictions of maximum climate changes in mid-continental locations strongly suggests that regions such as the BFTCS area are likely to be among the first to exhibit systematic responses to global change. Hence they may be considered "information-rich" systems with potentially large returns of knowledge for the research effort invested. BFTCS attempts to maximise this return by looking not only at the relatively stable central regions of the boreal biome, but also at the transitions with adjacent biomes to the north and south.

A working hypothesis is that these ecotones are likely to be particularly sensitive to change, since the vegetation is on the edge of transition from one quasi-stable state to another (Holling, 1992; Apps, 1993). A corollary to this hypothesis is that the first signs of change will manifest as "fragmentation" of the landscape at the biome boundaries. Due to the greater sensitivity of these ecosystems to changing conditions at the extremes of their geographic range and a likely increase in disturbance rates (Flannigan and van Wagner, 1991; Kurz *et al.*, 1995b), smaller and increasingly disjoint patches may be expected (C. S. Holling, University of Florida, pers. comm., 1993).

The *direction* of change will also be a very important indicator, e.g., are there signs that grasslands are replacing the aspen parkland (as suggested by Rizzo and Wiken, 1992)? Are krummholz woodlands in the boreal/tundra transition developing into upright seed-producing forms, resulting in increased northward seed dispersal and establishment of previously unrecorded woodland islands north of the current tree line? (e.g., MacDonald *et al.*, 1993). Have the proportions of the landscape covered by each ecosystem type within the transition zone changed in recent years? If so, is the mechanism hydrological (e.g., as reported for permafrost peatland area by Vitt *et al.*, (1994) and Zoltai and Vitt (1990)), or biogeochemical? (Oechel *et al.*, 1993; Gignac *et al.*, 1991; Rastetter *et al.*, 1991; Schindler and Bayley, 1994), or a change in disturbance frequency, or a combination of these?

The possible impacts of global change on boreal ecosystems outlined above warrant detailed investigation,. but economic and practical constraints make it impossible to subject the entire boreal region to intensive study. Some ecosystem processes may be investigated in the laboratory, or in field experiments, while others may be explored using simulation models. The problem is to be able to aggregate the effects of small-scale processes over an area extensive enough to assess their importance at the regional scale, when considering interactions with larger-scale processes such as ecosystem disturbances.

Under a changing environment, ecosystem disturbances can have significant effects on regional ecosystem dynamics, such as succession and productivity (Overpeck *et al.*, 1990). These effects are caused mainly by changes in disturbance frequency, which will lead to shifts in forest age-class distribution, and species composition. Disturbances should not alter regional dynamics when they occur at a constant frequency. Factors influencing the disturbance frequency, however, such as a warming climate leading to increased losses from fires and insect attacks (the latter due to longer summers and warmer winters), or protection programs leading to reduced losses, will almost certainly affect forest structure (Apps, 1993). In turn, these changes are likely to change the spatial integrals of biological processes such as net C assimilation (Kurz *et al.*, 1995b), and of physical climate processes including radiative and convective transfer, with consequences for the regional energy balance and hydrology (Shuttleworth, 1991; BOREAS Science Steering Committee, 1991; Hall *et al.*, 1993).

The latter point is not widely appreciated by earth system scientists attempting to assess the role of surface vegetation in the global climate (Sellers *et al.*, 1989; Dickinson *et al.*, 1993). The "soil–vegetation–atmosphere–transfer" (SVAT) surface parameterization schemes currently favoured by GCM modellers typically do not consider spatial and temporal vegetation dynamics. Usually, the landscape is broadly classified into static vegetation types such as "tundra", "old conifers" or "young hardwoods". Intensive studies of species ecophysiology should yield parameterizations appropriate for such classifications, but problems will be encountered when the proportions of these vegetation types, and their vertical structures (i.e., species composition, height, leaf area index etc.), alter in response to environmental changes.

The concept for BFTCS emerged as a means of integrating within a single study framework, both modelling and experiments (including remote sensing) carried out over a relatively large area and a time period of at least ten years. The objective was to gain maximum value from several large collaborative projects, both planned and underway, while facilitating independent studies by local researchers targeted at specific problems not adequately addressed by the larger scale projects.

3.2. Scaling Methodologies

In recent years, there has been an increasing trend toward large multidisciplinary field studies which encourage collaborative field work and data analysis. Access to wide-coverage multi-spectral remote sensing products has increased enormously, while low-cost, high-performance computing has resulted in increasingly sophisticated simulation models covering a range of paradigms. It is now feasible to link such models both "laterally" by coupling ecosystem models to physiologically-based canopy process models and to models of local area hydrology and microclimate; as well as "vertically" in the manner of nested hierarchies (leaf → canopy → patch → landscape → region). Indeed, large scale field experiments such as the Boreal Ecosystem–Atmosphere Study (BOREAS) are largely devoted to testing the hypotheses expressed in such linkages.

BOREAS is a joint Canada–U.S. study based on satellite, aircraft and ground observations, aimed at detailed understanding of canopy level processes governing boreal forest function under two distinct environmental regimes (Figure 1). There are two main components to the study: intensive measurements of physical and ecophysiological processes (with typical time-frames of seconds to weeks); and development of models to extend this understanding to the landscape scale (10–50 km) where aircraft and satellite remote sensing measurements may be used for validation (Hall *et al.*, 1993). Hence, an important goal of BOREAS is to develop and test spatial scaling methodologies. The nested experimental design (cuvette → canopy flux-measurement tower → aircraft-mounted sensors → satellite) allows simultaneous measurement of radiative transfers, and energy, mass and momentum fluxes, between forest and atmosphere at several scales. This will also be essential to validate and/or revise aggregation hypotheses (e.g., Gardner *et al.*, 1982; Shuttleworth, 1991), which can then be used in larger scale models of vegetation processes and vegetation–atmosphere exchanges. Remote sensing techniques are expected to be a key observational tool for longer term monitoring of ecosystem processes at the landscape and regional (transect) scales.

The BFTCS area encloses all the major BOREAS study sites (Figure 1), providing a logical step for testing aggregation techniques at the regional scale. The experimental component of BFTCS extends the spatial and temporal ranges of BOREAS from a few discrete sites studied intensively for one growing season, to a greater number of sites studied in much less detail over a much longer period. The modelling component supports this extension, by integrating models of the small-scale processes into models of ecosystem dynamics operating at larger scale and over longer durations (time-frames of years to decades), including disturbances, vegetation succession, and human influences (Price and Apps, 1993).

4. Studies in Progress

Experimental work in BFTCS covers many of the disciplines needed to assess vegetation/climate interactions, ranging from studies of ecophysiological processes affecting primary productivity and respiration, and accumulation and decomposition of organic matter in soils, to studies of disturbance history and its correlation with past climates. Collaborative studies with several Canadian universities include studies of C dynamics in forest soils and the application of a regional climate model (RCM) nested within the Canadian Climate Centre's GCM (Caya *et al.*, 1995, this volume).

4.1. Biomass Inventory and Productivity Assessments

As a component study of BOREAS, almost 100 forest sites have been surveyed within the BFTCS area (Halliwell *et al.*, 1995, this volume). At each site, standard measurements of stand density, basal area and standing volume, and soil variables have been made, together with estimates of detrital mass and understory vegetation composition. These data are being combined with studies of species allometry (S. T. Gower, University of Wisconsin, pers. comm., 1995) to provide estimates of vegetation biomass, C storage density, and annual production, as functions of site conditions.

4.2. CIDET

The ongoing Canadian Intersite Decomposition ExperimenT (CIDET) is a nation-wide comparison study of forest floor litter decomposition rate (Trofymow *et al.*, 1995, this volume). Sets of mesh bags containing litter collected from 21 forest ecoclimatic regions across the country, (including three paired lowland/wetland boreal forest sites in the

BFTCS area) have been staked out in groups at each site. Decomposition rates are being assessed and cross-compared annually over a 10-year period, to assess the interacting effects of substrate quality and climate. This study is designed to complement a similar LTER (Long Term Ecological Research) program in the USA (Harmon *et al.*, 1990).

4.3. Peatland studies

Peatlands are a major feature of the landscape for much of the BFTCS area, and represent significant accumulations of atmospheric carbon due to low decomposition rates, resulting from low average temperatures and anaerobic conditions. Permafrost is present across more than half the length of the BFTCS, with most of the forested regions lying in the discontinuous permafrost zone. These areas are likely to be sensitive to slight increases in temperature, hence climate warming could have major effects on the productivity and patch-scale dynamics of wetland–forest ecosystems. Dated peat deposits may be used to determine past changes in peat accumulation rates, and serve as proxies for past climatic conditions in the BFTCS area. In the southern parts of the boreal forest, patterns in the vegetated landscape are associated with melting of permafrost islands, and may be an indicator of warming within the last 150 years (Vitt *et al.*, 1994). To explore possible future changes, models of peat ;and forest interactions are required and both conceptual models (e.g., Zoltai, 1994) and simulation models (e.g., Apps *et al.*, unpublished ms.).

4.4. Disturbance histories

Most stands in the western Canadian boreal forest are of fire-origin, indicating that large-scale natural disturbances are a key factor influencing forest structure and successional development in this region. In order to assess the possible trends in disturbances due to wildfire and insect attacks under a warming climate (e.g., Flannigan and Van Wagner, 1991; see also Bergeron and Flannigan, 1995; Fleming and Volney, 1995, this volume), it is imperative that disturbance histories be established for stands under investigation within the BFTCS area. Such work is being undertaken by Stocks and colleagues (B. J. Stocks, Canadian Forest Service, pers. comm.,1993).

Some 3.28 Mha burned in Manitoba in 1989, one of the warmest years on record (Hirsch, 1991), illustrating the potential sensitivity of boreal ecosystems to a warming climate. The strong influence that such changes may have on future C budgets has been stressed by Kurz *et al.* (1995b). Reconstruction of past climates and their relationship to changes in disturbance frequency is possible from paleo-ecological data (e.g., Campbell and McAndrews, 1993; MacDonald *et al.*, 1993); from fossil charcoal deposition studies (e.g., MacDonald *et al.*, 1991; Clark, 1988, 1989); from dendrochronological studies (e.g., Van Wagner, 1978; Clark, 1989; Filion *et al.*, 1993); and from historical records (Kurz *et al.*, 1995a). Models are also being developed to explore the impacts of spatially connected disturbance processes on spatial and temporal vegetation dynamics in the landscape (Li and Apps, 1995, this volume) and to assess possible consequences of climate change on insect and tree host phenological interactions (Fleming and Volney, 1995, this volume).

4.5. Remote sensing

By providing large-scale time-series data, remote sensing is potentially very useful for monitoring regional changes in vegetation structure. U.S. and Canadian agencies are collaborating in data acquisition and analysis of remote sensing data as an integral component of BOREAS and BFTCS. Vegetation boundaries are being mapped and changes monitored. Over a period of several years to decades, vegetation dynamics monitored in this way may be compared to predictions derived from regional- and transect- scale model simulations.

For areas where change is manifest, ground-based studies can be deployed to test and refine the understanding of the causal mechanisms. As part of BOREAS, intensive field measurements took place during 1993 and 1994 in the two large study areas near the northern and southern ecotones of the boreal forest (Figure 1). A further intensive field season is planned for the summer of 1996. Abundant process data have already been obtained from these sites selected for stand uniformity and general suitability for energy balance and trace gas flux measurements, but there will be relatively little information about larger scale and longer term processes.

4.6. Carbon Budget Model

The results of field studies (e.g., see Halliwell *et al.,* 1995) will also be used for validation of a large-scale process-based Carbon Budget Model of the Canadian Forest Sector (CBM-CFS) developed from a national scale model (Kurz *et al.*, 1992, Apps and Kurz, 1993) as a component study of the BFTCS (Price and Apps, 1993). The major purpose of this model is to provide an integrating framework for estimating the contributions of vegetation productivity and other ecosystem processes (including soil and wetland dynamics, and large-scale disturbances) to regional and national C budgets (Kurz *et al.*, 1992). The CBM-CFS is a C-conserving mass balance model that tracks photosynthetic C uptake by living forest phytomass, and tracks the subsequent releases of C during transfers to detrital material, soils, peat deposition and forest products (Kurz *et al.*, 1992). The effects of disturbances on large-scale C dynamics are an important factor explicitly represented in this model, based on historical records. Linkage to studies of the relationships between climate and disturbance frequency mentioned earlier, will allow CBM-CFS to be used for assessing impacts of changes in disturbance regime resulting from anticipated future climate change on the regional C budget. Developed and applied originally at the national scale, it is now being applied and tested at the regional scale of BFTCS (Price and Apps, 1993) and the landscape/management scale of individual forest management units (Price *et al.*, 1994). These finer scale implementations of CBM-CFS are also being used to examine the sensitivity of the regional C budget to transient changes in climate and resource management.

4.7. Transect Ecosystem Models

Boreal ecosystems are typically organised in patches, with areas of homogeneous vegetation rarely exceeding 2 km^2 (although some lakes may be larger than this in total area, there are no large lakes in the BFTCS area). This means that most of the surface vegetation is organised at length scales of significantly less than 10 km, which in turn indicates that it may be appropriate to use simple area-weighted aggregations of the vegetation cover when estimating surface fluxes of water vapour and CO_2 as was found by Pinty *et al.* (1989) in the FIFE experiment (an earlier study comparable to BOREAS, but in the short-grass prairie of Kansas (e.g., see Shuttleworth *et al.*, 1989)).

The basic spatial unit for transect-scale simulations is the 10 km square, of which 1000 are needed to represent the entire BFTCS area. Data from BOREAS will allow scaling up from flux tower sites to 10 km (e.g., based on validations from BOREAS flux tower and flux aircraft measurements). Hence, BFTCS provides an opportunity to simulate interactions between vegetation and atmosphere at resolutions within a 10 km grid square combined with larger scale processes e.g., shifts in biome distribution and composition that can be simulated on a 10 km grid. Eventually these simulations will be combined with a nested regional climate model co-located over the transect area (Caya *et al.,* 1995). Interactions between 10 km squares will also be considered, particularly for those mechanisms affecting species migration and ecosystem succession: seed dispersal, management, and the spread of natural disturbances.

Using BFTCS as a source of data for development and testing, CBM-CFS can use metadata derived for the transect area from runs of several smaller scale sub-models, which can be tested and refined against field data (Price and Apps, 1993). These include: detailed models of ecophysiological responses; single-tree and stand-level growth models; ecosystem-level models of species competition and succession and of insect population dynamics; and regional hydrological and climate models. Future C budget assessments are not the only outcome: it will also be possible to develop maps of predicted changes in vegetation composition. The integrated transect model structure will be used to explore future responses of regional and national forest resources to anticipated changes in climate (Price and Apps, 1993), and to enable forest scientists to better determine the probable relative effects of possible climate change and various forest management options on rates of C sequestration and release by boreal forest ecosystems. It may also reveal further strategic research needs (i.e., gaps in knowledge) yet to be identified. With suitable parameterizations and further validation work, the model should be extensible to other forest ecosystems, and even to other major vegetation types.

5. Summary

In the boreal zone of central Canada, both natural and managed forest ecosystems appear particularly sensitive to global change, with potentially serious consequences both regionally (in vegetation composition and distribution) and globally (as feedbacks to the global climate). The Boreal Forest Transect Case Study (BFTCS) is an adaptive investigative approach that encapsulates within a single study area of some 100 000 km^2, the major ecosystems and ecological processes of the circumpolar boreal and tundra biomes likely to be affected by, and to feed back to, anticipated global change. Field studies of past dynamics and present-day processes will be used to develop and test a suite of predictive models, to better assess short-term (50–100 yr) transient ecosystem dynamics and their implications for management and policy.

A second benefit of this approach will be the development of appropriate environmentally-sustainable management practices, based on improved understanding of fundamental ecosystem processes and of their potential responses to environmental change. The co-ordinated investigation of ecosystem processes and human impacts within BFTCS is assisting researchers to provide answers to many policy-relevant questions concerning current and possible future interactions between boreal ecosystems and the global environment.

Acknowledgments

We would particularly like to recognise the very careful and considerate review, including many detailed editorial suggestions, made by an anonymous reviewer, which helped us to very significantly improve the organisation and content of this paper.

References

Apps, M. J. and Kurz, W. A., 1993. *In: Carbon Balance of World's Forested Ecosystems: Towards a Global Assessment,* Publ. Acad. Finland, Helsinki, Finland, pp. 14-28.

Apps, M. J., Kurz, W. A., Luxmoore, R. J., Nilsson, L. O., Sedjo, R. A., Schmidt, R., Simpson, L. G. and Vinson T. S.: 1993, *Water, Air, Soil Pollut.* **70**, 39-53.

Apps, M. J., Belland, R., Gignac, L. D., Hogg, E. H., Kurz, W. A., Seburn, D., Warner, B., Wein, R. W., Vitt, D. H. and Zoltai, S. C.: 1994, *A Proposed Carbon Dynamics Model of Canadian Peatlands*, Draft Design Document, Can. For. Serv., Northwest Region, Edmonton, Canada. (Unpublished manuscript).

Atmospheric Environment Service (AES): 1983, *Canadian Climate Normals, 1951-1980. Temperature and Precipitation, Prairie Provinces*, Environment Canada, Downsview, Canada, 429 pp.

Bergeron, Y. and Flannigan, M. D.: 1995, *Water, Air, Soil Pollut.* **82**, 437-444.

Bonan, G. B., Pollard, D. and Thompson, S. L.: 1992, *Nature* **359**, 716-718.

Bonan, G. B., Shugart, H. H. and Urban, D. L.: 1990, *Climatic Change* **16**, 9-29.

BOREAS Science Steering Committee: 1991, *EOS* **72**, 33-40.

Campbell, I. D. and McAndrews, J. H.: 1993, *Nature* **366**, 336-338.

Chapin III, F. S. and Matthews, E.: 1993, *In:* Kolchugina, T. P. and Vinson, T. S. (eds.), *Carbon Cycling in Boreal Forests and Subarctic Ecosystems: Biospheric Responses and Feedbacks to Global Climate Change.* U.S. Environmental Protection Agency Report EPA/600R-93/084, Office of Research and Development, Washington D.C., USA, pp. 9-20.

Caya, D., Laprise, R., Giguère, M., Bergeron, G., Blanchet, J.-P., Stocks, B. J., Boer, G. J. and McFarlane, N. A.: 1995, *Water, Air, Soil Pollut.* **82**, 477-482.

Clark, J. S.: 1988, *Nature* **334**, 233-235.

Clark, J. S.: 1989, *J. Ecology* **77**, 989-1004.

Clark, J. S.: 1990, *Can. J. For. Res.* **20**, 219-232.

Clayton, J. S., Ehrlich, W. A., Cann, D. B., Day, J. H. and Marshall, I. B.: 1977, *Soils of Canada, Volume 1: Soil Report.* Can. Dept. Agr. Res. Br., Supply and Services Canada, Ottawa, Canada, 243 pp.

D'Arrigo, R., Jacoby, G. C. and Fung I. Y.: 1987, *Nature* **329**, 321-323.

Dickinson, R. E., Henderson-Sellers, A. and Kennedy, P. J.: 1993, *Biosphere-Atmosphere Transfer Scheme (BATS) Version 1e as Coupled to the NCAR Community Climate Model.* NCAR/TN-387+STR, Climate and Global Dynamics Division, National Center for Atmospheric Research, Boulder, USA, 72 pp.

Ecoregions Working Group: 1989, Ecoregions Working Group of Canada Committee on Ecological Land Classification. *Ecological Land Classification Series*, no. **23**, Sustainable Development Br., Can. Wildlife Serv., Conservation and Protection, Environment Canada, Ottawa, Canada, 118 pp.

Filion, L, Jardon, Y. and Cloutier, C.: 1993, *In:* Ung, C-H. (ed.), *Forest Growth Models and their Uses.* Natural Resources Canada, Canadian Forest Service, Sainte Foy, Canada, pp. 235-239.

Flannigan, M. D. and Van Wagner, C. E.: 1991, *Can. J. For. Res.* **21**, 66-72.

Fleming, R. A. and Volney, W. J. A.: 1995, *Water, Air, Soil Pollut.* **82**, 445-454.

Gardner, R. H., Cale W. G. and O'Neill, R. V.: 1992, *J. Ecology* **63**(6), 1771-1779.

Gignac, L. D., Vitt, D. H., Zoltai, S. C. and Bayley, S. E.: 1991, *Nova Hedwigia* **53**, 27-71.

Gorham, E.: 1991, *Ecol. Appl.* **1**, 182-195.

Hall, F. G., Sellers, P. J., Apps, M., Baldocchi, D., Cihlar, J., Goodison, B., Margolis, H. and Nelson, A: 1993, *IEEE Geoscience and Remote Sensing Society Newsletter*, **March 1993**, 9-16.

Halliwell, D. H., Apps, M. J. and Price, D. T.: 1995, *Water, Air, Soil Pollut.* **82**, 257-270.

Hansen, J., Fung, I., Lacis, A., Rind, D., Russell, G. Lebedeff, S., Reudy, R. and Stone, P.: 1988, *J. Geophys. Res.* **93**, 9341-9364.

Harmon, M. E., Baker, G. A., Greene, S. and Spycher, G.: 1990, *For. Ecol. Manage.* **31**, 55-66.

Hirsch, K. G.: 1991, *For. Chron.* **67**, 358-365.

Hogg, E. H.: 1994, *Can. J. For. Res.* **24**, 1835-1845.

Hogg, E. H. and Hurdle, P. A.: 1995, *Water, Air, Soil Pollut.* **82**, 391-400.

Holling, C. S.: 1992, *Ecol. Monogr.* **62**, 447-502.

Jensen, M. E., Burman, R. D. and Allen, R. G.: 1990, *Evapotranspiration and irrigation water requirements. Amer. Soc. Civil Eng. Man. Rep. Eng. Pract.* No. **70**, New York, USA, 332 pp.

Kurz, W. A., Apps, M. J., Webb, T. M. and McNamee, P. J.: 1992. *The carbon budget of the Canadian forest sector: Phase 1*, Forestry Canada Information Report NOR-X-326, Northwest Region , Edmonton, Canada, 93 pp.

Kurz, W. A., Apps, M. J., Beukema, S. J., and Lekstrum, T.: 1995a, *Tellus* **47B**, 170-177.

Kurz, W. A., Apps, M. J., Stocks, B. J., and Volney, W. J. A.: 1995b, *In:* Woodwell, G. M. and Mackenzie, F. T. (eds.), *Biospheric feedbacks in the global climate system: will the warming speed the warming?* Oxford University Press, New York, USA, pp. 119-133.

Li, C. and Apps, M. J.: 1995, *Water, Air, Soil Pollut.* **82**, 429-436.

MacDonald, G. M., Larsen, C. P. S., Szeicz, J. M. and Moser, K. A.: 1991, *Quaternary Sci. Rev.* **10**, 53-71.

MacDonald, G. M., Edwards, T. W. D., Moser, K. A., Pienitz, R. and Smol, J. P.: 1993, *Nature* **361**, 243-246.

Oechel, W. C., Hastings, S. J., Vourlitis, G., Jenkins, M., Riechers, G. and Grulke, N.: 1993. *Nature* **361**, 520-523.

Overpeck, J. T., Rind, D. and Goldberg, R.: 1990, *Nature* **343**, 51-53.

Pinty, J.-P., Mascart, P. and Rosset, R.: 1989, *J. Appl. Meteorol.* **28**, 976-992.

Prentice, I. C., Cramer, W., Harrison, S. P., Leemans, R., Monserud, R. A. and Solomon, A. M.: 1992, *J. Biogeography*, **19**, 117-134.

Price, D. T., and Apps, M. J.: 1993, *World Resour. Rev.* **5**, 15-31.

Price, D. T., Apps, M. J., Kurz, W. A., Prentice, I. C. and Sykes, M.: 1993. *In:* Ung, C-H. (ed.), *Forest Growth Models and their Uses.* Natural Resources Canada, Canadian Forest Service, Sainte Foy, Canada, pp. 251-264.

Price, D. T. and Apps, M. J.: *Climatic Change* (Unpublished manuscript).

Price, D. T., Apps, M. J., Kurz, W. A., Wesbrook, M. and Curry, R. S.: 1994, *World Resour. Rev.* **6**, 461-476.

Rastetter, E. B., Ryan, M. G., Shaver, G. R., Melillo, J. M., Nadelhoffer, K. J., Hobbie, J. E., and Aber, J. D.: 1991, *Tree Physiology* **9**, 101-126.

Rizzo, B. and Wiken, E.: 1992, *Climatic Change* **21**, 37-55.

Schindler, D. W. and Bayley, S. E.: 1993, *Global Biogeochemical Cycles* **7**, 717-733.

Sellers, P. J., Shuttleworth, W. J. and Dorman, J. L.: 1989, *J. Appl. Meteorol.* **28**, 727-759.

Shuttleworth, W. J.: 1991, *Rev. Geophys.* **29**, 585-606.

Shuttleworth, W. J., Gurney, R. J., Hsu, A. Y and Ormsby, J. P.: 1989, *Intl. Assoc. Hydrol. Sci. Publ.* **186**, 67-74.

Sirois, L.: 1992, *In:* Shugart, H. H., Leemans, R. and Bonan, G. B. (eds.), A Systems Analysis of the Global Boreal Forest. Cambridge University Press, Cambridge, UK, pp. 196-215.

Solomon, A. M. and Bartlein, P. J.: 1992, *Can. J. For. Res.* **22**, 1727-1738.

Smith, T. M., Weishampel, J. F. and Shugart, H. H.: 1992, *Water, Air, Soil Pollut.* **64**, 307-326.

Tans, P. P., Fung I. Y. and Takahashi, T.: 1990, *Science* **247**, 1431-1438.

Tans, P. P., Berry, J. A. and Keeling, R. F.: 1993, *Global Biogeochemical Cycles* **7**, 353-368.

Townsend, A. R., Vitousek, P. M. and Holland, E. A.: 1992, *Climatic Change* **22**, 293-303.

Trofymow, J. A., Preston, C. M and Prescott, C. E.: 1995, *Water, Air, Soil Pollut.* **82**, 215-226.

Van Wagner, C. E.: 1978, *Can. J. For. Res.* **8**, 220-227.

Vitt, D. H., Halsey, L. A. and Zoltai, S. C.: 1994, *Arctic and Alpine Res.* **26**, 1-13.

Woodwell, G. M.: 1989, *Climatic Change* **15**, 31-50.

Woodwell, G. M. and Houghton, R. A.: 1993, *In:* Kolchugina, T. P. and Vinson, T. S. (eds.), *Carbon Cycling in Boreal Forests and Subarctic Ecosystems: Biospheric Responses and Feedbacks to Global Climate Change.* Environmental Protection Agency Report EPA/600R-93/084, Off. Res. Dev., Washington D.C., USA, pp. 1-8.

Woodwell, G. M. and Mackenzie, F. T. (eds.): 1995, *Biotic Feedbacks in the Global Climatic System: Will the Warming Feed the Warming?.* Oxford University Press, Oxford, New York, UK, 416 pp.

Zoltai, S. C.: 1994, *In:* Grünig, A. (ed.) *Mires and Man,* Swiss Fed. Inst. for Forest, Snow and Landscape Res., Birmensdorf, Switzerland, pp. 315-319.

Zoltai, S. C. and Vitt, D. H.: 1990, *Quaternary Res.* **33**, 231-240.

Zoltai, S. C.; Singh, T.; Apps, M. J., 1991: *In:* Navratil, S. and Chapman, P. B. (eds.), *Aspen Management for the 21st Century,* Forestry Canada, Northwest Region and Poplar Council of Canada, Edmonton, Canada, pp. 143-152.

LITTER QUALITY AND ITS POTENTIAL EFFECT ON DECAY RATES OF MATERIALS FROM CANADIAN FORESTS

J. A. TROFYMOW[1], C. M. PRESTON[1], and C. E. PRESCOTT[2]

[1]*Pacific Forestry Centre, Canadian Forest Service, 506 West Burnside Rd, Victoria, British Columbia V8Z 1M5, Canada*

[2]*Faculty of Forestry, University of British Columbia, Vancouver, British Columbia V6T 2A2, Canada*

Abstract. Decomposition is influenced by a wide array of factors including macroclimate, microclimate, soil biota, soil nutrients, substrate piece size and substrate quality. To separate the influence of some of these factors a 10-year study, the Canadian Intersite Decomposition Experiment, was established in 1992 to measure the decay of 11 standard litter types on a range of forest types at 21 sites across Canada. As part of the study we analysed the initial elemental contents (N, P, S, K, Ca, Mg) and carbon (C) fractions (extractables, cellulose, hemicellulose, lignin) by ^{13}C NMR and wet chemical proximate analysis in a total of 37 primarily foliar litter types representative of the range of species found at the different CIDET sites. Litter types especially non-conifer species varied greatly in their qualities. Principal component analyses showed that the litter types could be distinguished by the elemental macronutrient contents through the ratio of N+P+K:S, by proximate chemical analyses through the ratio of water soluble:acid fractions, and by NMR through the ratio of O-alkyl:alkyl C. Litter quality data was used in three simple models of litter decay to predict how the mass loss of the different litter types could vary. Two models using a linear or single exponential decay equation and litter lignin and N content predicted a 2–5 fold difference in total mass loss for the different litter types. A third model using a summed exponential decay equation for three chemical fractions and a ligno-cellulose index predicted that for all but one litter type, variation in mass loss between types would be less than a 20%.

Keywords. NMR, PRINCIPAL COMPONENTS, PROXIMATE ANALYSIS, DECAY MODELS, CIDET.

1. Introduction

The breakdown and decay of plant materials represents the major pathway by which carbon (C) is returned to the atmosphere in most forested ecosystems, especially the boreal forest, and has the potential to be greatly impacted by climate change.

Forestry practices can also affect site C balances, influencing the amounts of C exchanged with the atmosphere. Current attempts to quantitatively model the current and potential changes in the C budgets of Canadian forests have relied upon published results and expert advice to develop algorithms and provide input data for models, yet for many of the processes modelled, information was limited (Kurz *et al.*, 1992). In particular, data on soil processes such as decomposition rates were found lacking in extent and in length. Some forest types have been extensively studied while others have not and more significant long-term studies are rare: most studies have been for 2–3 years (Aber *et al.*, 1990). To overcome this deficiency a group of 18 researchers from across Canada have assembled to conduct a Canadian Intersite Decomposition Experiment (CIDET) with the purpose of investigating the long-term rates of litter decomposition and nutrient mineralization over a broad range of forested ecoclimatic regions in Canada. The study, established in fall 1992, involved the preparation and placement of close to 11 000 litterbags comprised of 10 sets of 11 litter types in four replicate plots at each of the 21 sites. Each year for ten years, one set of bags will be removed per plot and analysed for weight loss and C, N, P content. CIDET

Water, Air and Soil Pollution **82**: 215–226, 1995.

complements several other long-term litter decay studies underway in the United States (Harmon *et al.*, 1990; LIDET, in press) and Europe (Berg *et al.*, 1993).

One of the main objectives of CIDET is to examine the role litter quality and climate have on long-term decomposition rates as previous studies have shown that these can greatly influence decay rates (Bunnell *et al.*, 1977a, 1977b; Meentemeyer, 1978; Swift *et al.*, 1979). The most commonly used measures of litter quality include elemental concentrations and ratios such as N, C:N, C:P (Melin, 1930; Hunt *et al.*, 1988; Heal and French, 1974; Berg and Staaf, 1980); organic matter fractions such as lignin (Meentemeyer, 1978; Harmon *et al.*, 1990), a ligno-cellulose index (Aber *et al.*, 1990); a combination of lignin plus N (Berg, 1984; Taylor *et al.*, 1991) or a lignin:N ratio (Melillo *et al.*, 1982). A more recent review of litters and forest floor materials using nuclear magnetic resonance spectroscopy (^{13}C NMR), (Baldock and Preston, in press) suggests that litter alkyl C content (waxes and cutins) may affect the resistant of litter to decay.

Some measure of litter quality is commonly used in most models of litter decay as a modifier of the decay constant. Both linear and power functions have been used to describe mass loss, although the most common function used is the negative exponential model:

$$\% \text{ initial mass remaining} = 100\, e^{-kt} \qquad [1]$$

where t is time and k, the rate constant, is assumed to be some function of litter quality. Melillo *et al.* (1982) applied this model to a litter decay study in NE USA and found a good fit with k a power function of the lignin:N ratio:

$$-k = 8.35(\% \text{ initial lig} : \% \text{ initial N})^{-0.784} \qquad [2]$$

Taylor *et al.* (1991) examined the decay of a variety of substrate types in Kananaskis, Alberta and found, with litters of N-fixing species excluded, that three year mass loss was best described by a piece wise linear function with a hinge point defined at lignin concentrations of 28%:

$$\% \text{ mass loss at 3 years} = 73.50 - 2.49(\%\text{lignin}) + 2.37(\text{floor}) + 20.50(\%\text{N}) \qquad [3]$$

$$\text{with: floor} = (\%\text{lignin} - 28) \text{ for } \%\text{lignin} > 28 \qquad [4]$$

Another approach suggested by Minderman (1968) is that as a first approximation, long-term decomposition of fine litter can be described as the sum of negative exponential decay curves for two chemical fractions:

$$\%\text{initial mass} = 100\left[Ae^{-k_1 t} + (1 - A)e^{-k_2 t}\right] \qquad [5]$$

where k_1 and k_2 are the respective rate constants for the labile, A, and non-labile, 1-A, litter fractions. This approach has been extended by others (Paul and Voroney, 1980; Parton *et al.*, 1987) to model mass loss as the sum of three or more chemical fractions.

Aber *et al.* (1990) examined mass loss data from two previous studies and found mass loss could be predicted as the summed losses of three chemical fractions:

$$\%\text{initial mass} = 100\left[\text{extract}\left(e^{-k_1 t}\right) + \text{cellulose}\left(e^{-k_2 t}\right) + \text{lignin}\left(e^{-k_3 t}\right)\right] \quad [6]$$

where k_1, k_2, k_3 are the respective rate constants for the extractives (sum of the water-soluble and non-polar fractions), cellulose (acid-soluble fraction) and lignin (residue corrected for ash content). The k value for each fraction was found to be linearly related to the ligno-cellulose index [LCI – ratio of %lignin:(%lignin + %cellulose)], with a different relationship for each chemical fraction:

$$k_1 = 1.346 - 2.119(\text{LCI}) \quad [7a]$$

$$k_2 = 0.591 - 0.555(\text{LCI}) \quad [7b]$$

$$k_3 = -0.565 + 1.821(\text{LCI}) \quad [7c]$$

For this paper we report the elemental, proximate chemical and ^{13}C NMR characteristics of 37 litter types, the 11 litter types used in the CIDET study (bolded, Table I) and an additional 26 litter types being used in independent studies by CIDET site co-operators or as part of a shorter-term decay study (Prescott, pers. comm.) on forested and clearcut stands at a range of sites in British Columbia. The purpose of this paper is two-fold, firstly, to describe how the litter types used in CIDET span the range of litter quality variables measured and identify which variables distinguish the different litter types. Secondly, we examine how the initial potential rates of mass loss of the litter types could vary based on predictions of three litter decay models.

2. Methods

Leaf litters (Table I) were collected by several co-operators from a range of forest types across Canada (Figure 1). Most litter types were collected over a one to two month period in the autumn of 1991 as newly senescent leaves passively falling into mesh traps or tarps placed in the forest. The western hemlock wood used consisted of ca. 50 g uniformly grained blocks cut from the outer heartwood of a single ca. 1 m diameter log. Prior to chemical analyses, samples were dried at 70 °C in a force-air oven and Wiley milled to pass through a 0.2 mm mesh.

Total %N, %C, and %S were determined by combustion; N by a modified Dumas method on a LECO FP-228 N analyser, C on a LECO CR-12 carbon system and S on a LECO SC-132 sulphur system. Total P, Ca, Mg, and K were determined through wet oxidation of the sample in a block digester (Parkinson and Allen, 1975) with the P in the digest measured using a Technicon Autoanalyzer and Ca, Mg, K by atomic adsorption spectroscopy. Single composite samples were measured for each litter type along with quality control duplicates and reference standards.

TABLE I

Species, common name, litter elemental concentrations (mg g^{-1}) and chemical proximate analysis fractions (mg g^{-1}) of total dry litter mass.

N	FC[1]	Genus_species	Common_name	C	N	P	S	Ca	Mg	K	NPE[2]	WSE	ACID	LIG	ASH
1	Ccn	*Chamaecyparis_nootkatensis*	Yellow_cedar	487.9	5.6	0.41	1.97	29.24	0.81	0.48	154.6	247.1	308.6	222.2	72.6
2	Cdc	*Pseudotsuga_menziesii*	Douglas_fir_nw	493.5	5.7	0.81	2.13	14.43	0.95	1.35	117.9	122.2	391.5	307.3	65.1
3	**Cdc**	***Pseudotsuga_menziesii***	**Douglas_fir_cl**	**496.1**	**7.0**	**1.05**	**2.65**	**12.77**	**1.09**	**1.55**	**102.7**	**114.8**	**416.3**	**303.3**	**67.4**
4	Cdi	*Pseudotsuga_menziesii*	Douglas_fir_in	476.5	5.0	0.87	1.94	22.26	1.11	2.05	91.0	328.8	316.8	212.6	53.4
5	Cfa	*Abies_amabilis*	Amabilis_fir	558.8	3.9	0.65	1.74	12.88	0.54	0.54	127.6	302.8	281.8	253.6	35.6
6	Cfb	*Abies_balsamea*	Balsam_fir	554.9	5.3	0.31	1.77	9.71	0.93	1.31	132.2	264.6	301.5	272.2	30.5
7	Cfl	*Abies_lasiocarpa*	Alpine_fir	502.9	4.4	0.75	2.18	18.24	1.26	2.39	139.2	340.2	253.2	214.5	56.2
8	Chm	*Tsuga_mertensiana*	Mountain_hemlock_n	507.3	4.2	0.67	2.04	6.58	0.81	0.73	92.5	261.5	292.4	330.4	24.0
9	Chw	*Tsuga_heterophylla*	Western_hemlock_n	503.2	7.6	0.91	1.49	7.49	1.14	0.96	84.7	259.1	290.0	339.3	27.8
10	**Cll**	***Larix_laricina***	**Tamarack_s**	**487.5**	**5.9**	**0.23**	**3.15**	**6.56**	**2.48**	**3.09**	**93.5**	**311.0**	**300.5**	**239.6**	**58.9**
11	Cll	*Larix_laricina*	Tamarack_larch	474.7	7.6	2.32	2.10	8.34	2.44	2.86	126.5	281.8	288.2	261.8	43.6
12	Clo	*Larix_occidentalis*	Western_larch	456.8	3.9	2.03	1.61	4.18	1.31	4.12	95.1	321.0	325.1	200.2	62.3
13	**Cpj**	***Pinus_banksiana***	**Jack_pine**	**497.2**	**12.8**	**1.27**	**1.37**	**45.53**	**1.23**	**2.66**	**69.7**	**152.4**	**424.4**	**327.9**	**26.5**
14	Cpl	*Pinus_contorta*	Lodgepole_pine_c	490.3	4.7	0.39	2.21	8.43	1.51	0.48	163.5	258.6	330.5	207.8	41.4
15	Cpp	*Pinus_ponderosa*	Ponderosa_pine	483.8	6.0	0.96	1.94	4.50	1.35	1.68	112.5	265.6	315.2	276.9	30.7
16	Cpr	*Pinus_resinosa*	Red_pine	501.4	4.3	0.41	1.64	4.50	1.14	1.58	140.0	159.6	400.6	281.9	18.3
17	Cpw	*Pinus_monticola*	Western_white_pine	506.2	5.5	0.38	1.87	11.99	1.47	0.84	87.9	314.7	324.9	238.0	35.9
18	Csb	*Picea_mariana*	Black_spruce_bc	478.0	3.4	0.63	1.67	18.60	0.57	1.50	134.5	365.9	287.6	156.1	59.3
19	**Csb**	***Picea_mariana***	**Black_spruce_s**	**494.7**	**7.3**	**0.79**	**2.83**	**6.62**	**0.89**	**2.18**	**109.2**	**198.5**	**369.9**	**282.6**	**41.6**
20	Cse	*Picea_engelmanii*	Engelmann_spruce	473.7	5.8	0.64	1.65	17.37	0.60	2.34	132.3	345.1	274.3	172.3	82.3
21	Css	*Picea_sitchensis*	Sitka_spruce	491.0	8.4	0.87	1.69	14.92	1.25	2.12	100.0	191.0	380.6	269.5	62.9
22	Csw	*Picea_glauca*	White_spruce	464.4	7.7	1.28	1.67	19.72	0.69	2.87	81.3	297.3	347.3	201.0	68.6
23	Cto	*Thuja_occidentalis*	Eastern_white_cedar	520.2	6.0	0.34	1.50	19.69	2.36	0.91	112.8	167.1	377.3	289.1	56.9
24	**Ctp**	***Thuja_plicata***	**Red_cedar**	**496.7**	**6.4**	**0.47**	**1.20**	**16.84**	**0.90**	**1.09**	**107.2**	**105.1**	**365.1**	**355.5**	**72.0**
25	Cyp	*Taxus_brevifolia*	Pacific_yew_l	512.0	9.5	1.13	1.65	13.89	1.44	1.77	93.2	157.0	308.8	392.4	51.2
26	**Dba**	***Fagus_grandifolia***	**American_beech**	**470.0**	**7.1**	**0.36**	**2.03**	**9.89**	**2.52**	**0.82**	**72.5**	**129.0**	**452.9**	**279.9**	**70.5**
27	**Dbw**	***Betula_papyrifera***	**White_birch**	**480.0**	**7.2**	**0.38**	**0.99**	**8.50**	**2.44**	**2.62**	**65.2**	**359.4**	**303.3**	**239.6**	**33.8**
28	**Dpt**	***Populus_tremuloides***	**Trembling_aspen**	**468.3**	**6.7**	**1.27**	**1.55**	**20.52**	**1.55**	**12.34**	**87.5**	**354.2**	**337.4**	**143.7**	**83.8**
29	Dro	*Quercus_rubra*	Red_oak	488.4	6.2	0.99	1.15	10.41	2.06	1.75	56.0	234.8	392.6	275.9	42.6
30	Dsa	*Alnus_rugosa*	Speckled_alder	480.7	25.3	1.50	1.30	15.18	1.79	6.72	80.1	302.0	362.8	198.0	60.9
31	Dsm	*Acer_saccharum*	Sugar_maple	441.9	7.7	1.16	1.37	21.99	2.73	2.46	88.9	317.0	377.2	123.8	102.9
32	**Fbf**	***Pteridium_aquilinum***	**Bracken_fern**	**463.3**	**8.8**	**0.66**	**1.15**	**7.72**	**3.10**	**4.29**	**22.6**	**90.4**	**491.0**	**328.9**	**72.1**
33	**Gfh**	***Festuca_hallii***	**Plains_rough_fescue**	**437.9**	**7.1**	**0.58**	**1.51**	**3.69**	**1.27**	**5.04**	**90.6**	**128.6**	**584.9**	**111.6**	**92.2**
34	Hea	*Epilobium_angustifolium*	Fireweed	452.2	4.9	1.15	1.13	22.79	4.10	4.80	81.0	596.6	263.8	24.5	66.5
35	Lcs	*Cladina_stellaris*	Reindeer_lichen	464.5	3.9	0.30	2.97	0.31	0.19	0.74	48.7	69.2	828.9	44.9	9.0
36	Sgs	*Gaultheria_shallon*	Salal	486.8	4.9	0.32	2.00	19.77	4.28	2.66	80.1	258.6	340.9	256.0	64.4
37	**Whw**	***Tsuga_heterophylla***	**Western_hemlock_w**	**473.3**	**2.4**	**0.05**	**2.76**	**1.13**	**0.23**	**0.32**	**16.5**	**29.2**	**656.1**	**294.3**	**4.0**

[1]The FC code consists of three letters: the first indicates life-form, Conifer, Decidous tree, Fern, Grass, Herb, Lichen, Shrub, Wood; the second the latin name or common name of the genus; the third the species latin or common name. Highlighted types are included in CIDET.

[2]Proximate chemical fractions: NPE- Non-Polar Extractable; WSE- Water-Soluble Extractable; ACID- Acid soluble; LIG- Klason Lignin.

Proximate chemical analyses were done following the methods of McClaugherty *et al.* (1985) and Ryan *et al.* (1990). Non-polar extractables (NPE) (i.e., soluble fats, waxes and oils) were removed with dicholormethane; water soluble extractables (WSE) (i.e., simple sugars, water soluble phenolics) removed with hot water; acid soluble carbohydrates (ACID) (i.e., cellulose) removed by sulphuric acid hydrolysis. The ash content of the initial litter and the remaining mass of acid-insoluble residue, commonly referred to as Klason lignin (LIG), was determined by ashing in a muffle furnace at 450 °C for 8 hours. The amount of each extracted fraction was reported as a percentage of ash-free dry organic matter. Duplicate analysis samples of each composite litter were measured and values averaged.

Solid state ^{13}C CPMAS NMR characterises the specific chemical types of C in a material and has recently been applied to litter and forest floor materials (Norden and Berg, 1990; Preston *et al.*, 1994; Baldock and Preston, in press). One composite sample of each litter type was analysed on a Bruker MSL 100 spectrometer operating at 25.18 MHz for ^{13}C at 2.35 Tesla. Spectra were divided in chemical shift regions (Baldock and Preston, in press) according to chemical type of C as follows: alkyl (fats and waxes) 0–50 ppm; methoxyl (side chains of guaiacyl lignin, C2 and C3 of condensed tannins) 50–60 ppm; O-alkyl (cellulose and sugars) 60–92 ppm; aromatic (lignin and condensed

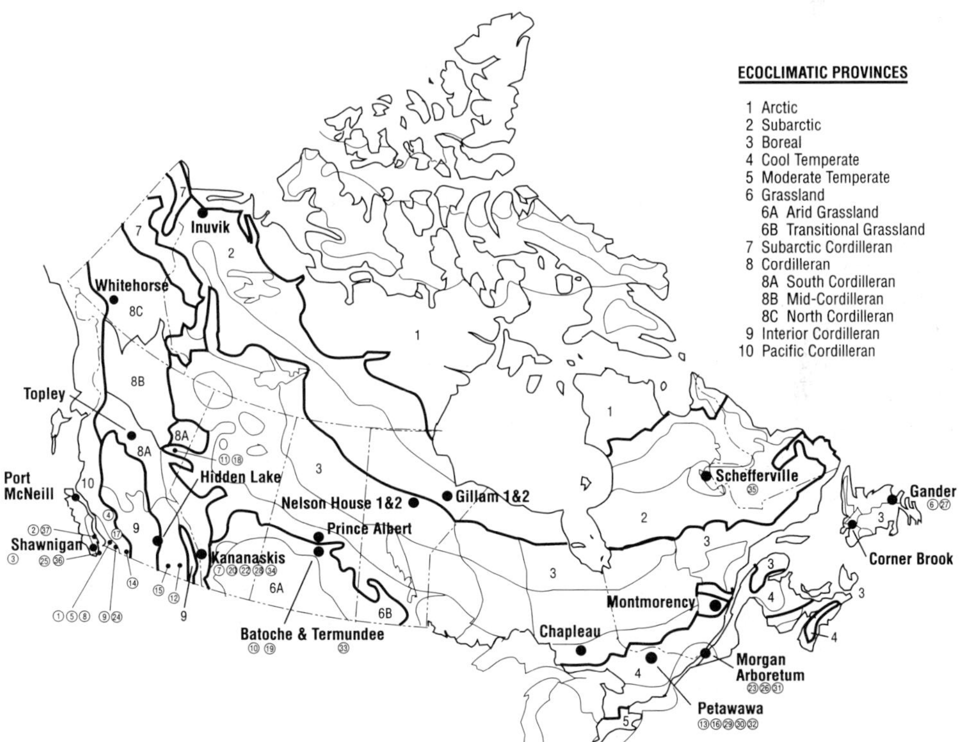

Fig. 1. Distribution of CIDET sites across the ecoclimatic provinces and regions of Canada (after Ecoregions Working Group, 1989) and approximate collection sites for the different litter types (circled numbers, refer to Table I). Sites at Batoche/Termundee, Nelson House and Gillam contain paired upland and wetland sites.

tannins) 92–140ppm; phenolic (tannins and lignin) 140–163 ppm; carboxyl (hemicellulose, amino acids) 163–185ppm. The relative area of the total spectrum in each region was determined by cutting and weighing the plotted spectra.

For further data analysis, the elemental and proximate chemical fraction data was expressed in mg g^{-1} of total dry litter mass and the NMR data as a relative percentage of total C. To better distinguish the most important factors differentiating litter types, a principal component analysis (PCA) (PRINCOMP procedure, SAS Institute Inc., 1985) was conducted for the elemental chemistry, proximate chemical and NMR data. PCA is useful for data with limited observations and many variables and has been used to identify the important chemical properties of different types of peat (Bohlin *et al.*, 1989), heterogeneous organic materials (Albano and Norden, 1988) and peat in different stages of decomposition (Axelson *et al.*, 1989). Comparisons of model predictions of mass loss for different litter types were done using the ranking and Spearman's rank correlation procedures as implemented in SAS.

3. Results

3.1. Chemical Analyses

Mean elemental concentrations (mg g^{-1} total dry mass) were C: 486.9, N: 6.65, P: 0.79, S: 1.83, Ca: 13.44, Mg: 1.53 and K: 2.38 with K, Ca, P and Mg showing the greatest range of variation amongst the different litter types (Table I). Wood and lichen had the lowest concentrations of N, P, Ca, Mg, K and highest S; alder the highest N and Mg concentrations; jack pine the highest Ca and second highest N concentration; western larch highest P; aspen highest K. The ACID, WSE and LIG fractions made up the majority of the mass of the various proximate chemical fractions (mean respective values 369.3, 243.3, 241.7 mg g^{-1} total dry mass) though significant amounts of NPE (mean 97.05 mg g^{-1}) and ASH (mean 52.6 mg g^{-1}) also occurred in all litter types (Table I). Wood and lichen had the lowest amounts of NPE, WSE and ASH and highest ACID fractions; lichen and fireweed the lowest amounts of LIG. The highest amounts of NPE were in lodgepole pine; highest WSE in fireweed; yew and western hemlock needles the highest LIG; sugar maple and fescue the highest ASH. As expected the NMR analysis (data not shown) indicated that most of the C was in the O-alkyl, carbohydrate, fraction (mean 45.7%) followed by the alkyl (20.47%) and aromatic (19.8%) fractions and least in the phenolic (7.3%), carboxylic (4.8%) and methoxyl (1.9%) region.

Results of the PCA for the elemental, proximate analysis and NMR data showed that the first 3 PCs explained close to or more than 70% of the variation in the data (Table II). The eigenvectors for each PC indicate the loadings each of the variables contribute to a PC; negative values indicate inverse relationships between the value and that PC. The first PC of the elemental data was heavily influenced by macronutrient content, N+P+K:S while amounts of Ca and C contributed to the second PC. For the proximate data the first PC related to ratio of total extractables (WSE+NPE):cellulose (ACID) while Klason lignin (LIG) related to the second PC. The ratio of O-alkyl:alkyl C contributed the strongest loading to the first PC of the NMR data while the sum, phenolics + aromatics, contributed to the second PC. This is illustrated in Figure 2, which shows the NMR spectra for six species spanning the range of values for the first and second PC. Engelmann spruce had the lowest ratio of O-alkyl:alkyl C and fescue the highest while lichen had the lowest amount of phenolic+aromatic C and tamarack the highest.

To examine whether the various litter types could be distinguished into groups based on broad botanical characteristics, the code for each type was plotted in the multivariate space defined by the first and second PC (Figure 3). In almost all graphs especially CHEM and PROX, conifer litters were clustered together with the non-conifer litters showing more spread. In the CHEM P1 *vs* P2 graph most conifers clustered in the upper portion with non-conifers spread in the lower portion of the graph, indicating that the conifers tended to have narrower ranges of macronutrient contents than the non-conifers. The exceptions were the reversed grouping of larches, salal and some of the spruces and separation of jack pine because of high Ca and N. Conifer species within the same or related genera tended to cluster with the exception of the pines. In the PROX P1 *vs* P2 graph conifers and non-conifers generally separated along the P2 axis, conifers clustered in the upper portion and non-conifers in the lower, spread along the P1 axis. Larches, cedars and yew, and hemlocks each clustered along the P1 axis. Conifers and non-conifers were less clearly separated in the

TABLE II

Summary of the Principal Component Analyses of elemental and proximate chemistries and NMR spectral data for the 37 litter types. For each of the top four principal components (PC) the eigenvalue and the proportional (PrVar), cumulative variance (CumVar) it explains are shown with the eigenvectors and loading each variable contributes to that PC.

Eigenvalues				Eigenvectors						
Elemental chemistry (mg g^{-1})										
PC	Value	PrVar	CumVar	C	N	P	S	Ca	Mg	K
1	2.53	0.36	0.36	-0.30	0.40	0.42	-0.40	0.28	0.32	0.48
2	1.23	0.18	0.54	0.63	0.24	-0.07	-0.34	0.57	-0.16	0.47
3	0.98	0.14	0.68	0.09	0.36	0.48	0.29	-0.16	-0.71	0.14
4	0.68	0.097	0.77	0.40	0.53	-0.16	-0.12	-0.65	0.32	0.02
Proximate chemical fractions (mg g^{-1})										
PC	Value	PrVar	CumVar	NPE	WSE	ACID	LIG	ASH		
1	2.21	0.44	0.44	0.44	0.58	-0.56	-0.25	0.31		
2	1.34	0.27	0.71	0.36	-0.17	-0.42	0.74	-0.34		
3	0.82	0.16	0.88	0.12	-0.42	0.02	0.25	0.86		
4	0.62	0.12	0.99	0.79	-0.32	0.34	-0.34	-0.17		
NMR spectral regions										
PC	Value	PrVar	CumVar	Alkyl	Methox	O-Alkyl	Arom	Pheno	Carbox	
1	2.44	0.41	0.41	-0.59	0.21	0.54	0.41	-0.09	-0.38	
2	1.73	0.29	0.70	0.02	-0.05	-0.38	0.54	0.72	-0.19	
3	0.95	0.16	0.85	0.13	0.96	-0.17	-0.16	0.06	-0.11	
4	0.77	0.13	0.98	-0.41	0.17	0.03	0.11	0.19	0.87	

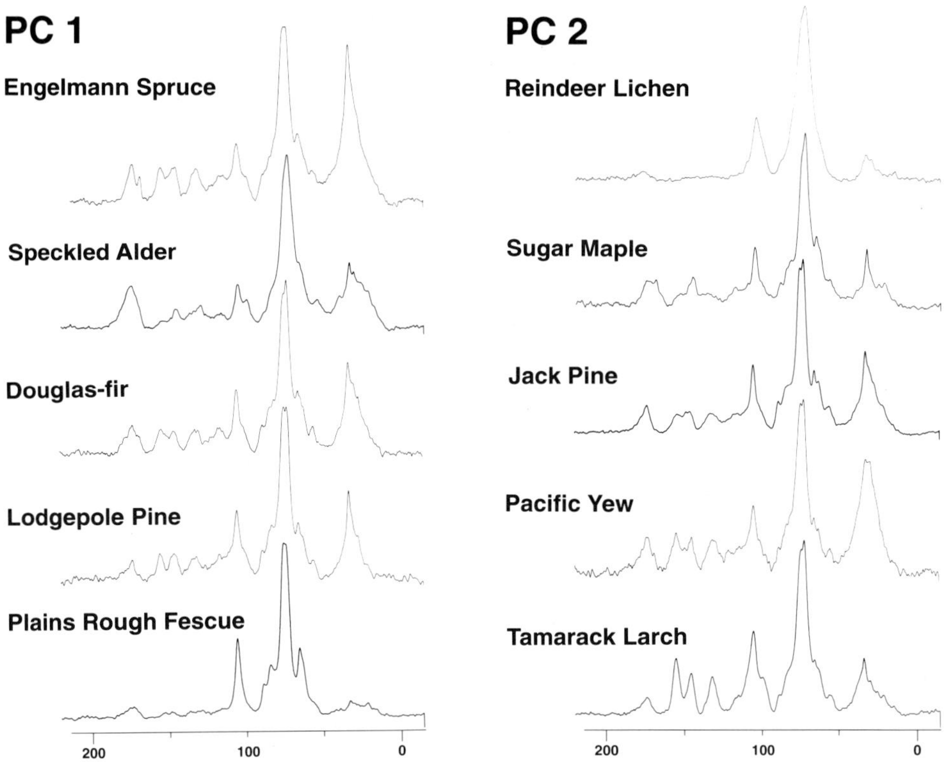

Fig. 2. Examples of NMR spectra of selected litter types as ranked by principal component (PC) analysis for PC 1 and PC2. Chemical shift regions: alkyl 0–50 ppm; methoxyl 50–60 ppm; O-alkyl 60–92 ppm; aromatic 92–140 ppm; phenolic 140–163 ppm; carboxyl 163–185 ppm.

the NMR P1 *vs* P2 graphs though clustering of some species of conifer genera did occur, the cedars and yew, Douglas-firs, hemlocks, and pines (P1 axis).

3.2. Model Rankings of Litter Quality

To examine the potential effects of litter quality on decay we examined three models, Melillo *et al.* (1982) Aber *et al.* (1990) and Taylor *et al.* (1991) and their predictions of mass percentage remaining (Figure 4). All models predicted that the non-conifer litter types would show the greatest range of mass loss, the lowest being wood and the highest fireweed. The actual range of predicted mass losses differed greatly from the model: the Aber model predicted 21–50% mass loss with the least variation with litter type, the Melillo model 15–90% mass loss, and Taylor 8–25%. The ranking of the different litter types were similar among all three models. Spearman's rank correlation coefficients for each pair of models was large (Aber/Taylor = 0.83, Melillo/Taylor = 0.56, Melillo/Aber = 0.78) and significant in all cases ($P < 0.001$).

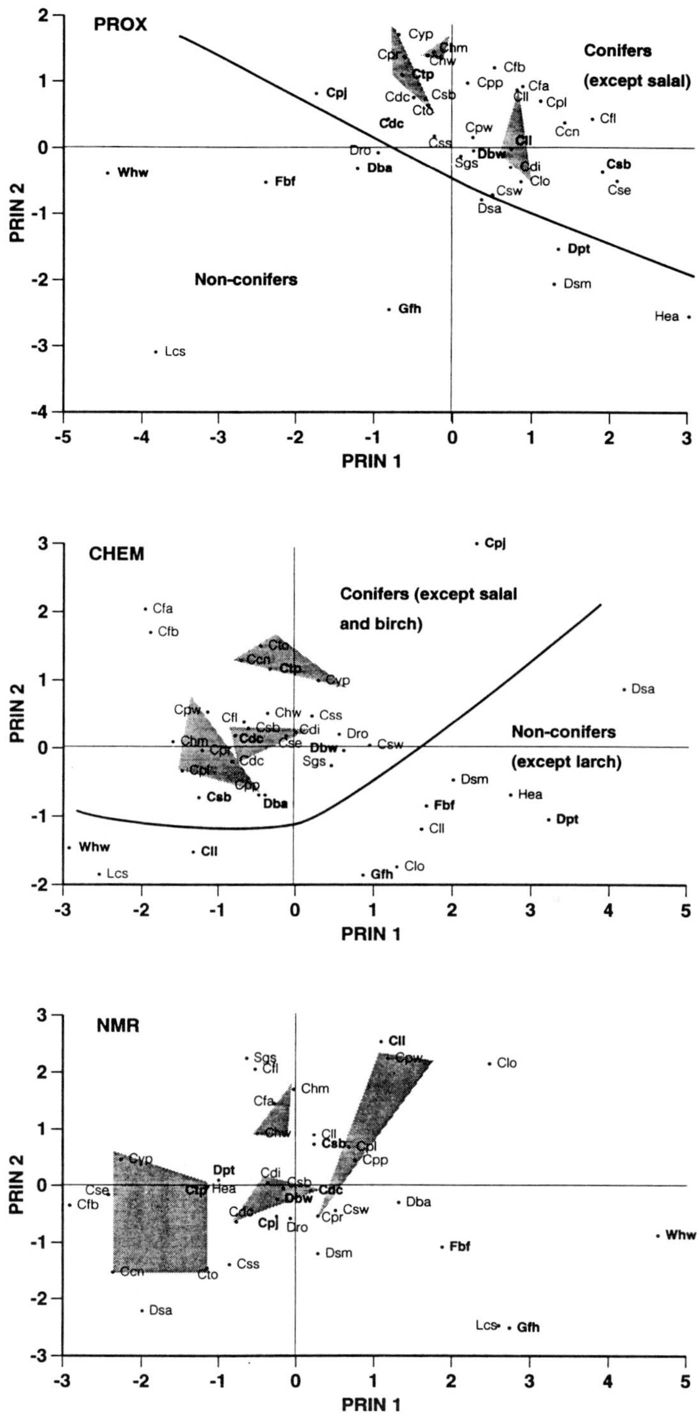

Fig. 3. All 37 litter types plotted (FC codes in Table I) with respect to principal component axes 1 and 2. Separate PC analyses were performed for the elemental (CHEM), proximate analysis (PROX) and ^{13}C CPMAS NMR (NMR) data.

4. Discussion

While the various measures of litter quality were useful in grouping coniferous and non-coniferous litters, they were less useful in grouping species within genera. Although Kiilsgaard *et al.* (1987) demonstrated that selected members of the Taxodiaceae and Cupressaceae families and *Pseudotsuga* could be distinguished from each other by litterfall elemental concentrations, the litters used were obtained from trees in the same arboretum. As leaf and litterfall nutrient concentrations are heavily influenced by site nutrient conditions (Stump and Binkley, 1993; Trofymow *et al.*, 1991),

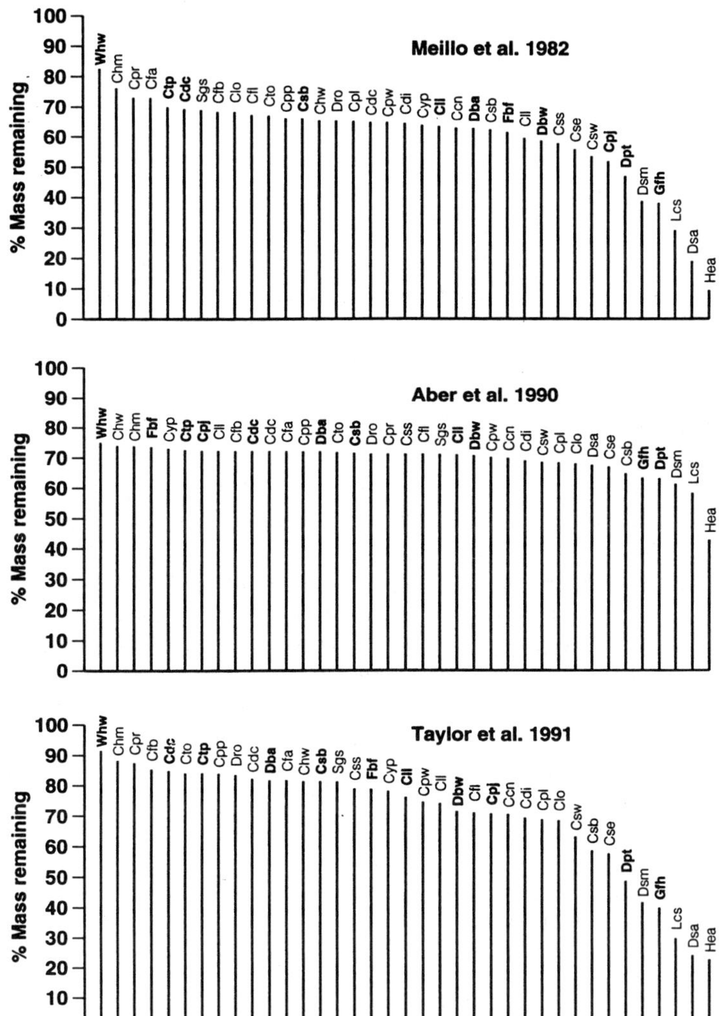

Fig. 4. Ranking of the 37 litter types (FC codes in Table 1) based on predictions of percent mass remaining from 3 models of litter decay, Meillo et al. 1982 (1 year decay), Aber et al. 1990 (1 year decay) and Taylor et al. 1991 (3 year decay).

Fig. 4. Ranking of the 37 litter types (FC codes in Table I) based on predictions of percent mass remaining from 3 models of litter delay. Melillo *et al.* (1982) (1 year decay) Aber *et al.* (1990) (1 year decay) and Taylor *et al.* (1991) (3 year decay).

it was not surprising the litters in this study were poorly grouped given the wide variety of sites from which litters were collected. As well, the separation of salal and larches for the elemental data give some indication that the conifer, non-conifer grouping may be more related to "evergreenness". Although some discrimination of botanical groupings was seen for the proximate and NMR data, less is known about the variation of organic fractions among genera and with site conditions. Certainly it has been observed that heather (Iason and Hester, 1993), willow (Waring *et al.*, 1985) and aspen (Bryant *et al.*, 1987) all change their lignin and tannin contents in response to changes in nutrient supply.

Examination of the three decay models does suggest that litter quality could have a large effect on short-term decay rates. It was surprising how differently the summed exponential model by Aber *et al.* (1990) behaved in comparison to the two other models. Considering the number of litter quality variables it used, the range of mass loss predicted for the different litter types was much less. Many of the recent studies of litter decomposition in Canadian forests confirm that that chemical measures of litter quality are important for predicting litter decay (Lousier and Parkinson, 1976; MacLean and Wein, 1978; Moore, 1984; Taylor *et al.*, 1991). Fyles and McGill (1987) do suggest that physical attributes of litter quality such as piece size and cuticle thickness may have as much or more importance than chemical measures. Their suggestions may be more applicable to the early stages of litter decay as their results were from a 35-week laboratory incubation and the chemical measures they used (lignin, N, tannin) are perhaps less appropriate to the initial stages of decomposition when low molecular weight compounds are lost first. They did observe that a double exponential model fitted their data best, compatible with litters composed of labile and non-labile fractions.

The CIDET litter types spanned a large range of litter quality characters and thus results from this experiment should provide a useful test of the effects of litter quality on decay rates. Based on its chemical characteristics, decay rates of wood may not be predictable from the same model used for other litter types. One test of the effects of litter quality that could be made as data from CIDET become available, would be to compare the observed and models' rankings of mass loss for the different litter types for all sites. Comparisons with model predictions of absolute mass loss would be inappropriate as the models were developed from sites with different climates and the decay rates were not temperature-compensated. Further work will use climate data for each site to develop specific functional relationships between k and site climate.

Acknowledgments

This study was supported with funding through the Canadian Forest Service's Green Plan Climate Change Program to the CIDET working group (Study Leader: J. Trofymow; Site Co-operators: D. Anderson, C. Camire, L. Duchesne, J. Fyles, L. Kozak, T. Moore, I. Morrison, C. Prescott, M. Siltanen, S. Smith, B. Titus, R. Trowbridge, S. Visser, R. Wein, D. White, S. Zoltai; Chemical Analyses Co-operators: C. Preston, A. van Niekerk, B. Sayer.) and through NSERC strategic grant No. STRIN-235 to C. Prescott. The chemical, proximate and NMR analyses were conducted under the direction of A. van Niekerk (Chemical Services Lab, Pacific Forestry Centre), M. Harmon (Forest Services lab, Oregon State University), and B. Sayer (Dept. Chemistry, McMaster University), respectively. M. Weber, C. Montreal and members of the CIDET working group collected the different litter types. Thanks to Eric Andersen, Stacey Dixon and Jennifer Kinnis for their help in preparing litterbags.

References

Aber, J. D., Melillo, J. M. and McClaugherty, C. A.: 1990, *Can. J. Bot.* **68**, 2201-2208.

Albano, C. and Norden, B.: 1988, *Fuel* **68**, 771-775.

Axelson, D. E., Dinel, H., Dudley, R. L., Levesque, M., Mathur, S. P. and Preston, C. M.: 1989, *Org. Geochem.* **14**, 393-403.

Baldock, J. A. and Preston, C. M.: *In: Proceedings of 8th N. American Forest Soils Conference*. August, 1993. Gainesville, USA, (In press).

Berg, B.: 1984, *Soil Biol. Biochem.* **16**, 609-617.

Berg, B. and Staaf, H.: 1980, *Structure and Function of Northern Coniferous Forests–An Ecosystem Study* **32**, 373-390.

Berg, B., Berg, M. P., Bottner, P., Box, E., Breymeyer, A., Calvan De Anta, R., Couteaux, M., Esudero, A., Gallardo, A., Kratz, W., Madeira, M., Malkonen, E., McClaugherty, C., Meentemeyer, V., Munoz, F., Piussi, P., Remacle, J. and Virzo De Santo, A.: 1993, *Biogeochemistry* **20**, 127-159.

Bohlin, E., Hamalainen, M. and Suden, T.: 1989, *Soil Sci.* **147**, 252-263.

Bunnell, F. L., Flanagan, P. W. and Tait, D. E. N.: 1977a, *Soil Biol. Biochem.* **9**, 41-47.

Bunnell, F. L., Tait, D. E., Flanagan, P. W. and Van Cleve, K.: 1977b, *Soil Biol. Biochem.* **9**, 33-40.

Bryant, J. P., Clausen, T. P., Reichardt, P. B., McCarthy, M. C. and Werner, R. A.: 1987, *Oecologia* **73**, 513-517.

Ecoregions Working Group: 1989, *Ecol. Land Classif. Ser. No. 23*, Sustainable Development Branch, Canadian Wildlife Service, Conservation and Protection, Environment Canada, Ottawa, Canada, 119 pp.

Fyles, J. W. and McGill, W. B.: 1987, *Can. J. For. Res.* **17**, 109-114.

Harmon, M. E., Baker, G. A., Greene, S. and Spycher, G.: 1990, *For. Ecol. Man.* **31**, 55-66.

Harmon, M. E. and Melillo, J. M.: 1990, *Protocol for Intersite Decomposition Experiments: I. Fine Root, Leaf Litter and Wooden Dowels*. Oregon State University. Unpublished Report.

Heal, O. W. and French, D. D.: 1974, *In:* Holding, A. J., Heal, O. W., MacLean, S. F. Jr. and Flanagan, P. W. (eds.) *Soil Organisms and Decomposition in Tundra*. IBP Tundra Biome Steering Committee, Stockholm, Sweden, pp. 227-248.

Hunt, H. W., Ingham, E. R., Coleman, D. C., Elliott, E. T. and Reid, C. P. P.: 1988, *Ecology* **69**, 1009-1016.

Iason, G. R. and Hester, A. J.: 1993, *J. Ecol.* **81**, 75-80.

Kiilsgaard, C. W., Greene, S. E. and Stafford, S. G.: 1987, *Plant and Soil* **102**, 223-227.

Kurz, W. A., Apps, M. J., Webb, T. M. and McNamee, P. J.: 1992, *The Carbon Budget of the Canadian Forest Sector: Phase I.* For Can. Inf. Rep. NOR-X-326, Forestry Canada, Northwest Region, Edmonton, Canada, 93pp.

Long-Term Intersite Decomposition Experiment Team (LIDET): *Bioscience* (In press)

Lousier, J. D. and Parkinson, D.: 1976, *Can. J. Bot.* **54**, 419-436.

MacLean, D. A. and Wein R. W.: 1978, *Can. J. Bot.* **56**, 2730-2749.

McClaugherty, C. A., Pastor, J., Aber, J. D. and Melillo, J. M.: 1985, *Ecology* **66**, 266-275.

Meentemeyer, V.: 1978, *Ecology* **59**, 465-472.

Melin, E.: 1930, *Ecology* **11**, 72-101.

Melillo, J. M., Aber, J. D. and Muratore, J. F.: 1982, *Ecology* **63**, 621-626.

Minderman, G.: 1968, *Ecology* **56**, 355-362.

Moore, T. R.: 1984, *Ecology* **65**, 299-308.

Norden, B. and Berg, B.: 1990, *Soil Biol. Biochem.* **22**, 271-275.

Parkinson, J. A. and Allen, S. E.: 1975, *Comm. in Soil Sci. Plant Analysis* **6(1)**, 1-11.

Parton, W. J., Schimel, D. S., Cole, C. V. and Ojima, D. S.: 1987, *Soil. Sci. Soc. Am. J.* **51**, 1173-1179.

Paul, E. A. and Voroney, P.: 1980,. *In:* Ellwood, D. C., Hedges, J. N., Latham, M. H. and Slater, J. H. (eds.) *Contemporary microbial Ecology*. Academic Press, London, UK, pp. 215-237.

Preston, C. M., Hempfling, R., Schulten, H. R., Schnitzer, M., Trofymow, J. A. and Axelson, D. E.: 1994, *Plant and Soil* **158**, 69-82.

Ryan, M. G., Melillo, J. M. and Ricca, A.: 1990, *Can. J. For. Res.* **20**, 166-171.

SAS Institute Inc.: 1985, *SAS User's Guide: Statistics, Version 5 Edition*. SAS Institute Inc. Cary, USA 956 pp.

Stump, L. M. and Binkley, D.: 1993, *Can. J. For. Res.* **23**, 492-502.

Swift, M. J., Heal, O. W. and Anderson, J. M.: 1979, *Decomposition in Terrestrial Ecosystems*. University of California Press, Berkley, USA, 362 p.

Taylor, B. R., Prescott, C. E., Parsons, W. F. J. and Parkinson, D.: 1991, *Can. J. Bot.* **69**, 2242-2250.

Trofymow, J. A., Barclay, H. J. and McCullough, K.: 1991, *Can. J. For. Res.* **21**, 1601-1615.

Waring, R. H., McDonald, A. J. S., Larsson, S., Ericsson, T., Wiren, A., Arwidsson, E., Ericsson, A. and Lohammar, T.: 1985, *Oecologia* **66**, 157-160.

DYNAMICS OF THE DEAD WOOD CARBON POOL IN NORTHWESTERN RUSSIAN BOREAL FORESTS

O. N. KRANKINA and M. E. HARMON

Department of Forest Science, Oregon State University, Corvallis, Oregon 97331, USA

Abstract. Our study examines dead wood dynamics in a series of permanent plots established in closed, productive second-growth forest stands of north-west Russia and in temporary plots that represent different successional stages and types of disturbance. Dead wood stores measured on 63 plots 0.2–1.0 ha in size range from 1–8 Mg C ha^{-1} in young to mature intensively managed stands, 17 Mg C ha^{-1} in an old-growth forest, 20 Mg C ha^{-1} on a clear-cut, and 21–39 Mg C ha^{-1} following a severe windthrow. A total of 122 logs, snags, and stumps aged by long-term plot records was sampled for decay rates and to develop a system of decay classes. Annual decomposition rates are: 3.3% for pine, 3.4% for spruce, and 4.5% for birch. Based on these decay rates the average residence time of carbon (C) in the dead wood pool is 22–30 years. The mortality input on the permanent plots was 23-60 Mg C ha^{-1} over 60 years of observation or 15–50% of the total biomass increment. This data suggests a dead wood mass of 10–22 Mg C ha^{-1} would be expected in these mature forests if salvage had not occurred. In old-growth forests, dead wood comprised about 20% of the total wood mass, a proportion quite similar to the larger, more productive forests of the Pacific Northwest (USA). If this proportioning is characteristic of cool conifer forests it would be useful to estimate potential dead wood mass for old-growth forests without dead wood inventories. However, the use of a single live/dead wood ratio across the range of successional stages, a common practice in C budget calculations, may substantially over- or under-estimate the dead wood C pool depending upon the type of disturbance regime. Intensive forest management including short harvest rotations, thinning and wood salvage reduces dead wood C stores to 5–40% of the potential level found in undisturbed old-growth forest. In contrast, natural disturbance increases dead wood C pool by a factor of 2–4.

Keywords. DEAD WOOD, CARBON CYCLE, RUSSIAN FORESTS, DECAY RATES, BIOMASS POOLS.

1. Introduction

One of the most important questions facing ecologists today is the role of the biota in the global carbon (C) cycle. The fact that 20–50% or 1–3 Pg yr^{-1} of the annual fossil fuel C can not be accounted for (Post *et al.*, 1990; Dixon *et al.*, 1994) raises serious questions about our current understanding of this cycle. Moreover, the recent decrease in atmospheric concentrations can not be fully explained; oceanic uptake cannot account for it (Sarmiento, 1993). Recent efforts to decrease uncertainty of global C cycling processes have focused upon improved estimates of land-use change in the tropics (Dale *et al.*, 1991), sequestration in soil humus (Schlesinger, 1990), increased production associated with CO_2 fertilisation (Keeling *et al.*, 1989), and sequestration by temperate forests (Tans *et al.*, 1990). Despite years of study it has recently become clear that we still do not understand the current role of forests in controlling the global C cycle (Dixon *et al.*, 1994). Over the past few years forests have been described as everything from a source (Houghton *et al.*, 1987) to a sink of C (Tans *et al.*, 1990).

We suggest there may be another fundamental problem with current global C budgets: dead trees and associated woody detritus are generally not considered (Kobak, 1988; Houghton and Woodwell 1989; Post *et al.*, 1990). At best this pool is included as a constant proportion to live biomass (Kolchugina and Vinson, 1993). There are many reasons why, until recently, there has been little research on this topic (Harmon *et al.*,

Water, Air and Soil Pollution **82**: 227–238, 1995.

1986). First, decomposition of woody detritus is a long-term process and, as with other long-term ecological processes (i.e., forest succession), there are few long-term observational studies (Franklin *et al.*, 1990). Second, inappropriate methodology has been used to study dead wood dynamics. For example, litter production is often estimated from a sample area <10 m^2 for 1–2 years, and although this is suitable for fine litter, it is inappropriate for woody detritus which requires many hectare-years for reliable estimates (Sollins, 1982). Finally, there are few estimates of dead tree stores because they have little commercial value and until recently were associated with "unhealthy" forest or bad management practices. Thus little empirical data exists upon which to base estimates or ground assumptions.

Ignoring dead trees has several profound impacts upon our understanding of global C dynamics. First, it excludes a flux at least as large as fossil fuel burning. Recent global estimates of tree mortality indicate 3–15 Pg C yr^{-1} of woody detritus is produced within intact stands, while natural catastrophic disturbances add another 1 to 2 Pg yr^{-1} globally (Harmon *et al.*, 1993). Second, the size of the terrestrial detrital pool has been underestimated by 85–295 Pg C (Harmon and Chen, 1991; Harmon *et al.*, 1993). If woody detritus was in steady-state this might not be a major concern, but land-use changes have placed woody detritus in disequilibrium. In the Pacific Northwest, for example, reduction of woody detritus accounted for 33% of the C flux to the atmosphere associated with the conversion of old-growth forests to plantations (Harmon *et al.*, 1990). If this is a general rule, then past calculations of C flux from forest clearing have been underestimated. It also implies that a large, unaccounted for C sink may be present in forests recovering from past harvest (Harmon and Chen, 1991). Third, excluding dead trees will lead to an unrealistic assessment of the response of detrital stores to projected climate change. Analysis of transient responses to projected climate change indicate a large pulse of C may be injected into the atmosphere as forests are disturbed or migrate (King and Neilson, 1992). To a large degree, the temporal dynamics of this transient pulse will be controlled by the decomposition of woody plants killed by catastrophic disturbances (i.e., fire), or increased stress. Fourth, estimated effect of intensive forest management on C sequestration (Sedjo and Solomon, 1991; Dixon *et al.*, 1991) would probably be reconsidered if woody detritus could be taken into account. The commonly held notion that young forests are active C sinks does not hold if the flux from decomposing dead wood is included in calculations (Cohen *et al.*, in review). We have found that in managed forests, when woody detritus removed by timber harvest, salvage operations, and firewood gathering is included total ecosystem C stores can be substantially decreased (Krankina and Harmon, 1994).

Clearly the time has come to begin including woody detritus in the global C cycle. However, with the currently available database it is extremely difficult to advance beyond order of magnitude estimates. To date the majority of work concerning the production, decomposition, and storage of C in dead trees has centred in the Pacific Northwest. Now the United States National Science Foundation and Earthwatch funding has allowed us to study the woody detritus production via tree mortality, as well as the dead wood decomposition and stores in the taiga forests of north-western Russia. Russian forests comprise about 50% of the globe's boreal forests and thus play a significant role in the global C cycle (Kobak, 1988). Russian taiga forests have a great potential in explaining the missing C sink yet our understanding of these forests is inadequate and their overall impact remains uncertain: some authorities consider it a major sink and others a source of CO_2 (Kolchugina and Vinson, 1993; Melillo *et al.*, 1988).

The goal of this paper is to examine the dynamics of woody detritus in relation to tree species, successional stage, management and disturbance regime in a Russian boreal forest based on preliminary results of our on-going research. To the best of our knowledge this is the first time that empirical data on long-term decomposition rates for Russia's major boreal tree species has been published.

2. Study Area and Methods

The study area is located at about 59°N and between 31°E and 32°E in the St. Petersburg and Novgorod Regions of north-western Russia. The climate is cool maritime with cool wet summers and long cold winters. The area is a part of the East-European Plain with elevations between 0 and 250 m above sea level; terrain is gentle and rests on ancient sea sediments covered by a layer of moraine deposits. Soils are mostly of podzol type on deep loamy or sandy sediments. Natural vegetation of the area belongs to southern taiga types; major dominant conifer species include Scots pine (*Pinus sylvestris* L.) and Norway spruce (*Picea abies* (L.) Karst.) both growing in pure and mixed stands. After disturbance, they are often replaced by hardwoods including birch (*Betula pendula* Roth.) and aspen (*Populus tremula* L.).

Experimental data for this study was collected in closed forest stands of medium to high productivity subject to intensive forest management. Although this data set does not represent the average for the region, it allows us to examine woody biomass distribution and the impact of management practices on C stores at stand level in intensively managed forest types.

Our work takes advantage of a permanent plot system started over 60 years ago to study effects of thinning upon forest growth in St. Petersburg (then Leningrad) region in north-western Russia. Most plots were set up in 1928–34 and supplemented by new plots in 1968–76 at the Leningrad (now St. Petersburg) Forestry Research Center (Sennov, 1984). The plots were set up in well stocked young stands (23–45 years old) and over their existence they were regularly remeasured at 4–11 year intervals so that by now an age span of 20 to 105 years is covered by the record. Each site consists of a control plot and 1 to 4 thinned plots with 15–50% of the growing stock removed. The size of each plot is 0.1–0.25 ha. The historical data on live tree measurements currently maintained by St. Petersburg Forest Academy is supplemented by one plot established in an old-growth spruce stand and 8 plots (including 3 controls) in a windthrow area by the St. Petersburg Forest Academy and by 4 temporary plots set up by the authors on recent clear-cuts. The scope of work includes: (1) estimation of decay rates; (2) live and dead wood inventory; (3) reconstruction of dead wood dynamics over succession.

A chronosequence approach was used to estimate decomposition dynamics of the three major species: birch, pine, and spruce. The time elapsed since the tree died was estimated based on plot records. A chronosequence is a substitution of space for time. Although not as precise as time series measures, chronosequences can give a valid first approximation of temporal dynamics. Two major problems associated with most chronosequences are avoided in this study. The first problem is uncertainty of the time of death, a minor problem in our case as the time of death will be known within ±3 years from the plot records. In the case of thinned trees, the exact year of death could be determined. The second problem involves lack of information concerning initial conditions of dead trees. This is an especially large problem when fragmentation has occurred; major underestimates of decay losses and nutrient flux can be made if fragmentation losses are ignored (Harmon and Chen, 1991). Fragmentation is not a major problem in our case because the diameter at the time of death was recorded, thus the original volume and mass can be accurately estimated. Each of the three species to be examined had a chronosequence sampled in the plots of at least three sites. Each chronosequence consists of 2–12 trees in each of 5 age/decay classes (Figure 1). The total of 122 logs, snags, and stumps was sampled to estimate average density for decay classes. Out of these, 75 were dated and used to develop the chronosequences.

For each log, snag or stump sampled, we recorded length, diameters, and bark cover, and noted external characteristics associated with the degree of decomposition. Cross-sections were then removed using a chainsaw to measure the density of the bark and wood. To account for the variation within logs, three to four cross-sections were taken systematically along the length of the bole. The maximum and minimum diameter and height above the root collar of each cross-section was recorded. For each cross-section the density and radial thickness of the bark, sapwood, and heartwood as well as the depth and type of decay (white rot *versus* brown rot) was measured. In the case of extremely decayed cross-sections, the external dimensions were recorded prior to removal to estimate sample volume. Mass was determined after oven drying at 55 °C until constant weight was reached. Sample density was calculated as dry mass divided by green volume.

To calculate the average density of each sampled log or snag we weighed sample densities by respective section volumes. Density values were then averaged by decay classes for each tree species. The result of this work was the decay class system that allows to convert the dead wood volume into biomass (Table I). To account for mass loss through fragmentation for dated logs and snags, residual density was calculated by dividing current biomass by the estimated bole volume at the time the tree died (Harmon *et al.*, 1987). Average decomposition rate of dead wood was estimated using regression of the residual density over time:

$$\text{density}_t = \text{density}_0 \, [\exp(-kt)] \qquad [1]$$

where t is the time elapsed since the tree died, density_t is the residual density, density_0 is the initial density, and k is the decomposition rate.

Live and dead aboveground tree biomass was measured in the 63 plots. Live tree biomass was calculated by standard procedures based on diameter and height measurements of trees in each plot and on local forest mensuration tables (Tretyakov *et al.*, 1952; Moshkalev *et al.*, 1984). A biomass to C conversion factor 0.5 was used. As dead wood is commonly overlooked in biomass assessments and C budget calculations we describe the methods for this component in greater detail.

Measurements of dead wood stores in the plots followed the methods procedures described in Harmon *et al.* (1986, 1987). Within each plot the end diameters and lengths of each piece of dead wood >10 cm diameter and >1 m in length was recorded. All forms of coarse dead wood were inventoried including snags (standing dead), logs (dead and downed), stumps (cut by harvest), and "blobs" of highly decayed wood. We did not inventory wood buried in the forest floor, although this can be a substantial proportion of that layer (Little and Ohman, 1988). Species and decay state of each piece were also noted. The mass was calculated by multiplying computed volume by the average bulk density of each decay class and species. Total organic mass was converted to C stores by assuming the C content of dead wood was 51% (Sollins *et al.*, 1987). Live tree and dead wood C stores were compared for old-growth and early successional forest including clear-cut, young (10–20 years after harvest), middle aged (30–70 years after harvest), and mature second growth forest (80–120 years after harvest). Where possible the effect of natural disturbance (windthrow), thinning and salvage was assessed. Successional stages beyond the limits of observations on permanent plots were reconstructed using local growth and yield tables (Tretyakov *et al.*, 1952; Moshkalev *et al.*, 1984) and live/dead wood ratios from the measured temporary plots (Figure 2).

Mortality input into the dead wood pool over time was calculated for several selected plots over the 60-year observation period based on the long-term records of the St. Petersburg Forest Academy (Sennov, unpublished data) and using the same biomass equations as for live trees (Tretyakov *et al.*, 1952; Moshkalev *et al.*, 1984). Decomposition rates applied to mortality inputs allowed us to reconstruct the dynamics of dead wood stores on the control and on the thinned plots (Figure 3). Comparison of reconstructed and measured dead wood stores provided an estimate of the impact of salvage.

TABLE I

Decay class system for Russian boreal tree species.

Species	Decay class	Density ± STD ($Mg\ C\ m^{-3}$)	Sample size (dated samples)	Mean age (yr)*
Pine	1	0.201 ±0.018	10 (6)	3
	2	0.160 ±0.037	14 (10)	8
	3	0.117 ±0.032	17 (12)	16
	4	0.071 ±0.042	12 (7)	29
	5	0.065 ±0.002	3 (2)	70
Spruce	1	0.169 ±0.030	6 (6)	4
	2	0.159 ±0.028	15 (11)	7
	3	0.106 ±0.029	10 (5)	12
	4	0.038 ±0.005	3 (0)	-
Birch	1	0.251 ±0.017	5 (5)	1
	2	0.236 ±0.038	5 (2)	4
	3	0.142 ±0.035	13 (5)	14
	4	0.063 ±0.036	9 (4)	20

*Based on 75 dated logs.

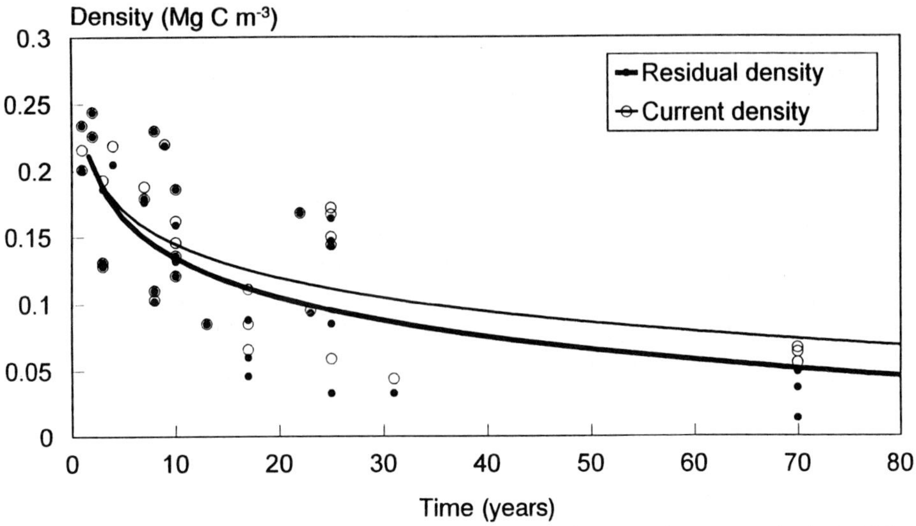

Fig. 1. Chronosequence of dead wood density for pine. Due to fragmentation current density is higher than residual density (current biomass of a dead tree divided by its initial volume).

3. Results

The decay class system relates the 5 decay classes for each of the 3 studied species to density values (Table I). The changes in density values with decay classes depend on initial density which is considerably higher for birch than for conifers and on the dynamics of decomposition process. Spruce and birch logs at final stages of decay were not found in studied plots resulting in missing density values in class 5. This may be due to the fact that these two species were decayed by white rot fungi, a functional group that does not leave lignin, the materials typically found in the decay class 5 boles (McFee and Stone, 1966).

The loss of density over time does not fully account for the loss of biomass by a dead tree because it does not include the loss of volume due to fragmentation. In the advanced stages of decay density values change little because fragmentation removes the most decayed material (Figure 1).

Average annual decay rates (±90% confidence interval) for pine, spruce and birch are 3.3 ±0.6%, 3.4 ±1.0%, and 4.5 ±1.3% respectively. Due to a significant variation in decay rates of individual logs, the correlation between the residual density and the time elapsed since the tree died is rather low, with r^2 ranging from 0.67 to 0.81. These decay rates generally agree with the common assumptions that birch decomposes faster than conifers, and that pine retains higher density in the advanced stages of decay because of the slow decomposition of its heartwood. Based on these decay rates the average residence time of C in the dead wood pool is 22–30 years depending on species.

In these forests, the position of a dead tree (log *versus* snag) is an important control at the early stages of decomposition. For example, the average density of pine snags during the first 10 years is equal to the sound wood density (0.205 Mg C m^{-3}) indicating

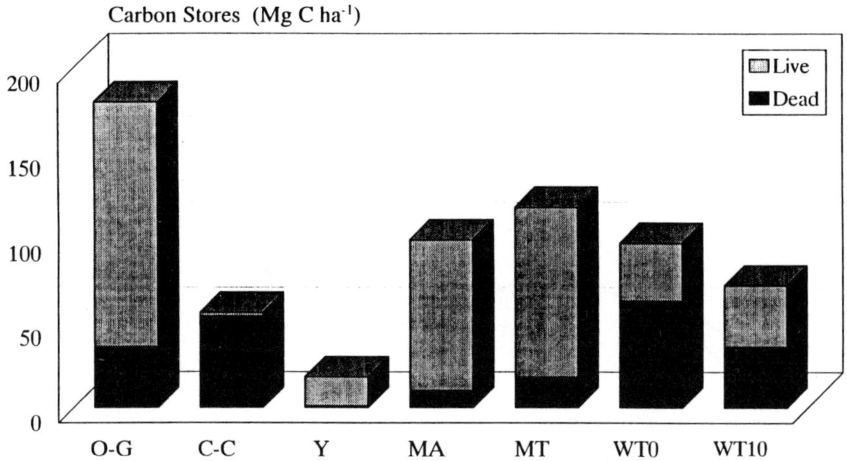

Fig. 2. Successional dynamics of live biomass and dead wood on a productive site (reconstruction based on local growth tables and live/dead wood ratio from measured plots): O-G: old-growth, C-C: clear-cut, Y: young, MA: middle age, MT: mature, WTO: following windthrow, WT10: 10 years after windthrow.

little decomposition occurred during this period. In contrast, the density of pine logs on the ground of the same time period is 0.145 Mg C m^{-3}, which represents a 27% biomass loss (the difference is significant at 0.0013 level). Similarly, spruce and birch logs lost more biomass compared to snags by 25% and 39%, respectively. As snags rarely appear to stand longer than 10 years, this decrease in decay rates is a temporary phenomenon.

Dead wood stores vary significantly over succession and do not necessarily parallel the dynamics of live biomass (Figure 2). The largest amount of dead wood was found in recently disturbed areas. Nearly all above-ground biomass present on a clearcut was in dead wood store which measured 20 Mg C ha^{-1}. Another site that was recovering for 10 years after a severe windstorm had 21–39 Mg C ha^{-1} in dead wood or 43–57% of the total biomass found on the site. Dead wood stores immediately following this windstorm, estimated by applying undecayed densities to class 2 and 3 log volumes, are even higher (39–52 Mg C ha^{-1} or 59–69% of total biomass). Dead wood stores are minimal in young stands (10–40 years old) regenerating after clearcut harvest and in older second-growth stands where dead wood is salvaged (1–8 Mg C ha^{-1}). In undisturbed old-growth forest, woody detrital stores measured 17 Mg C ha^{-1} or 20% of the total aboveground biomass.

The dynamics of woody detritus on permanent plots were reconstructed using the data on dead wood input by tree mortality adjusted for losses to decomposition (Figure 3). The mortality input on the control plots was 23–60 Mg C ha^{-1} over 60 years of observation or 15–50% of the total above-ground biomass increment. Thinning reduced tree mortality: on thinned plots it was 4–26 Mg C ha^{-1} or 7–12% of the total biomass increment. Our reconstruction of dead wood dynamics indicates significant accumulation of woody detritus with stand age. Predicted dead wood stores (10–22 Mg C ha^{-1}) were higher than those actually measured in the plots due to salvage operations.

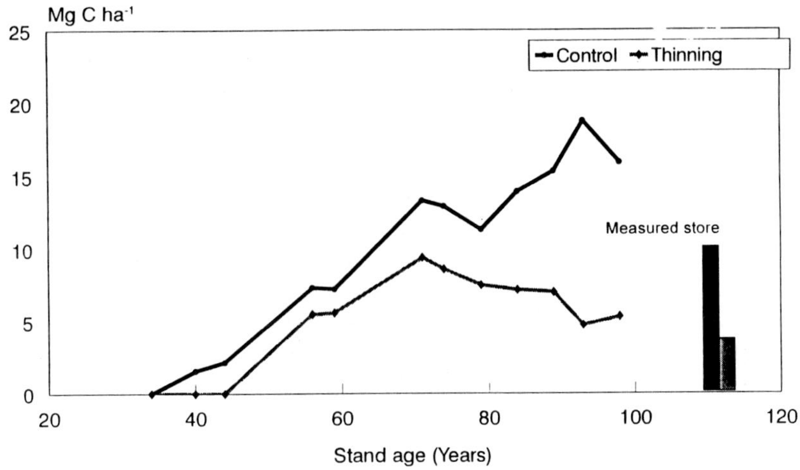

Fig. 3. Reconstructed dead wood dynamics in a pine stand over the period of observations and measured dead wood store at the end of the period (example).

4. Discussion

Comparison to other carbon pools

This study indicates a considerable pool of woody detritus can be found in boreal forests. The estimated range for undisturbed forests of 17–36 Mg C ha^{-1} is similar to the stores found in surface litter (10–32 Mg C ha^{-1}) measured on the same plots, but is less than the mass of organic matter stored in the very shallow but highly organic soil (57–128 Mg C ha^{-1}) (Sennov, 1984; Kobak, 1988). In forests disturbed by fire, which consumes much of the litter and organic soil, or by clearcut harvest, which removes much of the live biomass, woody detritus can become the dominant store of organic matter and C within the ecosystem, making 98% of above-ground biomass on a clearcut and 59–69% following a windthrow.

Our estimates of woody detritus for boreal forest ecosystems may be low for two reasons. First, we did not consider the wood associated with dead coarse roots and branches. This could increase our estimates up to 30% depending upon the degree of sprouting in top killed trees, the difference between above- and below-ground decay rates, and the effect of size on decay rates. Second, we did not consider wood buried within the soil and forest floor in our inventories. Including these other forms of woody detritus would increase our estimates of total woody detritus in these forests.

Comparison to other biomes

The mass of dead wood depends upon a number of factors including the productivity of the ecosystem, climate, substrate quality, and organisms degrading the material (Harmon *et al.*, 1986). As latitude increases the production rate of dead wood should decrease along with the overall productivity of the forest (Harmon *et al.*, in press). Although this should lead to lower dead wood stores, this trend is partially offset by a decrease in the decomposition rate caused by a decrease in temperature and macro-invertebrate abundance (e.g., termites), and by more frequent disturbances. Comparison of our study results to other major biomes indicates that dead wood C pool in undisturbed Russian boreal forest is generally within the range of that reported for some other forest regions: 9–21 Mg C ha^{-1} in undisturbed dry tropical forests (Harmon *et al.*, in press), 2–25 Mg C ha^{-1} in moister tropical forests (Kauffman *et al.*, 1988; Edwards and Grubb, 1977; Uhl and Kauffman 1990), 15–25 Mg C ha^{-1} in cold deciduous forests (Gore and Patterson, 1986; Harmon *et al.*, 1986; Tritton, 1980) or pine dominated conifer forests where 14–21 Mg C ha^{-1} of coarse woody debris has been reported (Fahey, 1983; Harmon *et al.*, 1986). In contrast, woody detrital stores in boreal forests are much smaller than those found in cool, moist conifer forests which range between 100 and 225 Mg C ha^{-1} (Agee and Huff, 1987; Grier and Logan 1977; Spies *et al.*, 1988), but exceed those reported for warm temperate deciduous forests where 10–12 Mg C ha^{-1} is typical (Muller and Liu, 1991; Onega and Eickmeir, 1991; Harmon *et al.*, 1986; Harmon and Chen, 1991). This comparison suggests there is no simple correlation between woody detrital mass and latitude.

Decomposition rates

The rates of decomposition that we estimated for the dominant conifer species in north-western Russia (0.033–0.034 yr^{-1}) are generally higher than for other northern ecosystems. Red spruce and balsam fir in the north-east USA decompose at rates of 0.0299 yr^{-1} (Lambert *et al.*, 1980) to 0.029–0.033 yr^{-1} (Foster and Lang, 1982). In western Canada lodgepole pine and Engelmann spruce had a wide range of decay rates (0.0025–0.0299 yr^{-1}) depending upon the age of forest (Johnson and Greene, 1991). Lodgepole pine had a rate of 0.027 yr^{-1} reported for central Oregon (Busse, 1994) and a lower rate of 0.016 yr^{-1} in a cool dry Wyoming site (Fahey, 1983). As these species are similar to those we studied, we would conclude that the overall environment in north-western Russia is slightly more favourable to wood decomposition than many other northern forests.

Decomposition rates of birch were lower than the 0.11 yr^{-1} reported for warmer hardwood dominated forests (Onega and Eickmeier, 1991). Although there been no reported studies on birch, another hardwood species, aspen, is reported to decompose at a rate of 0.05 yr^{-1} in Minnesota (Miller, 1983). Our observations indicate the aspen decomposes faster than birch in Russian forests, so we can not judge the degree to which climate is affecting hardwood decomposition rates.

Influence of disturbance and succession

Disturbance plays a major role in determining the amount and nature of the dead wood. The most obvious effect of disturbances is to increase the amount of woody detritus. In the forests we studied, clear-cutting and windthrow increased woody detrital levels 1.5–1.7 and 2.5–3.8 fold, respectively. This increase is similar to a fire killed *Pseudotsuga/Tsuga* forest in which woody detritus increased 2.7 fold (Agee and Huff, 1987) and a logged tropical forest which had woody detritus increase 3.4 fold (Uhl and Kauffman, 1990; Harmon *et al.*, in press).

Russian boreal forests are frequently disturbed and should contain a significant pool of disturbance related dead wood. Forest fires occur annually on about 3 million hectares of forest land in Russia (Dixon and Krankina, 1993); windthrow is also fairly widespread: in 1990 it was reported on 30 000 ha but the actual extent could be even greater (Krankina *et al.,* in press). Predicted climate change is expected to increase disturbance frequency and severity, which may result in a significant positive feedback to greenhouse gas accumulation in the atmosphere (King and Neilson, 1992). Understanding the dynamics of dead wood is critical for assessing the scope and timing of this feedback.

As the time since a disturbance increases, the mass of dead wood in disturbed forests should decrease dramatically. Given the decomposition rates observed here, we would expect the disturbance generated wood to largely disappear within 65–90 years depending upon the species composition of the disturbed forest. For that reason dead wood stores reach minimum in young stands. The average density of dead wood in a young stand (0.07–0.10 Mg C m^{-3}) is about one half that in an old-growth forest because harvest generated residues have lost most of their biomass while there is no measurable mortality input from the new generation of trees. The density values are the highest (0.15–0.18 Mg C m^{-3}) following disturbance that generates a pulse of tree mortality.

The post-disturbance recovery of dead wood stores to old-growth levels depends upon the decay rate and time required to restore input or mortality to old-growth rates (Harmon *et al.,* 1986; Harmon and Chen, 1991). The recovery of the woody detritus pool is delayed by thinning and dead wood salvage for firewood (Figure 3), which are widely practised in populated areas of Russia. While increasing biomass output and economic efficiency, which is the goal of traditional forestry, these intensive forest management practices significantly reduce C stores in forest ecosystems (Krankina and Harmon, 1994; Lamas and Fries, 1994).

Since experimental data on dead wood C stores and flux is very rare it is important to develop methods of estimation based on other more readily available parameters, such as live biomass. Our measurements show that in Russian old-growth forests, dead wood comprised about 20% of the total wood mass, a proportion quite similar to the larger, more productive forests of the Pacific Northwest (USA) (Krankina and Harmon, 1994). If this proportioning is characteristic of cool conifer forests it would be useful to estimate potential dead wood mass for old-growth forests without dead wood inventories. However, we also found tremendous variation in dead wood stores largely controlled by disturbance and management practices (Figure 2). On our plots the share of dead woody in total above-ground biomass stores ranged from 2 to 98%. The use of a single live/dead wood ratio across the range of successional stages, which is a common practice in C budget calculations, may substantially over- or under-estimate the dead wood C pool depending upon the type of disturbance regime.

Dynamics of dead wood can also be estimated if mortality input and decomposition rates are known. There is a substantial body of information regarding tree mortality associated with disturbance (e.g., Auclair, 1985) and values of tree mortality for undisturbed stands can be found in many growth-and-yield tables. However, the process of decomposition needs to be studied further to develop adequate modelling tools to dynamically estimate C pools and flux associated with woody detritus.

Acknowledgments

The research was sponsored by the National Science Foundation (BSR-9213187) and in part by Earthwatch Center for Field Research. The authors gratefully acknowledge helpful assistance of Drs. S. N. Sennov, A. V. Gryazkin, E. S. Melnikov, and A. V. Selikhovkin of St. Petersburg Forest Academy (Russia) who provided access and background information on their research plots and Dr. O. G. Chertov for facilitating co-operation between Russian and United States participants of the project. We also thank the Oregon State University Forest Science Data Bank (FSDB) and Gody Spycher, for assistance in managing and processing the data used in this paper.

References

Agee, J. K. and Huff, M. H.: 1987, *Can. J. For. Res.* **17**, 697-704.
Auclair, A. N. D.: 1985, *Can. J. For. Res.* **15**, 279-291.
Birdsey, R. A.: 1992, *Gen. Tech. Rep. WO*-59, 51 pp.
Busse, M. D.: 1994, *Soil Sci. Soc. Am. J.* **58**, 221-227.
Cohen, W. B., Wallin, D. O., Harmon, M. E. and Fiorella, M.: (Unpublished manuscript).
Dale, V. H., Houghton, R. H. and Hall, C. A. S.: 1991. *Can. J. For. Res.* **21**, 87-90.
Dixon, R. K. and Krankina, O. N.: 1993, *Can. J. For. Res.* **23**, 700-705.
Dixon, R. K., Schroeder, P. E. and Winjum, J. K.: 1991, EPA report 600/3-91/067.
Dixon, R. K., Brown, S., Houghton, R. A., Solomon, A. M., Trexler, M. C. and Wisniewski J.: 1994, *Science* **263**, 185-190.
Edwards, P. J. and Grubb, P. J.: 1977, *J. Ecol.* **65**, 943-969.
Fahey, T. J.:1983, *Ecol. Monogr.* **53**, 51-72.
Foster, J. R. and Lang, G. E.: 1982, *Can. J. For. Res.* **12**, 617-626.
Franklin, J. F., Bledsoe, C. S. and Callahan, J. T.: 1990, *Bioscience* **40**, 509-523.
Franklin, J. F., Shugart, H. H. and Harmon, M. E.; 1987, *Bioscience* **37**, 550-556.
Gore, J. A. and Patterson, W. A. III: 1986, *Can. J. For. Res.* **16**, 335-339.
Grier, C. C. and Logan, R. S.: 1977, *Ecol. Monogr.* **47**, 373-400.
Harmon, M. E. and Chen, H.: 1991, *Bioscience* **41**, 604-610.
Harmon, M. E., Franklin, J. F., Swanson, F. J., Sollins, P., Gregory, S. V., Lattin, J. D., Anderson, N. H., Cline, S. P., Aumen, N. G., Sedell, J. R., Lienkaemper, G. W., Cromack, K. Jr. and Cummins, K.: 1986, *Adv. Ecol. Res.* **15**, 133-302.
Harmon, M. E., Cromack, K. Jr. and Smith, B. G.: 1987. *Can. J. For. Res.* **17**, 1265-1272.
Harmon, M. E., Ferrell, W. K. and Franklin, J. F.: 1990, *Science* **247**, 699-702.
Harmon, M. E., Brown, S. and Gower, S. T.: 1993, *In:* Vinson, T. S. and Kolchugina, T. P. (eds.) *Carbon Cycling in Boreal and Sub-Arctic Ecosystems,* EPA Number 600-93/084, Corvallis, USA, pp. 167-178.
Harmon, M. E., Whigham, D. F. and Sexton, J.: *Biotropica,* (In press).
Houghton, R. A. and Woodwell G. M.: 1989, *Sci. Amer.* **260,** 36-44.
Houghton, R. A., Boone, R. D., Fruci, J. R.,. Hobbie, J. E, Melillo, J. M., Palm, C. A., Peterson, B. J., Shaver, G. R., Woodwell, G. M., Moore, B., and Skole, D. L.: 1987, *Tellus* B **39**, 122-139.
Johnson, E. A. and Greene, D. F.: 1991, *J. Veg. Sci.* **2,** 523-530.
Kauffman, J. B., Uhl, C. and Cummings, D. L: 1988, *Oikos* **53**, 167-175.
Keeling, C. D., Bacastow, R. B., Carter, A. F., Piper, S. C., Whorf, T. P., Heimann, M., Mook, W. G. and Roeloffzen, H. J.: 1989, *Geophysical Monogr.* **55**, 165-236.
King, G. and Neilson, R. P.: 1992, *Water, Air, and Soil Pollution* **64**, 365-383.
Kobak, K. I: 1988, *Biotical Compounds of Carbon Cycle.* Gidrometeoizdat Publishers, Leningrad, Russia.
Kolchugina, T. P. and Vinson, T. S.: 1993, *Water, Air, and Soil Pollution* **70**, 207-221.
Krankina O. N. and Harmon, M. E.: 1994, *World Resour. Rev.* **6**(2), 161-177.
Krankina, O. N., Dixon, R. K., Shvidenko, A. Z. and Selikhovkin, A. V.: *World Resour. Rev.*

Lämås, T. and Fries, C.: 1995, *In: Forest Survey Designs: 'Simplicity versus Efficiency' and the assessment of Non-timber Resources,* Proceedings of Conference, 2-7 May, 1994, Monte Verità, Switzerland, Swiss Federal Institute of Technology, Swiss Federal Institute of Forest, Snow and Landscape Research, and IUFRO S4.02/S4.11. (In press.)

Little, S. N. and Ohmann, J. L.: 1988, *Forest Science* **34**, 152-164.

Lambert, R. C., Lang, G. E. and Rieners, W. A.: 1980, *Ecology* **61**, 1460-1473.

McFee, W. W. and Stone, E. L.: 1966, *Soil Sci. Soc. Am. Proc.* **30**, 513-516.

Melillo, J. M., Fruci, J. R. and Houghton, H. A.: 1988, *Tellus* B **40**, 116-128.

Miller, W. E.:1983, *Forest Science* **29**, 351-356.

Muller, R. N. and Liu, Y.: 1991, *Can. J. For. Res.* **21**, 1567-1572.

Onega, T. L. and Eickmeier, W. G.:1991 *Bull. Torrey Bot. Club* **118**, 52-57.

Post, W. M., Peng, T. H., Emanuel, W. R., King, A. W., Dale, V. H., and DeAngelis, D. L.: 1990, *American Scientist* **78**, 310-326.

Sarmiento, J. L: 1993, *Nature* **365**, 697-698.

Schlesinger, W. H.: 1990, *Nature* **348**, 232-234.

Sedjo, R. A. and Solomon, A. M.: 1991, *In: Greenhouse Warming: Abatement and Adaptation. Proceedings for the Workshop, Washington, USA.* pp. 105-19.

Sennov, S. N.: 1984, *Tending the Forest: Ecological Basis.* Lesnaya Promyshlennost, Moscow, Russia, [In Russian]

Sollins, P.:1982, *Can. J. For. Res.* **12**, 18-28 1982

Sollins, P., Cline, S. P., Verhoven, T., Sachs, D. and Spycher, G.: 1987, *Can. J. For. Res.* **171** 1585-1595.

Spies, T. A., Franklin, J. F. and Thomas, T. B.: 1988, *Ecology* **69**, 1689-1702.

Tans, P. P., Fung, I. Y., and Takahashi, T.: 1990, *Science* **247**, 1431-1438.

Tretyakov, N. V., Gorskiy, P. V., and Samoilovich, G. G.: 1952, *Forest Inventory Reference Book.* Goslesbumizdat Publishers, Moscow-Leningrad, Russia. [In Russian]

Tritton, L. M.:1980, Ph.D. Dissertation, Yale Univ., New Haven, USA.

Uhl, C. and Kauffman, J. B.: 1990, *Ecology* **71**(2), 437-449.

CARBON POOLS AND FLUXES OF 25-YEAR OLD CONIFEROUS AND DECIDUOUS STANDS IN MIDDLE SIBERIA

E. F. VEDROVA

Institute of Forests, Russian Academy of Sciences, Siberian Branch, Akademgorodok, Krasnoyarsk 660036, Russia

Abstract. Between 72 and 88% of carbon (C) loss in forest litter decomposition returns to the atmosphere in the form of carbon dioxide. The share of water-soluble organic products does not exceed 3–4%. Between 8% under spruce and 25% under aspen and pine of the total C loss from litter organic matter goes to the formation of humus. Decomposition intensity of the dead organic matter on the soil surface is close to annual litterfall income (except under cedar). The specific rate of decomposition processes among the coniferous litters is minimum for cedar (167 mg C g^{-1} yr^{-1}) and maximum for larch (249 mg C g^{-1} yr^{-1}). The specific rate of decomposition of organic residues under aspen and birch canopies are 344 and 362 mg C g^{-1} yr^{-1}.

Keywords. LITTERFALL, LITTER, CARBON, DECOMPOSITION, MINERALISATION, HUMIFICATION, SPECIFIC RATE.

1. Introduction

A multi-year experiment including the six main forest forming species of Siberia—cedar, pine, larch, spruce, aspen and birch—was carried out from 1969 to 1972 by the Laboratory of Forest Soil Science of the Institute of Forests (Siberian Branch, Russian Academy of Sciences) on old arable dark-grey soils (Shugalei *et al.*, 1984). The stands under investigation grow in similar climate conditions and on a homogeneous soil substratum. The turnover of carbon, nitrogen and mineral elements was studied in the experiment both at the level of forest ecosystems and at the level of differences in stands of different tree species.

This paper presents the results of a C balance study of the transformation of plant residues into organic soil matter. The biological turnover constituting the sequence of tree litterfall–litter–soil humus is analysed. Each block of this sequence is characterised by organic matter storage per unit area (g C m^{-2}). Blocks are interrelated by exchange reactions (flows), which are characterised by the intensity of the change in organic matter storage per unit time (g C m^{-2} yr^{-1}). Flow intensity is represented by the specific rate showing the effectiveness of 1 g organic matter in a given exchange reaction (mg C g^{-1} C yr^{-1}).

2. Materials and Methods

The experimental plot is situated in the Kemchug Hills, which according to their geomorphological division, are part of the Chulymo–Enisei denudational plain in the south-west edge of the West-Siberian Lowland. Climatic conditions are characterised by continentality, sufficient humidity supply and insufficient heat quantity. Average temperatures are 18 °C in July, and -24.7 °C in January. Mean annual precipitation is 476 mm, although most falls during the summer (314 mm). The total heat sum for the growing season is 1500 degrees (i.e., sum of positive mean daily Celsius temperatures). The last spring frost is observed in late May to early June. The first autumn frost is

Water, Air and Soil Pollution **82**: 239–246, 1995.

recorded in early or mid-August. The frost-free period is generally 65–120 days. Snow cover lasts for 171 days; its maximum depth is 60–80 cm. Depth of freezing is up to 70–160 cm. Snowmelt begins in late March to early April. Soils thaw out completely by late June.

The central part of the Kemchug Hills is covered with fir and spruce forests admixed with cedar. Pine and birch dominate in the north. In the southern part in bogged plots spruce and fir forests are widespread, while pine grows in dry watersheds (Lyubimova, 1964). The soil cover of the experimental plot and of adjacent areas is represented by dark-grey weakly podzolised gleysolic forest soils. Pedogenic soils (rocks) are brown clays.

The experimental plot (1.7 ha in area) was periodically used as agricultural land after clearing. Immediately before the experiment began, this plot was left fallow for two years. After special preparation (deep breaking and grading of the surface and a soil-fertility bioassay with agricultural wheat) 2–3 year old seedlings of spruce (*Picea abovata*), cedar (*Pinus sibirica*), larch (*Larix sibirica*), pine (*Pinus sylvestris*), aspen (*Populus tremula*) and birch (*Betula fruticosa*) were planted in 1971–1972 (Shugalei *et al.*, 1984). Each species was planted on twenty four 10 m × 10 m plots.

The stands are presently 25 years old (Table I) with the following characteristic parameters under investigation: The formation of forest litter on the surface; the rate of its decomposition; the correlation of mineralisation and humification processes as plant residues decompose; and the role of water-soluble organic compounds in the humus forming process, were studied in the stands from 1987 to 1991. Litterfall was collected every two weeks during May–August and every week during heavy litterfall, until snow cover began (mid-October to early November). Litterfall samples were collected using ten 0.25 m^2 litter-traps for each stand. Litterfall samples of every collection were sorted into fractions (needles, foliage, branches of diameter less than 1 cm, bark, cones and others). Each fraction was weighed, and water content determined by drying at 100 °C for 5–8 hours. At the end of the year, an average sample was made for every fraction from subsamples of all the collections. The chemical composition was determined in this composite sample.

Litter was collected twice during the growing season: in mid-May and late August until heavy litterfall began, using fifteen to twenty 0.0625 m^2 litter-traps. The litters were sorted by sub-horizon in the laboratory: litter (L), fermentation (F) and humification (H). The L layer (01) is formed by litterfall from the previous (autumn) and current (spring) years and is weakly changed by the biochemical processes of transformation and levigation. F (02) is the brown mass of plant residues. H (03) is a dark brown organic material without visible plant residues. When moist, it is spreadable to a homogenous mass. L, F, and H samples were weighed and water content was determined for them. All the data presented here are reported on an oven-dry basis.

TABLE I

Characteristics of trial plantations of six boreal species at age 25 years.

	Cedar	Pine	Larch	Spruce	Aspen	Birch
Height (m)	5.2	8.5	7.6	3.8	9.7	5.7
Diameter (cm)	5.4	7.9	6.0	4.1	8.3	4.0
Stand density (stem ha^{-1})	11 700	10 500	1 080	9 900	4 200	8 800

Carbon content of plant samples was determined according to Anstet's method modified by Ponomareva and Nikolaeva (1959, in Ponomareva and Plotnikova, 1975). Carbon of soil organic matter and of solutions filtering through the litter were determined using Tyurin's method (Arinushkina, 1970). The plastic lysimeters of Schilova's system were used to obtain litter solutions (Schilova, 1955).

3. Results and Discussion

The specific biohorizon—the forest litter—is fully formed in the 25-year old plantations. Tree litterfall is the source of its formation. No essential differences between deciduous and coniferous species are observed in the mass of annual litterfall (Table II). Litterfall mass can be subdivided into groups. The leaf fall is 397 g m^{-2} yr^{-1} in aspen forest and 245 g m^{-2} yr^{-1} in birch forest. Appropriate tree species leaves constitute 80–88% of the leaf fall. Branches account for 9–16%. Among the conifers the maximum litterfall was observed in cedar stands (359 g m^{-2}). Two peaks are clearly seen: in winter and in early summer. The summer maximum is not expressed in the pine forest. In terms of litterfall mass the larch forest is similar to birch forest (282 g m^{-2}), but the spruce forest gives the least litterfall (94 g m^{-2} yr^{-1}) compared with other conifers. Spruce forest is similar to cedar forest in terms of the dynamics of litterfall input.

The grass layer is less important to litter formation: the grass share does not exceed 3% of the annual litterfall in coniferous stands; it is 3.7% in aspen forest and 6.7% in birch forest.

Litterfall of different tree species is characterised by similar C concentration, therefore C storage and annual turnover depend mainly on the annual mass of leaf-fall (Table III).

Fallen plant residues influenced by moisture, heat, soil mesofauna and microflora are subjected to complicated quantitative as well as qualitative changes. First of all it is reflected in litter structure. In coniferous litters, except spruce, the fermentation layer (F) is dominant in the spring sampling; in deciduous litters the relatively unaltered leaves and branches of the autumn litterfall dominate, i.e., the L layer (Table IV). Cedar and pine forest litters differ in the humification layer (H). An L layer is almost absent in cedar forest litter and the litter itself is easily separated from the humus accumulating in the soil horizon. The horizon surface is heavily penetrated by white fungal mycelia. The H layer constitutes 16% of the entire litter mass in pine litter and consisting of humified pine needle residues. Later it transforms gradually into the mineral

TABLE II

Litterfall mass and composition and (average for 4 years).

Stand	Litterfall components (g m^{-2})						
	Foliage	Twigs	Branches	Bark	Cones	Other	Total
Cedar	322.8	0.2	10.8	0.4	1.2	23.8	359.2
Pine	259.7	4.0	16.6	13.3	8.4	9.4	311.4
Larch	219.4	6.4	39.1	1.2	1.6	14.6	282.1
Spruce	72.5	2.4	4.9	0.6	10.0	8.0	93.5
Aspen	347.4	0.3	35.2	0.1	-	14.2	397.2
Birch	196.6	0.1	39.6	1.4	-	7.2	244.9

TABLE III

Carbon concentration of litterfall components under six 25-year old stands of different species (expressed as percentage of oven-dry mass).

Fraction	Cedar	Pine	Larch	Spruce	Aspen	Birch
Litterfall						
Foliage	60.1	60.1	55.7	55.1	53.9	56.4
Branches	59.3	60.4	51.0	57.8	51.6	58.1
Other	57.3	58.8	58.9	55.0	52.2	52.0
Litter, horizons						
L	54.2	52.0	-	48.3	46.6	47.5
F	45.1	44.7	35.5	39.8	28.2	28.2
H	36.0	29.2	27.3	-	23.9	27.4

horizon. An unusual horizon is formed under the larch: its litterfall is quickly occupied by fungi transforming it into the fragile dark-brown heavily compressed mass of the F layer. The L layer is absent in this litter. Spruce litter differs from other conifer litters in that it is characterised by 95% relatively unaltered fallen needles, while the H layer is absent. Litters of deciduous species are similar in their structure: all three layers occur in them.

Carbon content in the litters is less than in litterfall and differs between layers: it decreases as humification of plant residues increases. The fractionation of litter organic matter showed that most humic acids are extract from the humification layer (Ponomareva and Plotnikova, 1975). As a rule, the lowest ratios of C contained in humic and fulvic acid complexes (C_{ha}:C_{fa}) occur here. The litters of different stands are clearly divided into two groups according to the ratios of these humic acid fractions: litters of deciduous species (birch, aspen, larch) and litters of evergreen conifer species (cedar, pine, spruce). In the first group the humic acids are dominated by the fraction which readily forms peptide bonds when heated in 0.02 normal NaOH. The availability of this fraction is characteristic for organic mineral soil horizons. This fraction is a rough humus, the most stable of humic substances, and shows more deep humification of plant residues in deciduous stands. In litters of the second group, the main humic substances are extracted in the first stages of fractionation, and after the decalcification transfer to 0.1 normal NaOH.

TABLE IV

Litter structure components under six 25-year old stands of different species (expressed as percentage of oven-dry mass).

	Litter storage (g m^{-2})							
	Spring estimate				Autumn estimate			
Horizons	L	F	H	Total	L	F	H	Total
Cedar	649	439	22	1110	192	797	77	1166
Pine	973	377	158	1508	185	952	76	1213
Larch	0	1298	169	1467	0	1200	246	1446
Spruce	224	12	0	236	0	223	0	223
Aspen	433	176	102	711	0	512	63	575
Birch	338	53	92	483	0	240	56	301

TABLE V

Parameters for the ratio of mineralisation and humification in litter decomposition carbon (C) loss (g m^{-2} yr^{-1}).

		C - CO_2		C - H_2O			C - CO_2
Stand	Total	g m^{-2}	% Total	g m^{-2}	% Total	C-humus	C-humus
Cedar	122	88	72	3.6	3.0	30	3.0
Pine	193	158	82	5.4	2.8	30	5.2
Larch	145	125	86	4.7	3.2	15	8.1
Spruce	61	54	88	2.5	4.0	4.7	11.5
Aspen	311	238	76	7.1	2.3	66	3.0
Birch	228	192	84	9.2	4.0	27	7.3

The pattern of litter storage in coniferous and deciduous stands is similar to that of the annual litterfall mass. In spite of small differences in annual litterfall income it was found that litter mass in coniferous forests (except spruce) is 1.5 and 3 times more than in aspen and birch forests, respectively. Differences in litter storage, in their structure, and in the qualitative composition of organic matter, show different directionality and rates of biochemical transformation processes from plant residues.

In litter decomposition, part of the C-bearing substances are mineralised to gaseous products and return to the atmosphere, while another portion is transferred to the soil in the form of complex humification products—organic compounds of low molecular weight and non-specific nature. Annual sampling of litter stores, of litterfall income dynamics, and of the quantity and composition of solutions which are filtered from litter to soil, allowed the estimation of the intensity of decomposition processes of litter plant residues as well as the role of water soluble organic products in humus formation. The three-year model field experiments made it possible to determine the correlation of two main processes of litter organic matter decomposition in different stands: mineralisation and humification (Table V).

The intensity of plant decomposition processes within every stand depends on the total reserve of organic residues being decomposed. The specific rate of the processes is mainly conditioned by their qualitative (chemical) composition. Between 72 and 92% of the C lost in decomposition returns to the atmosphere in the form of carbon dioxide. The share of water soluble organic products—one of the humus sources in the soil—does not exceed 3–4%. Between 8% (under spruce) and 25% (under aspen and pine) of total litter organic matter C lost is spent in humus formation. Experimentally derived parameters for the correlation of mineralisation and humification processes in litter decomposition in different stands were used to calculate the balance in the C cycle sequence: litterfall → litter → soil → atmosphere (Table VI).

According to the data in Table VI, the organic matter of plant residues is decomposed in leafy stands 1.5–2 times faster than in coniferous stands: when expressed as a fraction of the total stock of decomposed material, C losses were 34–36% in deciduous stands, and 17–25% in coniferous stands, during the year. The large litter store in coniferous stands, as compared to deciduous stands, is explained by different the decomposition rates. In terms of the rate of destruction, they rank as follows: larch > pine > spruce > cedar. Mineralisation is the main process by which litter loss of organic matter occurs in all stands: under coniferous stands, the proportion

TABLE VI

Carbon (C) balance for litter decomposition in 25-year old stands.

Parameter		Cedar	Pine	Larch	Spruce	Aspen	Birch
Litter C store (g m^{-2})							
At beginning of 1987	(A)	565	643	565	134	343	160
At end of 1990	(B)	800	702	502	169	429	256
C in litterfall							
1987–1988		185	197	178	40	275	158
1988–1989		189	167	136	42	194	137
1989–1990		248	236	162	67	201	127
Total (1987–1990)	(C)	622	594	476	149	670	422
Decomposition	(A+C)	1187	1237	1041	283	1013	582
C loss	(A+C)–B	381	535	539	114	584	326
Rate (g m^{-2} yr^{-1})							
Decomposition		129	178	80	38	195	109
Mineralisation		97	150	160	35	154	96
Humification		32	28	20	3	41	13
Specific rate (mg g^{-1} yr^{-1})							
Decomposition		167	212	249	206	344	362
Mineralisation		126	178	222	191	271	319
Humification		41	34	27	15	73	43
Soil humus store (0–10 cm) (g m^{-2})							
In 1972		2864	2464	3262	2838	2850	2748
In 1990		3144	2745	3809	2983	3323	2890
C-humus accumulation		280	281	547	151	473	142
Rate of humus accumulation (mg m^{-2} yr^{-1})		22	22	42	12	36	11

of total C lost as CO_2 ranges from 75% (under cedar) to 92% (under spruce) and in deciduous forests from 77% (under aspen) to 88% (under birch) of the total C loss.

Humification processes are strongest under cedar and in aspen forest during the transformation of C-containing plant products: the newly formed humic substances constitute a quarter of the decomposed product here. The ratios of mineralisation to humification in coniferous litter decomposition are: 3.0, 5.2, 8.1 and 11.5 under cedar, pine, larch, and spruce stands, respectively. Destruction processes in deciduous stands are built in different ways: mineralisation flow is 3 times greater under aspen and 7.3 times greater under birch than is humification. Taking into account that all the plantations grow in similar soil and climatic conditions, the observed differences in rate and directionality of decomposition processes may be explained by qualitative (chemical) litter composition.

Water-soluble organic products of the litter, as well as newly formed humus, are fixed partly on the surface of the mineral soil and transform to stable humic substances, and partly mineralised to CO_2 returning to the atmosphere from soil respiration. Humus reserves increased in the upper 10 cm soil layer in all stands during the growing season (Table VI). Maximum humus accumulation rates were observed under larch and aspen, whereas humus stores under spruce and birch changed very little. Assuming that the active interaction of decomposing litter products with the mineral soil lasted 13 years (i.e., from the period of full crown closure in 1978–1980) the mean rate of accumulation may be calculated. Calculated data for the rates of humus accumulation as a whole correspond to those characteristics obtained from studying the decomposition processes of plant residues. The significant increase of humus stores under larch is related, obviously, to humification of underground mortmass. Carbon stores in small roots(d < 2 mm) in the 0–6 cm soil layer under larch are 1.7 times greater than in aspen and 2.5–4.8 times greater than other species.

We consider finally how C turnover is structured using the 25-year old pine stand as an example (Figure 1). Approximately 364 g C m^{-2}, (i.e., 4.3% of total phytomass), accumulates in the current increment of the aboveground phytomass. The net accumulation is 186 g m^{-2} yr^{-1} in the tree stand with an annual litterfall of 178 g C m^{-2}. The release of C as CO_2 to the atmosphere during litter decomposition amounts to 149 g C m^{-2} yr^{-1}. Approximately 28 g C m^{-2} yr^{-1} is spent in new humus formation. In root decomposition (specific rate is 184 mg C g^{-1} yr^{-1}) 10 g C returns to the atmosphere and 0.5 g C is transformed to soil humus. The total C accumulation in humus is about 22 g C m^{-2} yr^{-1}.

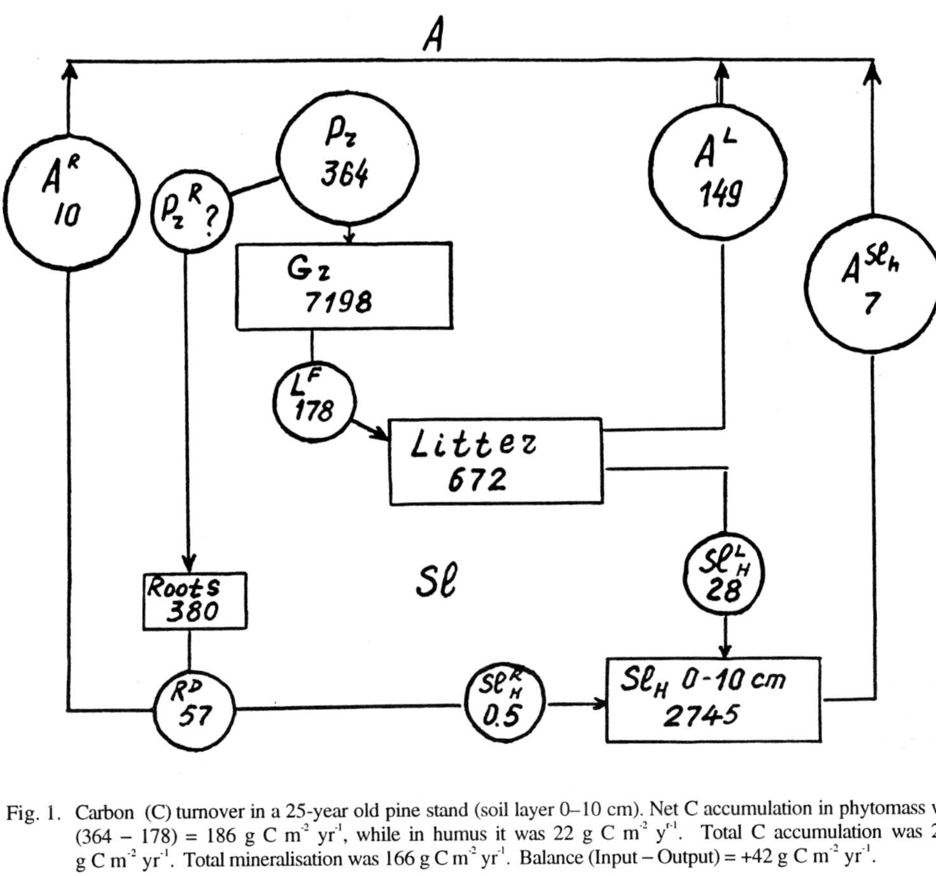

Fig. 1. Carbon (C) turnover in a 25-year old pine stand (soil layer 0–10 cm). Net C accumulation in phytomass was (364 – 178) = 186 g C m^{-2} yr^{-1}, while in humus it was 22 g C m^{-2} y^{r-1}. Total C accumulation was 208 g C m^{-2} yr^{-1}. Total mineralisation was 166 g C m^{-2} yr^{-1}. Balance (Input – Output) = +42 g C m^{-2} yr^{-1}.

Humus mineralisation amounts to 6.5 g C m^{-2} yr^{-1} as CO_2. Total C accumulation in the 25-year old pine forest is 208 g C m^{-2} yr^{-1}. The input–output balance is therefore: (208 – 166) = +42 g C m^{-2} yr^{-1}. This means the rates of production and destruction of dead organic matter are not balanced, because the pine forest is in a stage of active growth.

4. Conclusions

The rate of decomposition of dead organic matter on the soil surface (which is equivalent to carbon) is approximately equal to annual litterfall inputs in 25-year old deciduous and coniferous stands. For the species studied here, only cedar is an exception: 64% of the annual input of C containing products are decomposed under its canopy. The specific rate of destruction which reflects the qualitative composition of the matter being decomposed is least for deciduous litters: 344 and 362 mg C g^{-1} yr^{-1} under aspen and birch stands respectively. Organic residues under larch canopies are characterised by the highest specific rate of decomposition of all the coniferous litters: 249 mg C g^{-1} yr^{-1}. Litters under pine and spruce are decomposed at a similar specific rates: 206–212 mg C g^{-1} yr^{-1}.

The differences among species in the rates of destruction are the cause of accumulation of dead organic matter on the soil surface which is 1.5–3 times higher under coniferous species than under deciduous species with similar annual rates of soil C inputs from litterfall.

The main C loss in decomposition derives from mineralisation of organic products to CO_2: 75–92% of C loss is released to the atmosphere in the form of CO_2 from coniferous plantations and 77–88% of C is lost through decomposition in deciduous stands annually. New formation of humic substances (humification) is greatest under cedar and aspen where 25 and 23% of C is lost in decomposition.

Carbon stores in humus increased by 17% in the upper 10 cm soil layer during stand growth under aspen and larch. Practically no changes occurred in the soil under birch and spruce. Humus C stores increased by 10% under pine and cedar stands in comparison with the primary stores.

Acknowledgments

The author wishes to acknowledge Russian and Canadian colleagues who assisted with this research and the preparation of this paper, specifically Dr A. Z. Shvidenko and C. Campbell. She also greatly appreciates financial assistance from the U.S. International Science Foundation (Washington D.C., Moscow) and from the Canadian Forest Service, to attend the IBFRA '94 Boreal Forests and Global Change conference.

References

Arinushkina, E. V.: 1970, *Methods of Soil Chemical Analyses*. MGU, Moscow, Russia. 487 pp.

Lyubimova, E. L.: 1964, *Vegetation cover: Middle Siberia*. Nauka, Moscow, Russia, pp. 226-276.

Ponomareva, V. V. and Nikolaeva, T. A.: 1959, Methods for studying organic matter in bogs. *In: Modern Soil Processes in the Forest Zone of European USSR*, Moscow Academy of Science, Moscow, USSR. pp. 170-173.

Ponomareva V. V. and Plotnikova, T. A.: 1975, *In: Methods for Determination of Soil Humus Content and Composition*. Leningrad, Russia, 105 pp.

Schilova, E. I.: 1955, Methods for obtaining soil solutions under natural conditions. *Pochvovedenie*. **11**, 86-90.

Shugalei, L. S., Semechkina, M. G., Yaschihin, G. I. and Dmitrienko, V. K.: 1984, Modelling the evolution of biogeocenoses. *Natura*, Novosibirsk, Russia, 151 pp..

CARBON STOCK AND DEPOSITION IN PHYTOMASS OF THE RUSSIAN FORESTS

A. ISAEV[1], G. KOROVIN[1], D. ZAMOLODCHIKOV[1], A. UTKIN[2], A. PRYAZNIKOV[1]

[1]*Centre for the Problems of Ecology and Forest Productivity of the Russian Academy of Sciences, Novocheriomuskinskaia str., 69, Moscow, 117418, Russia*
[2]*Institute of Forestry Science of the Russian Academy of Sciences, Uspenskoe, Moscow Region, 143030, Russia*

Abstract. Russian forests occupy an area of 771 Mha with a wood volume of 81.6×10^9 m^3. In this paper we estimate the carbon (C) storage and annual C deposition in living vegetation for stands of different age groups according to information in the Russian National Forest Inventory (NFI). Conversion of root C storage into phytomass was based on phytomass/storage ratios calculated from data from 1900 test areas in the different forest regions of Russia. We developed these conversion ratios for different phytomass fractions, namely: stems, branches, roots and foliage, for different forest forming species. Of the total forest area (771 Mha), C storage in the living phytomass is 35.07 Pg C. Total annual C deposition in forest vegetation is estimated at 213.2 Tg C. We considered the role of the main forest forming species in the C cycle of the forest vegetation taking into account the actual structure of the Russian forests.

Keywords. RUSSIA, PHYTOMASS, CARBON, CONIFERS, CO_2, BOREAL FOREST.

1. Introduction

Russian forests occupy a considerable part of Eurasia and constitute about 21% of the global wood volume, including 57% of the most valuable coniferous species. The importance of the boreal forests has grown sharply with intensification of human pressure on the environment and deforestation in the tropical zone. Boreal forests are considered to be one of the most important components of the biosphere, exercising an influence on global ecosystem health and biodiversity. Boreal forests are comparable with the oceans in terms of global biosphere processes. Thus did G. F. Morozov, the founder of Russian forest science, call forests "Oceans of the Land".

Concern over global climate change, related to increasing atmospheric CO_2 and the enhanced greenhouse effect, has prompted detailed study of the carbon (C) balance in forest ecosystems. Russian boreal forests are an enormous repository of C in living phytomass, and in dead phytomass at different levels of decay including humus and peats.

According to current predicitons, global change will first and most strongly affect the Northern regions of the Taiga zone, which is mainly in Russia. It is impossible to solve the problems of global ecology without evaluating the potential role of Russian forests in the C cycle, both through quantitative description and mathematical simulation of the main C fluxes.

A large number of studies devoted to estimating C balance have been published in recent years. These studies can be divided into two groups: (1) studies based on the biological productivity of forest ecosystems and the areas of different forest types in ecoregions or other territorial complexes (Kolchugina and Vinson, 1993a,b,c; Kolchugina *et al.*, 1993); and (2) studies using the National Forest Inventory (NFI,

Water, Air and Soil Pollution **82**: 247–256, 1995.

1988) and ecosystem productivity data (Sedjo, 1992; Isaev, 1993; Isaev *et al.*,1993; Kolchugina and Vinson, 1993c).

This paper presents calculations based on data in the Russian NFI. The NFI is a unique data base including various information about Russian forests, including their age structures, species compositions, productivity and condition, at the local, regional and federal levels. The NFI does not contain information about phytomass and other components of C balance. For this reason we have developed several approaches to estimating the forest C cycle from the NFI. The purpose of this paper is to develop one approach suggested in the preceding paper (Isaev *et al.*, 1993) for estimating the C balance of Russian forests.

2. Methodology and Data Base

2.1. Data of the National Forest Inventory

Lands of the Russian National forest fund are divided into forest and non-forest lands (bogs, plough lands, pastures, glaciers, sands and other). The area of forest lands is 884.1 Mha; the area of non-forest lands is 298.5 Mha (Forest fund of the RSFSR, 1990). This paper considers only the forest lands.

Forest lands are divided into forested lands (771.1 Mha) and unforested lands (113.0 Mha), where forest is temporarily absent due to fire, harvest, sparse trees, or other causes.

In this paper we use information from the NFI relating to the distribution of areas and volumes of wood by age groups (young stands, middle aged, premature, mature and overmature) of the main forest forming species. The main forest forming species are: coniferous - pine (*Pinus sylvestris*), spruce (*Picea abies, P. obovata, P. ajanensis*), fir (*Abies sibirica, A. holophylla, A. nephrolepis, A. nordmanniana*), larch (*Larix sibirica, L. gmelinii, L. cajanderi, L. olgensis, L. kurilensis*), cedar pine (*Pinus sibirica, P. koraiensis*), dwarf siberian pine (*Pinus pumila*); hardwood - high stem (stands) and low stem (coppice) oak (*Quercus robur, Q. mongolica*), beech (*Fagus orientalis*), ash (*Fraxinus excelsior, F. mandshurica, F. sieboldiana, F. rhynchophyla*), stone birch (*Betula ermanii*); softwood - birch (*Betula pendula, B. alba* and other), aspen and poplar (*Populus tremula, P. nigra, P. alba, P. suaveolens, P. maximoviczii*), lime (*Tilia cordata, T. amurensis* and others).

The area and stocking data for stands of different age groups allows the reconstruction of the dynamics of the average stocks (m^3 ha^{-1}) and to determine the sizes of annual changes in the total stock of rootwood. In other words, it is possible to estimate C pools and fluxes taking into account the actual age structure and species composition of forests. This is particularly important in the Russian forest, with its enormous territories and different landscapes and ecological conditions.

2.2. Phytomass Data

We used the computer data base "Biological productivity of forest ecosystems" (Utkin *et al.*, 1994) as the second information source. This data base includes information about more than 1900 test plots established in the forests of Russia and neighbouring foreign countries. It includes the following main blocks: geographical situation, taxation characteristics of stands, phytomass, detritus production, annual litterfall and attrition. This data base is different from the Bazilevich (1993) data base in that it has a forestry orientation, so is easily compared with data in the NFI.

To calculate conversion ratios, we used stand stocking, age, and values of the different fractions of the stand phytomass.

2.3. Estimating Phytomass Carbon Storage

Phytomass and C storage estimates were derived from wood volume data and phytomass/rootwood ratios. Similar methods are widely used in forest inventories and to calculate C storage in different geographical zones (e.g., Johnson and Sharpe, 1983; Brown and Lugo, 1984; Brown *et al.*, 1989; Makarevsky, 1991; Sampson, 1992; Nabuurs and Mohren, 1994). Conversion ratios were calculated for different age groups of the main forest forming species for stems, branches, roots, and foliage.

Conversion ratios for dwarf siberian pine were calculated differently. We ascribed V. G. Kolischyk (1968) observations to *Pinus mugo*, and the observations in the mountains of central Japan of T. Kajimoto (1989, 1992, 1993) to dwarf siberian pine. This resulted in a new estimate of the C storage importance of dwarf siberian pine, lower than that of Isaev *et al.* (1993).

The total phytomass of each forest forming species was calculated by summing the wood stocks multiplied by the corresponding conversion ratios for each phytomass fraction for each age class.

The phytomass of lower layers was calculated from the stand species and age structure and average values (Mg ha^{-1}) of bushes and soil phytomass.

We equated average phytomass storage on the unforested lands with the phytomass storage of forest zone grasslands (Guricheva *et al.*, 1975). We excepted only areas designated as sparse wood, which in the Russian forest fund is mainly northern and montane sparse woods; its phytomass was assumed to be 20% of the phytomass of closed stands. We calculated C storage from the phytomass assuming 1 kg of dry mass of stems, branches and roots contains 0.5 kg of C, and 1 kg of dry mass of pine-needles, foliage and vegetation of lower layers contains 0.45 kg. Table I shows the ratio of phytomass C (including stems, branches, roots, foliage and vegetation of lower layers) to the wood stock for the age groups of stands of the main forest forming species.

2.4. Estimating Annual Carbon Deposition

The calculation of annual C deposition in the living phytomass of the forest ecosystems was based on the dynamics of areas and the average stocks of the main forest forming species. We assumed that the areas of young, middle aged and premature stands of each forest forming species were evenly distributed by age. For each species we determined fixed age group boundaries, corresponding to normal ages of cutting (Zagreev *et al.*, 1992). Then we calculated average stocks (m^3 ha^{-1}). We calculated annual accumulation of wood stocks (Q) for each forest forming species as follows:

$$Q = \sum_{j=0}^{m-1} \frac{S_j}{T_j} \left(W_{j+1} - W_j \right) \qquad [1]$$

TABLE I

Conversion ratios for phytomass carbon (Mg m^{-3}).

Species	Age groups			
	Young stands	Middle aged	Premature	Mature/Overmature
Pine	0.47	0.34	0.35	0.34
Spruce	0.55	0.37	0.36	0.38
Fir	0.39	0.32	0.27	0.29
Larch	0.51	0.48	0.52	0.51
Cedar	0.42	0.35	0.34	0.42
Dwarf Siberian pine	0.98	0.98	1.03	1.07
High stem oak	0.58	0.53	0.45	0.46
Low stem oak	0.89	0.54	0.62	0.73
Beech	0.47	0.51	0.49	0.49
Ash	0.51	0.50	0.42	0.43
Birch	0.48	0.41	0.37	0.37
Aspen, poplar	0.48	0.42	0.31	0.35
Lime	0.41	0.34	0.32	0.32
Other softwood	0.37	0.39	0.33	0.34

where m = number of age groups (j = 0 corresponding to the unforested area); S_j = area of age group j; T_j = temporary interval of group j; and W_j = average stock of the age group j.

Annual phytomass and phytomass C deposition was calculated with the applicable conversion ratios (Table I).

Unforested areas were divided into three groups for calculations of C depositon. The first group includes nursery, plantations, glades, and barrens. These areas will not transfer to the forested category. Annual C deposition for these areas was considered to be zero.

The second group includes areas where vegetation is a net C sink during stand regeneration. These are cuttings, burned areas, dead stands, etc; areas that will transfer to the forested category in the future. We considered the period of transition to the forested category to be 20 years, and that 30% of burned areas do not regenerate.

Sparse forests are in the third group. Calculation of C deposition for this group of forests is similar to that of stands with a crown closure density of 20%.

3. Results and Discussion

3.1. Carbon Storage and Deposition

The resulting estimates of C storage and deposition for Russian forests is given in Table II. The average C storage for Russian forest land is 41.1 Mg ha^{-1}, and the average annual deposition is 0.24 Mg ha^{-1} yr^{-1}. Coniferous and softwood species play the largest role in the C cycle in the Russian forests (Tables III and IV).

TABLE II

Carbon (C) storage and deposition in Russia by forest region.

Indicator	Covered with forests	Not forest-covered	Total
Area, Mha	771.1	113.0	884.1
C storage, Pg	35.07	1.25	36.32
C deposition, Tg	184.8	28.4	213.2

The distribution of the areas and volumes of wood of the main forest forming species (without bushes) is shown in Figure 1. Larch forest covers the largest area (42.4%). Together with other conifers (pine, spruce, cedar, fir) larch forests constitute 80.5% of the forested area. Softwood species cover 16.9% of the area. Hardwood species (oak, maple, ash and other) cover 2.6% of the area.

The distribution of wood volumes by species is similar to its distribution by area (Figure 1). Note that the per centage of cedar increases and of larch decreases. Cedar in Russia is mainly in old stands with large volumes (m^3 ha^{-1}) of rootwood. Larch forests grow mainly in the permafrost zone and have lower productivity.

The main C storage in living vegetation is concentrated in the larch, pine, spruce and birch forests (Figure 2). These forests account for 81% of the total C storage of the main forest-forming species. Consequently, the distribution of C storage in living forest vegetation corresponds to the distribution of these areas and stocks (Figure 1).

TABLE III

Indicators of phytomass stock and carbon (C) in forest vegetation.

Groups of species and land categories	Area		Wood volume		Total phytomass		C storage	
	Mha	%	10^9 m^3	%	Pg	%	Pg	%
Coniferous	568.0	74	65.78	81	55.77	79	27.79	79
Hardwood	18.5	2	1.99	2	2.42	3	1.21	3
Softwood	118.9	15	12.40	15	9.45	13	4.71	13
Other species	0.2	0	0.02	0	0.02	0	0.01	0
Bushes	65.6	9	1.45	2	2.74	4	1.36	4
Lands covered with forest	771.1	100	81.64	100	70.40	100	35.07	100
Lands not covered with forest	113.0	-	1.30	-	2.52	-	1.25	-
Forest lands	884.1	-	82.95	-	72.91	-	36.32	-

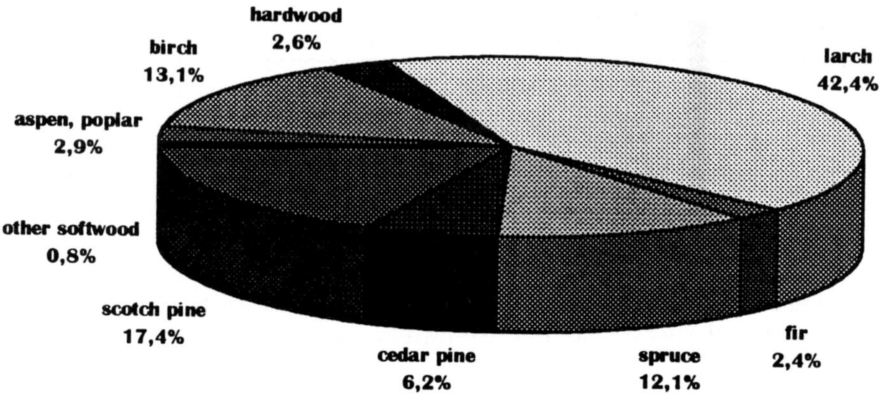

(a) Areas of forests (705.4 Mha)

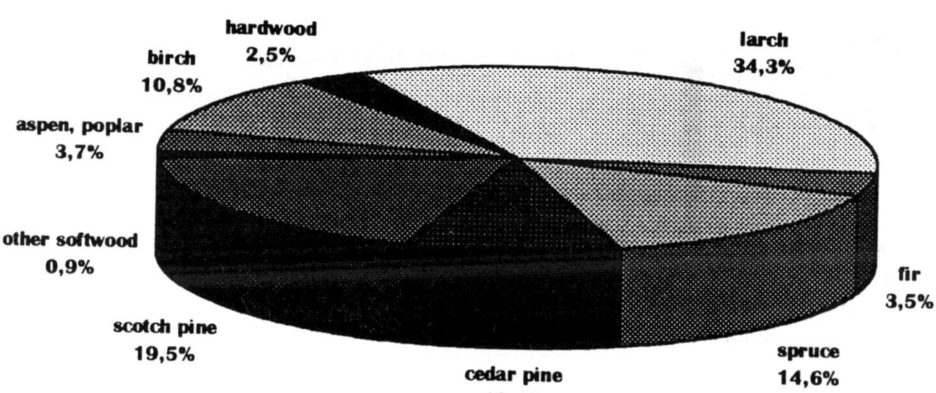

(b) Stem volume (80.17×10^9 m^3)

Fig. 1. Distribution of total area of Russian forests (a) and wood volumes (b) by main forest species.

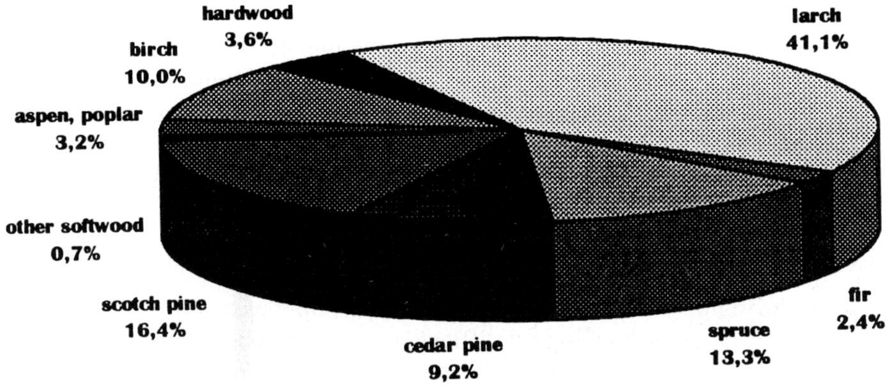

(a) Carbon stock (33.7 Pg)

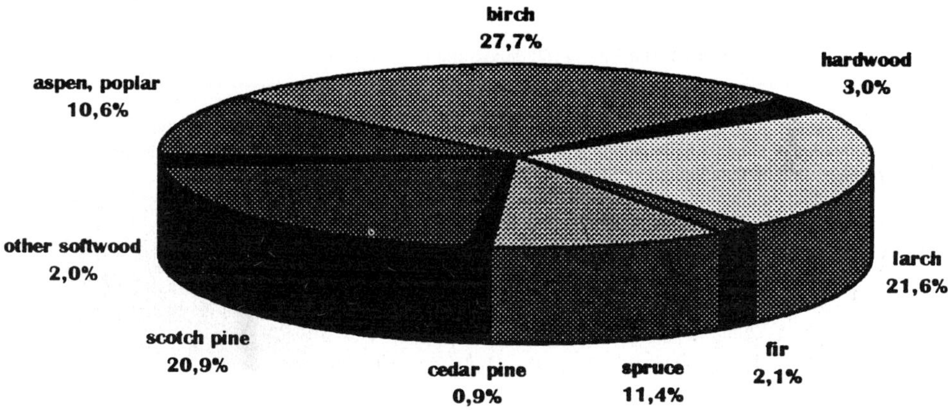

(b) Carbon deposition (183.0 Tg yr^{-1})

Fig. 2. Distribution of total carbon (C) storage in living phytomass (a) and average C deposition (b) for main forest species (without bushes).

The greatest annual C deposition is in birch stands, followed by aspen, then larch and spruce (Figure 2). The importance of aspen is greater, and that of conifers other than pine is lower. This is because deposition depends largely on the age structure of stands. Thus, prevalence of mature and overmature stands produces large C storage but low rates of deposition, whereas, prevalence of young and middle aged stands increases C deposition and reduces C storage. Thus the importance of aspen and birch increased in C deposition because of the prevalence of young and middle aged stands, formed on the enormous areas burned and harvested over the last fifty years.

Since, according to the NFI, conifers store less C in mature stands than in premáture stands, then when these last will transfer to the older age group we will see negative values of deposition (Table IV).

3.2. Comparison of Different Estimates

The process of the regional and global estimation for the components of the C cycle is fraught with considerable methodological difficulties, whereas the significance of these estimates is very high. The only criteria to prove reliability of these estimates is coincidence with estimates by different investigators using different methods.

Comparison of the estimates of phytomass C storage (Table V) shows relatively close values for the forests of Russia. The differences can be explained by the use of different approaches and different conversion ratios. For instance, Kolchugina and Vinson (1993c) used the value of 0.53 Mg C m^3 to estimate a C pool of 50.2 Pg from the NFI. This value is recommended by R. Sampson (1992) for forests of the USA. According to our estimates, taking into account the species and age structure of the Russian forests, this value should be 0.43 Mg C m^3.

TABLE IV

Annual change in wood stock and carbon (C) deposition by groups of the main forest forming species.

Species group	Age group	Area ha	Change of stock 10^6 m^3 yr^{-1}	Deposition Tg C yr^{-1}
Conifers	Young	91.9		
	Middle-aged	107.5	185.3	82.0
	Premature	54.0	67.9	25.9
	Mature and overmature	314.6	-13.7	-4.0
Hardwood	Young	2.3		
	Middle-aged	4.6	5.1	3.4
	Premature	2.1	3.1	1.5
	Mature and overmature	9.5	1.1	0.6
Softwood	Young	25.4		
	Middle-aged	38.8	92.7	40.0
	Premature	13.6	62.9	24.8
	Mature and overmature	41.0	24.6	8.9

It is more difficult to compare annual values of C accumulation by the forests of Russia to those of the former Soviet Union (Table VI). Direct comparison of published values is impossible because they are significantly different in their content. The value of annual deposition is the increase in C storage in living phytomass. This value is a function of the current species and age structures of stands. The value "net ecosystem production (NEP)" is usually defined as the difference between "net primary production (NPP)" and "heterotrophic respiration (Rh)". Kolchugina and Vinson (1993b) considered NEP to be the total change of C storage in phytomass, mortmass, and soil organic matter, assuming this change to be effected only by natural processes. They considered "net increment of vegetation (NIV)" as the result of change in the quantity of C in phytomass and mortmass. The concept "net sink" is the net increase of C in all pools in the forest cover of fixed areas, excluding all emissions (fires, cuttings, etc.).

TABLE V

Estimates of carbon storage in forest phytomass.

Region	Area, Mha	Carbon pool, Pg	Source
Russia	771	39.8	Isaev *et al.*, 1993
Russia	771	35.1	This paper
Former USSR	800	46.3	Kolchugina and Vinson, 1993b
Former USSR	800	50.2	Kolchugina and Vinson, 1993c

Sedjo calculated C sequestration for the Former Soviet Union using the change in timber volume for the period 1973–1984. According to Sedjo's estimate, the annual change in timber volume was 400×10^6 m^3. According to our estimate, the corresponding value for the forested territory of Russia is 435×10^6 m^3. Sedjo's estimate of C sequestration is higher than ours essentially because it includes C deposition in mortmass and soil organic matter.

4. Conclusion

We consider the estimates of carbon (C) storage and deposition in the Russian forest vegetation given in this article as the completion of the first stage of research. In the future, based on this approach, we plan to develop estimates of NPP, mortmass, soil C, and other components of the C balance of the forest cover of Russia.

At present, calculations for separate zones, subzones and other large territorial landscape units, as well as for administrative/economical regions of Russia, are been done using similar methods. This approach will allow not only increased reliability of the estimates but also the standardization of estimates for different geographical regions in Eurasia. Carbon storage and deposition estimates for different parts of Russia are interesting from the perspective of "carbon capacity", and potential CO_2 emmissions. Studies relating to "carbon credits" are of current interest.

The approach demonstrated in this paper opens perspectives for long-term prediction of carbon fluxes taking into account the natural and anthropogenic dynamics of the forest fund. Building on this approach, we have begun to create complex mathematical models for forecast calculations.

TABLE VI

Estimates of different indicators of carbon accumulation. NEP is Net Ecosystem Production; NIV is Net Increment of Vegetation (see text).

Region	Parameter	Tg yr^{-1}	Source
Russia	Deposition	211	Isaev *et. al.*, 1993
Russia	Deposition	213	This paper
Former USSR	NEP	825	Kolchugina and Vinson, 1993b
Former USSR	NIV	780	Kolchugina and Vinson, 1993b
Former USSR	Net sink	485	Kolchugina and Vinson, 1993b
Former USSR	NIV	880	Kolchugina and Vinson, 1993c
Former USSR	Carbon sequestration	416	Sedjo, 1992

Acknowledgment

This study was supported by the project N 9404-13775-B of the Russian Fund of Fundamental Investigations.

References

Bazilevich, N. I.: 1993, *Bioproductivity of Northern Eurasian Ecosystems*, Nauka, Moscow, Russia, 295 pp.

Brown S., Lugo A. E.: 1984, *Science* **223**, 1290-1293.

Brown S., Gillespie A. J. R., Lugo, A. E.: 1989, *Forest Science* **35**(4), 881-902.

Forest Fund of the RSFSR (under accounting on the January 1, 1988): 1990, *Ministry of Forestry of the RSFSR*, 738 pp.

Guricheva N. P., Demina O. M., Kozlova G. I. *et al.*: 1975, *L. Nauka. Is.* **1**, 96-127.

Isaev A. S., Korovin G. N., Utkin A. I., Pryaznikov A. A., Zamolodchikov D. G.: 1993, *Forest Science* **6**, 3-10.

Isaev A. S.: 1993, *Piroda*, **7**, 18-21.

Johnson W. C, Sharpe D. M.: 1983, *Canadian Journal of Forest Research* **13**(3), 372-383.

Kajimoto T.: 1989, *Ecological Research* **4**(1), 55-69.

Kajimoto T.: 1992, *Ecological Research* **7**(3), 333-339.

Kajimoto T.: 1993, *Ecological Research* **9**(2), 193-204.

Kolchugina T. P., Vinson T. S.: 1993a, *Canadian Journal of Forest Research* **23**(1), 81-88.

Kolchugina T. P., Vinson T. S.: 1993b, *Global Biogeochemical Cycles* **7**(2), 291-304.

Kolchugina T. P, Vinson T. S.: 1993c, *Water, Air, and Soil Pollut.* **70**(1-4), 207-221.

Kolchugina T. P., Vinson T. S., Shwidenko A. Z., Dixon R. K., Kobak K. I., Botch M. S.: 1993, *Publications of the Academy of Finland.* **3**, 52-62.

Kolicshuk V. G.: 1968, *Forest Science* **4**, 28-38.

Makarevsky M. F.: 1991, *Ekologia* **3**, 3-10.

Nabuurs G. J., Mohren G. M. J.: 1994, Koolstofvoorraden En-vastlegging in Het Nederlandse Bos Nederlands Bosbouw Tijdschrift. **66**(4), 144-157.

National Forest Inventory: 1988, *VNIITZlesresurs M., CEPL RAN.*

Sampson R. N.: 1992, *Water, Air and Soil Pollut.* **64**, 83-120.

Sedjo R. A: 1992, *Ambio* **21**(4), 274-277.

Zagreev V. V., Sukhikh V. I., Shvidenko A. Z., Gusev N. N., Moshkalev A. G.: 1992, *In:* Kolos M.(eds.), Generally Union Norms for Forest Taxation, 495 pp.

Utkin A. I., Gylbe A. *et al.*: 1994, *In:* Bioproductivity of Forest Ecosystem Computer Data Base, Institute of Forestry Science of RAN- CEPL RAN, Moscow.

A SURVEY OF THE FOREST SITE CHARACTERISTICS IN A TRANSECT THROUGH THE CENTRAL CANADIAN BOREAL FOREST

D. H. HALLIWELL, M. J. APPS and D. T. PRICE

Canadian Forest Service, Northern Forestry Centre, 5320-122 St., Edmonton, Alberta T6H 3S5, Canada

Abstract. As a contribution to the BOReal Ecosystem Atmosphere Study (BOREAS), a total of 97 sites have been surveyed in the boreal forest regions of Manitoba and Saskatchewan. The sites extend across an eco-climatic transect through the boreal forest of central Canada, from aspen parkland to sub-arctic woodland. Sites were selected using satellite, air photograph, and forest cover map data. Each site represents a minimum 100 m by 100 m homogeneous stand, suitable for satellite imaging. Data collected at each site covers overstory (species composition, DBH, pathogens, age, etc.), understory (percent cover by species), debris, and soil characteristics. These data provide a background database for the diverse measurements of terrestrial ecology, land surface climatology, tropospheric chemistry, and remote sensing. Such data provide an important baseline for ongoing studies of the boreal forest in a region sensitive to global change. The results of this study aid integration of a variety of more detailed studies being carried out in the region and also allow comparison with other international studies of a similar nature. Sites are concentrated in two regions: north of Prince Albert, Saskatchewan, and west of Thompson, Manitoba. Additional sites are located in the area of the Boreal Forest Transect Case Study (BFTCS) between Batoche, Saskatchewan and Gillam, Manitoba. The surveyed sites were classified according to three criteria: dominant species (*Picea mariana, Populus tremuloides, Pinus banksiana*, or "mixed"), age (mature, immature, or young/recently disturbed), and productivity (low, medium, or high). Overstory and understory measurements were usually taken at three points within each site. Debris measurements were taken at two points, and soil measurements at one point per site. This paper focuses on overstory and debris data taken at 97 sites in 1993 and 1994. The sites are predominantly in medium and high productivity stands. Overstory data, primarily from point-sampling, have been used to estimate stand basal areas, stem densities, volumes, and aboveground biomass (excluding foliage). Stand basal areas ranged from 5 m^2 ha^{-1} to over 60 m^2 ha^{-1}. Stem densities ranged from a few hundred to over 40 000 stem ha^{-1}. Estimates of total aboveground biomass range from a few Mg ha^{-1} to over 250 Mg ha^{-1}. The Prince Albert sites tend to have greater values than Thompson for basal area, volume and biomass, but stem densities are highly variable. Detritus data show values from zero to 60 Mg ha^{-1}. Generally, these values are small compared to aboveground biomass, but in recently disturbed sites the detrital mass can exceed aboveground biomass.

Keywords. BOREAL FOREST, OVERSTORY, DETRITUS, BIOMASS, FIELD METHODS.

1. Introduction

The work presented in this paper is the result of field studies carried out for two projects: the BOReal Ecosystem Atmosphere Study (BOREAS), and the Boreal Forest Transect Case Study (BFTCS). BOREAS is a large, multidisciplinary collaborative project examining the interaction between the boreal forest and the atmosphere. It entails diverse measurements of terrestrial ecology, land surface climatology, tropospheric chemistry, and remote sensing, with a focus on short-term (seasonal) processes. More extensive descriptions of the project are found in either Apps *et al.* (1991), or BOREAS (1990). The measurements discussed in this paper are a Canadian Forest Service contribution to the project, providing background information on forest characteristics. BFTCS is a study which extends the BOREAS forest measurements to a wider geographic region and a longer time scale. More detail can be found in Price and Apps (1995). A major goal of the Canadian Forest Service studies is an examination of regional carbon (C) budgets, in the context of global change.

Water, Air and Soil Pollution **82**: 257–270, 1995.

The work in this study also ties in with several other studies in this volume. Alexeyev *et al.* (1995), Isaev *et al.* (1995) and Krankina and Harmon (1995) are all concerned with C storage within boreal forests. Prince *et al.* (1995) examines remote sensing of forest characteristics. Hogg and Hurdle (1995) use sites that are included in this study. This study provides information about C storage for use in modelling work and provides spatial information about forest characteristics required for remote sensing validation.

The nature of BOREAS and BFTCS influences the methods used in this study. These projects extend over a large area of the boreal forest in Manitoba and Saskatchewan, Canada. The forest surveys cover a diverse and complex ecosystem. The data collected provide information suitable for a variety of tasks, as demanded by the diverse interests of the collaborators, yet are simple enough that the necessary measurements can be carried out at a large number of sites. For remote sensing purposes, each BOREAS site was to be located in an homogeneous forest stand at least 100 m by 100 m. (This allows each site to fill a 30 m × 30 m pixel on a satellite image, allowing a one-pixel positioning error.) The sites selected also encompass the range of productivity, age and species dominance represented by the study area.

Originally, approximately one thousand potential sites were identified using satellite, air photograph, and forest cover map data. After eliminating logistically unsuitable sites, several hundred were visually examined either from the air or the ground. Eventually, the number of sites was narrowed down to less than a hundred. The current Canadian Forest Service database[1] contains information on 97 sites in the region, with data collected in the summers of 1993 and 1994. Of these sites, nine are BOREAS tower sites (locations instrumented to measure energy balance and trace gas exchange), 59 are BOREAS "Auxiliary" sites (locations with less comprehensive measurement schemes, primarily established for remote sensing purposes), 25 are BOREAS/BFTCS "Transect" sites (similar to Auxiliary sites, but generally outside the two main study areas) and four are sites related to a non-BOREAS/BFTCS study.

2. Sample Design

The locations of the sites are shown in Figure 1. The BOREAS project is concentrated in two study areas, selected to cover the southern and northern limits of the boreal forest. The northern limit is one where there is sufficient precipitation, but tree growth is thought to be limited by low temperature. The southern limit is one where conditions are warmer, but tree growth is thought to be limited by growing-season soil moisture deficits. The Southern Study Area (SSA), Northern Study Area (NSA), and limits of the BFTCS are indicated in the figure. A cluster of transect sites can be seen in Figure 1 along the northern edge of the BFTCS region, in Saskatchewan, near the Manitoba border. However, most of the transect sites are grouped within the clusters representing the SSA and NSA. This geographical bias must be considered in interpreting the data.

After preliminary investigations, largely based on remote sensing or forest cover maps, each BOREAS Auxiliary site was assigned a site code indicating three characteristics of the site: dominant tree species, age, and productivity. The details of

[1] These data will be available for public distribution through the BOREAS Information System (BORIS) and from the Northern Forestry Centre.

this coding are given in Table 1. Not all age or productivity classes were present for each species, but an attempt was made to select sites covering the range of possibilities.

BOREAS tower sites and BOREAS/BFTCS transect sites were only classified on the basis of species dominance and age. For all sites, these preliminary classifications will be subject to re-evaluation based on the more detailed information collected in the course of the study.

The site characterisation data can be subdivided into five groups: site locations and descriptions, overstory, understory, detritus, and soils. The site location information includes geographic co-ordinates, forest cover map and air photograph lists, brief site descriptions, and directions for finding sampling points within the site. This information is contained in the report by Apps *et al.* (1994b). The other four data groups constitute the scientific data. Two further reports outline this information: Apps *et al.* (1994a) examines overstory and understory data, and Apps *et al.* (1995) examines

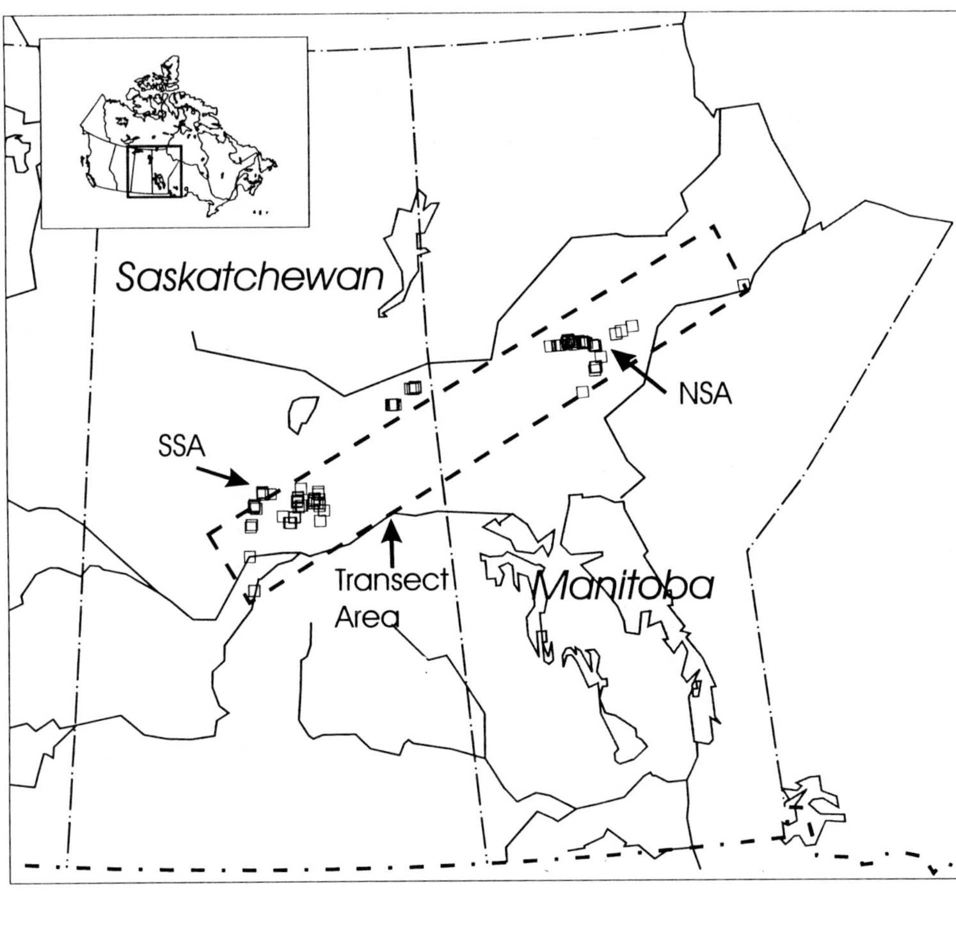

Fig. 1. Map of study area, showing site locations. Inset map shows location of main map within Canada.

soils and detritus data. The referenced versions of these reports are based largely on 1993 data and all three reports are under revision to include 1994 data.

At each site, three sampling points were established, usually prior to field visits, using air photographs and forest cover maps. These three points were typically 100–400 m apart, and were located in the field using a compass and hip chain. On the basis of either air photographs or field examination, twelve of the sites (seven in Manitoba, five in Saskatchewan) were expected to show greater non-homogeneity and were subdivided into a/b or a/b/c sampling strata. As a result, 84 distinct geographical locations contain 97 sampled sites, with the extra 13 sites representing these additional sampling strata. Each sampling stratum contained three sampling points (with a few exceptions where sites were too small), for a total of 285 sampling points (119 in Manitoba; 166 in Saskatchewan). Close to half of these points were visited in the summer of 1993, and the remainder in the summer of 1994.

For overstory measurements, a sample of trees was selected (at each sampling point) using either point-sampling methods or fixed-area plots. Generally, point-sampling was the preferred technique, using either a relascope or a wedge prism. The Basal Area Factor (BAF) ranged from 0.394 to 3 $m^2\ ha^{-1}$, and was typically selected to provide a minimum sample size (when possible) of about 15 trees. Point-sampling does not provide good results in young stands dominated by small trees. In 1994, such stands were sampled using fixed-area plots ranging from 25 to 100 m^2 (5 m or 10 m square plots).

For each tree in the sample, variables recorded include diameter at breast height (DBH, at 1.3 m), species, canopy class (dominant, codominant, intermediate, or suppressed), evidence of pathogens, and whether the tree was dead or alive. At each sampling point, several trees were selected for additional measurements: height, crown diameter, and height from the base of the tree to the live crown, plus coring using an increment borer. The cores provide samples for age determination, sapwood and bark thickness, and detailed dendrochronological analysis. Typically, these additional measurements were done on ten trees at each site—three or four at each sampling point. The trees were selected to cover a range of canopy classes and species (if more than one was present).

The overstory data are used directly to calculate sample values of stand basal area and stem density. Details on sampling theory and data analysis can be found in any basic text on forest measurement, e.g., Avery and Burkhart (1983) or Husch *et al.* (1982). Other calculations, such as stand volume or biomass,

TABLE I

Site classification hierarchy.

Code	Description
Species dominance	
A	Aspen (*Populus tremuloides* Michx.)
B	Black spruce (*Picea mariana* (Mill.) B.S.P.)
J	Jack pine (*Pinus banksiana* Lamb.)
M	Mixed (no dominant species; typically *Populus tremuloides* and *Picea glauca* (Moench) Voss. common)
Age	
D	recently disturbed; young
I	Immature
M	Mature
Productivity	
L	Low
M	Medium
H	High

require further assumptions which are discussed later.

Understory measurements used fixed-area plots of 4 or 25 m^2 (2 m or 5 m square plots). The sample plot was established a few metres away from the point selected as plot centre for overstory sampling. For each plot, percent cover was estimated visually for all species present. Estimation of understory biomass from these measurements is problematic, and requires regression equations not yet developed. However, the data serve as ecological indicators for the sites. Plants were identified to the species level where possible, but some are only recorded to the genus level.

Detritus sampling was based on the line intersect method described in Van Wagner (1982). At two of the three sampling points within each site, a 30 m equilateral triangle was set out a few metres away from the overstory and understory plots, and woody debris was tabulated on each side of the triangle. Tabulations were done in six diameter classes: 0–5, 5–10, 10–30, 30–50, 50–70, and >70 mm. For the first five classes, a simple count of debris items was recorded. For the >70 mm class, each diameter was measured individually, along with the species and physical condition of the item (e.g. sound, rotten, etc.) The smallest class was tabulated over the first 5 m of the line, the second over 10 m, etc., with only the last class being recorded over the full 30 m. The triangular orientation guards against any bias introduced by non-random orientation of the fallen debris. Van Wagner (1982) discusses error sources, and methods of estimating detrital mass using the frequency counts. In addition to the debris tallies, duff samples were collected from the mid-point of each leg of the triangle, to determine bulk density and for analysis of C content. Duff depth was also measured at five points (5 m intervals) on each side of the triangle.

Soil sampling was carried out at one of the three sampling points within each site. Each soil pit was excavated to approximately 1 m, and the soil described using the Canadian Soil Classification System (Canada Soil Survey Committee, 1978; Land Resource Research Institute, 1983). Samples were collected from each identifiable horizon, for bulk density, carbon, and other chemical analyses.

3. Overstory Results

This paper will focus on two aspects of the data collected: overstory characteristics, and detritus. Overstory calculations are presented for stand basal area, stem density, stem volume, and total aboveground biomass (excluding foliage). Detritus data is used to calculate total detrital mass. As mentioned earlier, the overstory data allow direct calculation of stand basal area and stem density using standard formulae. All overstory calculations were carried out for live trees only. Stem volume and total aboveground biomass require information not measured directly, so several assumptions were made. The most important of these involves tree heights.

Heights were only recorded for trees that were cored (approximately 1000 trees, or 20% of the total). Estimation of heights for the other trees used a procedure which accounted for both species and site factors. Initially, all trees with measured heights were plotted on graphs of height *versus* DBH, using log-log axes. Data was sorted by province (Manitoba or Saskatchewan) and by species.

Figure 2 shows the graph for Saskatchewan *Picea mariana*. There are two distinct linear sections: trees of small DBH show a more shallow slope. As a result, two linear segments were used to approximate the data. For most species, a single linear relationship was deemed sufficient. Analysis of all graphs indicated several features. First, there was no noticeable difference between the two provinces, so subsequent analysis was based on species alone. Second, there was little difference between species within the same genus, so these were also pooled. The final equations used for height estimation are given in Table II, along with the number of trees in the database with measured heights for each group. Sample sizes for some of the equations are small, but the three main genera—*Picea*, *Pinus*, and *Populus*—all have reasonable sample sizes. These three genera include 5 808 of the total 6 097 trees in the database. The height-DBH relationships were compared to similar ones from Huang *et al.* (1992). Significant deviations were noted, demonstrating the importance of using local forms.

For stem volume and aboveground biomass calculations, the equations in Table II were applied to all trees without measured heights. However, productivity is not the same from site to site, so a site adjustment was devised to account for local effects. The procedure involved taking trees with measured heights within each sample, and comparing those heights with the height estimated from DBH. A mean calculated/estimated height ratio was determined for each species, and this ratio was then used to adjust the estimated heights. For example, if a sample had 15 *Picea mariana* trees, and three had measured heights that were 0.8, 0.95, and 0.65 times the height estimated from their DBH, then a site adjustment factor of 0.8 was used as a multiplier for the other 12 heights calculated from DBH. Figure 3 is a scatter diagram of estimated *versus* measured height, for all trees (with measured heights) in the database. The agreement is generally good, with a Mean Bias Error of 0.18 m, and a Root Mean Square Error of 1.47 m.

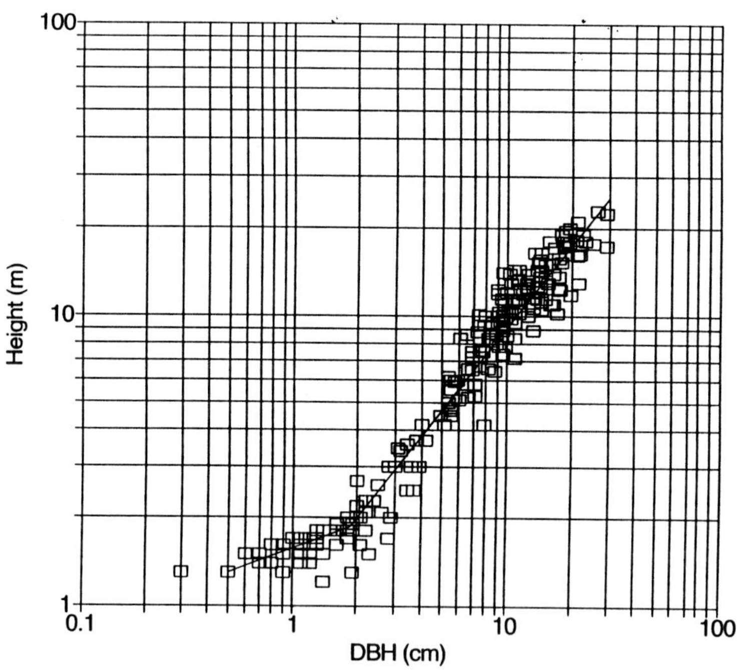

Fig. 2. Height to DBH relationship for *Picea mariana* from the Saskatchewan sites. The solid line represents the fitted equations.

TABLE II

Height (m) *versus* DBH (cm) relationships. Equation fitted is of form *ln*(height) = a *ln*(DBH) + b.

Species	N	Coefficients a	Coefficients b	
Abies balsamea	5	0.82	0.27	
Betula papyrifera	10	0.67	0.74	
Larix laricina	20	1.06	-0.38	
Picea glauca, mariana	526	0.94	0.04	(DBH>1.86)
		0.28	0.45	(DBH<1.86)
Pinus banksiana	310	0.86	0.19	(DBH>3.51)
		0.39	0.78	(DBH<3.51)
Populus balsamifera, tremuloides	234	0.80	0.52	

Tree volumes were calculated from estimated height (or the measured value, when available) and measured DBH, assuming a conical stem shape. The tree volume was then multiplied by the number of stems per hectare the tree represents[2], and this value was summed for all live trees to provide an estimate of stand stem volume.

Total aboveground biomass (excluding foliage) was estimated using biomass regression equations developed by Singh (1982). Singh provides equations in several forms, using combinations of height (H), DBH, DBH^2, DBH^3, and DBH^2H. Polynomial forms show odd characteristics when extrapolated outside the range of DBH found in Singh, so for simplicity a linear form with mass = a + b DBH^2H was selected. This also has the advantage of explicitly accounting for the varying height-DBH relationships, as indicated in the stand volume calculation procedures.

Figures 4 and 5 show the results of the stand basal area, stem density, stem volume, and total biomass calculations. Figure 4 is for sites in Manitoba, and Figure 5 is for Saskatchewan. The figures require some explanation, as the organisation is quite complex. However, each panel in the figures is similar, so we can focus on a single panel. Consider the basal area graph, in the upper left panel. Stand basal area is on the vertical axis, and the different sites are arranged in order along the horizontal axis. For each site, the vertical bar represents the range of basal area (each site usually has three sampling points), and the square symbol represents the mean. The ordering of sites along the x-axis is hierarchical, based on three criteria: species dominance, age, and basal area. Species dominance is derived from the preliminary site classification presented in Table I, and the graph is labelled for each grouping. Within each species group, the sites are grouped according to the three age classes (again from Table I). These classes are indicated by the horizontal bar just above the X axis, and are ordered —from left to right—recently disturbed, immature, and mature. (The mixed sites are not classified by age.) Finally, within each age class, sites are sorted by ascending mean stand basal area.

The other three graphs within each figure are organised in exactly the same fashion, with the final sorting criterion always being basal area (i.e., the site order is exactly the same in each of the four graphs in the figure). Vertical axes for all the graphs except

[2]For a fixed-area plot, each tree in the sample represents one stem per Xm^2, where X is the plot area. For point samples, each tree represents a different number of stems per hectare, dependent on DBH. Details can be found in the references previously cited.

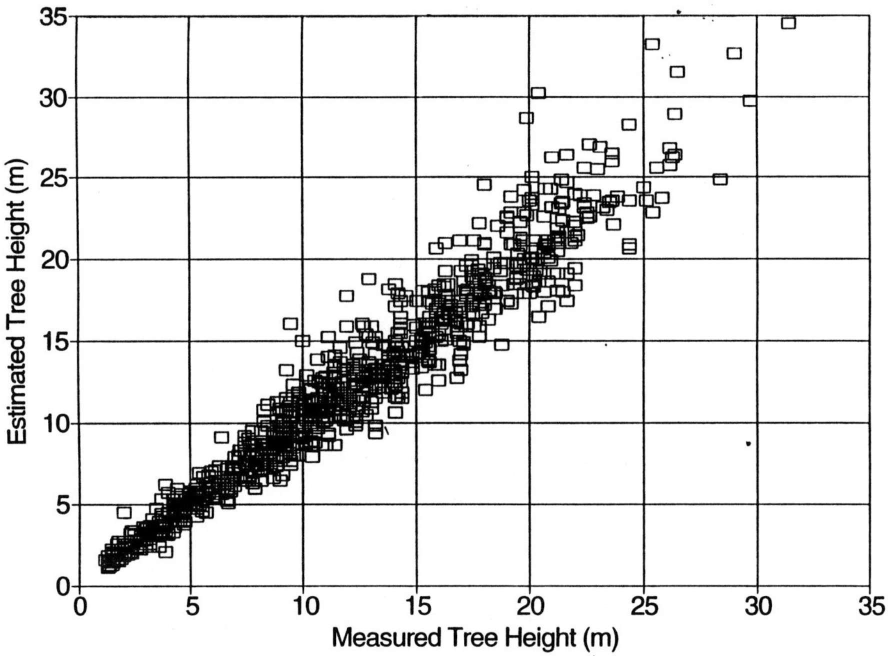

Fig. 3. Estimated tree height *versus* measured tree height, for all trees with measured heights from all sites.

stem density are the same in Figures 4 and 5, to facilitate comparison between regions. The graphs give an indication of variability between and within provinces, species, age classes, and sites. The graphs also give an indication of the relative frequency of sites of a given species and age class.

4. Detritus Results

Analysis of the detritus data follows the methods described in Van Wagner (1982) and McRae *et al.* (1979). Briefly, the detrital volume per hectare is calculated from:

$$V = \frac{\pi}{4L}\sum_{i=1}^{n} A_i \qquad [1]$$

where L is the length of the sample line, A is the cross-sectional area of each piece at intersection, and π/4 is a probability factor derived from sampling theory. The final result was converted to $m^2\ ha^{-1}$. Volume was then converted to mass using wood density. The densities used in the analysis varied with size class: 0.48 $Mg\ m^{-3}$ (5–10 mm), 0.46 $Mg\ m^{-3}$ (10–30 mm), 0.44 $Mg\ m^{-3}$ (30–50 mm) and 0.42 $Mg\ m^{-3}$ (50–70 mm). For items >70 mm diameter, density ranged from 0.4 $Mg\ m^{-3}$ for sound material to 0.3 $Mg\ m^{-3}$ for rotten material.

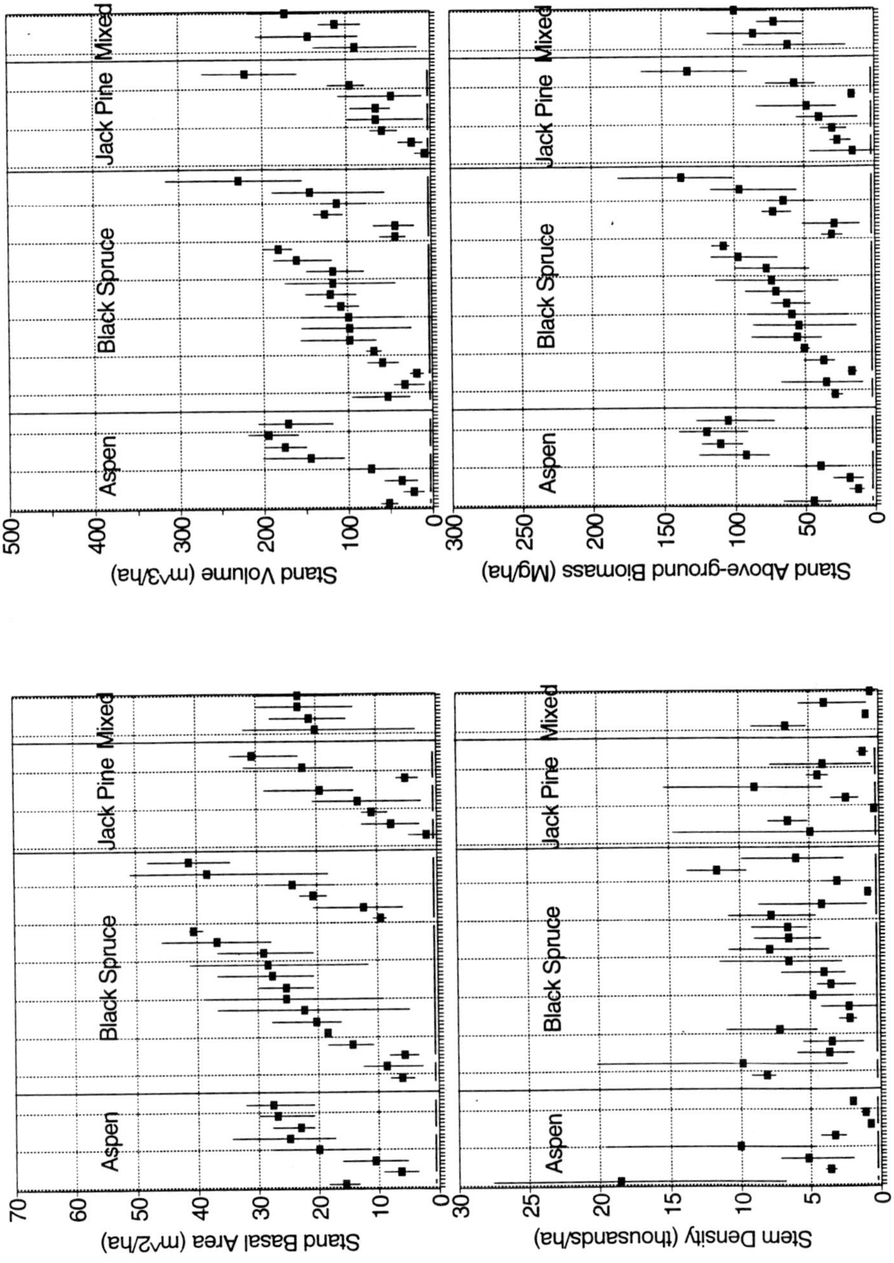

Fig. 4. Stand basal area (m^2 ha^{-1}), stem density (ha^{-1}), stem volume (m^3 ha^{-1}), and above-ground biomass, excluding foliage, (Mg ha^{-1}) for the Manitoba sites. Symbol and vertical line represent mean and range for each site. Ordering of sites is explained in the text.

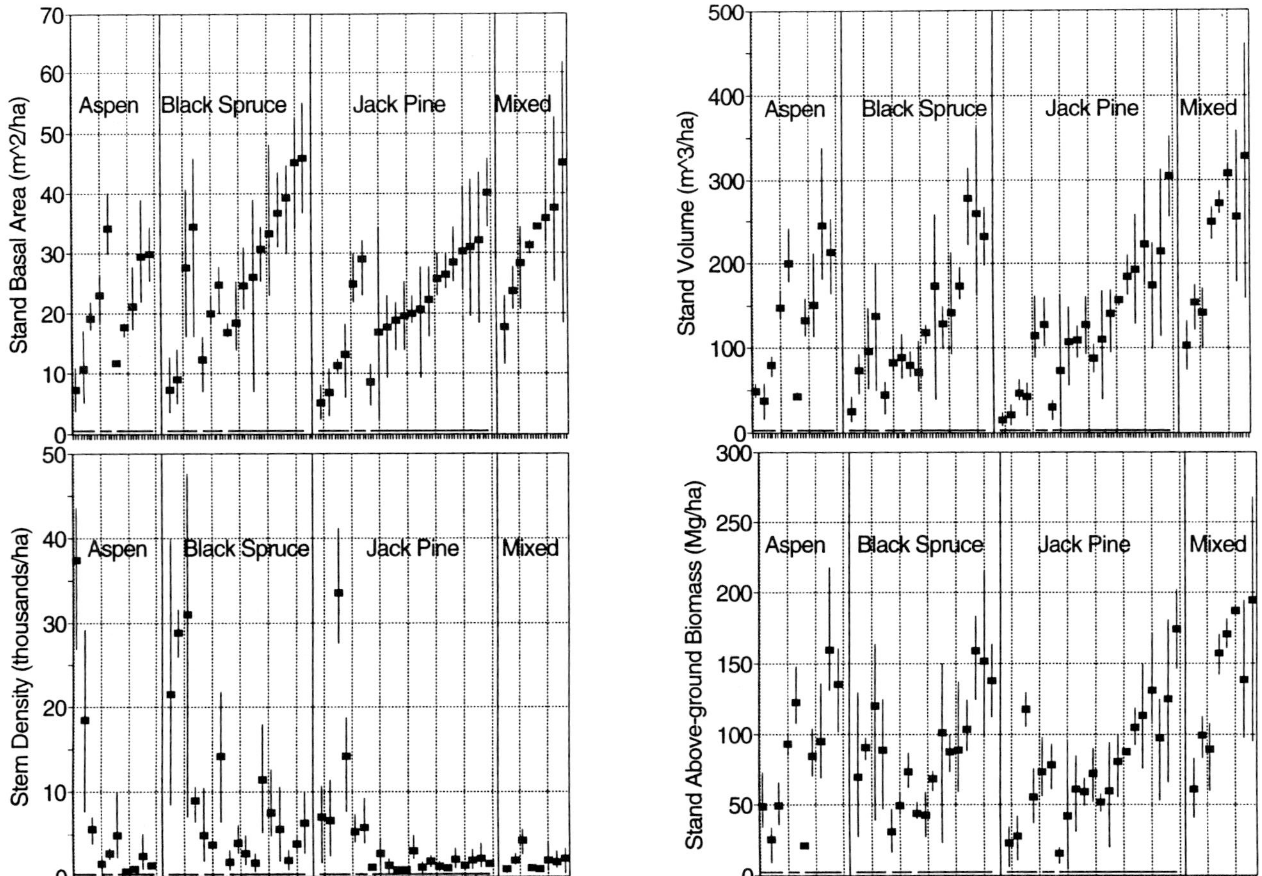

Fig. 5. Stand basal area ($m^2\ ha^{-1}$), stem density (ha^{-1}), stem volume ($m^3\ ha^{-1}$), and above-ground biomass, excluding foliage, ($Mg\ ha^{-1}$) for the Saskatchewan sites. Symbol and vertical line represent mean and range for each site. Ordering of sites is explained in the text and is the same as in Figure 4.

For this study, all but the largest size class were recorded as counts without diameters, so equation 1 is applied using a mean area term (A) within each class. Van Wagner (1982) suggests an area based on the Quadratic Mean Diameter for the size class, but the value would depend on assumptions about diameter distributions. In the >70 mm class, individual areas were used. For the smaller classes, the midpoint value was used (e.g., 40 mm for 30–50 mm class). The smallest size class can be problematic if an inappropriate mean area is selected, so this paper presents data for detritus larger than 5 mm diameter only. Figure 6 shows the detrital mass values. Two graphs are given, one for each province. The sites are in the same order as in Figures 4 and 5.

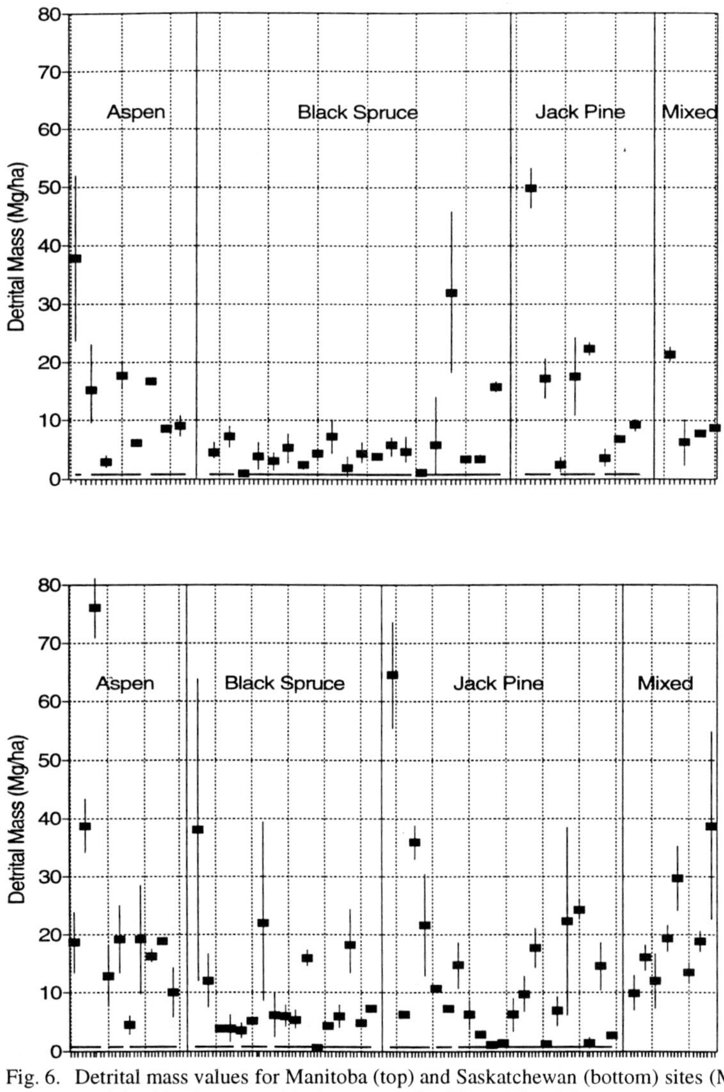

Fig. 6. Detrital mass values for Manitoba (top) and Saskatchewan (bottom) sites (Mg ha^{-1}). Ordering of sites is explained in the text and is the same as for Figures 4 and 5.

5. Discussion

Generally, the Saskatchewan sites (SSA, Figure 5) show greater productivity than the Manitoba sites (NSA, Figure 4), measured in terms of stand basal area, volume, or aboveground biomass. Basal areas reach 45 m^2 ha^{-1} in Saskatchewan, compared to 40 m^2 ha^{-1} in Manitoba. Stand volumes approach 300 m^3 ha^{-1} in Saskatchewan, but rarely exceed 200 m^3 ha^{-1} in Manitoba. Aboveground biomass has a maximum close to 200 Mg ha^{-1} in Saskatchewan, and remains less than 150 Mg ha^{-1} in Manitoba. However, both regions show some low-productivity stands, and there is often a significant range of productivity between sampling points within the stands. There is also a large range of values within a single age class for a species, e.g., mature black spruce in either province. Volume and biomass are strongly correlated with basal area, as expected. Stem counts can be very high in recently disturbed sites, although the small individual tree size keeps basal areas and volumes relatively small. The biomass values look reasonable in comparison to values reported for the areas by Bonnor (1985).

Species comparisons between provinces are interesting. Aspen and black spruce sites tend to show similar ranges in values, but jack pine and mixed-wood sites are clearly more productive in Saskatchewan. However, one must keep in mind that the graphs can mislead: there is a greater number of mature jack pine sites in Saskatchewan, so a direct comparison is not straightforward.

Detrital mass estimates (Figure 6) can be compared to the biomass estimates in Figures 4 and 5. Generally, the detrital values are much smaller (1–30 Mg ha^{-1}) than the aboveground values (50–150 Mg ha^{-1}) in immature and mature sites. Recently-disturbed sites can have relatively high detrital mass values (e.g. >60 Mg ha^{-1} in aspen and jack pine sites in Saskatchewan) that exceed the aboveground biomass. Detrital mass can be an important component of the ecosystem C storage.

The sample design was based on coverage of all stand types, without explicit consideration of the fraction of the boreal forest each type represents. If remotely-sensed data can be used to identify stands and provide estimates of overstory characteristics, the aerial extent can be assessed. This is part of the overall BOREAS project work, but is beyond the scope of this paper. The number of sites within each species class may also mislead, since subdivision of preselected stands into sampling strata leads to multiple sites at one location. This is the explanation for the large number of black spruce sites in Manitoba, and of jack pine sites in Saskatchewan. This division into sampling strata was based on perceived variability within stands, either from air photograph analysis before field work, or at the time of field measurements. Even after division into sampling strata, the within-site variability is evident in Figures 4, 5 and 6.

The stand volume and aboveground biomass estimates presented here can be improved with better allometric data. The volume calculations assumed a simple conical shape. The biomass estimates used equations from the literature (Singh, 1982). Harvesting of selected trees was carried out in August 1994, by personnel from Canadian Forest Service and University of Wisconsin, in order to develop allometric equations specific to the study areas. Once analysis of these data are complete, more accurate taper and biomass relationships will be available, and the stand volume and biomass estimates can be improved. However, the values reported here give a good indication of the range of values found within this portion of the boreal forest.

6. Conclusion

This paper presents an overview of the site characterisation data collected for the BOREAS/BFTCS research projects. Analysis has been presented for the overstory and detritus portions of this data, giving estimates of stand basal area, stem density, volume, aboveground biomass (excluding foliage), and detrital mass. The data presented is useful in two contexts: as an indication of the range of values that can be expected in the study areas (which represent two important limits for the North American continental boreal forest), and as an indication of the characterisations that can be derived from a field program designed to cover a large number of sites over a wide area (with a necessarily abbreviated measurement time for each site). The database that is now available provides numerous possibilities for spatial analysis of various forest characteristics.

Acknowledgments

This study has been carried out with the financial assistance of the CFS Climate Change Green Plan Initiative, the Canada-Saskatchewan Partnership Agreement in Forestry (PAIF), and the Canadian Federal Panel on Energy Research and Development (PERD) through the ENFOR (ENergy from the FORest) program of Canadian Forest Service. Field work and analysis has been carried out with the assistance of many people, including Paul Albu, Colin Borys, Dean Gauthier, Norman Fraser, Forrest Hall, Chris Higgs, Maneesh Kalra, Liz Kruit, Sandy Mah, Joe Niederleitner, Tamara Oishi, Karen Oldershaw, Jason Proche, Dave Seburn, Piers Sellers, Marty Siltanen, Gavin Sulewski, Thierry Varem-Sanders, Tammy Verhaeghe, and Jason Vogel.

References

Alexeyev, V. A., Korotkov, I. A., Stakanov, V. D. and Birdsey, R. A.: 1995, *Water, Air, Soil Pollut.* **82**, 271-282.

Apps, M. J., Albu, P., Halliwell, D., Niederleitner, J., Seburn, D. and Siltanen, M.: 1994a, BOREAS biometry and auxiliary sites: overstory and understory data, (Version 1.0). Can. For. Serv. Rep., Northwest Region, Edmonton, Canada, 124 pp.

Apps, M. J., Niederleitner, J., Seburn, D., Varem-Sanders, T. and Siltanen, M.: 1994b, BOREAS biometry and auxiliary sites: locations and descriptions, (Version 1.0). Can. For. Serv. Rep., Northwest Region, Edmonton, Canada, 33 pp.

Apps, M. J., Albu, P., Halliwell, D., Price, D. T., Siltanen, M. and Varem-Sanders, T.: 1995, BOREAS biometry and auxiliary sites: soils and detritus data, (Version 1.0). Can. For. Serv. Rep. Northwest Region, Edmonton, Canada, 73 pp.

Apps, M. J., Goodison, B., Cihlar, J. C., Hall, F., Leckie, D., LeDrew, E., Matson, P., Running, S. and Sellers, P. J.: 1991, *BOREAS: Global Change and Biosphere-Atmosphere Interactions in the Boreal Forest Biome, Science Plan*, NASA, Goddard Space Flight Center, Greenbelt, USA, September 1991, 67 pp.

Avery, T. E. and Burkhart, H. E.: 1983, *Forest Measurements*, 3rd edn., McGraw-Hill, New York, USA, 331 pp.

Bonnor, G. M.: 1985, *Inventory of Forest Biomass in Canada*, Can. For. Serv. Petawawa National Forestry Institute, Chalk River, Canada, 63 pp.

BOREAS Science Steering Committee: 1990, *Eos* **72**, 33-40.

Canada Soil Survey Committee: 1978, *The Canadian System of Soil Classification*, Agric. Can. Publ. **1646**, Ottawa, Canada, 164 pp.

Hogg, E. H. and Hurdle, P. A.: 1995, *Water, Air, Soil Pollut.* **82**, 391-403.

Huang, S., Titus, S. J. and Wiens, D. P.: 1992, *Can. J. For. Res.* **22**, 1297-1304.

Husch, B., Miller, C. I. and Beers, T. W.: 1982, *Forest Mensuration*, 3rd edn., Wiley, New York, USA, 402 pp.

Isaev, A., Korovin, G., Zamolodchikov, D., Utkin, A. and Pryaznikov, A.: 1995, *Water, Air, Soil Pollut.* **82**, 247-256.

Krankina, O. N. and Harmon, M. E.: 1995, *Water, Air, Soil Pollut.* **82**, 227-238.

Land Resource Research Institute: 1983, Hay J. H. (ed.). *The Canadian Soil Information System (CanSIS): Manual for Describing Soils in the Field.* LRRI Contrib. No **82-52**, Agric. Can., Ottawa, Canada, 164 pp.

McRae, D. J. Alexander, M. E. and Stocks B. J.: 1979, *Measurement and Description of Fuels and Fire Behavior on Prescribed Burns: A Handbook,* Can. For. Serv. Inf. Rep. **O-X-287**, Great Lakes Region, Sault Ste. Marie, Canada, 58 pp.

Price, D. T. and Apps, M. J.: 1995, *Water, Air, Soil Pollut.* **82**, 203-214.

Prince, S. D., Goetz, S. J. and Goward, S. N.: 1995, *Water, Air, Soil Pollut.* **82**, 509-522.

Singh, T.: 1982, *Biomass Equations for Ten Major Tree Species of the Prairie Provinces*, Can. For. Serv. Inf. Rep. **NOR-X-242**, Northern Forest Research Centre, Environment Canada, Edmonton, Canada, 35 pp.

Van Wagner, C. E.: 1982, *Practical Aspects of the Line Intersect Method*, Can. For. Serv. Inf. Rep. **PI-X-12**, Petawawa National Forestry Institute, Chalk River, Canada, 11 pp.

CARBON IN VEGETATION OF RUSSIAN FORESTS: METHODS TO ESTIMATE STORAGE AND GEOGRAPHICAL DISTRIBUTION

V. ALEXEYEV[1], R. BIRDSEY[2], V. STAKANOV[1], and I. KOROTKOV[1]

[1] *V.N. Sukachev Institute for Forests Research (Siberian Branch of Russian Academy of Sciences), Russia,*
[2] *USDA Forest Service, Northeastern Forest Experiment Station, United States*

Abstract. Estimates of carbon (C) storage for the forest vegetation of Russia vary to a considerable extent. In this paper we describe methods of estimating C storage of forest ecosystems and present estimates of C storage in the forest vegetation of administrative territories and ecoregions of Russia. Estimates of C storage are based on National Forest Inventory data, estimates of the phytomass of trees and other vegetation of different forest ecosystems in various ecoregions of Russia, and mapped data for natural ecoregions. The forest ecosystems of Russia (771.1 Mha) have 118.8 Pg of C with 23.6% of the total in forest vegetation, 14.1% in the mortmass, and 62.3% in the soils. The estimates from this study are comparable to the estimates from other countries with temperate and boreal forest ecosystems.

Keywords. RUSSIAN FORESTS, CARBON STORAGE, METHODOLOGY. FOREST ECOSYSTEMS, FOREST INVENTORY, BOREAL FORESTS, RUSSIA.

1. Introduction

Lands classified as forest areas of Russia occupy more than 1182 Mha. According to National Inventory data as of January 1, 1988 (Forest Fund, 1990), these lands include 297.6 Mha of non-forest land (peatlands, water, rocks, wastelands, etc.), and 884.1 Mha of forest land, including 771.1 Mha of tree stands and 113.0 Mha of non-stocked lands (woodlands, clear-cuttings, burned stands of trees, etc.). The forest lands comprise 22% of the Earth's forest area or 43% of the Earth's temperate and boreal forest area (Anonymous, 1992).

To estimate carbon (C) storage for such a huge territory, it is necessary to use a variety of information sources, some not easily accessible. It is therefore not surprising that the estimates of C storage for the forest vegetation of Russia (or for the former USSR) vary to a considerable extent. For example, Kolchugina and Vinson (1993a) estimate that forests of the former USSR contain 88 Pg, Isaev *et al.* (1993) estimate 38 Pg, and Dixon *et al.* (1994) estimate 74 Pg.

The purposes of this paper are: (1) to describe methods of estimating C storage of forest ecosystems; and (2) to estimate C storage in the forest vegetation of administrative territories and ecoregions of Russia.

2. Methods

Estimates of C storage are based on (1) National Forest Inventory data; (2) data from many sources about phytomass of trees and other vegetation of different forest ecosystems in various ecoregions of Russia, (3) mapped data for natural ecoregions, and (4) professional experience. The basic inventory is the published data of the National Forest Inventory as of 1 January, 1988 (Forest Fund of the USSR, 1990, 1991; see also Backman and Waggener, 1991) for 71 administrative units of Russia.

Water, Air and Soil Pollution **82**: 271–282, 1995.

Additionally, we used unpublished data from the forest inventory database for 98 forestry enterprises in Krasnoyarsk and the Republic of Sakha (Yakutia).

Russian forest inventories are designed to estimate the volume of growing stock for all trees (commercial, non-commercial and damaged trees). Since the published forest inventories do not include estimates of the biomass of trees and stands, it is necessary to convert estimates of volume to estimates of biomass.

2.1. EQUATIONS FOR CONVERTING VOLUME OF GROWING STOCK TO PHYTOMASS OF TREE STANDS

The volume of growing stock includes both wood and bark. Bark is different from wood in density, chemical composition, and rate of decay. These properties of bark make it necessary to account for this fraction separately from timber:

$$V_{tb} = V_t + V_b, \qquad [1]$$

where V_{tb} is the volume of growing stock, V_t is the underbark volume of growing stock, and V_b is bark volume.

Volume of growing stock can be converted to mass by:

$$V_{tb} = M_t/P_t + M_b/P_b \qquad [2]$$

where M_t and M_b are the mass of wood and bark of the stem, and P_t and P_b are the density of wood and bark in Mg m^{-3}or kg m^{-3}.

Formula Eq. [2] has two unknowns: M_t and M_b. To exclude one of them, we can introduce a bark conversion factor K_b, which is the ratio of the bark mass to the timber mass:

$$K_b = M_b/M_t = V_b \cdot P_b / V_t P_t \qquad [3]$$

Transforming Eq. [2], we have:

$$M_t = V_b \frac{P_t P_b}{P_b + (K_b P_t)} \qquad [4]$$

Thus, to determine the mass of timber from the volume of growing stock of the various tree species we need to know the density of timber and bark of these tree species.

Stand phytomass (M_{st}) is equal to the sum of its constituent fractions:

$$M_{st} = M_t + M_b + M_{cr} + M_r \qquad [5]$$

where M_t is the mass of stem timber, M_b is the mass of bark, M_{cr} is the mass of crown, and M_r is the mass of roots.

Expressing the mass of fractions of Eq. [5] through the mass of stem timber:

$$M_{st} = M_t + K_b M_t + K_{cr} M_t + K_r M_t \qquad [6]$$

where K_{cr} is the conversion factor of the crown, equal to the ratio of the mass of crown to the mass of stem timber ($K_{cr} = M_{cr}/M_t$) and K_r is the conversion factor of roots, equal to the ratio of the mass of root to the mass of stem timber ($K_r = M_r/M_t$).
Transforming equation 4.6, we have:

$$M_{st} = V_{tb} \frac{P_t P_b}{P_b + K_b P_t}(1 + K_b + K_{cr} + K_t) \qquad [7]$$

Designate the second multiplier of formula Eq. [7] as K_{ph}:

$$K_{ph} = \frac{P_t P_b}{P_t + K_b P_t}(1 + K_b + K_{cr} + K_r) \qquad [8]$$

Equation [7] takes a form more convenient for many calculations:

$$M_{st} = K_{ph} V_{tb} \qquad [9]$$

The adopted K_{ph} coefficient reflects the relationship between the volume of growing stock and the phytomass of tree stands. Clearly it can be calculated only when all its constituent indices are calculated. Thus, to estimate the stand phytomass and C by using the volume of growing stock it is necessary to know: (1) density of timber and bark of the forest tree species; (2) conversion factors of bark, crown and roots of species (ratios of their masses to the stem timber mass), and (3) content of C in phytomass.

To include all living parts of forest plant communities, we need data about undergrowth and other live vegetation. To include the dead part of forest vegetation it is necessary to use data about coarse woody debris and litter.

2.2. DENSITY OF TIMBER AND BARK

Timber density varies in rather wide limits not only for different species, but for regions and for age of trees. Therefore it is appropriate to use the regional values of timber density for every tree species and for every age class. We used a conventional definition of timber density, the mass of oven dry timber in volume unit of living tree. Estimates of the bark density were obtained from Poluboyarinov (1976).

2.3. COEFFICIENTS FOR CONVERSION OF GROWING STOCK INTO PHYTOMASS AND CARBON

It is well known that allocation of tree phytomass between stem, roots, bark, twigs and leaves is different for different ages of trees and habitat conditions. For determination of conversion factors K_b, K_{cr}, and K_r, we used our own data and data from the literature for 2290 sample plots of forest ecosystems throughout Russia (detailed estimates are documented in Alexeyev and Birdsey, 1994). For example, the phytomass data used to determine conversion factors for Siberian fir (at 41 years of age, oven dry mass in Mg ha^{-1}) are: main stem 21.2, bark 5.6, crown 9.6, and roots 7.5; this produces values for K_b = 0.26, K_{cr} = 0.45, and K_r = 0.35. The values of the conversion factor K_{cr} vary significantly by tree species, age class and ecoregion. For example, the estimate of K_{cr} for young stands of Scots pine in the northern taiga is 1.22

and in the middle taiga, 0.84. For mature stands the respective values are 0.18 and 0.11. For young stands of larch in the northern taiga, K_{cr} is estimated to be 0.26. For some studies of C balance and dynamics it is necessary to have separate estimates for C in bark, roots, and crowns (leaves and twigs). For this analysis it is sufficient to have information about the total C of stands. Therefore it is possible to use the K_{ph} conversion factor, derived after evaluation of the foregoing coefficients (Table I). For the forests of European part of Russia the K_{ph} values are close to the estimates for the Asian part of Russia shown in Table I because species, age classes, and ecoregions are similar.

For transforming phytomass of tree stands and understory to C we used a coefficient of 0.5 since the C content of most forest tree species and understory plants is sufficiently close to this value.

2.4. Vegetation of the Lower Layers of Forest Communities

The phytomass of the lower layers of forest communities includes data on understory vegetation (seedlings, shrubs, dwarf-shrubs, herbs, forbs, grasses), mosses and lichens. The lower layers of phytomass are most abundant in open northern forests and woodlands. In the northern larch forests the lower layers of phytomass include up to 16–19 Mg ha^{-1} of oven dry matter (Ignatenko *et al.*, 1973; Gorchakovsky and Andreashkina, 1975; Sofronov and Volokitina, 1990). Northern spruce forests contain up to 8–15 Mg ha^{-1} of phytomass in the lower layers (Kazimirov and Morozova, 1973; Bobkova, 1987), and Scots pine forests contain up to 11 Mg ha^{-1} (Kazimirov and Volkov, 1977).

In the middle and southern taiga with closed canopy, the phytomass of lower vegetation layers is substantially less. Some examples of generalised data for the Asian part of Russia are shown in Table II.

2.5. Coarse Woody Debris and Litter

The mass of standing and fallen dead trees and branches can be derived from available tables of growth for even-aged stands with a single dominant species (Tretyakov *et al.*, 1952; Anonymous,1962, 1974; Kozlovski and Pavlov, 1967). Yet, since even-aged, single species stands are rare in nature, and the uniform structure and persistence of standing dead trees are frequently altered by various factors, data derived from such growth tables do not agree with the real values estimated in forest inventories. Nevertheless, because of a lack of comprehensive and representative data on coarse woody debris from forest inventories, estimates were partly taken from even-aged growth tables.

The standing time of dead trees was determined from published (Molchanov, 1971) and unpublished data (Sofronov and Volokitina, private communications). For the forests of Krasnoyarsk and the Republic of Sakha (Yakutia), volumes of coarse woody debris were taken from the unpublished forest inventory data base for forest enterprises. To account for the lower density of partially decayed wood, the density of coarse woody debris was reduced 10% in comparison with healthy live trees.

To estimate C in litter we used the methods and data of soil scientists and we reference their methods rather than describe them here (Shugalei *et al.*, 1994).

TABLE I

Factors to convert volume of growing stock to stand phytomass (K_{ph}) for main tree species in different ecoregions of the Asian part of Russia, by broad age class.

Region/Species	Regeneration	Pole-timber	Middle-aged	Maturing	Mature
Northern taiga					
Pinus sylvestris	0.835	0.661	0.666	0.648	0.590
Picea sp.	0.984	0.723	0.710	0.704	0.650
Larix sp.	0.806	0.762	0.768	0.802	0.795
Betula sp.	1.102	0.834	0.828	0.886	0.858
Middle taiga					
Pinus sylvestris	0.654	0.528	0.545	0.587	0.557
Picea sp.	0.857	0.670	0.640	0.620	0.602
Abies species	0.660	0.604	0.544	0.596	0.468
Larix sp.	0.726	0.714	0.708	0.738	0.724
Pinus sibirica	0.714	0.710	0.668	0.646	0.600
Betula	1.026	0.740	0.752	0.778	0.771
Populus tremula	0.779	0.504	0.534	0.548	0.484
Southern taiga, forest-steppe					
Pinus sylvestris	0.735	0.638	0.545	0.585	0.557
Picea sp.	0.847	0.630	0.621	0.632	0.584
Abies sp.	0.714	0.660	0.576	0.546	0.510
Larix sp.	0.726	0.714	0.704	0.738	0.724
Pinus sylvestris	0.610	0.522	0.528	0.506	0.488
Betula sp.	1.026	0.735	0.744	0.775	0.762
Populus tremula	0.779	0.504	0.534	0.548	0.484

2.6. Estimation of Carbon Storage in Natural Ecoregions

Estimates of C storage in the administrative territories is useful for planning forest management strategies. But for better understanding of the C balance and C dynamics, it is necessary to know the distribution of C storage in natural ecoregions. Information about forest lands in ecoregions has been taken from Korotkov (1994). The C storage of tree stands, vegetation of lower layers of forests, and coarse woody debris have been estimated on the basis of data derived for administrative territories.

In the cases when the areas of an ecoregion and an administrative unit coincide (for example, Kaliningrad, Kamchatka, Sakhalin, Baltic, Kamchatka and Kuril-Sakhalin provinces), estimation of C stocks needed no additional calculation. It was simple and accurate to make calculations for provinces and districts which completely enclosed administrative territories (for example, Leningrad, Novgorod, Pskov, Tver, Kostroma, and so on). Carbon storage of the forests in the ecoregions that covered parts of administrative territories was evaluated in two ways. For the largest administrative territories—Krasnoyarsk and Republic of Sakha (Yakutia)—we used the 1988 forest inventory databases. In other cases, additional statistical and literature data were used (Forest fund of USSR, 1990, 1991; Atlas of the USSR Forests, 1973; Anonymous, 1975, etc.) for determination of stocked areas and model stand characteristics in those provinces.

TABLE II

Phytomass (oven dry, Mg ha^{-1}) of lower layers of vegetation in forests of main tree species in different ecoregions of Russia

Region/Species	Regeneration	Pole-timber	Middle-aged	Maturing	Mature
Northern taiga					
Pinus sylvestris	3.0	3.2	3.4	3.5	4.1
Picea obovata	1.4	1.5	1.7	2.2	2.3
Larix sp.	1.2	6.0	9.2	12.0	16.0
Middle taiga					
Pinus sylvestris	1.2	0.8	2.2	2.5	2.8
Picea obovata	0.3	0.4	0.8	1.2	1.6
Abies sibirica	0.1	0.5	1.2	1.7	2.8
Larix sp.	0.5	2.0	3.2	4.1	5.0
Pinus sibirica	6.0	3.0	3.6	3.6	4.0
Betula sp.	4.0	1.2	2.5	2.5	2.0
Populus tremula	1.5	1.2	2.3	2.3	1.2
Southern taiga and forest-steppe					
Pinus sylvestris	1.2	1.0	2.0	2.4	2.5
Picea obovata	0.5	0.6	1.0	1.2	2.0
Abies sibirica	0.2	0.3	1.1	1.5	2.5
Larix sp.	0.2	1.8	3.0	4.4	6.0
Pinus sibirica	8.0	4.0	3.5	2.2	2.5
Betula sp.	1.0	0.8	0.6	0.6	0.7
Populus tremula	1.0	0.8	0.6	0.6	1.0

3. Results and Discussion

We estimated C storage of the forest vegetation in 71 administrative units and 63 ecoregions of Russia (Alexeyev and Birdsey, 1994). Here we include information mainly for all administrative units together and for ecoregion groups; only some illustrations have been included for administrative territories.

3.1. Carbon of Forest Vegetation

According to our calculations the C pool of forest vegetation of Russia equals 28 Pg, including 25.6 Pg C in tree stands, 0.6 Pg C in krummholz and shrub communities, and 1.9 Pg C in lower layers of forests. The weighted mean C density in forest vegetation is 36.3 Mg ha^{-1}.

The composition of a hypothetical average forest community of Russia expressed in C storage is:

Larch	35%	Spruce	14%	Fir	2%
Scots pine	18%	Siberian pine	8%	Oak	2%
Birch	16%	Aspen	4%	Other species	1%

Krummholz and shrubs are not included in the composition estimate.

More than 3/4 of the C (76.8%) that has accumulated in the forest communities is contained in coniferous forests (Table III), with larch forests accounting for almost half the C of all other conifers together. The C storage of the hardwood forests is very small, consisting of only 4% of the total. For all tree species the highest portion of accumulated C is in mature and overmature stands.

The Asian part of Russia is characterised not only by the largest amount of C storage but also the largest areas of mature and overmature conifer forests. The European part of the country has a much more balanced distribution of areas by age classes. For example, the areas of mature and overmature forests in Voronezh, Smolensk, Ivanovo, Pskov, Yaroslavl and other nearby administrative territories ranges from only 7–12% of the total forest area. Deciduous tree species are widely represented by young and middle-aged stands which have succeeded on cutover conifer forests.

3.2. Distribution of Carbon in Vegetation by Ecoregion Groups

3.2.1. Forest-tundra and mountain subarctic forests

The stocked area of the forest-tundra and mountain subarctic forests is more than 108 Mha (14% of total stocked area), with the mountain forests totalling 66.3 Mha. A relatively simple composition of the stands, low productivity and open stands (relative density is 0.3–0.4 of the standard density) produce a low C accumulation. The average density of C in vegetation is 15 Mg ha^{-1} (Table IV). Carbon of lower layers of plants in subarctic forests amounts to 20% of the vegetation C.

3.2.2. Boreal forests

The stocked area of the boreal (taiga) zone within Russia is 521.9 Mha, of which 157.2 Mha are covered by mountain forests. This zone makes the largest contribution to the C pool of forest vegetation in Russia, containing 19.6 Pg (70%) of C, an average density of 37.5 Mg ha^{-1}.

It is well known that the northern, middle, and southern subzones of boreal forests and the subzone of hardwood-conifers forests (sub-taiga) are essentially different in structure, productivity, and composition of stands. These parts of the boreal zone also have varying C density (Table IV). The lower layers of the taiga forests are less developed under the closed stand canopy and add up to no more than 1–3% of the C storage of the stands.

The relative productivity of different geographic ecoregions is uneven, but productivity and C of forests is determined at present not only by natural factors but also by anthropogenic activity. Areas little affected by clear-cuttings such as the southern taiga forests of middle Siberia contain the maximum amount of C in the boreal zone—62 Mg ha^{-1} (Table IV). In the European part of the country the C density in the same subzone amounts to 14 Mg ha^{-1} less.

One of the factors determining the amount of C storage of the forests in the Asian part of the country is permafrost. About half (46%) of the forests in Russia grow on permafrost soils. The forest communities of middle, eastern Siberia and Yakutia growing on permafrost contain 9.03 Pg of C with an average density of 26 Mg ha^{-1}.

TABLE III

Carbon in forest vegetation of various age classes, Tg.

Dominant species	Regeneration	Pole-timber	Middle-aged	Maturing overmature	Mature	Total
Conifers						
Pinus sylvestris	95.51	245.21	1 083.08	731.25	2 535.20	4 690.25
Picea	76.67	182.44	634.91	433.77	2 184.09	3 511.88
Abies	6.21	17.32	119.58	98.59	440.02	681.72
Larix	72.68	253.77	1 260.43	1 346.57	5 696.26	8 629.72
Pinus sibirica	8.46	52.72	340.88	468.44	1 254.60	2 125.12
Subtotal	259.53	751.47	3 438.88	3 078.62	12 110.18	19 638.68
Deciduous hardwoods						
Quercus	10.28	35.41	208.86	99.10	212.31	565.96
Fagus	0.46	3.17	41.96	14.11	36.98	96.68
Carpinus	0.11	0.58	10.90	2.98	7.56	22.13
Ulmus	0.01	0.06	0.64	0.28	0.92	1.91
Betula ermanii	0.38	3.22	16.72	23.23	283.08	326.63
Subtotal	11.24	42.44	279.08	139.70	540.85	1 013.31
Deciduous softwoods						
Betula	52.62	141.07	1 187.95	682.77	1 719.89	3 784.29
Populus tremula	14.03	33.51	305.27	165.68	456.75	975.24
Populus	0.12	0.57	7.51	2.93	13.49	24.62
Tilia	1.00	4.30	40.83	18.99	44.92	110.04
Alnus	0.30	0.84	11.11	4.81	3.79	20.85
Subtotal	68.06	180.29	1 552.67	875.19	2 238.84	4 915.04
Total all species	338.84	974.20	5 270.63	4 093.51	14 889.87	25 567.04

In most climatic sectors, an increasing amount of accumulated C is observed from forest-tundra (and mountain subarctic forests) to the forest-steppe and mountain subboreal forests, with productivity decreasing in the zone of steppes and deserts. Exceptions are the plain regions of middle and western Siberia (Kemerovo for example) where birch and aspens have replaced conifers as a result of intensive fellings.

An increasingly continental climate in Siberia and Yakutia precludes growth of high-yield hardwood deciduous species, and therefore the C density culminates in the southern taiga (Table IV).

3.2.3. Hardwood deciduous forests (zone of forest-steppes) and mountain sub-boreal forests

The hardwood and mountain sub-boreal forests cover an area of 134.6 Mha (17.5% of the stocked territory). The major vegetation community is mountain forest. The total storage of C accumulated in vegetation is 6.5 Pg, with an average density of 48 Mg ha^{-1}. In the plains the C stock is somewhat higher, but is substantially less than it might have been without the impact of human disturbance of the forests. The forests of this zone have not only hardwood tree species, but as result of anthropogenic activity, conifers and deciduous softwoods. The total C storage in hardwood tree stands is 1.0 Pg.

The regional C density is the highest in the Caucasus mountains (82 Mg ha^{-1}). The mountain belt of these forests is traditionally classed as sub-boreal, even though the region has warmer climate and much higher C accumulation.

3.2.4. *Forests of steppes, deserts and mountain forests of subarid and arid zonality*

The forests of steppes, deserts and mountain forests of subarid and arid zonality cover a total of 0.8% of the stocked area and do not make any considerable contribution to the total storage of C in the forests of Russia (Table IV). The forests of the steppe biome generally grow on lands unsuitable for agricultural crops.

3.3. Carbon Storage in the Forest Ecosystems of Russia

Carbon in vegetation forms only part of the C in forest ecosystems. Other components are mortmass (coarse woody debris and litter), soil organic matter, and C of consumers of organic matter (animals, fungi and micro-organisms).

The forest ecosystems of Russia (771.1 Mha) have 118.8 Pg C (Alexeyev *et al.* 1994). The C portion of the forest vegetation equals 23.6 % (Table V), the mortmass 14.1%, the soils 62.3%. Carbon of the forest consumers is 0.2% (according our approxïmate estimation the group of forest consumers amounts to about 1% of C in tree stands).

TABLE IV

Distribution of storage and density of carbon in forest vegetation in ecoregions of Russia.

	European Part		Western Siberia		Middle Siberia		Eastern Siberia		Russian Far East		Total	
Ecoregions	Pg	Mg ha^{-1}	Pg	Mg ha^{-1}	Pg	Mg ha^{-1}	Pg	Mg ha^{-1}	Pg	Mg ha^{-1}	Pg	Mg ha^{-1}
Plains												
Zone of Forest-Tundra	0.04	12	0.14	12	0.41	15					0.59	14
Subz. of north. taiga	0.99	28	0.66	32	0.79	24					2.43	27
Subz. of middle taiga	1.60	43	1.75	42	1.09	45	1.95	29			6.39	37
Subz. of south. taiga	1.70	48	1.50	50	1.55	62					4.75	52
Subz. of mixed forests	0.64	49									0.64	49
Zone of Forest-Steppe	0.49	51	0.34	50	0.17	43			0.20	34	1.20	46
Zone of Steppe	0.06	36	0.07	40							0.13	38
Mountains												
Subarctic zonality		10			0.13	15	0.67	17	0.25	14	1.05	16
Boreal zonality	0.37	46			0.89	39	1.78	28	2.33	37	5.37	34
Sub-boreal zonality	10.28	48			2.29	50	1.14	43	1.32	49	5.04	48
Sub-boreal (Caucasus)	0.27	82									0.27	82
Subarid zonality					0.10	39					0.10	39
Total	6.44		4.46		7.42		5.54		4.10		27.96	36

We estimate that the C pool in the forest vegetation of Russia equals 28 Pg. This is less than 10% different from the estimate of Isaev *et al.* (1993) when the C estimate for shrubs is excluded.

Other estimates of the C in the vegetation of forest ecosystems of the former USSR exceed ours (Kolchugina and Vinson, 1993b,c; Dixon *et al.*, 1994). Lack of detailed descriptions of methods in these publications prevents a complete discussion of the reasons for the discrepancies. Some of the difference may be explained by the use of conversion factors from other countries and from vegetation types that may not be accurate for Russian forests because of differences in standards for estimating the inventory of growing stock, and differences in average forest composition.

TABLE V

Carbon of forest ecosystems in European and Asian parts of Russia (Alexeyev and Birdsey, 1994). Area of forest in European part of the country is 164 Mha, in the Asian part 607 Mha.

	European Part		Asian Part		Russia Total	
Ecosystem Component	Pg	%	Pg	%	Pg	%
Tree stands	6 605	27.3	19 499	20.6	26 103	22.0
Lower layers	520	2.1	1 357	1.4	1 877	1.6
Forest vegetation total	7 125	29.4	20 855	22.0	27 980	23.6
Coarse woody debris	587	2.4	2 250	2.4	2 837	2.4
Litter	3 068	12.7	10 781	11.4	13 849	11.7
Mortmass total	3 655	15.1	13 031	13.8	16 686	14.1
Soils	13 348	55.3	60 554	64.0	73 902	62.1
Forest consumers	65	0.2	195	0.2	260	0.2
Forest ecosystems total	24 193	100.0	94 328	100.0	118 828	100.0

3.4. Uncertainties and Errors

Estimation of the C storage for national and subcontinent territories is based on various sources of data. Below are listed the most significant uncertainties and errors contained in the C estimates.

- There are some inaccuracies in the National Inventory data. According to Alexeyev *et al.* (1994) the volume of timber (overbark) is underestimated by 10%.
- There is insufficient information on the density of timber and bark for different tree species, for different ages of trees, and for various parts of the country. The use of available handbooks and other published data makes it possible to estimate regional timber density with an accuracy of ±10–20% (the least accuracy is for the Asian part of the country).
- A lack of information about the distribution of decay in living trees and losses of phytomass and C induced by decay may overestimate the C storage by 10%.
- Published information is insufficient to give an exhaustive characterisation of the fractional phytomass of forest tree stands in all age groups under different growth

conditions. Least accurate are conversion coefficients for young forests with broadest natural variability due to difference in composition and initial density of trees. The error of K_{cr} for age class I in individual cases may be more than ±50%. For other age groups it is within ±10–25%. An incomplete accounting of physiologically active roots less than 1 mm in diameter may be the cause of underestimated K_r. The conversion coefficient K_r values are assumed to be underestimated by 10–25%.

- Lack of sufficient data in some territories, especially the regions of Siberia and the Russian Far East.
- Some errors may arise due to more complicated composition, age and structure of the forests than was reflected in sample plots used for producing conversion factors. It is not possible to evaluate the magnitude of this source of error.

Taking into consideration various uncertainties, it is our opinion that the accuracy of estimation of the phytomass and C storage are within the limits of ±10–20%. Errors in estimating the C storage in the ecoregions are approximately in the same range.

4. Conclusions

The estimates from this study are comparable to the estimates from other countries with temperate and boreal forest ecosystems. Birdsey (1990) estimated that forests of the United States contain an average of 177 Mg ha^{-1}, similar to the estimate of 154 Mg ha^{-1} for Russia. About 32% of the C is in live vegetation in the U.S., versus 24% in Russia, a logical difference considering the larger proportional area of Russia covered by boreal forests with high amounts of soil organic matter. For Canada, Kurz *et al.* (1992) estimated an average C density of 219 Mg ha^{-1}, with 14% of the total in live vegetation. This significantly higher estimate of C density in Canada is the result of higher estimates of C in soils. Dutch forest ecosystems are estimated to contain 191 Mg ha^{-1}, with 31% of the total in live vegetation (Nabuurs and Mohren 1993).

Acknowledgments

This study was supported by the International Science Foundation, the USDA Forest Service Northeastern Forest Experiment Station (Grant 23-817) and the Krasnoyarsk Science Foundation (Grant 3F0038).

References

Alexeyev, V. A. and Birdsey, R. A.: 1994, *Carbon in Ecosystems of Forests and Peatlands of Russia*, Sukachev Institute for Forest Research, Krasnoyarsk, Russia 224 pp. [In Russian]

Alexeyev, V. A. Birdsey, R. A., Korotkov, I. A., Stakanov, V. D., Yefremov, S. P., Yefremova, T. T., Melentyeva, N. V., Shugalei, L. S. and Popova, E. P.: 1994, *In:* Alexeyev, V. A. and Birdsey, R. A., (eds.) Carbon in Ecosystems of Forests and Peatlands of Russia, Sukachev Institute for Forest Research, Krasnoyarsk, Russia, pp. 40-145. [In Russian].

Anonymous: 1962, *The Timber, Physical-Mechanical Characteristics, Guiding Technical Information.* Standartgis, Moscow, Russia, 48 pp. [In Russian].

Anonymous: 1974, *Tables of Increment for Main Tree Species Stands of the USSR*, Lesproect, Forest Industry, Moscow, Russia, 128 pp. [In Russian].

Anonymous: 1975, *Growth Rate of Main Forest Forming Tree Species, Part 2*, Institute of Forest, Siberian Branch of the Russian Academy of Sciences. Krasnoyarsk, Russia, 196 pp. [In Russian].

Anonymous: 1992, *The Forest Resources of the Temperate Zones, 1992.* Main findings of the UN-ECE/FAO 1990 Forest Resource Assessment, United Nations, New York. 32 pp.

Atlas of the USSR's Forests: 1973, *Main Administration of Geodesy and Mapping at Ministry Council*, Moscow, Russia, 222 pp. [In Russian].

Backman, C. A. and Waggener, T. R.: 1991, *Soviet Timber Resources and Utilization: Interpretation of the 1988 National Inventory*, CINTRAFOR Working Paper **35**, Seattle, USA, 296 pp.

Birdsey, R. A.: 1990, *In:* Research in Forest Inventory, Monitoring, Growth and Yield (IUFRO Proceedings), Virginia Polytechnic Institute, Blacksburg, Virginia, USA Publication No. FWS-*3-90*, pp. 24-31

Birdsey, R. A., 1992, *Carbon Storage and Accumulation in United States Forest Ecosystems*. USDA For. Serv. Gen. Tech. Rep. WO-**59,** 51 pp.

Bobkova, K. S.: 1987, *Biological Productivity of Conifer Forests of European North-East*, Nauka, Leningrad, Russia, 156 pp. [In Russian].

Dixon, R. K., Brown, S., Houghton, R. A., Solomon, A. M., Trexler, M. C. and Wisniewski, J.: 1994, *Science* **263**, 185-190.

Forest Fund of the USSR: 1990, *In:* Drozhalov, M. M (ed.), *Statistical Collection, State Forestry Committee,* Vol. **1**, Moscow, Russia, 1006 pp. [In Russian].

Forest Fund of the USSR: 1991, *In:* Drozhalov, M. M. (ed)., Statistical Collection, *State Forestry Committee,* Vol. **2**, Moscow, Russia, 1021 pp. [In Russian].

Gorchakovsky, P. L., Andreyashkina, N. I.: 1975, Study of the Primary Production in Plant Communities of Forest-Tundra and in Field Station "Harp", *Resources of Biosphere,* Nauka, Leningrad, Russia, Vol. **1**, pp. 25-33. [In Russian].

Ignatenko, I. V., Knorre, A. V., Lovelius, N. V., Norin, B. N.: 1973, *Ecologia*, No. **3**, 36-43. [In Russian].

Isaev, A. S., Korovin, G. N., Utkin, A. I., Pryanishnikov, A. A., Zamolodchikov, D. G.: 1993, *Lesovedenie*, No **5**, 3-10. [In Russian].

Kazimirov, N. I., Morozova, R. M.: 1973, *Biological Turnover of the Substances in the Spruce Forests of Karelia*, Nauka, Leningrad, Russia, 175 pp. [In Russian].

Kazimirov, N. I., Volkov, A. D.: 1977, *Mass and Energy Exchange in Scots Pine Forests of the European North*, Nauka, Leningrad, Russia, 303 pp. [In Russian].

Kolchugina, T. P., Vinson T. S.: 1993a, *Can. J. For. Res.* **23**, 81-88.

Kolchugina, T. P. and Vinson, T. S.: 1993b, *Global Biogeochemical Cycles* **7**, 291-304.

Kolchugina, T. P. and Vinson, T. S.: 1993c, *Water, Air, Soil Pollut.* **70**, 207-221.

Korotkov, I. A.: 1994, In: Alexeyev, V. A. and Birdsey, R. A., (eds.), *Carbon in Ecosystems of Forests and Peatlands of Russia*, Sukachev Institute for Forest Research, Krasnoyarsk, Russia, pp. 140-145. [In Russian].

Kozlovsky, V. B. and. Pavlov, V. M: 1967, *Growth Rate of Main Forest Tree Species*, Lesnaya Promyshlennost, Moscow, Russia, 327 pp. [In Russian].

Kurz, W. A., Apps, M. J., Webb T. M. and McNamee, P. J..: 1992, *The Carbon Budget of the Canadian Forest Sector, Phase I*, Forestry Canada, Inf. Rep. NOR-X-326, Northwest Region, Edmonton, Canada, 93 pp.

Molchanov, A. A.: 1971, *Productivity of the Organic Mass in the Forests of Different Zones*, Nauka, Moscow, Russia, 276 pp. [In Russian].

Nabuurs, G. J. and Mohren, G. M. J.: 1993, *J. Agric. Sci.* **41**, 309-326 pp.

Poluboyarinov, O. I.: 1976, *Timber Density*, Forest Industry, Moscow, Russia, 160 pp. [In Russian].

Shugalei, L. S., Popova, E. P., Alexeyev, V. A.: 1994, *In: Carbon in Ecosystems Forests and Peatlands of Russia*, Krasnoyarsk, pp. 27-44. [In Russian].

Sofronov, M. A., Volokitina, A. V.: 1990, *Fire Regionalization in Taiga Zone*, Nauka, Novosibirsk, 203 pp. [In Russian].

Tretyakov, N. P., Gorsky, P. V., Samailovich, G. G.: 1952, *Handbook for Forest Inventorists*, Goslesbumisdat, Moscow, Russia, 853 pp. [In Russian].

SIMULATING CARBON DYNAMICS OF THE BOREAL FOREST IN PUKASKWA NATIONAL PARK

I. A. NALDER[1] and H. G. MERRIAM

Biology Department, Carleton University, 1125 Colonel By Drive, Ottawa, Ontario K1S 5B6, Canada
[1]*Department of Renewable Resources, 751 General Services Building, University of Alberta, Edmonton, Alberta T6G 2H1, Canada*

Abstract. The development of forests in Pukaskwa National Park, Ontario, Canada, was simulated over 150 years to investigate boreal carbon dynamics and to test the feasibility of simulating large tracts of heterogeneous boreal forest. Pukaskwa National Park, located on the north shore of Lake Superior, encompasses 1835 km^2 of the Superior Section of the boreal forest. We developed a patch model, called BOPAS (BOreal PAtch Simulator), to simulate the development of carbon pools as a function of environmental parameters. Using GIS techniques, we divided the park into patches defined by a unique combination of forest type, age, climatic variables, soil type and topography, then used a forest gap model to develop biomass-over-time relationships for each patch type. BOPAS uses these relationships to simulate the development of carbon pools for trees, moss and litter/humus. We report results for constant climate, but BOPAS can be easily adapted to changing climate scenarios. Good results were obtained for predictions of carbon storage in trees. The initial value was 3.61 kg C m^{-2}, which agrees closely with literature values. With no disturbance, tree carbon increased to a maximum of 3.97 kg C m^{-2} at 30 years then slowly declined. Carbon storage was stabilized by introducing fire as a disturbance with a return interval of 100 years. Predicted forest floor carbon density, however, was much lower than expected, being less than half that of trees. It was anticipated to be substantially higher than tree carbon density based on a preliminary survey in the park and values reported in the literature. Published data, however, are very limited in coverage and give such a wide range of values that it was impossible to draw any firm conclusions about the validity of the model. BOPAS also showed that the forest floor carbon pool was relatively constant over the timescales of the simulation, but no published data were available to test this prediction. In summary, this work has demonstrated the feasibility of the BOPAS approach, but has high-lighted the necessity for more extensive data on forest floor carbon storage and dynamics.

Keywords. GAP MODELS, CARBON DYNAMICS, BOREAL FOREST, PUKASKWA NATIONAL PARK, DISTURBANCE, FIRE REGIMES.

1. Introduction

Boreal forests store such vast amounts of carbon that changes in this pool can have a significant effect on atmospheric levels of carbon dioxide. Modelling of global carbon dioxide fluxes has indicated a large, terrestrial sink in the northern hemisphere (Tans *et al.*, 1990), and recent estimates of carbon sequestration indicate that the boreal forest may, in part, account for this sink (Apps *et al.*, 1993). If these estimates are correct, boreal forests are currently playing an important role as a negative feedback on climate change. There is concern, however, that in a warming world this may change to a positive feedback and accelerate the warming. Nearly one quarter of the world's boreal forests lie in Canada (Rowe, 1992). The potential for mitigating or exacerbating climate change needs to be considered when formulating resource policy for Canada's boreal forests.

An appreciation of the role of boreal forests requires an understanding of their carbon dynamics, i.e., the long term balance between photosynthesis, respiration and decomposition. Several researchers have used forest gap models to determine carbon

Water, Air and Soil Pollution **82**: 283–298, 1995.

dynamics for a limited range of site conditions (Pastor and Post, 1988; Bonan, 1990b; Kellomaki and Kolstrom, 1992; Prentice *et al.*, 1993). Others have taken a broader scale view, examining dynamics at the ecoregion or biome level (Kurz *et al.*, 1992; Rastetter *et al.*, 1991). Yet the boreal forest is extremely heterogeneous: forest cover and environmental conditions change substantially over small distances. Consequently, to adequately represent the carbon dynamics of large areas it may be necessary to simulate at a much higher spatial resolution, such as the forest stand, and integrate across the larger area of interest. In the past, this has been impracticable because of the vast number of stands that would need to be manually assessed. Now, forest managers increasingly use Geographic Information Systems (GIS) and the necessary computer databases are becoming available. In this paper, we test the feasibility of modelling the carbon dynamics of large tracts of heterogeneous boreal forest by combining forest inventory data stored on a GIS with the ecological knowledge of forest development inherent in gap models. We selected one of Canada's larger national parks, Pukaswka National Park, as a test site because of the wide range of environmental conditions, stand compositions and ages within the park and the availability of inventory data on GIS.

2. Methods

2.1. Site

Pukaskwa National Park is located on the north shore of Lake Superior (see Figure 1) and covers an area of 183,500 ha. It is underlain by Precambrian bedrock and is cut by a number of valleys formed from eroded diabase dikes. These and other valleys have received deposits of glacial till and fluvial sediments, and lacustrine deposits in the lower areas, to give a pattern of rocky uplands interspersed with variable soil types in the valleys (Agriculture Canada, 1983). Lakes are numerous but none are large. Elevation ranges from 183 m AMSL along the lake shore to 640 m in the central

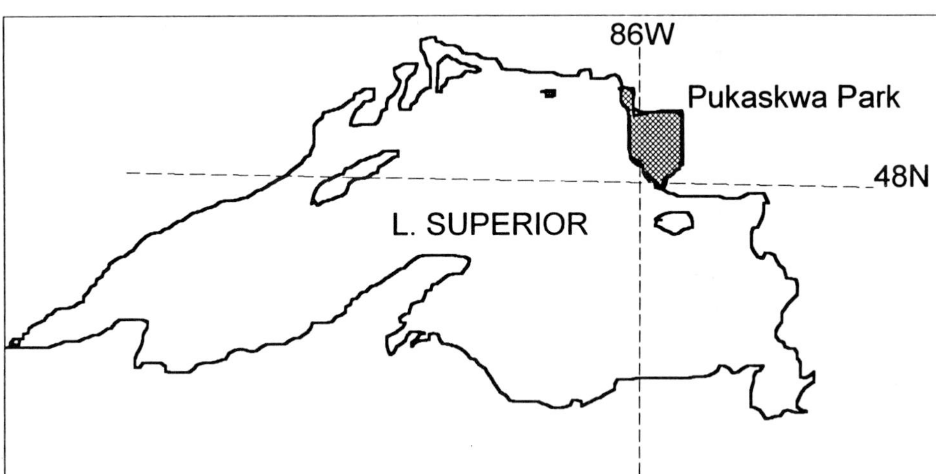

Fig. 1. Location of Pukaskwa National Park, Ontario, Canada.

portion of the park. In terms of vegetation, the park lies on the southern edge of the boreal forest, falling within the Superior Section as defined by Rowe (1959) and the Superior Highlands Ecoregion defined by Wickware and Rubec (1989). The dominant trees are *Betula papyrifera* Marsh., *Populus tremuloides* Michx., *Picea mariana* (Mill.) B.S.P., *Abies balsamea* (L.) Mill. and *Pinus banksiana* Lamb.

2.2. General

We hypothesized that the following variables were significant to the development of carbon pools in the park: initial forest composition, structure and age of stands, growing degree days, precipitation, cloudiness, soil type, slope and aspect.

To simulate carbon dynamics, we created a digital layer for each environmental variable, and overlayed these with layers for forest cover and forest age to create a unique-conditions map. We refer to each area defined by a unique set of conditions as a "patch" and simulate carbon dynamics at this spatial resolution. For each patch, we generated biomass-over-age relationships, which we term "growth curves", using a gap model. Finally, we simulated the development of carbon pools in each patch based on the appropriate growth curve, initial conditions and defined disturbance intervals, and summed across patches to determine total carbon storage. A TYDAC Technologies SPANS GIS™ system was used to manipulate data layers which were stored as raster images (TYDAC, 1993). The spatial resolution of the images was 38 m.

In the limit, each patch could have been a 38 × 38 m cell, but this would have led to an unacceptably high number of patches. Consequently, we restricted the number of patches created by aggregating classes and by categorising each continuous variable as described in the next section.

2.3. Data layers

A forest inventory of the park was carried out in 1983 (Hunter and Associates, 1989) which became the base year for our simulation. From this inventory we created an image of forest composition and structure utilizing the 14 classes defined in Table I. The 1983 inventory did not assess forest age so we used a forest age image from a 1973 biophysical inventory of the park (Gimbarzevsky *et al.*, 1978) and assumed that there were no disturbances in the intervening decade. The image had five classes, representing ages of 5, 20, 45, 70 and 100 years.

Climate data were obtained from Findley (1973). In the absence of climate stations within the park, Findley interpolated data from surrounding stations (principally Marathon, Wawa and White River) while accounting for the effects of topography and the influence of Lake Superior, and produced contour maps of mean monthly values for temperature, rain and snowfall. We digitized these maps to create overlays, but defining patches in terms of monthly temperatures and precipitation created far too many patches to characterise with the gap model. Instead, we created one overlay each for temperature and precipitation using annual growing degree days (GDD) and annual precipitation (P) as integrating variables where:

$$P = \sum_{\text{MONTH}=1}^{12} \left(\text{mean\_monthly\_rainfall} + \frac{\text{mean\_monthly\_snowfall}}{10} \right) \qquad [1]$$

$$GDD = \sum_{MONTH=1}^{12} \left(\max\{(mean\_monthly\_temperature - 5), 0\} \times days\_in\_month \right) \quad [2]$$

GDD values for the park ranged from 1106 to 1254. We reclassified the image into three classes (1106–1156, 1157–1207 and 1208–1254) to produce the GDD overlay. The classes were initially set to be equal width, but were adjusted slightly to give a more even distribution of areas in each class. The number of classes was determined by carrying out sensitivity runs with the gap model and recording the GDD increment necessary to give a significant change in model output as judged subjectively by examining the graphical output shown in Figure 3. There was a significant change in model output for an incremental change associated with three classes but not for four or more classes, therefore three classes were chosen.

Precipitation ranged from 760 to 1196 mm yr^{-1}. Sensitivity runs over this range with the gap model showed no changes in its output, therefore this image was not reclassified or utilized in the final overlay.

In the absence of spatially explicit cloudiness data, we averaged the monthly values from six climate stations around Lake Superior (Environment Canada, 1984) to give the values shown in Figure 2 and assumed these values applied equally to all areas of the park.

TABLE I

Forest classes utilized for the simulation showing composition in terms of dominant tree species and structure as percent coverage. Classes and coverages derived from ELS classifications defined in Hunter and Associates (1989).

Class	Description	Coverage	ELS classification	Area (km^2)
1	Birch	>60%	White birch, birch-redmaple, and riparian birch forest	423
2	Birch	30–60%	Birch woodland	182
3	Aspen/Birch	>60%	Aspen/birch forest and aspen forest	138
4	Aspen/Birch	30–60%	Aspen/birch forest and aspen woodland	11
5	Mixed wood	>30%	Mixedwood woodland and forest, riparian balsam fir-mixedwood	192
6	Mixed conifer	>60%	Mixed conifer forest, mixed conifer-cedar forest	90
7	Mixed conifer	30–60%	Spruce-fir mixed woodland, spruce-fir woodland, mixed conifer woodland, mixed conifer-cedar woodland	323
8	Black spruce	>60%	Spruce forest, spruce-tamarack forest	119
9	Black spruce	30–60%	Spruce woodland	4
10	Black spruce/ Jack pine	>60%	Spruce-jackpine mixed forest, woodland and parkland	224
11	Jack pine	>60%	Jack pine forest	22
12	Jack pine	30–60%	Jack pine woodland	9
13	Water or aquatics	N/A	Undefined aquatics	56
14	Undefined	N/A	Various classes defining non-forested areas	40
			TOTAL AREA	1835

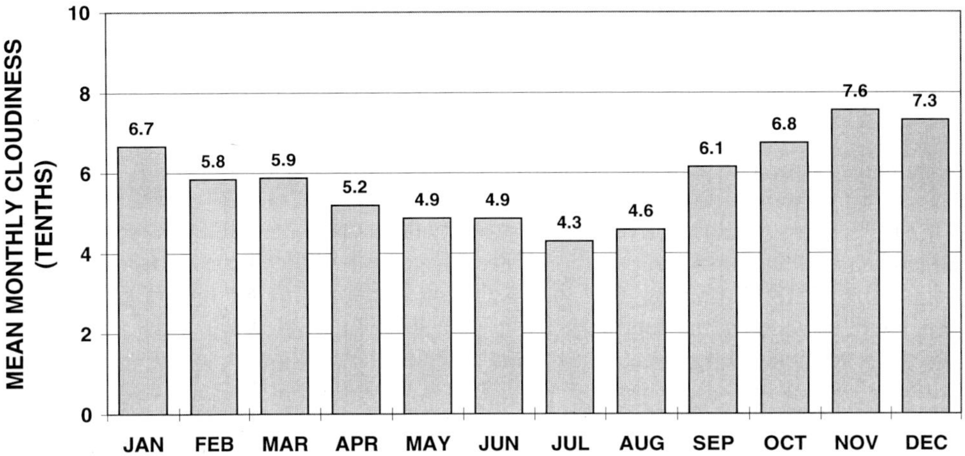

Fig. 2. Mean monthly cloudiness values assumed to apply to Pukaskwa National Park based upon the average of data from climate stations at Thunder Bay, Kapuskasing, Caribou Island, Atikokan, Armstrong Lake and Sault St. Marie.

Soils data were taken from a 1:100 000 soils map of the park (Agriculture Canada, 1983). The map delineated thirty soil classes and qualitatively described each class in sufficient detail that we could define three categories for each of the input soil variables required by the gap model (water holding capacity, soil depth and organic content). Although this gave a potential for 27 combinations, only 10 combinations actually occured. We assigned a class to each combination, then digitised the soils map and reclassified the image into these ten classes to produce the soils overlay. The definition of each class is shown in Table II.

TABLE II

Soil classes utilized for the simulation to define the water holding capacity (WHC), soil depth and organic content. Poor WHC corresponds to rockland and well-drained gravel and sand. Moderate WHC corresponds to well drained sand with some loam. Good WHC corresponds to poorly drained or very poorly drained soil of any sort.

Soil Class	WHC	Soil Depth	Organic Content	Area (km^2)
1	Good	≥100cm	Deep organics	6
2	Poor	≥100cm	Significant organics	33
3	Moderate	≥100cm	Significant organics	97
4	Good	≥100cm	Significant organics	78
5	Poor	≥100cm	No organics	14
6	Moderate	≥100cm	No organics	99
7	Good	≥100cm	No organics	21
8	Good	≅ 50 cm	Deep organics	17
9	Moderate	≅ 50 cm	No organics	96
10	Poor	≅ 20 cm	No organics	1343

To generate slope and aspect data, we used an elevation image of the park derived from 1:250 000 scale topographical maps. This image gave elevation for each pixel in 30.5 m increments. We smoothed this image with a 5-pixel by 5-pixel Gaussian filter to interpolate between the 30.5 m intervals, then combined the effect of slope and aspect by calculating the sun's angle of incidence at a nominal time, taken as midsummer noon. The resultant image was reclassified into three functionally distinct classes: south-facing, flat and north-facing.

Finally, we overlayed the five images—forest class, stand age, GDD class, soil class and slope/aspect class—to create a combined image with 1706 patches. After grouping patches that differed only in age and excluding patches smaller than 1 km^2 (these accounted for 15% of the total area), we were left with 134 patch types for which growth curves had to be generated using the gap model.

2.4. Gap Model

We surveyed gap models that have been applied to boreal regions and selected the one developed for the Alaskan boreal (Bonan, 1990a,b; Bonan and Korzuhin, 1989; Bonan, 1988). This is an individual tree simulator derived from the JABOWA (Botkin *et al.*, 1972) and FORET (Shugart and West, 1977) line of models, but goes further to include soil climate as well as the dynamics of moss/lichen and the forest floor. The model calculates aboveground tree biomass, moss and lichen biomass and forest floor biomass in yearly time steps, and has given reasonable predictions of biomass at several locations across the North American boreal (Bonan, 1990b). Root, shrub or herb biomass is not simulated.

For each of the 134 patch types, we transformed the patch variables into input variables required by the gap model. For example, the model requires monthly mean temperatures rather than GDD: in this case, we averaged the values in the mean monthly temperature image for each month over the area covered by the appropriate GDD class. Climate was held constant rather than being allowed to vary stochastically and each run was initialised with a fire of high intensity and severity. We ran the model with the five major species identified in the 1983 forest inventory: *Betula papyrifera*, *Populus tremuloides*, *Picea mariana*, *Abies balsamea* and *Pinus banksiana*. Species, canopy and litter parameters were taken from Bonan and Korzuhin (1989) and Bonan (1990b).

After modifiying the output routines to provide a listing of biomasses at 10-year intervals, we ran the model for 300 years for each patch type to provide 134 growth curves. Patches less than 1 km^2 in area were assigned to one of these growth curves by assuming they were soil class 10 and slope/aspect class 2, which were the most common of these classes. Figure 3 shows one of the growth curves.

The successional sequence shown in Figure 3 occurred for nearly all conditions in the park which caused problems in recreating the stands actually existing. Jack pine, for instance, would only become dominant if it was the only species allowed in the run. We assumed, therefore, that the existence of jack pine or young conifer stands indicated an absence of competing species to begin with and removed these from the simulation where necessary to recreate stands.

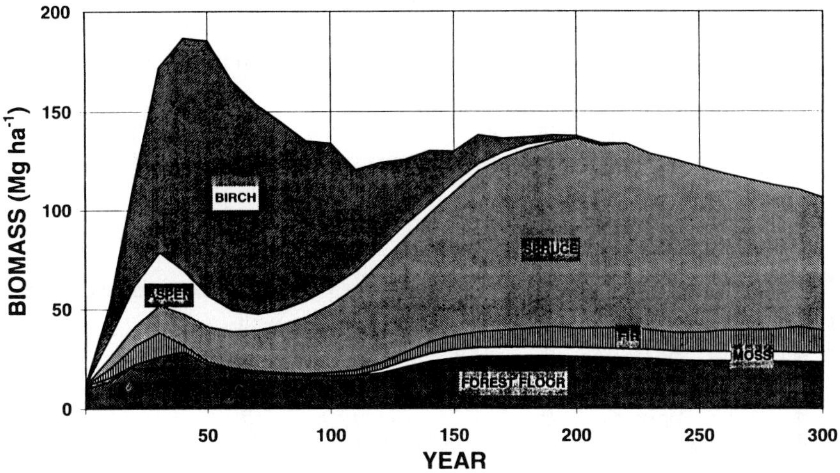

Fig. 3. Growth curve with all species present for GDD class 2, flat terrain, and shallow soil with low water holding capacity.

2.5. BOPAS

We next developed a PC-based model, termed BOPAS (BOreal PAtch Simulator), to apply the growth curves to each patch and thus simulate development of carbon pools in the park over 150 years. This approach assumes that there is no interaction among patches. BOPAS takes as inputs: (1) a file listing the patch IDs, area, forest class, age, and ID of the growth curve applicable to that patch, (2) a file containing the growth curves, and (3) a file of initialization variables that can be adjusted for different runs or different assumptions.

The principal functions of BOPAS are shown in Figure 4. For each patch, the model selects the correct growth curve and determines the age based upon initial age, simulation year and a disturbance function. When invoked, the disturbance function resets the patch age whenever a randomly-generated number between 0 and 1 is less than the probability of disturbance in a 10-year period (defined as ten divided by the disturbance return interval in years). Consequently, it simulates the effect of a high intensity/severity fire which is the initializing condition in generating each growth curve. From the patch age, BOPAS looks up the biomasses on the growth curve, calculates the new forest class (dependant on the dominant species) then determines carbon density for that patch. Carbon density is determined in two steps. First, the biomasses for all tree species are summed and this value is reduced by a factor that depends on the original forest density as defined in Hunter and Associates (1989). For forests that had greater than 60% cover we use a factor of 0.8 and for forests with 30–60% cover we use 0.5. Second, tree, moss and forest floor biomasses are converted to carbon assuming 45% carbon content. The model then multiplies carbon density for each patch by the area of that patch and sums across patches to obtain the total carbon content of the forest. Two outputs are provided at specified intervals for the duration of the simulation. The first output gives the total amount of carbon stored in the park in each of three pools (trees, mosses and forest floor), and the second output gives the total amount of carbon for each forest class.

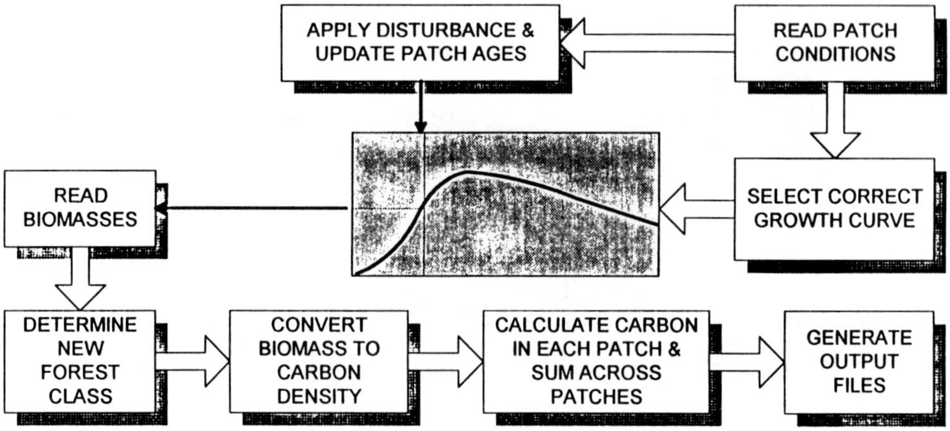

Fig. 4. Main functions of the BOreal PAtch Simulator.

2.6. Field Survey

A carbon survey was carried out as part of a field course in Pukaswka National Park in September 1993. Eleven forested patches were surveyed within a distance of approximately five km at the north end of the park. Patch sizes varied considerably, but were in the order of one hectare. Each patch was sampled with appropriately sized quadrats. Aboveground carbon was determined by measuring the dbh of each tree, basal diameter of each shrub and the percent cover of herbs, and using standard allometric relationships to convert to biomass (Stanek and State, 1978; Evert, 1985; Smith and Brand, 1983). Forest floor carbon density (kg C m^{-2}) was estimated by measuring the depth of the moss, lichen, litter and humus layers and multiplying by their carbon content (kg C m^{-3}). Based upon loss-on-ignition measurements of typical samples of moss, litter and humus, and a 45% carbon:organic matter ratio, we assumed a carbon content of 14 kg C m^{-3} for moss and lichen, 26 kg C m^{-3} for litter and 54 kg C m^{-3} for humus.

3. Results

3.1. Simulation Outputs

We first simulated carbon dynamics without disturbances for 150 years starting in 1983. Figure 5 charts the changes for the three pools—trees, mosses and forest floor—and Figure 6 breaks the total carbon storage down by the dominant species in each forest cover type. Total carbon content of forests in the park (excluding root carbon which was not simulated) rises to a peak of nearly 8.8 Tg by the year 2013 and declines slowly thereafter. The variation occurs mainly in tree carbon. Forest floor carbon is relatively constant. Moss carbon is a small component, but increases slightly over the period. In terms of species, aspen and birch dominated patches account for nearly half the storage in 1983 but gradually decline over time and are replaced by black spruce. Jack pine shows an initial increase but it too then declines. After 150 years, nearly all carbon is stored in black spruce patches.

To facilitate comparison with other data, Table III lists the carbon densities in the park for each 10-year interval of the simulation. These densities were obtained by dividing the total carbon storage reported by BOPAS by the area actually simulated (1652.43 km^2), i.e., excluding water and other non-forested areas.

TABLE III

Simulated carbon densities (kg C m^{-2}) without disturbance in Pukaskwa National Park.

Year	1983	1993	2003	2013	2023	2033	2043	2053	2063	2073	2083	2093	2103	2113	2123	2133
Trees	3.61	3.88	3.95	3.97	3.86	3.77	3.68	3.61	3.56	3.50	3.45	3.41	3.34	3.30	3.25	3.19
Forest-floor	1.18	1.26	1.27	1.25	1.23	1.20	1.19	1.19	1.19	1.19	1.20	1.21	1.22	1.22	1.22	1.21
Moss	0.08	0.09	0.10	0.11	0.12	0.13	0.14	0.14	0.15	0.16	0.17	0.18	0.19	0.19	0.20	0.20
Total	4.87	5.23	5.32	5.32	5.20	5.10	5.01	4.94	4.90	4.86	4.83	4.80	4.75	4.71	4.66	4.60

The simulation was also run for three disturbance regimes: a 50-year return interval, a 100-year return interval and a 300-year return interval. The effect of the disturbance depended on patch type, but in general it removed most of the aboveground biomass as well as a portion of the forest floor. The total carbon storage by simulation year is plotted in Figure 7 along with no-disturbance data. The effect of disturbance is to reduce carbon storage initially, but to stabilise it over the long term. A 100-year return interval provides the most stable result with storage being approximately 7.8 Tg (4.8 kg C m^{-2}). With this regime, storage is approximately 10% less than in an undisturbed system for the first 50 years, but becomes greater after 120 years. Very high disturbance rates, i.e., 50-year return interval, give substantially less storage over the 300 years of simulation.

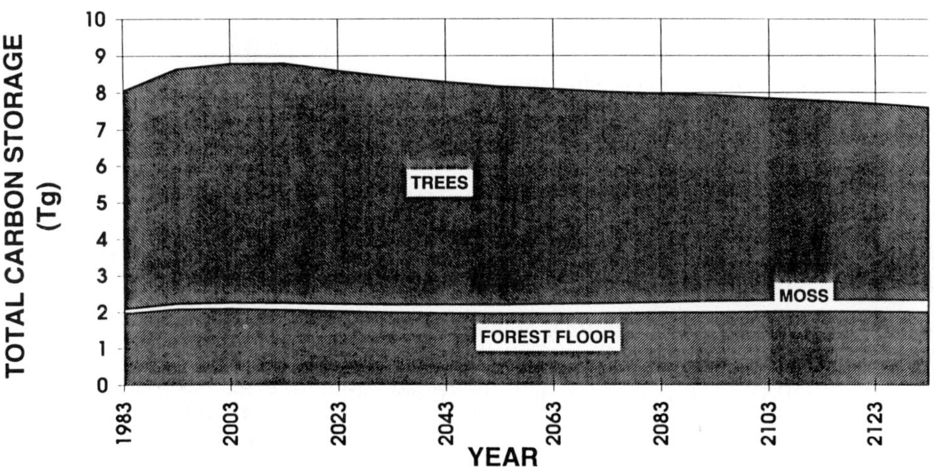

Fig. 5. Predicted carbon storage in the forests of Pukaswka National Park for 150 years from 1983 without disturbance. Total storage segregated by strata.

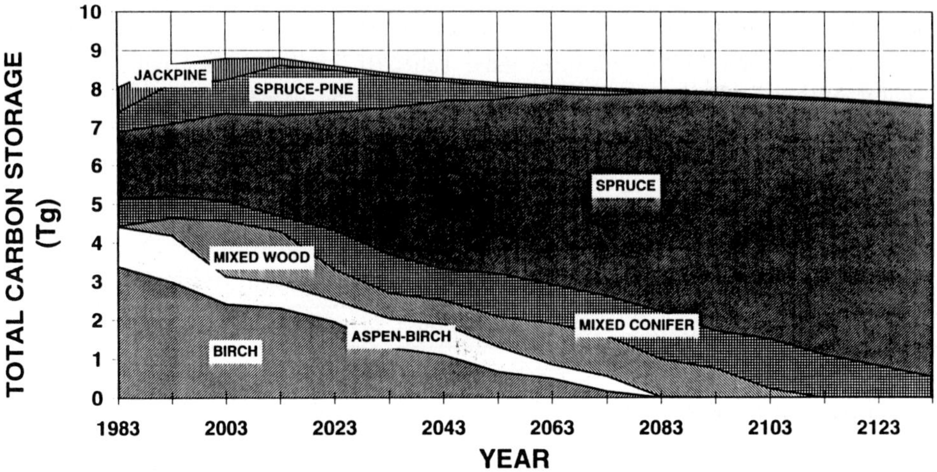

Fig. 6. Predicted carbon storage in the forests of Pukaswka National Park for 150 years from 1983 without disturbance. Total storage segregated by dominant tree species in each patch.

3.2. Field Survey

The carbon density measurements from the field survey at the north end of the park are summarized in Figure 8. Although neither scientifically rigorous nor necessarily representative of the park as a whole, this survey did provide a set of expectations for comparison with the simulation outputs. It indicated that total carbon storage in trees and the forest floor varies widely across patches, but is on average 16.9 kg C m^{-2}. In contrast, the initial value predicted by BOPAS is 4.87 kg C m^{-2} (see Table III—we use the 1983 no-disturbance value for comparison since we have no data on actual disturbance rates). The major discrepancy is in the forest floor pool: the survey found an average of 11.0 kg C m^{-2} but BOPAS predicted only 1.18 kg C m^{-2}. The value for trees agrees reasonably well—the survey average was 5.1 kg C m^{-2} versus 3.61 kg C m^{-2} from BOPAS. For moss and lichen, the survey found 0.8 kg C m^{-2} compared to 0.08 kg C m^{-2} from BOPAS. The survey also found that shrubs and herbs accounted for 0.16 kg C m^{-2} or less than 1% of the total storage.

4. Discussion

4.1. Initial Storage

In an attempt to validate the BOPAS predictions, we carried out a literature survey to determine expected values based on other research. Table IV summarises these data. The initial BOPAS values tabulated in Table III are within the range of those reported. There is, however, a tremendous spread in reported values: 1.65 to 9.50 kg C m^{-2} for aboveground carbon, 0.003 to 0.43 kg C m^{-2} for moss and 0.50 to 29.44 kg C m^{-2} for forest floor carbon. This reflects not only the heterogeneity of the boreal forest but also the variation in methods of measurement.

For aboveground carbon, the best data for comparison comes from a study by Botkin and Simpson (1990) which sampled 152 randomly selected sites with 5 plots at each site across the North American boreal. They measured biomass at every plot regardless of vegetation cover—only plots that fell in water were excluded—and obtained an overall figure for aboveground carbon of 1.88 ±0.45 kg C m^{-2}. This was much lower than previous estimates, but is very close to the value of 1.96 kg C m^{-2} achieved for the Canadian boreal by Kurz *et al.* (1992) based on simulation with the Canadian Forest Inventory as an input database. The value for Pukaskwa, however, should be higher based upon Botkin and Simpson's regional analysis. The value for Region XI, which includes the Lake Superior north shore, was 2.82 ±0.90 kg C m^{-2}. We would expect the values for Pukaskwa National Park to be somewhat higher for two reasons. First, Pukaswka has been protected from the harvesting that has occurred in Region XI over the past two decades. Second, biomass densities generally increase from north to south across the boreal and Pukaskwa is on the southern edge of Region XI. Values of approximately 4 kg C m^{-2} would seem reasonable for Pukaskwa National Park. The BOPAS initial value of 3.61 kg C m^{-2} is in good agreement with these expectations. BOPAS does not account for shrubs and herbs, but as indicated by our field survey, these would only add approximately 3% to aboveground carbon.

Botkin and Simpson (1990) did not measure forest floor carbon, nor are there any comparable surveys for forest floor biomass in the North American boreal zone. BOPAS predicts an initial value of 1.18 kg C m^{-2} for the forest floor, a value that is very low compared with published estimates for boreal regions. The most comprehensive global survey of soil carbon, based on more than 3500 soil profiles collected from undisturbed ecosystems since 1946, is that of Zinke *et al.* (1986). Their value for North American wet boreal forest, the Holdridge Life Zone that includes northern Ontario, is 19.8 ±8.2 kg C m^{-2}. This is, however, not directly comparable to BOPAS because it includes mineral soil carbon and excludes litter. Also, the samples

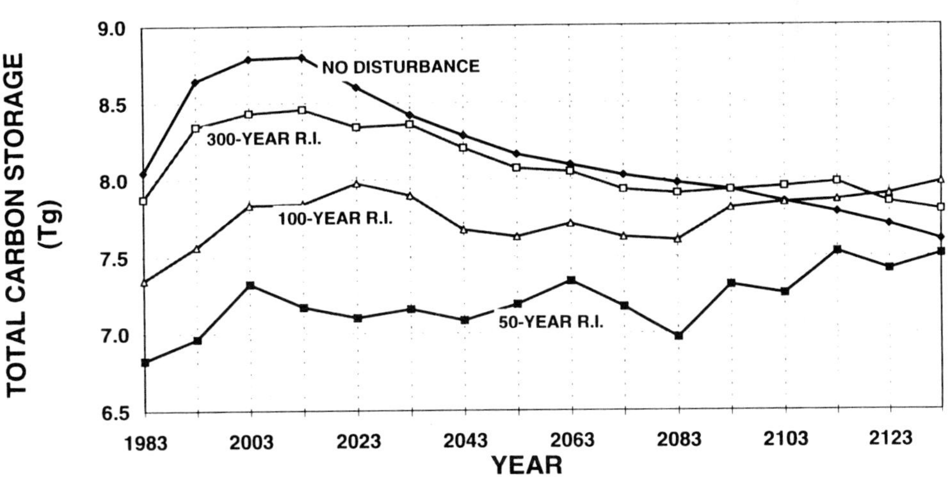

Fig. 7. Simulated total carbon storage in Pukaskwa National Park for 150 years from 1983 for four disturbance regimes of varying return intervals (R.I.).

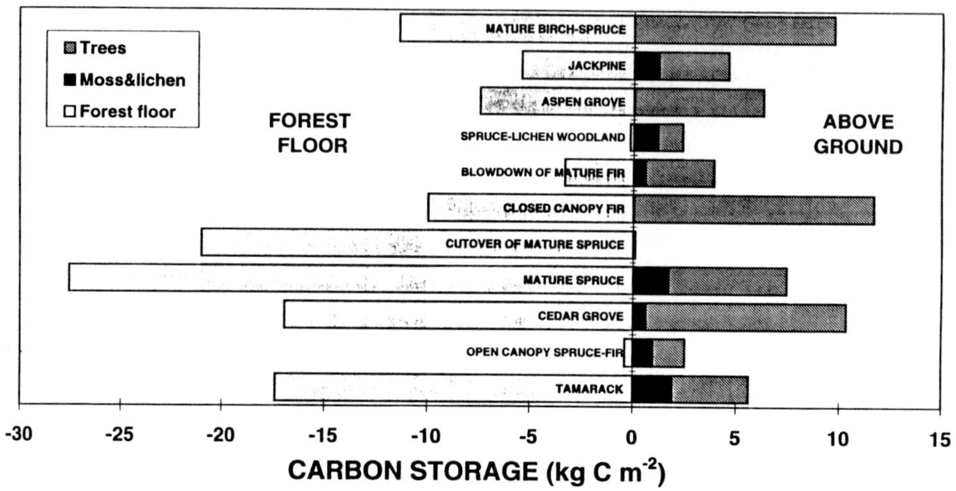

Fig. 8. Results of a survey of carbon density in eleven forest patches at the north end of Pukaswka National Park.

taken in Canada were strongly spatially clustered. We conclude that there are insufficient published data to characterise forest floor carbon in Pukaswka, or for most regions of the Canadian boreal.

Nevertheless, the pattern of carbon storage shown by Table IV and Figure 8 is that forest floor carbon is much greater than that stored above ground. BOPAS gives the opposite pattern, with forest floor carbon being nearly a third of that in trees. BOPAS seems to seriously underestimate forest floor carbon.

4.2. Carbon Dynamics

The BOPAS predictions for no disturbance provide a surprising view of long term carbon dynamics. We expected that carbon would continue to increase over time as slow-growing species, such as black spruce, continued to accumulate biomass and as the humus layer built up. In contrast to these expectations, tree carbon declines substantially after reaching a peak around the year 2013 and forest floor carbon remains relatively constant.

The decline in tree carbon can be explained by two factors. First, the peak biomass of birch is much greater than that attained by spruce, a pattern evident in Figure 3. Consequently, as succession occurs, birch dies and is replaced by lower density spruce. Second, old spruce stands tend to lose biomass as the first successional cohort begins to die off. Consequently, the increase and subsequent decline in total tree carbon is a result of not allowing any disturbance. When disturbances are introduced, the pattern changes as shown in Figure 7. A fire return interval of 100 years provides a relatively stable carbon storage over the 150 years of the simulation. Interestingly, this approximates the observed fire return interval in unmanaged southern boreal forests (Heinselman, 1973).

TABLE IV

Summary of carbon density values (kg C m^{-2}) in boreal forests reported in the literature. Where possible values have been put on a common basis for comparison. Biomasses have been converted to carbon assuming 45% carbon content. Unless otherwise noted "aboveground" refers to phytomass excluding roots and "forest floor" refers to the litter and humus layers not including mineral soil carbon or live roots.

Source	Forest type	Location	Above-ground	Moss, lichen	Forest floor
Global/regional measures					
Whittaker and Likens (1975)	Boreal forest	North America	9.0		
Schlesinger (1977)	Boreal forest	Global			15.9[2]
Atjay *et al.* (1979)	Boreal forest	Global	7.89		18.77
Olson *et al.* (1983)	Boreal forest	North America	5.6		
Post *et al.* (1982)	Boreal wet forest (Holdridge Life Zone system)	Global			13.1 ±7.1[1]
Botkin and Simpson (1990)	Boreal forest	North America	1.88 ±0.45		
Kolchugina and Vinson (1993)	Taiga	Former Soviet Union	6.11		29.44
Stand-level measures					
Van Cleve *et al.* (1983) Oechel & Van Cleve (1986)	Mature black spruce	Interior Alaska	2.29 ±0.95	0.16	3.44 ±0.27
Van Cleve *et al.* (1983) Oechel & Van Cleve (1986)	Mature paper birch	Interior Alaska	5.02 ±0.80	0.003	2.61 ±1.51
Van Cleve *et al.* (1983) Oechel & Van Cleve (1986)	Mature trembling aspen	Interior Alaska	4.96 ±2.89		2.13 ±0.55
Van Cleve *et al.* (1983) Oechel & Van Cleve (1986)	Mature white spruce	Interior Alaska	7.85 ±1.93	0..20	3.34 ±0.67
Auclair and Rencz (1982)	Black spruce-lichen woodland	Central Quebec	1.65[3]	0.43	2.06
Auclair (1985)	Mature black spruce-lichen woodland	Central Quebec	2.26	0.22	2.28
Damman (1971)	Black spruce	Newfoundland		0.09	3.92[4]
Damman (1971)	Balsam Fir	Newfoundland		0.10	2.93[4]
McLean and Wein (1976, 1977)	Jack pine (average of 12 mixed age stands)	N.E. New Brunswick	2.46		0.50[5]
Stocks (1987)	Immature jack pine	Central Ontario			0.58
Simulations					
Kurz *et al.* (1992)	Canadian boreal	Eastern region	2.688		13.65
Kurz *et al.* (1992)	Canadian boreal	Western region	1.946		12.85
Kellomaki and Kolstrom (1992)	N-S transect of Finnish boreal		6.75		0.90
Pastor and Post (1988)	Spruce	Schefferville	3.38		8.55
Pastor and Post (1988)	Mixed wood	Kapuskasing	6.39		12.28
Pastor and Post (1988)	Mixed wood	N.E Minnesota	9.50		14.49

1 Value is for depth to 1m or to bedrock, but excludes litter. Based on 15 samples.
2 Value is for surface litter and detritus to at least 50 cm.
3 Value includes roots.
4 Value is for humus only.
5 Value is for litter only.

The constancy of forest floor carbon is more difficult to explain. We could find no empirical evidence to confirm or deny this trend, but it is not consistent with our observations in the boreal forest where the depth of the humus layer generally increases with increasing forest age. More field data would be required to validate the dynamics of the model.

4.3. Pukaswka as a Carbon Store

In addition to their legislated role to preserve and protect natural ecosystems, Canada's National Parks can be considered to have a role as a carbon store. According to BOPAS predictions, Pukaskwa National Park is storing around 8.0 Tg of carbon (excluding non-forested areas). Although this is a small part of the 89 Pg stored by Canada's boreal forests (Kurz *et al.*, 1992), it is significant in relation to Canada's commitment to stabilise CO_2 emissions at 1990 levels by the year 2000. Emissions of fossil fuel carbon in Canada are around 128 Tg C yr^{-1} (Environment Canada, 1991). Harvesting or burning an area the size of Pukaskwa National Park could release a significant percentage of the national annual carbon budget. The role of national parks in sequestering forest carbon should not be ignored.

This is not to imply that park managers should prevent disturbances such as fire. On the contrary, the simulation shows that suppressing disturbances to prevent release of carbon can be counter-productive in the long run. With no disturbances for 150 years, the forests would end up storing less carbon than at present (Figure 7). Stabilisation of carbon storage requires disturbances.

5. Conclusions

We have shown that it is possible to model the carbon dynamics of large tracts of boreal forest at the patch level under the simplifying assumption that there are no interactions between patches. Predictions of aboveground biomass are in good agreement with expectations and the approach could be reasonably extended to other areas and to changing climates.

Our work, however, raises a major question about the amount of carbon in the forest floor. The BOPAS model predicts forest floor carbon density in Pukaskwa to be around 1.2 kg C m^{-2} which is substantially less than documented in field surveys. Also unexpected was the result that forest floor carbon is relatively constant over time under a no-disturbance regime. Further research is required to validate the model's characterisation of forest floor carbon dynamics.

This work has also highlighted the lack of data available on forest floor carbon pools in Canadian boreal forests. These pools may be very important for climate change predictions. Field measurements are urgently required to determine their magnitude and to investigate the long term dynamics.

Acknowledgments

This project would not have been possible without the assistance of Frank Burrows and Henry Schryber of the Canadian Parks Service. We would also like to acknowledge the work of Anna Gibson, Laura Tomat, Kara Webster, and Andrea Wilson who carried out the carbon survey in Pukaswka National Park. We are grateful for the constructive comments provided by two anonymous reviewers.

References

Agriculture Canada: 1983, *Soils of Pukaskwa National Park, Ontario, 1:100,000*, Report No. 53 Ontario Institute of Pedology, 1 p.

Apps, M. J., Kurz, W. A., Luxmoore, R. J., Nilsson, L. O., Sedjo, R. A., Schmidt, R., Simpson, L. G., Vinson, T. S.: 1993, *Water, Air, Soil Pollut.* **70**, 39-53.

Atjay, G. L., Ketner, P., and Duvigneaud, P.: 1979, *In:* Bolin B. (ed.) *The Global carbon cycle. Workshop on the Carbon Cycle*. Scope #13, Chichester, UK pp. 129-165.

Auclair, A. N. D.: 1985, *Can. J. For. Res.* **15**, 279-291.

Auclair, A. N. D. and Rencz, A. N.: 1982, *Can. J. For. Res.* **12**, 947-968.

Bonan, G. B.: 1988, *Ecol. Model.* **45**, 275-306.

Bonan, G. B.: 1990a, *Biogeochem.* **10**, 1-28.

Bonan, G. B.: 1990b, *Can. J. For. Res.* **20**, 1077-1088.

Bonan, G. B. and Korzuhin, M. D.: 1989, *Vegetatio* **84**, 31-44.

Bonan, G. B. and Shugart, H. H.: 1989, *Ann. Rev. Ecol. Sys.* **20**, 1-28.

Botkin, D. B., Janak, J. F. and Wallis, J. R.: 1972, *J. Ecology* **60**, 849-872.

Botkin, D. B. and Simpson, L. G.: 1990, *Biogeochem.* **9**, 161-174.

Damman, A. W. H.: 1971, *Ecol. Monogr.* **41**, 253-270.

Environment Canada: 1984, *Principal Station Data* Report PSD/DSP-29, 48 pp.

Environment Canada: 1991, *The State of Canada's Environment*. Government of Canada, Ottawa, Canada, 688 pp.

Evert, F.: 1985, *Systems of Equations for Estimating Ovendry Mass of 18 Canadian Tree Species*. Information Report PI-X-59. Petawawa National Forestry Institute, Canadian Forestry Service. Chalk River, Canada, 50 pp.

Findley, B. F.: 1973. *Climatology of Pukaskwa National Park, Ontario*. Canadian Parks Service, Ottawa, Canada, 80 pp.

Gimbarzevsky, P., Lopoukhine, N. and Addison, P.: 1978, *Biophysical Resources of Pukaskwa National Park*. Forest Management Institute Information Report FMR-X-106, Can. For. Serv., Ottawa, Canada, 129 pp.

Heinselman, M. L.: 1973, *Quaternary Res.* **3**, 329-382.

Hunter and Associates: 1989, *Pukaswka National Park Vegetation Re-inventory and GIS database creation*. Report to Parks Canada under contract CR87-36, Canadian Parks Service, Ottawa, Canada, 160 pp.

Kellomaki, S. and Kolstrom, M.: 1992, *Vegetatio* **102**(1), 47-68.

Kolchugina, P. and Vinson, T. S.: 1993, *Can. J. For. Res.* **23**, 81-88

Kurz, W. A., Apps, M. J., Webb, T. M., McNamee, P. J.: 1992, *The Carbon Budget of the Canadian Forest Sector: Phase I*. For. Can. Inf. Rep. NOR-X-326, Forestry Canada, Northwest Region, Edmonton Canada. 93 pp.

MacLean, D. A. and Wein, R. W.: 1976, *Can. J. For. Res.* **6**, 441-447.

MacLean, D. A. and Wein, R. W.: 1977, *Can. J. For. Res.* **7**, 562-578.

Oechel, W. C. and Van Cleve, K.: 1986, *In:* Van Cleve, K. *et al.* (eds.) *Forest Ecosystems in the Alaskan Taiga*. Springer-Verlag, New York, USA. pp. 121-137.

Olson, J. S., Watts, J. A. and Allison, L. J.: 1983, *Carbon in live vegetation of major world ecosystems*. United States Department of Energy, Report TR004, 152 pp.

Pastor, J. and Post, W. M.: 1988, *Nature* **334**, 55-58.

Post, W. M., Emanuel, W. R., Zinke, P. J., and Stangenberger, A. G.: 1982, *Nature* **298**, 156-159.

Prentice, I. C., Sykes, M. T., and Cramer, W.: 1993, *Ecol. Model.* **65**, 51-70.

Rastetter, E. B., Ryan, M. G., Shaver, G. R., Melillo, J. M., Nadelhoffer, K. J., Hobbie, J. E. and Aber, J. D.: 1991, *Tree Physiology* **9**, 101-126.

Rowe, J. S.: 1972, *Forest Regions of Canada*. Dept. of Northern Affairs and National Resources, Ottawa, Canada. 172 pp.

Rowe, J. S.: 1992. Boreal Forest in the Global Context. *In:* Sherman, J. (ed.), *Boreal Forest Conference Proceedings Oct. 4-6, 1991. Athabasca, Alberta.* Athabasca University, Canada, 206 pp.

Schlesinger, W. H.: 1977, *Ann. Rev. Ecol. Syst.* **8**, 51-81.

Shugart H. H. and West, D. C.: 1980, *Bioscience* **30**(5), 308-313.

Shugart H. H. and West, D. C.: 1977, *J. Env. Man.* **51**, 161-179.

Smith, W. B. and Brand, G. J.: 1983, *Allometric Biomass Equations for 98 Species of Herbs, Shrubs, and Small Trees*. USDA Forest Service Research Note NC-299, ^^^. 8 pp.

Stanek, W. and State, D.: 1978, *Equations Predicting Primary Productivity (Biomass) of Trees, Shrubs and Lesser Vegetation Based on Current Literature*. Report BC-X-183, Pacific Forest Research Centre, Canadian Forestry Service, Victoria, Canada, 58 pp.

Stocks, B. J.: 1987, *Can. J. For. Res.* **17**, 80-86.

Tans, P. P., Fung, I. Y. and Takahashi, T.: 1990, *Science* **247**, 1431-1438.

TYDAC, 1993. *SPANS GIS Reference Manual*. TYDAC Technologies Inc., Nepean, Canada, 510 pp.

Van Cleve, K., Viereck, L. A., Dyrness, C. T.: 1983, *In:* Wein, R. W., Riewe, R. R., and Methven, I. R. (eds.), *Resources and Dynamics of the Boreal Zone: Proceedings of a Conference held at Thunder Bay, Ontario, August 1982.* Association of Canadian Universities for Northern Studies, Ottawa, Canada, pp. 148-166.

Weetman, G. F. and Harland, R.: 1964, *For. Sci.* **10**, 80-88.

Whittaker, R. H. and Likens, G. E.: 1975, *In:* Lieth, H. and Whittaker, R. H. (Eds.) *Primary Productivity of the Biosphere*. Springer-Verlag, Berlin, Germany, pp. 305-328.

Wickware, G. M. and Rubec, C. D. A.: 1989. *Ecoregions of Ontario*. Ecological Land Classification Series # 26, Environment Canada., 50 pp.

Zinke, P. J., Stangenberger, A. G., Post, W. M., Emanual, W. R. and Olson, J. S.: 1986, *Worldwide Organic Soil Carbon and Nitrogen Data* NDP-018 Carbon Dioxide Inf. Centre, Oak Ridge Natl. Lab., Oak Ridge, USA, 117 pp.

SIMULATING CARBON STORAGE IN FORESTS OF EASTERN RUSSIA

P. BRADLEY, G. GASTON, T. KOLCHUGINA and T. S. VINSON

Department of Civil Engineering, 202 Apperson Hall, Oregon State University, Corvallis, Oregon 97331, USA

Abstract. In a pilot study investigating a method for estimating carbon (C) in forested ecosystems of Russia, a forest growth model was selected and applied to a test area. BORFOR, a forest gap model which simulates growth of trees, mosses and forest floor decay was used to model a mountainous area in the Russian Far East. Vegetation and landcover were identified using an unsupervised classification of Global Vegetation Index (GVI). The continental scale image class descriptions required a supplemental description of the species-specific to the area before they could be used as input data. Site level data was provided for soils and disturbances using published literature and thematic maps. Carbon values for forest ecosystems estimated with BORFOR were compared to other estimates. Good comparisons were noted for living tree C in all forest types, except mountain deciduous forests. Forest floor C was both overestimated or underestimated depending on forest type. Spatial resolution of ecosystem identification data was found to be very important to the estimation of C in mountainous regions. The BORFOR model could be improved to provide better C estimates with species-specific biomass calculations and improved simulation of decay of dead trees.

Keywords. FOREST MODELLING, CARBON BUDGETS, RUSSIA, FOREST GAP MODEL, BORFOR, GLOBAL VEGETATION INDEX

1. Introduction

Forests play a crucial role in the global carbon (C) cycle, storing an estimated 80% of terrestrial C in forest biomass, and 70% of terrestrial soil C (Post *et al.*, 1982; Olson *et al.*, 1983). While present global forest practices result in a C loss (Dixon *et al.*, 1994), changes in forest management could increase C stored in forests (Marland, 1988).

Carbon budgets and climate change mitigation strategies require estimates of C pools (or total C stored) and fluxes (rates of C stored and released annually) for forest ecosystems. Estimating C in forest ecosystems is challenging because growth and decay depends on weather (which varies from year to year), local site conditions (including past climatic conditions, permafrost, slope, aspect, elevation and accumulation of soil C), and disturbance regimes (such as fire, insect attack and harvest). Forest ecosystems in mountainous regions present further challenges because landscape changes over relatively small areas can cause diversity in ecosystem types (Swanson *et al.*, 1988).

Forest growth models can provide information on C pools and fluxes based on actual site conditions and local climate regimes (Solomon, 1993). By simulating forest growth and forest floor accumulation and decay over time, a model can predict C pools and fluxes in terrestrial C. The purpose of the research reported herein is to (1) evaluate the suitability of a forest growth model to estimate C in Russian forests, (2) compare model results to other C estimates, and (3) recommend a method to model the forested areas of Russia. The results of modelling forests across Russia can be used to estimate C storage in Russian forests and utilised in selection of forest management strategies to respond to the challenges of global climate change.

Water, Air and Soil Pollution **82**: 299–308, 1995.

2. Background

A forest ecosystem C budget includes C pools in living trees and other forest vegetation, dead trees, forest floor, and soil organic layers. It also requires estimates of C fluxes associated with forest growth, and decay of dead trees and forest floor (Kolchugina and Vinson, 1993).

Most Russian forests are classified as boreal and contain a limited number of tree species. In most forest ecosystem types, one or two tree species predominate, with productivity dependent upon a number of interconnected environmental factors (Bonan and Shugart, 1989). Soil conditions such as moisture, temperature, depth of active layer over permafrost and available nutrients from forest floor decomposition influence the ecosystem type and productivity.

Of the well over 100 forest growth models available (Waring and Schlesinger, 1985), two models evaluated were developed specifically for boreal forests. These models include BORFOR (Bonan, 1990a,b), an ecosystem model which emphasises below ground dynamics, and FORSKA (Leemans, 1992), a dynamic forest stand model that emphasises canopy structure.

The BORFOR forest gap model was selected to estimate C pools of forest ecosystems in Russia. The rationale for choosing BORFOR over FORSKA and other available models was the necessity to include forest floor dynamics in the estimates of C pools, and the inclusion of a permafrost component in BORFOR. BORFOR was also chosen because of a previous history of validation at a number of sites in boreal forests (Bonan, 1990a,b) and because a number of authors have recommended it as an appropriate model for boreal forests (e.g., Urban *et al.*, 1991; Solomon, 1993).

BORFOR simulates the development of some components of a forest ecosystem (i.e., live trees, mosses and forest floor organic layer) by assuming a maximum growth rate which is then modified by environmental factors such as light availability, soil moisture, length of growing season, nutrient availability, and depth of the active layer. Forest floor decomposition is simulated following the approach of Aber *et al.* (1982), and Pastor and Post (1986).

The input data necessary for BORFOR includes site, climate and tree species information. Site input data include slope, aspect, elevation, soil water holding capacity, fire frequency, and moss productivity. Climate input data include average monthly values of cloudiness, temperature and accumulated precipitation. Tree species input data consists of biomass constraints (such as maximum height and diameter at breast height), environmental tolerances and reproduction information (Bonan, 1990a).

In previous validation efforts (Bonan, 1990b), forest biomass values predicted with BORFOR were in good agreement with actual site data. Living biomass was correctly simulated on most sites, but forest floor biomass was underestimated by about 20% for half of the sites.

3. Methods and Materials

A mountainous forest area was selected in the Amur region of the Russian Far East (Figure 1) to evaluate BORFOR's capability to estimate carbon pools. The area is located in the Sikhote–Alin mountain range, bounded between 47°N and 48°N latitude and 138°E and 139°40'E longitude, in the transition zone between boreal and temperate forests. Forest ecosystem types in the area include both spruce–fir forests, (typical of boreal regions) and mixed deciduous forests (more common to the temperate regions) (Suslov, 1987).

3.1. Ecosystem Classification

Carbon estimates for forest ecosystems show a range of values which is partially based on ecosystem type (Bazilevich, 1986, 1993). While BORFOR can provide a simulation of a site with all available species of trees being modelled, (and the remaining trees after a period of time used as the determinant of an ecosystem type), other factors not included in the model also determine surviving tree species. Because a classification of ecosystem type was provided, the model's ability to produce an appropriate ecosystem type was not tested.

Of the available continental scale vegetation maps (e.g., Ryabchikov, 1988; Gaston *et al.*, 1994) the classification by Gaston *et al.* (1994) derived from Global Vegetation Index (GVI) provided the most detailed ecosystem identification. GVI provides a record of temporal changes in greenness for a 15 × 15 km grid cell. Four year average monthly GVI observations (1985–1988) were used to classify regions with similar timing, productivity, and duration of seasonal photosynthesis. In contrast to the Ryabchikov vegetation map which classifies the entire study area as one ecosystem type (mountainous light crown coniferous), the GVI image classes differentiated the study area into two non-forested and four forested ecosystem types. The GVI database has an advantage when compared to thematic maps in that it is timely, and can be updated, while thematic maps are static in time.

While GVI image classes are used to identify forest ecosystem types, BORFOR requires a list of tree species to include in simulations. To translate each ecosystem type into a list of tree species, image classes were compared to regional forest species maps and descriptive information (Suslov, 1987; G.U.G.K., 1960; Knystautas, 1987) (Table I). A significant difference in GVI data and the thematic forest species map (G.U.G.K., 1960) was noted in "sparse trees". While several grid cells in the forest species map are defined as "sparse trees", the GVI data set simply identifies an ecosystem type.

3.2. Model Parameterisation

Tree species-specific input parameters were obtained from published literature (Nikolov and Helmisaari, 1992; Suslov, 1989; Prentice and Helmisaari, 1991). An input file of tree species was completed for each of the four forested ecosystems identified in the study area.

The 40 grid cells identified as Spruce–Fir forests occupied most of the northern portion of the area and ranged in elevation from 115 to 742 m. Tree species selected to model this ecosystem include Yeddo spruce (*Picea jezoenis*), Khingan fir (*Abies nephrolepis*), and Japanese stone pine (*Pinus pumila*). Four cells of Larch-Pine forests, located in lowlands near the coast, were modelled as Yeddo spruce, Dahurian larch (*Larix dahurica*), Korean pine (*Pinus koraiensis*), and Erman's birch (*Betula ermanii*). The 98 cells of Mixed (conifer/deciduous) forest ranged in elevation from 168 to 1065 m. Species selected to model this ecosystem include Yeddo spruce, Khingan fir, Dahurian larch, Erman's birch and Korean pine. Mountain Deciduous forest occupied

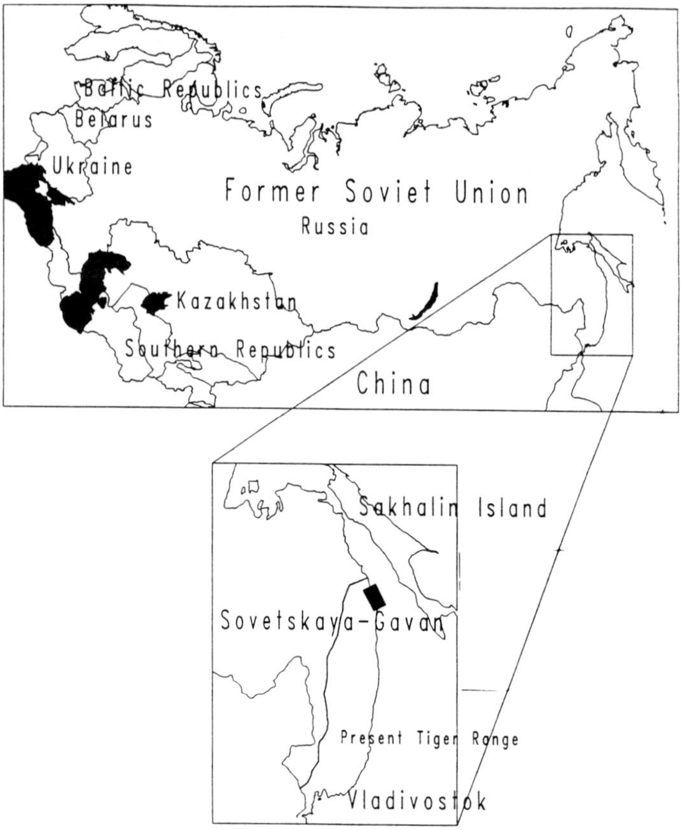

Fig. 1. The shaded study area is located in the Amur region of the Russian Far East. The study area of nearly 1 Mha is bounded by 47°N and 48°N latitude and 138°E and 139°40'E longitude.

26 grid cells in the study area, mostly in the lower latitudes. Tree species selected included Manchurian oak (*Quercus mongolica*), Yeddo spruce, Erman's birch, Khingan fir and Korean pine.

Monthly climate data for cloudiness, precipitation, and temperature were obtained from three weather stations in the study area. Cloudiness for each grid cell was assigned from the nearest weather station. Temperature was extrapolated from the nearest station using a wet adiabatic rate of -6.5° C for each 1000 m increase in elevation.

Monthly precipitation was estimated using PRISM, an analytical model that distributes point measurements of monthly and annual precipitation to regularly spaced grid cells (Daly *et al.*, 1994). PRISM was applied to a region spanning 44°N to 52°N and 132°E to 141°E. Input data were monthly average precipitation from 81 weather stations in the region. An estimate of elevation for each grid cell was derived from ETOPO5, a global digital elevation grid (Kineman and Ohrenschall, 1992), by inverse distance weighting the closest five grid cells. The finer resolution of ETOPO5 reduced the 15 × 15 km (GVI) grid cell size to approximately 7.5 × 7.5 km, and differentiated the area into 179 grid cells (each approximately 5625 ha).

Site data of fire frequency, soil properties, and physical geography were provided for each 7.5 × 7.5 km grid cell. The fire disturbance frequency for simulation was set at 300 years based on historical information on the age of regional stands (from 60 to 120 years) and a national average interval of 320 years (i.e., 2.5 Mha of 800 Mha burned annually) (Kolchugina and Vinson, 1993). Three soil parameters (water holding capacity, saturation point and wilting point of the underlying mineral soil) were interpolated from a digitised soils map (Fridland, 1988). Soil texture information (limited to "fine" or "granular") was obtained by correlating mapped soil data types to published soil texture data (Kachinskii, 1966).

Because grid cells often contained several slopes and aspects, each was modelled as horizontal at the average elevation of the cell. The lack of slope and aspect for each grid cell was perceived to be a significant weakness of the input data set so the sensitivity of the horizontal assumption was tested as part of the modelling effort.

TABLE I

Image classes from GVI satellite data are compared to the Forest Species Map (G.U.G.K., 1960) for the same grid cells. The percentages of area from the Forest Species Map are noted. The tree species selected to model each forest ecosystem type were obtained from regional forest species descriptions (Suslov, 1987; Knystautas, 1987).

Image class (GVI)	Forest Species Map	Percentage	Species modelled
Conifer Forests	Spruce	43	Yeddo spruce
	Spruce-fir	13	Khingan fir
	Larch	18	Japanese stone pine
	Sparse trees	19	
Larch-Pine Forests	Spruce	9	Yeddo Spruce
	Larch	3	Dahurian Larch
	Cedar	12	Korean pine
	Birch	4	Erman's birch
	Sparse trees	72	
Mixed Forests	Spruce	15	Yeddo spruce
	Spruce-Fir	7	Khingan fir
	Larch	12	Dahurian larch
	Cedar	10	Korean pine
	Birch	9	Erman's birch
	Sparse trees	42	
Mountain Deciduous	Spruce	2	Yeddo spruce
	Spruce-fir	15	Khingan fir
	Cedar	12	Korean pine
	Birch	21	Manchurian oak
	Other	3	Erman's birch
	Sparse trees	34	

3.3. FOREST SIMULATIONS

Eleven grid cells in the area identified as non-forest (e.g. meadows or cultivated land and tundra or wetlands) were not included in forest modelling. A 500 yr simulation of 30 plots (1/12 ha) was completed for each of the 168 forested grid cells. Following Bonan (1990b), BORFOR was initialised to represent conditions following a severe fire in a mature forest. Carbon values for each grid cell were obtained by averaging the output values simulated for the entire 500 years. All pixels in each class were then used to provide a range for that class in the study area.

4. Results

4.1. CARBON POOLS

The range of carbon pools obtained from 500 year simulations are summarised for each of the four ecosystem types modelled (Table II) and compared with previous estimates of similar forest ecosystems (Bazilevich, 1986,1993).

Carbon pools simulated for Spruce-Fir forests ranged from 15 to 27 Mg C ha^{-1} for aboveground phytomass and 3 to 8 Mg C ha^{-1} for the forest floor. The species in the simulation consisted of 60% spruce and 40% fir. The literature data for similar ecosystem type of 17 Mg C ha^{-1} was reported for Japanese stone pine forests of Okhotsk to the north of the study area (Bazilevich, 1993). This corresponded to the high elevation model results in which Japanese stone pine predominated. The literature data of 31.5 Mg C ha^{-1} from Spruce-Fir forests of mountainous regions near Okhotsk (Bazilevich, 1993), were slightly higher than simulated data. The forest floor was reported as 15 Mg C ha^{-1} (Bazilevich, 1986) while BORFOR results (3.25–8.5 Mg C ha^{-1}) were substantially less. This is consistent with Bonan (1990b) indicating the spruce forest floor is underestimated by BORFOR.

In Larch–Pine forest simulations, larch predominated with secondary species of pine and birch comprising less than 10% of the forest. Simulated C values ranged from 29 to 59 Mg C ha^{-1} for aboveground phytomass and from 13 to 22 Mg C ha^{-1} for the forest floor. Literature data for similar ecosystems ranged from 24.3 Mg C ha^{-1} in Okhotsk to 80 Mg C ha^{-1} in forests with a higher percentage of birch (Bazilevich, 1993). Literature estimates of forest floor of 8 to 14 Mg C ha^{-1} (Bazilevich, 1986) were lower than simulated forest floor. The forest species map identifies 72% area of sparse trees in the Larch-Pine forest, but this was not modelled in BORFOR. It appears that the potential areas of wetlands are not identified by the modelling process.

In mixed forest simulations, larch trees predominated in most cells until later succession when hardwoods such as pine and birch predominated. The mixture of tree species was partially dependent upon the elevation, with grid cells over 800 m being much less productive than lower elevation cells. This successfully modelled some areas of sparse trees that the GVI classes could not identify. Simulated C values ranged from 22 to 45 Mg C ha^{-1} for aboveground phytomass. Literature data for similar ecosystems ranged from 24.3 Mg C ha^{-1} in Okhotsk to 70 Mg C ha^{-1} in the Sikhote–Alin mountain range (Bazilevich, 1993). Simulated forest floor from 10 to 22 Mg C ha^{-1} was comparable to literature data of 12.5 Mg C ha^{-1} (Bazilevich, 1986).

TABLE II

The GVI class/ETOPO5 overlay identified 179 grid cells of six ecosystem types within the study area. Eleven non-forested grid cells included Class 4 (tundra or wetlands) and Class 21 (meadows and agricultural lands). The remaining 168 grid cells included four types of forested ecosystems which were modelled with BORFOR. The results of ecosystem modelling are expressed as ranges for all grid cells of that class, and then compared to literature data (Bazilevich, 1986, 1993) for similar ecosystem types.

GVI Class	Forest type	No. cells	BORFOR	Bazilevich, 1993	BORFOR[1]	Bazilevich, 1986[2]
			Biomass ($Mg\ C\ ha^{-1}$)	Phytomass ($Mg\ C\ ha^{-1}$)	Forest floor ($Mg\ C\ ha^{-1}$)	Forest Floor ($Mg\ C\ ha^{-1}$)
4	Tundra or Wetland	4				
21	Agricultural	7				
31	Spruce-Fir forest	40	15–27	17–31.5	3.25–8.5	15
36	Larch-Pine forest	4	29–59	24.3–80	13–22	8–14
40	Mixed forest	98	22–45	24.3–70	10–22.5	12.5
42	Mountain Deciduous forest	26	52–80	100–124	18–25	4.5–5

1 BORFOR results include dead trees.
2 Forest Floor does not include dead trees.

In Mountain Deciduous forests, simulated late successional forests included predominant oak and birch as predicted in ecosystem descriptions (Suslov, 1987; Knystautas, 1987). Simulated C values ranged from 52 to 80 Mg C ha^{-1} for aboveground phytomass, significantly lower than literature data for similar ecosystems in the Sikhote-Alin mountain range of 100 to 124 Mg C ha^{-1} (Bazilevich, 1993). The simulated forest floor from 18 to 25 Mg C ha^{-1} was higher than literature data of 4.5–5 Mg C ha^{-1} (Bazilevich, 1986). The lesser value of phytomass might be due to literature reporting C in undisturbed forests while BORFOR results include fire.

4.2. Summary Carbon Estimates

To compare C pool estimates with previous estimates of the same area, average C pool values for each grid cell were multiplied by cell area. Totals for all grid cells were then summed to obtain area wide C estimates (Table III). Previous C estimates classified the entire area as mountainous light crown coniferous forest with a living phytomass pool of 26.7 Mg C ha^{-1} (Kolchugina and Vinson). The GVI-based forest modelling differentiated more ecosystems within the study area, with C pool estimates of living phytomass from 15 Mg C ha^{-1} in less productive Spruce-Fir forests to 80 Mg C ha^{-1} in Mountain Deciduous forests.

When compared to Kolchugina and Vinson (unpublished database, 1993), the living tree biomass predicted by BORFOR was 19% greater and forest floor was 12% less. Total C for the area was 26% less in BORFOR simulations, primarily due to a much reduced dead tree pool in BORFOR model predictions.

4.3. Slope and Aspect Sensitivity Analysis

A sensitivity analysis tested the validity of assuming a horizontal surface as an average for multiple slopes and aspects within a grid cell. The model was executed for different slopes (10°, 20°, 30° and 40°), and aspects (north, south, east and west) for each of the four forest classes. When the output for all the slopes and aspects modelled are averaged, the C estimate varied 2 to 9% from the values obtained when modelling the same surface area as horizontal. A higher resolution data set could provide slope, aspect and an ecosystem differentiation of tree species between northern and southern slopes. The assumption of a horizontal surface when taken over a landscape average may not be as significant to C estimates as the need for a higher resolution data set for accurate ecosystem identification within a mountainous landscape.

5. Discussion

The 15 × 15 km scale resolution of GVI appears adequate for relatively homogeneous terrain. However, the 15 × 15 km resolution may not allow vegetation and land cover regions to be accurately identified in heterogeneous areas such as mountainous regions with several ecosystem types within each grid cell. Satellite system imagery at 1.1 × 1.1 km (AVHRR LAC) or 80 × 80 m (LANDSAT MSS) may be required for more accurate identification of vegetation and land cover in mountainous regions.

The occurrence of sparse trees is simulated in BORFOR results for high elevation areas but not for wetlands and possibly other sparse tree areas. The GVI ecosystem classification uses the "greenest pixel" to identify the entire area, which tends to overestimate by assigning each pixel its most productive portion. Higher elevation grid cells reflected reduced productivity or areas of "sparse trees", but BORFOR did not predict the sparse trees of the Mountain Deciduous or Larch-Pine forests. The potential inability for unsupervised classification of GVI to identify wetland areas would also point to the need to use a wetlands map overlay when modelling the forests across Russia. The difference in classification for the Mountain Deciduous area points to a potential problem in using satellite data for vegetation and the possibility of confusing forests for deciduous brush. Any continental scale estimation of forests with GVI data must take into account the possibility of this overestimate in deciduous forests.

The BORFOR model could be modified to better estimate the dead tree C pool. In the current model, all dead trees decay at a constant 6% per year rate, even though actual data of decay rates is both species-specific and size dependent (Harmon, 1993). Dead trees have been noted to be a significant C pool in forest ecosystems (Harmon, 1993; Kolchugina and Vinson, 1993), and could be improved in the model to provide species-specific decay rates that are also varied according to size.

BORFOR uses regression equations based on diameter at breast height of a single tree species (*Liriodendron tulipefira*) to predict biomass of stems and branches for all species. By using species-specific regression equations for each type of tree modelled, C estimates might be significantly improved.

6. Conclusion

BORFOR, a forest model that simulates growth of live trees, mosses and forest floor decay, was used to estimate carbon pools in a mountainous area of the Russian Far East. The forest ecosystem types for the BORFOR simulations were identified from unsupervised classification of GVI spectral data at 15 × 15 km resolution. A higher resolution data set to identify ecosystems in mountainous regions could improve the

TABLE III

Results of forest modeling include average carbon (C) pool values over the 500 year simulation. Each grid cell was individually modeled and the averaged C was summed for all grid cells identified as that forest ecosystem type. Results are compared to previous C estimates (Kolchugina and Vinson, unpublished database based on Bazilevich (1986)). The principle difference between estimates is the reduced value of dead trees in BORFOR.

GVI Class	No. cells	Tree, moss & foliage	Phytomass	Forest floor	Forest floor and dead trees (Forest floor	Dead trees)	Total	Total
		BORFOR	Kolchugina and Vinson	BORFOR	Kolchugina and Vinson	Kolchugina and Vinson	BORFOR	Kolchugina and Vinson
		(Tg C)	(Tg C)	(Tg C)	(Tg C)	(Tg C)	(Tg C)	(Tg C)
Spruce-Fir	40	4.54		1.34			5.88	
Larch-Pine	4	9.20		0.571			1.49	
Mixed	98	2.08		10.9			31.7	
Mountain Deciduous	26	9.41		3.33			12.7	
Total	168	35.6	28.9	16.2	18.4	16.9	51.8	64.9
Difference			-19%		+12%			+26%

modelling effort. Overall, the differences between ecosystem types (e.g., Spruce–Fir or Mountain Deciduous forest), may be more significant than the difference in the methods used to estimate carbon for similar ecosystems.

BORFOR estimates of C were lower than previously published estimates, partially due to an underestimated dead tree pool, and perhaps to lower productivity of actual conditions in the northern mountainous region. Forest floor C values were underestimated in Spruce-Fir and overestimated in Mountain Deciduous forests and warrant investigation of model assumptions of decomposition and dead tree decay.

The procedure of using GVI to classify forested ecosystems, and then modelling the areas identified with BORFOR should provide an estimate of C in live vegetation and forest floor for Russian forests, with the limitation noted for mountainous areas if the 15 × 15 km resolution is used for GVI. The dead tree pool and forest floor should be considered as potential problems considering the differences noted with published data.

Acknowledgments

The authors would like to thank two anonymous reviewers for their helpful comments and suggestions. Although the research described in this article has been funded under U.S. Environmental Protection Agency Agreement CR 17682-01 to Oregon State University, it has not been subjected to the EPA's review and, therefore does not necessarily reflect the views of the EPA, and no official endorsement should be inferred.

References

Aber J. D., Mellilo, J. M. and Federer, C. A.: 1982, *Forest Sci.* **28**, 31-45.

Bazilevich, N. I.: 1986, *Bulletin of the Academy of Sciences of the USSR, Geographical Series*, **2**, 49-66.

Bazilevich, N. I.: 1993, *Biological Productivity of Ecosystems of Northern Eurasia*, Naukapress, Moscow, Russia.

Bonan, G. B.: 1990a, *Biogeochemistry* **10**, 1-28.

Bonan, G. B.: 1990b, *Can. J. Forest Res.* **20**, 1077-1088.

Bonan, G. B. and Shugart, H. H.: 1989, *Ann. Rev. Ecol. Syst.* **20**, 1-28.

Daly, C. R., Neilson, R. P., and Phillips, D. L.: 1994, *J. Appl. Meteorol.*, **33**, 140-157.

Dixon, R. K., Brown, S., Houghton, R. A., Solomon, A. M., Trexler, M. C. and Wisniewski, J.: 1994, *Science* **263,** 185-190.

Fridland, V. M.: 1988, Map: Soils of the USSR, G.U.G.K., Moscow, Russia.

Gaston, G. G., Jackson, P. L., Vinson, T. S., Kolchugina, T. P., Botch, M., and Kobak, K: 1994, *Int. J. Remote Sensing* **15**(16), 3199-3221.

G.U.G.K.: 1960, Editor unspecified. Map: Forest Tree Species of the USSR, Moscow State University, Moscow, Russia.

Harmon, M.: 1993, *In:* Turner, D. P., Lee, J. J., Koerper, G. J., and Barker, J. R., (eds.), *The Forest Sector Carbon Budget of the United States: Carbon Pools and Flux Under Alternative Policy Options*, EPA/600/3-93/093, USEPA-ERL, Corvallis, USA.

Isachenko, A. G.: 1988, (ed.) Landscape Map of the U.S.S.R., Institute of Geography, Leningrad State University, Leningrad, Russia.

Kachinskii, N. A.: 1966, *Mechanical and Micro Aggregate Composition of Soils,* Israel Program for Scientific Translations, Jerusalem, Israel.

King, A. W.: 1991, *In*: Vinson, T. S. and Kolchugina, T. P. (eds.), *Proceedings of the International Workshop on Carbon Cycling in Boreal Forest and Sub-Arctic Ecosystems,* USEPA-ERL-C, Corvallis, USA, pp. 219-232.

Kineman, J. J. and Ohrenschall, M. A.: 1992, Global Ecosystems Data Base—Version 1.0 Documentation *Manual.* National Geophysical Data Center, Boulder, USA.

Knystautas, A.: 1987, *The Natural History of the USSR*, McGraw-Hill, New York, USA.

Kolchugina, T. P. and Vinson, T. S.: 1993, *Global Biogeochemical Cycles* **7**(2), 291-304.

Leemans, R.: 1992, *In:* Shugart, H. H., Leemans, R. and Bonan, G. B. (eds.), *A Systems Analysis of the Global Boreal Forest,* Cambridge University Press, Cambridge, UK, pp. 428-445.

Marland, G.: 1988, *The Problem of Solving the CO_2 Problem through Global Reforestation.* Rep. DOE/NBB - 0082, U. S. Dept. Energy, Off. Energy Res., Washington, USA.

Nikolov, N and Helmisaari, H.: 1992, *In:* Shugart, H. H., Leemans, R. and Bonan, G. B. (eds.), *A Systems Analysis of the Global Boreal Forest,* Cambridge University Press, Cambridge, UK, pp. 13-84.

Olson, J. S., Watts, J. A. and Allison, L. J.: 1983, *Carbon in Live Vegetation of Major World Ecosystems*. Rep. ORNL-5862, Oak Ridge Natl. Lab., Oak Ridge, USA.

Pastor, J. and Post, W. M.: 1985, *Development of a Linked Forest Productivity-Soil Process Model*, Oak Ridge National Laboratory, ORNL/TM-9519, Oak Ridge Natl. Lab, Oak Ridge, USA.

Pastor, J. and Post, W. M.: 1986, *Biogeochemistry* **2**, 3-27.

Post, W. M., Emanuel, W. R., Zinke, P. I. and Stangenberger, A. G.: 1982, *Nature* **298**, 156-159.

Prentice, I. C. and Helmisaari, H.: 1991, *Forest Ecol. Manage.* **42**, 79-93.

Ryabchikov, A. M.: 1988, Map: Geographical Belts and Zonal Types of Landscapes of the World. School of Geography, Moscow State University, Moscow, Russia.

Solomon, A. M.: 1993, *In:* Alden, J. *et al.* (eds.), *Forest Development in Cold Climates*, Plenum Press, New York, USA.

Sochava, V. B.: 1960, Map: Vegetation of the USSR. Ministry of the Interior, Moscow, Russia.

Suslov, S. P.: 1989, *Physical Geography of Asiatic Russia,* Freeman, San Francisco, USA.

Swanson, F. J., Kratz, T. K., Craine, N. and Woodmansee, R. G.: 1988, *Bioscience* **38**(2), 92-98.

Urban, D. L, Bonan, G. B., Smith, T. M. and Shugart, H. H.: 1991, *Forest Ecol.* **42**, 95-110.

Waring, R. H. and Schlesinger, W. H.: 1985, *Forest Ecosystems Concepts and Management,* Academic Press, New York, USA.

SIMULATION OF FOREST AND WOOD PRODUCT CARBON BUDGET UNDER A CHANGING CLIMATE IN FINLAND

T. KARJALAINEN and S. KELLOMÄKI

Faculty of Forestry, University of Joensuu, P.O. Box 111, FIN-80101 Joensuu, Finland

Abstract. Flows of carbon (C) in the forest ecosystem were simulated with a gap-phase dynamics type model, while flows of C in wood products were simulated using a model that processes raw material into final products. In southern Finland, the ratio between gross production and total storage for the 500 year period was 3052–3572 Mg C ha^{-1} : 192–223 Mg C ha^{-1} under the current climate and 4257–5096 Mg C ha^{-1} : 260–318 Mg C ha^{-1} under the predicted changing climate. In northern Finland, the respective ratios were 1721–2021 Mg C ha^{-1} : 103–134 Mg C ha^{-1} and 3409–4475 Mg C ha^{-1} : 212–244 Mg C ha^{-1}. The average total C storage in southern Finland over the 500 year period was 174–181 Mg C ha^{-1} under the current climate, and 206–217 Mg C ha^{-1} under the changing climate. In northern Finland, average total storage was 101 Mg C ha^{-1} under the current climate, and 191–198 Mg C ha^{-1} under the changing climate. The average C storages in unmanaged forest ecosystems under the current climate and under changing climate were 200 and 191 Mg C ha^{-1} respectively in southern Finland, and 142 and 193 Mg C ha^{-1} in northern Finland. Approximately 27–43% of total C was stored in wood products over a 500 year period. Wood products contributed 15–22% of the total emissions to the atmosphere. Over short periods, C sequestration potentials are much greater than over longer periods.

Keywords. CARBON BUDGET, FOREST ECOSYSTEM, WOOD PRODUCTS, FOREST MANAGEMENT, CARBON SEQUESTRATION, CLIMATE CHANGE.

1. Introduction

In Finland, the mean annual temperature has been projected to increase by 2–4 °C during the next 100 years and the mean annual precipitation to increase by 10–15% (Kettunen *et al.*, 1987; Holopainen and Carter, 1992; Carter *et al.*, 1993). These kinds of changes will influence the structure and productivity of forest ecosystems (Kellomäki *et al.*, 1988; Kellomäki and Kolström, 1992).

The increasing concentration of atmospheric carbon dioxide (CO_2) contributes over 50% to the enhanced greenhouse effect (Houghton *et al.*, 1990, 1992). This increase could be mitigated by reducing emissions by cutting down use of fossil fuels and deforestation or by intensifying terrestrial carbon (C) sinks by increasing vegetation cover on the current land area or by increasing the area covered by vegetation. The role of forest ecosystems is crucial in this context (UNCED, 1992; Helsinki Ministerial Conference Resolutions, 1993), as their share of the terrestrial C storage is about 60% (Winjum *et al.*, 1992). The C storage represented by forests in the temperate and boreal zone has increased during the past few decades (Tans *et al.*, 1990; Sedjo, 1992; Kauppi and Tomppo, 1993; Birdsey *et al.*, 1993) possibly replacing part of the "missing C" in the perturbed global C budget.

In C balance assessments, attention should be paid not only to the sequestration potential of stemwood, but also to ground vegetation, litter, and soil organic matter. As forests are extensively utilised to produce fuelwood or raw material for forest industry, the utilisation of forests should be included in assessments, not forgetting wood-based products until the C contained in them is released back to the atmosphere.

The aim of this study was to assess the potential of forest ecosystems and wood-based products to sequester atmospheric C. This study covers the dynamics and state of

Water, Air and Soil Pollution **82**: 309–320, 1995.

C sequestration in managed, naturally regenerated mixed stands under the current and changing climates with different thinning and final felling options. The production of wood-based products is based on current conversion figures for the wood-processing industry. Simulations were made covering periods of 500 years under the current and changing climate in medium fertile southern and northern boreal forest sites in Finland.

2. Method

The simulation of C flows in the forest ecosystem is based on a gap-type model developed by Kellomäki *et al.* (1992). The site conditions in this model are described in terms of availability of light, growing season degree days, available soil moisture, and available soil nitrogen. Consequently, it is also possible to simulate the effects of changing climate (changes in temperature and precipitation) on the forest ecosystem. Forest management is possible through regeneration, thinning, clearfelling and fertiliser application. The model utilises a Monte Carlo simulation technique in such a way that the birth and death of trees are stochastic events. Plot size in the model is 100 m^2. The model was run on an annual timestep.

The simulation of C flows in wood-based products is based on a model that processes harvested timber into products and follows the products until they are removed from use and the C contained in them is released back into the atmosphere (Karjalainen *et al.,* 1994, 1995; Karjalainen, 1994).

2.1. BOUNDARY CONDITIONS

2.1.1. Location and site

The simulations represent medium fertile sites in southern and northern Finland with a climate similar to that of the regions in Tampere (61°28'N, 23°44'E, elevation 50 m), and Rovaniemi (66°34'N, 25°50'E, elevation 199 m). The soil texture in all simulations was assumed to be that of sandy loam with a water retention capacity of the top 30 cm soil layer of 2.6 cm at the wilting point and of 5.7 cm at field capacity. The amount of soil organic matter in the humus layer (SOM) at the beginning of simulations was 85.4 Mg dry weight ha^{-1}, containing 46.9 Mg C ha^{-1} and 1.82 Mg nitrogen (N) ha^{-1}. Similarly, the amount of litter was 68.4 Mg dry weight ha^{-1}, containing 35.8 Mg C ha^{-1} and 0.202 Mg N ha^{-1}. Nitrogen deposition on the sites was assumed to be 10 kg $ha^{-1} yr^{-1}$ over the whole simulation period. Simulations were repeated 15 times for each scenario.

2.1.2. Climate

The current climate conditions applied in the simulations were monthly mean temperatures and precipitation for the period 1961–1980 (Ilmatieteen laitos, 1983; Kellomäki *et al.,* 1992). In the changing climate conditions, temperature and precipitation increases were based on a hybrid of two scenarios (Kettunen *et al.,* 1987, Holopainen and Carter, 1992), which are in line with the IPCC predictions (Houghton *et al.,* 1990, 1992). In southern Finland, temperature was assumed to increase linearly at 0.4 °C per decade during the first one hundred years. After one hundred years, the rise in temperature was assumed to slow during the next 50 years to half of the previous

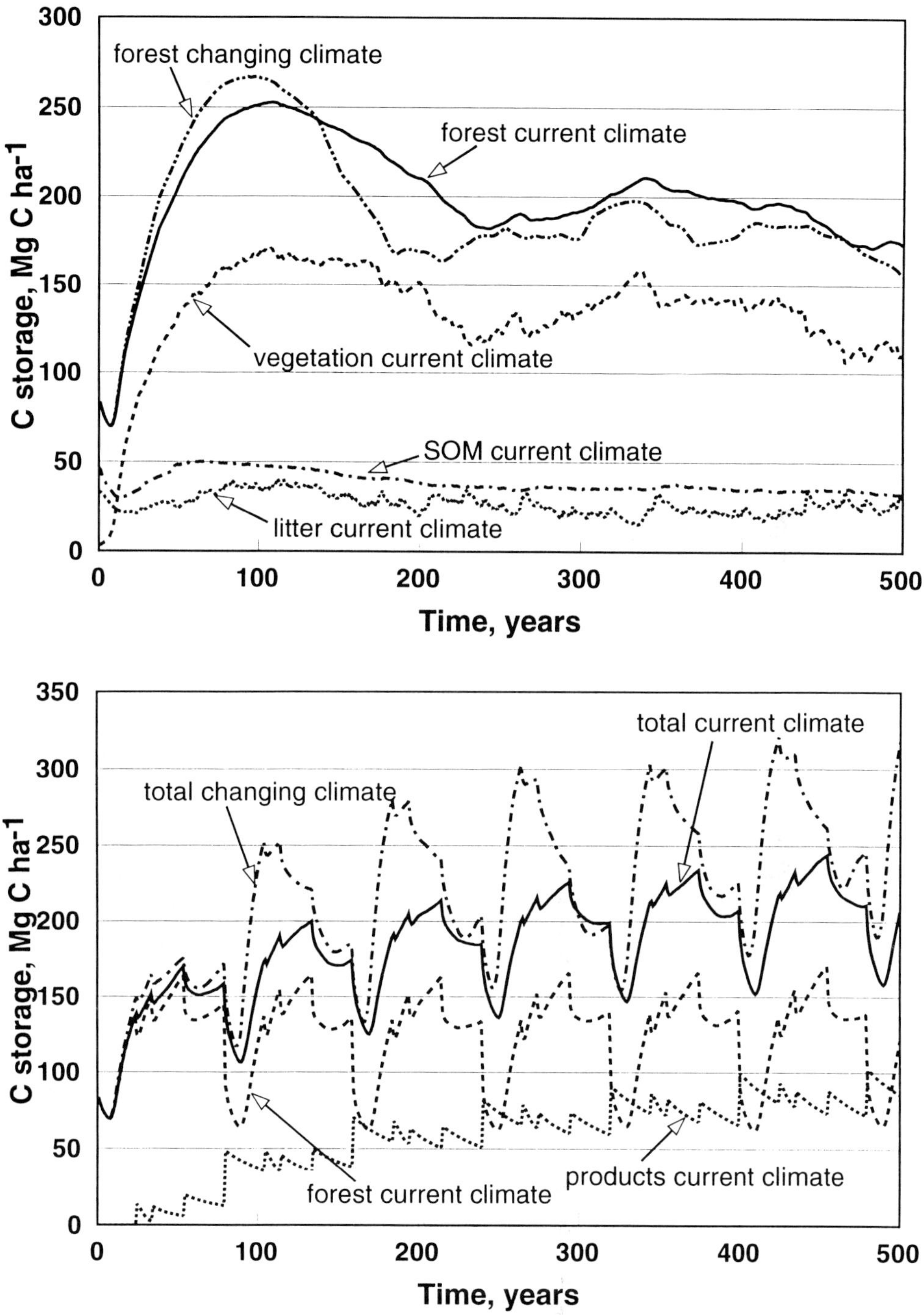

Fig. 1. Development of C storage under current and changing climates for unmanaged stands (upper pane) and for managed stands under Traditional Management, as described in text (lower pane).

rate and to stay at this level from there on. Temperatures were assumed to increase by 4.3–6.2 °C in winter (October–March) and by 1.6–4.4 °C in summer (April–September) during the first 100 years. Warming in northern Finland was assumed to be 5.0–6.4 °C during the winter and 2.6–4.8 °C during the summer. Precipitation was assumed to increase in both locations during the first one hundred years, 5% per decade during the winter months, and no change during the summer months. After the first one hundred years, the increase in precipitation was assumed to slow down during the next 50 years to half of the earlier rate and to stay at that level thereafter.

2.1.3. *Management*

The simulations were done for unmanaged and managed stands for 500 years. Unmanaged stands were naturally regenerated, mixed stands of Scots pine (*Pinus sylvestris*), Norway spruce (*Picea abies*), pendula birch (*Betula pendula*) and pubescent birch (*B. pubescens*).

Managed stands were also naturally regenerated, mixed stands, to which three thinning programs were applied. In Traditional Management and in More Natural Management, intermediate and final fellings were based on the age of the stands. In southern Finland thinnings (30% of the basal area) were done after 25, 35 and 55 years and final felling after every 80 years, while in northern Finland thinnings were done after 35, 55 and 75 years and final felling after every 100 years. The difference between Traditional Management and More Natural Management was that final felling in Traditional Management was clearcut, while in More Natural Management it amounted to 70% of the basal area. In Development Management, the time for thinnings was defined by the basal area of the stand exceeding a given thinning limit, and stands were thinned to a given level (Metsänhoitosuositukset, 1989). Moreover, the time for regeneration was based on diameter limits: 26 cm in southern Finland, and 23 cm in northern Finland. All the stands were low-thinned, and only stemwood was harvested.

2.1.4. *Production and use of products*

The conversion of timber into wood products was based on product/timber units typical for the current wood-processing industry in Finland (Yearbook of Forest Statistics 1992). The products were divided into four lifespan categories. Both export and import of timber and products were excluded. When products were removed from use, one third were assumed to be recycled, one third used to generate energy, and one third disposed of in landfills. The decay rate in landfills was assumed to be 0.5% yr^{-1}.

3. Results

3.1. Carbon sequestration in southern Finland

In terms of the vegetation's gross production, C sequestration from the atmosphere under the current climate was 685–747 Mg C ha^{-1} during the first 100 years, depending on the management option applied. At the same time, forests and wood products were able to sequester 146–166 Mg C ha^{-1}. Under the changing climate, gross production was 809–913 Mg C ha^{-1}, and storage 182–233 Mg C ha^{-1}. During the entire 500 year simulation, the ratio between gross production and storage under the current climate was 3052–3572 Mg C ha^{-1} : 192–223 Mg C ha^{-1}, and under the changing climate 4257–5096 Mg C ha^{-1} : 260–318 Mg C ha^{-1}.

Gross production in unmanaged stands was lower than in managed forests when the time-frame exceeded 100 years. The difference was greater under the changing climate than under the current climate. The actual C storages after the first 100 years in the managed stands and wood products were able to exceed those of unmanaged stands, regardless of climate and forest management applied (Figure 1).

In unmanaged stands, the annual net C balance over the first 100 years was larger under the changing climate than under the current climate. Over 200–500 years, however, it was smaller. After about 100 years in the simulation, the C balance in unmanaged stands was in equilibrium, whereas in managed stands the C sequestration phase followed thinnings and clearfelling (Figure 2). In the managed stands and wood products, the annual net C balance was bigger under the changing climate regardless of the time frame and forest management (Figure 3). Under the current climate, the biggest total annual C balance was achieved when the forests were managed applying Development Management. More Natural Management provided a smaller annual C balance than did Traditional Management. Under the changing climate, Traditional Management gave the biggest annual C balance.

The differences in the average total C storages between the forest management options were bigger under the changing than under the current climate, and tended to increase with time. The forest management option of Development Management provided the biggest average total C storage, particularly under the changing climate. This implies that forest management based on the development phase of the stand is more beneficial from a C storage point of view than is management with a fixed cutting regimen. Under the current climate, the average C storage in unmanaged stands was greater than in the managed stands and wood products, but under the changing climate it was smaller over simulation periods of 300–500 years (Figure 3). When the time frame exceeded 200 years, average C storage decreased in unmanaged stands, while in managed stands and wood products average storage increased. Over the first 100 years, 45–47% of the average C storage was distributed in the vegetation, 17–20% in the litter, 23–27% in SOM, 7–8% in products in use, and 3% in products in landfills; i.e., 89–90% in the forest and 10–11% in products. Over a longer time frame, the forest storage share of total storage decreased while that of products increased. Over the entire 500 year period, the share of vegetation was 31–36%, that of litter 13–18%, that of SOM 13–20%, that of products in use 13–16%, and that of products in landfills 19–23% of the average C storage; i.e., the share of the forests was 60–68% and that of products 32–40%.

The C storage represented by vegetation, litter, products in use and products in landfills was greater under the changing climate than under the current climate, whereas the C storage of SOM was smaller under the changing climate (Table I). Of the emissions into the atmosphere, respiration by the vegetation contributed 28–32%, decay of litter 23–27%, decay of SOM 26–29%, process emissions (burning of by-products) 8–11%, burning of products 5–8%, and decay of products in landfills 2–3%, depending on the management options and climate. In unmanaged stands, respiration by the vegetation was 34–37% of the emissions into the atmosphere, decay of litter contributed 32–33%, and decay of SOM 31–33%.

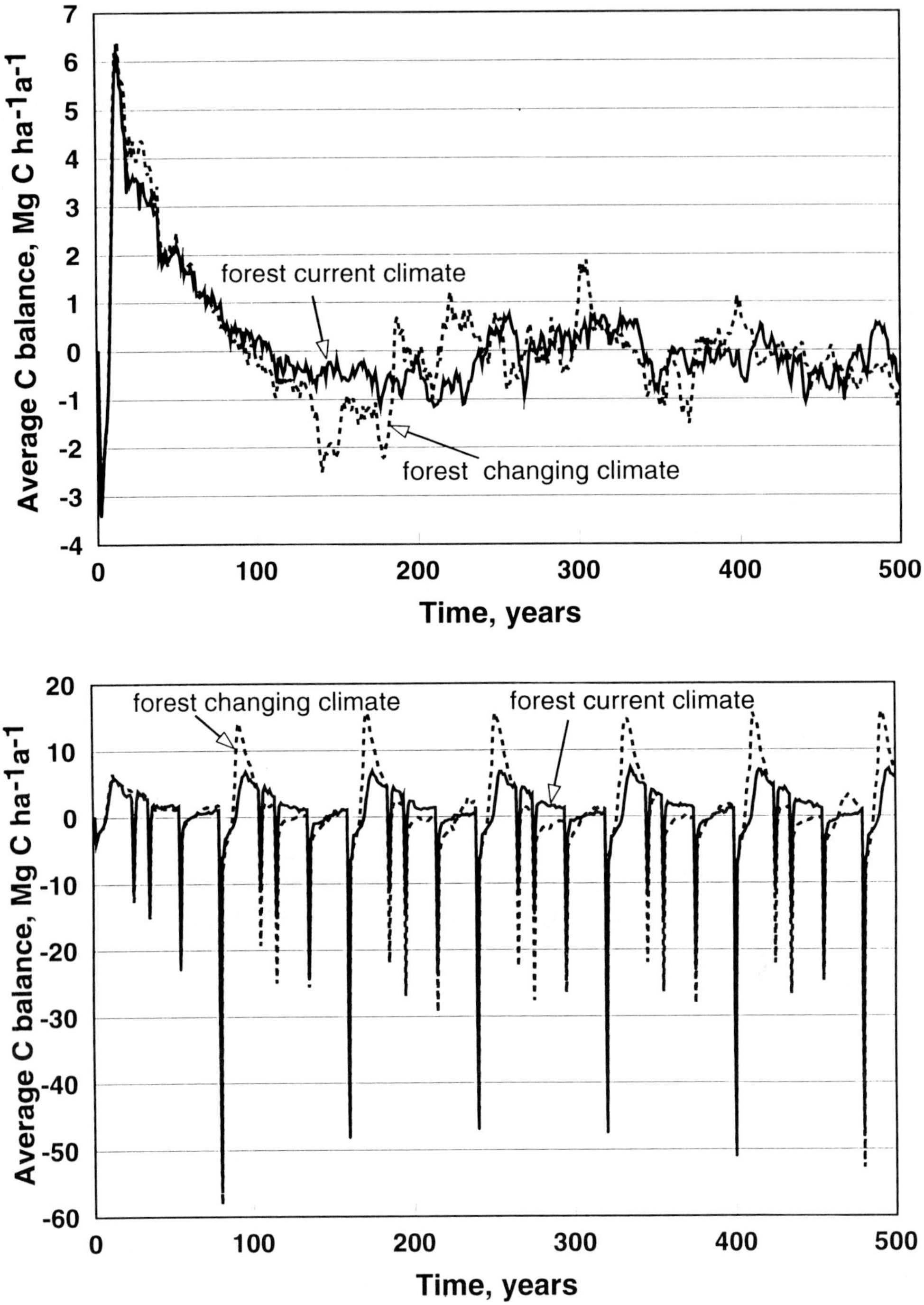

Fig. 2. Development of C balances under the current and changing climate for unmanaged stands (upper pane) and managed stands under Traditional Management, described in text (lower pane).

3.2. Carbon sequestration in northern Finland

The effect of a changing climate on C sequestration in northern Finland was much more pronounced than in southern Finland. Gross production in managed stands was 1721–2021 Mg C ha^{-1} over the whole 500 year simulation under the current climate, but under the changing climate gross production was 3409–4475 Mg C ha^{-1} depending on the management option applied. The highest gross production was achieved under Development Management.

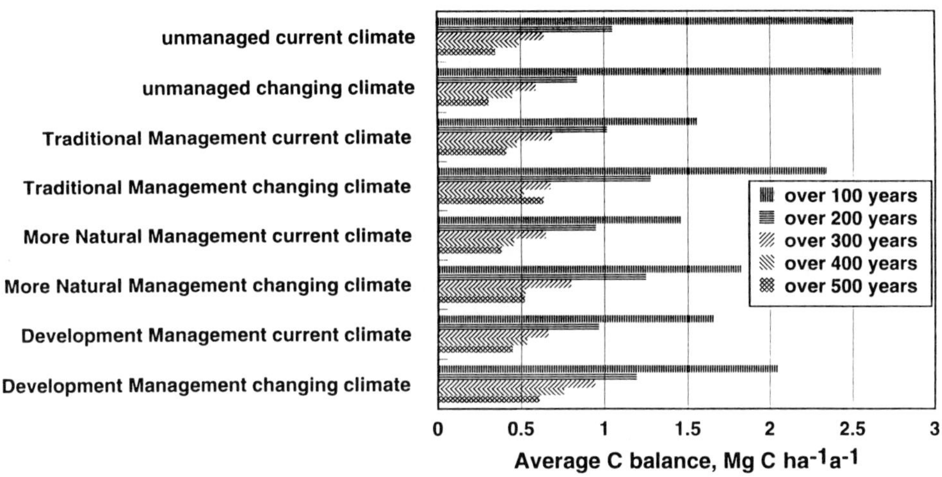

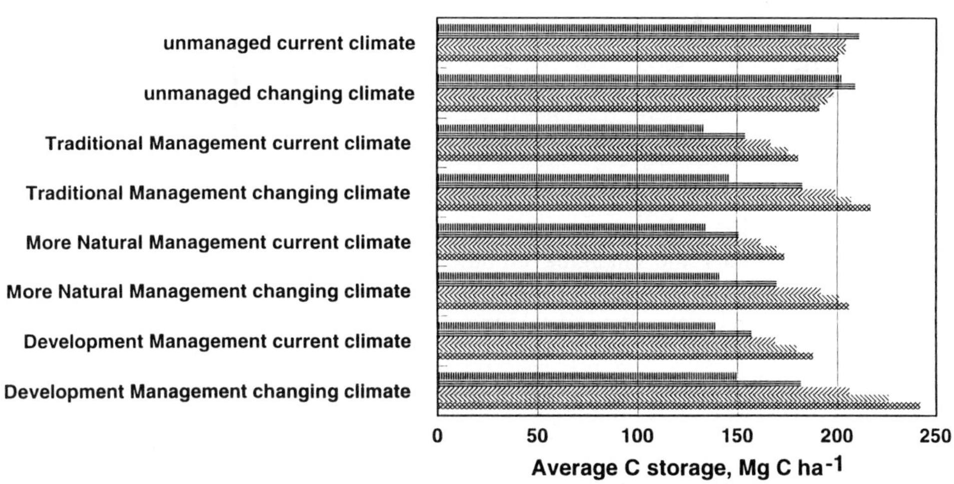

Fig. 3. Average C balances and storage for unmanaged stands (upper pane) and managed stands and wood products (lower pane) when applying different forest management options in southern Finland.

TABLE I

Distribution of average C storage over 500 years simulation applying different management under current and changing climate. Figures above for southern Finland and figures below for northern Finland.

Applied management and climate	Average C storage, Mg C ha^{-1}							
					Products			
	Vegetation	Litter	SOM	Forest	in use	in landfills	total	Total
Unmanaged								
Current	134	27	39	200	-	-	-	200
	95	12	36	143	-	-	-	143
Changing	128	31	32	191	-	-	-	191
	137	22	33	193	-	-	-	193
Traditional Management								
Current	61	26	36	124	23	35	58	181
	35	6	34	75	11	16	27	102
Changing	79	39	31	148	28	41	69	217
	73	33	35	141	23	34	56	198
More Natural Management								
Current	61	23	36	119	22	33	54	174
	34	5	35	73	12	16	28	101
Changing	78	34	30	141	26	39	65	206
	76	28	34	138	22	32	54	191
Development Management								
Current	58	26	36	121	26	41	67	188
	37	8	33	77	13	19	32	109
Changing	75	40	32	147	39	56	95	242
	47	31	34	112	33	52	86	198

In unmanaged stands, gross production was 1423 Mg C ha^{-1} under the current climate and 2110 Mg C ha^{-1} under the changing climate. The vegetation's C storage fluctuated more under the changing climate than it did under the current climate. Unmanaged stands were C sinks for about 220 years under the current climate and for about 180 years under the changing climate; from there on, the C balance was in equilibrium.

Under Development Management, the total C balance under the changing climate was about double that under the current climate. The differences were smaller under Traditional Management or More Natural Management (Figure 4). The C storages were also distributed differently under the changing climate than they were under the current climate (Table I). The C storages in the vegetation and litter increased while the SOM storage was influenced less; in fact, the latter decreased proportionally. Under the current climate, the vegetation's C storage was 33–34% of the total C storage, that of the litter was 5–7%, that of the SOM 30–34%, that of products in use 11–12%, and that of products in landfills 27–30%. Under the changing climate, the vegetation accounted for 24–40%, the litter for 14–17%, the SOM for 17–18%, products in use for 11–17%, and products in landfills for 17–27% of the total C storage.

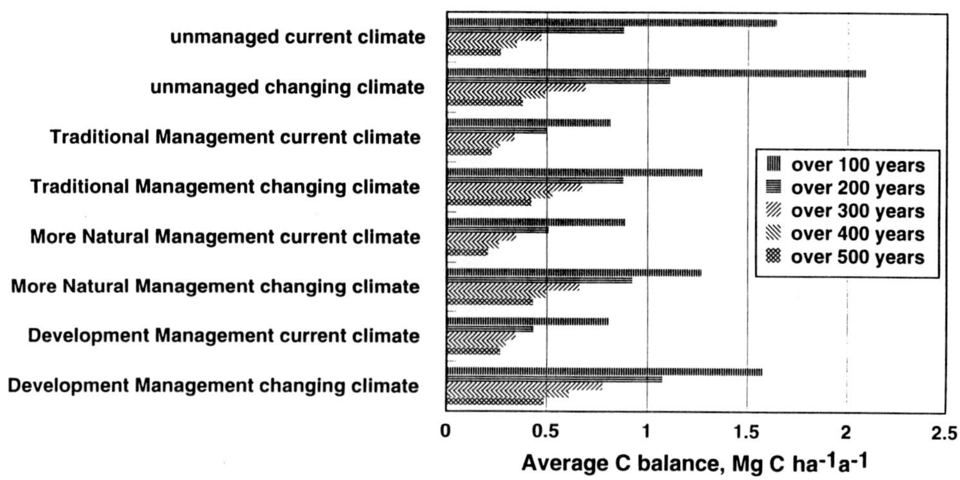

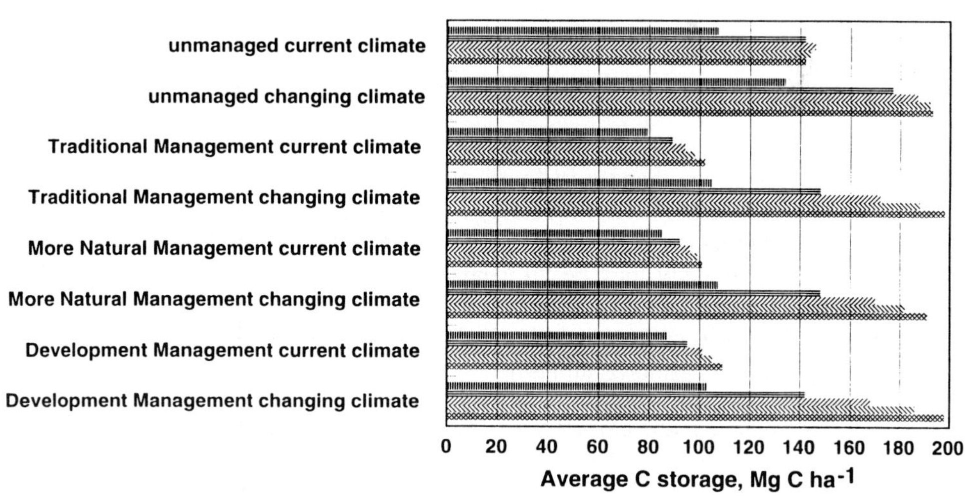

Fig. 4. Average C balances and storage for unmanaged stands (upper pane) and for managed stands and wood products (lower pane) under different forest management options in northern Finland.

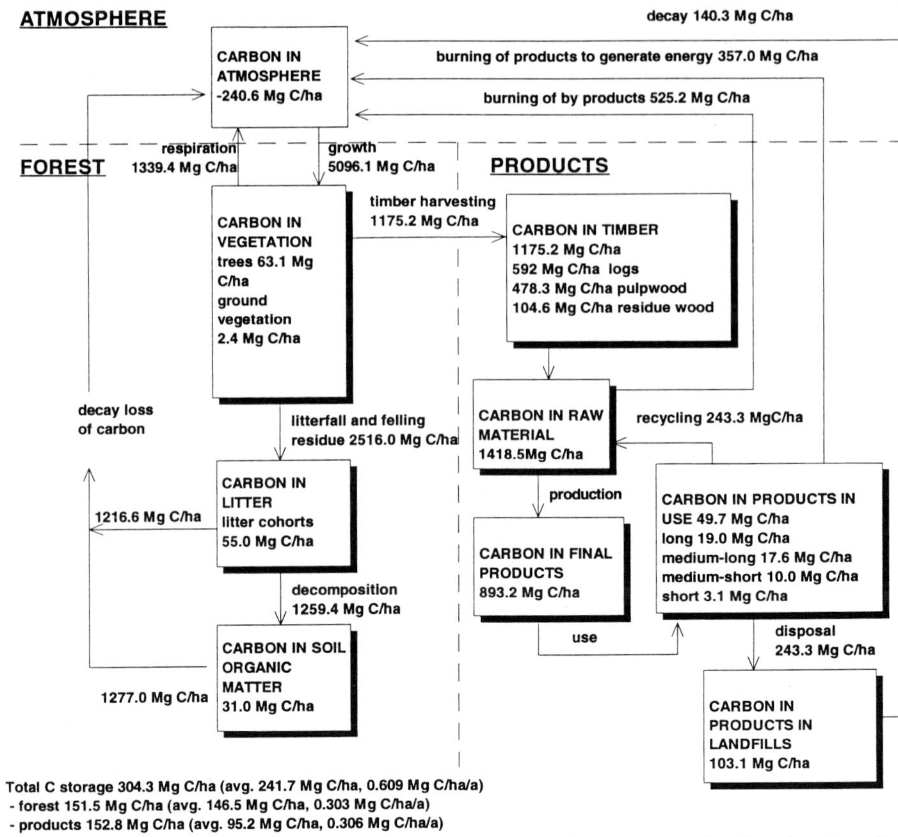

Fig. 5. Example of C sequestration applying Development Management under the changing climate in southern Finland. C storage after 500 years simulation, C flows are cumulative.

4. Discussion

The potential of forests and wood products to sequester atmospheric C at the stand level taking into account changes in forest management and climate in Finland were outlined in this study. It is clear that forest management has a considerable influence on the forest's C storage as well as on the forest sector's C storage. The climate change scenario used in this study may be an extreme one, but it favoured C sequestration more in managed stands and wood products than in unmanaged stands. It also indicated that changes in C sequestration were more pronounced in northern Finland than in southern Finland. Furthermore, thinnings should be done earlier and perhaps more intensively under the changing climate than under the current climate, and rotation lengths could also be shorter under the changing climate. Over short time frames (less than 200 years), the forest's C balances and C sequestration potentials are much higher than over long time frames. However, it seems to be important also to include wood products in C balance assessments since about 27–43% of the forest and wood product C was in either products in use or products in landfills (Figure 5). Wood products' share of the total C emissions into the atmosphere was 15–22%.

Nabuurs and Mohren (1993) assessed C sequestration potentials of selected forest types. In Norway spruce stands and wood products in boreal Russia, the long-term average amount of C storage was 192 Mg C ha^{-1}, which is close to figures presented in this study. The highest long-term average C storage was found for Douglas-fir in northwest USA, 339 Mg C ha^{-1}, while the highest net annual C fixation in the first rotation was achieved with Caribbean pine in Brazil, 5.1 Mg C ha^{-1} yr^{-1}. However, the long-term average storage for Caribbean pine was 179 Mg C ha^{-1}. Dewar and Cannell (1992) have assessed C storages for different types of plantations and wood products in the UK. They have reported 93–254 Mg C ha^{-1} equilibrium C storages, depending on tree species and yield class.

In boreal conditions, stand-destroying disturbances other than timber harvesting could substantially affect the forest's C storage (Auclair and Carter, 1993; Kurz and Apps, 1993, 1994). These can be forest fires, insects, fungal diseases and wind damage. However, forest fires for example have had only a small impact on national forest C balance in Finland, since burned forest area during the last 20 years has been only about 500 ha yr^{-1}, which is magnitudes less than the annual cutting area of 422 000 ha yr^{-1} (Yearbook of Forest Statistics 1993–94). The contribution of natural mortality on the national C balance has also been small compared to timber harvesting. However, if forest resources are utilised at the current level in the future, natural mortality has been predicted to increase due to ageing and the dense forest structure (Report of the Revision Committee of the Forest 2000 Programme, 1992).

The amount of C in the stemwood of the growing stock in Finland has increased from 320 Tg to 400 Tg between 1953 and 1990 due to intensive silviculture and growth exceeding removal since 1971 (Karjalainen and Kellomäki, 1993). The total amount of C on 23.3 Mha of forest and scrub land in 1990 was estimated to be 2800 Tg, of which 74% was in the soil. In the late 1980s, harvested timber contained approximately 9.9 Tg C yr^{-1}, process emissions amounted to 3.3 Tg C yr^{-1}, and 6.6 Tg C yr^{-1} was bound in the final products (Karjalainen *et al.,* 1994). Moreover, at that time the forest C balance in terms of growth of the stemwood stock (16.9 Tg C yr^{-1}) and losses (11.2 Tg C yr^{-1} including 1.3 Tg C yr^{-1} natural mortality and felling residues) was 5.7 Tg C yr^{-1}, which was 38% of the C emission from energy production and consumption. Assuming the use of forest resources remains at the late 1980s level in the future, the size of the forest sector sink in the late 2030s could be 18.9 Tg C yr^{-1}, of which the forest sink will be 16.3 Tg C yr^{-1} and the wood product sink will be 2.6 Tg C yr^{-1} (Karjalainen *et al.,* 1995). If the use of forest resources equals annual growth, the size of the forest sector sink could be 5 Tg C yr^{-1}. These figures also include the C in exported products. The effect of changing climate on C sequestration was excluded in those assessments, but according to this study, C sequestration could be enhanced under the predicted changing climate especially in northern Finland. Additionally, recent recommendations for forest management and increasing the area of protected forests seem to enlarge forest C storages at least for some time.

On the global scale, increasing amounts of wood products have been produced during this century. Some of these products are still in long-term storages (either in use or disposed of to permanent disposal storages), and may have contributed to the global C balance, providing possibly one additional solution for the missing C sink.

The emissions of C from timber harvesting and transportation were excluded in this study, but Karjalainen *et al.* (1994) have estimated that they were about 1% of the amount of C contained in the annual timber harvest in late 1980s. Moreover, the consumption of external energy, which was excluded in this study, could be high in some production processes (mechanical pulp processes, paper production), while some processes could produce more energy than they consume (chemical pulp). Furthermore, emissions caused by manufacturing machines and the construction of mills may be considered, but then emissions should be split over the time the machines and mills are in use. Some of the carbon emissions during the production process could be delayed if more of the by-products could be used as raw material for particle and fibre boards instead of burned. Particle and fibre boards could have long lifespans if used to manufacture furniture or used as construction material.

References

Auclair, A. N. D. and Carter, T. B.: 1993, *Can. J. Forest Res.* **23**:1528-1536.

Birdsey, R. A., Plantinga, A. J. and Heath, L. S.: 1993, *Forest Ecol. Manage.* **58**, 33-40.

Carter, T., Holopainen, E. and Kanninen M. (eds.): 1993, *Publ. Acad. Finland* **2**/93, 63 pp.

Dewar, R. C. and Cannell, M. G. R.: 1992, *Tree Physiol.* **11**, 49-71.

Helsinki Ministerial Conference Resolutions: 1993, *Resolutions of the Ministerial Conference on The Protection of Forests in Europe*, 22 pp.

Holopainen, E. O. and Carter, T.: 1992, *In:* Kanninnen, M. (ed.), *Muuttuva ilmakehä.* Ilmasto, luonto ja ihminen. Vapk-Kustannus, Finland, pp. 51-56.

Houghton, J. T., Jenkins, G. J. and Ephraums, J. J. (eds.): 1990, *Climate Change. The IPCC Scientific Assessment*, Intergovernmental Panel on Climate Change, Cambridge University Press, Cambridge, UK, 364 pp.

Houghton, J. T., Callander, B. A. and Varney, S. K. (eds.): 1992, *Climate Change 1992. The Supplementary Report to the IPCC Scientific Assessment*, Intergovernmental Panel on Climate Change, Cambridge University Press, Cambridge, UK, 200 pp.

Ilmatieteen laitos: 1983. Tilastoja Suomen ilmastosta 1961-1980. *Suomen Meteorologinen vuosikirja* **80**, 560 pp.

Karjalainen, T.: 1994, Thesis for Licentiate Degree in Forest Sciences, University of Joensuu, Joensuu, Finland, 123 pp.

Karjalainen, T. and Kellomäki, S.: 1993, *Publ. Acad. Finland* **3**/93, 40-51.

Karjalainen, T., Kellomäki, S. and Pussinen, A.: 1995, *Climatic Change*, 28 pp. (In press.)

Karjalainen, T., Kellomäki, S. and Pussinen, A.: 1994, *Silva Fenn.* **28**(2), 67-80.

Kauppi, P. E. and Tomppo, E.: 1993, *Water, Air, and Soil Pollut.* **70**, 187-196.

Kellomäki, S., Hänninen, H. and Kolström, T.: 1988, *Silva Fenn.* **22**(4), 293-305.

Kellomäki, S. and Kolström, M.: 1992, *Vegetatio* **102**, 47-68.

Kellomäki, S., Väisänen, H., Hänninen, H., Kolström, T., Lauhanen, R., Mattila, U. and Pajari, B.: 1992, *Silva Fenn.* **26** (1), 1-18.

Kettunen, L., Mukula, J., Pohjonen, V., Rantanen, O. and Varjo, U.: 1987. In: Parry, M. L., Carter, T. R. and Konijn, N. T. (eds.), *The Impact of Climatic Variations on Agriculture. Vol. 1. Assessments in cool temperate and cold regions.* IIASA, Vienna, Austria, 90 pp.

Kurz, W. A. and Apps, M. J.: 1993, *Water, Air, and Soil Pollut.* **70**, 163-176.

Kurz, W. A. and Apps, M. J.: 1994, *Environ. Pollut.* **83**, 55-61.

Metsänhoitosuositukset: 1989, Keskusmetsälautakunta Tapio, 55 pp.

Nabuurs, G. J. and Mohren, G. M. J.: 1993, *IBN Research Report* **93**/4, 205 pp.

Report of the Revision Committee of The Forest 2000 Programme.: 1992, Komiteamietintö 1992:**5**, 112 pp.

Sedjo, R. A.: 1992. *Ambio* **21**(4), 274-277.

Tans, P., Fung, I. Y. and Takahashi, T.: 1990, *Science* **247**, 1431-1438.

UNCED: 1992, *Adoption of Agreements on Environment and Development*, 97 pp.

Winjum, J. K., Dixon, R. K. and Schroeder, P. E.: 1992, *Water, Air, and Soil Pollut.* **64**, 213-227.

Yearbook of Forest Statistics 1992, *Official Statistics of Finland*: 1993:**5**, 317 pp.

Yearbook of Forest Statistics 1993-94. *Official Statistics of Finland*: 1994:**7**, 348 pp.

AN ANALYSIS OF FUTURE CARBON BUDGETS OF CANADIAN BOREAL FORESTS

W. A. KURZ[1] and M. J. APPS[2]

[1] *ESSA Technologies Ltd., 1765 West 8th Avenue, Vancouver, British Columbia V6J 5C6, Canada*
[2] *Department of Natural Resources Canada, 5320-122nd Street, Edmonton, Alberta T6H 3S5, Canada*

Abstract. The Canadian boreal forest covers over 300 Mha of land area. Its dynamics are largely influenced by fires and insect-induced stand mortality and to a much lesser extent by forest management. This paper analyses six scenarios of future (1990–2040) carbon (C) budgets of the Canadian boreal forest, each based on different assumptions about natural disturbances, rates of reforestation of disturbed land, and conversion of non-stocked to productive forest stands. The objective of these scenarios is to explore the range of responses to different management options. The results indicate an overall inertia of a system whose dynamics are strongly influenced by a recent 20-year period (1970–1989) of large-scale forest disturbances by fire and insects. The 50-year C budget of the six scenarios ranges from an estimated net source of 1.4 Pg C to a net sink of 9.2 Pg C. These estimates indicate the range of response to the management of the Canadian boreal forest. Although a full-scale implementation of the management activities examined here is not likely given ecological and economic realities in the Canadian boreal forest, the analyses explore the relative merits of reducing forest disturbance rates, regeneration delays, and the area of non-stocked forest land.

Keywords. CARBON BUDGET, BOREAL FOREST, CBM-CFS, CANADA, DISTURBANCE, FUTURE PROJECTIONS.

1. Introduction

Anthropogenic tampering with the global carbon (C) cycle has reached levels that have led to considerable concern about future concentrations of C in the atmosphere. Numerous efforts are therefore underway to elucidate the many processes that are influencing the global C cycle. Forest ecosystems are significant stores of C and contribute to both gross and net fluxes in the global C budget (Smith *et al.*, 1993; Apps *et al.*, 1993a). Forest ecosystems are dynamic systems that respond to environmental changes and to forest management actions. It is increasingly recognised that forest dynamics could significantly affect the future C content of the atmosphere. In this paper we explore a series of alternative scenarios of forest dynamics in the Canadian boreal forest over the next 50 years. Each scenario is analysed in terms of its implication for the net exchange of C between Canadian boreal forest ecosystems and the atmosphere.

This paper examines how the C budget of the Canadian boreal forest will change over the next 50 years assuming different:

1. scenarios of natural disturbance rates;
2. policies aimed at the regeneration of disturbed areas; and
3. policies aimed at increasing the area of productive forests.

Although we recognise that in the next 50 years, changes in climatic conditions (Houghton *et al.*, 1990) could further influence forest dynamics and disturbance regimes (Kurz *et al.*, 1995b; Kurz and Apps, 1994), we prescribe scenarios for growth and decomposition rates that are independent of climatic change for the purpose of this analysis. We also limit the scope of the analysis to assessments of the C-budget implications of different scenarios without exploration of the ecological or economic constraints of each scenario.

Water, Air and Soil Pollution **82**: 321–331, 1995.

2. Methods

The Carbon Budget Model of the Canadian Forest Sector (CBM-CFS) is a national-scale model of forest biomass, soil, and forest product sector C dynamics that has been used to assess the net C budget of Canadian forests (Apps and Kurz, 1994; Kurz *et al.*, 1992). A preliminary assessment of the model's sensitivity to changes in growth and decomposition rates and in disturbance regimes has been undertaken (Kurz and Apps, 1994). More recently, a revised version of the model (CBM-CFS2) has been used to conduct a retrospective assessment of the C budget of Canadian forests for the period 1920 to 1989 (Kurz *et al.,*1995a; Kurz and Apps, 1995). A comparison of the C budget of a 1.0 Mha forest management unit in the boreal forest with and without forest management has been completed using the same model at a higher spatial resolution (Price *et al.*, 1995).

The projective scenario analyses described here are for the three ecoclimatic provinces (Ecoregions Working Group, 1989) that comprise the Canadian boreal forest: boreal east, boreal west, and subarctic. The boreal forest area represented in the model is 303 Mha (Kurz *et al.*, 1992; Apps *et al.*, 1993b) and is similar to the area used in a recent 70-year retrospective analysis (Kurz and Apps, submitted). The total forest area is assumed to remain constant for the duration of the 50-year projective model runs.

Ecosystem dynamics of the Canadian boreal forest are affected by several factors and processes whose time dynamics introduce considerable memory and inertia into the system. Understanding the recent past of this dynamic system is therefore a useful prerequisite to assessing its behaviour in the near future. For example, the forest age-class distribution reflects past episodes of fluctuations in the disturbance rates (Kurz and Apps, 1993) as does the amount of decomposing organic matter in the four soil pools represented in the model. In the CBM-CFS2, detritus and coarse woody debris are included in the soil pools. During a disturbance, only some of the ecosystem C is released to the atmosphere (as in the case of fire) or transferred to the forest product sector (in the case of harvesting). A portion of the biomass C is transferred from the living biomass to the decomposing soil C pools. The composition and magnitude of this transfer differs among disturbances: fires primarily leave boles that are added to the soil C pools, harvesting primarily leaves foliage and branches, and in insect-killed stands, most of the biomass C is added to soil C pools. The CBM-CFS2 accounts for the total annual transfer of biomass to soil C pools, including litterfall, stem mortality associated with stand break-up, and the transfer associated with disturbances.

All scenarios described in this analysis commence in 1990 with simulated initial ecosystem conditions that are the endpoint of the 70-year retrospective model run for the period 1920–1989 (Kurz *et al.,* 1995a; Kurz and Apps, 1995). The forest age-class distribution and the C content of biomass and soil C pools are all affected by the forest dynamics of the 70-year period prior to 1990. The principal findings of the retrospective analysis are summarised below and further details on the model and the assumptions underlying the retrospective analysis can be found elsewhere (Kurz and Apps, 1994; Kurz *et al.*, 1995a; Kurz and Apps, 1995).

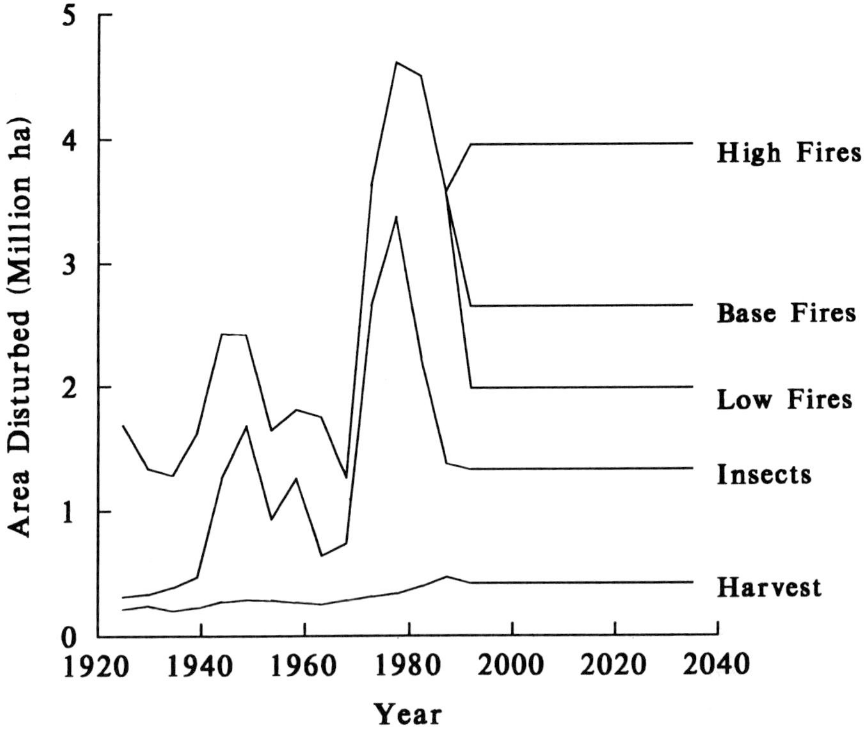

Fig. 1. The observed 5-year average of the area annually disturbed in the Canadian boreal forest for the period 1920–1989 (Kurz and Apps, 1995) and the assumed disturbance regimes for the projective base run (1990–2039), and the low and high fire scenarios. The area annually burned in each scenario is the difference between the insect and the fire lines.

The purpose of the six projective scenarios is to explore the net C budgets resulting from different assumptions about forest management policies as expressed in rates of natural disturbances, rates of replanting of disturbed areas, and deliberate increases in the area of productive forest through conversion of non-stocked forest land to productive forest land. All scenarios assume that ecosystem-specific growth and decomposition rates are similar to those in the retrospective analysis. Growth curves used to simulate forest biomass dynamics are parameterized using data derived from the national biomass inventory (Bonnor, 1985), as described in more detail elsewhere (Kurz and Apps, 1994). As noted above, the analysis does not assess the possible effects of changes in temperature or precipitation regimes on growth or decomposition.

For all projective scenarios, we assume that annual disturbance rates are constant during the 50 years of the analysis. The magnitude of disturbance rates is expressed relative to that of the last 5-year period in the retrospective analysis, i.e., 1985–1989. Figure 1 summarises historic disturbance rates in the boreal forest (1920–1989, Kurz and Apps, 1995) and the disturbance rates assumed for the projective base run. To reflect recent decreases in harvest activities (Canadian Council of Forest Ministers, 1994), we assume that harvest and planting rates are at 90% of the 5-year average of the period 1985–1989. Insect-induced stand mortality is assumed to be similar to the average of the period 1985–1989. Our assumptions about fire regimes in the base run are guided by the observation that the decade of the 1980s was the warmest decade on record in Canada (Gullett and Skinner, 1992) during which the area annually burned was also the highest since the beginning of the statistical record. The average area annually burned in the period 1990–1993 in Canada was about half of that during the 1980s (Canadian Council of Forest Ministers, 1994; Canadian Forest Service, unpublished data). For the base run, the area burned annually is assumed to be 60% of the average area burned annually in 1985–1989, or 143% of the annual average in the period 1920–1969. We assume that the distribution of all disturbances by spatial units (i.e., ecoprovinces within Provinces and Territories) is similar to that in the period 1985–1989.

Table I summarises the assumptions of the six projective scenarios. The impacts of different natural disturbance regimes and fire suppression policies are addressed in scenarios 2 ("Low Fire") and 3 ("High Fire"). Natural regeneration is assumed to take place following disturbance. The fourth scenario ("Plant All") explores the implications of planting all areas that are disturbed through harvesting, fires, or insect-induced stand mortality. In the model, planting as a management intervention eliminates the 10-year regeneration delay that is assumed for all natural regeneration (Figure 2). Column 3 of Table I summarises the total area planted during the 50-year projective runs.

The fifth and sixth scenario runs analyse the impacts on the C budget of converting areas that are currently considered non-stocked to stocked stands with high growth rates (a "best case" scenario). Non-stocked stands do not meet stocking standards (which differ between Provinces) but may contain some forest. To simulate stand conversion, the sites are first disturbed with impacts on biomass and soil C pools that are equivalent to a wildfire. The sites are then planted with trees with the highest growth rate of any forest ecosystem in the spatial unit of the non-stocked area (Figure 3). Thus, regional constraints on growth rates are incorporated in the model: for example, a site in the northern boreal forest will not be assigned growth rates from a southern site. Scenario five assumes that in every year of the 50-year analysis, 1% of

TABLE I

Summary of assumptions for 6 scenarios of forest dynamics of Canadian boreal forests. For scenarios two to six, only the differences from the base run are listed.

Scenario	Assumptions	Area planted (Mha)
1. Base Run	Harvest and planting at 90% of 1985–89 levels; fire at 60% of 1985–89 levels; insect-induced mortality at 100% of 1985–89 levels.	4.6
2. Low Fire	Fire at 50% of Base Run	4.6
3. High Fire	Fire at 200% of Base Run	4.6
4. Plant All	All disturbed areas are planted	132.0
5. Conversion	Non-stocked areas converted to high yield stands at 1% of eligible area each year	58.2
6. Selective Conversion	Non-stocked areas with less than 2 Mg C ha^{-1} converted to high yield stands at 5% of eligible area each year	51.8

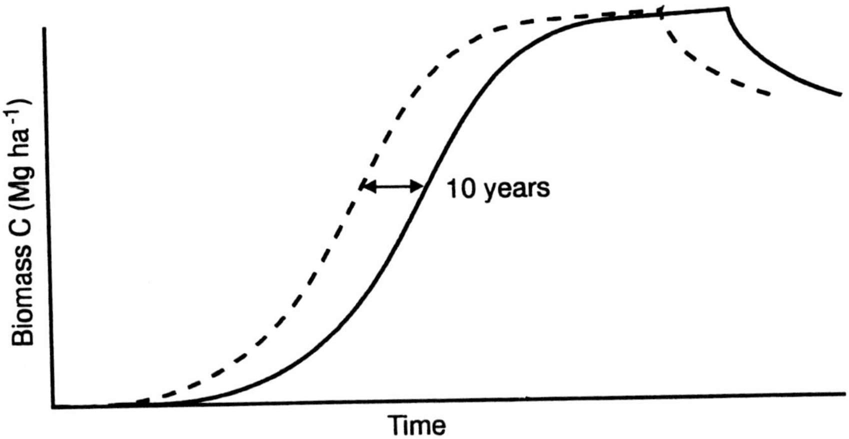

Fig. 2. The impact of planting on stand biomass C dynamics, as simulated in the model. The solid line shows stand development under the assumption of natural regeneration. The dashed line shows the assumptions for planted stands.

all non-stocked area in each spatial unit is converted to productive stands. In scenario six, each year 5% of the non-stocked area that at the time contains less than 2 Mg C ha^{-1} in total forest biomass is converted to productive stands. The low biomass values of the selected sites can be the result of either a recent disturbance or poor site conditions. As discussed below, many ecological and economic factors may constrain the actual implementation of these scenarios.

3. Results

The retrospective analysis of the period 1920–1989 demonstrated that changes in the area annually disturbed (primarily by wildfire and insects) greatly influence biomass and soil C dynamics and therefore the annual net C exchange with the atmosphere (Kurz and Apps, submitted). From 1920 to 1969, the five-year average of the area annually disturbed in the Canadian boreal forest averaged 1.7 Mha yr^{-1}. In the period 1970–1989, the disturbance rate was more than twice as high: 3.9 Mha yr^{-1} were disturbed (Figure 1).

This increase in disturbances resulted in a large increase in the annual transfer of biomass to the soil C pools containing above- and below-ground detritus and coarse woody debris. For the period 1920–1969, the annual transfer averaged 57 Tg C yr^{-1}. This increased to 156 Tg C yr^{-1} in the last 20 years of the analysis. This recent addition of large quantities of decomposable material to the soil C pools affects the amount of C released through decomposition in the years following 1989.

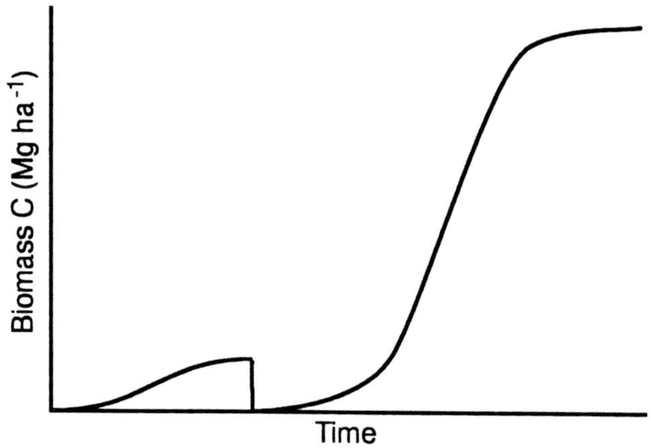

Fig. 3. The impact on stand biomass C dynamics of converting non-stocked stands to stands with high biomass productivity.

The retrospective analysis also indicated that the changes in disturbance regimes resulted in shifts in the forest age-class structure and in the average forest age. The average age of the boreal forest increased from 60.9 years in 1920 to 82.5 years in 1970, and subsequently decreased to 76.4 years in 1989 (Kurz and Apps, submitted). The recent large increase in the area annually disturbed decreased forest age and added recently-disturbed area to the youngest age class.

The rate of net biomass accumulation for the different stand growth curves in the model is age-dependent and is low during the initial stages of stand development, in particular during the 10-year regeneration delay assumed for natural regeneration (Figure 2). The recent increase in the area in the youngest age-class therefore also affects biomass accumulation in the future.

The annual net changes in biomass, soil, and total ecosystem C for the retrospective run (Kurz and Apps, submitted) and for the base run of the projective analysis, are shown in Figure 4. In the 50-year projective run, net ecosystem productivity (NEP, the net balance of ecosystem C uptake and losses) switches from a small C loss (source) in the first 20 years to an increasing gain (sink) in the last 30 years. The total net C accumulation over 50 years was 0.9 Pg C (Table II). The C release during the initial years is a consequence of the 20-year period of high disturbances from 1970–1989.

Reducing the rate of disturbances ("Low Fire" scenario) accelerates the rate at which the forests switch from a C source to a C sink (Figure 5). In contrast, doubling the area annually burned maintains the boreal forests as a C source for most of the 50-year period and results in a net loss of ecosystem C of 1.4 Pg C (Table II).

When compared to the base run, the scenario run in which all areas disturbed are planted shows greater C uptake throughout the 50-year period (Figure 6). The two scenarios in which we convert non-stocked to stocked stands with high growth rates

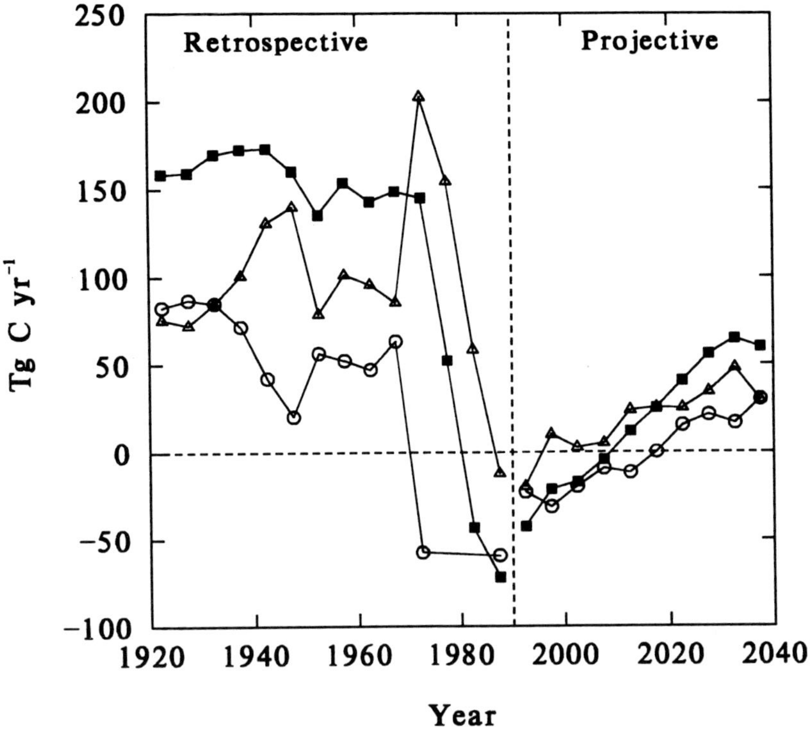

Fig. 4. Annual net changes in biomass, soil, and total ecosystem C pools for the retrospective analysis (Kurz and Apps, submitted) and for the projective base run. The lines show annual change in ecosystem C (■), biomass C (O) and soil C (Δ) pools.

TABLE II

Comparison of the change in ecosystem C of Canadian boreal forests for the six projective scenarios over 50 years (1990–2040).

Scenario	Change in ecosystem C (Pg C)
1. Base Run	0.86
2. Low Fire	2.3
3. High Fire	-1.4
4. Plant All	1.8
5. Conversion	0.92
6. Selective Conversion	9.2

have very different impacts on the forest C budget although the total non-stocked area converted in the two scenarios differs by only 11%. The conversion of 1% of all non-stocked area results in less C uptake than the base run for the first 30 years, and results in a total C uptake which is only slightly higher, at 0.92 Pg C (Table II). In contrast, converting 5% of the non-stocked area with less than 2 Mg C ha^{-1} yields more C uptake than the base run in all but the first 5-year time step. This scenario also has the highest total C uptake, at 9.2 Pg C. The impacts of site preparation and stand conversion on the biomass C pools and the subsequent release of C from the soil organic matter pools are such that, for some time, the ecosystem C losses exceed the gains from the improved stand growth. Thus, limiting stand conversion to those areas with low initial biomass (i.e., scenario six) greatly enhances the regional C uptake.

4. Discussion

Forest management policies aimed at increasing net C storage in forest ecosystems by controlling the rates of natural disturbances, accelerating the rate of regeneration, or converting non-stocked forest land to productive stocked forests, will require a sustained long-term commitment and investment. This analysis demonstrates an upper bound to additional atmospheric C that could be sequestered by Canadian boreal forest ecosystems under various scenarios. The actual net uptake of atmospheric C is likely to be smaller because: (1) this analysis does not account for the additional release of fossil C required to implement some scenarios; (2) this analysis assumes 100% success in all planting and site conversion projects; and (3) the ecological factors responsible for low initial stocking levels in some areas may also reduce the high growth rates assumed after conversion of these areas to stocked stands. Two main factors contribute to non-stocked status of sites: failure to regenerate following disturbance, and site constraints on tree growth (e.g., edaphic, climatic, or hydrologic factors). Scenarios five and six assume that site preparation and species selection can overcome site constraints in selected areas. The extent to which this is realistic cannot be assessed in the present analysis because the reasons for the non-stocked status are not included in the national biomass inventory (Bonnor, 1985).

One policy option which has not been addressed directly in this analysis is the afforestation of previously non-forested and marginal agricultural land, as has been suggested in a number of analyses (Dixon *et al.*, 1993). The difficulty in these assessments is that they ought to make an assumption about the contribution to the C budget of the land area prior to afforestation. Many studies have assumed that afforestation would occur on marginal agricultural land, with a zero net contribution to the C budget prior to afforestation. There is little agricultural land use in the Canadian boreal forest. For the afforestation scenarios in this analysis, we therefore chose non-stocked land that was already included in the national forest biomass inventory (Bonnor, 1985) and for which we can simulate biomass and soil C dynamics prior to afforestation.

The non-stocked stands included in the conversion contain little forest biomass. We therefore have assumed that none of this biomass C would be utilised for energy (e.g., substituting fossil energy) or forest products. Such uses would reduce the net C release to the atmosphere and further increase the effective carbon sequestration by the Canadian forest sector.

The results of the projective analysis demonstrate the importance of the system's memory and inertia: the 20-year period of high disturbances (1970–1989) has affected ecosystem conditions such that not one of the six scenarios assessed in this analysis was able to produce a net C sink in the period 1990–2000. Moreover, all scenarios show a common trend of increasing C uptake over the 50-year period. This further supports our hypothesis that, in the largely unmanaged boreal forest of Canada, natural forest dynamics will continue to be the dominant force determining future C budgets.

This analysis highlights the importance of disturbance regimes to the future C budget of the Canadian forest sector. We assumed that the 5-year average of disturbance rates will be constant throughout the 50-year period. In reality, year-to-year variations in disturbance regimes will introduce additional variation in boreal forest C fluxes. Periods with higher disturbance rates will generally result in reduced net C uptake. Large-scale outbreaks of insect populations periodically occur in the Canadian boreal forest (Blais, 1983; Volney, 1988). For example, during the 70 years of the retrospective analysis, outbreaks of spruce budworm occurred several times, particularly in the 1940s and in the 1970s. For the 50-year projective analysis, we assume an intermediate and constant level of insect activity. The likely reality will be periods of low insect activity interspersed by some major outbreaks. While both temporal patterns could generate similar cumulative fluxes over the 50-year period, the assumptions employed in this analysis do not capture the anticipated temporal variation in C fluxes associated with major insect outbreaks.

The scenarios explored here assumed that growth and decomposition rates of boreal ecosystems would remain unchanged for the next 50 years. We have previously demonstrated that the effects on the C budget of increased growth rates are opposite to those of increased decomposition rates and that the two should therefore not be assessed independently of each other (Kurz and Apps, 1994). Changes in environmental conditions in the boreal forest that enhance decomposition rates also result in higher nitrogen availability and therefore improved tree growth (Bonan,

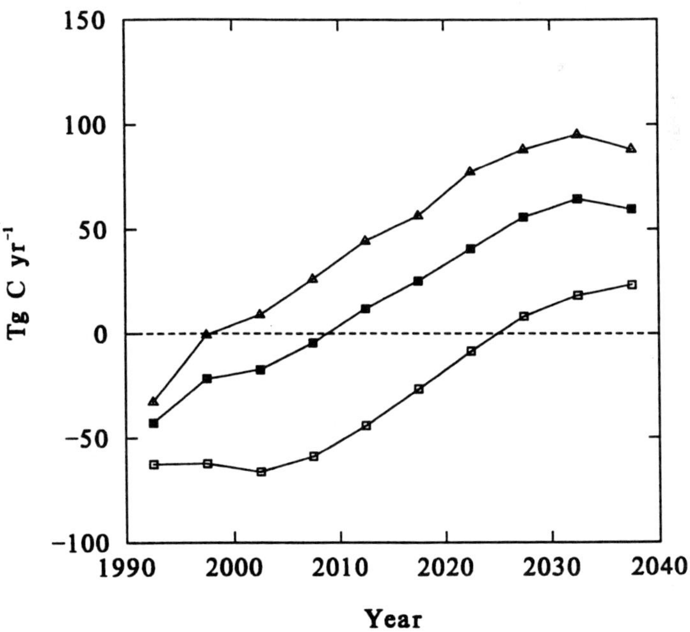

Fig. 5. Net ecosystem productivity for the base run (■), and the low (Δ) and high (□) fire scenarios.

1990). The combined net effect of the two changes, however, is difficult to project and could cause either increased or decreased C storage.

The impact of the six scenarios on the 50-year C budget of the Canadian boreal forest ranges from a net source of 1.4 Pg C (28 Tg C yr^{-1}) to a net sink of 9.2 Pg C (184 Tg C yr^{-1}) (Table II). A comparison of scenarios four, five and six emphasises that selective conversion is much more efficient than the two alternatives. The forest management and investment efforts required to achieve the largest potential C sink identified in this analysis are so intensive that the implementation of such a scenario is highly unlikely. It is important to note, however, that large-scale efforts would be required if boreal forest ecosystems were to offset the annual national fossil C emissions of ~114 Tg C in 1990 (Marland *et al.*, 1994) or ~5.7 Pg C for the 50-year period (assuming no change in annual emissions).

It has been demonstrated previously (Marland and Marland, 1992), that there is a large potential for affecting net emissions of fossil C by meeting some of society's energy needs through bioenergy derived from forest biomass. In our base run scenario, the annual forest harvest yields 14.4 Tg C. If all of this biomass was used for energy production, 7.4 to 15.6 Tg C from fossil sources could be substituted, depending on the assumptions about the energy content of wood and fossil fuels (Hall and Scurlock, 1993). Forest products, such as building materials, meet important needs of our society. Substituting these forest products with alternative materials, for example plastic, steel, or aluminium, has significant impacts on fossil C use elsewhere in the economy. Thus, the net impact on the regional C budget of using all forest biomass for bioenergy can only be assessed after also accounting for the fossil C cost of substituting the services previously provided by forest products.

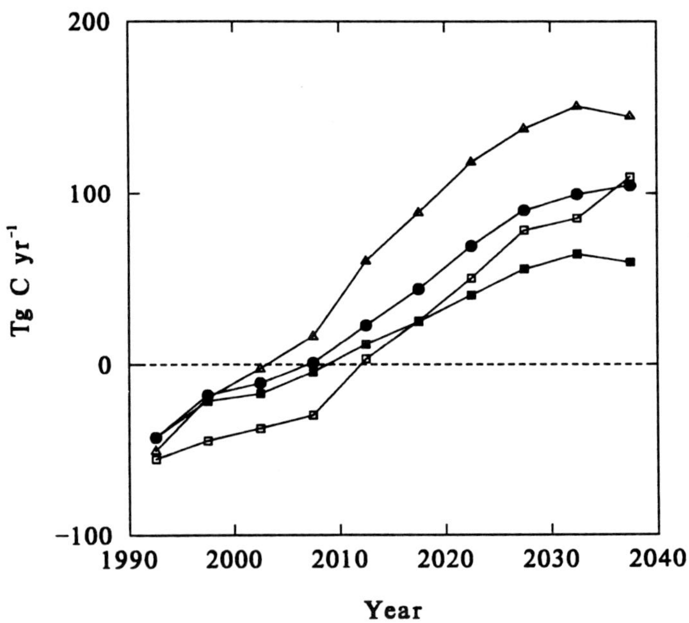

Fig. 6. Net ecosystem productivity for the base run (■), the scenarios that assume planting of all disturbed areas (●), converting 1% of all non-stocked area per year(□), and converting 5% per year (Δ) of all non-stocked area with less than 2 Mg C ha^{-1}.

Acknowledgments

This work was supported in part by the ENFOR (ENergy from the FORest) program of the Canadian Forest Service. We thank Dr. J. Greenough and two anonymous reviewers for comments on the draft manuscript and S. Beukema for executing the scenario analyses runs.

References

Apps, M. J., Kurz, W. A., Luxmoore, R .J., Nilsson, L. O., Sedjo, R. A., Schmidt, R., Simpson, L. G. and Vinson, T. S.: 1993a, *Water, Air, Soil Pollut.* **70**, 39-53.

Apps, M. J., Kurz, W. A. and Price, D. T.: 1993b, *In*: Vinson, T. S. and Kolchugina, T. P. (eds.), *Carbon Cycling in Boreal Forest and Subarctic Ecosystems*. U.S. EPA. Washington, EPA/600R-93/084, pp. 243-252.

Apps, M. J. and Kurz, W. A.: 1993, *In*: *Carbon Balance of World's Forested Ecosystems: Towards a Global Assessment*, Academy of Finland No. 3/1993, Helsinki, Finland, pp. 14-28.

Blais, J. R.: 1983, *Can. J. For. Res.* **13**, 539-547.

Bonan, G. B.: 1990, *Biogeochem.* **10**, 1-28.

Bonnor, G. M.: 1985, *Inventory of Forest Biomass in Canada.* Canadian Forestry Service, Petawawa National Forestry Institute. Chalk River, Canada, 63 pp.

Canadian Council of Forest Ministers: 1994, *Compendium of Canadian Forestry Statistics 1993.* National Forestry Database. Ottawa, Canada, 151 pp.

Dixon, R. K., Andrasko, K. J., Sussman, F. G., Lavinson, M. A., Trexler M. C. and Vinson, T. S.: 1993, *Water, Air, Soil Pollut.* **70**, 561-577.

Ecoregions Working Group: 1989, *Ecoclimatic Regions of Canada, First Approximation.* Ecological Land Classification Series, No. 23, Environment Canada, Ottawa, Canada. 119 pp. and map at 1:7,500,000.

Gullett, D. W. and Skinner, W. R.: 1992, *The State of Canada's Climate: Temperature Change in Canada 1895-1991.* Environment Canada, Atmospheric Environment Service, SOE Report No. 92-2., Ottawa, Canada, 36 pp.

Hall, D. O. and Scurlock J. M. O.: 1993, *In*: Hall, D. O., Scurlock, J. M. O., Bolhàr-Nordenkampf, H. R., Leegood, R. C. and Long, S. P. (eds.), *Photosynthesis and Production in a Changing Environment: A Field and Laboratory Manual*, Chapman and Hall, London, UK, pp. 425-444.

Houghton, J. T., Jenkins, G. J. and Ephraums, J. J.: 1990, *Climate Change: The IPCC Scientific Assessment.* Cambridge University Press, Cambridge, UK, 365 pp.

Kurz, W. A. and Apps, M. J.: 1993, *Water, Air, Soil Pollut.* **70**, 163-176.

Kurz, W. A. and Apps, M. J.: 1994, *Environ. Pollut.* **83**, 55-61.

Kurz, W. A. and Apps, M. J.: 1995, *In:* Apps, M. J., Price, D. T. (eds.), *Role of Forests, Forest Management in the Global Carbon Cycle.* Springer-Verlag, Heidelberg, Germany. (In press).

Kurz, W. A., Apps, M. J., Beukema, S. J., and Lekstrum, T.: 1995, *Tellus* **47B**, 170-177.

Kurz, W. A., Apps, M. J., Stocks, B. J. and Volney, W. J. A.: 1995, *In*: Woodwell, G. M. and Mackenzie, F. (eds.), *Biotic Feedbacks in the Global Climatic System: Will the Warming Feed the Warming?* Oxford University Press, Oxford, UK, pp. 119-133.

Kurz, W. A., Apps, M. J., Webb, T. M. and McNamee, P. J.: 1992, *The Carbon Budget of the Canadian Forest Sector: Phase I.* Forestry Canada, Information Report NOR-X-326, Northwest Region, Northern Forestry Centre, Edmonton, Canada, 93 pp.

Marland, G., Andres, R. J., and Boden, T. A.: 1994, *In*: Boden, T. A., Kaiser, D. P., Sepanski, R. J., Stoss, F. W. (eds.), *Trends '93: A Compendium of Data on Global Change.* ORNL/CDIAC-65. Carbon Dioxide Information Analysis Center, Oak Ridge National Laboratory, Oak Ridge, USA, pp. 505-584.

Marland, G. and Marland, S.: 1992, *Water, Air, Soil Pollut.* **64**, 181-195.

Price, D. T., Mair, R. M., Kurz, W. A. and Apps, M. J.: 1995, *In:* Apps, M. J., Price, D. T. (eds.), *Role of Forests, Forest Management in the Global Carbon Cycle.* Springer-Verlag, Heidelberg, Germany. (In press).

Smith, T. M., Cramer, W. P., Dixon, R. K., Leemans, R., Neilson, R. P. and Solomon, A. M.: 1993, *Water, Air, Soil Pollut.* **70**, 19-37.

Volney, W. J. A.: 1988, *Can. J. For. Res.* **18**, 1152-1158.

A SYSTEM FOR EVALUATION OF GROWTH AND MORTALITY IN RUSSIAN FORESTS

A. SHVIDENKO, S. VENEVSKY, G. RAILE and S. NILSSON

International Institute for Applied Systems Analysis, A-2361 Laxenburg, Austria

Abstract. A modelling system (MS) for evaluation of current increment (net and gross growth) and mortality in Russian forests is presented. The informational basis of the MS are the data of the Russian forest state account (growing stock by dominant species, age, site indices, stocking) and some additional indicators (types of age structure, average species composition, type and intensity of disturbance regimes by ecoregions and/or landscapes. The MS includes model blocks generated with available information including yield tables (general, regional, full-stocked, with different stocking and types of age structure), growth tables, models, and experimental data for managed and unmanaged, disturbed and undisturbed forests. Two-dimensional (by age and site index) and three-dimensional (age, site index and stocking) models by species and ecoregions, based on regularities of coefficients of Mitscherlich growing functions, were used as the nuclei of the MS. The preliminary results, received for several Russian forest forming species gave a good fit to experimental data.

Keywords. NET GROWTH, GROSS GROWTH, MORTALITY, PRODUCTIVITY, RUSSIAN FORESTS, YIELD TABLES, RUSSIA.

1. Introduction

Russia has a long tradition and outstanding results in investigations of growth (increment) of forest stands, and the results have been employed as a basis for the theory development on forest stand productivity and forest management (Turskii, 1925; Orlov, 1926, 1928; Tovstoles, 1938; Matveev-Motin, 1962; Tjurin, 1945; Naumenko, 1941; Dvoretzkij, 1964; Zakharov, 1966; Nikitin, 1966; Antanaitis and Sagreev, 1981). A large number of scientific conferences on increment studies, held in the 1970s in the former USSR (Anonymous, 1967a, 1968, 1972, 1975), kept momentum in the area of forest research in the former USSR.

However, for each primary inventory unit, the indicators of growth (net and gross) are not measured by the current forest inventory system. The existing instructions do not require direct measurements of growth in the field inventory. The increment estimation is limited to net increment calculations for periods between inventories and are aggregated for all species by the forest enterprises inventories (Anonymous, 1986, 1991; Goscomles SSSR, 1990, 1991). Thus, the so called average increment is presented in the form of an accumulative characteristic of the growing stock. The official estimates on net or gross increment for the total Russian forests are not reported. Different expert estimates exist, and vary from 500 to 1500 million m^3 yr^{-1} (all wood volumes are stem wood overbark for trees taller than breast height) per year for the net growth and from 1400 to 2600 million m^3 yr^{-1} for the gross growth for all forests in Russia (V.V. Sagreev, 1991, pers. comm.; Shvidenko, 1991). Official data on the average increment for forested areas (see the section Definitions) under state forest management (676.2 Mha or 76.9% of all Russian forested areas), was reported in 1988 to be 822.5 million m^3 of which 268.0 million m^3 were located in European Russia (Goscomles SSSR, 1990, 1991). Natural mortality is estimated to vary from 400 to 1200 million m^3 yr^{-1} (Shvidenko, 1991; V.V. Sagreev, 1991, pers. comm.; Kusmitchev, 1991).

Water, Air and Soil Pollution **82**: 333–348, 1995.

The necessity of improved growth calculations, as one of the most important indicators on forest productivity and for implementation of a rational forestry management, is recognised in Russia. Additionally, the net and gross growths are important components of the global carbon (C) cycle and also as an integrated indicator of sustainable development.

Current revisions of the forest state inventory system in Russia (Strakhov *et al.*, 1994) suggest an improvement of the evaluation of the state and the productivity of the forests, by carrying out direct measurements in the field of the growth by the inventory crews. However, the implementation of a new inventory system in Russia, with respect to the vast area, will require at least 20 years.

At this stage, there are possibilities to carry out aggregated estimates on the net and gross increment, for relatively large territories in Russia, using data from the Forest State Account combined with specific information from models of forest growth developed in Russia. Such an approach leads to the design of a modelling system for increment evaluation with elements of expert systems, and in combination with databases. Although the proposed system is designed to generate the "large picture" of forest growth, specifically for general ecological conclusions (e.g., for productivity or C budget evaluation), it is also obvious that the system can provide useful information necessary for the development of an efficient future Russian inventory system.

2. General Structure of the Modelling System

There exists numerous yield tables for different forest types (so called normal and modal, over-all (general) and regional, pure and mixed stands, etc.) for the major forest forming species in Russia, as well as a number of different models designed for increment estimation. These models have been approved officially by the former State Committee on Forest in the USSR, and published in special reference books to be used by the forest inventory (Sagreev *et al.*, 1992; Moshkalev, 1984; Shvidenko *et al.*, 1987; Voinov, 1987; Korjakin, 1990; Gusev, 1993). The basic objective from our work has been to generate a computer system which would make it possible to estimate the gross and the net increments and the natural mortality, by employing the models and tables mentioned above, supplemented and verified with data from the forest inventory and direct measurements on test plots. The modelling system (MS) uses both aggregations of data from individual stands and aggregated data from the forest state account.

The system developed is able to estimate the gross and net increment, as well as the natural mortality for the major forest species, and for ecoregions which are defined and employed by IIASA's Siberian Forest Study. The system is also capable with the help of additional auxiliary models, to estimate the Net Ecosystem Productivity (NEP). A simplified structure of the system developed is shown in Figure 1.

2.1. Basic Principles for the System Development

The modelling system is developed as a system of spatial and parameter extension. It means that (a) for any type of initial data and for different locations in Russia, the MS can chose a way to calculate gross and net increments and natural mortality, with the employment of regional models, in order to minimise errors; and (b) the MS can be

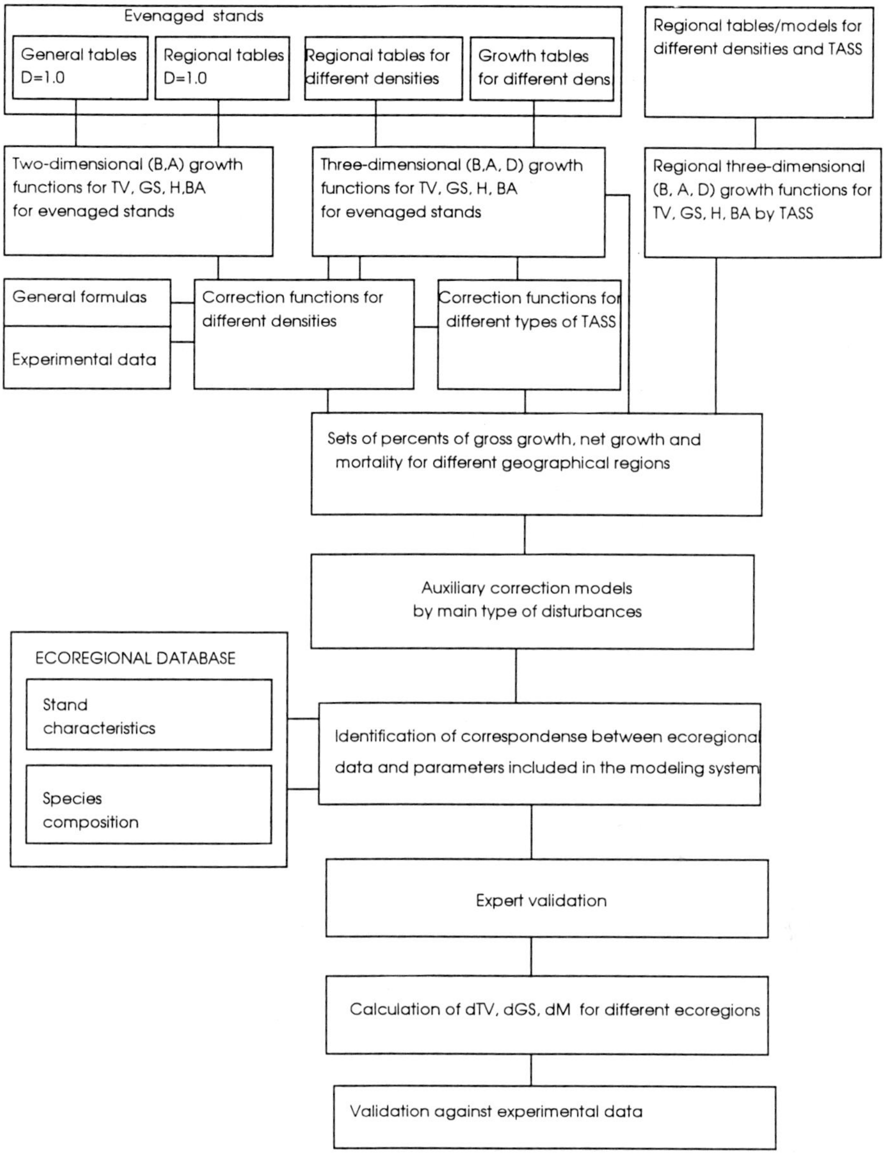

Fig. 1. Simplified block-diagram of the modelling system for evaluation of growth and mortality in Russia.

implemented as a component of a geographical information system (GIS) and knowledge databases.

The basic mathematical principle of the MS is the use of a set of growth functions (Shvidenko and Juditsky, 1983; Kiiviste, 1988) based on different stand characteristics. The growth functions employed are simple and aggregated. These kinds of functions are able only to describe the growth processes of even-aged stands up to a certain age.

2.2. The Objects of the Modelling

The final objects for estimation are sets of sample stands of dominant forest species from different ecoregions. The ecoregions have been defined by using the following criteria:

1. the biospheric features and biogeochemical fluxes and pools should be of comparable orders in the selected ecoregions;
2. each ecoregion should have a relative climatic homogeneity in order to describe the main forest species in terms of average productivity and other average characteristics;
3. mountainous and plain areas should be divided into separate ecoregions;
4. permafrost regions should be treated separately;
5. the level of anthropogenic impact or disturbances should be of similar magnitude in each ecoregion;
6. the division into ecoregions should coincide with the borders of the administrative territories.

According to these criteria, 161 ecoregions have been defined for Russia. Sizes of ecoregions are different in European Russia (average total area is 4.4 Mha, forested area is 1.7 Mha) and in Siberia and Far East (average total area is 20.3 Mha, forested area is 9.6 Mha).

2.3. Initial Data for Growth Evaluation

The minimal information set of attributes for modelling of the growth includes, at regional level, species, age, site index, density, age structure of stands, regime of forest management (regularly thinned or unmanaged stands), and for specified stands, the radial increment (Nikitin and Shvidenko, 1976; Antanaitis and Sagreev, 1981).

The Forest State Account (FSA) contains the distribution of areas over growing stock by age and site indices, and the densities aggregated by dominant species within a specific utilisation category for each forest enterprise in Russia.

The FSA does not contain the distribution of forests over the types of structures of stand age. However, this information is very important for growth estimation. A unified classification scheme has been developed for age structure based on different Russian publications (e.g., Shanin, 1965, 1967; Semetchkin, 1963, 1970):

1. Even-aged stands. In which 70% and more of the growing stock is constituted by trees belonging to one age class (the age classes for coniferous in Russia and for hard deciduous species of seed regeneration is 20 years (for cedar - 40 years). For the other species with some exclusions is 10 years). The variation coefficients for diameter frequency distribution of age is $Va < 8\%$ and for tree diameters is $Vd < 28\%$;
2. Relative uneven-aged stands. In which 70% or more of the growing stock is constituted by trees belonging to more than one, but within two, age classes ($Va = 8 \ldots 15\%$, $Vd = 28 \ldots 35\%$);
3. Uneven-aged stands. The growing stock constituted by trees of more than two, and usually of all, age groups (young, middle-aged, premature, mature, and overmature). ($Va > 15\%$, $Vd > 35\%$);
4. Gradually uneven-aged stands. The growing stock is comprised by trees of more than one canopy layer in which dominant species belong to different age groups.

Numerous investigations (e.g., Gusev, 1964; Shanin, 1967; Anonymous, 1967b; Verkhunov, 1975; Semetchkin, 1973; Svalov 1968, 1979; Sokolov *et al.*, 1993) report that an essential part of the taiga forests of Russia are uneven-aged. Data compiled from available data by ecoregions (collected by Sedykh, Sokolov, Efremov, Khomentovsky, in 1994, under the auspices of the IIASA Siberian Forest Study) show that in Siberia and the Far East, about 40% of the coniferous forests belong to uneven-aged and gradually uneven-aged groups.

The FSA usually presents the collected information in the form of two-dimensional tables. This simplification means that model applications will decrease the accuracy of the data. The only way to improve the accuracy conditions is to employ multidimensional tables for coniferous, hard deciduous and soft deciduous species, which provide the distribution of areas by ages, densities and site indices. Therefore, a special algorithm for preparation of multidimensional tables of the FSA data, for growth estimation, is required.

3. Structure and Current State of the Modelling System

The system, which is under development, includes a hierarchical and spatially distributed set of modules for:

1. Fully stocked stands, both general and aggregated for ecoregions.
2. Productivity under different stocking.
3. Regional models and correction coefficients based on observations in real stands.
4. Auxiliary models of disturbance regimes impact (at this stage we only consider influences of main types of air pollutants).

The modelling activity so far has been oriented with available information on species composition, age, site indices and density, and basic models were developed for even-aged stands. A set of auxiliary models modify, the growth of stands for the different types of age structure mentioned above. Aggregated diameter distributions for trees and aggregated schemes of mortality, dependent on species and climatic zones, are employed in the models.

The Mitscherlich growth functions were chosen as the basic mathematical functions for implementing experimental data, yield tables, and model outputs into a simple, but adequate modelling system. Height, diameter, basal area, growing stock, total volume and their derivatives, such as annual gross increment, annual net increment and natural mortality, are dependent on the stand age A and three different coefficients in the Mitscherlich functions (c_1, c_2, c_3). For example, total volume, *TV*, can be calculated as:

$$TV = c_1 \cdot (1 - \exp(-c_2 \cdot A))^{c_3} \qquad [1]$$

The coefficients represent different biological processes (c_1 is the maximal value of a growth function (the asymptote), $c_1 c_2 (1 - 1/c_3)^{(c_3 - 1)}$ is the maximal increment value (slope of the curve), and $\ln(c_3)/c_2$ is the turning point of a growth function).

With the estimates of the coefficients of the Mitscherlich functions, there are possibilities to calculate annual increments and percentages of annual increments. For example, for the gross increment, and percentage of gross increment, P_{TV}, we get:

$$dTV = c_1 c_2 c_3 (1 - \exp(-c_2 A))^{(c_3 - 1)} \exp(-c_2 A) \qquad [2]$$

and

$$P_{TV} = c_2 \cdot c_3 \cdot \exp(-c_2 \cdot A) / (1 - \exp(-c_2 \cdot A)) \qquad [3]$$

The difference between *dTV* and net increment, *dGS*, both calculated as the derivatives from the Mitscherlich functions, gives us the natural mortality, *dM*:

$$dM = dTV - dGS \qquad [4]$$

The coefficients of the Mitscherlich functions vary for different forest species and geographical locations.

The implementation of the experimental data is carried out using the following approach:

1. Firstly, a non-linear regression with the Levenberg-Marquart method is executed in order to evaluate coefficients of the Mitscherlich functions by age, site index and density variations. The computer exercises have shown that the estimation of the exponential coefficient is particularly important, as all the functions designed are very sensitive to variation of this coefficient.
2. Then, quadratic regressions give us a statistical dependence of the three coefficients by site indices. This stage allows us to design a two-dimensional growth function for our stand characteristic by age and site index. There are possibilities to use other non-linear expressions for the regression calculations. However, more sophisticated regressions can generate considerable calculation errors and do not agree with the simplicity and intergrality of the system developed. The analysis so far showed that linear regressions cannot be used due to the low accuracy of the results, while quadric regressions seem to give adequate results.
3. The next step is to design a three dimensional growth function by age, site index and density, using the results obtained by the first two steps and discussed above. From the experimental data, the density can be calculated by using basal areas from fully-stocked stands evaluated in the previous steps. However, usually the density variations for a separate region are very small and one is forced to find separate individual methods for constructing the final three-dimensional approximated surface. From a mathematical point of view, this is the problem of the multidimensional approximation of the plane with irregular points by an analytical function. Such a problem can only be solved in some cases with a "good" configuration of experimental data sets. Nevertheless, one can try to design such a function by using additional data from other locations and projections from already designed three-dimensional growth planes for similar forest species. Computer analyses for two coniferous species (larch and pine) allow to state that such manipulations can be done within in the limits of the accuracy of the function coefficients.
4. Finally all growth functions designed are evaluated by experts with respect to the stand variables employed in the functions (height, basal area, diameter, growing stock, total volume, gross and net increments, the consequent percentage, natural mortality). The iterative procedure described can be restarted if the results are not

explicit enough or are in contradiction with for results from the forest taxation (inventory).

The basic functional relationships in the whole modelling system are based on a variety of fully stocked yield tables for the major forest species. Each coefficient of the Mitscherlich function (as mentioned above) is presented in the form of a quadric function of the site index N (N = 1 to 7 for site indices from Ia to Va in line with the Russian site classification system).

For example, the equation for larch (fully stocked stands), the growing stock for the former USSR, totally calculated as a two dimensional non-linear regression function, is the following:

$$GS(A,N) = (1.882 \cdot N^2 - 122.287 \cdot N + 931.045) \times (1 - \exp(-(0.044 \cdot N^2 - 0.666 \cdot N + 3.859)/100 \cdot A))^{((0.468 \cdot N^2 - 4.377 \cdot N + 26.829)/10)} \quad [5]$$

Fully stocked stands at the regional level are described by the same growth functions employing the two variables, age, A, and site index, N. However in addition, sets of polynomial correction functions for the Mitscherlich coefficients are also employed, depicting the local density variations. Calculations for regional fully stocked and non fully stocked stands for the territory of the former USSR showed that the methodology chosen generates adequately sound results.

For stands with different density (stocking), d, the growth functions can be proposed as three-dimensional functions, where each of the three coefficients of the Mitscherlich function will be dependent on site index N and density d (d has a range from 1.0 to 0.3). The simplest form of such functions for coefficient estimations are systems of polynomial non-linear equations. For example, for the calculation of the total volume for pine, these coefficients are estimated to be:

$$c_1 = 4.174N^2 - 66.982N - 193.159d^2 + 1778.316d - 174.997dN + 207.467 \quad [6a]$$

$$c_2 = \left(0.034N^2 - 0.430N + 0.362d^2 - 2.136d + 0.113dN + 4.285\right)/100 \quad [6b]$$

$$c_3 = \left(0.389N^2 - 2.872N + 8.230d^2 - 12.963d + 0.776dN + 27.684\right)/10 \quad [6c]$$

Thus, a computer system has been developed with a set of multidimensional matrixes of coefficients determining the growth functions as elements of an interactive software system allowing to evaluate gross and net increment and natural mortality by a large number of ecoregions of Russia with input data from the Russian forest inventory.

4. Some Preliminary Results of Modelling

For the current stage of the MS development, the results obtained so far will be summarised in the following. The general models for even-aged stands, including the parameter height, diameter, basal area, growing stock and total volume for the major forest forming species (pine, spruce, larch, oak by classes of origin, beech, aspen, black alder), have been constructed. The upper limits of the modelling indicators have been taken from general yield tables. In Table I, a set of nine coefficients for two-dimensional growth functions, for the total volume estimate of different species are presented as an example.

Table I shows that the set of the three coefficients, determining the maximal total volume, has the most predictable behaviour (alder being an exception), while other coefficients have wide variation and may have different signs.

All of the calculations done for general tables and aggregated increment models have a good accuracy. These tables were prepared by using an averaging procedure for a large sets of local tables and widely distributed experimental data. The results of the calculations for the growing stock of larch are presented in Figure 2 as an example.

The analyses of the local fully-stocked tables, which do not exist for all regions, showed that there are possibilities to generate sufficient models for the growth indicator estimation. Such tables reflect the most representative stands for the regions investigated, and consequently also reflect stands with different age structures (TASS).

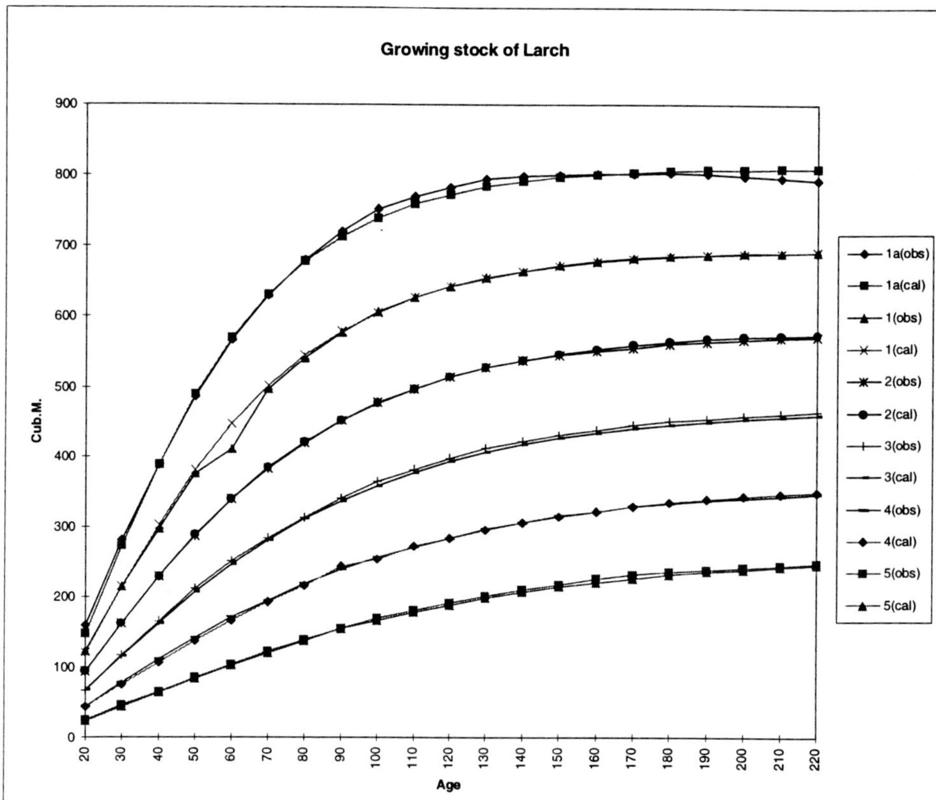

Fig. 2. Growing stock of larch for the FSU, observed and calculated from two-dimensional growth function. Odd-numbered series belong to yield table values, whereas the even-numbered series belong to calculated values, for site indices 1a, 1, 2, 3, 4 and 5, respectively.

TABLE I

Sets of volume growth function coefficients[a] for different species from territory of the former USSR.

Species	C_{11}	C_{12}	C_{13}	C_{21}	C_{22}	C_{23}	C_{31}	C_{32}	C_{33}
Pine	5.65	-261.2	2095.5	-0.0001	0.0018	-0.0223	0.0239	-0.160	2.16
Spruce	1.87	-299.2	2597.8	0.00001	0.0003	-0.0218	0.0123	0.077	2.71
Oak H[b]	13.38	-288.3	1200.1	0.0002	-0.0026	-0.0184	0.0088	0.073	1.58
Oak L[b]	8.41	-306.8	2021.2	-0.0001	0.0007	-0.0172	0.0128	-0.022	1.81
Beech	28.25	-489.2	2473.2	0.0000	-0.0005	-0.0148	-0.0050	0.200	1.32
Birch	7.49	-187.2	1080.4	0.00002	-0.0002	-0.0219	-0.0002	0.216	1.35
Aspen	13.26	-230.3	1206.9	-0.0001	0.0011	-0.0319	0.0091	-0.052	1.95
Alder	-93.60	313.9	736.7	-0.0039	0.0230	-0.0490	0.1000	0.564	2.23

a *TV*, total volume, is given by the equation

$$TV = c_1 \cdot (1 - \exp(c_2 \cdot A))^{c_3}$$

where *A* is stand age, and the coefficients c_1, c_2 and c_3 are given by

$$c_1 = c_{11} \cdot N^2 + c_{12} \cdot N + c_{13}$$

$$c_2 = c_{21} \cdot N^2 + c_{22} \cdot N + c_{23}$$

$$c_3 = c_{31} \cdot N^2 + c_{32} \cdot N + c_{33}$$

where *N* is site index.

b Oak H, Oak L indicate high and low productivity oak, which correspond generally to stands of seed and vegetative origin, respectively.

A complete algorithm using data for regional fully stocked stands and a three-dimensional dependency for the total volume estimate (and for the gross increment as well as for the natural mortality) by age, site index and density, has so far been developed for larch and pine, which are the most common forest species in the former USSR.

A three-dimensional dependency for the total volume estimate by age, site index and density has primarily been estimated for pine. The table of gross increment for different densities of pine (Sagreev *et al.*, 1992) has been used for statistical estimation of the non-linear regression coefficients in the expression:

$$dTV = c_1 \cdot c_2 \cdot c_3 \cdot (1 - \exp(-c_2 \cdot A))^{(c_3 - 1)} \cdot \exp(-c_2 \cdot A) \qquad [7]$$

Hence, we obtained complex two-dimensional surfaces, describing the dependence of each Mitscherlich function coefficient by site index and density. The surfaces were approximated by paraboloids generated by the expressions shown above. Figure 3, illustrates the approximation of the three coefficients c_1, c_2 and c_3 for the growth function on total volume for pine.

Given approximations for the coefficients, total volume, gross increment and the percentage of gross increment for every age, the density and site index can be calculated. The gross growth for site index 1a (observed and calculated), the total volume and the percentage of gross growth by age and density, are illustrated in

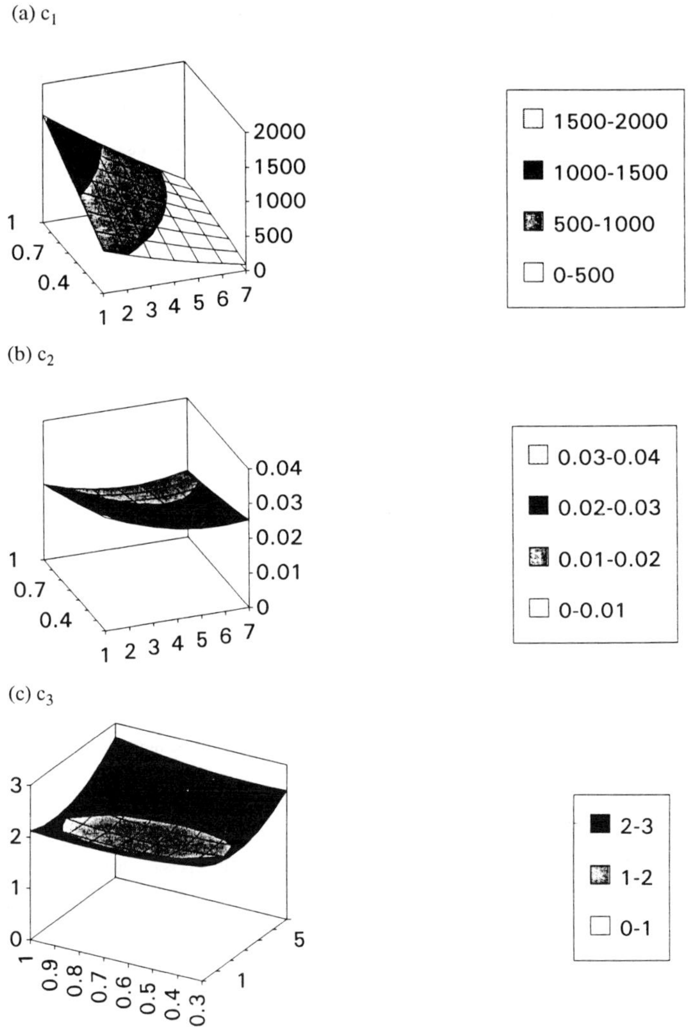

Fig. 3. Calculated coefficients for a 3-dimensional growth function for total volume of pine in former USSR.

Figures 4 and 5. It is interesting that despite a complex non-linear character of the three-dimensional growth function for total volume, gross increment and percentage of gross increment, there is practically a linear dependence for density by age and site index.

The two-dimensional growth function for fully- and partially-stocked stands of larch was prepared, based on local yield tables (Far East, Magadan region and Angara river basin), and taking into account data from the different Siberian ecoregions. Unfortunately, availability of local data on the volume variation over age is limited. However, we constructed a three dimensional dependence of total volume by age, density and site index, based on the local tables available and by using estimates on the coefficients for Pine. By this, there is the possibility to estimate the total volume, the gross increment and the natural mortality for larch at local Siberian conditions.

TABLE II

Calculated net growth (m^3 yr^{-1}) of partially stocked larch stands in the Magadan region for site indices 3, 4 and 5.

Age	3	4	5
10	2.51	1.28	0.87
30	2.05	1.75	1.08
50	1.62	1.52	0.91
70	1.27	1.17	0.69
90	0.99	0.85	0.50
110	0.77	0.60	0.35
130	0.59	0.42	0.25
150	0.46	0.29	0.17
170	0.35	0.20	0.12

TABLE III

Calculated growing stock (m^3) of partially stocked larch stands in the Magadan region for site indices 3, 4 and 5.

Age	3	4	5
10	25.3	8.0	5.7
30	71.0	40.4	26.4
50	107.6	73.5	46.5
70	136.4	100.5	62.6
90	158.8	120.7	74.5
110	176.2	135.1	83.0
130	189.8	145.3	88.9
150	200.2	152.2	93.0
170	208.3	157.0	95.8

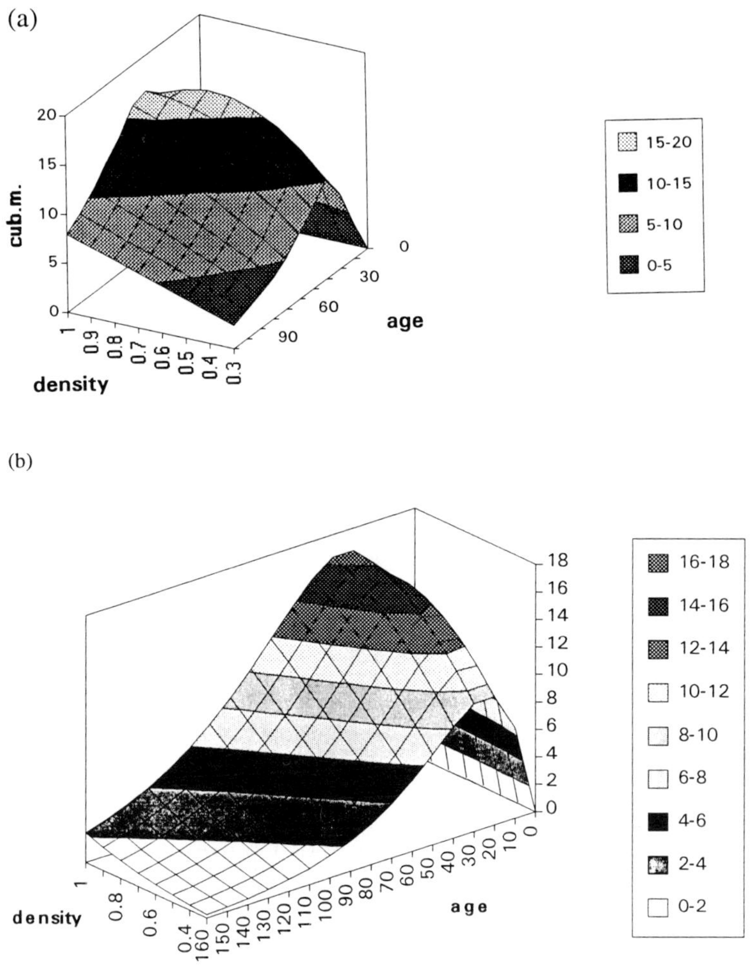

Fig. 4. Gross increment of pine at site index 1a: (a) observed, (b) calculated.

(a)

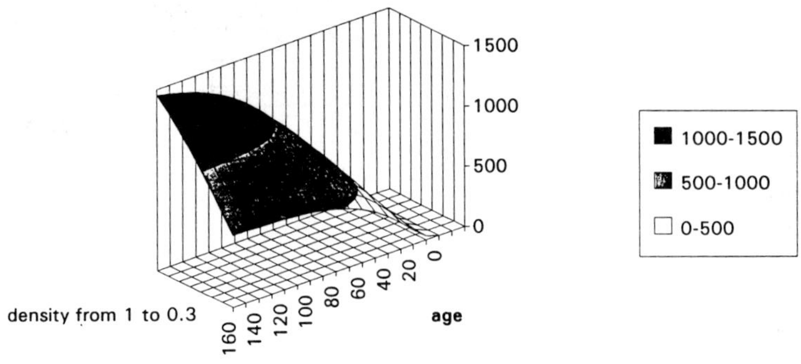

(b)

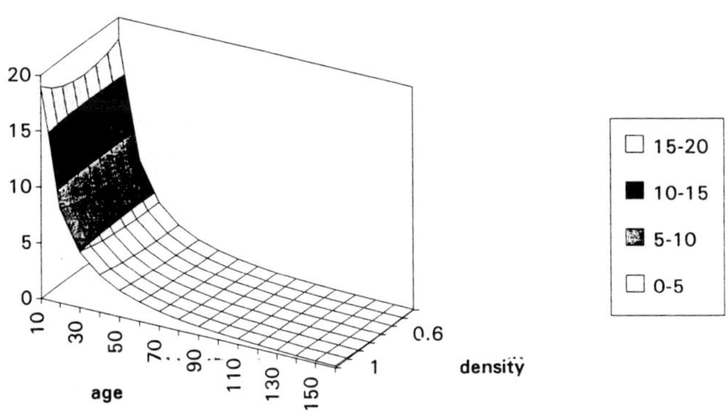

Fig. 5. Total volume (a) and percentage of gross increment (b) of pine at site index 1a.

TABLE IV

Total volume (m^3) of partially stocked larch stands in the Magadan region for site indices 3 and 4. Observed respectively calculated by a three-dimensional function, dependent on age, site index and density.

(a) Observed

Age	3	4
10	33	23
30	128	75
50	227	140
70	294	202
90	347	249
110	392	291
130	429	321
150	464	342
170	481	353

(b) Calculated

Age	3	4
10	36.0	15.3
30	131.2	80.0
50	213.9	151.1
70	292.2	205.2
90	351.5	243.6
110	395.8	280.6
130	431.3	309.8
150	457.9	328.9
170	478.9	337.5

TABLE V

Calculated gross increment (m^3 yr^{-1}) of partially stocked larch stands in the Magadan region, for site indices 3 and 4.

Age	3	4
10	4.57	2.44
30	4.68	3.69
50	3.90	3.48
70	3.28	2.76
90	2.61	2.07
110	2.02	1.60
130	1.56	1.23
150	1.20	0.91
170	0.91	0.65

TABLE VI

Calculated natural mortality (m^3 yr^{-1}) of partially stocked larch stands in the Magadan region, for site indices 3 and 4.

Age	3	4
10	2.06	1.15
30	2.63	1.95
50	2.28	1.95
70	2.01	1.59
90	1.62	1.22
110	1.25	1.00
130	0.97	0.81
150	0.74	0.62
170	0.56	0.46

Tables II, III, IV, V and VI show a complete example of the calculations for the Magadan region for two site indices. It can be concluded that there are considerable differences between the calculated growing stock for this region and the available general tables.

5. Conclusions

The analyses done so far show that the proposed modelling system could generate quite satisfactory results and allow to estimate the gross increment, the net increment and the mortality in Russian forests. The next stages of research will deal with the finalization of the set of three dimensional growth functions for non-fully stocked stands for all of Russia, identification of the parameters for the auxiliary models on the growth transformation for different TASS, development of a matrix of the correction coefficients for mixed forest, and validation of the estimates against measurements on about 10 000 sample plots in the territories of the former USSR.

References

Antanaitis, V. V. and Sagreev, V. V.: 1981, *Growth of Forest*. Forest Industry, Moscow, USSR, 201 pp. [In Russian.]

Anonymous: 1967a, *Problems of Forest Increment in the Forest Inventory and Planning*. Kaunas, USSR, 270 pp. [In Russian.]

Anonymous: 1967b, *Uneven-aged Forests of Siberia, Far East and Ural*. Institute of Forest and Wood, Siberian Division, Academy of Sci. of the USSR, Krasnojarsk, USSR 278 pp. [In Russian.]

Anonymous: 1968, *Materials of the All-Union Scientific Conference on Problems of Dendrochronology and Dendroclimatology*. Vilnius, USSR, 247 pp. [In Russian.]

Anonymous: 1972, *Current Increment of Stands and Its Application in Forestry*. Riga, USSR, 243 pp. [In Russian.]

Anonymous: 1975, *Current Increment of Stands*. Short reports of the All-Union Scientific Conference. Minsk, USSR, 287 pp. [In Russian.]

Anonymous: 1986, 1991, *Instructions on the Forest Inventory and Planning in the State Forest Fund of the USSR*. The State Committee on Forest of the USSR, Moscow, USSR, Part I (1986) 133 pp., Part II (1991) 327 pp. [In Russian.]

Dvoretzkij, M. L.: 1964, *Growth of Trees and Stands*. Forest Industry, Moscow, USSR, 162 pp. [In Russian.]

Goscomles SSSR: 1990, 1991, *Forest Fund of the USSR*. State Committee on Forest of the USSR, Moscow, USSR, **1**, 1005 pp, **2**, 1020 pp. [In Russian.]

Gusev, I. I.: 1964, *Structure and Peculiarities of Inventory in Spruce Stands of the North*. Forest Industry, Moscow, USSR, 75 pp. [In Russian.]

Gusev, N. N. (ed): 1993, *Forest Inventory Reference Book for Central and Southern Regions of European Russian Federation*. Russian Forest Federal Service, Moscow, Russia, 418 pp. [In Russian.]

Kartashov Ju., G. (ed): 1986, *Normatives for Forest Inventory in Sakhalin and Kamchatka*. Far Eastern Forestry Inst., Jusno-Sakhalinsk, USSR, 814 pp. [In Russian.]

Kiiviste, A. A.: 1988, *Growth Functions for Forests*. Estonia Agricultural Academy, Tartu, USSR, 171 pp. [In Russian.]

Korjakin, V. N. (ed): 1990, *Forest Inventory Reference Book for the Far East*. Far Eastern Forestry Inst., Khabarovsk, USSR, 526 pp. [In Russian.]

Kusmitchev, A. S. (ed): 1991, *Impact of Climate Change on Structure, State and Dynamics of Forests of the USSR*. Unpublished Report, Moscow, USSR, **1-3**. [In Russian.]

Matveev-Motin, A. S.: 1962, *Growth and Productivity of Forests*. Forest Industry, Moscow, USSR, 188 pp. [In Russian.]

Moshkalev, A. G. (ed): 1984, *Forest Inventory Reference Book for the Northwestern USSR*. Forest Technical Academy, Leningrad, USSR, 319 pp. [In Russian.]

Naumenko, I. M.: 1941, *Growth of Forests on Watershed Zones*. Voronjesh, USSR45 pp., [In Russian.]

Nikitin, K. E.: 1966, *Larch in Ukraine*. Uroshaj Publ., Kiev, USSR, 331 pp. [In Russian.]

Nikitin, K. E. and Shvidenko, A. Z.: 1976, *Methods and Techniques of Forest Information Handling*. Lesnaja prom. Publ., Moscow, USSR, 286 pp. [In Russian.]

Orlov, M. M.: 1927-1928, *Forest Inventory and Forest Management*, **1**, 428 pp; **2**, 326 pp; **3**, 348 pp. Books publ. by the Journal, Forest Management and Forest Industry, Leningrad, USSR. [In Russian.]

Sagreev, V. V., Sukhikh, V. I., Shvidenko, A. Z., Gusev, N. N. and Moshkalev, A. G.: 1992, *All-Union Standards and Normatives for the Forest Inventory*. Kolos Publication, Moscow, Russia, 495 pp. [In Russian.]

Semetchkin, I. V.: 1963, *Structure of Uneven-aged Cedar Stands and Specific Features of their Measurement*. In: Materials on the Study of the Forests in Siberia and the Far East. Siberian Division, Academy of Science of the USSR, Krasnojarsk, USSR, pp. 217-224. [In Russian.]

Semetchkin, I. V.: 1970, *Dynamic of Stand Age Structure and Methods of its Investigation*. In: Problems of Forestry, **1**, Inst. of Forest and Wood, Siberian Division, Academy of Science of the USSR, pp. 422-445. [In Russian.]

Shanin, S. S.: 1965, *Structure of Pine and Larch Stands in Siberia*. Krasnojarsk, USSR, Forest Industry, Moscow, Russia, 104 pp. [In Russian.]

Shanin, S. S.: 1967, *Management of the Structure of Pine and Larch Stands in Siberia*. Krasnojarsk, USSR, 40 pp. [In Russian.]

Shvidenko, A .Z., Savich Ju., N., Strochinskij, A. A. and Kashpor, S. N.: (eds): 1987, *Forest Inventory Reference Book for Ukraine and Moldavia*. Uroshaj, Kiev, Ukraine, 559 pp. [In Russian.]

Shvidenko, A. (ed): 1991, *Concept of a Forest Monitoring System for the USSR*. Moscow, VNIIZ Lesresurs, Unpublished Manuscript, USSR, 35 pp. [In Russian.]

Shvidenko, A. and Juditsky Ja., A.: 1983, *Application of Non-linear Functions to Investigations of Growth Processes*. Unr. Agr. Academy, Kiev, USSR, 12 pp. [In Russian.]

Sokolov, V. A., Atkin, A. S., Ziganshin, R. A., *et al.*: 1993, *Structure and Growing of Stands in Siberia*. Inst. of Forest and Wood, Sub. Div., Russ. Acad. Sci., Krasnojarsk, USSR, 173 pp. [In Russian.]

Strakhov, V. V., Filipchuk, A. N. and Shvidenko, A. Z.: 1994, *Development of a New System for the Forest Inventory in Russia*. Moscow, Russia, Unpublished Manuscript.

Svalov, N. W.: 1963, *Basic Problems of Forest Management and Forest Use*. Gosleskomizdat, Moscow, USSR, 208 pp. [In Russian.]

Svalov, N. W.: 1979, *Modelling of Productivity of Stands and Theory of Forest Use*. Forest Industry, Moscow, Russia, 216 pp. [In Russian.]

Tjurin, A. V.: 1945, *Forest Taxation*. Goslestekhizdat, Moscow, USSR, 376 pp. [In Russian.]

Tovstoles, D. I.: 1938, *A New Method for Forest Growth Estimation*. Kiev, USSR, 78 pp. [In Russian.]

Turskii, S.: 1925, *Essays on Increment Theory*. Novaj Derevnja, Moscow, USSR, 147 pp. [In Russian.]

Verkhunov, P. M.: 1975, *Variety and Interconnections of Taxation Indicators in Uneven-aged Pine Stands*. Nauka Publ., Novosibirsk USSR, 205 pp. [In Russian.]

Voinov, G. S. (ed): 1986, *Forest Inventory Reference Book for the North East of European Russia*. Institute of Forest and Forest Chemistry, Archangelsk, USSR, 358 pp. [In Russian.]

Zakharov, V. K.: 1967, *Forest Taxation*. Visshaja Shkola, Moscow, USSR, 359 pp. [In Russian.]

Appendix

In the following we present definitions of some terms central to the article.

Growing stock, GS(A), is total volume of all living trees in a stand at age A, in $m^3\ ha^{-1}$.

Total volume, TV(A), (total production at age A), the total volume produced by all stem wood by a stand up to age A, in $m^3\ ha^{-1}$.

Accumulated mortality, M(A), the accumulated volume of trees which died of natural causes, up to age A, in $m^3\ ha^{-1}$. Hence:

$$TV(A) = GS(A) + M(A) \qquad [D1]$$

Gross increment per year at age A, $dTV(A)$, is defined as:

$$dTV(A) = TV(A) - TV(A\text{-}1) \qquad [D2a]$$

or,

$$dTV(A) = f'(A) \qquad [D2b]$$

where $TV(A) = f(A)$ is the functional expression for the total volume by age.

Net increment per year at age A, $dGS(A)$, is defined as:

$$dGS(A) = GS(A) - GS(A\text{-}1) \qquad [D3a]$$

or,

$$dGS(A) = g'(A) \qquad [D3b]$$

where $GS(A) = g(A)$ is the functional expression for the growing stock by age.

Natural mortality per year for age A, $dM(A)$, is the difference between gross and net increment:

$$dM(A) = dTV(A) - dGS(A) \qquad [D4]$$

Gross increment per year can be calculated approximately as a percentage of total production to date:

$$P_{TV}(A) = \frac{dTV(A)200}{TV(A) + TV(A-1)} \qquad [D5]$$

or explicitly as:

$$P_{TV}(A) = \frac{f'(A)100}{f(A)} \qquad [D6]$$

Net increment per year can be calculated in the same manner as a percentage of growing stock to date.

Average diameter, D, is DBH, calculated as the average quadratic value of DBH of the trees constituting a stand.

Average height, H, is calculated from the regression equation as the height of the tree with diameter D.

Basal area, BA, is the sum of basal areas of all trees of the stand at breast height, in m^2 ha^{-1}.

Density or stocking is determined as the ratio of the sum of the basal areas of the measured stand at breast height, to the sum of the basal areas of an ideal stand according to yield tables.

Site indices are determined in Russia from average height at a specified age of the stand.

Type of age structure (TASS) is a classification system for stands, which reflects the variation of age inside of separate stand.

General tables are yield tables for fully stocked stands for a whole area, dominated by a certain species and are valid for all of the former USSR. They are used as general standards for different comparisons, and for forest inventories and forest management, if regional tables are absent.

Regional tables are yield tables generated for a definite part of areas dominated by a certain species. There are different types of regional tables: for normal (or fully stocked stands); for modal (or not fully stocked stands), for different TASS; for specific goals of the forest management (so-called goal programs of forest regeneration); for forest plantations with different initial densities, etc. They are used for the local forest inventory and by the forest management.

Growth tables (or models) are regional tables (or models) of dependency of net and/or gross growth of different stand characteristics.

Forested Area is a category of lands which was used in the former USSR and is used in the countries of CIS. These are lands covered by forests with stocking >0.3 for young stands and stocking >0.2 for the rest of the age groups. Stocking 1.0 is defined by yield tables or special "tables of basal area and growing stock under stocking 1.0". Temporary non-stocked forest land is separately accounted as "unforested area". Lands covered by shrubs are included in the Forested Area only if closed forests cannot grow in this territory due to severe climatic conditions (extreme north, sub-alpine mountain zone, etc.).

PART III

ADVANCED TECHNOLOGIES AND THE IMPACTS OF GLOBAL CHANGE

BOREAL FOREST CATCHMENTS: RESEARCH SITES FOR GLOBAL CHANGE AT HIGH LATITUDES

C. W. SLAUGHTER, V. Y. E. GLOTOV[2], L. A. VIERECK[3], and V. M. MIKHAILOV[4]

[1]Northwest Watershed Research Center, USDA-Agricultural Research Service, Boise, Idaho 83712, USA,
[2]Northeast Interdisciplinary Research Institute, Russia Academy of Sciences, Far East Branch, Magadan 685000, Russia,
[3]Institute of Northern Forestry, USDA Forest Service, Fairbanks, Alaska 99775 USA,
[4]Permafrost Institute, Northeastern Department, Laboratory of General Permafrostology, Magadan 685000, Russia

Abstract. Circumpolar subarctic boreal forest ecosystems are subject to change from a variety of agents and processes. Climate warming predicted by many GCMs indicates that regions north of 60°N may be subjected to major warming in coming decades, producing increased permafrost thaw, altered vegetation distribution and biological productivity, and perhaps release of large quantities of stored organic carbon into the global carbon cycle. Research into change in ecosystems can entail use of ecosystem "samples," i.e., sectors of landscape such as catchments (watersheds) which are representative of the larger ecoregion and available for repeated, long-term measurement and analysis. Boreal forest research and monitoring programs have been established in hydrologically-defined catchments in discontinuous-permafrost regions at 65°N, 148°W in the Yukon-Tanana Uplands of central Alaska, and at 62°N, 158°W in the Kolyma River headwaters of Magadan Oblast, north-eastern Russia. These sites are available for sustained research into global change.

Keywords. HYDROLOGY, CATCHMENT, WATERSHED, PERMAFROST, GLOBAL CHANGE, BOREAL FOREST.

1. Introduction

The circumpolar northern boreal forest extends to north of 70°N in Siberia, and south of 50°N in Mongolia, north-east China adjacent to south-east Siberia, and south of Hudson Bay in eastern Canada (Larsen, 1980). About two-thirds of the circumpolar boreal forest lies in Eurasia, and one-third in North America. Throughout its circumpolar range the boreal forest is characterised by relatively few tree species (Larsen, 1980). Conifers include spruce (*Picea* sp.), fir (*Abies* sp.), pine (*Pinus* sp.), and larch (*Larix* sp.); hardwoods typically include aspen or cottonwood (*Populus* sp.), birch (*Betula* sp.), alder (*Alnus* sp.), and willow (*Salix* sp.). Many shrubs and forbs of the boreal forest exhibit similar circumpolar commonality at the genus and species level (Hulte'n, 1968; Bliss and Matveyeva, 1992).

Much of the northern boreal forest in Eurasia, Alaska and Northwest Canada is underlain by permafrost (Figure 1). Permafrost is earth material which is perennially frozen (Pe'we', 1982; Brown and Pe'we', 1973). The proportion of the landscape underlain by permafrost increases with latitude, with increasingly northern exposure, and with greater soil moisture. Permafrost in the northern boreal forest is commonly associated with north-aspect slopes, low-lying areas and wetlands. In Fennoscandia and western Russia permafrost is largely lacking; however, in central and eastern Siberia large sectors of boreal forest occur on continuous permafrost landscapes. The northern boreal forest of North America lies largely in the zone of discontinuous or sporadic permafrost.

Water, Air and Soil Pollution **82**: 351–361, 1995.

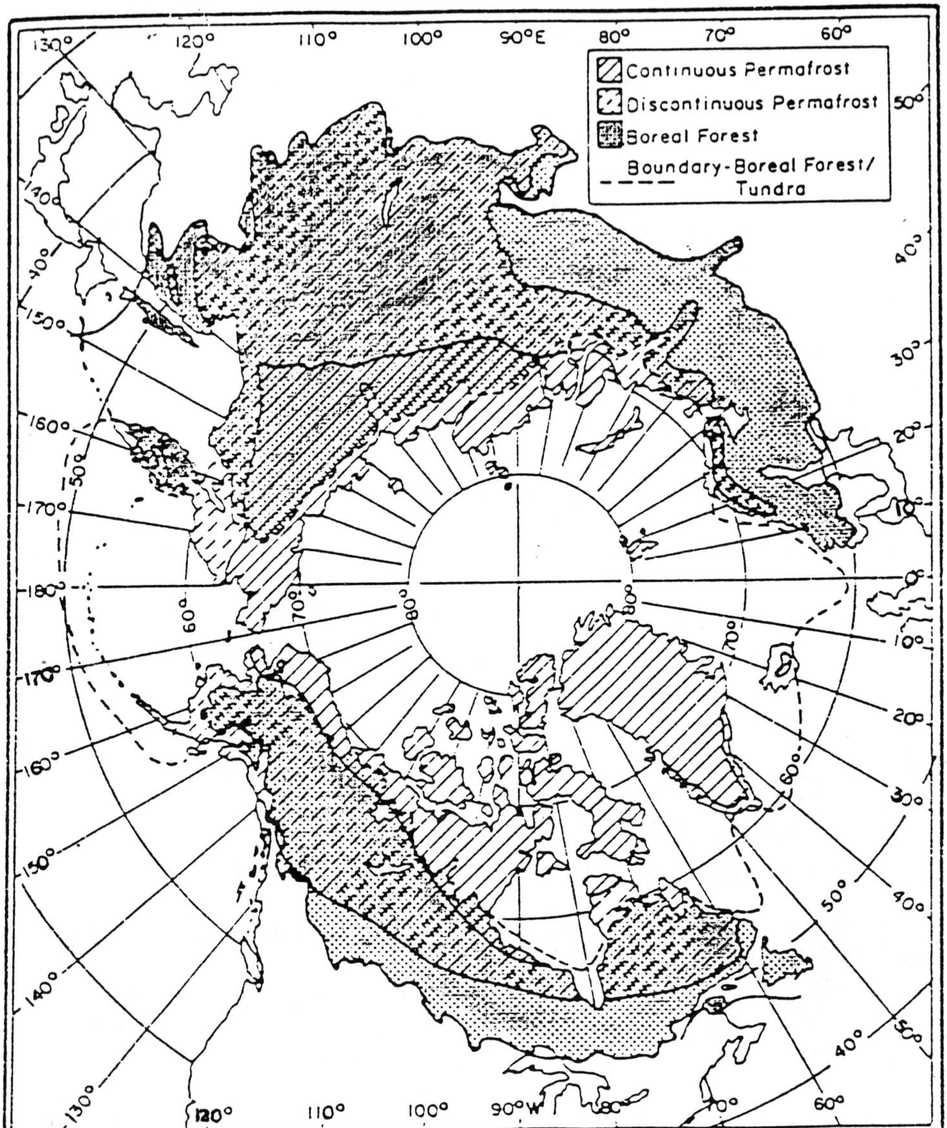

Fig. 1. The circumpolar northern boreal forest is associated with permafrost-affected landscapes through much of its range. (Figure courtesy Van Cleve *et al.* (1986).)

The northern boreal forest is a strongly fire-influenced ecosystem; the vegetation patterns of the taiga landscape are inextricably linked to fire history (Viereck, 1973; Dyrness *et al.*, 1986). The area burned annually in Alaska alone has exceeded 400 000 ha (one million acres) at least five times since 1955, almost entirely affecting boreal forest lands. The recurrence interval for wildfire in Alaska's boreal forest has been estimated to range from 50 to 200 years (Heinselman, 1981; Viereck and Schandelmeier,

1980). Wildfire regime has major implications for evaluating stability and biological productivity of ecosystems, and for management for sustained production of commodity and non-commodity forest resources.

1.1. CONSEQUENCES

The circumpolar boreal forest ecosystem is closely linked to global change issues (Post, 1989; Apps *et al.*, 1993). The carbon (C) store in northern boreal forests and peatlands is a major factor in controlling or forcing global change; CO_2 and CH_4 sequestration or release in taiga ecosystems has the potential to drastically alter rates of global warming (Harriss and Frolking, 1991; Gorham, 1991). The total C pool of boreal forests, including associated peatlands, is significant globally. Estimates cited by Armentano (1980) varied from 135 to 247 Pg C stored in boreal forests. Apps *et al.* (1993) suggest ca. 800 Pg C total storage in northern hemisphere boreal forest biomes. This constitutes perhaps 10–20% of the global C pool, which is estimated at ca.1395 Pg (Post *et al.*, 1982). General Circulation Models (GCMs) have been utilised to study the consequences of increasing atmospheric concentrations of greenhouse gases. These models, while differing in detail, converge to indicate that global warming will be detected earlier and will be of greater magnitude at 50–70°N latitude (Hansen *et al.*, 1988; Hall, 1988; Mitchell, 1989), i.e., the boreal forest zone, and will directly affect this major C store.

The landscapes of the zone of discontinuous permafrost—largely northern boreal forest —are extremely sensitive to thermal disruption: ground temperatures range from perhaps 0.5 °C to -2 °C in central Alaska, while GCMs predict as much at 8 °C warming over the next 50 to 100 years at these northern latitudes. Warming may result in widespread and long-lasting thermal and physical degradation of terrain, altered rates and patterns of vegetation succession and species composition shift, and damage to engineered facilities (Pe'we', 1982). Disturbance regimes may change in boreal forests under an altered climate (Franklin *et al.*, 1991). A warming climate might result in expanding ranges for insects and pathogens deleterious to the maintenance of boreal forests. Change in fire severity and frequency is a primary concern, given the ubiquitous role of fire in northern forest ecosystems (Rowe and Scotter, 1973; Viereck, 1973; Auclair and Carter, 1993).

The northern boreal forest landscape is occupied by seasonal snow and ice for six to nine months of the year (Slaughter and Benson, 1986). Seasonal snow alters landscape albedo while holding water in "detention storage" prior to its release at spring melt. The soil mantle is seasonally or perennially frozen affecting surface and groundwater regimes (Kane *et al.*, 1992). Presence of permafrost in watersheds increases the "flashiness" of water yield: permafrost-dominated catchments have higher peak flows, lower base flows, and steeper recessions (on a unit-area basis) than adjacent permafrost-free basins. These effects increase with increasing proportion of permafrost (Slaughter *et al.*, 1983; Hilgert and Slaughter, 1987). A similar pattern has been demonstrated for sediment yield from undisturbed upland catchments; a permafrost-dominated catchment yields higher loads of suspended sediment than does an adjacent non-permafrost basin (Slaughter and Collins, 1981). Warming which alters the proportion and distribution of permafrost in boreal landscapes will consequently affect hydrologic behaviour, including sediment and nutrient flux, within that landscape.

Biological attributes of boreal stream systems will be affected by global warming. Freshwater stream ecosystems are integral components of boreal forest landscapes and stream system composition and functions are strongly influenced by processes and events occurring within the contributing watershed (Meyer and Pulliam, 1991). For example, altered thermal regimes of watersheds and changes in C input to streams will affect periphyton primary production, leading to trophic changes within the freshwater systems (Oswood *et al.*, 1991). Despite the importance of hydrologic processes in global change, Rind *et al.* (1992) have shown that models currently being utilised to evaluate hydrological consequences of global warming yield inadequate forecasts of hydrological cycle changes. Better understanding of basic hydrological processes and modelling techniques is needed.

1.2. SUSTAINED FIELD RESEARCH

Evaluation of change at local, regional or global scales is facilitated by using reliable long-term data sets. Likens (1983), Callahan (1984), and Franklin (1989) are among many authors who cite the need for long-term, sustained and documented monitoring and research programs. Few dedicated sites for such research are available to scientists in the circumpolar north. Examples are provided here of two high-latitude boreal forest research sites, one in Northeast Russia and one in central Alaska. Each was established on the basis of hydrologically-defined landscape units, has long-term documented data, a useful research history, and provides a dedicated, protected setting for sustained boreal forest ecosystem research.

2. Northern Research Sites

2.1. BONANZA CREEK EXPERIMENTAL FOREST AND CARIBOU-POKER CREEKS RESEARCH WATERSHED

The Bonanza Creek Experimental Forest/Caribou-Poker Creeks Research Watershed is the northernmost forested site in the United States Long-Term Ecological Research (LTER) network (Figure 2). Bonanza Creek Experimental Forest (BCEF) (Figure 3) covers 5000 ha about 30 km west of Fairbanks, Alaska. BCEF lies within the Tanana Valley State Forest of central Alaska, in the southern sector of the Yukon–Tanana Uplands. BCEF includes environments of the floodplain of the Tanana River (120 m amsl), and adjacent uplands up to 450 m amsl Vegetation is typical of the principal boreal forest types of central Alaska. Occurrence of black spruce (*Picea mariana*), larch (*Larix laricina*) and bogs generally indicates the presence of permafrost. Aspen (*Populus tremuloides*) and white spruce (*P. glauca*) occur on permafrost-free sites. Paper birch (*Betula papyrifera*) is common on permafrost-underlain and permafrost-free sites. Balsam poplar (*Populus balsamifera*) develops in extensive stands on permafrost-free floodplain sites. The glacially-fed Tanana River carries heavy sediment loads throughout the summer months, and occasionally undergoes rapid lateral channel migration, bank erosion, and bar aggradation within its flood plain.

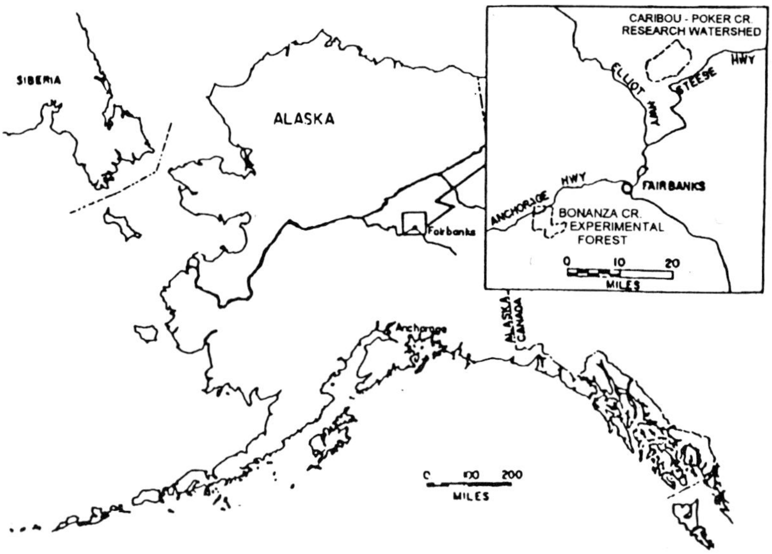

Fig. 2. Bonanza Creek Experimental Forest and Caribou-Poker Creeks Research Watershed are located near Fairbanks, Alaska, in the discontinuous permafrost zone of the northern boreal forest.

LTER research at BCEF focuses on a long-term study of ecosystem structures and functions affected by forest successional processes. Long-term climate monitoring at several locations supports experimental research into resource allocation within various ecosystem components, C availability in the forest floor (soil and plant communities), and herbivory by invertebrates and mammals. The emphasis is on the floodplain where primary succession is initiated by periodic flooding and on uplands where secondary succession is strongly influenced by periodic wildfire.

The Caribou-Poker Creeks Research Watershed (CPCRW) (Figure 4) is a 10 400 ha upland 45 km north of Fairbanks. Research at CPCRW started in 1969 to address hydrologic and environmental questions in the discontinuous-permafrost boreal forest (taiga) of the Yukon–Tanana Uplands of central Alaska. CPCRW encompasses more than a dozen first-, second-, and third-order sub-drainages over an elevation range from 210 to 826 m amsl, allowing analysis of a stream system continuum from headwaters through fourth-order streams. The watershed is in a region of uplifted Precambrian Birch Creek schist, which is overlain by a thin mantle of loess derived from glacial outwash floodplains on the north flanks of the Alaska Range. CPCRW lies in the zone of discontinuous permafrost. South-facing slopes are generally free from permafrost; valleys and north-facing slopes are underlain by frozen ground at shallow depth. The vegetation reflects typical subarctic patterns. North slopes are characterised by black spruce stands, with ericaceous shrubs, a thick feather moss/sphagnum moss/lichens ground cover and a deep organic matter layer. South-facing slopes support forest stands of varying age classes of aspen, birch and lesser amounts of white spruce and black

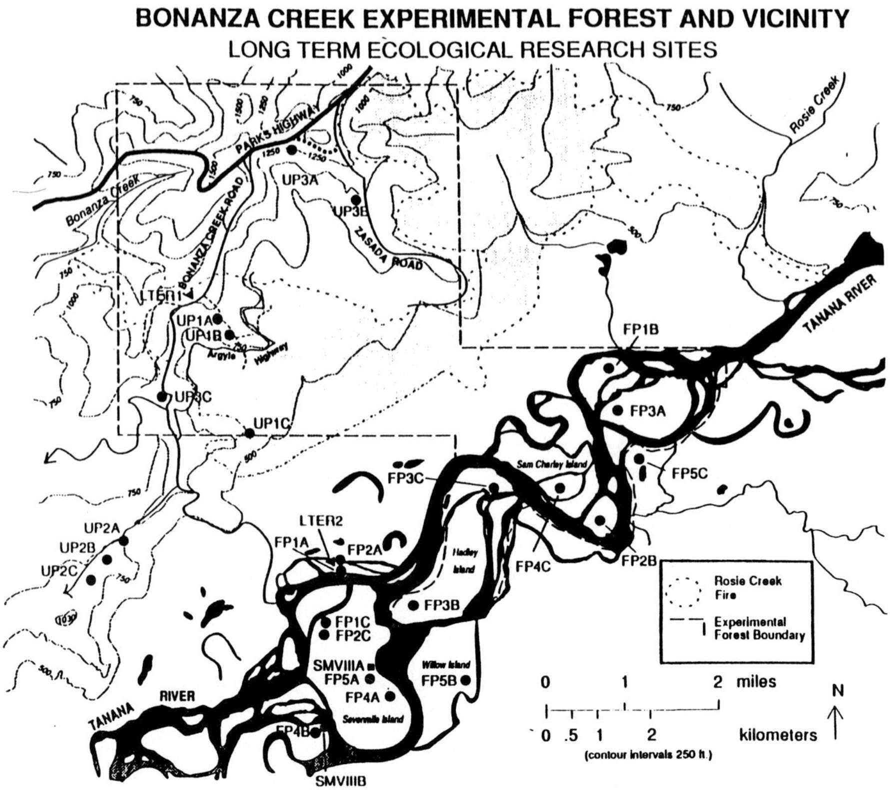

Fig. 3. The 50 km^2 Bonanza Creek Experimental Forest occupies portions of the Tanana Valley floodplain and adjacent upland slopes. Research and monitoring sites of the Long-Term Ecological Research Program (LTER) are located in the Experimental Forest and in adjacent sectors of the Tanana Valley State Forest.

spruce. Alder is found throughout the watershed. Valley floors are generally poorly drained and underlain by permafrost, and support dense stands of shrub birch (*B. glandulosa*), blueberry (*Vaccinium uliginosum*), willows (*Salix* spp.) and associated shrubs and forbs. The stream valleys and lower slopes support sedge fens and sedge tussock communities.

Early research in CPCRW addressed the hydrologic regime of permafrost-affected headwaters catchments, identification of hydrologic processes that are important in this high latitude environment, and the determination of factors determining patterns of streamflow and water quality. Research has been completed on seasonal snowpack processes, development of aufeis (nalyds or "icings") in stream channels and floodplains, bedrock dissolution and geochemistry of base flow in streams, and documentation of long-term soil temperature trends in permafrost and non-permafrost forests. Research monitoring includes a basin-wide system of hydrologic and climatology stations with many records dating back more than two decades. Four 25 ha Long-Term Forest Reserve Stands have been established for permanent baseline ecosystem monitoring in four major environmental settings: treeline; coniferous woodland on a permafrost-underlain north slope; a permafrost-underlain valley and

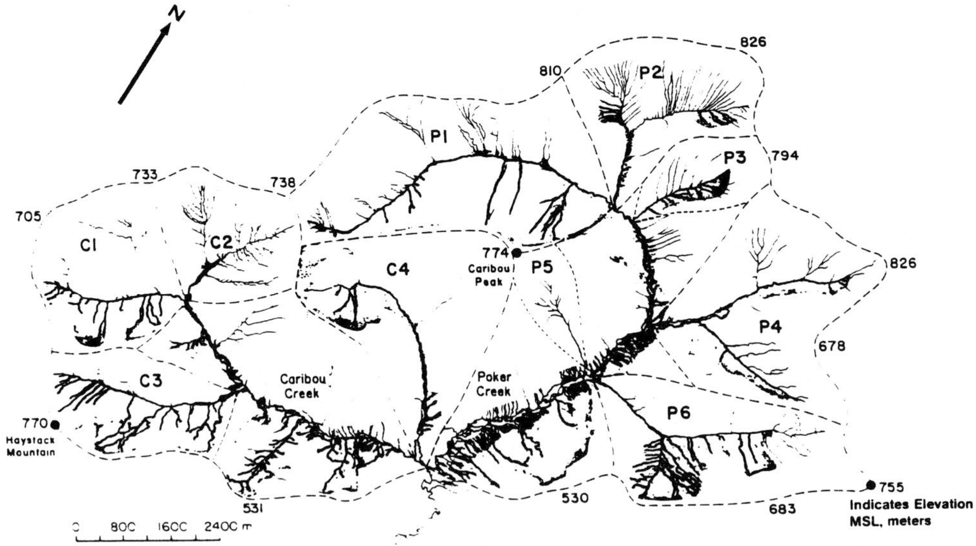

Fig. 4. The 104 km^2 Caribou-Poker Creeks Research Watershed lies in the Yukon–Tanana Uplands north of Fairbanks, Alaska.

riparian zone; and mixed deciduous/coniferous forest on a permafrost-free south slope . These stands complement a BCEF network of 24 LTER sites established to study forest succession on floodplain and uplands. Precipitation chemistry is continuously monitored in the Caribou Creek valley (230 m elevation) in collaboration with the National Atmospheric Deposition Program (NADP).

The composite BCEF/CPCRW LTER SITE encompasses slightly more than 150 km^2, and includes highly productive floodplain forests, permafrost-free coniferous and hardwood forest stands on south-facing slopes, and low-productivity permafrost-underlain coniferous woodland on north-facing slopes and in poorly drained cold valley settings. The composite BCEF/CPCRW research site offers opportunities for scaling ecosystem research from plot to slope to watershed, over an elevational gradient from floodplain to treeline and over a range of soil, vegetation and permafrost conditions.

2.2. Kolyma Water Balance Station

The Kolyma Water Balance Station (KWBS) is located on Kontaktovy Creek, a tributary to the Kulu River 200 km from Susuman in the headwaters of the Kolyma River in eastern Russia. KWBS (Figure 5) was established in 1948 to evaluate hydrologic processes and water supply. KWBS is located in intensively dissected mountainous terrain and lies within the continuous permafrost zone. Climate is strongly continental, with long cold winters and short, relatively warm and dry summers. Dominant tree species are larch (*Larix dahurica*) and pine shrub (*Pinus pumila*). Higher elevations are occupied by tundra and bare talus slopes. Talus and stony mountainous tundra occupies 39% of the whole drainage area, and stony moss tundra occupies 32%. Trees and shrubs occupy 25% of the area, while 4% is recently burned wildfire sites.

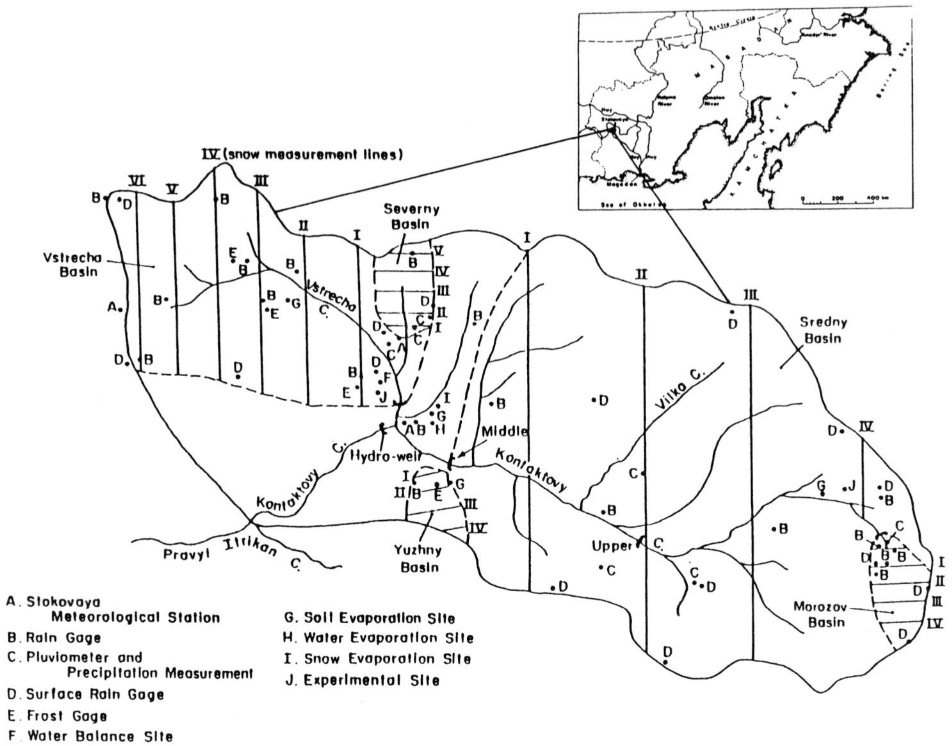

Fig. 5. The 21 km^2 Kolyma Water Balance Station lies in the headwaters of the Kolyma River in north-eastern Russia.

The drainage area of KWBS at the lowermost (Nizhny) gauging station is 21.2 km^2. Discrete catchments within KWBS range from 0.7 to 15 km^2. Elevation varies from 790 m amsl at the mouth of the main watercourse, to almost 1700 m amsl at its extreme headwaters. The watershed is asymmetrical, with some 70% of the total drainage area found on the right (northern) side of the stream. River channel density is 2.5 $km\,km^{-1}$; sinuosity is 1.24 $km\,km^{-1}$. All streams have well-defined channels abutting steep valley slopes. North-facing slopes are mostly concave; south-facing slopes tend to be convex.

Bedrock consists mainly of Permian sandy-clay shales and granites. Valleys of the largest sub-basins, Kontaktovy and Vstrecha, coincide with the contact of granitic batholith with shales. The whole area was subjected to tectonic uplift: batholiths, having lower density, rose faster than the surrounding shale bedrock. This situation is common in the larger Kolyma River basin and is accompanied by widespread and numerous faults. Alluvial Quaternary deposits are mainly coarse-grained and highly permeable. On the lower parts of slopes and in flat, poorly drained settings, lenses and interlayers of fine materials with low permeability are often formed. In the Kontaktovy valley, alluvial deposits reach a maximum thickness of 6.5 to 8 m near the Nizhny gauging station. The underlying shale is weathered to a depth of 20 m.

Permafrost thickness varies from 200 to 300 m at upper elevations, to 130 m in the lower reaches of the Kontaktovy Creek valley. The thickness of the seasonal thaw layer (active layer) is 30 to 50 cm on north-facing slopes, 40 to 60 cm on flat sites with

TABLE I

Mean annual precipitation and air temperature, Kolyma Water Balance Station.

Site	Precipitation (mm)			Air temperature (°C)
	Total	Rain	Snow	
Nizhnyaya	318	151	167	-11.5
Verkhnyaya	331	166	165	-10.4

low permeability, and reaches 100 cm to 250 cm on dry south-facing slopes. A supra-permafrost talik exists in the middle and lower reaches of Kontaktovy Creek. The talik width near the Nizhny gauging station is 40 to 50 m; its depth under the stream channel is 8.5 to 10 m, and 4 to 5 m near the lateral boundaries.

Three types of groundwaters occur in KWBS. They are: (1) groundwater of the seasonal thaw layer (active layer), (2) groundwaters of the supra-permafrost *talik*, and (3) sub-permafrost groundwaters. The first exists from the beginning of snowmelt in early May until November–January (depending on the active layer thickness). Sub-permafrost waters have no connection with the other types except near the creek mouth, where the talik extends upward to intercept the seasonally thawed active layer. During the winter, a small nalyd (aufeis) develops upstream from the Sredny gauging station due to forcing of groundwater from the freezing talik.

Comprehensive climatic and hydrologic measurements are conducted year-round. Climatic characteristics from KWBS are given in Table I. These are derived from meteorological sites at 860 m amsl (Nizhnyaya) and 1100 m amsl (Verkhnyaya). Winter temperatures and total precipitation (mostly rain) increase with elevation while summer temperatures decrease with elevation. At the upper meteorological site the warm period is shorter, and snowfall is possible throughout the summer.

Snow cover usually lasts in the stream valleys from late September/early October to late May. On the water divides (ridgelines) average snow duration is longer by about three months. Maximum snow depth varies from zero on the windward, upper sectors of slopes to as much as 3 to 4 m on lower slopes with dense shrub vegetation. Snow depth may exceed 5 m in small, V-shaped valleys of steep ephemeral streams. Average snow depth is 0.7 to 1.2 m; average snowpack density is 0.2 g cm^{-3}, with a wide range of spatial and temporal heterogeneity.

Surface streamflow is usually observed from early May to mid-October. In minor creeks, flow ceases 2–3 weeks earlier in the fall. Snowmelt causes a rapid growth of stream discharge in spring beginning with water flowing over ice. Daily fluctuations in this period are distinctive. In accordance with the spatial variability of snow cover, characteristics of spring runoff change throughout the basin and from year to year. The flow is lowest in the Yuzhny Creek watershed (from 23 to 125 mm) and highest in the Morozova Creek watershed (134 to 345 mm). Maximum stream discharge rates are usually observed in June, when snowmelt is accompanied by rainfall.

The Kolyma Water Balance Station is unique in that there is no similar research site currently operated in the continuous permafrost zone in Russia. This basin provides good representation of upland environments in the vast territory of the inner parts of the Northeastern Russia. The existence in KWBS of a dense hydrometeorological network with a long period of observations provides a solid basis for conducting many research programs. Hydrogeological observations were initiated in 1968, twenty years after starting hydrologic and climatic measurements. A comprehensive study was completed in the course of the International Hydrological Decade and a permanent hydrogeological post was established. Related ecological investigations (hydrobiological, geobotanical, soil cover, etc.) were conducted at KWBS by the Institute of Biological Problems of the North, including study of the interrelation between vegetation cover and surface flow. This research contributes to the suitability of KWBS for monitoring long-term climatic and ecological changes in relation to global change.

3. Summary

Two long-established field research sites, one in central Alaska and one in north-east Russia, are available for sustained interdisciplinary research into boreal forest ecosystems under current conditions and under conditions of altered climate including potential warming. Each site includes hydrologically defined catchments and has a long-term baseline of climatic and hydrologic monitoring and has an established infrastructure and instrumentation network. Each site is available to investigators from institutions of the circumpolar north.

The Bonanza Creek Experimental Forest/Caribou-Poker Creeks Research Watershed and the Kolyma Water Balance Station offer opportunities for comparative research and monitoring in subarctic boreal forest ecosystems of north-western North America and north-eastern Eurasia. Results already available from these sites provide a quantitative baseline for assessing physical and biological changes in response to regional pollution and global warming. Continued research and monitoring at KWBS and BCEF/CPCRW can contribute to an improved understanding of long-term ecological functioning of boreal forest ecosystems, and to an improved ability to anticipate consequences of global change in boreal ecosystems.

Acknowledgments

This work was supported by the Institute of Northern Forestry, Pacific Northwest Research Station, USDA Forest Service, and the Office of International Co-operation and Development (OICD), US Department of Agriculture.

References

Apps, M. J., Kurz, W. A., Luxmoore, R. J., Nilsson, L. O., Sedjo, R. A., Schmidt, R., Simpson, L. G. and Vinson, T. S.: 1993, *In:* Wisniewski, J. and Sampson, R. N. (eds.), *Boreal Forests and Tundra*, Kluwer, Boston, USA, pp. 39-53.

Armentano, T. V.: 1980, *Bioscience* **30**(12), 825-830.

Auclair, A. and Carter, T. B.: 1993, *Can. J. For. Res.* **23**, 1528-1536.

Bliss, L C. and Matveyeva, N. V.: 1992, *In:* Chapin, F. S. III, Jefferies, R. L., Reynolds, J. F., Shaver, G. R., Svoboda, J., Chu, E. W. (eds.), *Arctic Ecosystems in a Changing Climate: an Ecophysiological Perspective*, Academic Press, New York, USA, pp. 59-89.

Brown, R. J. E. and Pe'we', T. L.: 1973, *In: Permafrost, the North American Contribution to the Second International Conference*, Yakutsk, Siberia, USSR. National Academy of Sciences, Washington D.C., USA, pp. 71-100.

Callahan, J. T.: 1984, *Bioscience* **34**(6), 363-367.

Dyrness, D. T., Viereck and L. A., Van Cleve, K.: 1986, *In:* Van Cleve, K., Chapin, F. S. III, Flanagan, P. W., Viereck, L. A. and Dyrness C. T. (eds.) *Forest Ecosystems in the Alaskan Taiga: A Synthesis of Structure and Function*, Springer-Verlag, New York, USA, pp. 74-86.

Franklin, J. F.: 1989, *In:* Likens, G. E. (ed.) *Long-Term Studies in Ecology: Approaches and Alternatives* Springer-Verlag, New York, USA, pp. 3-19.

Franklin, J. F., Swanson, F. J., Harmon, M. E., Perry, D. A., Spies, T. A., Dale, V. H., McKee, A., Ferrell, W. K., Means, J. E., Gregory, S. V., Lattin, J. D., Schowalter T. D. and Larsen, D.: 1991 *Northwest Environmental Journal* **7**, 233-254.

Gorham, E.: 1991 *Ecological Applications* **1**(2), 182-195.

Hall, D. K.: 1988, *Rev. Geophysics* **26**, 26-39.

Hansen, J., Fung, I., Lacis, A., Lebedeff, S., Rind, D., Ruedy, R., Russell, G. and Stone, P.: 1988, *J. Geophys. Res.* **93**(D8), 9341-9364.

Harriss, R. C. and Frolking, S. E.: 1991, *In:* Firth, P. and Fisher, S. G. (eds.), *Global Climate Change and Freshwater Ecosystems*, Springer-Verlag, New York, USA, pp. 48-67.

Heinselman, M. L.: 1981, *In:* Mooney, H., *et al.* (eds.), *File Regimes and Ecosystem Properties.* Proc. Conf., USDA For. Serv. Gen. Tech. Rep. W0-26, Washington D.C., USA, pp. 7-57.

Hilgert, J. W. and Slaughter, C. W.: 1987, *Water Quality and Streamflow in the Caribou-Poker Creeks Research Watershed, Central Alaska, 1979.* Research Note PNW-RN-463, USDA Forest Service, Pacific Northwest Research Station, Portland, USA, 34 pp.

Hulte'n, E.: 1968, *Flora of Alaska and Neighboring Territories.* Stanford Univ. Press, Stanford, USA, 1008 pp.

Kane, D. L., Hinzman, L. D., Woo, M. and Everett, K.: 1992, *In:* Chapin, F. S. III, Jefferies, R. L., Reynolds, J. F., Shaver, G. R., Svoboda, J. and Chu, E. W. (eds.), *Arctic Ecosystems in a Changing Climate: An Ecophysiological Perspective*, Academic Press, New York, USA, pp. 35-57.

Larsen, J. A.: 1980, *The Boreal Ecosystem.* Academic Press, New York, USA, 500 pp.

Likens, G. E.: 1983, *Bull. Ecol. Soc. Amer.* **64**(4), 234-243.

Meyer, J. L., Pulliam, W. M.: 1991, *In:* Firth, P. and Fisher, S. G. (eds.), *Global Climate Change and Freshwater Ecosystems.* Springer-Verlag, New York, USA, pp. 177-191.

Mitchell, J. F. B.: 1989, *Rev. Geophysics* **27**, 115-139.

Oswood, M. W., Milner, A. M. and Irons, J. G. III.: 1991, *In:* Firth, P. and Fisher, S. G. (eds.), *Global Climate Change and Freshwater Ecosystems.* Springer-Verlag, New York, USA, pp. 192-210.

Pe'we', T. L.: 1982, *Geologic Hazards of the Fairbanks Area, Alaska.* Special Report **15**, Div. of Geological and Geophysical Surveys, Alaska Department of Natural Resources, Fairbanks, USA,. 109 pp.

Post, W. M. (ed.): 1989, *Report of a Workshop on Climate Feedbacks and the Role of Peatlands, Tundra, and Boreal Ecosystems in the Global Carbon Cycle.* U.S. Department of Energy, Washington D.C., USA, 28 pp.

Rind, D., Rosenzweig, C. and Goldberg, R.: 1992, *Nature* **358**, 119-122.

Rowe, J. S. and Scotter, G. W.: 1973, *Quaternary Research* **3**(3), 444-464.

Slaughter, C. W. and Collins, C. M.: 1981, *J. Hydrology (New Zealand)* **20**, 39-48.

Slaughter, C. W., Hilgert, J. W. and Culp, E. H.: 1983, *In: Proceedings, 4th International Conference on Permafrost.* National Academy of Sciences, Polar Research Board, Washington D.C., USA, pp. 1172-1177.

Slaughter, C. W. and Benson, C. S.: 1986, *In:* Kane, D. L. (ed.), *Proceedings, Cold Regions Hydrology*, American Water Resources Association, Bethesda, USA, pp. 101-109.

Van Cleve, K., Chapin, F. S. III, Flanagan, P. W., Viereck, L. A. and Dyrness, C. T. (eds.): 1986, *Forest Ecosystems in the Alaskan Taiga: A Synthesis of Structure and Function.* Springer-Verlag, New York, USA, 230 pp.

Viereck, L. A.: 1973, *Quaternary Research* **3**(3), 465-495.

Viereck, L. A. and Schandelmeier, L. A.: 1980, *Effects of Fire in Alaska and Adjacent Canada—A Literature Review.* BLM-Alaska Technical Report **6**. U.S. Department of the Interior, Bureau of Land Management, Anchorage, USA, 124 pp.

THE NASHWAAK EXPERIMENTAL WATERSHED PROJECT: ANALYSING EFFECTS OF CLEARCUTTING ON SOIL TEMPERATURE, SOIL MOISTURE, SNOWPACK, SNOWMELT AND STREAM FLOW

F.-R. MENG, C. P.-A. BOURQUE, K. JEWETT,
D. DAUGHARTY and P. A. ARP

Faculty of Forestry and Environmental Management, University of New Brunswick, Fredericton, New Brunswick E3B 6C2, Canada

Abstract. A forest disturbance such as clearcutting affects local climate conditions by affecting surface reflectance, amount of snow catch, amount and timing of snowmelt and stream water discharge, soil moisture, soil temperature and soil frost. In this paper, snowpack water equivalents, soil moisture and soil temperature are simulated for pre- and post-harvest conditions at the Nashwaak Experimental Watershed Project. This Project involved a paired watershed experiment, with one of the forest-covered basins cut by way of a conventional harvesting operation, and the other basin left as a control. The computer simulations involved the application of a forest hydrology model (ForHyM) and a soil temperature model (ForSTeM). Both models used monthly amounts of rain and snow and mean monthly air temperatures records as input. Forest-atmosphere energy balances were calculated for year-round conditions. In this report, special attention was given to the energy balance when the ground is covered with snow. Doing so generated a process-oriented approach for calculating snowmelt throughout the winter and at the beginning of spring. Year-round model simulations for stream discharge were compared with pre- and post-harvest discharge observations. Also compared were simulated and observed snowpack water equivalents. Simulated results were in good agreement with field observations, thereby supporting the general calculations made for pre-and post-harvest soil moisture, temperature and frost conditions. Altogether, it was found that clearcutting produced positive as well as negative contributions to the water budget of the harvested basin. Positive contributions were likely due to reduced evapotranspiration. Negative contributions were likely due to reduced snow and fog water catch. Increased ground-level insolation advanced the snowmelt season for the cut basin by about two weeks.

Keywords. SNOWMELT, FOREST HYDROLOGY, SOIL TEMPERATURE, CLEARCUTTING, COMPUTER MODELS.

1. Introduction

This paper reports on effects of a conventional clearcutting operation on snow hydrology, soil moisture, soil temperature and stream discharge, as observed and simulated for the forest-covered Narrows Mountain Brook (NMB) and Hayden Brook (HB) basins in Central New Brunswick. These watersheds formed a paired watershed experiment (Hewlett and Nutter, 1969), with NMB cut in 1978-79, and HB serving as the control (Dickison *et al.*, 1981; Daugharty 1984). Experimental data obtained from this experiment pertain to stream water yield (Daugharty and Dickison, 1988), snow cover (Dickison and Daugharty, 1980; Daugharty, 1984), albedo (Daugharty and Dickison, 1988; Bourque *et al.*, 1995), stream and soil water chemistry (Bacon, 1981; Krause, 1982), and vegetative responses (MacDonald and Powell, 1983).

Hydrological simulations for the before and after cutting situation were performed with the forest hydrology model (ForHyM; Arp and Yin, 1992), and with its affiliated forest soil temperature model (ForSTeM; Yin and Arp, 1993). These models use monthly meteorological data (monthly amounts of snow and rain, mean monthly air temperatures) and simple site descriptions (depth of the forest floor and the rooted part of the mineral soil, overall soil texture, coarse fragment content, dominant tree species

Water, Air and Soil Pollution **82**: 363–374, 1995.

at the site, canopy closure) as driving variables (Figure 1). Monthly solar radiation input was simulated with a solar radiation module (ForRad, a forest radiation module). With ForHyM, we were able to simulate interception, evapotranspiration, throughfall, percolate volumes through the forest floor and the mineral soil, stream discharge rates (stream flow), snowpack, and soil water contents (Figure 2). ForHyM-simulated soil moisture was used as input for simulating soil and snowpack temperatures with ForSTeM (Figure 2). Field data regarding snowpack water equivalents and stream water data were used for model verification.

Comparing field data with model simulations provided some relevant benchmark information about the performance of the models used, and about the extent of conventional clearcutting effects on soil moisture, soil temperature, snowpack accumulations, time and amount of snowmelt, and stream discharge rates. This information, in turn, was useful for assessing post-harvest effects of clearcutting on the water and heat budgets of the forested landscape from winter to summer, as affected by the successional regrowth of the forest vegetation.

2. Study Area

The Nashwaak Experimental Watershed is located in west central New Brunswick approximately 70 km north of Fredericton (46°18'N, 67°02'W; Figure 3). The NMB basin spans 391 ha, and the HB Brook basin spans 660 ha. Elevations in NMB range from 220 to 420 m asl, with a mean of 296 m, an average slope of 7.6% and a predominantly south-eastern aspect. Elevations in HB range from 197 to 478 m, with a mean of 317 m, an average slope of 11.4%, and a predominantly southern aspect (Daugharty, 1984).

Forest cover was similar for both basins: 50% was deciduous forest (located mainly at higher elevations), 25% was coniferous forest (located mainly in the lower elevations), and 25% was mixed wood (Figure 3). Dominant tree species were beech (*Fagus grandifolia* Ehrh.), sugar maple (*Acer saccharum* Marsh.), red spruce (*Picea rubens* Sarg.) and balsam fir (*Abies balsamea* (L.) Mill) (MacDonald and Powell, 1983).

The climate, according to the Koeppen Classification system (Lutgens and Tarbuck, 1982), is humid continental, with at least four months of mean temperatures greater than 10°C. Precipitation is distributed fairly evenly, with lowest monthly rates in August and highest monthly rates in December (Palmer, 1988). Mean annual temperatures from 1972 to 1983 were 3.5 °C. Mean annual precipitation from 1972 to 1985 was 1322 mm (Palmer, 1988). Approximately one-third of precipitation fell as snow. Snowpacks were generally persistent, and lasted from late November to mid-April (Daugharty and Dickison, 1988).

Both watersheds are underlain by argillite which is covered with till (usually ablation till on top of basal till), with a depth varying from less than 50 cm to several metres. The till is derived from the underlying bedrock, but also contains granite, diorite and gabbro coarse fragments from areas north-west of the watersheds (Krause, 1982). Soils of the area are: (1) Orthic Humo-Ferric and Ferro-Humic Podzols on hill tops and gentle to strong slopes, (2). Gleyed Ferro-Humic Podzols on very gently sloping and level areas, and (3) Orthic Gleysols in small areas next to streams (Krause, 1982).

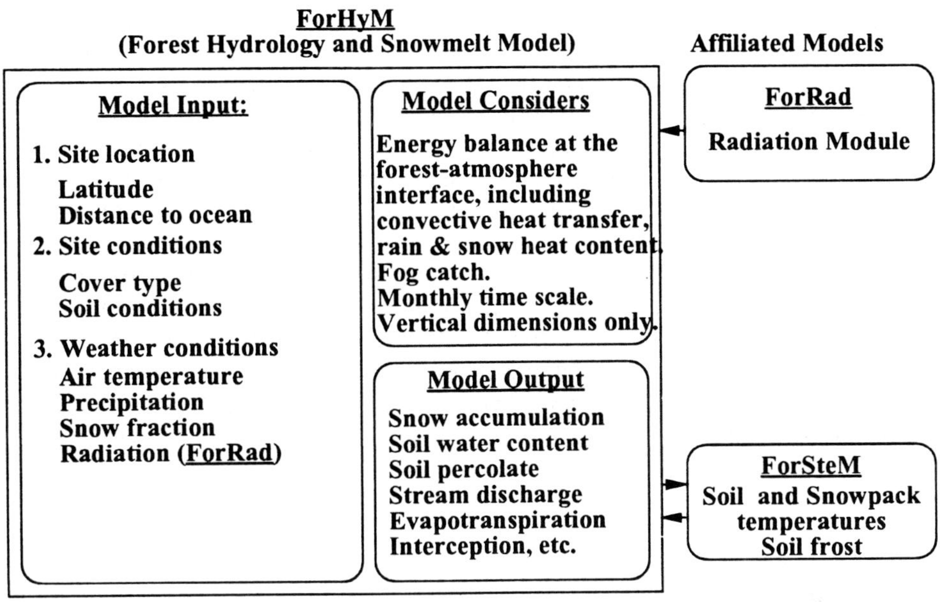

Fig. 1. Overview and data requirements of the modelling effort. ForHyM: forest hydrology model; ForSTeM: forest soil temperature model; ForRad: forest radiation module.

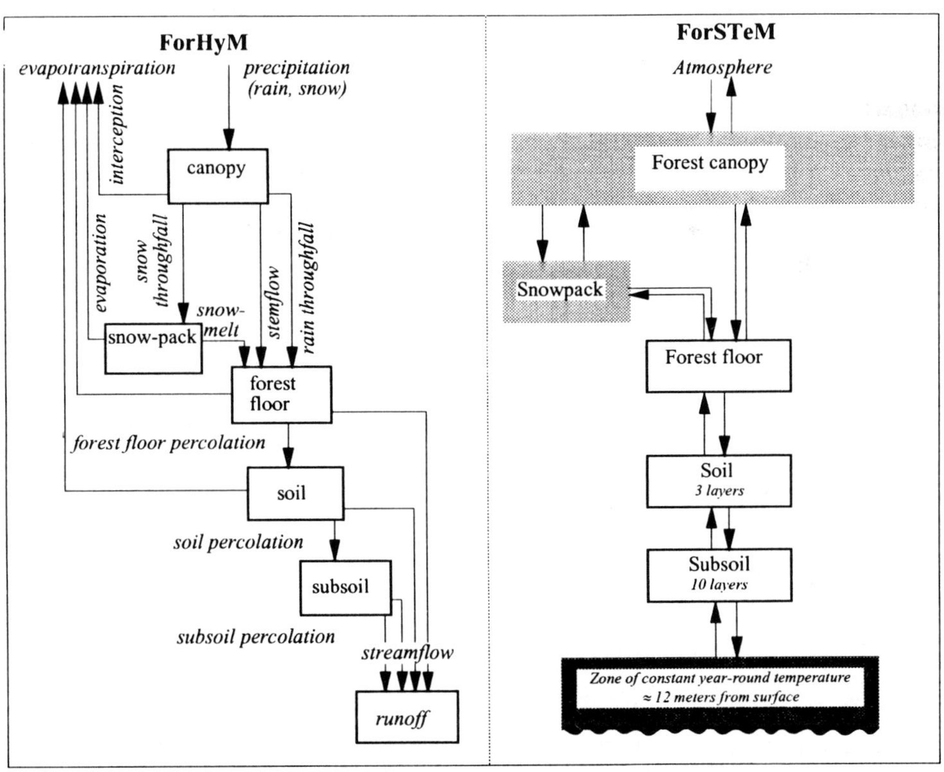

Fig. 2. Flowchart of water fluxes as considered in the Forest Hydrology Model (ForHyM, left), and of heat fluxes as considered in the Forest Soil Temperature Model (ForSTeM, right).

3. Methods

Snow depth and snowpack water equivalents were monitored within the HB and NMB basins from 1973 to 1982. Snowpack water content was measured 4 to 7 times during each winter. Before 1978, sampling began after the first significant snowfall. Sampling ended in the middle of April. After 1978, snow sampling was done at three-week intervals until the time of maximum accumulation. At that time, the surveying interval was reduced to two weeks in order to observe snowpack ablation dynamics. Snow survey data were collected from permanent sample plots located along parallel lines transecting the narrow axes of the basins, at a grid spacing of 316×316 m and 283×283 m for HB and NMB, respectively. Additional snow depths were measured at two equally spaced intervals between plots (Daugharty, 1984). Standard Canadian Atmospheric Environment Service (AES) snow samplers were used to determine both depth and snow water equivalent in terms of 0.25 cm increments (Daugharty, 1984). Snow survey data were averaged when sampled more than once a month.

Stream flow was monitored from 1972 to 1993. A natural control stream gauge was used to monitor stream water discharge rates near the outlet of HB into the Nashwaak River. A concrete 90° V-notched weir was constructed for NMB approximately 0.75 km from its outlet into the Nashwaak River. Both streams were monitored with Stevens-type A-35 automatic water level recorders.

Harvesting the NMB basin started in May 1978 and ended in February 1979. Logging roads were constructed before the cut. Harvesting was done with chainsaws. Trees with breast height diameter greater than 10 cm were cut and delimbed. Tree-length logs were transported to road-side with cable skidders on rubber tires. A 75 m wide riparian strip was left on each side of the main NMB branch (Figure 3). The residual stand [mainly eastern white cedar (*Thuja occidentalis* L.)] had a wood volume of 5 m^3 ha^{-1} and a stem density of 450 trees ha^{-1} (Krause, 1982).

4. Modelling Aspects

Hydrological processes such as canopy interception, evapotranspiration, surface runoff, forest floor infiltration, and soil percolation were simulated with the forest hydrology model (ForHyM, Figure 2; Arp and Yin, 1991) according to the monthly weather records for the location, and according to a generalised site description of the area (Table I). Soil and snowpack temperatures were calculated with the forest soil temperature model (ForSTeM, Figure 2; Yin and Arp, 1993).

Water and energy balances at the soil-atmosphere interface were evaluated by considering the related water and heat transfer processes at the snow surface in accordance with the vegetative cover. For this surface, we replaced the originally empirical snowmelt formulation of ForHyM with a set of surficial energy balance equations. With this balance, snowmelt was related to the net amount of incoming short wave and long wave radiation, to convective heat transfer, and to rain-carried energy, i.e.,

$$W_m = (R_s + R_l + H_c + H_r - H_s)/h_f \qquad [1]$$

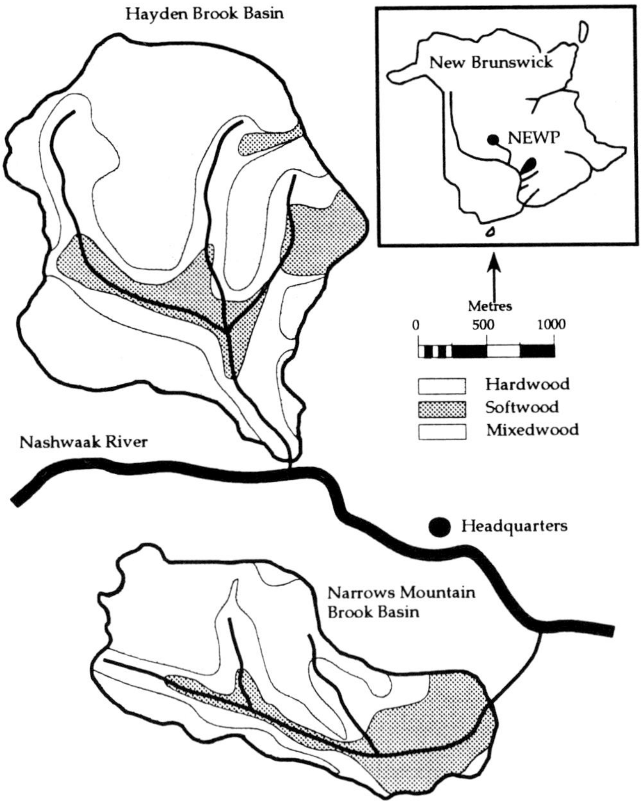

Fig. 3. The Nashwaak Experimental Watershed Project: pre-harvest cover types. The Narrows Mountain watershed was cut in 1978/1979.

TABLE I

General site descriptions and model input for the Hayden Brook Basin (HB) and the Narrows Mountain Brook (NMB) Basin of the Nashwaak Experimental Watershed Project (Jewett, 1995).

Parameter	Value
Mean annual snow depth (cm)	279
Annual temperature amplitude (°C)	15.2
Mean annual temperature (°C)	4.3
Mean July temperature (°C)	18.4
Deciduous (%)	70
Coniferous (%)	30
Forest floor	9
Mineral soil (cm)	50
Subsoil depth (cm)	100
Mineral soil texture	silt loam
Subsoil texture	sandy loam
Distance to coast (km)	170

where

h_f = heat of fusion,
R_s = short wave radiation absorbed by snow surface (W m^{-2}),
R_l = long wave radiation absorbed by snow surface (W m^{-2}),
H_c = convective sensible heat exchange between snow surface and the atmosphere (W m^{-2}),
H_r = rain caused energy input to the snowpack (W m^{-2}), and
H_s = heat transfer from snow surface to snowpack (W m^{-2}).

The amount of short wave radiation that reaches and is absorbed by the snow surface is given by

$$R_s = (1 - \alpha)e^{-k\ VPI} \qquad [2]$$

where

k = extinction coefficient,
VPI = vegetation area index, and
α = snow surface albedo.

Underneath the forest canopy, the amount of longwave radiation transferred between the ground surface and the atmosphere is given by (Bonan, 1991; Black *et al.*, 1991)

$$R_l = \sigma\left[(1 - a/2)T_{air}^4 - T_{ss}^4\right] \qquad [3]$$

where

T_{air} = air temperature under canopy,
T_{ss} = snow surface temperature, and
σ = Stefan-Boltzmann constant.

The amount of sensible heat transferred between the snow and the atmosphere is assumed to be proportional to the temperature differences between the snow surface and the atmosphere. This amount is calculated from (Yin and Arp, 1993):

$$H_c = k_c[1 - e^{\min(V_c,\ V_v)-6.8}](T_{air} - T_{ss}) \qquad [4]$$

where

k_c = heat transfer coefficient,
V_c = vegetation surface area index, and
V_v = maximum vegetation surface area.

Rain falling on snowpack affects the rate of snowmelt. Firstly, rain temperatures are higher than snow temperatures, and the related energy input contributes directly to snowmelt. Secondly, rain water increases the heat conductivity of the snowpack. Thirdly, small amounts of rain falling into snow with temperatures < 0 °C are converted into ice. Finally, incoming rain that freezes, and incoming water vapour that condenses on or into the snowpack, release heat through the process of fusion and condensation. Altogether, all of this facilitates heat transfer into and through the snowpack, thereby affecting snowmelt. A completely mechanistic description of all potential rain effects on snowmelt, however, is difficult. Instead, we assume that the rain effect on snowmelt is proportional to the difference between the mean monthly rain temperature, T_r, and the mean monthly snowpack temperature, T_{sp}, i.e.,

$$R_r = k_r(T_r - T_{sp}) \qquad [5]$$

Heat transfer from snow surface to snowpack was simulated with the following heat conductance equation (Yin and Arp, 1993): i.e.,

$$H_s = k_{ss}(T_{ss} - T_{sp}) \qquad [6]$$

where

k_{ss} = snow surface conductance (= 2 k_{sp}/D_{sp}),
k_{sp} = snowpack conductance [= $7537(0.358 + 26.4\, d_s^2)$],
d_s = snow density (g cm^{-3}), and
D_{sp} = snowpack depth (cm).

Note that $T_{ss} = 0$ when the snow melts from the top. For $T_{ss} < 0$,

$$T_{ss} = \frac{R_s(1-\alpha) + (4 - \alpha/2)\sigma(T_{air} + 273)^3 T_{air} + k_c + k_{ss}T_{sp} + k_r T_r}{4\sigma(T_{air} + 273)^3 + k_c + k_{ss} + k_r} \qquad [7]$$

The downward temperature profiles within the snowpack and within the soil were calculated with ForSTeM, as described by Yin and Arp (1993).

5. Results and Discussion

Observed and simulated snowpack water equivalents for the NMB and HB basins were well synchronised as they varied from year to year (Figure 3). This agreement was particularly good for 1975, 1979, 1980, 1981 and 1982, but fell somewhat short in 1973, 1974, 1976, 1977 and 1978. Both natural variations and sampling errors could have caused these inconsistencies. For example, basin-wide snowpacks were not easily surveyed, and were neither uniform nor consistent from year to year. In addition, snow accumulations in cutover areas were strongly affected by wind-induced drifting. Such drifting varied according to topography, patches of remaining vegetation, slash pile patterns, and changes in wind speed and direction. Rainfall during winter further complicated the situation by causing variable patterns of water flow and subsequent icing within the snow accumulations (Daugharty, 1984). In reference to all these complications, we note that the model calculated average snow water equivalents across each basin as they would vary throughout each winter season. Given the fact that recorded and simulated cumulative stream discharge were in good agreement with each other (Figure 5), it follows that the model simulated average pre-and/or post-harvest snow water equivalents quite well for both basins.

Figure 5 was constructed as follows:

1. In Part A, we examined the stream discharge situation that would have occurred if the NMB basin had not been cut, and compared this with the actual pre- and post-harvest records. Evidently, the cumulative stream discharge rate for the uncut situation (as simulated) would have exceeded the cumulative stream discharge rate for the cut situation (as recorded). The total post-harvest difference 12 years after harvesting, however, was quite small at about 1000 mm, or 100 mm per year on average. Research results compiled by MacGregor (1994) indicate that clearcutting increases water yield in general, with amounts varying from +3 to +300% according to site-specific conditions.
2. In Part B, we superposed a vegetational post-harvest growth function on the simulated and recorded stream discharge rates. This post-harvest growth function was constructed from observed forest-atmosphere reflectance values (Bourque *et al.*, 1995), and was incorporated into the Forest Hydrology Model as a means to simulate positive and negative influences of the regrowing vegetation on, e.g., evapotranspiration, rain and snow interception, and fog deposition (Jewett, 1995). Good agreements between recorded and simulated stream discharge rates and cumulated observations were achieved by adjusting the evapotranspiration coefficient of the regrowth function.
3. In Part C, we show recorded and simulated stream water discharge observations for the uncut HB basin. Evidently, the agreement was good, thereby confirming that the ForHyM model, as modified, simulated snow water equivalents, snowmelt and related flow processes with a reasonable degree of accuracy.

In Figures 4 and 6, we further illustrate the ForHyM and ForSTeM simulations by showing:

1. plots of simulated soil temperatures at various depths from 1978 to 1992 for the cut situation (Figure 6A);
2. plots of simulated mineral soil temperatures (10 cm depth) for the cut and uncut situation (Figure 6B);
3. plots of simulated depth of frost penetration into the soil for the cut and uncut situation (Figure 4).

The detailed snowpack water equivalent calculations in Figure 6 show that snow, on average, should have accumulated slightly faster on post-harvest ground than on non-harvested ground. This difference was directly related to differences in pre- and post-harvest snow interception by the vegetation (or lack thereof). This difference would be particularly important at the beginning of the winter when air temperatures drop sharply, and when frost may enter into the ground without much resistance depending on the depth of the snowpack. Accordingly, and on average, early-winter frost (as simulated) should enter the mineral soil more often and more deeply on uncut areas than on cut areas. Extensive snow drifting on cut-over areas, however, should reverse this situation, because soil frost would penetrate deeply into wind-exposed soils.

The results of the soil moisture calculations for the topsoil (rooted mineral soil, top 60 cm) and the subsoil (next 100 cm), for the cut and uncut situation are shown in Figure 7. These plots indicate that:

1. Post-cut summer soil moisture levels would generally exceed the uncut summer soil moisture levels for a number of years.

2. Peak soil moisture levels would occur in spring, at the time of snowmelt.
3. Secondary soil moisture peaks would generally occur in the fall when the evapotranspirational demand diminishes.
4. Mid-winter soil moisture levels should remain at or near field capacity because of reduced water flux into the soil and low to near-zero evapotranspiration rates. Thereafter, soil moisture should gradually increase until the final snowmelt

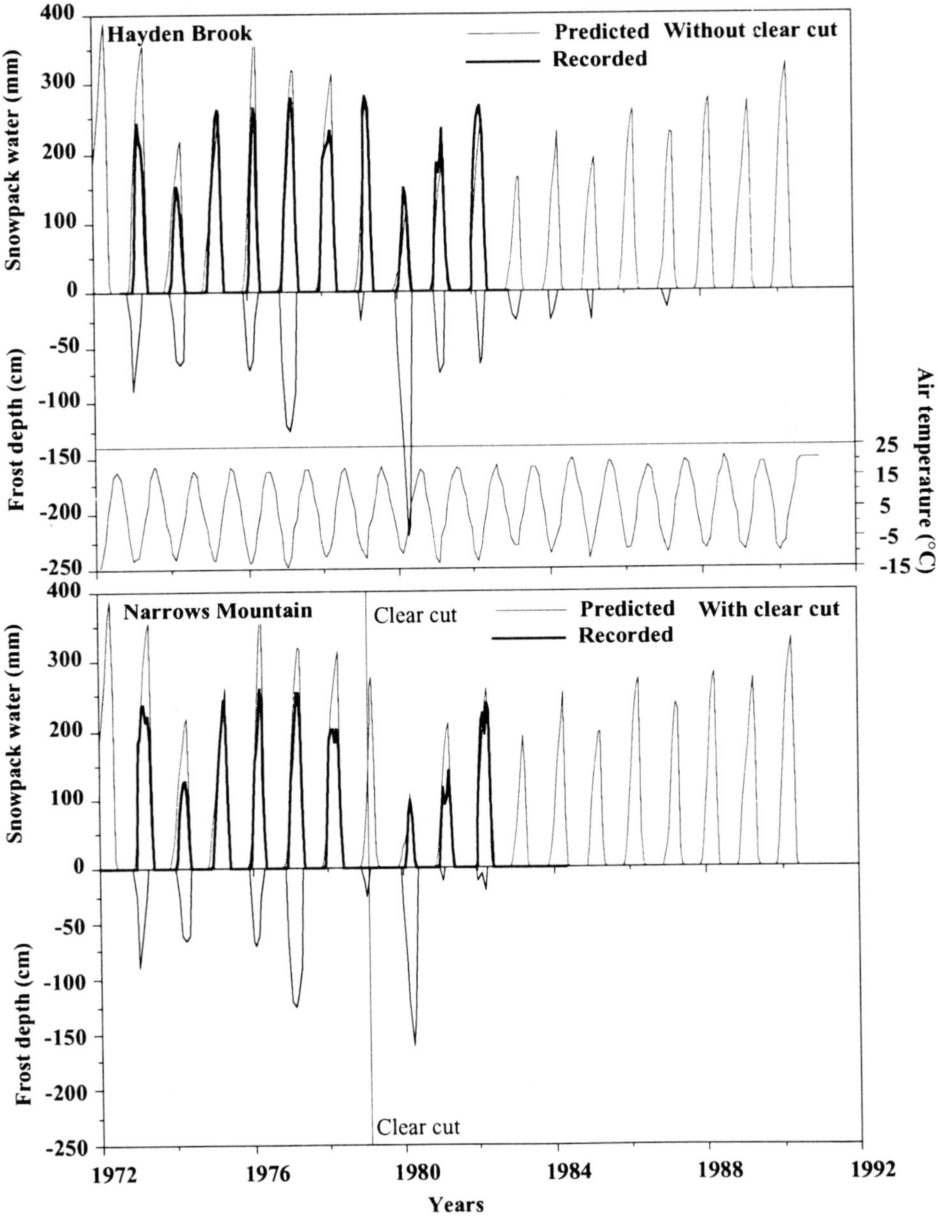

Fig. 4. Simulated (thin lines) and observed snowpack water contents (thick lines) for the Narrows Mountain Watershed basin (top) and Hayden Brook basin (bottom) from 1972 to 1992. Also shown are the depths of frost penetration into the mineral soil, and a trace of the mean monthly air temperatures.

event because of gradually increasing rates of snowmelt.

5. Clearcutting should advance the snowmelt season by about 2 weeks. Simulation results obtained by Leaf and Brink (1972) suggested that snowmelt should occur earlier for a hypothetical partial cut (diameter of canopy opening close to tree height), but the total snowmelt period should not change.

6. Concluding Remarks

By accounting for the forest-atmosphere energy balance in winter, we were able to simulate conventional clearcutting effects on basin-wide snowmelt and stream discharge with reasonable accuracy. We found that these effects on forest hydrology and soil temperatures were relatively small. In particular, calculated post-harvest summer and winter soil moisture levels and temperatures were higher in cut areas than in uncut areas for about the first 5 post-harvest years. The simulations further indicated that clearcutting should advance the spring snowmelt season by about 2 weeks.

The model calculations regarding soil moisture, soil temperature and soil frost penetration were not checked with actual data (such data were not available), but similar calculations are in good agreement with actual observations at other locations (Arp and Yin, 1992; Yin and Arp, 1993).

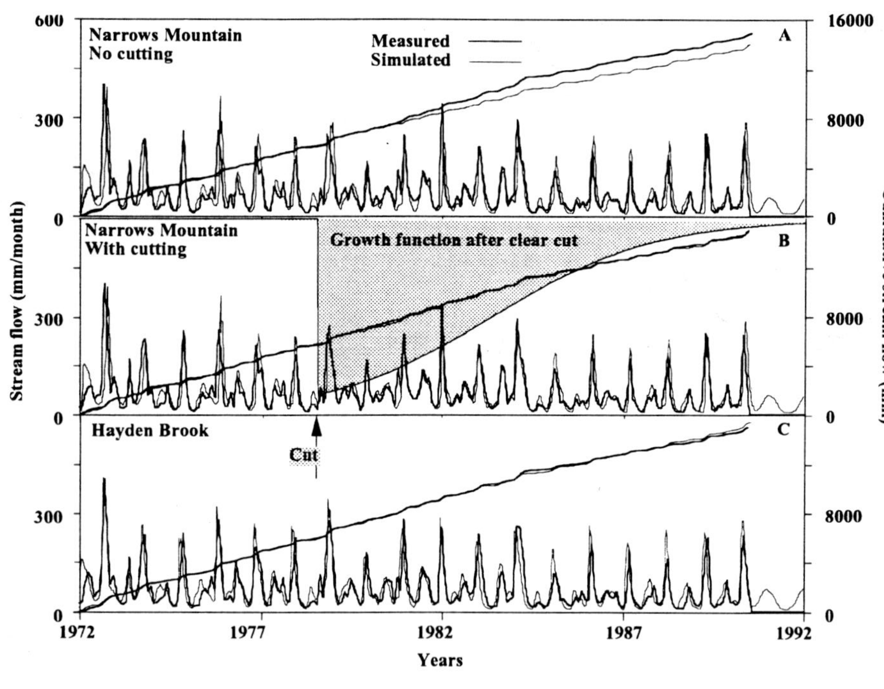

Fig. 5. Plots for observed (heavy lines) and simulated (thin lines) stream discharge rates: monthly rates (mm month^{-1}) as well as cumulative discharge (mm). Top: Narrows Mountain Brook calculations with no corrections made for harvesting. Middle:Narrows Mountain Brook calculations with regrowth corrections to account for the vegetation-induced changes following harvesting. The post-harvest shading relates to a vegetative regrowth function that provides a lumped index for leaf area index as well as the depth development of the re-emerging canopy. Bottom: Hayden Brook (control).

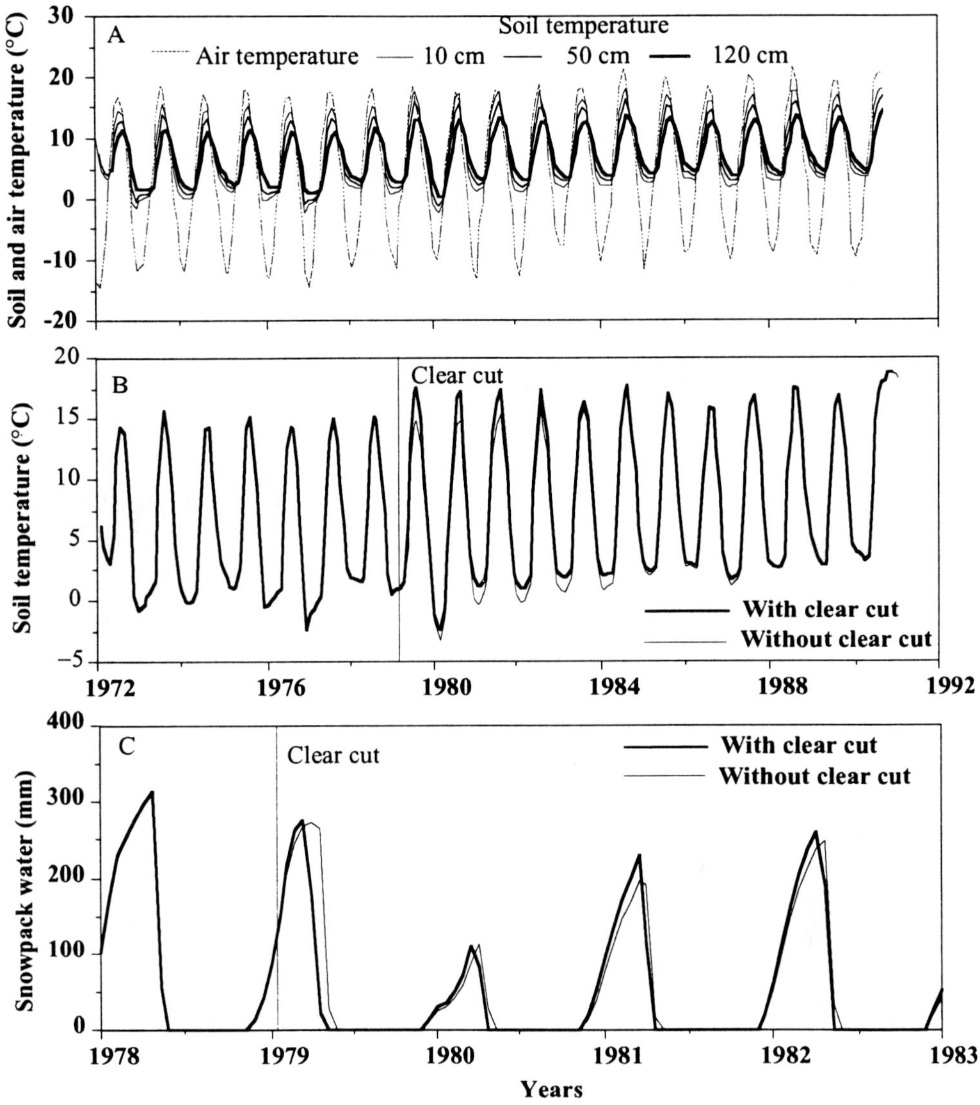

Fig. 6. Top: Predicted soil temperature at various depths in the mineral soils of the Narrows Mountain Brook, from 1972 to 1992 without harvesting. Middle: Predicted soil temperature at 10 cm in the mineral soil of the Narrows Mountain Brook, from 1972 to 1992 with and without out harvesting in 1978/1979. Bottom: Snowpack water equivalents from 1978 to 1983, with or without harvesting in 1978/1979.

References

Arp, P. A. and Yin, X.: 1992, *Can. J. For. Res.* **22**, 864-877.

Bacon, G. B.: 1981, New Brunswick Research and Productivity Council Rep. No. AS/82/164.

Black, T. A., Chen, J.-M., X. Lee and Sagar, R. M.: 1991, *Can. J. For. Res.* **21**, 1020-1028.

Bonan, G. B.: 1991, *Water Resour. Res.* **27**, 767-781.

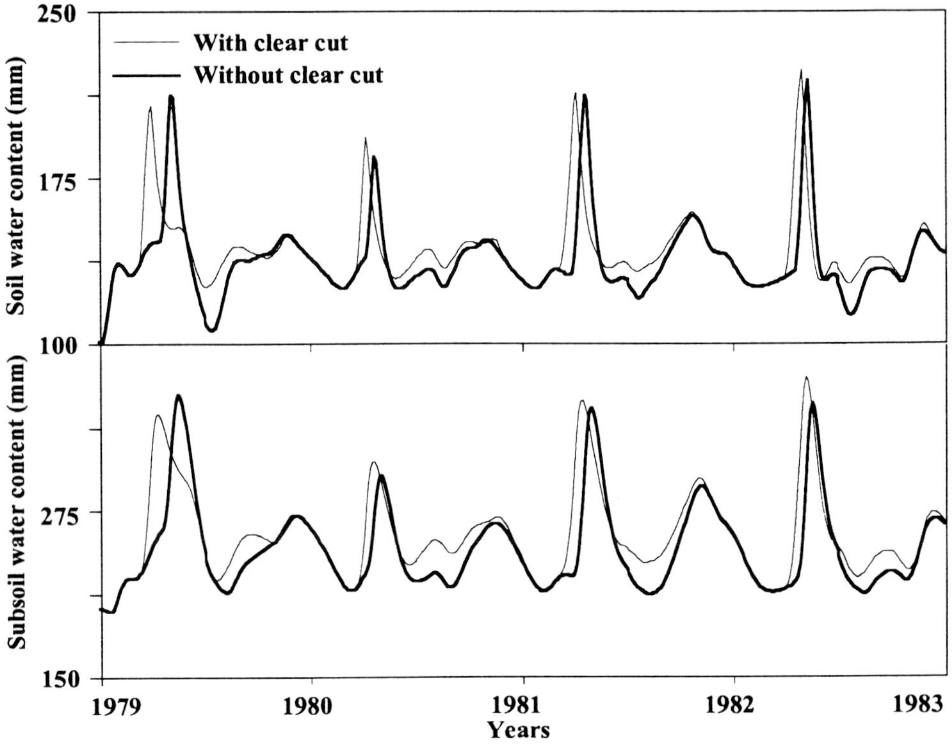

Fig. 7. Simulated soil moisture values for the top 60 cm of the rooted mineral soil, and for 200 cm of subsoil with (thin line) and without (thick line) clearcutting for the uplands soil situation within the Narrows Mountain Brook Basin.

Bourque, C. P.-A., Daugharty, D. A., Dickison, R. B. B. and Arp, P.A.: 1995, *For. Sci.* (In press.)

Daugharty, D. A.: 1984, M.Sc. Thesis. Dept. Forest Resources, University of New Brunswick, Fredericton, Canada, 256 pp.

Daugharty, D. A. and Dickison, R. B. B.: 1988, Proc. Can. Hydrol. Symp. May 9-11. Banff, Canada, pp. 289-296.

Dickison, R. B. B. and Daugharty, D. A.: 1980, Int. Assoc. Sci, Hyd. Proc. of the Helsinki Symp., 23-26 June, 1980. Publ. No. 130. pp. 329-335

Dickison, R. B. B., Daugharty, D. A. and Randall, D. R.: 1981, Proc. 5th Can. Hydrotech. Conf. pp. 59-74.

Hewlett, J. D. and Nutter, W. L.: 1969, *An Outline of Forest Hydrology*. University of Georgia Press, Athens, USA, 137 pp.

Jewett, K.: 1995, M.S. Thesis, University of New Brunswick, Fredericton, Canada. 250 pp.

Krause, H. H.: 1982, *Can. J. For. Res.* **12**, 922-930.

Leaf, C. F. and Brink, G. E.: 1972, *In:* Csallany, S. C., McLaughlin, T. G., and W. D. Striffler, (eds.) *Watersheds in Transition*. Proc. Natl. Symp., Ft. Collins, USA, 1972. A.W.R.A. Illinois, USA. pp. 191-196.

Lutgens, F. K. and Tarbuck, E. J.: 1982, The Atmosphere. 2nd ed. Prentice-Hall, Englewood Cliffs, USA.

MacDonald, J. and Powell, G. R.: 1983, *Can. J. Bot.* **63**, 819-828.

MacGregor, H. G. : 1994, Can. For. Serv. R&D Report No. 10. Canadian Forest Service, Maritimes Region, Fredericton, Canada, 119 pp.

Palmer, D. C.: 1988, M.Sc. Thesis. Dept. Forest Resources, University of New Brunswick, Fredericton, Canada, 109 pp.

Yin, X. and Arp, P. A.: 1993, *Can. J. For. Res.* **23**, 2521-2536.

TEMPORAL AND SPATIAL VARIATIONS OF TERRESTRIAL BIOMES AND CARBON STORAGE SINCE 13 000 YR BP IN EUROPE: RECONSTRUCTION FROM POLLEN DATA AND STATISTICAL MODELS

C. H. PENG[1,2], J. GUIOT[1], E. VAN CAMPO[3] and R. CHEDDADI[1]

[1]*Laboratoire de Botanique Historique et Palynologie, UA CNRS 1152, Faculté de St Jérôme, Boîte 451, 13397 Marseille cedex 20, France*
[2]*Canadian Forest Service, 5320 - 122 Street, Edmonton, Alberta T6H 3S5, Canada (present address)*
[3]*Laboratoire de Géologie du Quaternaire, CNRS Luminy, 13288 Marseille cedex, France*

Abstract. Statistical models calibrated from field measurement data are used to reconstruct the past terrestrial carbon (C) storage from pollen data for the last 13 000 yr BP in Europe. The pollen-based climatic and biome reconstructions provide the input data for these statistical models, i.e., annual mean temperature, total annual precipitation, annual actual evapotranspiration, annual potential evapotranspiration and biome type with a spatial resolution of 0.5° × 0.5° longitude/latitude. Our reconstructions indicate that the last 13 000 yr BP were characterized in Europe by variations of terrestrial biome and net primary productivity (NPP) at various temporal and spatial scales. For the considered region, our results also suggest that changes in climate have significantly altered the distribution of terrestrial biomes and affected the uptake of CO_2 for NPP. However, these changes did not translate into significant C storage change in potential terrestrial biosphere during the Holocene. The largest decrease of terrestrial C storage (compared to modern levels) is found during the late-Glacial period mainly due to the persistence of ice sheets and the small extension of forest.

Keywords. BIOME, POLLEN DATA, CARBON STORAGE, STATISTICAL MODEL, EUROPE.

1. Introduction

An understanding of past and possible future climate changes and the global carbon (C) cycle will require a clear picture of how vegetation changed in the past and may change in the future (Prentice *et al.*, 1991; Overpeck, *et al.*, 1992). The distribution of terrestrial biomes responds to changes in summer and winter temperatures and also moisture balance (Woodward, 1987; Prentice *et al.*, 1992). Late Quaternary climatic changes produced large changes in the distribution of vegetation types. Sets of ^{14}C-dated pollen diagrams provide records of vegetation patterns at various spatial and temporal scales in the past (Huntley and Birks, 1983).

Patterns of primary productivity and of C storage in vegetation and soil also respond to climatic changes. The assumption is that C storage in global terrestrial biomass was relatively low during the full glacial time, increasing considerably to a maximum between 9500 and 4500 years ago and then declining to an intermediate amount by the present time was drawn by Grove (1984). On the basis of the empirical Miami regression model (Lieth, 1975) and a climatic model (Kutzbach and Guetter, 1986), Meyer (1988) has shown that the net primary productivity (NPP) for the past 18 000 years was sensitive to annual temperature and precipitation. Foley (1994) found that total C storage in the terrestrial biosphere did not change significantly over the last 6000 yr BP, which corresponds to the results of Peng *et al.*, (1994a,b). These problems challenge our ability to identify the location and magnitude of terrestrial C sinks and sources during periods of climate change.

Water, Air and Soil Pollution **82**: 375–390, 1995.

Previous studies (Peng *et al.*, 1994a, 1994b) have demonstrated the ability of statistical biospherical models to provide reconstructions of terrestrial C storage from pollen data. The weak point of the Osnabrück Biosphere Model (OBM) is the soil submodel as was reported previously (Peng *et al.*, 1994b). Soil C storage was formerly expressed as a constant percentage of litter production. We now express this quantity as a function of actual evapotranspiration (AET), annual soil moisture deficit (which itself depends on precipitation and potential evapotranspiration (PET)) and of the site disturbance due to human land-use effect (Meentemeyer *et al.*, 1985), the latter being neglible at the palaeo-scale. The calculation of NPP is also improved by the use of AET (Montreal model, Lieth and Box, 1972). NPP then is determined by the most limiting climatic factor amongst the annual temperature, annual precipitation and AET, the latter introducing a climatic seasonal effect in NPP.

The objective of this paper is to reconstruct biome variations and the corresponding climatic changes since the end of the last glaciation and to study their influence upon C storage. This will enable us to better understand the role of the temperate and boreal forests of Europe in the natural global C cycle during large scale climatic changes.

We use the pollen data of Huntley and Birks (1983) (which, for Europe, are available for the last 13 000 years) to calculate the vegetation and climatic parameters needed by the model and to provide maps of C storage in time steps of 1000 years. Dating of the pollen record is not perfect, but is acceptable, given the well known errors in the calculations of the different components of the C cycle. Better data will soon be available from the European Pollen Database.

For each time-slice, the calculations follow four steps:

- attribution of a biome ("biomization") to each pollen site;
- deduction of the four climatic variables (annual temperature, annual precipitation, PET and AET);
- interpolation of the biomes and climatic parameters to a $0.5° \times 0.5°$ latitude/longitude grid;
- calculation of C storage in vegetation and soil.

The models are validated using independent observations of NPP and C storage in vegetation and soil.

2. Data and Methods

The use of pollen data to reconstruct climate and C storage is based on the hypotheses that modern analogues exist for the past and that the equilibrium of C storage depends on the vegetation structure and climate. These hypotheses are acceptable for the time-period studied. The main limitation lies in the C storage equilibrium hypothesis which nevertheless remains acceptable, even for soil, when time steps of 1000 years or more are taken (Schlesinger, 1990).

2.1. Pollen Data

The modern pollen data set consists of 1719 surface samples collected in Europe, *i.e.*, those already used in previous work (Guiot, 1990; Peng *et al.*, 1994a).

The pollen sum used in the calculations is the sum of 26 pollen types (*Abies, Alnus, Betula, Buxus, Carpinus, Cedrus, Corylus, Fagus, Fraxinus, Juniperus, Larix, Olea,*

Picea, Pinus, Pistacia, Quercus dec., Quercus ilex, Salix, Tilia, Ulmus, Artemisia, Chenopodiaceae, Ephedra, Ericaceae, Poaceae and *Hedera*). These taxa are rarely recognized by the palynologist to the species level, more frequently to the genus level and sometimes to the family level, which limits the precision.

Fossil pollen data spanning the last 13 000 yr BP in Europe were derived from Huntley and Birks (1983). These data consist of relative pollen counts of major taxa in lake or mire sediments for *ca.* 360 sites. The area extends from 40°N to 75°N and from 10°W to 60°E. Sixty-five per cent of the sites are ^{14}C-dated. The rest were dated by pollen-correlation with ^{14}C-dated nearby sites, or by comparison with standard ^{14}C-dated regional pollen stratigraphies. The same 26 taxa retained in the modern pollen data were used, and all percentages were calculated relative to the sum of these 26 taxa.

2.2. Pollen-based biomes reconstruction

Prentice *et al.* (unpubl. ms.) have developed a method to attribute a biome to each pollen assemblage. Each pollen taxon is assigned to one of the plant functional types such as defined in the BIOME model (Prentice *et al.*, 1992). Because each pollen taxon is not identified to the species level, it is sometimes impossible to do a unique classification. For example, a given taxon such as *Pinus* can be a cool temperate conifer, a boreal conifer or a warm temperate evergreen tree. A likelihood index is calculated for each plant functional type and translated in terms of biomes according to the combinations defined for the BIOME model. Finally for each biome, we obtain an index defined as the sum of the percentage square root of all the taxa potentially present in the biome. These indices are compared and the biome for which the index is maximum is attributed to the spectrum.

A particular interpolation scheme is applied to maintain the influence of topography on vegetation through the pollen record, even with a sparse pollen coverage. This scheme is based on the fact that the deviations between past and modern indices constitute a more spatially homogenous field than the a single index. This is because that index variations over time are relatively insensitive to topography. We calculate the anomalies by subtracting the modern indices from palaeo-indices. The anomalies are then smoothly interpolated to the $0.5° \times 0.5°$ latitude/longitude grid, using a standard method of weighted averaging according to the inverse space distance (Guiot, 1991). Simultaneously, the modern pollen database is used to interpolate the modern biome indices to the same grid (Figure 1b). Because the modern coverage is much better than the palaeo-coverage, this grid has higher quality than one based on direct interpolation grid of the palaeodata. The modern grid values are then added to the values of the grid of palaeo-anomalies to provide gridded palaeo-biome indices. Finally the biome with the higest index is attributed to each grid point.

2.3. Map comparison

The biome map reconstructed from pollen and that simulated by the global BIOME model of Prentice *et al.* (1992), are compared numerically using the Kappa statistic, which measures the grid cell by grid cell agreement between these maps (Cohen, 1960; Monserud, 1990; Monserud and Leemans, 1992).

Monserud (1990) and Prentice *et al.* (1992) used the following qualitative descriptors to characterize the degree of agreement suggested by the Kappa statistic:

very poor to poor agreement if $\kappa < 0.4$, fair agreement if $0.4 < \kappa < 0.55$, good agreement if $0.55 < \kappa < 0.7$, very good agreement if $0.7 < \kappa < 0.85$, and excellent agreement if $\kappa > 0.85$.

A visual comparison of the biomes reconstructed from pollen with the biomes predicted by the BIOME model indicated good agreement for Europe (Figure 1) except for some fuzzy boundaries, e.g., taiga/tundra and cool conifer/cool mixed forests. This result is supported by the overall value of the Kappa statistic for the two maps ($\kappa = 0.57$).

For individual modern biomes, the agreement was very good for steppe (warm/cool grass and shrub) ($\kappa = 0.85$) and semidesert ($\kappa = 0.72$); and was good for cool mixed forest ($\kappa = 0.56$), taiga ($\kappa = 0.62$), temperate deciduous forest ($\kappa = 0.56$), and xerophytic woods/shrub biome ($\kappa = 0.62$). But there was poor agreement for tundra, cool conifer forest, cold deciduous forest, cold mixed forest, and evergreen/warm mixed forest ($\kappa < 0.4$), mainly because: (1) pollen samples integrate a variable area due to the long distance transport of some pollen types and (2) there is anthropogenic disturbance of the vegetation.

2.4. Pollen-based Climatic Reconstruction

The climatic parameters needed for calculating the vegetation and the soil C storage (i.e., annual temperature, annual precipitation, AET and PET) are deduced directly from the biomes to ensure a good correspondence between climate and biome. The average value of these climatic parameters is calculated for each biome and is attributed

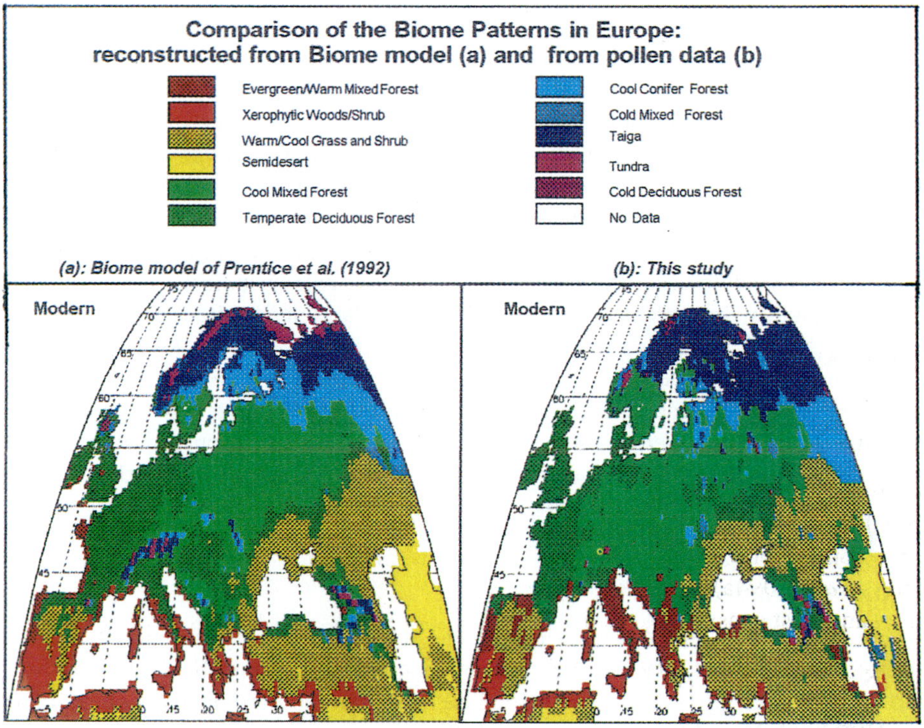

Fig. 1. Comparison of two biome maps, (a) reconstructed from the BIOME model of Prentice *et al.* (1992) and (b) from modern pollen data in Europe. The Kappa statistic (κ = 0.57) indicates a good agreement between two maps.

to each fossil pollen spectrum according to its biome. We have done the same for the geographically corresponding modern pollen spectra, so that the climatic anomalies have been calculated. These anomalies have been interpolated to the 0.5° × 0.5° grid cells using the weighted averaging method with a large radius (5°) and these interpolated anomalies have been added to the corresponding modern climate provided by Leemans and Cramer (1991). This stepwise procedure is again developed to preserve the topography of Europe.

The main consequence of this climate reconstruction procedure is that a point located in the central part of a given biome zone is attributed the mean climate of the biome and a point located in a transition between two biomes is attributed a climate at the midpoint of the two corresponding climates.

2.5. Description of Model

2.5.1. Net Primary Productivity (NPP)

NPP, which is the rate at which vegetation in an ecosystem fixes C from the atmosphere (gross primary productivity) minus the rate at which it returns to the atmosphere (plant respiration), represents net C input from the atmosphere into the biosphere. The NPP (g m^{-2} yr^{-1}) can be expressed as a function of annual temperature T_a (°C), annual precipitation P_a (mm) and actual evapotranspiration AET (mm). We use three empirical functions based on the original MIÀMI model of Lieth (1975) (equations (1) and (2)) and on the Montreal model of Lieth and Box (1972) (equation (3)).

$$NPP_T = 3000\,\{1 + \exp(1.315 - 0.119\,T_a)^{-1}\} \qquad [1]$$

$$NPP_P = 3000\,\{1 - \exp(-0.000664\,P_a)\} \qquad [2]$$

$$NPP_{AET} = 3000\,\{1 - \exp[-0.0009695(AET - 20)]\} \qquad [3]$$

According to the limiting factors principle, the NPP is the minimum of these three values:

$$NPP = \mathrm{Min}\,(NPP_T, NPP_P, NPP_{AET}) \qquad [4]$$

2.5.2. Vegetation carbon storage

The relationship between vegetation biomass and NPP has been discussed by Whittaker and Woodwell (1971), by Esser (1984, 1991) and by Peng *et al.*, (1994a). Vegetation C storage (VCS, in g m^{-2} yr^{-1}) is expressed as a non-linear function of NPP (Peng *et al.*, 1994a) and mean stand age, calibrated on 106 measurement sites in European forest ecosystems (Cannell, 1982; Rodin *et al.* 1975; Lossaint and Rapp, 1978).

$$VCS = 0.45(0.0147\,A^{0.775}\,NPP^{0.579}) \qquad [5]$$

Here NPP is expressed in g m^{-2} yr^{-1} dry matter. Mean stand age (A, expressed in years) of each vegetation type is derived from Esser (1991). The equation assumes a biomass C content of 0.45 (Olson *et al.*, 1985).

2.5.3. *Soil carbon storage*

The geographic and climatic model of global soil C has been proposed by Meentemeyer *et al.*, (1985). This model predicts soil C storage (SCS) from the annual actual evapotranspiration (AET), the site disturbance (DIS) due to land-use, and annual soil moisture deficit (D). SCS is expressed as the following function:

$$SCS = \exp\{-1.05989DIS + 0.00156(400[\ln(AET + 1)] - AET) - 0.00102D - 0.2269\} \quad [6]$$

The disturbance factor (DIS) splits the land coverage into undisturbed (DIS = 0) and disturbed (DIS = 1). Soil moisture deficit (D) is defined as the difference between annual potential evapotranspiration (PET) and AET. An AET which falls below PET implies a soil moisture deficit. The AET and PET are calculated by using a simple bucket model and assuming a soil water capacity of 150 mm, which is a reasonable estimate for deeply rooted natural vegetation in Europe (Harrison *et al.*, 1993). In any case, we have tested the value of 300 mm used by Meentemeyer (1985) for the globe, without significant changes in AET for Europe.

2.6. MODEL VALIDATION

Simulations of NPP are compared to 20 independent observations reported by Raich *et al.* (1991), McGuire *et al.* (1992) and Gauquelin *et al.* (1994), which represent 11 major ecosystems. The simulations and observations of NPP are highly correlated (r = 0.93) (Table I). It should be noted that19 of these observations have already been used to calibrate the TEM model (Raich *et al.* 1991; McGuire *et al.*, 1992; Melillo *et al.*, 1993), the CASA model (Potter *et al.*, 1993), and the DEMETER model (Foley, 1994).

The same dataset is used to validate the vegetation C and soil C estimates. In both cases high correlations are obtained: r = 0.91 for vegetation C and r = 0.81 for soil C (Table I).

These independent tests provide a means of estimating errors associated with these calculations. For this the standard deviation of the differences between the 20 observations and the 20 estimates have been calculated for NPP, vegetation and soil C density. Table I shows that the errors range between 19% and 29%, which is relatively small when considering that they are calculated from independent observations.

TABLE I

Verification statistics of the carbon parameters: correlation between estimates and observations, mean value, standard error and percentage of the mean represented by the standard error. These parameters are calculated from 20 observations.

Parameter	Correlation	Mean	Standard error	% Mean
NPP (g C m^{-2} yr^{-1})	0.93	477	140	29
Soil carbon (kg C m^{-2})	0.81	12	2.3	19
Vegetation carbon (kg C m^{-2})	0.91	9.3	2.7	29

3. Results and Discussion

We will successively present the reconstructions and discuss the results for the biomes, the NPP, vegetation and soil C storage. We will present maps only for the 12, 9, 6, 3 and 0 ka BP time periods, but the synthetic results will be discussed for each 1000 year time-slice.

3.1. Spatial and Temporal Variations in Biome Patterns

The reconstructed 12 000 yr BP (12 ka) biomes are dramatically different from the modern ones (Figure 2). In southern Europe, the typical Mediterranean biomes, e.g., xerophytic woods/shrub and evergreen/warm mixed forests were replaced by temperate deciduous forests and by warm grass/shrub (steppes). In central Europe, modern cool mixed and temperate forests were replaced by cold deciduous forests and tundra. Taiga was only present in eastern Europe northwest of the Black Sea. Most of northern Europe was covered by ice sheet (Denton and Hughes, 1981). No data were available east of 30°E.

At 9 ka, the most dramatic changes are in northern Europe (between 60°N and 70°N) where ice cover was strongly reduced and replaced by tundra, cold deciduous and mixed forests. Temperate deciduous forests extended over most of western Europe. Dry vegetation (xerophytic wood/shrubs and steppes) was reduced to a few spots in southern Europe, due to a more humid climate in summer.

For 6 ka, the most pronounced changes in biome cover are seen in Fennoscandia, where taiga and tundra were considerably decreased. This indicates a warmer climate especially in summer. Between 9 ka and 6 ka, the northern limit of the temperate deciduous and cold mixed forests shifted North. In southern Europe, we do not note any significant changes.

The reconstructed 3 ka biomes distribution are broadly similar to those at 6 ka. The only large differences are found in northern Scandinavia, where the cold mixed forest was replaced by cool conifer forest and taiga. The limit of the cool temperate deciduous mixed forest was already occupying its present position and the Mediterranean vegetation started to extend into southern Europe.

Climatically speaking, these biome area variations show a cold and dry climate predominant over much of Europe during the Late-Glacial period (between 13 and 10 ka), and warm and wet climatic conditions in most of Europe (especially in northern Europe) during the mid-Holocene (approximately between 4000 and 8000 yr BP). This is in agreement with the climatic reconstructions of Huntley and Prentice (1988, 1994), and Guiot *et al.* (1993); and is also consistent with the reconstruction of past moisture conditions based on lake-level records (Harrison *et al.*, 1991, 1993).

3.2. Spatial and Temporal Variations in Net Primary Productivity (NPP)

For 12 ka, Figure 3 shows the largest decrease in NPP compared to the modern situation because of the cold climate especially due to the proximity of the ice sheet. The only small regions with large positive anomalies of NPP are found in southern Europe, especially in Spain, because of increased precipitation over this area at 12 ka (as already reconstructed in COHMAP, 1988). Total NPP for 12 ka is approximately 39% lower than today.

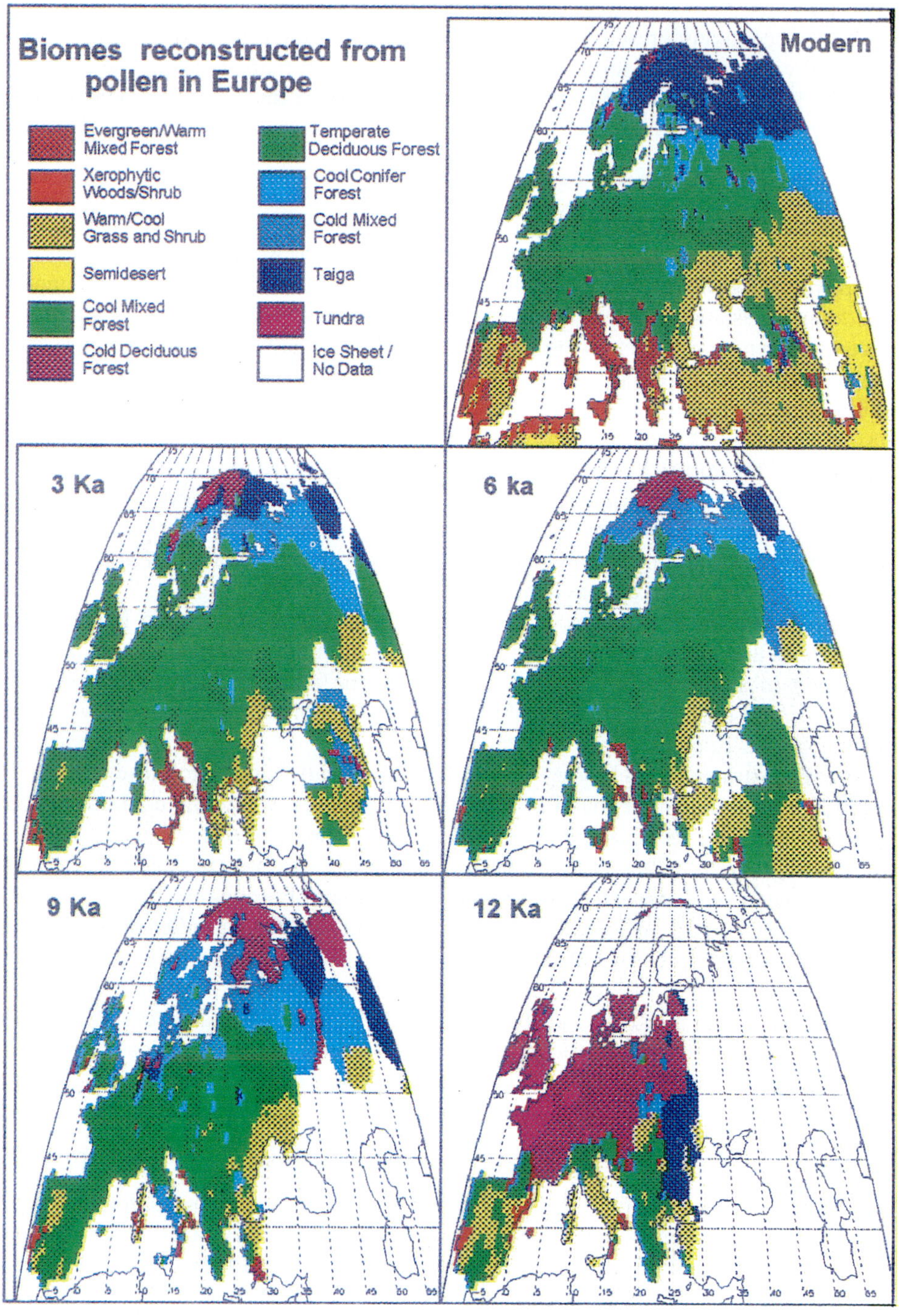

Fig. 2. Spatial patterns of terrestrial biomes at 3000 yr intervals as reconstructed from pollen data since 12 000 yr BP in Europe. The ice sheet extent follows Denton and Hughes (1981).

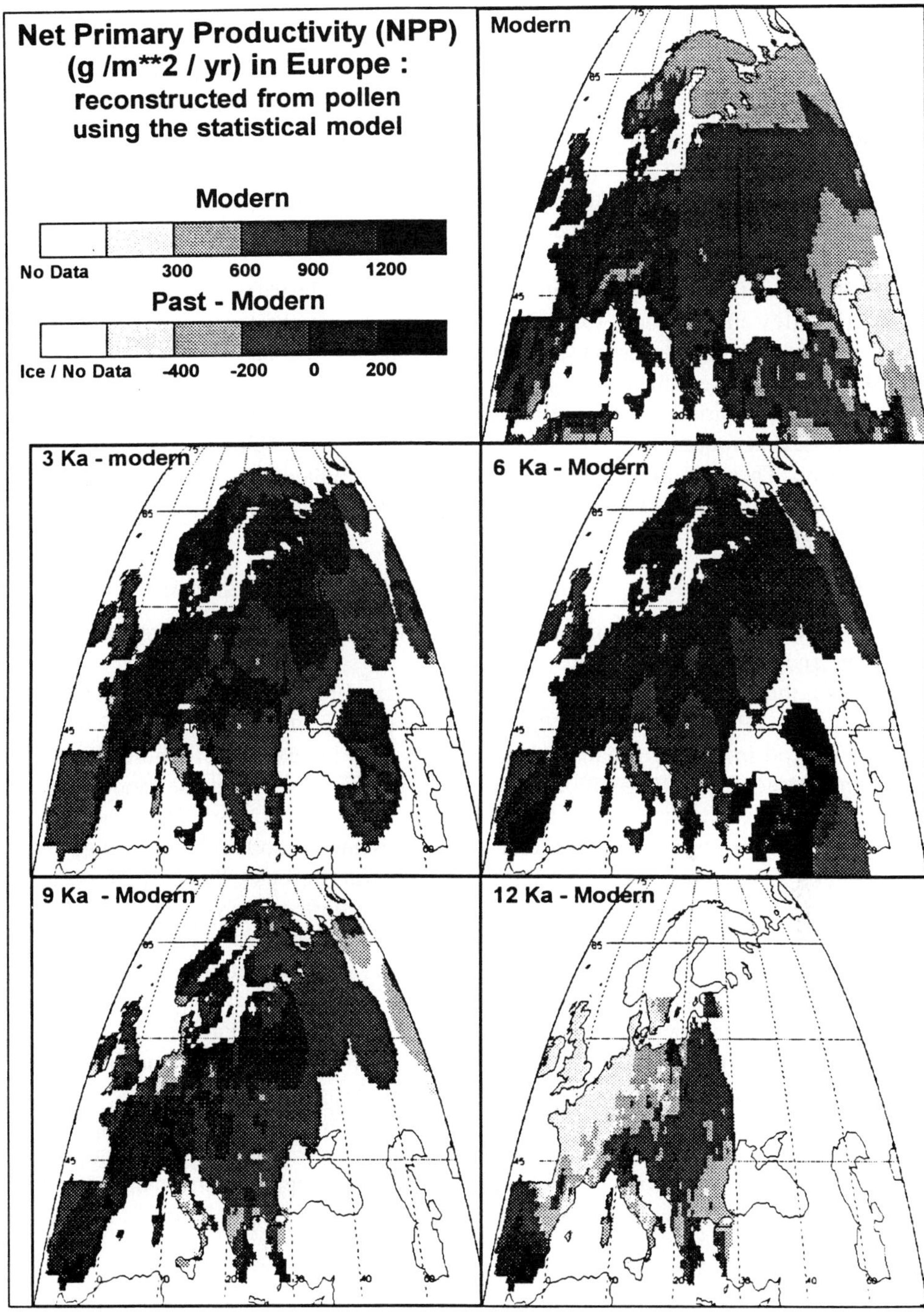

Fig. 3. Spatial variation of Net Primary Productivity (NPP, in g m^{-2} yr^{-1}) (Past *minus* Modern) at 3000 yr intervals since 12 000 yr BP in Europe: reconstructed from pollen data using the statistical model.

At 9 ka (Figure 3), the regions of positive NPP anomalies are found in Fennoscandia, France and Spain, where temperature was more than 2 °C warmer than now. Total NPP is about 3% lower than at the pesent time.

At 6 ka, warmer summers over northern Europe and warmer winters over central and northern Europe (Huntley and Prentice, 1988, 1994), where the temperature is the main limiting factor, explain the positive anomalies of NPP. In southern Europe, the positive anomalies must be explained by high precipitation. Total NPP over the common area covered by available pollen data in Europe for 6 ka is about 7% larger than today, which is qualitatively consistent with the results of the global simulations of Meyer (1988) and of Foley (1994).

The NPP variations at 3 ka are similar to those at 6 ka because of a similar climatic regime (Huntley and Prentice, 1994). Some differences are observed in Spain, Portugal and Italy, where the precipitation was higher at 6 ka than at 3 ka. The total NPP for this period is only 1% larger than the modern value.

The temporal variations of NPP show that maximum NPP is found at 6 ka, minimum NPP is found at 13 ka, and only small changes in NPP are observed during the Holocene. These results are very similar to those simulated by Meyer (1988) on the basis of the Miami Model and the global climate model simulations of Kutzbach and Guetter (1986).

3.3. Temporal and Spatial Variations in Vegetation Carbon Storage

The spatial variability of modern vegetation C density reflects spatial heterogeneity of the biome and climate distributions (Figure 4).

At 12 ka, the spatial variations of vegetation C densities (Figure 4) are correlated to variations in NPP (Figure 3). The largest decrease is found in the area of western Europe covered in tundra. Small regions of C storage increase are found in southern Spain and Greece. Total vegetation C storage was 59% lower than the modern value (Figure 5).

For 6 ka, changes of vegetation C densities (compared to modern levels) range from -5 to +5 kg C/m^2 over northern Europe and central Europe. The largest positive changes are found in Spain, southern Italy, and on the northeastern shore of the Black Sea, mainly due to the extension of the temperate deciduous forests relative to the less productive steppes and xerophytic woods and shrub. Total vegetation C storage was only 5% larger than the modern value (Figure 5).

The overall spatial patterns of vegetation C densities for 3 ka and for 9 ka are similar to those of 6 ka. Nevertheless, total vegetation C storage is lower than at the present time by 4.2% and 12.3% respectively (Figure 5).

Based on the common area (5.64×10^6 km^2) covered by available pollen data, Figure 5 shows the temporal variation of vegetation C storage. There are no significant changes in terms of vegetation C storage after 8 ka. If we apply the errors calculated in section 2.6, we find that vegetation C storage varies from 18 (±5) to 70 (±20) Pg C.

3.4. Temporal and Spatial Variations in Soil Carbon Storage

For 12 ka (Figure 6), apart from glaciated areas of northern Europe, much of central Europe and southern Europe shows no dramatic change in soil C densities (ranging from -3 to +3 kg C/m^2). Some southern regions have increased by more than 3 kg C/m^2. In summary, the total soil C storage was reduced by 23% (Figure 5).

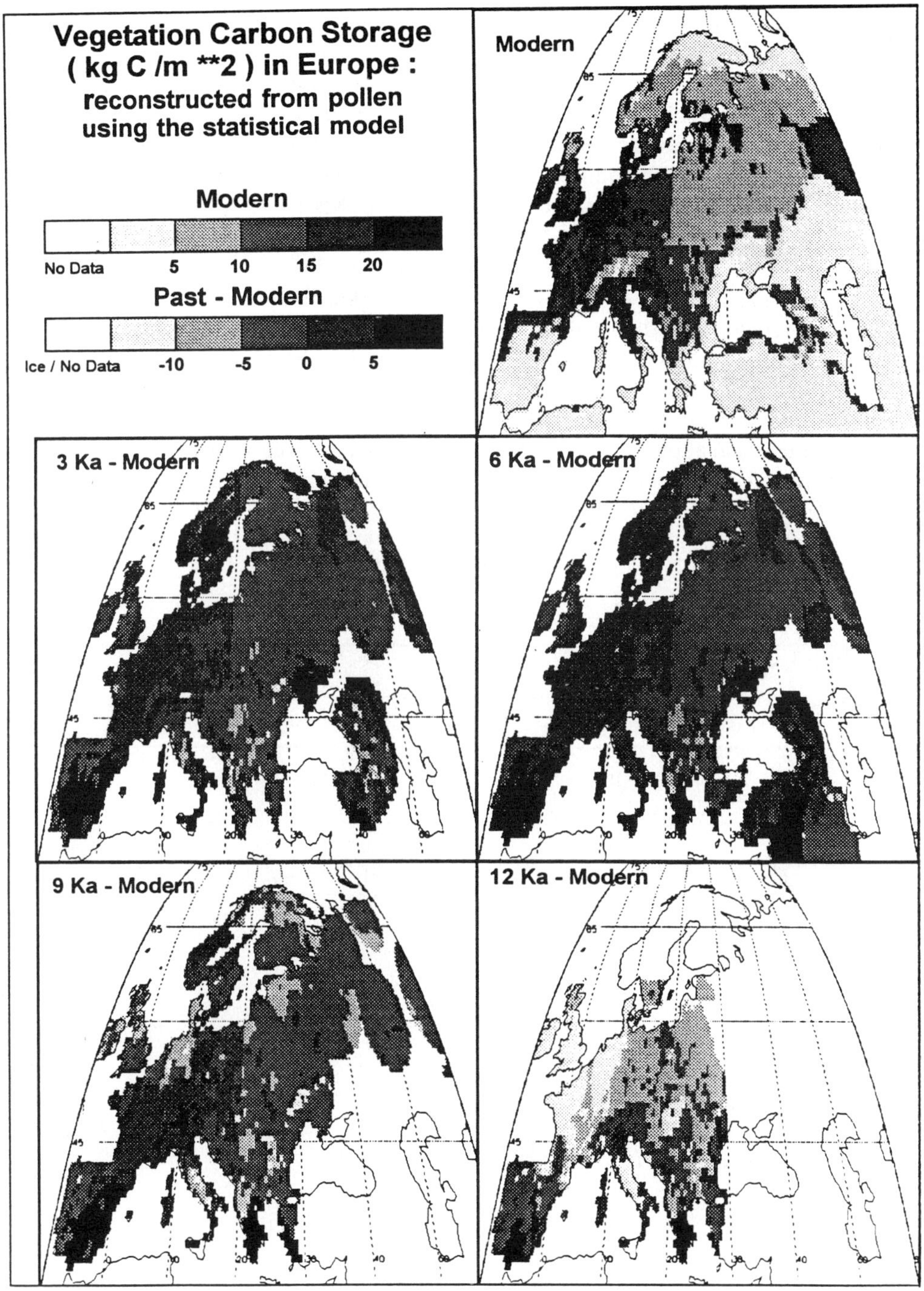

Fig. 4. Spatial variation of vegetation carbon densities (in kg C m^{-2} yr^{-1}) (Past *minus* Modern) at 3000 yr intervals since 12 000 yr BP in Europe: reconstructed from pollen data using the statistical model.

At 6 ka, the changes in soil C densities (compared to modern levels) range from -3 to +3 kg C/m^2 over much of Europe, except in the far eastern part. This is because of the reduction in steppe (Figure 2). Total soil C storage is not different from the modern value for the considered area (Figure 5).

Although the spatial patterns of soil C densities of 3 ka and 9 ka were similar to those of 6 ka, the total soil C storage anomalies were respectively -1% and -6% (Figure 5).

The temporal variation of soil C storage is roughly parallel to that of vegetation C storage (Figure 5) but with a much smaller amplitude. If we apply the errors calculated in section 2.6, we find that soil C storage varies from 66 (±13) to 92 (±17) Pg C.

3.5. Spatial Variations in Terrestrial (Vegetation+Soil) Carbon Storage

Figure 5 shows that after 9 ka, changes in total terrestrial C storage range from -5% to +3% compared with modern values, and between -31% and -47% during the Late-Glacial periods, mainly due to the persistence of the ice sheet (accounting for a value of -28%), and cold and dry climatic conditions (accounting for a value between -4% and -19%). Gliemeroth (pers. comm.) found a similar result when examining changes in belowground and aboveground biomass in Europe since Late-Glacial times.

In conclusion, if we add the errors calculated in sections 3.3 and 3.4, we find that the total C storage varies from 84 (±18) to 162 (±37) Pg C.

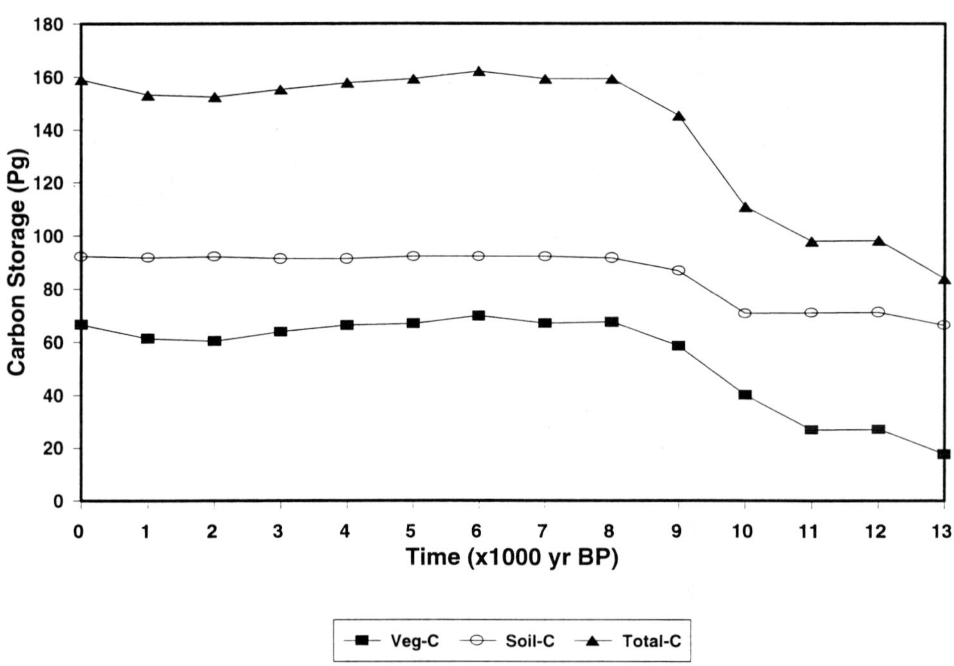

Fig. 5. Temporal variations of total carbon storage (in Pg C) at 1000 yr intervals since 13 000 yr BP based on the common area (5.64×10^6 km^2) covered by pollen data.

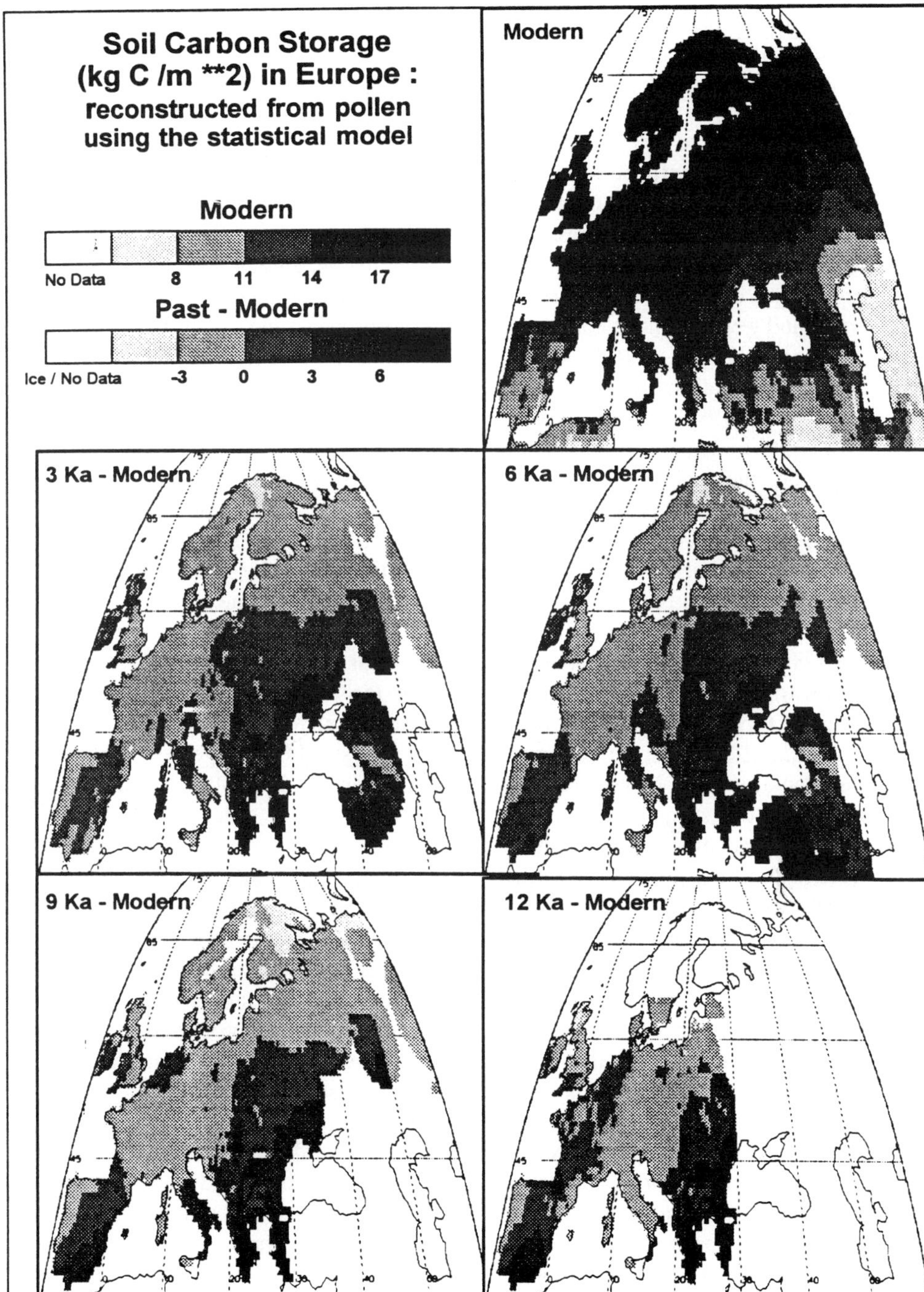

Fig. 6. Spatial variation of soil carbon densities (in kg C m^{-2} yr^{-1}) (Past *minus* Modern) at 3000 yr intervals since 12 000 yr BP in Europe: reconstructed from pollen data using the statistical model.

4. Summary and Conclusions

The distribution of biomes reconstructed from pollen is in good agreement with results obtained from the modern climate using the BIOME model of Prentice *et al.* (1992), excluding some boundaries that are still fuzzy (e.g., the taiga/tundra transition and the cool conifer/cool mixed forests transition). The pollen-based biome patterns have changed significantly during the last 13 ka. The most striking changes of reconstructed biome (compared to the present) are found during the Late-Glacial period (13–10 ka) with a large part of northern Europe covered by ice and with reduction of the temperate forests and their replacement by colder forests and tundra. At 6 ka, we observe a northward shift of forest zones and an eastward shift of the temperate deciduous forest. These biome changes suggest a cold and dry climate over much of Europe during the Late-Glacial period, and warmer and wetter climatic conditions in most of Europe during the mid-Holocene.

NPP, vegetation and soil C storage calculations have been validated using 20 independent observations reported by Raich *et al.* (1991), McGuire *et al.* (1992) and Gauquelin *et al.* (1994). These tests show that the errors associated with the models range from 19% to 29% of the estimates.

For Europe, the regional scale patterns of NPP are strongly affected by climatic changes. Maximum NPP is found at 6 ka mainly due to warmer summer over northern Europe and warmer winters over central and northern Europe, while minimum NPP is found at 13 ka because of the cold climate resulting from the proximity of the ice sheet. Small changes in NPP are observed before the mid-Holocene.

On the basis of the common area covered by pollen data, we do not find any significant changes in total C storage of the terrestrial biosphere during the Holocene. This result appears to be in agreement with the ice core evidence that shows that atmospheric CO_2 concentration did not change dramatically over the last 10 000 yr BP (Barnola *et al.*, 1987). The major changes in total terrestrial C storage are found during the Late-Glacial period. Nevertheless, changes in litter and in peatlands are not taken into account. Our estimates are qualitatively comparable to other studies based on paleoclimatic simulations (Foley, 1994).

In conclusion, the results presented here suggest that changes in climate can significantly alter the distribution of the terrestrial biomes and affect net primary production. In Europe, however, these changes did not translate into significant changes of C storage during the Holocene. The largest decreases of terrestrial C storage (between -31% and -47%) are found during the Late-Glacial period due to the persistence of the ice sheet (accounting for -28% reduction), and cold and dry climatic conditions (accounting for a value ranging between -4% and -19%). These changes are significant as the standard error is around 23% of the estimate's value. Of course, our understanding of past vegetation changes and modelling ability for terrestrial C budgets of the past are rudimentary and are limited not only by available pollen data, but also by the accuracy of pollen-based climatic reconstructions and statistical models (Peng *et al.*, 1994b). Our results only give a snapshot of the selected periods and the considered region assumed to be in equilibrium. However, these records of the past could help to improve our understanding of future climatic change and its potential impact on vegetation and on the global C cycle. Moreover, the most noticeable improvements in this field of research will be realized using the European Pollen Database which contains pollen data from all over Europe.

Acknowledgments

We are grateful to F. Saadi, J. Belmonte, V. Ruis-Vasquez and S. Bottema who have kindly provided pollen surface samples for Spain, Morocco and the Near East and B. Huntly who has kindly provided fossil pollen data. Thanks also to T. Gauquelin, and G. Jalut who provided some field data for Spain to validate the statistical model. Valuable comments have been given by J. C. Duplessy and R. Leemans. The Programme Environnement of the French Centre National de Rechèrche Scientifique, and the EPOCH and ENVIRONMENT programs of the European Community have funded this study.

References

Barnola, J. M., Raynaud, D., Korotkevitch, Y. S., and Lorius, C: 1987, *Nature* **329**, 408-414.

Cannell, M. G. R. (ed.): 1982, *World Forest Biomass and Primary Production Data*. Academic Press, London, UK, pp. 52-82.

Cohen, J.: 1960, *Educat. Psychol. Measur.* **20**, 37-46.

COHMAP Project Members: 1988, *Science* **241**, 1043-1052.

Denton, G. H. and Hughes, T. J.: 1981, *The Last Great Ice Sheets*. Wiley, New York, USA. pp. 484-488.

Esser, G.: 1984, *Progress in Biometeorology* **3**, 253-294.

Esser, G.: 1991, *In:* Esser G. and Overdieck D. (eds.), *Facets of Modern Ecology*, Elsevier, Amsterdam, The Netherlands, pp. 679-705.

Foley, J. A.:1994, *Global Biogeochemical Cycles* **8**, 505-525.

Gauquelin, T., Iglesias, M., Jalut, G., Formard, F. and Dedoubat J. J.: 1994, *Journal of Arid Environment* (Unpublished manuscript).

Grove, A. T.: 1984, *Progress in Biometeorology,* **3**, 5-10.

Guiot, J.: 1990, *Palaeogeography, Palaeoclimatology, Palaeoecology* **80**, 49-69.

Guiot, J.: 1991, *In:* Guiot, J. and Labeyrie, L. (eds.), *Quantification des Changements Climatiques: Méthodes et Programmes* - Monographie N° 1, Centre Universitaire d'Arles, Arles, France, pp. 217-218.

Guiot, J., Harrison, S. P. and Prentice, I. C.: 1993, *Quaternary Research* **40**, 139-149.

Harrison, S. P., Saarse, L. and Dogerfeldt, G.: 1991, *Paläoklimaforschung* **6**, 159-179.

Harrison, S. P., Prentice, I. C. and Guiot, J.: 1993, *Climate Dynamics* **8**, 189-200.

Huntley, B.'and Birks, H. J. B.: 1983, *An Atlas of Past and Present Pollen Maps for Europe 0–13 000 years ago*. Cambridge University Press, Cambridge, UK, pp. 10-35.

Huntley, B. and Prentice, I. C.: 1988, *Science* **241**, 687-690.

Huntley, B. and Prentice, I. C.: 1994, *In:* Wright, Jr. H. E., Kutzbach, J. E., Webb III, T., Ruddiman, W. F., Street-Perrott, F. A. and Bartlein, P. J. (eds.) *Global Climates since the Last Glacial Maximum*, University of Minnesota Press, USA, pp. 136-168.

Kutzbach, J. E. and Guetter, P. J.: 1986, *J. Atmos. Sci.* **43**, 1726-1759.

Leemans, R. and Cramer, W. P.: 1991, *The IIASA Database for Mean Monthly Values of Temperature, Precipitation, and Cloudiness on a Global Terrestrial Grid*; IIASA Research Report RR-91-18, International Institute for Applied System Analysis, Laxenburg, Austria, pp. 5-61.

Lieth, H. and Box E.: 1972, *Climatology* **25**, 37-46.

Lieth, H.: 1975, *In:* Lieth, H. and Whittaker, R. H. (eds.), *Primary Productivity of the Biosphere*, Springer-Verlag, New York, USA, pp. 237-263.

Lossaint, P. and Rapp, M.: 1978, *In*: Lamotte, M. and Boulière, F. (eds.), *Problèmes d'Ecologie: Structure et Fonctionnement des Ecosystèmes Terrestres*, Masson, Paris, France, pp. 129-185.

McGuire, A. D., Melillo, J. M., Joyce, L. A., Kicklighter, D. W., Grace, A. L., Moore III, B. and Vorosmarty, C. J.: 1992, *Global Biogeochem. Cycles* **6**, 101-124.

Meentemeyer, V., Gardner, J. and Box, E.: 1985, *Earth Surface Processes and Landforms* **10**, 557-567.

Melillo, J. M., McGuire, A. D., Kicklighter, D. W., Moore III, B., Vorosmarty, C. J. and Schloss, A. L.: 1993, *Nature* **363**, 234-240.

Meyer, M. K.: 1988, *Net Primary Productivity Estimates for the Last 18 000 Years Evaluated from Simulations by a Global Climate Model*, M.S. Thesis, University of Wisconsin, Madison, USA, pp. 11-36.
Monserud, R. A.: 1990, *Methods for Comparing Global Vegetation Maps*, Working Paper WP-90-40, International Institute of Applied Systems Analysis, Laxenburg, Austria, pp. 5-21.
Monserud, R. A. and Leemans, R.: 1992, *Ecological Modelling* **62**, 275-293.
Olson, J. S., Watts, J. A. and Allison, L. J.: 1985, *Major World Ecosystem Complexes Ranked by Carbon in Live Vegetation: A Database*. NDP-017, Carbon Dioxide Information Centre, Oak Ridge National Laboratory, Oak Ridge, USA, pp. 3-41.
Overpeck, J. T., Webb, R. S. and Webb III, T.: 1992, *Geology* **20**, 1071-1074.
Peng, C. H., Guiot, J., Van Campo, E. and Cheddadi, R.: 1994a, *Journal of Biogeography* **21,** 19-31.
Peng, C. H., Guiot, J., Van Campo, E. and Cheddadi, R.: 1994b, *Global Ecology and Biogeography Letters* (in press).
Potter, C. S., Randerson, J. T., Field, C. B., Matson, P. A., Vitousek, P. M., Mooney, H. A., and Klooster, S. A.: 1993, *Global Biogeochemical Cycles* **4**, 811-841.
Prentice, I. C., Bartlein, P. J. and Webb III, T.: 1991, *Ecology* **72**, 2038-2056.
Prentice, I., Cramer, W., Harrison, S. P., Leemans, R., Monserud, R. A. and Solomon, A. M.: 1992, *Journal of Biogeography* **19**, 117-134.
Prentice, I. C., Guiot, J., Jolly, D., Huntley, B. and Cheddadi, R.: *Quaternary Sciences Review,* (Unpublished manuscript).
Raich, J. W., Rastetter, E. B., Melillo, J. M., Kicklighter, D. W., Steudler, P. A., Peterson, B. J., Grace, A. L., Moore III, B. and Vorosmarty, C. J.: 1991, *Ecological Applications* **1**, 399-429.
Rodin, L. E., Bazilevich, N. I. and Drozdov, N. N.: 1975, *In:* Reichle, D. E. (ed.) *Productivity of the World's Ecosystems*, Proc. Seattle Symp. 1974, National Academy of Science, Washington D.C., USA, pp.12-78.
Schlesinger, W. H.: 1990, *Nature* **348**, 232-234.
Whittaker, R. H. and Woodwell, G. M.: 1971, *In:* Duvigneaud, P. (ed.) *Productivity and Forest Ecosystems*, Ecology and Conservation **4**, UNESCO, Paris, France, pp. 159-175.
Woodward, F. I.: 1987, *Climate and Plant Distribution*. Cambridge Univ. Press, Cambridge, UK, pp. 62-80.

THE ASPEN PARKLAND IN WESTERN CANADA: A DRY-CLIMATE ANALOGUE FOR THE FUTURE BOREAL FOREST?

E. H. HOGG and P. A. HURDLE

Canadian Forest Service, 5320-122 Street, Edmonton, Alberta T6H 3S5, Canada

Abstract. Predicted future changes in regional climate under a doubling of atmospheric CO_2 concentrations were applied to the 1951–80 normals of 254 climate stations to examine future impacts on the boreal forest of western Canada. Previous analyses have indicated that in this region, the southern boreal forest is presently restricted to areas where annual precipitation (P) exceeds potential evapotranspiration (PET). The present analysis suggests that a predicted 11% increase in P would be insufficient to offset the increases in PET resulting from a predicted warming of 4–5°C. As a result, half of the western Canadian boreal forest could be exposed to a drier climate similar to the present aspen parkland zone (P < PET), where conifers are generally absent and aspen is restricted to patches of stunted trees interspersed with grassland. Future changes could result in permanent losses of forest cover following disturbance and an increase in the proportion of exposed edge habitat in remaining stands, where environmental conditions might induce additional stresses on tree growth. Thus if the predicted warming and drying occurs, productivity of aspen and other commercial species in the southern boreal forest would be greatly reduced.

Keywords. CLIMATE CHANGE, DROUGHT, EVAPOTRANSPIRATION, *Populus tremuloides*, PRAIRIE-FOREST BOUNDARY.

1. Introduction

Most of the recent General Circulation Models (GCMs) predict relatively large changes in the climate of the western Canadian interior as a consequence of increases in the levels of atmospheric CO_2 and other greenhouse gases (e.g., Houghton *et al.*, 1990). Existing climate records indicate that there has already been a significant warming of between 0.9 and 1.7 °C in this region over the past century (Gullet and Skinner, 1992). One of the greatest concerns, however, is that the small predicted increases in precipitation will be insufficient to offset the much higher evapotranspiration rates expected under a warmer climate (Zoltai *et al.*, 1991). This could lead to significantly drier soils and more severe droughts over much of this region in the future (e.g., Houghton *et al.*, 1990; Manabe *et al.*, 1992).

In western Canada, the boreal forest forms its southern boundary with the aspen parkland (Figure 1), a transitional vegetation zone characterised by patches of trembling aspen (*Populus tremuloides* Michx.) interspersed with grassland. Farther south, the aspen parkland grades into the true grassland or prairie, where trees are absent except along water-courses. Vegetation of the region is discussed in relation to ecological and historic factors by Rowe and Coupland (1984).

A recent analysis of climate gradients in the southern Prairie Provinces of western Canada (Hogg, 1994) supports previous suggestions that forest distribution is primarily limited by soil moisture deficits and recurring severe drought (e.g., Looman, 1979; Zoltai *et al.*, 1991). In that analysis, regional gradients in moisture were mapped using a climate moisture index that was based on the difference between mean (1951–80) annual precipitation (P) and annual potential evapotranspiration (PET), using the Jensen–Haise method. The southern boundary of the boreal forest was found to

Water, Air and Soil Pollution **82**: 391–400, 1995.

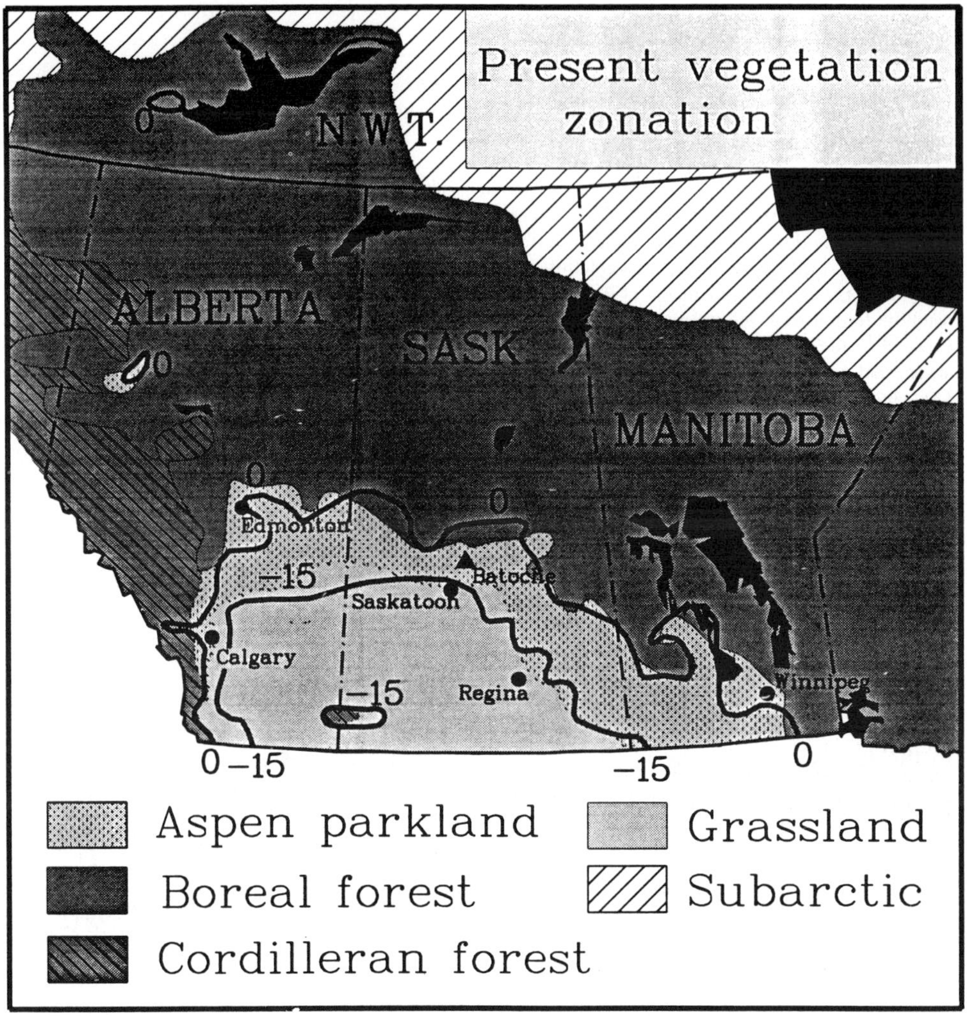

Fig. 1. Present distribution of vegetation zones in the western interior of Canada (after Ecoregions Working Group, 1989), in relation to the zero and -15 cm isolines of the P - PET climate moisture index for the period 1951–80 (after Hogg, 1994). Dry climate outliers (P - PET < 0) in northwestern Alberta (Peace River) and the Northwest Territories (Providence) are also shown.

correspond closely to the isoline where P = PET; i.e., a climate moisture index (P – PET) of zero. The analysis also showed a good correspondence between the -15 cm isoline, i.e., where P – PET = -15 cm, and the southern boundary of the aspen parkland with the grassland zone (Figure 1). The importance of moisture to the present vegetation zonation is also suggested by regional gradients in hydrological processes. In the boreal forest, the quantity of annual runoff typically ranges from 50–200 mm per year, but sharply decreases to < 25 mm per year over most of the aspen parkland and grassland zones (Figure 2). As a consequence, lake levels in the boreal forest tend to be constant from year to year, but show high interannual variability in the drier zones to the south, where lakes may dry up completely during periods of drought (Campbell *et al.*, 1994, pers. obs.). Hydrological gradients can also account for the abrupt change that occurs in a variety of ecosystem processes and characteristics at the boundary between boreal forest and aspen parkland (summarised in Table I). However, the GCM climate predictions indicate that the climate of the southern boreal forest may become warmer and drier in the future, similar to that found in the present aspen parkland. Such a change, accompanied by a northward shift in the zero P – PET isoline, would cause severe stresses on ecosystem functioning in the southern boreal forest.

The first objective of this study was to examine which areas of the western Canadian boreal forest are likely to be most sensitive to severe moisture stress under climate change. This was achieved by imposing a $2{\times}CO_2$ climate regime and mapping the equilibrium zonation of vegetation based on future predicted values of the P – PET index. Several GCM predictions are available; here we apply a generalised prediction for the region based on the Canadian Climate Centre GCM (Boer *et al.*, 1992). The second objective was to identify processes and characteristics of the aspen parkland zone, especially stresses on aspen under the present climate, that might serve as a climate change analogue for the most sensitive areas of the western Canadian boreal forest.

2. Materials and Methods

The predicted future climate of the western Canadian interior was based on the North American data subset of the CCC General Circulation model (Boer *et al.*, 1992), provided by the Canadian Climate Centre of Environment Canada, Downsview, Ontario. For the region bounded by 50°06'N to 61°12'N and 93°45'W to 116°15'W (28 GCM grid points), the average predicted changes in annual climate characteristics under a doubling of atmospheric CO_2 levels include a 4.2 °C increase in mean daily maximum temperature, a 4.9 °C increase in mean daily minimum temperature, and an 11% increase in precipitation. To estimate the future climate of the region, we applied these changes to the monthly climate normals (1951–1980) of 254 year-round stations in the region. Climate data sources and station locations are given in Hogg (1994).

TABLE I

Summary of differences in processes and characteristics between the boreal forest and aspen parkland in western Canada.

Characteristic	Boreal forest	Aspen parkland	Reference
Climate	P > PET (moist)	P < PET (dry)	Hogg, 1994
Coniferous trees	4–6 species	0 (1) species	Zoltai, 1975
Aspen	productive stands	stunted patches	Zoltai *et al.*, 1991
Peatlands	common	rare or absent	Vitt *et al.*, 1994
Annual runoff	50 to > 200 mm	< 50 mm	see Figure 2
Lake levels	constant	variable	Campbell *et al.*, 1994

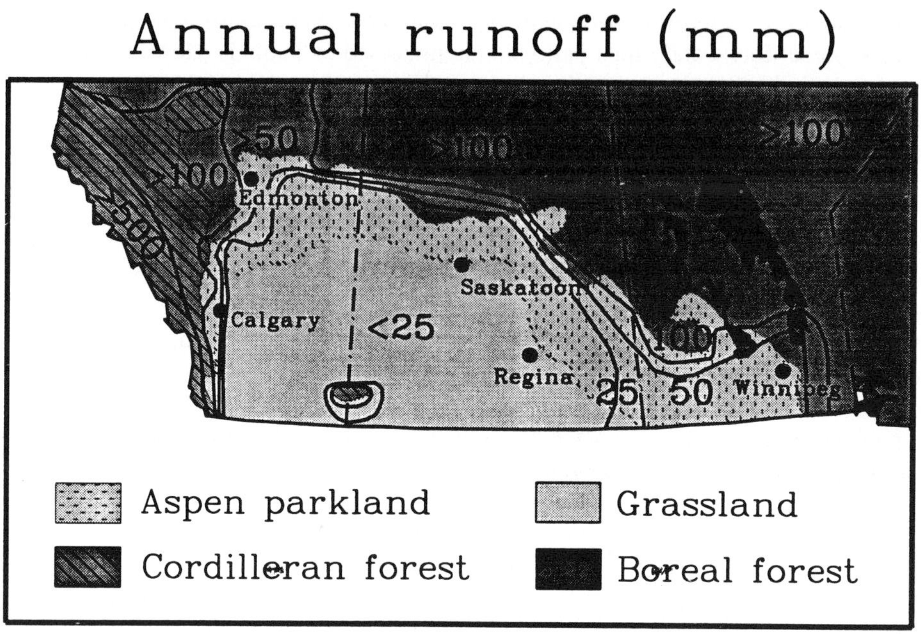

Fig. 2. Estimated annual runoff in the southern Prairie Provinces of western Canada (Alberta, Saskatchewan and Manitoba), based on Fisheries and Environment (1978). The 25, 50, 100 and 500 mm isolines are shown.

For each station, mean PET was estimated for the predicted future climate using the Jensen-Haise method (Bonan, 1989; Jensen *et al.*, 1990) as described by Hogg (1994). In the calculations of PET, it was assumed that future change in mean monthly solar radiation will be negligible. The P - PET climate moisture index under the predicted $2\times CO_2$ climate was determined for each station by subtracting PET from annual precipitation (P), with units in cm. The equilibrium distribution of future vegetation zonation under the $2\times CO_2$ predicted climate was then mapped, using the zero and -15 cm isolines of the future P - PET moisture index as follows: forest (P - PET > 0); aspen parkland (P - PET from 0 to -15); grassland (P - PET < -15). For purposes of presentation only, we used the 15 °C mean July isotherm as the approximate future position of the northern limit of the boreal forest.

3. Results and Discussion

The projected changes in the zero and -15 cm isolines of the P - PET moisture index in the western Canadian interior under a $2\times CO_2$ climate, as reflected by the vegetation zonation shown in Figure 3, suggest that a 4–5 °C increase in mean temperature would lead to much drier conditions in the region, despite the 11% predicted increase in precipitation (Figure 3). Based on this analysis, fully half of the present boreal forest in western Canada could develop a dry climatic moisture regime similar to the current aspen parkland (P - PET of 0 to -15 cm). The distribution of areas vulnerable to vegetation change (Figure 3) is similar to that suggested by others (e.g., Schwarz and Wein, 1990; Smith *et al.*, 1992), but our analysis takes topographic variation into account in greater detail. In general, we found that a 100 m increase in altitude resulted in a 5 cm increase in the P - PET moisture index, both because of cooler temperatures and greater precipitation (Hogg, 1994). Thus most areas of upland boreal forest (altitudes of ca. 600–1400 m) are expected to remain sufficiently moist to support forest cover under the 2×CO_2 climate. Parts of the boreal forest at lower altitudes in northern Manitoba are also relatively moist. In contrast, much of the lowland boreal forest elsewhere, might become dry enough to degrade into a mixture of forest remnants and grassland (i.e., an aspen parkland-type climate). These potentially sensitive lowlands occur over large areas adjacent to the major lakes and rivers (shown in Figure 3) across portions of northern Alberta, Saskatchewan and the Northwest Territories. The analysis suggests that climate impacts on these areas might eventually lead to an east-west fragmentation of the boreal forest in the region.

The map of future forest zonation (Figure 3) should only be regarded as an estimate of the equilibrium condition under a regionally generalised 2×CO_2 scenario, if it were to remain fixed for several centuries. The actual vegetation responses would likely be much more complex under the rapid, transient change in climate that is predicted. Schwarz and Wein (1990) point out that under climate change, rapid conversion of boreal forest to grassland may not occur until there is a major disturbance such as fire. Also, differences in plant species responses, e.g., dispersal and regeneration capacity following disturbance, may result in future vegetation types that have no modern analogue (MacDonald and Ritchie, 1985). This analysis does not consider the potential benefits of atmospheric CO_2 levels on plant water-use efficiency, which could partially offset the impacts of drought in the future (Long and Hutchin, 1991), nor does it take into account the possible negative impacts on vegetation caused by continuing increases in UV-B radiation from atmospheric ozone depletion (Kerr and McElroy, 1993; Yakimchuk and Hoddinott, 1994). Nevertheless, under the predicted future climate regime, moisture limitations could still lead to serious reductions in tree productivity within the next few decades in the more accessible, southern portions of the boreal forest in western Canada.

In the more remote, northern boreal forest and in other areas at high elevation, warmer temperatures and longer growing seasons, coupled with moist climatic conditions, might be expected to improve forest productivity in the future. However, a large proportion of these potentially favourable areas are located on the Canadian shield, where edaphic conditions are likely to remain as a constraint to tree growth in the future (Zoltai *et al.*, 1991).

Trembling aspen is the major hardwood species in the southern boreal forest of western Canada, where it has become an important commercial species for the forest industry since the late 1980s (Peterson and Peterson, 1992). For this species, the climatically warmer and drier aspen parkland zone can serve as an excellent analogue for probable growth responses under a 2×CO_2 climate. This approach forms the basis for an ongoing study (Hogg and Hurdle, unpublished) of the aspen parkland at Batoche, near Saskatoon, Saskatchewan (shown in Figure 1), as part of the Boreal Forest Transect Case Study (Price and Apps, 1995, this volume).

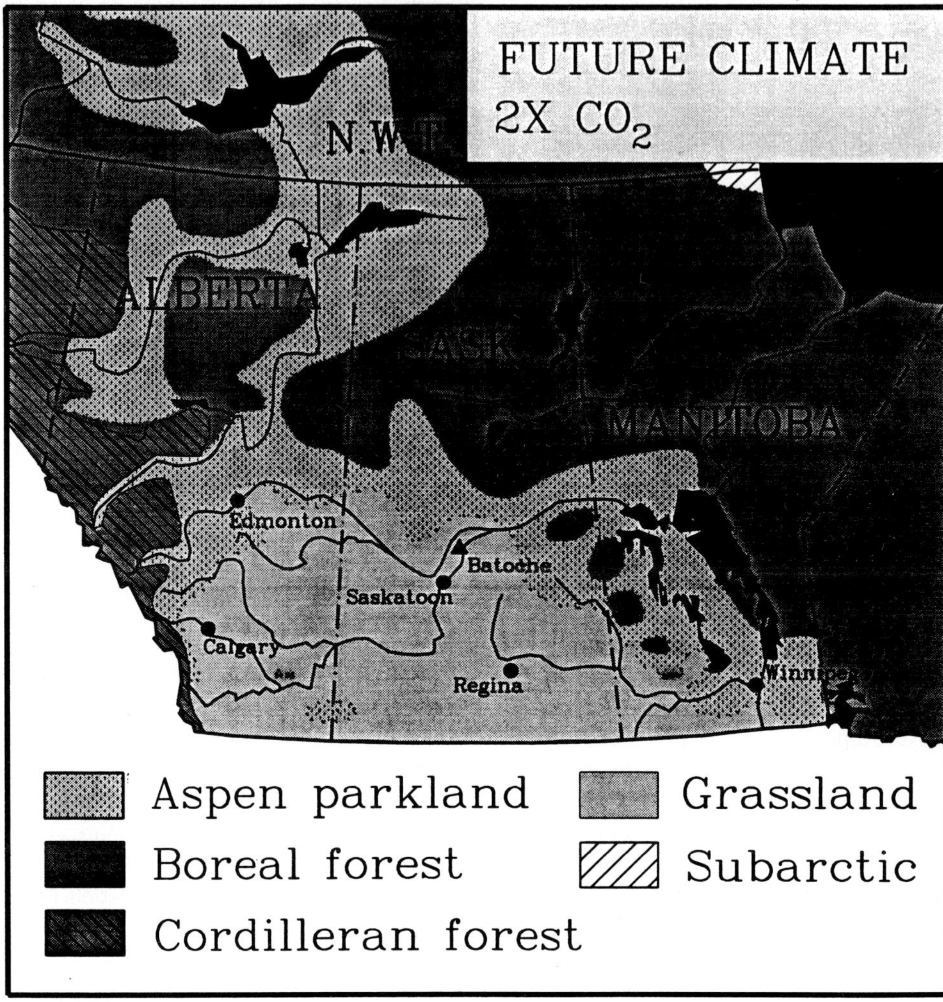

Fig. 3. Projected equilibrium vegetation, based on estimated future changes in the zero and -15 cm isolines of the P - PET climate moisture index, under 4.2 °C and 4.9 °C increases in mean daily maximum and minimum temperature, respectively, and an 11% increase in annual precipitation. New position of the northern limit of the boreal forest is based on the 15 °C mean July isotherm.

Surveys conducted at the Batoche site indicate that in this dry climate (P - PET = -13), aspen has a strongly stunted growth form and a reduced biomass productivity. For a given diameter, height growth at Batoche (Figure 4) was only about half to two-thirds that expected in the boreal forest (based on Huang *et al.*, 1992). The difference in growth form is shown in Figure 5. At Batoche, 75-year old aspen had achieved a height of only 12–14 m and estimated stand basal area of 15 m^2 ha^{-1} (mean of 24 prism sweeps), compared to the ca. 18 m and 25 m^2 ha^{-1} reported by Kirby (1975) for "poor sites" in the boreal forest of Saskatchewan. Similar comparisons have been made by Maini (1968) and by Johnstone and Peterson (1980), giving an indication of the impacts on aspen productivity that might be

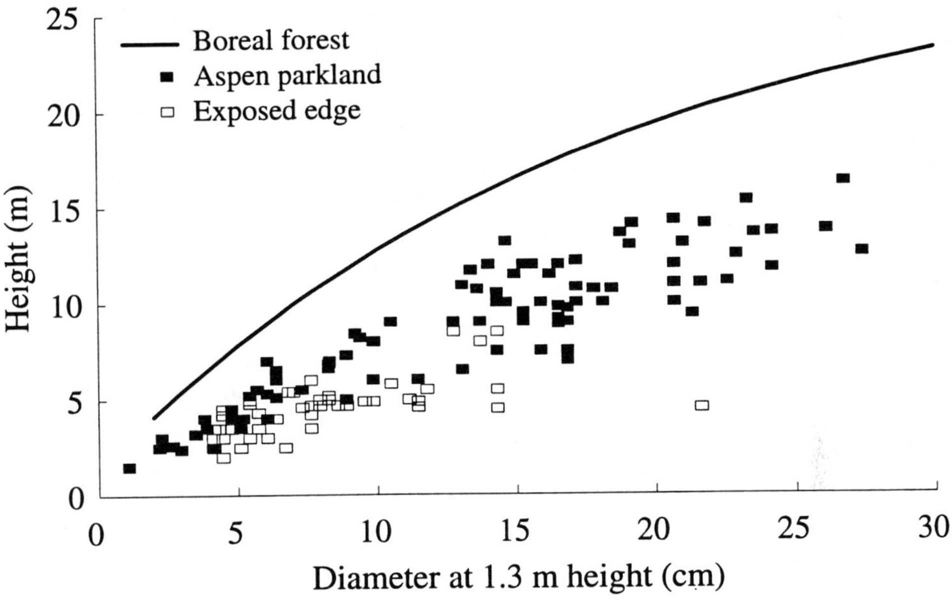

Fig. 4. Height-diameter relationships for trembling aspen (*Populus tremuloides*) in the boreal forest (equation 4 of Huang *et al.* 1992); and the aspen parkland at Batoche, Saskatchewan (authors' measurements) for forest interior sites (solid symbols) and exposed sites within 20 m of a forest-grassland edge (open symbols).

expected in the southern boreal forest under climate change. Furthermore, tree ring analyses on aspen at Batoche (Hogg, unpublished) indicate that radial growth is reduced by up to 90% during severe drought periods. Insect defoliators, such as the forest tent caterpillar (*Malacosoma disstria*), and numerous fungal species impose additional stresses on aspen (Peterson and Peterson, 1992). Some are promoted by warm, dry conditions (Ives, 1981) and thus their negative impacts on productivity and wood quality might be expected to increase under a 2×CO_2 climate.

Severe drought, coupled with insect or disease infestations, can also lead to extensive mortality of aspen groves, as occurred over large areas of the aspen parkland in 1961 (Zoltai *et al.*, 1991) and again following the 1988 drought (pers. obs.). However, the clonal nature of aspen, and its ability to resprout vegetatively following fire or drought would make this species well-suited to persist under a drier future climate in the boreal forest. In contrast, conifers lack such an effective means of vegetative regeneration. If climate change leads to increased fire frequencies in the future boreal forest (Flannigan and Van Wagner, 1991), conifers could be gradually eliminated if they are burned before reaching a seed-bearing age (ca. 30 years), or if soil conditions become too dry for successful seedling establishment (Hogg, 1994). The area affected by fire in Canadian forests increased dramatically under the dry conditions of the 1980s, notably in 1989 when fires burned over 3 Mha of boreal forest in northern Manitoba (Hirsch, 1991). In the Nisbet Forest, located 100 km north of Saskatoon, Saskatchewan, poor regeneration of jack pine (*Pinus banksiana* Lamb.) following one of the 1989 fires has lead to stand conversion to aspen and grassland in some areas (S. Hyde, Saskatchewan Forestry Branch, pers. comm.).

One of the main characteristics of the aspen parkland is that the trees are typically restricted to numerous small groves or "bluffs", often interspersed with large expanses of grassland. As a result, a high fraction of the aspen stems are located in exposed, edge habitat where environmental conditions can be substantially more severe than in the interior of larger areas of forest. This often results in even more pronounced stunting of aspen height in relation to diameter as shown in Figure 4 for exposed aspen at Batoche. In these situations, the aspen often develop crooked, leaning or forked stems (Figure 5). Measurements taken from June to September 1994 at the Batoche site indicate that aspen crowns at a grassland edge experienced 5 times greater mean wind speed than in the forest interior, located 300 m away from the edge. Furthermore, in-canopy vapour pressure was 6% lower and vapour pressure deficit was 11% greater at the edge compared to the forest interior during the same period (Hogg and Hurdle, unpublished). These and other differences in microclimate appear to control the differences in aspen growth form, although the precise mechanisms are not yet understood.

4. Conclusions

The present analysis suggests that climate change could have a major impact on moisture regimes and the future distribution of the southern boreal forest in western Canada. In areas where the forest becomes fragmented into small patches, either through natural processes or as a result of forest harvesting, a higher proportion of trees would occupy edge habitats with an even greater exposure to environmental stress, leading to stunted growth.

Acknowledgments

We thank S.C. Zoltai at the Northern Forestry Centre, and two anonymous reviewers for useful comments on the manuscript. This study has been supported by Green Plan funding (NOR-1683) within the Department of Natural Resources Canada.

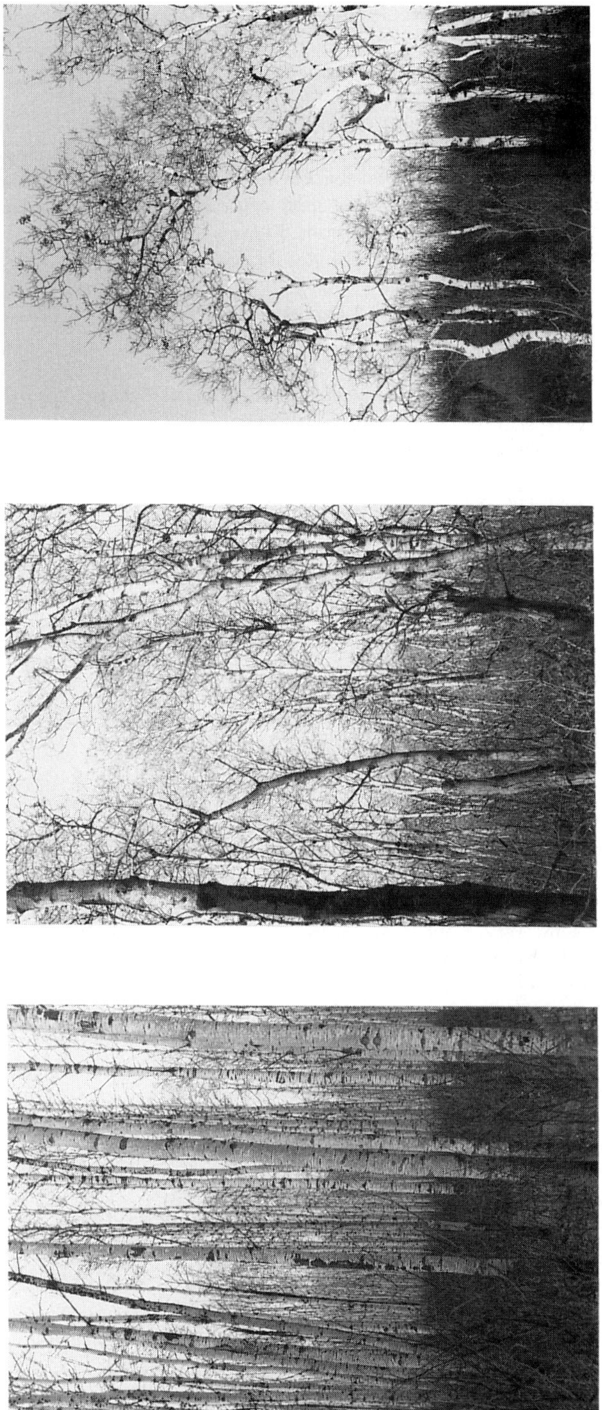

Fig. 5. Growth form of 75-year old aspen stands in (left) the boreal forest in Prince Albert National Park (53°38'N, 106°12'W) and (centre) in the aspen parkland at Batoche (52°44'N,106°09'W). Also shown (right) are 30-year old aspen in an exposed, grassland edge site at Batoche. All photos were taken in mid-October following leaf-fall.

References

Boer, G. J., McFarlane, N. A. and Lazare, M.: 1992, *J. Climate* **5**, 1045-1077.

Bonan, G. B.: 1989, *Ecol. Model.* 45, 275-306.

Campbell, C., Campbell, I. D. and Hogg, E. H.: 1994, *Geogr. Phys. Quatern.* **48**, 207-212.

Ecoregions Working Group: 1989, *Ecol. Land Classifi. Series No. 23,* Sustainable Development Branch, Canadian Wildlife Service, Conservation and Protection, Environment Canada, Ottawa, Canada, 119 pp.

Fisheries and Environment Canada: 1978, *Hydrological Atlas of Canada*, Surveys and Mapping Branch, Department Energy, Mines and Resources, Ottawa, Canada, 34 maps.

Flannigan, M. D. and Van Wagner, C. E: 1991, *Can. J. For. Res.* **21**, 66-72.

Gullett, D. W. and Skinner, W. R.: 1992, *The State of Canada's Climate: Temperature Change in Canada 1895-1991*, Environment Canada SOE Report No. 92-2, Dept. Supply and Services, Ottawa, Canada, 36 pp.

Hirsch, K. G.: 1991, *For. Chron.* **67**, 358-365.

Hogg, E. H.: 1994, *Can. J. For. Res.* **24**, 1835-1845.

Houghton, J. T., Jenkins, G. and Ephraums, J. J. (eds.), 1990, *Climate Change, The IPCC Scientific Assessment*, Cambridge University Press, Cambridge, UK, 365 pp.

Huang, S. and Titus, S.J.: 1992, *Can. J. For. Res.* **22**, 1297-1304.

Ives, W. G. H.: 1981, *Environmental Factors Affecting 21 Forest Insect Defoliators in Manitoba and Saskatchewan, 1945-69.* Can. For. Serv. Inf. Rep. NOR-X-233, Canadian Forestry Service, Edmonton, Canada, 142 pp.

Jensen, M. E., Burman, R. D. and Allen, R. G. (eds.).: 1990, *Evapotranspiration and Irrigation Water Requirements*, American Society of Civil Engineers, Man. Rep. Eng. Pract. **70**, Am. Soc. Civ. Eng., New York, USA 332 pp.

Johnstone, W. D. and Peterson, E. B.: 1980, *Above-Ground Component Weights in Alberta Populus Stands.* Can. For. Serv. Inf. Rep. NOR-X-226, Canadian Forestry Service, Edmonton, Canada, 18 pp.

Kerr, J. B. and McElroy, C. T.: 1993, *Science* **262**, 1032-1033.

Kirby, C. L., Bailey, W. S. and Gilmour, J. G.: 1957, *The Growth and Yield of Aspen in Saskatchewan*, Tech. Bull. **3**, Sask. Dep. Nat. Resour., For. Branch, Prince Albert,.Canada, 67 pp.

Long, S. P. and Hutchin, P. R.: 1981, *Ecol. Appl.* **1**, 139-156.

Looman, J.: 1979, *Phytocoenologia* **5**, 347-366.

MacDonald, G. M. and Ritchie, J. C.: 1986, *New Phytol.* **103**, 245-268.

Maini, J. S.: 1968, *In:* Maini, J. S. and Cayford, J. H. (eds.), *Growth and Utilization of Poplars in Canada*, Publ. 1205, Dep. For. Rural Dev., Ottawa, Canada, pp. 20-69.

Manabe, S., Spelman, M. J. and Stouffer, R. J.: 1992, *J. Climate* **5**, 105-126.

Peterson, E. B. and Peterson, N. M.: 1992, *Ecology, Management, and Use of Aspen and Balsam Poplar in the Prairie Provinces, Canada*, Forestry Canada, Northwest Region, Special Report 1, Edmonton, Canada, 252 pp.

Price, D. T. and Apps, M. J.: 1995, *Water, Air, Soil Pollut.* **82**, 203-214.

Rowe, J. S. and Coupland, R. T.: 1984, *Prairie Forum* **9**, 231-248.

Schwarz, A. G. and Wein, R. W.: 1990, *In:* Smith, D. D. and Jacobs, C. A. (eds.), Proc. 12th North Am. Prairie Conf., 5-9 August 1990, University of Northern Iowa, Cedar Falls, USA, pp. 1-4.

Smith, T. M., Leemans, R. and Shugart, H. H.: 1992, *Clim. Change* **21**, 367-384.

Vitt, D. H., Halsey, L. A. and Zoltai, S. C.: 1994, *Arctic and Alpine Res.* **26**, 1-13.

Yakimchuk, R. and Hoddinott, J.: 1994, *Can. J. For. Res.* **24**, 1-8.

Zoltai, S. C.: 1975, *Southern Limit of Coniferous Trees on the Canadian Prairies*, Can. For. Serv. Inf. Rep. NOR-X-128, Canadian Forestry Service, Edmonton, Canada, 12 pp. + maps.

Zoltai, S. C., Singh T. and Apps, M. J.: 1991, *In:* Navratil, S. and Chapman, P. B. (eds.), *Aspen Management for the 21st Century*, Forestry Canada, Northwest Region, Edmonton, Canada, pp. 143-152.

POTENTIAL EFFECTS OF CLIMATIC CHANGE ON SOME WESTERN CANADIAN FORESTS, BASED ON PHENOLOGICAL ENHANCEMENTS TO A PATCH MODEL OF FOREST SUCCESSION

P. J. BURTON and S. G. CUMMING

Department of Forest Sciences, University of British Columbia, #270 - 2357 Main Mall, Vancouver, British Columbia V6T 1Z4, Canada

Abstract. We enhanced the forest patch model, Zelig, to explore the implications of $2 \times CO_2$ climate change scenarios on several forest regions in British Columbia and Alberta, Canada. In addition to the processes and phenomena commonly represented in individual-based models of forest stand dynamics, we added some species-specific phenology and site-specific frost events. The consideration of bud-break heat sum requirements, growing season limits, and chilling requirements for the induction of dormancy and cold hardiness slightly improved the ability of Zelig to predict the present composition of B.C. forests. Simulations of the predicted effects of future climatic regimes (based on the averaged predictions of four general circulation models) include some major shifts in equilibrial, forest composition and productivity. Lowland temperate coastal forests are predicted to be severely stressed because indigenous species will no longer have their winter chilling requirements met. High-elevation coastal forests are expected to increase in productivity, while interior subalpine forests are expected to remain stable in productivity but will gradually be replaced by species currently characteristic of lower elevations. Dry, interior low-elevation forests in southern B.C. are likely to persist relatively unchanged, while wet interior forests are expected to support dramatic increases in yield, primarily by western hemlock. Northern interior sub-boreal forests are likewise expected to increase in productivity through enhanced growth of lodgepole pine. Conversely, the precipitous collapse of spruce stands in the true boreal forests of northeastern B.C. is expected to be associated with reduced productivity as they are replaced by pine species. Boreal-Cordilleran and Moist Boreal Mixedwood forests in Alberta are less likely to undergo compositional change, while becoming somewhat more productive. We believe these model enhancements to be a significant improvement over existing formulations, but the resulting predictions must still be viewed with caution. Model limitations include: (1) the current inability of climate models to predict future variation in monthly temperature and precipitation; (2) sparse information on the phenological behaviour of several important tree species; and (3) a poor understanding of the degree to which growth is constrained by different suboptimal climatic events.

Keywords. ALBERTA, BRITISH COLUMBIA, CLIMATE CHANGE, CHILLING REQUIREMENT, CONIFEROUS FORESTS, FROST TOLERANCE, GAP MODEL, TEMPERATURE RESPONSE, SUCCESSION, ZELIG

1. Introduction

Empirical data and the predictions of global circulation models (GCMs) suggest that elevated levels of CO_2 in the atmosphere (due to the combustion of fossil fuels and global declines in forest biomass) are in the process of warming the surface of the earth (Schneider, 1989). Individual-based simulations of forest succession in canopy gap-scaled homogeneous patches have been widely used to predict forest responses to climatic change (Solomon and Shugart, 1993). Such models have been successfully applied to forests around the world, including a number of boreal and other conifer-dominated ecosystems (Shugart and Prentice, 1992). These models, following an approach originally devised by Botkin *et al.* (1972), simulate the regeneration, growth and mortality of individual trees on an annual time step. Growth of individual trees is deterministic, with optimal growth (a sigmoid function of age) scaled down according to resource availability in each plot. Regeneration and mortality have stochastic components, so the simulation of many trees in several plots is required to describe stand dynamics. Competition is primarily for light, but site constraints of moisture, temperature and nutrients can also limit the growth of different species (Shugart, 1984; Botkin, 1993). Output consists of average densities, basal area, and biomass or wood volume produced by each species over time; stand tables describing size class distributions can also be generated at desired intervals. Unlike most forest growth and yield models, patch models include multiple species, natural regeneration, and uneven-aged stand structures. Patch models exhibit their greatest reliability over long (several century) simulation times, and quantitative predictions cannot be relied upon without extensive local calibration. Model execution requires estimates of a number of species' attributes, and description of the site (soil water retention, fertility, and monthly norms for temperature and precipitation). Simulation can be initiated from bare ground or from descriptions of existing stands.

We here report on further enhancements and applications of a patch model of the JABOWA-FORET genre. Zelig++ is a C language re-implementation of Zelig 1.0 (Smith and Urban, 1988; Urban, 1990), fully embedded within a command shell and graph plotting system derived from 'gnuplot' (Cumming and Burton, 1993).

2. Methods

2.1. Patch Model Enhancements: ZELIG++

A severe limitation to the use of patch models in predicting forest response to climate change has been the inadequate portrayal and calibration of temperature response functions (Austin, 1992; Prentice *et al.*, 1993; Urban *et al.*, 1993). A simple parabolic response function calibrated by the generic "growing degree-days" (heat sums above a threshold of 5.6 °C) found at the northerly (or high elevation) and southerly (or low elevation) limits of a species' range seriously confounds numerous factors limiting survival, growth, reproduction and dispersal (Figure 1a). As proposed by Cannell and Smith (1986), we have developed an alternative portrayal of the mechanisms by which the temperature regime influences tree growth, through a series of phenological events as well as species-specific heat sums which are used primarily to determine the onset of the annual growing season (Figure 1b). A similar approach has been independently developed by Sykes and Prentice (1995).

A year in the life of a tree simulated by Zelig++ basically follows the degree growth stage model of cold acclimation in temperate woody plants devised by Fuchigami *et al.* (1982):

1. Trees are physiologically inactive until daily temperatures reach a species-specific threshold temperature, for which we found published data on 9 species (T_{base} column in Table I and Table II);
2. Degree-days above that threshold are accumulated until a species-specific heat sum prompts bud-burst, or initiation of the growing season (data obtained for 10 species, BB column in Table I and Table II); during this time, minimum survivable temperature increases linearly from its deep winter value (T_{min}, known

for 28 species) to -2° C; if frosts colder than current frost-tolerance strike during this interval, annual growth is curtailed to 20% of its potential;

3. The end of the growing season occurs when quasi-daily temperatures fall below the species-specific threshold; the ratio of the length of the growing season experienced by each tree during the growing season, relative to the maximum length of growing season experienced by any established species in that plot, is used as a scalar to reduce realized growth in recognition of different periods of carbon assimilation;
4. Starting half-way through the growing season of each species, a record is kept of the number of days in which mean temperature is between -5° and +5° C; the degree to which this chilling history meets a species-specific chilling requirement (known for 8 species; CR column in Table I and Table II) determines the degree to which a tree attains its potential frost-tolerance, and constrains the subsequent year's growth to the same degree;
5. Any frost colder than the realized species-specific maximum frost tolerance (data found for 12 species; T_{min} column in Table I and Table II) is fatal to the tree. The likelihood of frost was determined as an empirical function of monthly mean temperatures for each zone (e.g., Figure 2).

Phenological parameters of species for which we could find no published values were estimated from the mean number of frost-free days weighted across their range of B.C. habitats, based on regressions which have an r^2 of 0.84 for T_{base}, $r^2 = 0.58$ for BB, and $r^2 = 0.55$ for CR. We tried to use other species parameters as used by published patch models with the following modifications: (1) light and moisture response categories are based on Klinka *et al.* (1990); and (2) inseeding rates (number of saplings to attempt establishment per plot per year) are based on the logarithm of published numbers of seeds per gram (Young and Young, 1992). Parameters for all natural hybrids were defined as the mean values of the parental species. Twenty-eight tree species (including varieties and hybrids) were assumed to be available in all zones west of the Rocky Mountains (Table I), while 15 species (including hybrids) were assumed available in all zones east of the Rocky Mountains (Table II), with no additional constraints to species migration among zones.

2.2. Climate Change Simulation

High-latitude areas, including the boreal and sub-boreal regions of Canada, are expected to be especially susceptible to atmospheric warming, with mean annual temperatures predicted to increase by 3 to 7° C (Singh and Wheaton, 1991). Different GCMs have predicted different degrees and directions of change in both temperature and precipitation at many northern locations under equilibrial $2\times CO_2$ scenarios. Having no conclusive information to believe the predictions of one GCM over another, we chose to explore the implications of the average level of change predicted by four different GCMs: those of the General Fluid Dynamics Laboratory (GFDL; Wetherald and Manabe, 1987), the Goddard Institute for Space Studies (GISS; Hansen *et al.*, 1988), Oregon State University (OSUM; Schlesinger and Zhao, 1988), and the United Kingdom Meteorological Office (UKMO; Mitchell, 1983). GCM output was interpolated to a grid of 0.5° × 0.5°, as described in Smith *et al.* (1992).

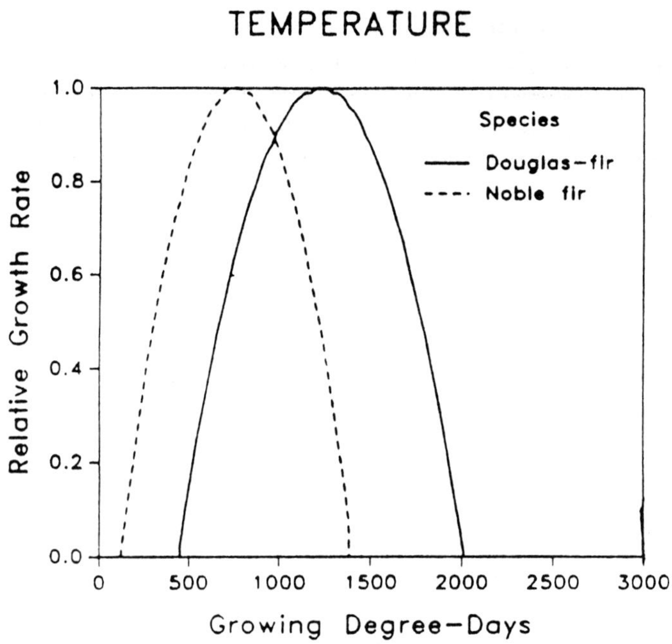

(a) Temperature response function in ZELIG Version 1.0

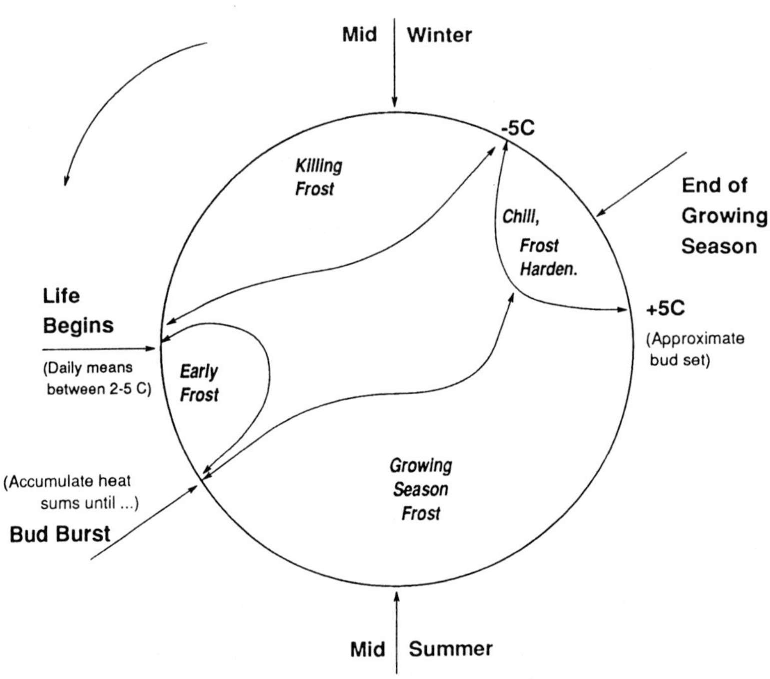

(b) Phenological events in ZELIG++ which determine temperature responses.

Fig. 1. Alternative portrayals of the temperature response behaviour of trees, as modelled in (a) traditional patch models (from Urban 1990), and in (b) Zelig++.

TABLE I

Species parameters used for simulations of forest regions west of the Rocky Mountains (B.C. only).

Code	Species	A_{max}	D_{max}	H_{max}	G	T_{base}	BB	CR	T_{min}	L	M	S
ABAM	*Abies amabalis*	600	200	75	180	4.3[a]	176[a]	16*	-30[b]	1	2	10
ABGR	*Abies grandis*	440[c]	182[c]	62[c]	190	4.3*	307*	13*	-35[b]	2	4	13
ABLA	*Abies lasiocarpa*	240[c]	104[c]	41[c]	150	2.6[a]	119[a]	10*	-67[b]	2	1	13
ACMA	*Acer macrophyllum*	300	250	40	100	5.0*	405*	17*	-20[b]	2	2	10
ALRU	*Alnus rubra*	100	150	40	225	5.3*	451*	18*	-46[b]	4	2	25
ARME	*Arbutus menziesii*	200	150	25	60	5.4*	468*	18*	-15[d]	4[d]	5	21
BEPA	*Betula papyrifera*	140[e]	76[e]	30[e]	187[e]	3.7*	231*	11*	-80[b]	3[e]	3[e]	29
CHNO	*Chamaecyparis nootkatensis*	3500	300	40	80	4.2*	295*	16*	-40[b]	2	1	18
LALA	*Larix laricina*	350	75	20	100	2.9*	111*	6*	-76[b]	5	2	22
LAOC	*Larix occidentalis*	450[f]	250[f]	69[f]	210[d]	3.4*	180*	10*	-40[g]	4	4	18
PIEN	*Picea engelmannii*	460[c]	217[c]	55	100	3.1*	145[h]	7*	-45[i]	2	2	18
PIGL	*Picea glauca*	400[d]	125[d]	46[d]	145[d]	2.7*	147[j]	6[k]	-70[l]	3	3	21
PIEGx	*P. engelmannii × glauca*	430	171	51	123	2.9*	146*	6.5*	-58	2	2	19
PIMA	*Picea mariana*	250[e]	46[e]	27[e]	94[e]	3.0*	123*	8*	-69[b]	2	3[e]	23
PISI	*Picea sitchensis*	750	400	90	210	5.0[m]	400[m]	20[m]	-34[b]	2	1	23
PICOc	*Pinus contorta contorta*	500	200	30	100	5.0*	409*	17*	-57[d]	5	4[c]	17
PICOl	*Pinus contorta latifolia*	335[c]	130[c]	40	145	2.9*	116*	9*	-85[b]	4	4[c]	18
PIMO	*Pinus monticola*	600	200	75	190	4.4[n]	468[n]	14[n]	-85[b]	3	3	13
PIPO	*Pinus ponderosa*	575[c]	225[c]	70[c]	175	3.9*	250*	11*	-41[b]	4	5	10
POBA	*Populus balsamifera*	200[d]	120[d]	35[d]	195[d]	2.1[o]	93[o]	7*	-80[b]	5	3	29
POTE	*Populus tremuloides*	200[d]	95[d]	39[d]	205[d]	3.5*	189*	10*	-80[b]	5	4	32
POTI	*Populus trichocarpa*	250	200	45	200	4.6[o]	175[o]	10[p]	-60[b]	5	2	29
PSMEg	*Pseudotsuga menziesii glauca*	525[c]	230[c]	54[c]	190	3.4*	255*	15[q]	-37[b]	3	5	15
PSMEm	*P. menziesii menziesii*	1100	425	85	190	2.8[r]	578[r]	13[s]	-23[b]	4	4	13
QUGA	*Quercus garryana*	500	225	25	65	5.6*	490*	19*	-18[b]	3	5	10
THPL	*Thuja plicata*	1500	350	60	150	4.1*	282*	13*	-35[b]	2	3	23
TSHE	*Tsuga heterophylla*	500	225	80	190	4.1*	277*	8[p]	-29[b]	1	3	21
TSME	*Tsuga mertensiana*	800	225	35	90	3.6*	214*	13*	-22[b]	1	3	18

Parameters are: A_{max}, maximum age (yr); D_{max}, maximum diameter (cm); H_{max}, maximum height (m); G, scalar of potential growth; T_{base}, basal temperature (°C) for counting heat sums and ending growing season; BB, heat sum (degree-days above T_{base}) required for bud break; CR, chilling requirement (weeks at -5 °C to +5 °C); T_{min}, maximum frost tolerance (°C); L, light response class (5 = most responsive); M, drought tolerance class (x10 = % of growing season which can be survived under a water deficit); S, sapling establishment rate ($plot^{-1}\ yr^{-1}$).

Parameter sources: Values as used by Urban *et al.* (1993) unless otherwise noted, where * denotes estimation from mean number of frost-free days in current B.C. habitat; [a]Worrall (1983); [b]mean of Sakai and Okada (1971) and Sakai and Weiser (1973); [c]mean of Keane *et al.* (1990) and Urban *et al.* (1993); [d]estimated on the basis of unpublished B.C. or Alberta data; [e]Bonan (1990); [f]Keane *et al.* (1990); [g]mean of Sakai and Weiser (1973) and Sakai (1983); [h]Nienstaedt (1967); [i]Becwar *et al.* (1981); [j]Nienstadt (1966); [k]Owens *et al.* (1977); [l]Sakai (1983); [m]Cannell and Smith (1983); [n]Steinhoff and Hoff (1972); [o]J. Worrall (unpublished data); [p]Nelson and Lavender (1979); [q]van den Driessche (1975); [r]Thomson and Moncrieff (1982); [s]McCrearey *et al.* (1990).

TABLE II

Species parameters used for simulation of forest regions east of the Rocky Mountains (Alberta and northeastern B.C.).

Code	Species	A_{max}	D_{max}	H_{max}	G	T_{base}	CR	BB	T_{min}	L	M	S
ABBA	*Abies balsamea*	200[a]	86[a]	25[b]	103[a]	2.8	7	121	-62[c]	1	2	22
ABLA	*Abies lasiocarpa*	240	104	41	150	2.6	10	119	-67	2	1	13
ABBLx	*A. balsamea* × *lasiocarpa*	220	95	33	127	2.7	8.5	120	-65	1	2	18
BEPA	*Betula papyrifera*	140	76	30	155	3.7	11	231	-80	4	3	29
LALA	*Larix laricina*	350	75	20	100	2.9	6	111	-76	5	2	22
PIEN	*Picea engelmannii*	460	217	55	100	3.1	7	145	-64	2	2	18
PIGL	*Picea glauca*	400	125	46	147	2.7	6	147	-69	3	3	21
PIEGx	*P. engelmannii* × *glauca*	430	171	51	124	2.9	6.5	146	-67	2	2	19
PIMA	*Picea mariana*	250	46	27	94	3.0	8	123	-69	2	3	23
PIBA	*Pinus banksiana*	150[a]	64[a]	30[a]	174[a]	2.8	8	108	-85[c]	5	5	20
PICOl	*Pinus contorta latifolia*	335	130	40	145	2.9	9	116	-85	4	4	18
PIBCx	*P. banksiana* × *contorta*	242	97	35	160	2.7	12	100	-85	5	5	19
POBA	*Populus balsamifera*	200	120	35	195	2.1	7	93	-80	5	3	29
POTE	*Populus tremuloides*	200	95	39	205	3.5	10	189	-80	5	4	32
PSMEg	*Pseudotsuga menziesii glauca*	525	230	54	190	3.4	15	255	-37	3	5	15

Parameters and their sources as in Table I, unless otherwise noted; [a]Bonan (1990); [b]modified from Solomon et al. (1984) and Bonan (1990) on the basis of unpublished Quebec and Alberta data; [c]mean of Sakai and Okada (1971) and Sakai and Weiser (1973).

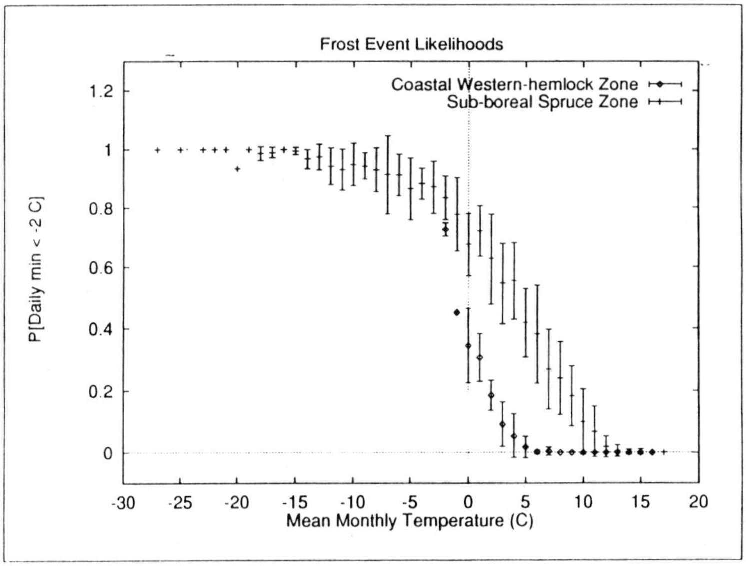

Fig. 2. Frost event likelihood as a function of mean monthly temperature for two regions of British Columbia, as estimated from long-term daily climate records. Likelihood is measured by the conditional expectation and standard deviation of the proportion of frost days in a month, given specific mean monthly temperatures.

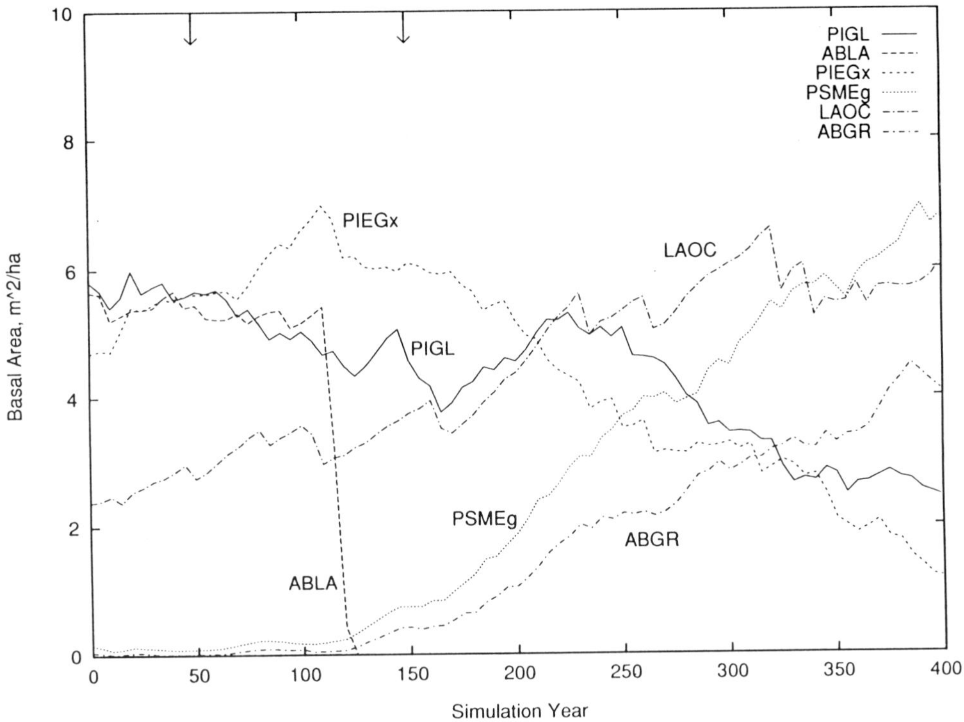

Fig. 3. Predicted composition of subalpine Engelmann spruce-subalpine fir forests in central B.C. under conditions of climatic warming. Climatic conditions were linearly ramped from current norms at simulation year 50 (first arrow) to equilibrial $2\times CO_2$ conditions (see Table III) at simulation year 150 (second arrow), after which those conditions were held constant. Only the three dominant species under the current climate and the three dominant species predicted under the future climate are shown; species abbreviations are defined in Table I.

The spatial resolution of GCM predictions span several degrees of latitude and longitude, so predictions of current and future climate in topographically complex regions are typically poor. We therefore chose to simulate forest composition for several locations in western Canada which exhibit climatic norms which are currently close to the values documented on a 0.5° grid in the International Institute for Applied Systems Analysis (IIASA) database (Leemans and Cramer 1990). Current climate attributes (based on climate station records) and mean $2\times CO_2$ GCM changes predicted for representative locations in 11 western forest regions are summarized in Table III. Input climate data consist of monthly means and standard deviations for temperature and precipitation; mean monthly temperatures are linearly interpolated to generate quasi-daily temperature values. GCMs do not presently provide estimates of variance in monthly precipitation and temperature, so we did not incorporate any variation in precipitation under current or future climates. However, sensitivity to low temperatures depends strongly on the incidence of rare, severe events, so stochastic temperature variation was used in all model runs, and the coefficient of variation in current temperatures was used to estimate variance associated with $2\times CO_2$ scenarios. Applying the same technique to future precipitation regimes resulted in uninterpretable unstable behaviour in many forest types, so no variance in precipitation was considered in any

TABLE III

Current[a] and predicted climatic attributes for different ecological regions in British Columbia and Alberta. The first value indicates climate station norms over recent 15 to 30 year periods, and the value after the slash indicates changes predicted by the mean of GFDL, GISS, OSUM and UKMO scenarios for the corresponding location under equilibrial $2\times CO_2$ conditions.

Ecological zone	Abbrev.	Climate station	Annual precip. (cm)	Annual temp. (°C)	Cold months (< 5 °C)
British Columbia					
Coastal Douglas-fir	CDF	Victoria	85.6/+9.3	9.6/+4.5	3/-3
Coastal western hemlock	CWH	Ethelda Bay	326.8/+58.5	7.8/+4.4	4/-4
Mountain hemlock	MH	Hollyburn	288.4/+31.6	4.9/+4.6	6/-2
Engelmann spruce-subalpine fir	ESSF	Allison Pass	147.5/+17.8	1.9/+4.6	8/-3
Interior Douglas-fir	IDF	Grand Forks[b]	38.9/+4.9	8.7/+4.7	5/-1
Interior cedar-hemlock	ICH	Rossland	76.3/+10.8	5.9/+4.8	5/-1
Sub-boreal spruce	SBS	Vanderhoof	39.8/+6.8	2.4/+5.0	7/-2
Boreal white and black spruce	BWBS	Ft. St. John	45.5/+8.0	1.5/+5.3	7/-2
Alberta					
Boreal mixedwood, dry	BMXD	Peace River	35.6/+6.9	0.7/+5.4	7/-2
Boreal-cordilleran	BCOR	Whitecourt	57.0/+10.3	2.5/+5.3	7/-2
Boreal mixedwood, moist	BMXM	Ft. McMurray	44.9/+9.5	0.1/+5.8	7/-2

[a] Data from Environment Canada, Atmospheric Environment Service, Canadian Climate Centre; "Climate Normals" on file as of March, 1994.

[b] Station actually in the adjacent Ponderosa Pine-Bunchgrass zone, but shows good agreement with typical IDF climate and generalized IIASA description of this region.

of the simulations reported here, pending the development of an improved soil water balance submodel.

Zelig++ simulations were conducted for site characteristics typical of deep sandy loam soils (capable of retaining 24.8 cm of water at field capacity, 9.7 cm at wilting point) at each location. All tree species found in the region were included in the simulation, as in the baseline runs. Forest stand development was simulated at each location, run from bare ground to equilibrial conditions over a period of 300 years in order to portray the composition of old-growth forests under the assumptions of the model as they interact with current climate. Comparisons of the presence and absence of different tree species in each region were then made with categorical descriptions in Meidinger and Pojar (1990). Results of Zelig++ simulations were compared with Zelig Version 1.0 (having a parabolic heat sum temperature response and no phenological features) predictions, incorporating degree-day limits used by Solomon *et al.* (1984), Bonan (1990) and Urban *et al.* (1993).

Climate change was introduced to these simulated mature forests as linearly gradual changes to the monthly mean temperature and precipitation values over a 100 year period; climate then remained constant at predicted $2\times CO_2$ equilibrial conditions for another 250 years of simulation. We present results primarily in terms of basal area

TABLE IV

Comparison of the abilities of Zelig Version 1.0 (first value) and Zelig++ (second value, after the '/' symbol) to correctly predict the current distribution of tree species from a single flora of 28 species west of the Rocky Mountains B.C.and 15 species east of the Rocky Mountains.

Ecological Zone	Correctly excluded	Incorrectly excluded	Incorrectly included	Correctly included	Total correct (%)
British Columbia					
Coastal Douglas-Fir (CDF)	6/4	6/7	8/10	8/7	50.0/39.3
Coastal Western Hemlock (CWH)	1/0	5/3	9/10	13/15	64.3/53.6
Mountain Hemlock (MH)	2/4	1/2	15/13	10/9	42.9/46.4
Engelmann Spruce-Subalpine Fir (ESSF)	7/7	4/1	6/6	11/14	64.3/75.0
Interior Douglas-Fir (IDF)	10/12	10/5	3/1	5/10	63.6/78.6
Interior Cedar-Hemlock (ICH)	3/3	5/1	8/8	12/16	53.6/67.9
Sub-Boreal Spruce (SBS)	9/9	10/10	6/6	3/3	42.9/42.9
Alberta and northeastern British Columbia					
Boreal White and Black Spruce (BWBS)	4/0	2/1	3/6	7/8	80.0/53.3
Boreal Mixedwood, Dry (BMXD)	5/6	7/7	3/2	0/0	33.3/40.0
Boreal-Cordilleran (BCOR)	1/1	2/0	2/2	10/12	73.8/86.7
Boreal Mixedwood, Moist (BMXM)	4/4	6/6	3/3	2/2	40.0/40.0
All species × zone combinations	52/50	57/43	65/67	82/96	52.3/57.0

(m^2 ha^{-1}) by species, because these predictions do not depend on allometric estimates of biomass or timber volume. Even so, the absolute values of basal area predictions cannot be relied upon, because the severe phenological constraints tested here have not been calibrated against growth and yield data in the many forest types simulated.

3. Results

3.1. Verification: Zelig++ Prediction of Current Forests

The ability of Zelig++ to predict the species composition of current forests was marginally superior to the baseline formulation of temperature response used in most patch models. Zelig++ predictions for the current floristics of southern interior forests (ESSF, IDF, ICH, and BCOR zones) in B.C. and Alberta were 64% to 87% reliable (in terms of the proportion of correct species inclusions plus correct species exclusions; Table IV). The model exhibited only a poor or moderate ability (39% to 54% accuracy) to predict the floristics of coastal B.C. forests (CDF, CWH, and MH zones) and of boreal and sub-boreal forests (SBS, BWBS, BMXD, and BMXM zones) with more complex temperature/moisture dynamics (40% to 53% floristic accuracy).

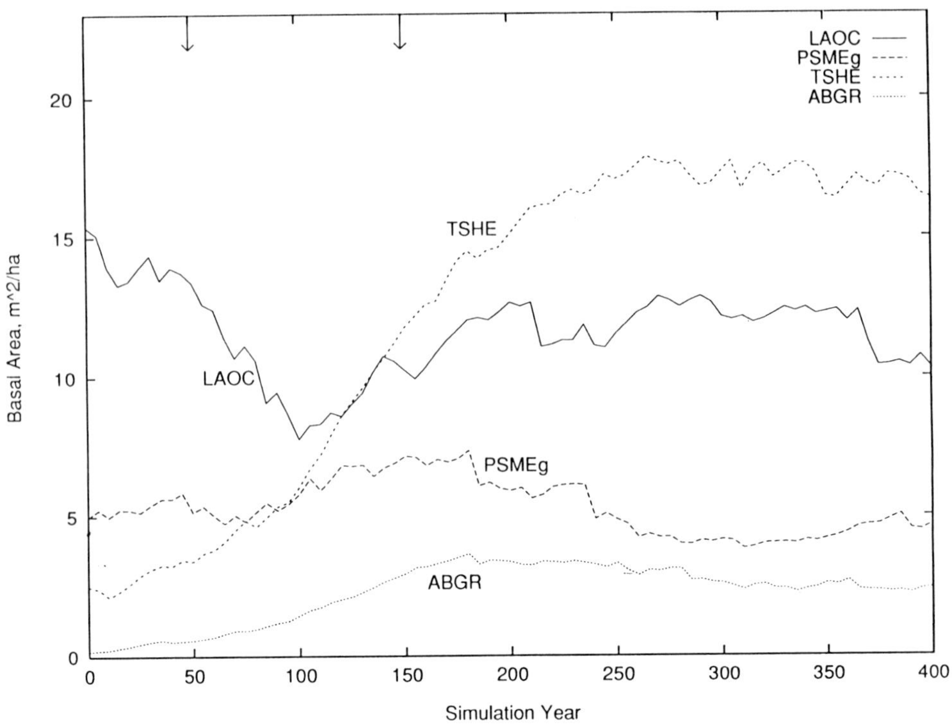

Fig. 4. Predicted composition of Interior Cedar-Hemlock forests in south-central British Columbia under conditions of climatic warming. Species abbreviations are defined in Table I; other details as in Figure 3.

3.2. Predictions of Future Forest Composition

Under the warmer conditions predicted for a 2×CO_2 climate, Zelig++ suggests that Coastal Douglas-Fir (CDF) and Coastal Western Hemlock (CWH) forests might undergo catastrophic collapse (results not illustrated). The monitoring of mortality during the simulations indicates that local species would suffer from a lack of sufficient chilling to induce winter cold hardiness, and would then succumb to frequent frost events. This is consistent with the fears articulated by Kimmins and Lavender (1992). We did not consider the potential for different genotypes and species from further south being able to offset these effects; such migrations would have to be actively promoted. In addition, the inability of Zelig++ to accurately predict the current composition of coastal forests indicates that we are failing to consider some important processes.

Simulations (results not illustrated) indicate that coastal subalpine Mountain Hemlock (MH) forests would undergo a large increase in productivity (on the order of 30 to 40%). They would also be able to support low elevation species such as *Tsuga heterophyllya.* The interior Engelmann Spruce-Subalpine Fir (ESSF) forests might likewise experience a marginal (5% to 10%) increase in productivity, led by the

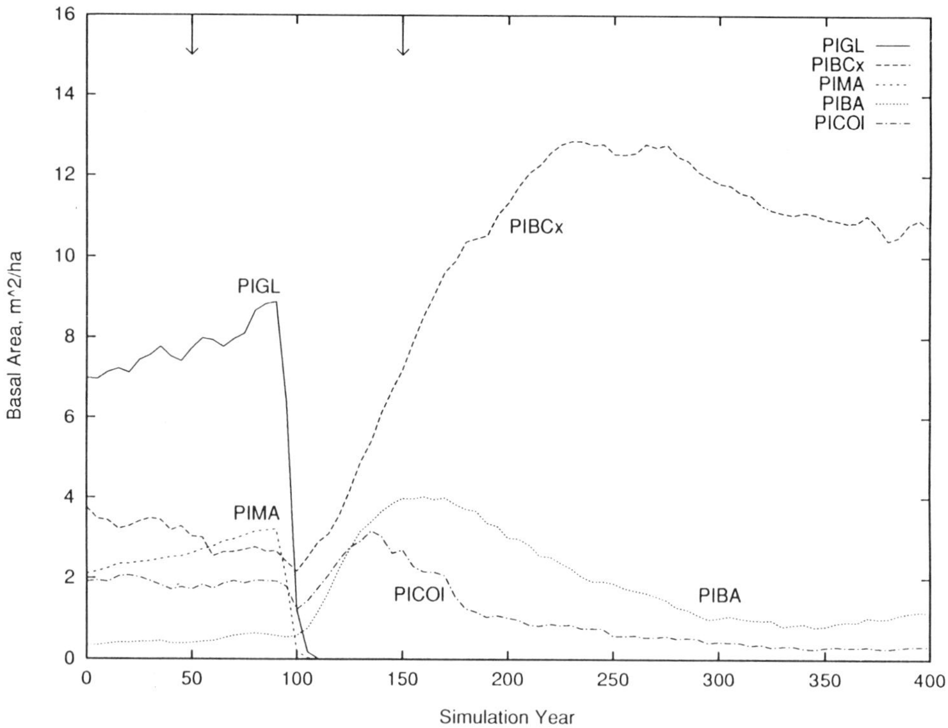

Fig. 5. Predicted composition of Boreal White and Black Spruce forests in northeastern B.C. under conditions of climatic warming. Species abbreviations are defined in Table II; other details as in Figure 3.

upward migration of western larch (*Larix occidentalis*), interior Douglas-fir (*Pseudotsuga menziesii* var. *glauca*), and grand fir (*Abies grandis*). Subalpine fir (*Abies lasiocarpa*) is rapidly lost, and members of the interior spruce complex (*Picea glauca, P. engelmannii*, and their hybrids) undergo a general decline as they are gradually replaced by the low elevation species (Figure 3).

In the low-elevation interior rain shadows, Interior Douglas-Fir (IDF) forests are expected to remain relatively unchanged in composition and productivity (results not illustrated), though ponderosa pine (*Pinus ponderosa*) may become marginally more abundant. Adjacent Interior Cedar-Hemlock (ICH, or "wet-belt") forests are predicted to increase in productivity by a phenomenal 55 to 65%, primarily through increased prevalence of western hemlock, *Tsuga heterophylla*; Figure 4). This enhanced level of forest productivity is predicted to persist indefinitely (at least for an additional 600 years of simulation time).

Forests of the Sub-Boreal Spruce (SBS) zone are likewise expected to undergo a 55 to 65% increase in productivity, primarily through increased growth of lodgepole pine (*Pinus contorta* var. *latifolia*). Trembling aspen (*Populus tremuloides*) may also become more vigorous and abundant in this central plateau region (results not illustrated). In the true boreal zone of British Columbia, Boreal White and Black

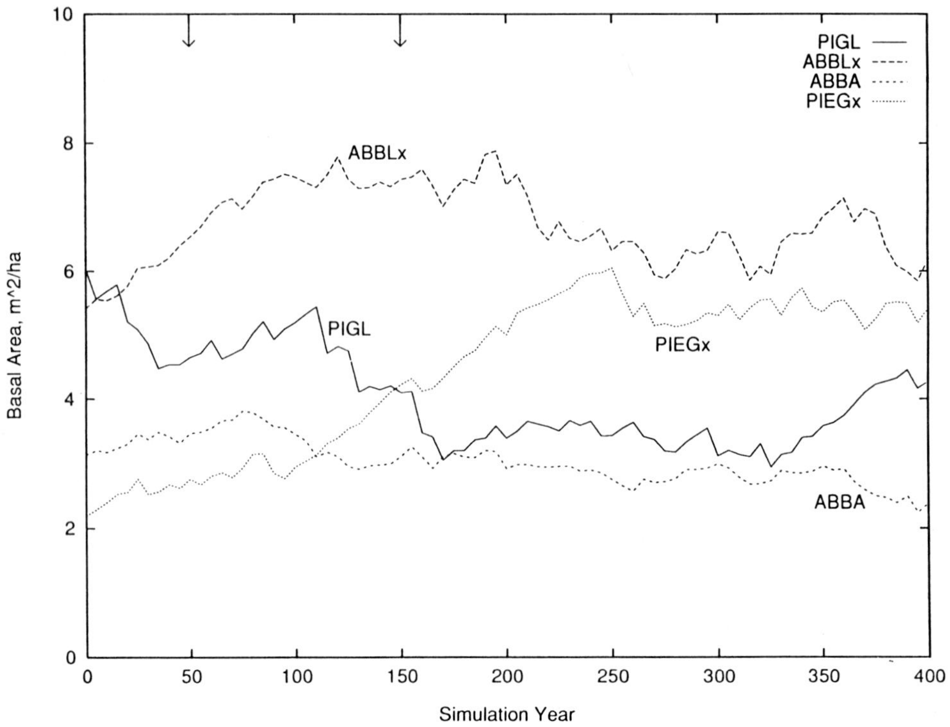

Fig. 6. Predicted composition of Boreal-Cordilleran forests in west-central Alberta under conditions of climatic warming. Species abbreviations are defined in Table II; other details as in Figure 3.

Spruce (BWBS) forests will undergo dramatic changes as *Picea glauca* and *P. mariana* stands rapidly collapse and are replaced by *Pinus* spp., with overall productivity decreasing by as much as 30 to 40%. Simulations suggest that hybrids of jack pine (*Pinus banksiana*) and lodgepole pine (*Pinus contorta*) may have a competitive advantage in this region under a warmer climate (Figure 5).

Zelig++ suggests that no trees should persist under present or future climates in the dry subregion of the Boreal Mixedwood region (BMXD) of Alberta (results not illustrated), probably reflecting an inadequate representation of the soil water balance and its influence on *Populus tremuloides* (which currently grows there). Boreal-Cordilleran (BCOR) forests of *Picea glauca* and *Abies* spp. would not collapse, but are expected to gradually become more dominated by *Picea engelmannii* × *glauca*, while becoming slightly (5 to 10%) more productive (Figure 6). The moist subregion of the Boreal Mixedwood (BMXM) is expected to become 30 to 35% more productive, though Zelig++ predicts dominance by *Pinus* spp. under both current and future climates (results not illustrated).

4. Discussion

Predicted levels of climate change may dramatically alter the composition and productivity of some western Canadian forests, but would have little effect on others. Expected responses are highly variable and regionally specific, but forest productivity is likely to increase more than decrease, so long as a wide array of tree species are available for forest regeneration. The potential for decreased productivity in the Boreal White and Black Spruce zone of B.C. is especially alarming. Furthermore, some areas (notably the SW coast of B.C.) may not be able to support forest cover at all. Some forest types, especially those including *Pseudotsuga menziesii* var. *glauca*, *Larix occidentalis*, and *Pinus* spp., may thrive and expand, but a number of other forest types could be lost.

The phenological considerations incorporated in Zelig++ somewhat improve our ability to understand and predict forest responses to climate, being especially valuable in demonstrating some of the more complex implications of climate change. Nevertheless, it must be emphasized that this model remains a very crude abstraction of reality. Data for a number of necessary parameters were unavailable for many species, and the degree to which growth is penalized under suboptimal conditions is inferred largely from anecdotal, unquantified observations. Simulations of future forests were initiated from (often unrealistic) simulated stands. All of these weaknesses point to the need for caution in interpreting model predictions, but can be overcome through a coordinated program of sampling, experimentation, and model refinement.

Acknowledgments

Ongoing Zelig++ development is supported by a grant from Alberta-Pacific Forest Industries Ltd. Thanks to T. M. Smith and the Global Change and Terrestrial Ecosystems (GCTE) project of the International Geosphere-Biosphere Programme (IGBP) for providing the IIASA database and GCM predictions. SGC is supported by a Canada Tri-Council Eco-Research Fellowship. We thank R. Guy, D. Hilbert, D. Lavender, A. Leduc, and J. Worrall for their assistance in formulating the model enhancements described here. The suggestions of two anonymous reviewers are gratefully acknowledged.

References

Austin, M. P.: 1992, *Aust. J. Bot.* **40**, 615-630.

Bonan, G.: 1990, *Can. J. For. Res.* **20**, 1077-1088.

Becwar, M. R., Rajashekar, C., Jansen Bristow, K. J. and Burke, M. J.: 1981, *Plant Physiol.* **68**, 111-114.

Botkin, D. B.: 1993, *Forest Dynamics, An Ecological Model.* Oxford Univ. Press, Oxford, United Kingdom.

Botkin, D. B., Janak, J. F. and Wallis, J. R.: 1972, *J. Ecol.* **60**, 849-872.

Cannell, M. G. R. and Smith, R. I.: 1983, *J. Appl. Ecol.* **20**, 951-963.

Cannell, M. G. R. and Smith, R. I.: 1986, *J. Appl. Ecol.* **23**, 177-191.

Cumming, S. G. and Burton, P. J.: 1993, *Environ. Software* **8**, 219-230.

Fuchigami, L. H., Weiser, C. J., Kobayashi, K., Timmis, R. and Gusta, L. V.: 1982, *In: Plant Cold Hardiness and Freezing Stress,* Li, P. H. and Sakai, A., (eds.) *Mechanisms and Crop Implications, Volume 2,* Academic Press, New York, USA, pp. 93-116.

Hansen, J., Fung, I., Lacis, A., Rind, D., Russell, G., Lebedeff, S., Reudy, R. and Stone, P.: 1988, *J. Geophys. Res.* **93**, 9341-9364.

Keane, R. E., Arno, S. F. and Brown, J. K.: 1990, *Ecology* **71**, 189-203.

Kimmins, J. P. and Lavender, D. P.: 1992, *Environ. Toxic. and Chem.* **11**, 1061-1068.

Klinka, K., Feller, M. C., Green, R. N., Meidinger, D. V., Pojar, J. and Worrall, J.: 1990, *In: Regenerating British Columbia's Forests*, Lavender, D. P., Parish, R., Johnson, C. M., Montgomery, G., Vyse, A., Willis, R. A. and Winston, D., (eds.) Univ. of British Columbia Press, Vancouver, Canada, pp. 55-72.

Leemans, R. and Cramer, W.: 1990, *The IIASA Database for Mean Monthly Values of Temperature, Precipitation, and Cloudiness of a Global Terrestrial Grid.* Working Paper WP-90-41, International Institute for Applied Systems Analysis, Laxenburg, Austria.

Manabe, S. and Wetherald, R. T.: 1987, *J. Atmos. Sci.* **44**, 1211-1235.

McCreary, D. D., Lavender, D. P. and Herman, R. K.: 1990, *Ann. Sci. For.* **47**, 325-330.

Meidinger, D. and Pojar, J.: 1991, *Ecosystems of British Columbia.* Special Report Series 6, Research Branch, B.C. Ministry of Forests, Victoria, Canada.

Mitchell, J. F. B.: 1983, *Quart. J. Royal Met. Soc.* **109**, 113-152.

Nelson, E. A. and Lavender, D. P.: 1979, *For. Sci.* **25**, 485-490.

Nienstaedt, H.: 1966, *For. Sci.* **12**, 374-383.

Nienstaedt, H.: 1967, *Silv. Gen.* **16**, 65-68.

Owens, J. N., Molder, M. and Langer, H.: 1977, *Can. J. Bot.* **55**, 2728-2745.

Prentice, I. C., Sykes, M. T. and Cramer, W.: 1993, *Ecol. Modelling* **65**, 51-70.

Schlesinger, M. and Zhao, Z: 1988, *Seasonal Climatic Changes Induced by Doubled* CO_2 *as Simulated by the OSU Atmospheric GCM/Mixed Layer Ocean Model.* Oregon State Univ., Climate Research Institute, Corvallis, USA.

Schneider, S. H.: 1989, *Global Warming.* Sierra Club Books, San Francisco, USA.

Shugart, H. H.: 1984, *A Theory of Forest Dynamics: The Ecological Implications of Forest Succession Models.* Springer-Verlag, New York, USA

Shugart, H. H. and Prentice, I. C.: 1992, *A Systems Analysis of the Global Boreal Forest*, Shugart, H. H., Leemans, R. and Bonan, G. B., (eds.) Cambridge Univ. Press, Cambridge, United Kingdom, pp. 313-333.

Sakai, A.: 1983, *Can. J. Bot.* **61**, 2323-2332.

Sakai, A. and Okada, S.: 1971, *Silv. Gen.* **20**, 91-97.

Sakai, A. and Weiser, C. J.: 1973, *Ecology* **54**, 118-126.

Singh, T. and Wheaton, E. E.: 1991, *For. Chron.* **67**, 342-348.

Smith, T. M., Leemans, R. and Shugart, H. H.: 1992, *Clim. Change* **21**, 367-384.

Smith, T. M. and Urban, D. L.: 1988, *Vegetatio* **74**, 143-150.

Solomon, A. M., Tharp, M. L., West, D. C., Taylor, G. E., Webb, J. M. and Trimble, J. C.: 1984, *Response of Unmanaged Forests to* CO_2*-Induced Climate Change. Available Information, Initial Tests, and Data Requirements.* U.S. Dept. of Energy, Washington, USA.

Steinhoff, R. J. and Hoff, R. J.: 1972, *Chilling Requirements for Breaking Dormancy of Western White Pine Seedlings.* U.S. For. Serv. Res. Note INT-153. Intermountain Forest and Range Experiment Station, Ogden, USA.

Sykes, M. T. and Prentice, I.C.: 1995, *Water, Air, Soil Pollut.* **82**, 413-426.

Thomson, A. J. and Moncrieff, S. M.: 1982, *Can. J. For. Res.* **12**, 448-452.

Urban, D. L.: 1990, *A Versatile Model to Simulate Forest Pattern, A User's Guide to ZELIG 1.0.* Univ. of Virginia, Dept. of Environmental Sciences, Charlottesville, USA.

Urban, D. L., Harmon, M. R. and Halpern, C. B.: 1993, *Clim. Change* **23**, 247-266.

van den Driessche, R.: 1975, *Flushing Response of Douglas-fir Buds to Chilling and to Different Air Temperatures After Chilling.* B.C. Forest Service Research Note #21, Victoria, Canada.

Worrall, J.: 1983, *Silv. Gen.* **32**, 203-209.

Young, J. A. and Young, C. G.: 1992, *Seeds of Woody Plants in North America.* Dioscorides Press, Portland, USA.

BOREAL FOREST FUTURES: MODELLING THE CONTROLS ON TREE SPECIES RANGE LIMITS AND TRANSIENT RESPONSES TO CLIMATE CHANGE.

M. T. SYKES and I. C. PRENTICE

Global Systems Group, Department of Ecology, Lund University, Ekologihuset, S-223 62 Lund, Sweden

Abstract. Range limits and broad-scale geographic variations in the productivity of boreal and northern deciduous tree species in Europe are simulated with a bioclimatic model (STASH). STASH is based on a small number of distinct mechanisms by which climate is thought to affect the survival, regeneration and growth of trees. Survival is limited by summer warmth requirements and winter cold tolerance; regeneration, additionally by winter chilling requirements; and growth rate by net assimilation, which in turn is related to photosynthetically active radiation, growing-season length, temperature (relative to species-specific optima) and soil moisture supply (relative to evaporative demand). These mechanisms are quantified either as thresholds (for survival and regeneration) or multipliers (for growth), based on bioclimatic variables computed from monthly climate normals interpolated three-dimensionally to a 10-minute grid. Growing-season and growing degree day calculations take into account the effects of chilling. The drought calculations also take into account the effects of soil-moisture storage by means of a physically based evapotranspiration calculation coupled to a one-layer soil hydrology model. STASH is used to examine changes in potential range limits under a 2×CO_2 climate-change scenario. Associated transient responses at selected sites in the boreal and boreo-nemoral zones of Sweden are also simulated, using the forest gap model FORSKA2. The species-specific survival and regeneration constraints and growth responses of STASH modify the growth, establishment and mortality of trees in FORSKA2. The results obtained in this way differ sharply from the results of conventional forest gap models, where growth rates are assumed to decline to zero at minimum and maximum growing degree day limits. For example, towards the southern limit of *Picea abies* (Norway spruce), STASH correctly shows no decline in productivity, but rather an abrupt cut-off corresponding to a chilling requirement during regeneration that is not met further south or west. In transient warming scenarios, this mechanism has the effect that natural regeneration can be blocked due to the warm winters even as yield is increasing due to the longer and warmer summers. STASH predicts drastic changes in species distributions in response to the large climate changes (especially winter warming) expected for northern Europe. Some of the common boreal species (e.g., *Picea abies*; *Pinus sylvestris; Alnus incana*) are unlikely to survive in much of their present range, withdrawing to the far north. Other species already widespread may be able to occupy some of the few sites that are today unavailable to them (e.g., *Betula* spp.; *Corylus avellana*). Other temperate deciduous species such as *Fagus sylvatica* could have dramatic range expansions, potentially occupying large tracts of the present boreal zone. FORSKA2 transient simulations illustrate some of the possible routes towards different types of forest in a changed climate. Some sites in the north show little change in species composition, but sites towards the southern boundary of the boreal zone could develop a new suite of dominants. The degree of sensitivity of a particular site depends both on the climate change prediction and on the transient dynamics of the forest community. Many types of transient behaviour are shown to be possible. Coupled with uncertainties about the future role of dispersal and changes in disturbance rate, the complexity and variety of these transient responses imply a highly uncertain future for the north European boreal forests.

Keywords. BIOCLIMATOLOGY, CHILLING REQUIREMENTS, CLIMATE CHANGE, FOREST DYNAMICS, FOREST MODEL, RANGE LIMIT, TRANSIENT EFFECTS.

1. Introduction

Boreal forests cover large areas of land in the northern high latitudes. They could respond in dramatic ways to future climate change. The greatest degree of climate change, in the form of large and rapid increases in winter temperatures, is predicted for just these areas. Modelling approaches make it possible to assess the range of potential responses by the affected ecosystems.

Water, Air and Soil Pollution **82**: 415–428, 1995.

Two models are used in this paper to assess possible responses of forest tree species to the climate change induced by a doubling of CO_2. The first, STASH (Sykes *et al.*, in press), uses a small number of bioclimatic limits each representing a particular, known type of constraint (growing-season heat requirement, winter cold tolerance, drought tolerance, chilling requirement) to characterise the distributions of northern European tree species under the current climate (Sykes *et al.*, in press). The success of STASH in representing the current distributions accurately with a small number of parameters is a consequence of its mechanistic basis, which also makes it meaningful to use the model to simulate future distributions under a different climate. Such simulations provide insights on the possible responses of boreal tree species' *potential* distributions to future climate change.

Further analysis of the forest dynamics, transient responses of species and overall biomass changes in affected regions can be assessed by transient modelling dynamics on a local landscape scale. This is done here using a version of the FORSKA2 model (Prentice *et al.*, 1993). FORSKA2 is a non-conventional (i.e., not derived from JABOWA) forest gap model. Bioclimatic limits for survival and regeneration, and the multipliers for growth as described in Prentice *et al.* (1993), are modified so as to be consistent with those used in STASH.

Model projections are naturally beset with uncertainties. Many hypotheses and assumptions are embodied in the models, which in any case can only give a partial picture of what may happen. For example, we have not tried to include "higher-order" effects such as possible increases in pest outbreaks due to climatic warming, nor have we considered likely management responses in any detail; and we have not included any representation of the possible modifying effects of CO_2 as a physiological (as opposed to climatic) agent. Nevertheless, our results indicate some key areas of concern and point to specific research needs (notably for more experimental information on the chilling requirements of a wide range of species) that are fundamental, even for the generalised and approximate scenarios that we are trying to develop.

2. Methods

2.1. CLIMATE DATA

The climate data set is based on weather station records of long-term mean monthly values for temperature, rainfall and sunshine (as proportion of possible sunshine hours, which is approximately the complement of average cloud cover). The data set is an updated version (W. Cramer, pers. comm., 1994) of the original dataset of Leemans and Cramer (1991). This update uses a surface fitting technique to interpolate to a 10' grid (Hutchinson, 1989). The entire window (25°E to 50°E, 35° to 75°N) was used for the STASH calculations. FORSKA2 runs were based on interpolated climate data for a few selected grid cells.

The $2\times CO_2$ scenario was created by using four general circulation models (GCMs) (GFDL (Manabe and Wetherald, 1987); OSU (Schlesinger and Zhao, 1989); UKMO (Mitchell, 1983), GISS (Hansen *et al.*, 1988)). Temperature and precipitation anomalies (differences between simulated $2\times CO_2$ and control climates) were interpolated to each 10' grid cell. These interpolated anomalies were then added to the current climate. The mean of the four scenarios was calculated to give "mean GCM" $2\times CO_2$ datasets of monthly mean temperature and precipitation. Sunshine anomalies were not available, so the current sunshine dataset was used. A constant 150 mm soil water holding capacity was assumed.

2.2. Modelling Tree Species Range Limits

There is still surprisingly little experimental information on the controls of the natural ranges even of the most economically important of the northern tree species. However, general observations of plant physiology and biogeography serve to identify the primary controlling variables (e.g., see Woodward, 1987). *Minimum temperatures* in winter are critical for most woody species (Sakai and Weiser, 1973), and extreme low temperatures are known to cause damage or death. The key variable is the extreme temperature, but mean coldest-month temperatures are well correlated with absolute recorded minimum temperatures (Prentice *et al.*, 1992). Some level of integrated *summer warmth* (usually expressed as growing degree days) is required by all species. *Drought* is a third major limiting factor. Drought is not directly related to precipitation, but rather to soil moisture supply—or more precisely, to the extent to which soil moisture supply satisfies evaporative demand. Thus, effective drought can be indexed by the ratio of actual to potential evapotranspiration, which in turn can be estimated using soil moisture accounting models (Prentice *et al.*, 1992, 1993). STASH therefore estimates mean coldest-month temperature, growing degree days above a 5 °C base (GDD) and the drought index α (strictly, the ratio of actual evapotranspiration to the equilibrium evapotranspiration computed from temperature and net radiation, which in turn is computed from temperature, insolation at the top of the atmosphere, and sunshine) and assumes that these three factors constitute separate and independent constraints on species' potential ranges.

In addition, northern woody species require a *period of chilling* to induce budburst in the spring. Just as specific cold tolerance and winter dormancy mechanisms confer protection against the lowest temperatures encountered during dormancy, so delayed budburst is an insurance against the possibility of low temperature damage occurring after dormancy is broken. This applied to evergreen conifers as well as to deciduous trees. The possibility of future warming in northern regions has stimulated experimental work on the subject of chilling requirements. It has been found that the chilling response is species-specific (Cannell and Smith, 1986; Murray *et al.*, 1989) and that the thermal time (i.e., the number of GDD) to budburst decreases exponentially with the chilling period. Some species (e.g. *Fagus sylvatica*) are especially sensitive to late spring frosts, and accordingly have a very high chilling requirement that implies very late budburst in regions with mild winters. Such species are however not well adapted to the most oceanic climates. STASH incorporates this trade-off by calculating an "effective" GDD (GDD°). GDD° is defined as the minimum number of required GDD, reduced by the number of GDD to reach budburst. This in turn is a function of the chilling period length. The form of the chilling response function (relation between GDD° and chilling period) was taken from Murray *et al.* (1989). For species tested by Murray *et al.* (1989), the parameters of this function were also constrained by the experimental data; Sykes *et al.* (in press) showed that including the known chilling response in this way substantially improved the range-limit predictions in western and northwestern Europe. For other species, chilling parameters were inferred from the range limits together with the other bioclimatic limits.

STASH further imposes a maximum value of the mean coldest-month temperature on a few continental species such as *Picea abies*, to reflect their notoriously poor regeneration in oceanic climates.

2.3. Modelling Transient Responses and Forest Dynamics

Forest dynamics were simulated at sites through the boreal and boreo-nemoral zone of Sweden. [The boreal (coniferous) forest zone extends southwards to *ca.* 60°N. The nemoral (temperate deciduous) forest zone is confined to the far south and the southwest coast; the boreo-nemoral (mixed) forest zone lies in between (Sjörs, 1965; Hutisch, 1960).] The simulations were done using the forest gap model FORSKA2 (Prentice *et al.*, 1993). FORSKA2 has at its core the FORSKA model of Prentice and Leemans (1990). FORSKA2 has been developed as a general forest landscape model for use in climatic change experiments. It has been used to simulate boreal and boreo-nemoral forest dynamics and transient effects under a number of climate scenarios (Prentice *et al.*, 1991; Price *et al.*, 1993; Sykes and Prentice, unpubl. ms.; Sykes, 1994) as well as providing estimates of carbon storage in boreal forested regions under present and future climates (Price *et al.*, 1993; Sykes and Prentice, 1995).

Five sites were selected for the simulations (Figure 3), from Reivo in the north of Sweden to Boa Berg in the south, on the southwestern boundary of the boreo-nemoral zone. At each of the five sites FORSKA2 simulated 1000 yr of forest dynamics using 200 independent 0.1 ha patches, scattered over a simulated landscape. Each patch had a 100 yr root-mean-square return time for disturbance, using a formulation that gives an increasing probability of disturbance for the patch as it ages (Prentice *et al.*, 1993). Nineteen of the most common northern European tree species were "available" for establishment and growth. At one site, Tyresta in the Stockholm region, an additional simulation was done in which immigration of "new" species was disallowed.

Monthly temperature, precipitation and sunshine values for the 10' cell containing the selected site were used. In the first 400 yr of the simulation the model was allowed to come to an equilibrium under the present climate. The climate was then changed linearly over 100 years so that it reached the "mean GCM" temperature and precipitation by year 500. Once the new climate was achieved, monthly temperature and precipitation were held at the new level for a further 500 yr .

3. Results

3.1. Species Distribution Limit Changes: Common Trees of the Boreal and Boreonemoral Zone

Picea abies (Norway spruce) is one of the major boreal forest trees in northern Europe. Its distribution is simulated well by STASH (Sykes *et al.*, in press). In particular, its complex northwestern and western boundary is described accurately: STASH shows where *Picea* is excluded from the coast in western and northern Norway and in the far south of Sweden (Figure 1). The simulation also suggests that *Picea* responds to increased temperatures towards the south, being at its most productive near its southern boundary.

Under the "mean GCM" climate scenario, where winters become substantially warmer, *Picea* establishment is predicted to fail in much of western Europe. Over time this would force a retreat eastwards, leaving only a few outposts in mountain regions (Figure 1). *Picea* would be removed from Fennoscandia except for parts of northern Sweden and Finland.

Pinus sylvestris (Scots pine), another major component of the present-day boreal forest, has a similar simulated present and future distribution to *Picea abies* (Figure 1). However the mechanisms (and past histories) behind these shifts are evidently different (Huntley and Prentice, 1993). *Pinus sylvestris* can regenerate at present in many places on the oceanic fringe of Europe where it is no longer found as a native species. Deciduous species may be outcompeting *Pinus* in the natural vegetation of these areas (Prentice *et al.*, 1991). Under a changed climate of the magnitude used here, *Pinus* declines in the boreal zone, as does *Picea*; but the implied mechanism involves competition instead of regeneration failure.

Acer platanoides (Norway maple) reaches its present northern boundary in the southern boreal zone of Sweden and Finland (Figure 1). Under the changed climate its western limit shifts, so that it is excluded from most of western Europe. *Acer platanoides* is shown at its most productive in central Europe and could still be found throughout Finland (except for the coastal strip) and in other northern or mountain regions of Norway and Sweden.

Alnus incana (grey alder) is found throughout the boreal zone today, and reaches its southern limit in the boreal-nemoral zone of southern Sweden (Figure 1). Under the changed climate it retreats both northwards and eastwards to become restricted almost entirely to northern Fennoscandia and Russia. *Betula pendula* (silver birch) occurs throughout the boreal zone except in the upper montane and northernmost regions of Fennoscandia. Under the changed climate the range of *B. pendula* increase to cover all the present boreal zone. *B. pendula* is also likely to become more productive, particularly in southern boreal regions (Figure 1).

Corylus avellana (hazel) is today distributed through many parts of the boreal zone, although it is excluded from montane and northernmost areas. Under the changed climate *Corylus* shows a considerable range expansion northwards, upwards, and eastwards into Russia. In Fennoscandia only the highest elevations remain outside its range (Figure 1).

Fraxinus excelsior (ash) reaches its northern limit in central Sweden and Finland in the southern boreal zone (Figure 1). Under the changed climate it is shown extending much further north, though still excluded from the far north. It is also likely to be excluded from coastal areas in the present day southern boreal zone, where it is found at present.

Populus tremula (aspen) is common throughout the boreal and boreo-nemoral zone though excluded from the far north. Under the changed climate *Populus tremula* moves to occupy these areas too, but at the same time its range is restricted in southern coastal areas (Figure 1).

Quercus robur (pedunculate oak) has its northern limit today at the border of the southern boreal and boreo-nemoral zones (Figure 1). Under the changed climate there is a strong range expansion northwards into most of the present boreal zone.

Tilia cordata (small-leaved lime) reaches its northernmost limit in the southern boreal zone (Figure 1). Under the changed climate it expands northwards, but retreats from southern and coastal regions.

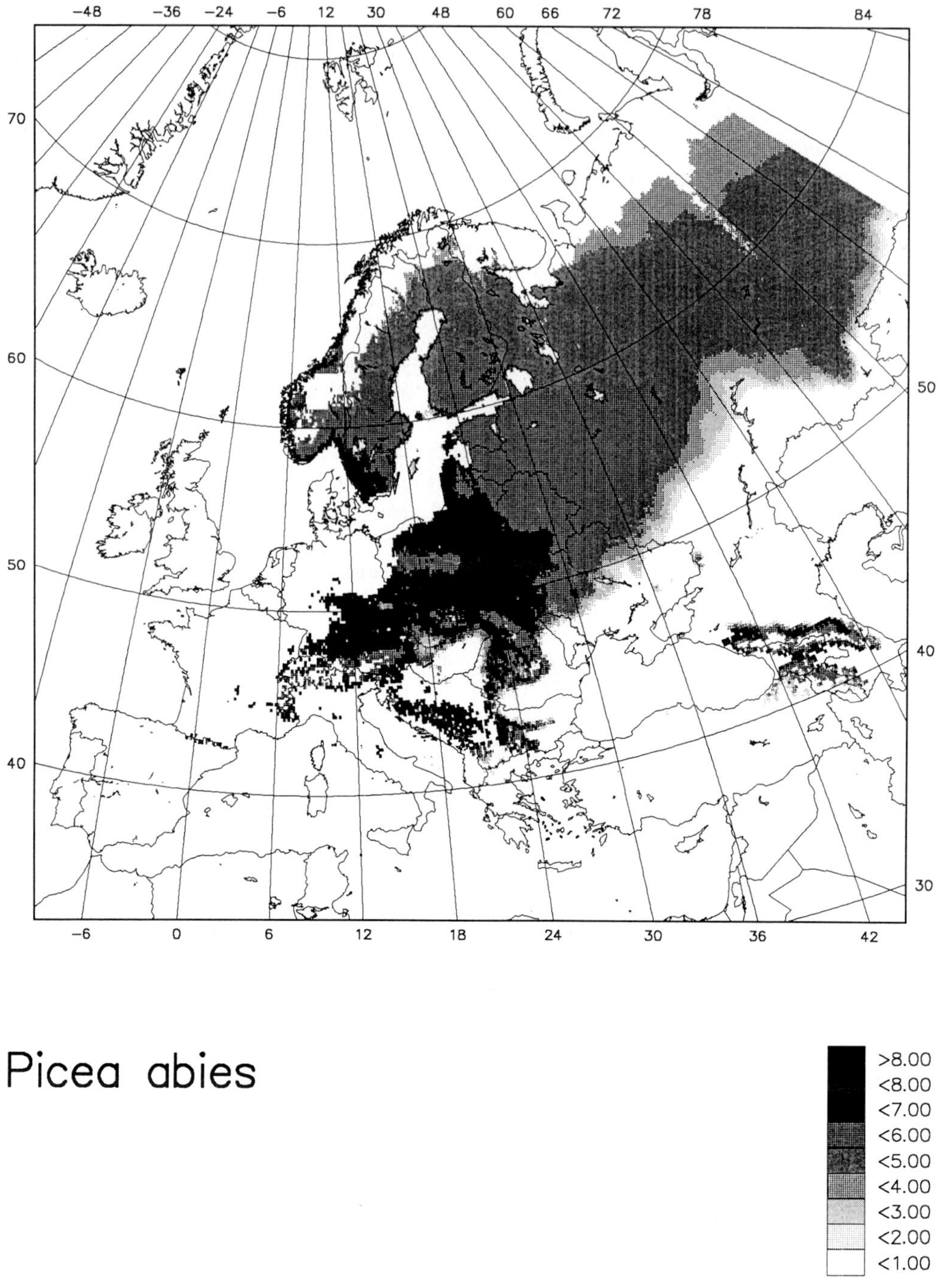

Fig. 1. Simulated distributions of 11 of the most common boreal or boreo-nemoral zone trees of northern Europe: under present climate, and under a changed climate derived from the mean of four $2\times CO_2$ scenarios.

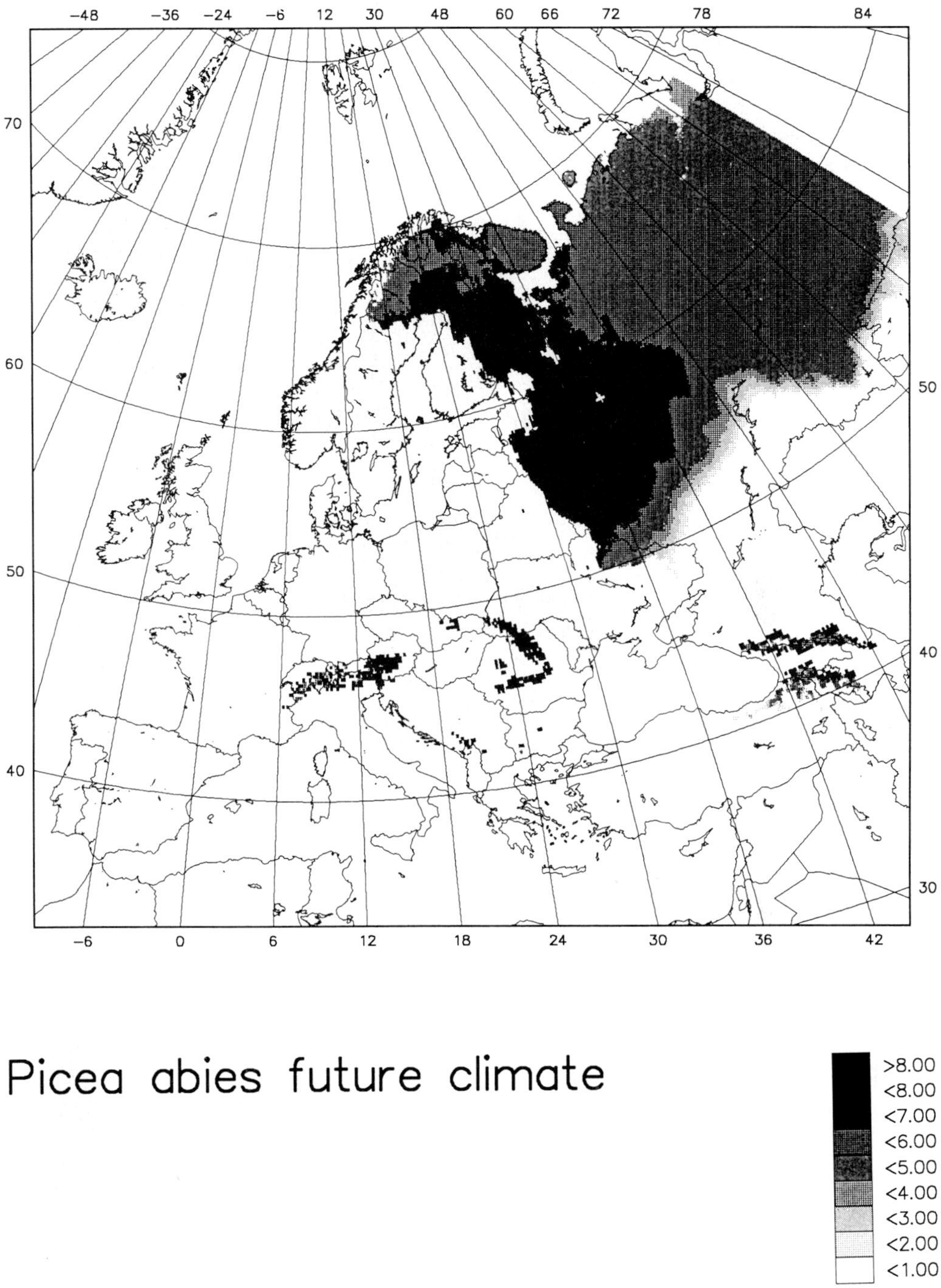

Fig.1. (continued)

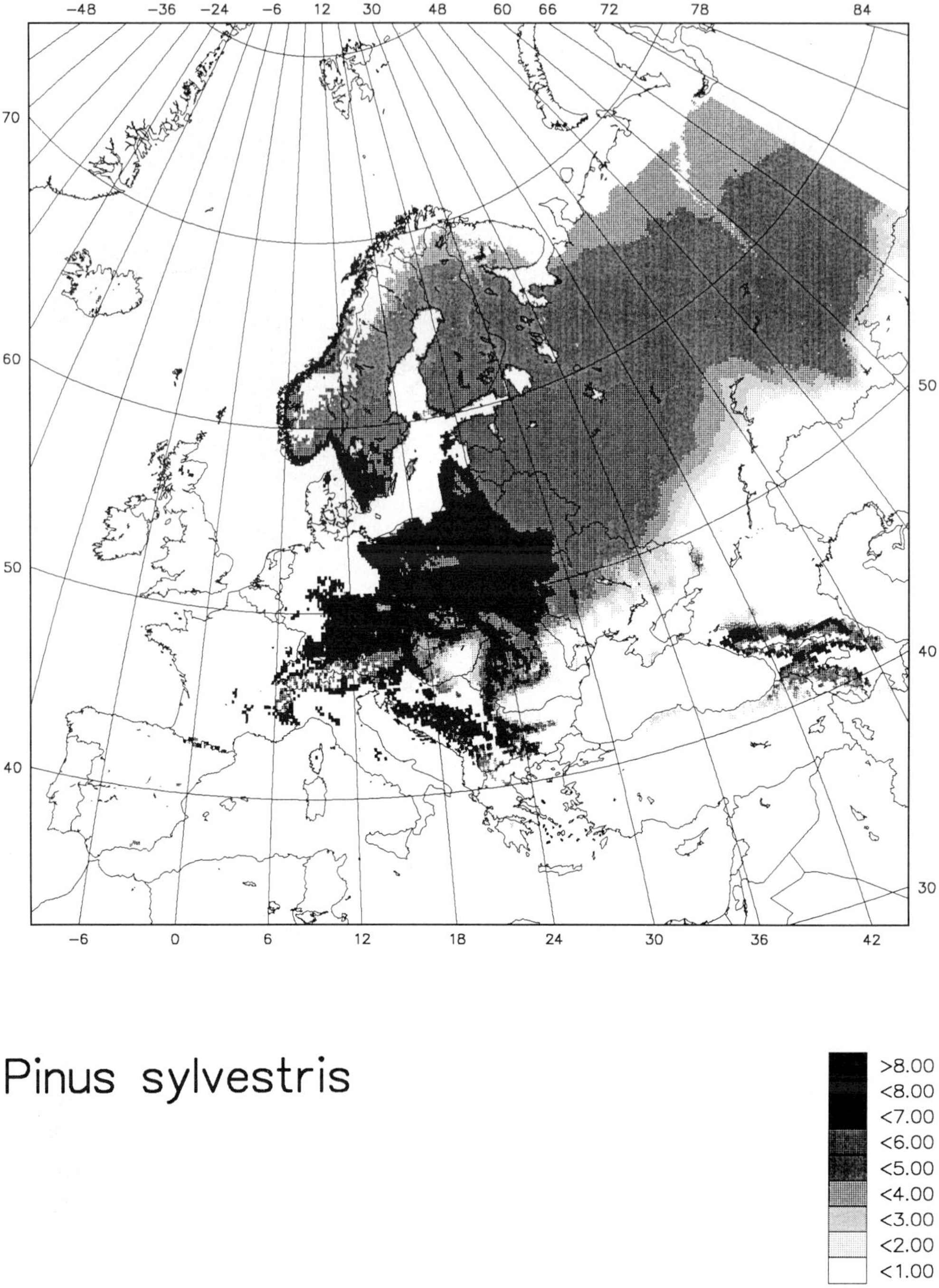

Fig. 1. (continued)

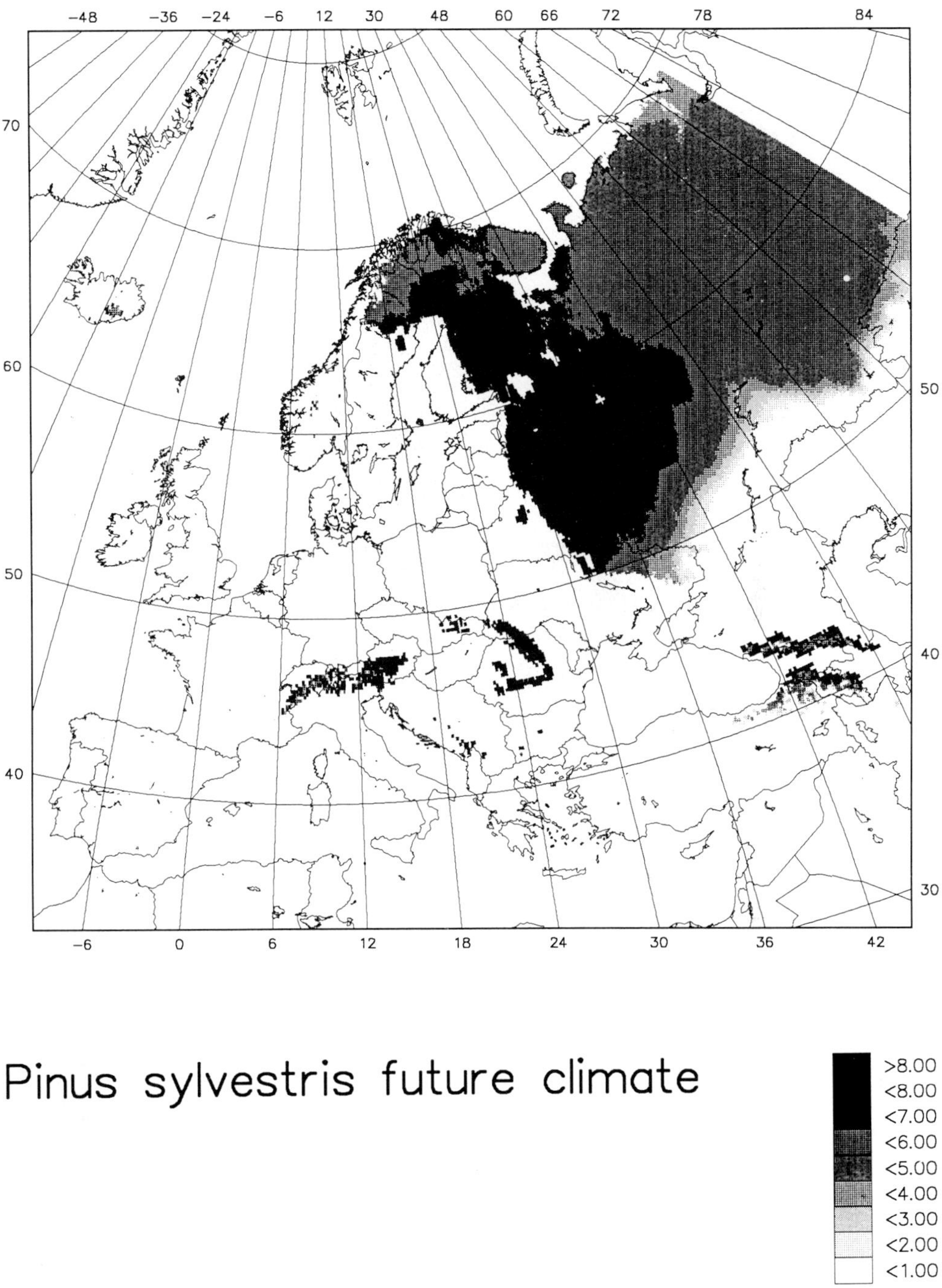

Fig. 1. (continued)

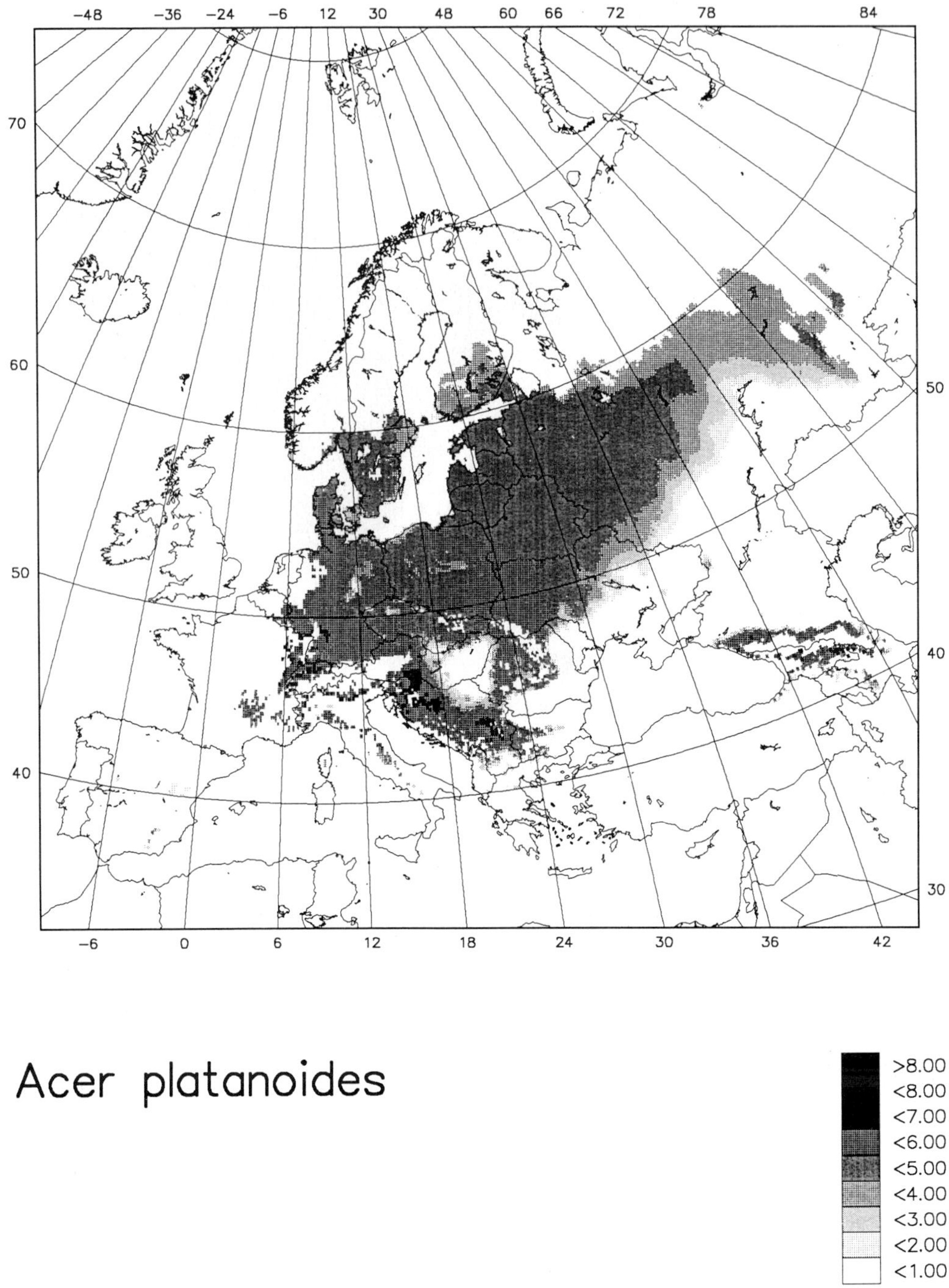

Fig. 1. (continued)

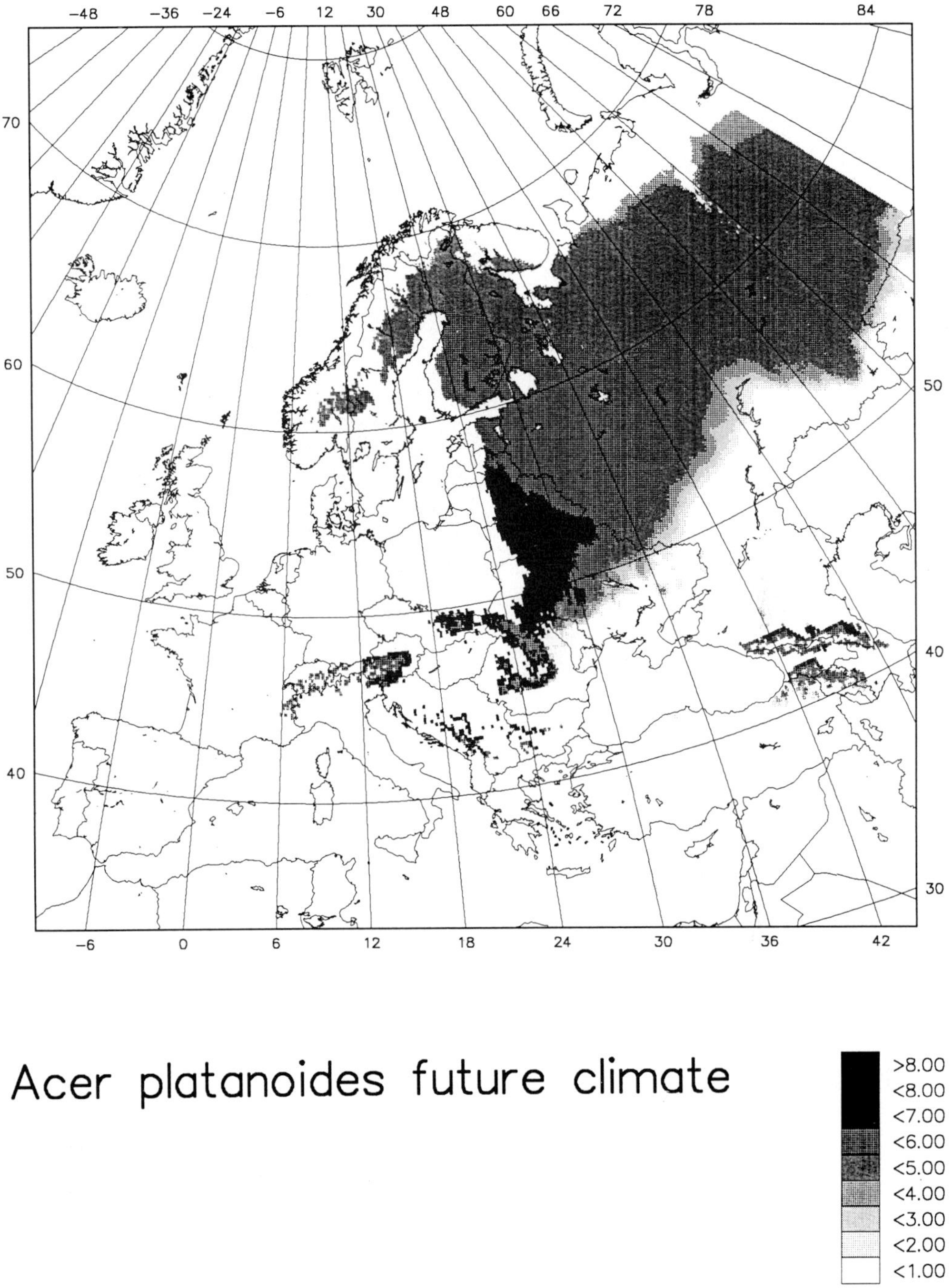

Fig. 1. (continued)

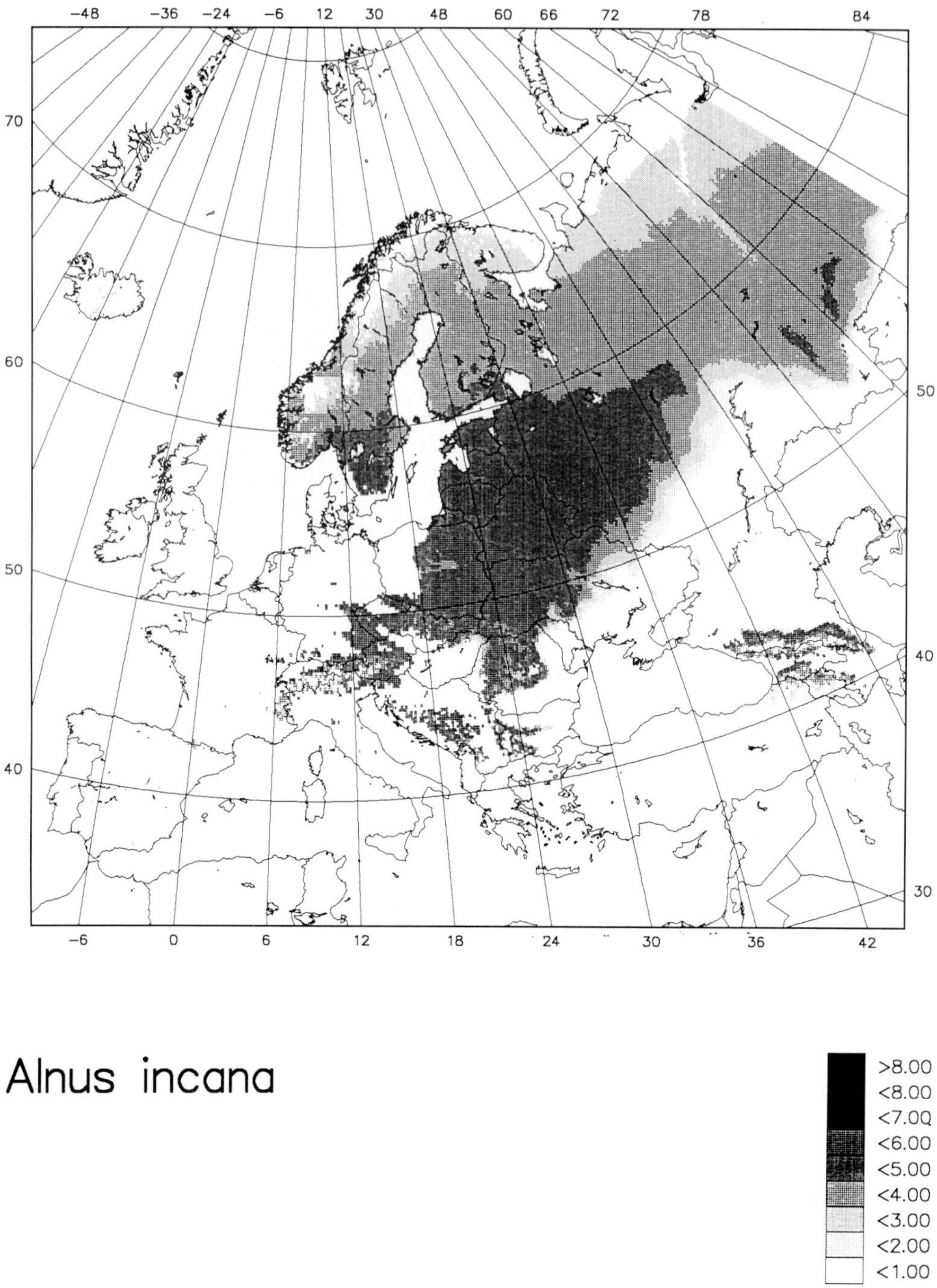

Fig. 1. (continued)

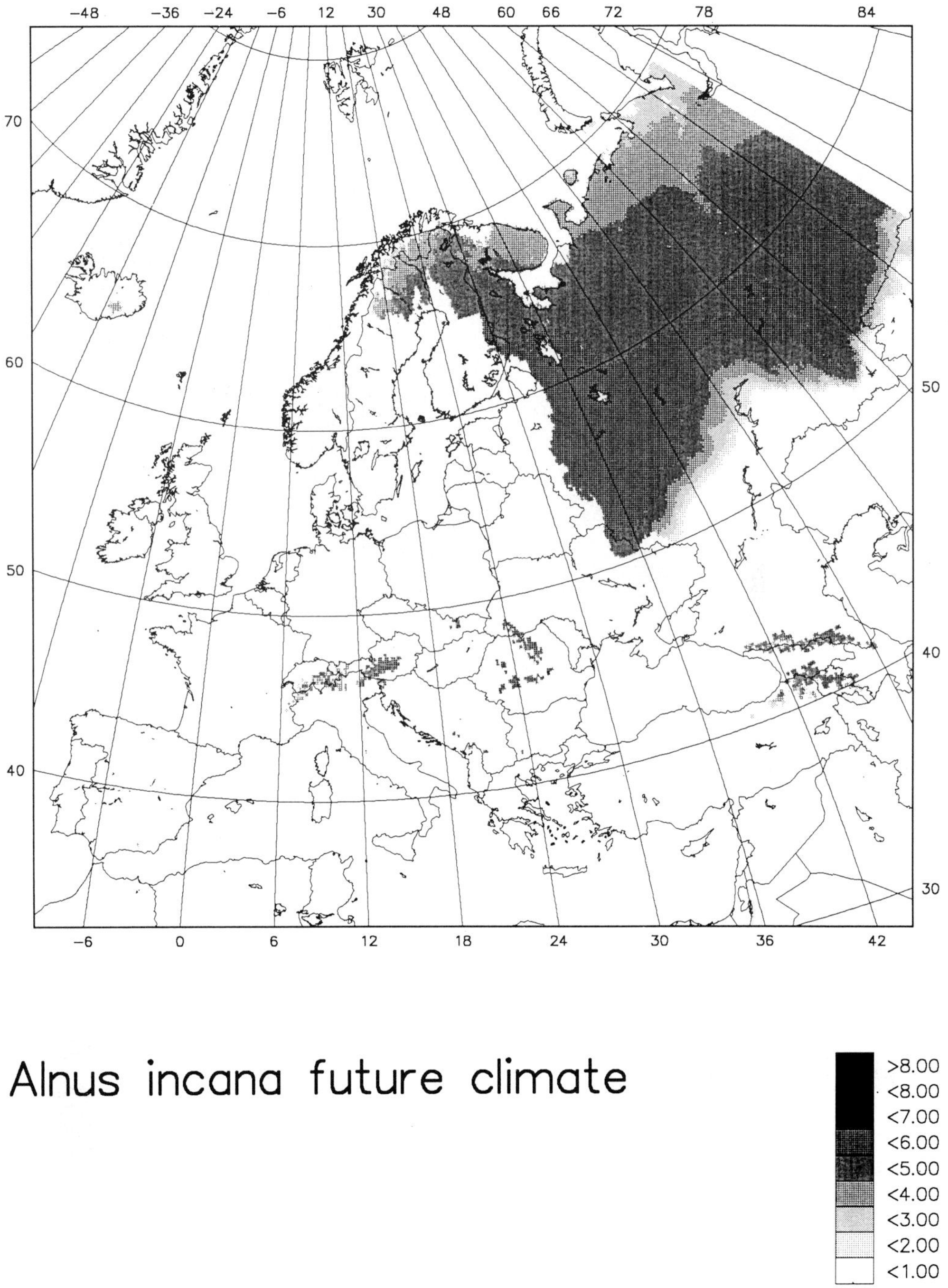

Fig. 1. (continued)

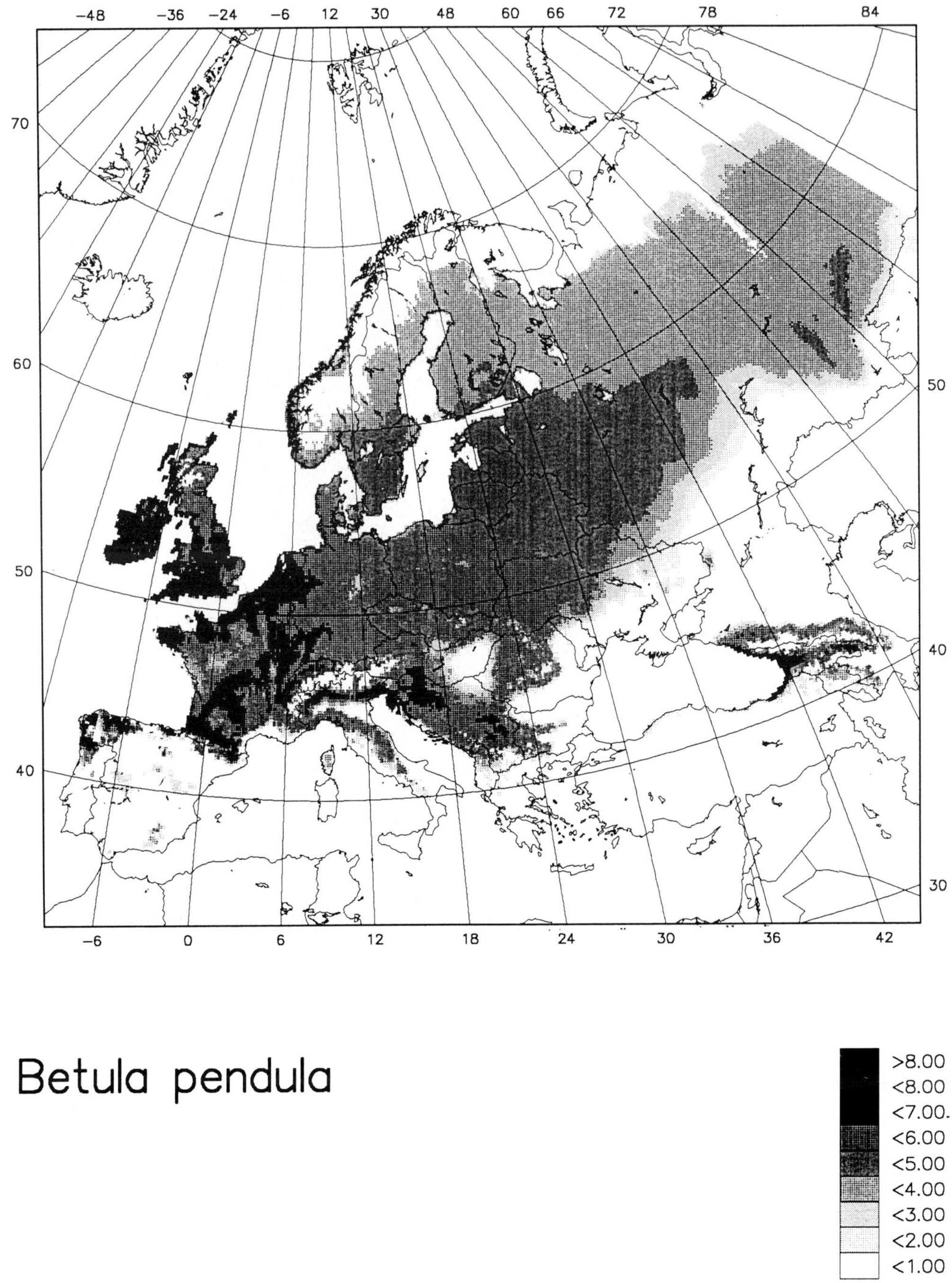

Fig. 1. (continued)

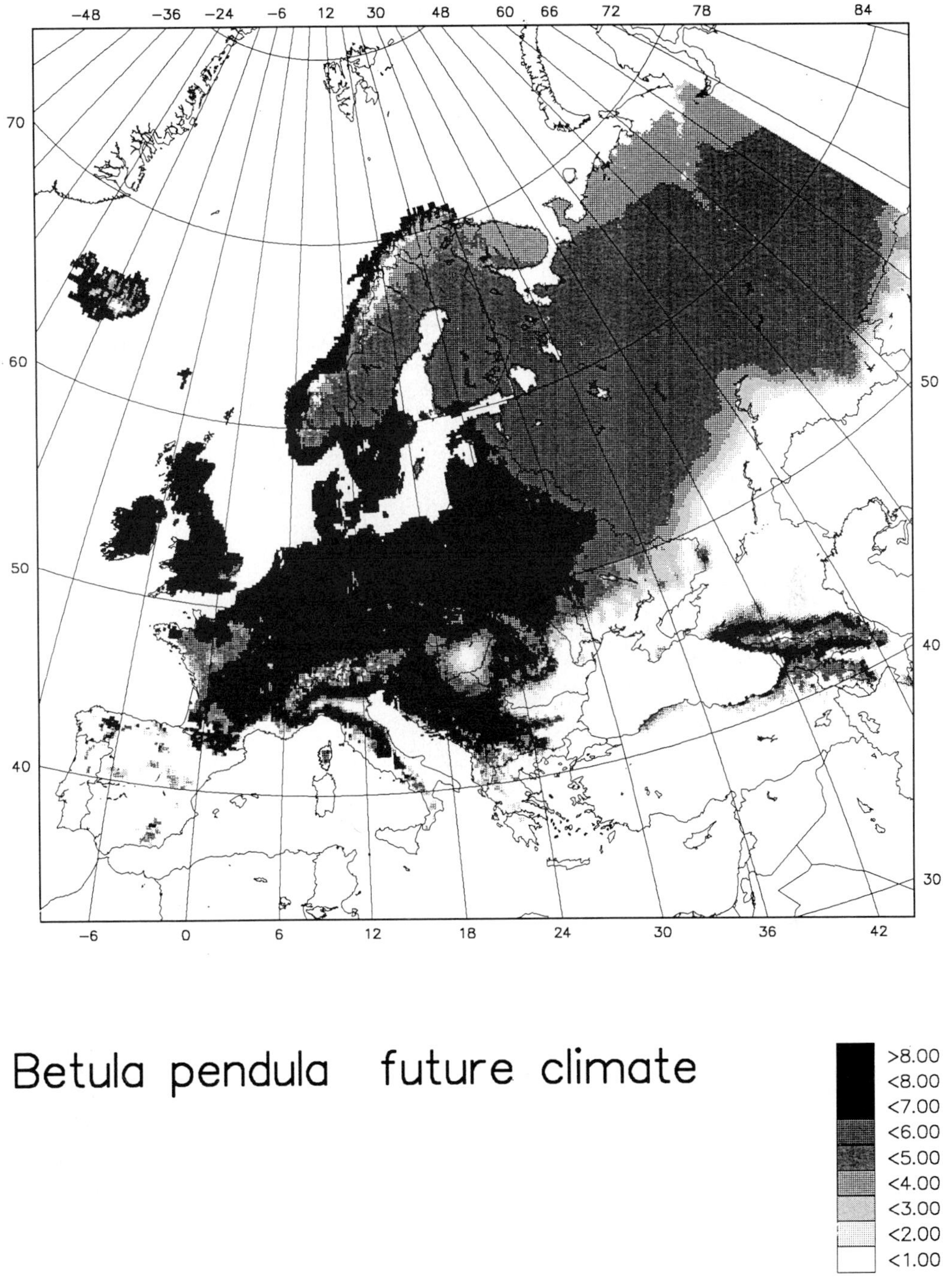

Fig. 1. (continued)

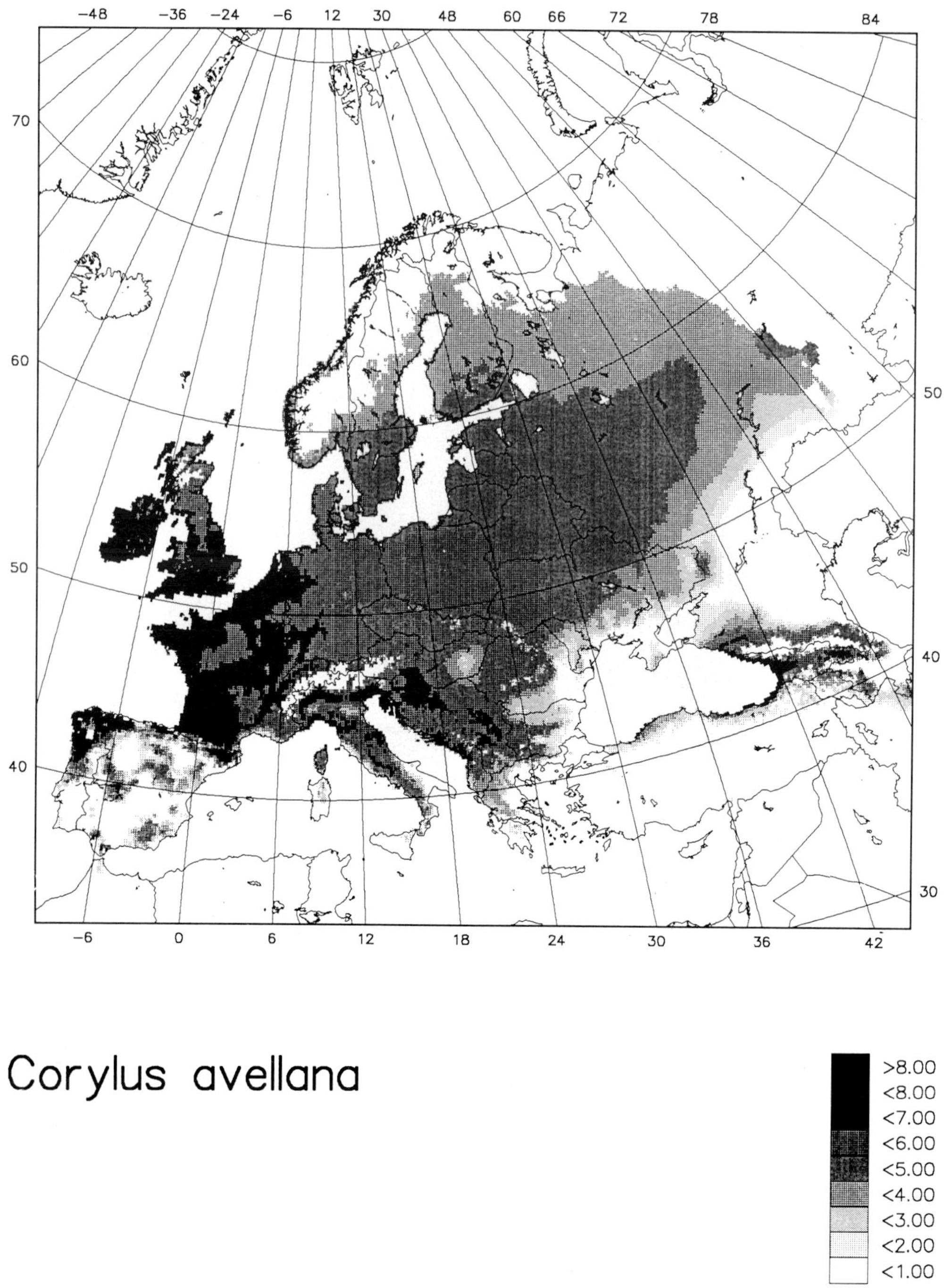

Fig. 1. (continued)

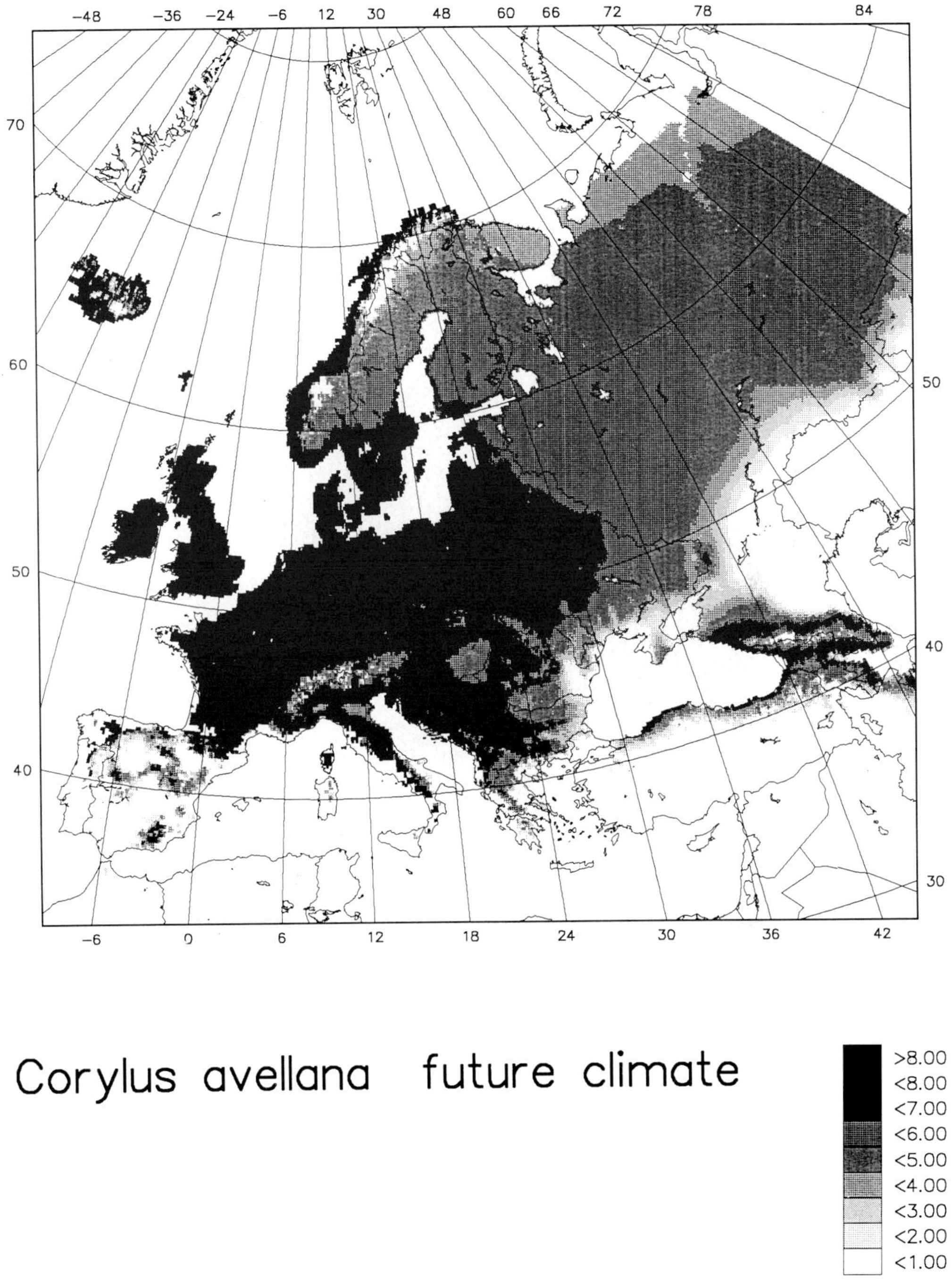

Fig. 1. (continued)

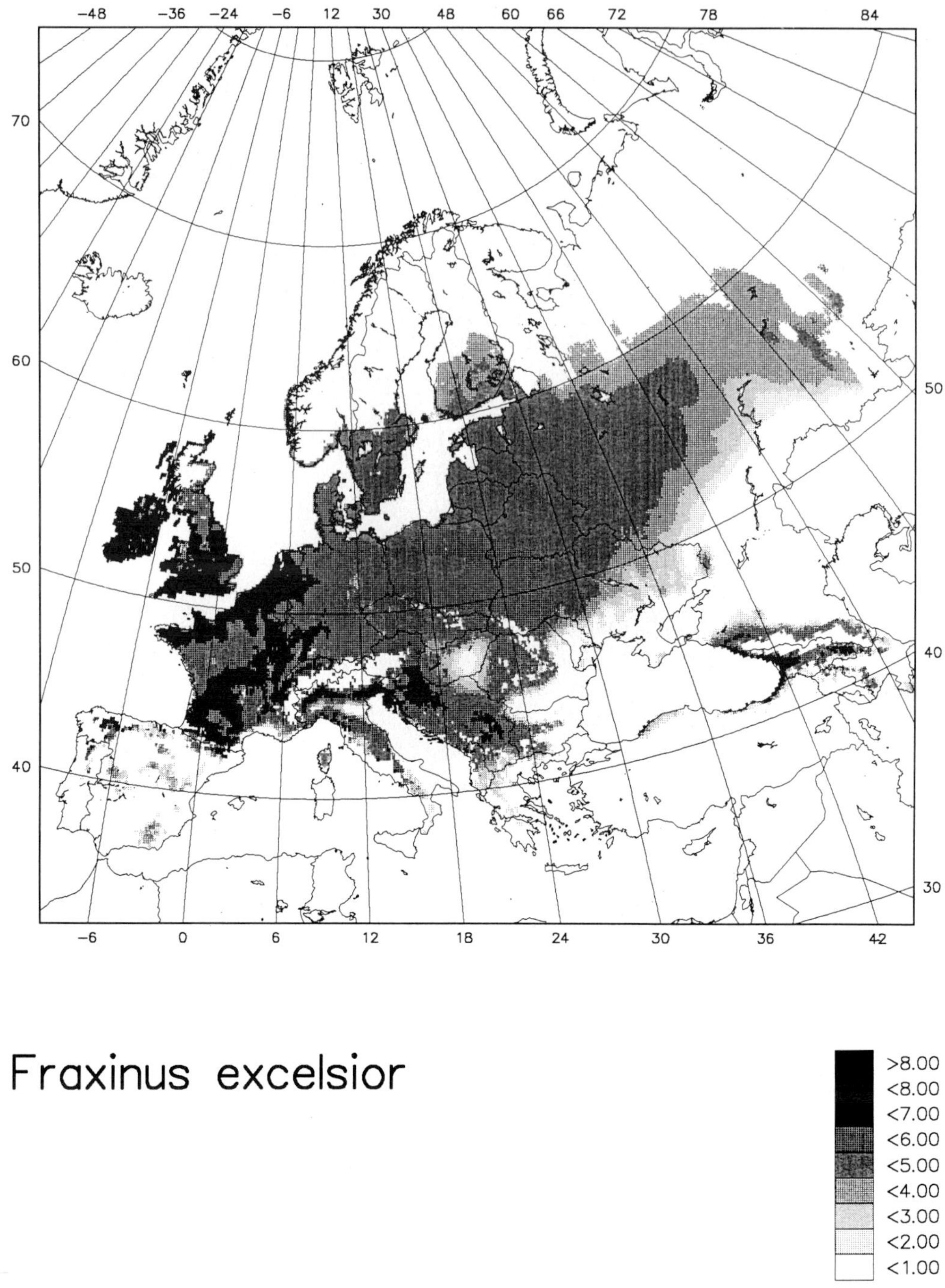

Fig. 1. (continued)

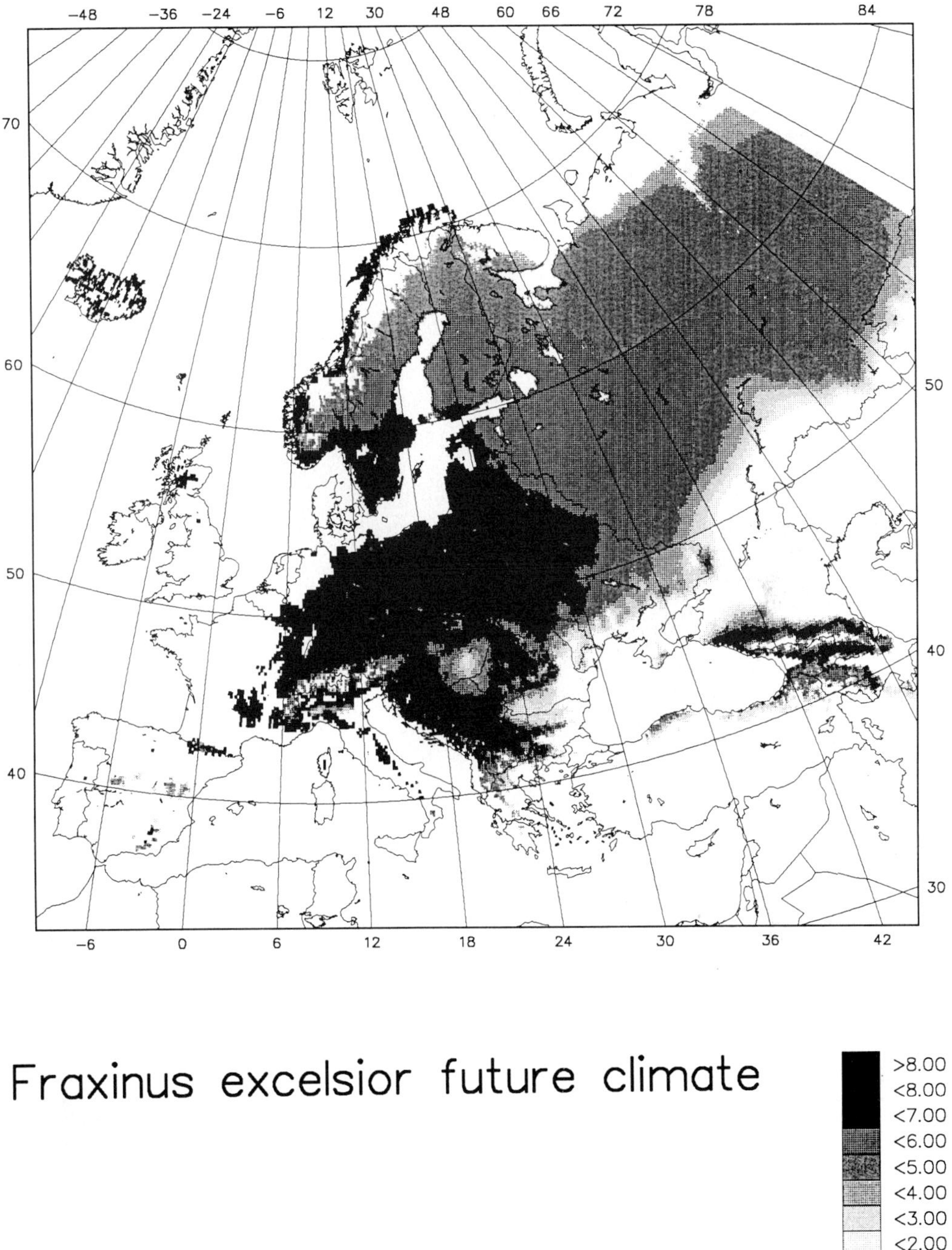

Fig. 1. (continued)

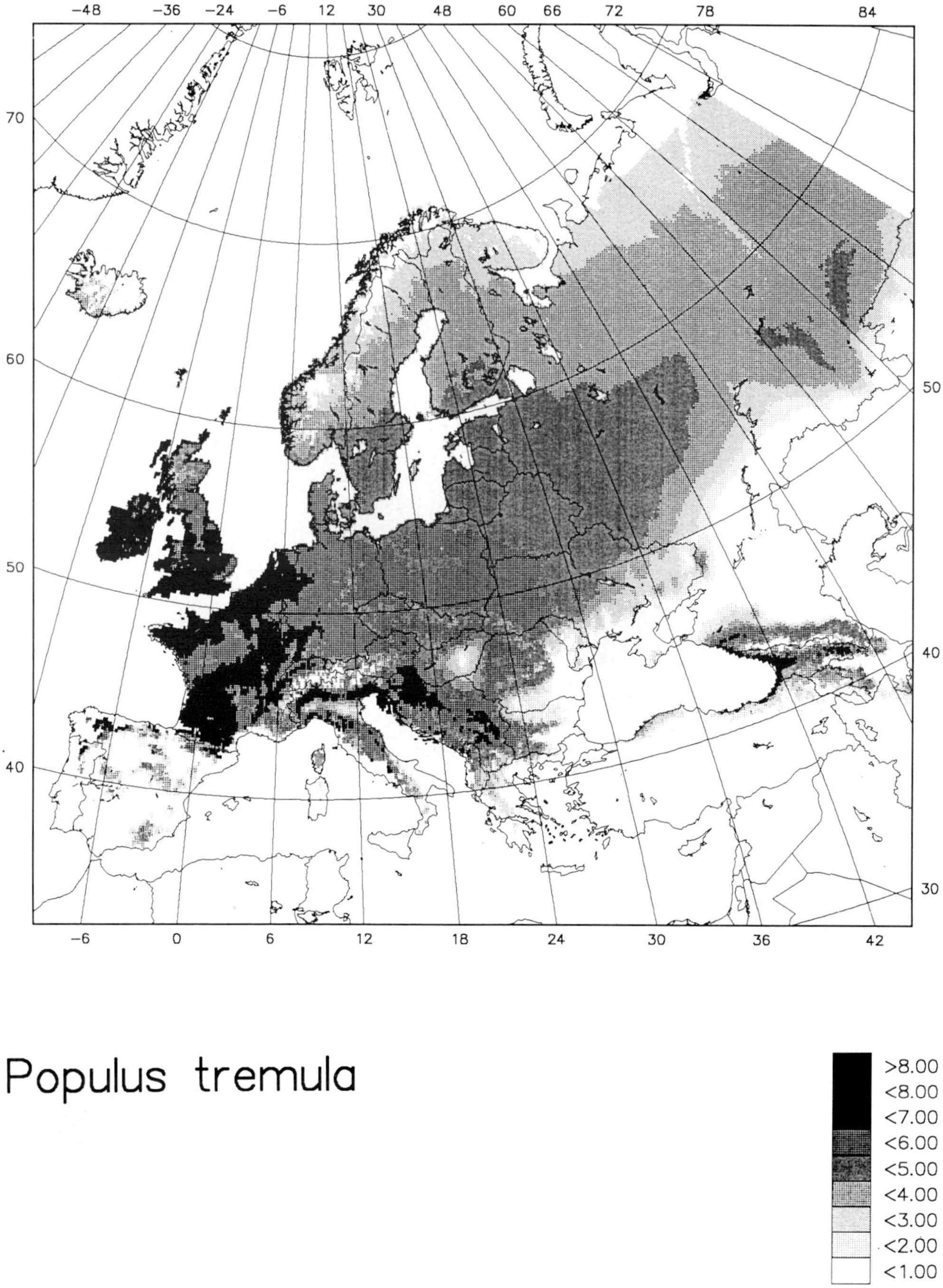

Fig. 1. (continued)

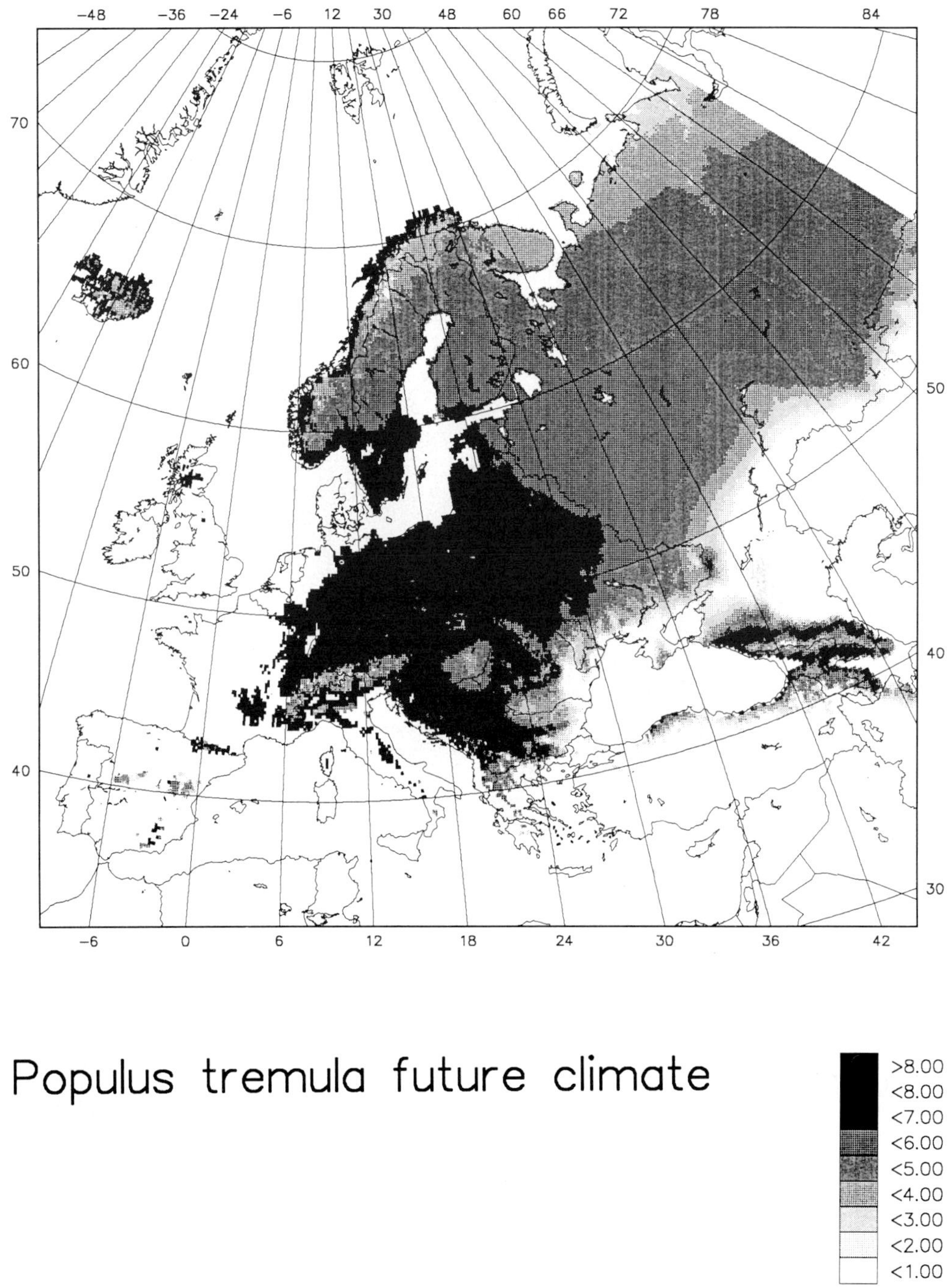

Fig. 1. (continued)

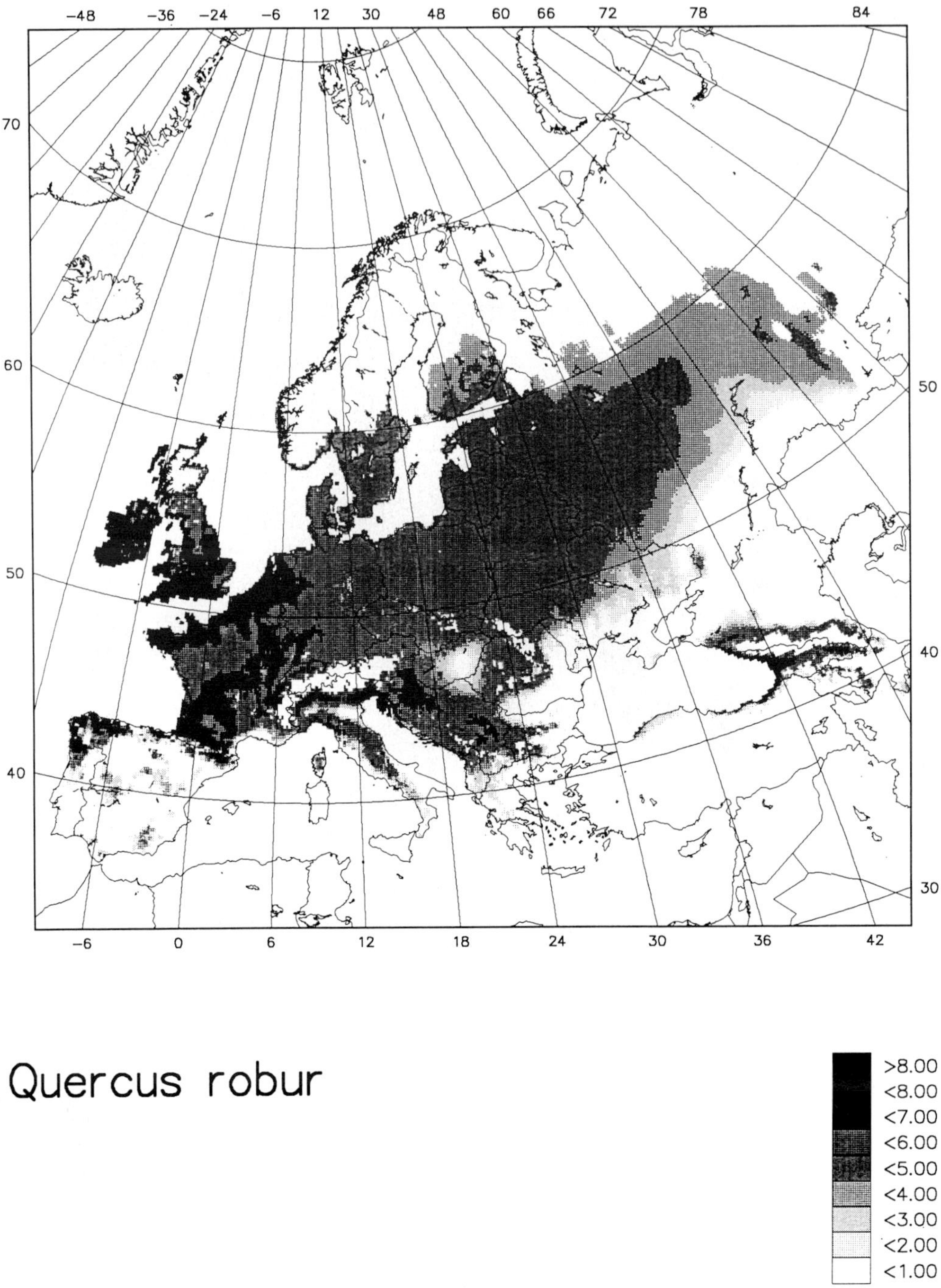

Fig. 1. (continued)

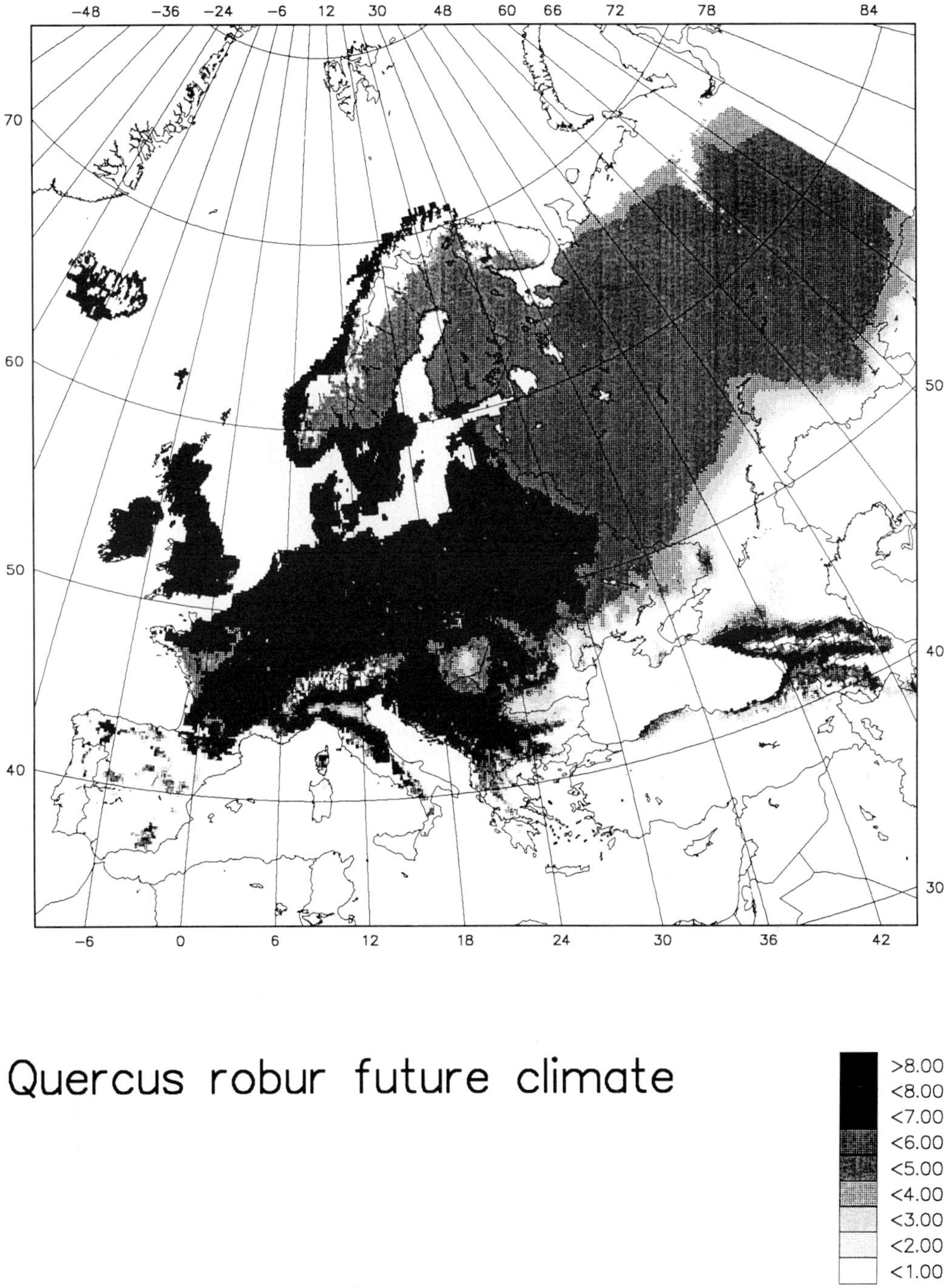

Fig. 1. (continued)

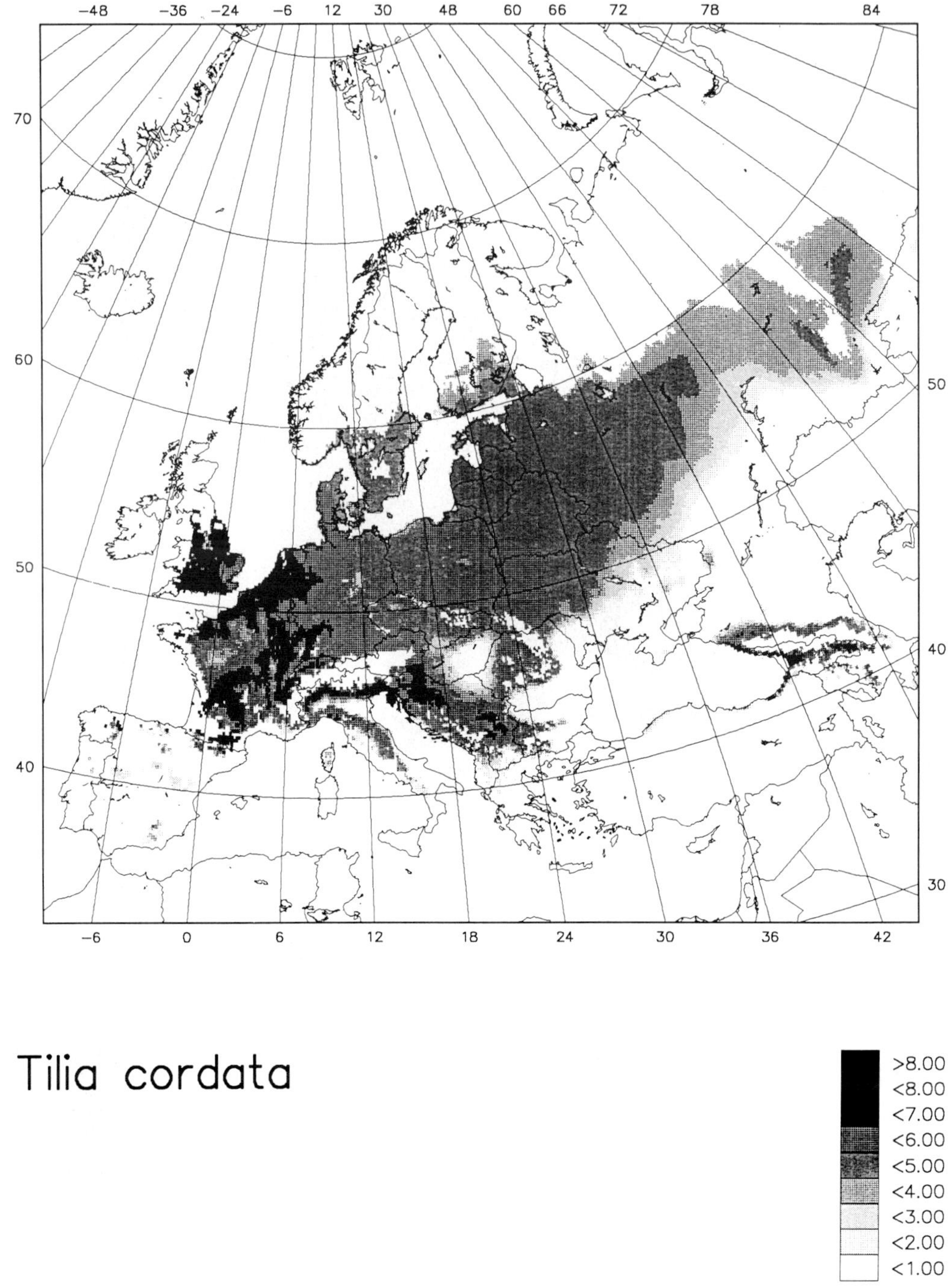

Fig. 1. (continued)

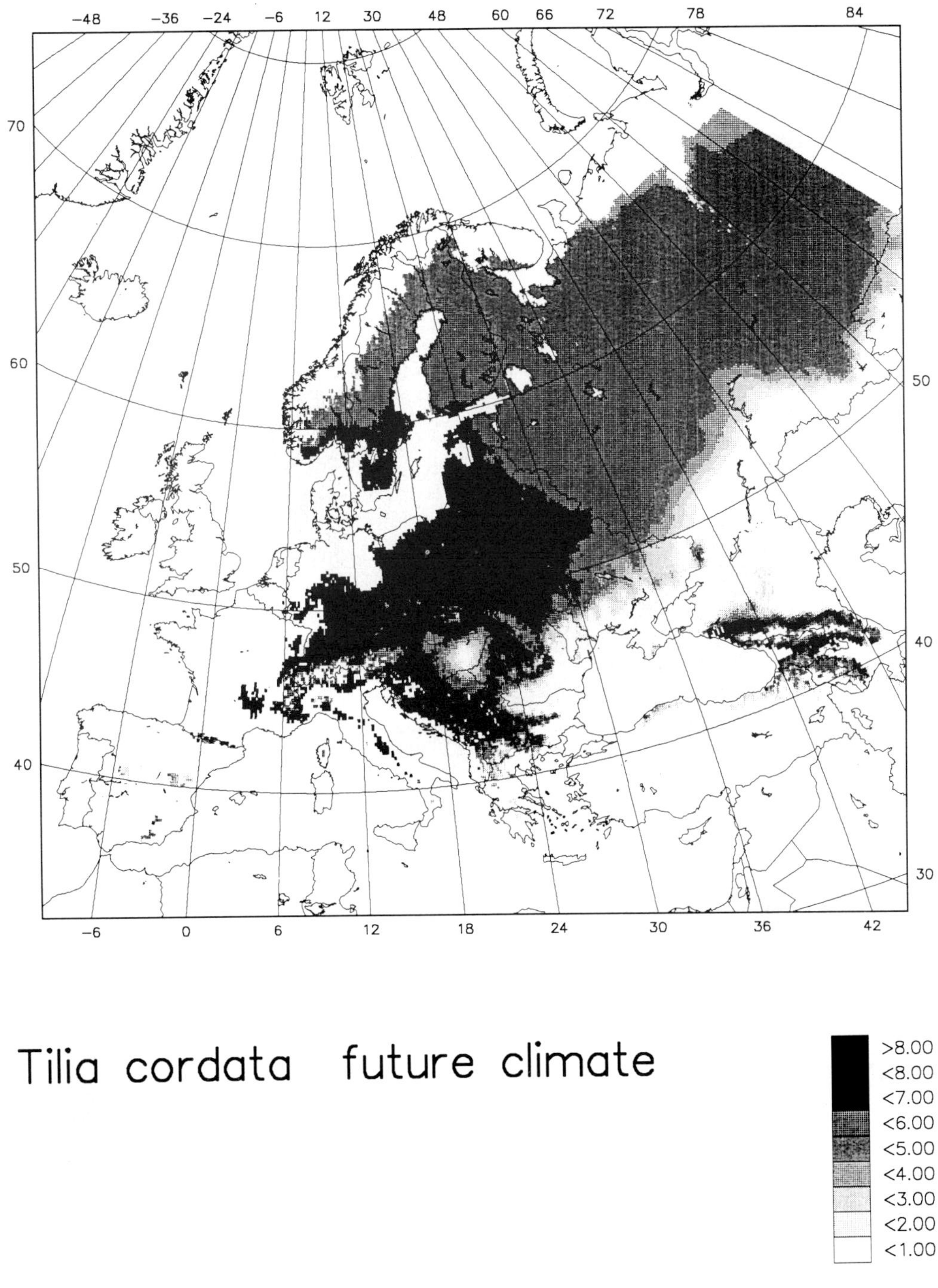

Fig. 1. (continued)

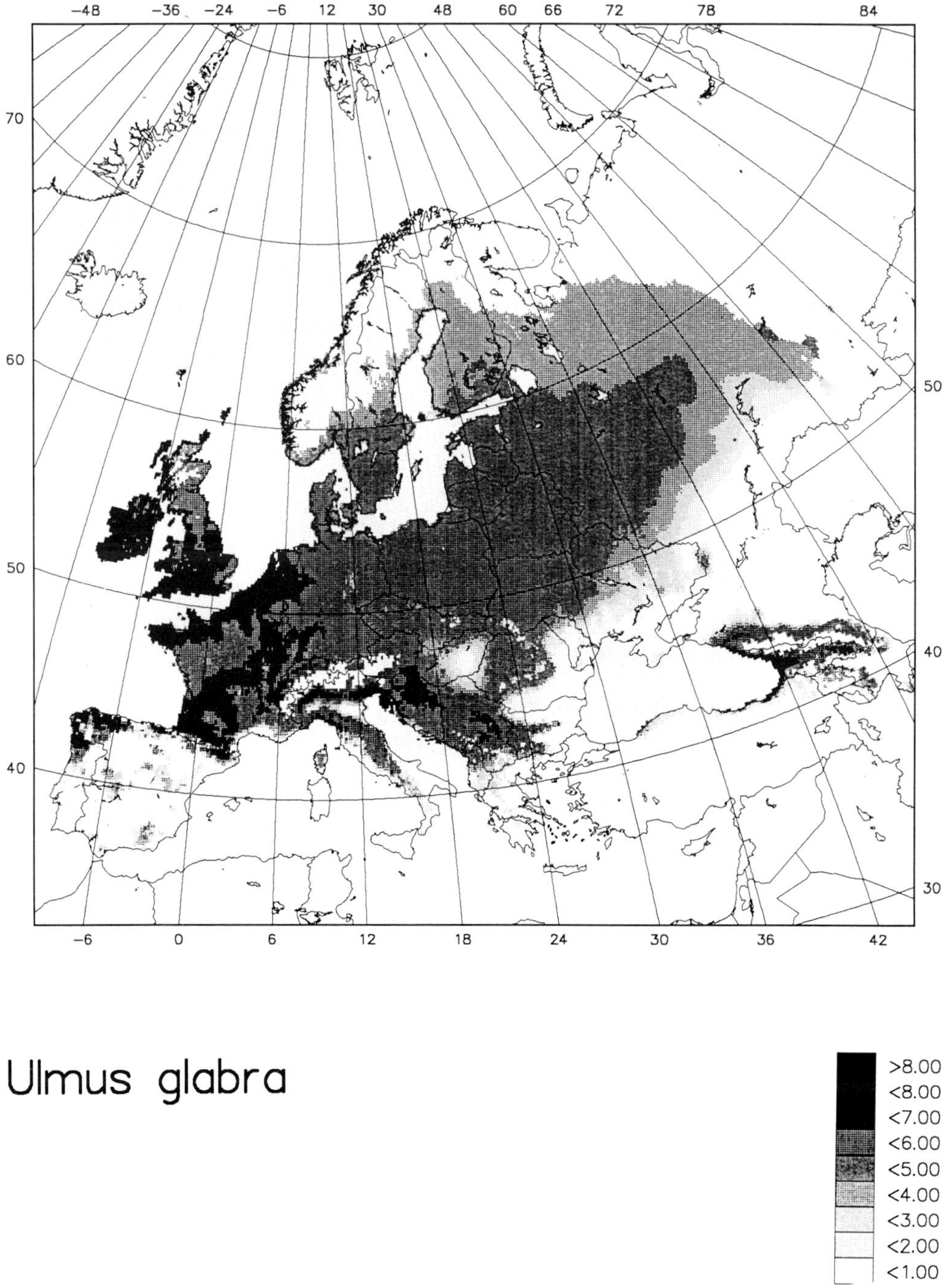

Fig. 1. (continued)

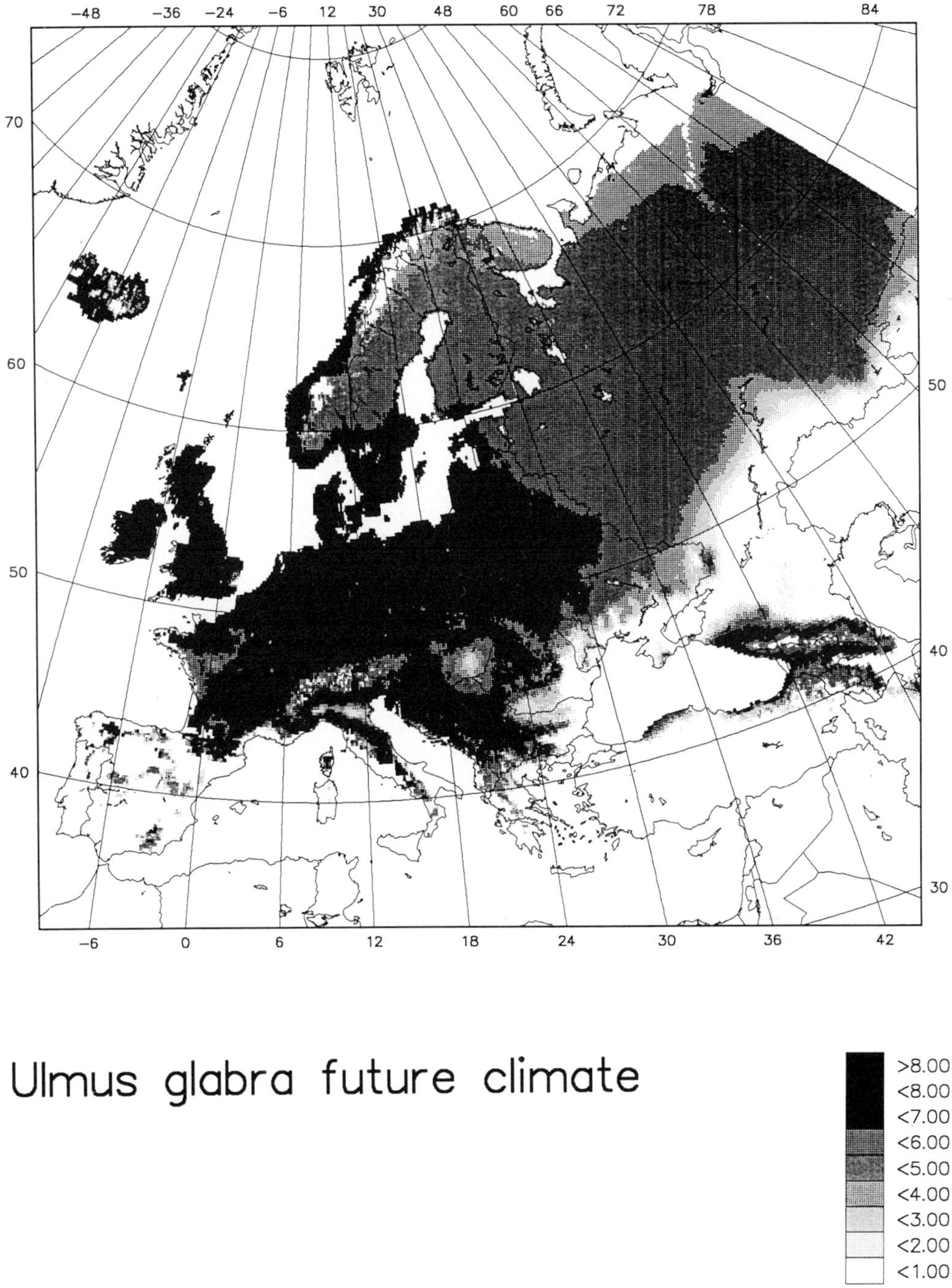

Fig. 1. (continued)

Ulmus glabra (Wych elm) is found throughout the southern and coastal central boreal zones (Figure 1). Under the changed climate it expands northwards to cover most of the boreal zone, except for some of the montane areas.

3.2. Species Distribution Limit Changes: Common Trees of the Nemoral Zone

Fagus sylvatica (beech). The northwestern distribution limit of *Fagus sylvatica* can be explained by its high chilling requirement (Figure 2). Without this adaptation to moderately continental conditions, *Fagus* would extend through Ireland and Scotland. Under the changed climate, *Fagus* extends a long way north into the boreal zone, while retreating from areas it at present occupies in the south and west.

Quercus petraea (sessile oak) occurs in part of southern Sweden, coastal Norway and Denmark (Figure 2). Under the changed climate its range is shown expanding to cover much of Fennoscandia, though it would be excluded from the western coastal fringes.

The degree and breadth of the possible changes in species ranges as described in these equilibrium simulations show the extent to which some the projections for climate changes in the boreal zone could dramatically alter the forest composition in northern latitudes. However, as these simulations are for equilibrium distributions, they give no information on the dynamics of change. Analysis of this is done at more local scale, using FORSKA2, as described in the following section.

3.3. Dynamic Simulations

3.3.1. Reivo, Norrbotten

At this most northerly site (Figure 3), FORSKA2 simulates a typical Swedish boreal zone landscape in which *Picea abies* and *Pinus sylvestris* co-dominate (Figure 4a). Increased warming in the transient period initially favours both species equally. *Pinus*, however, in the longer term responds less favourably to the increased competition from *Picea* so that *Picea* comes to dominate. Boreal deciduous species also decline. Total biomass increases as *Picea* grows faster in response to warming.

3.3.2. Tandövala, Kopparberg

Tandövala lies in the boreal zone of central west Sweden (Figure 3) where the equilibrium vegetation in the present climate is again simulated as a *Picea–Pinus* forest (Figure 4b). During the transient period an increase in biomass occurs as *Picea* responds favourably to increased growing-season warmth. At the same time *Pinus* responds negatively to the increased competition from *Picea*. However, the warming continues beyond the limits for establishment of *Picea*, leading to local extinction within 250 yr. The major replacement is *Pinus sylvestris*. Here, as *Picea* declines, *Pinus* comes to dominate over new deciduous arrivals such as *Fagus sylvatica* (Figure 2). Forest composition changes dramatically from one climate phase to the next, but total site biomass seems to be little affected, even during the transient phase.

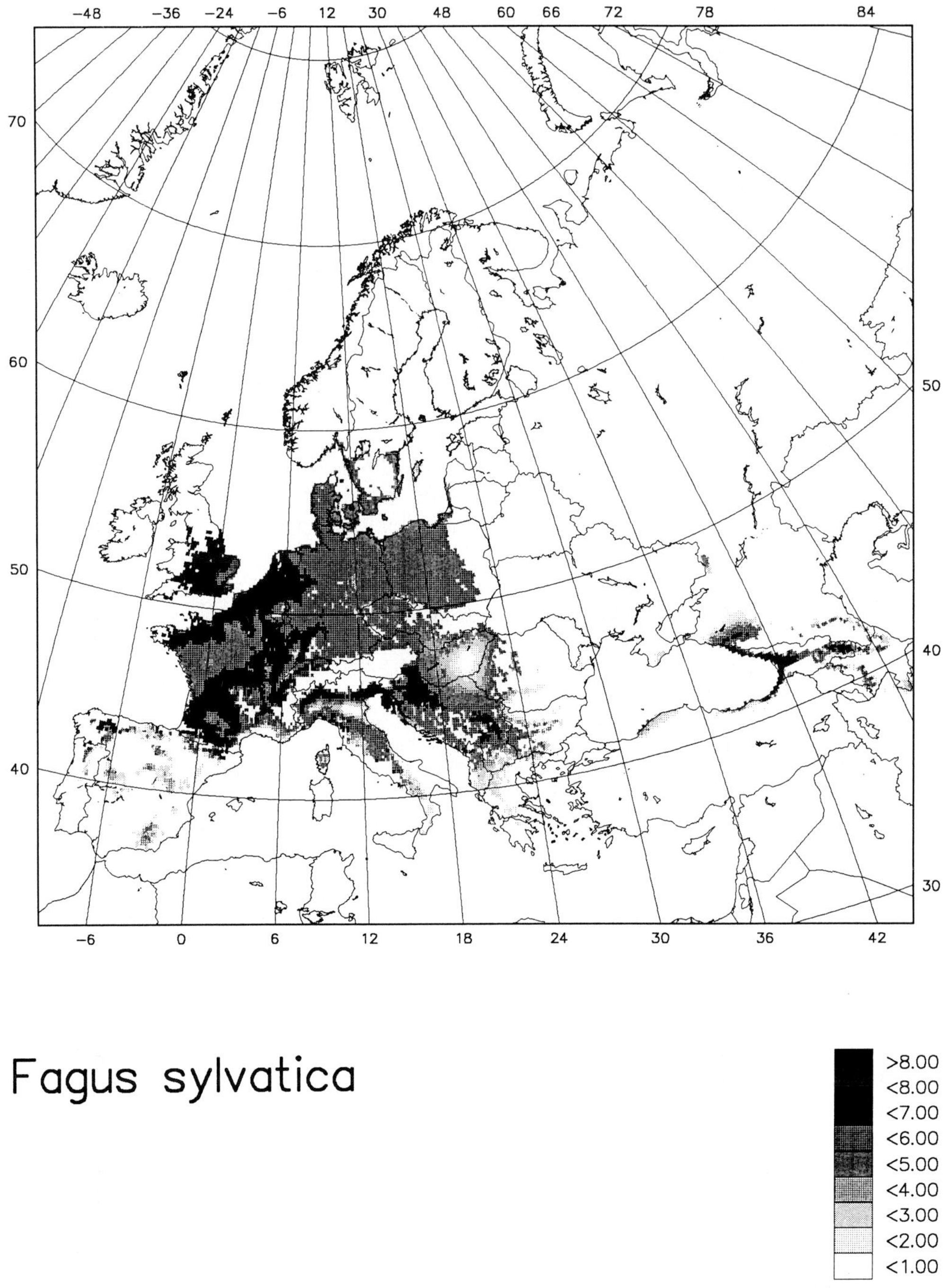

Fig. 2. Simulated distributions of two common nemoral zone trees from northern Europe under present and future climates (see Figure 1 for details).

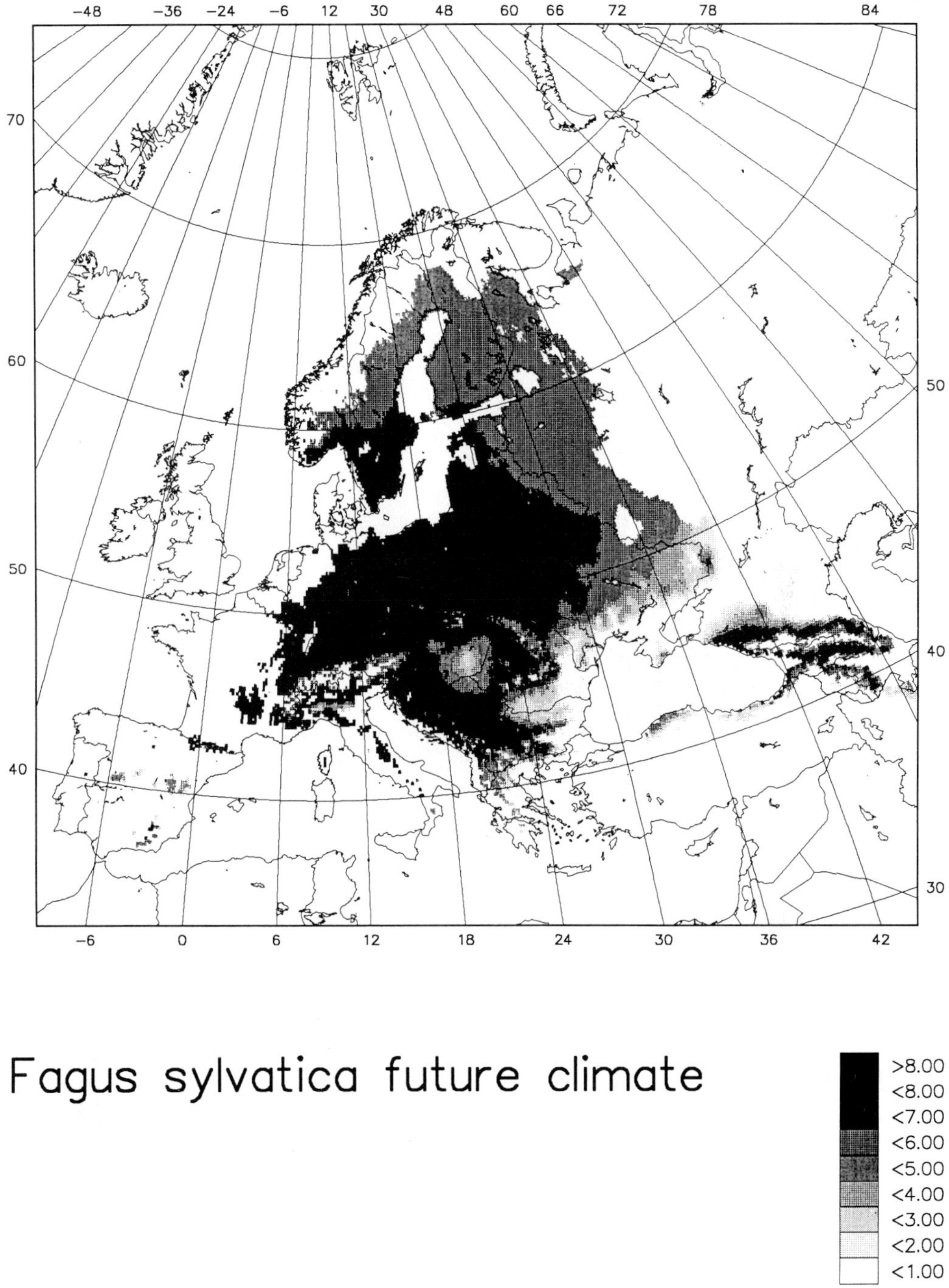

Fig. 2. (continued)

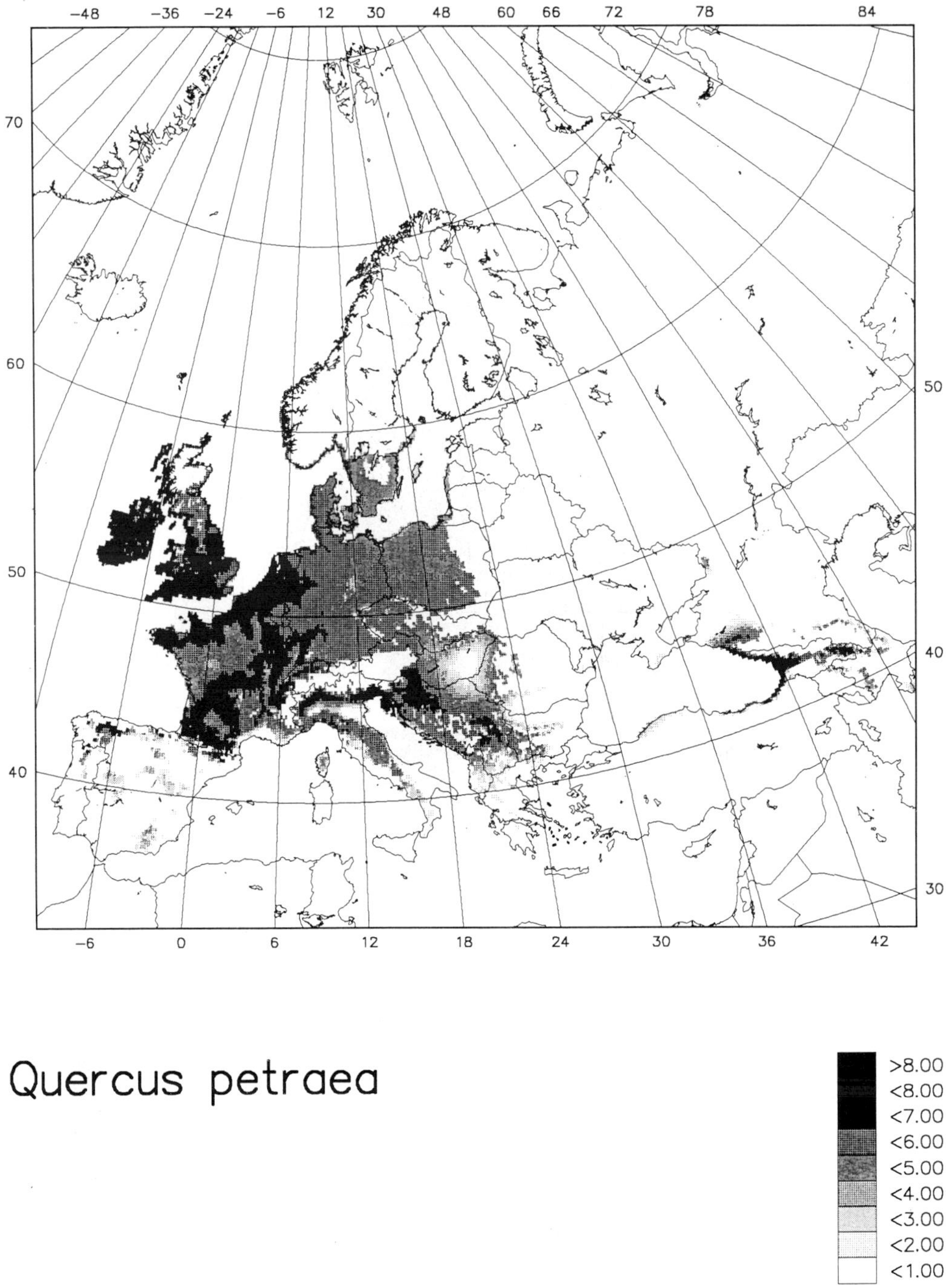

Fig. 2. (continued)

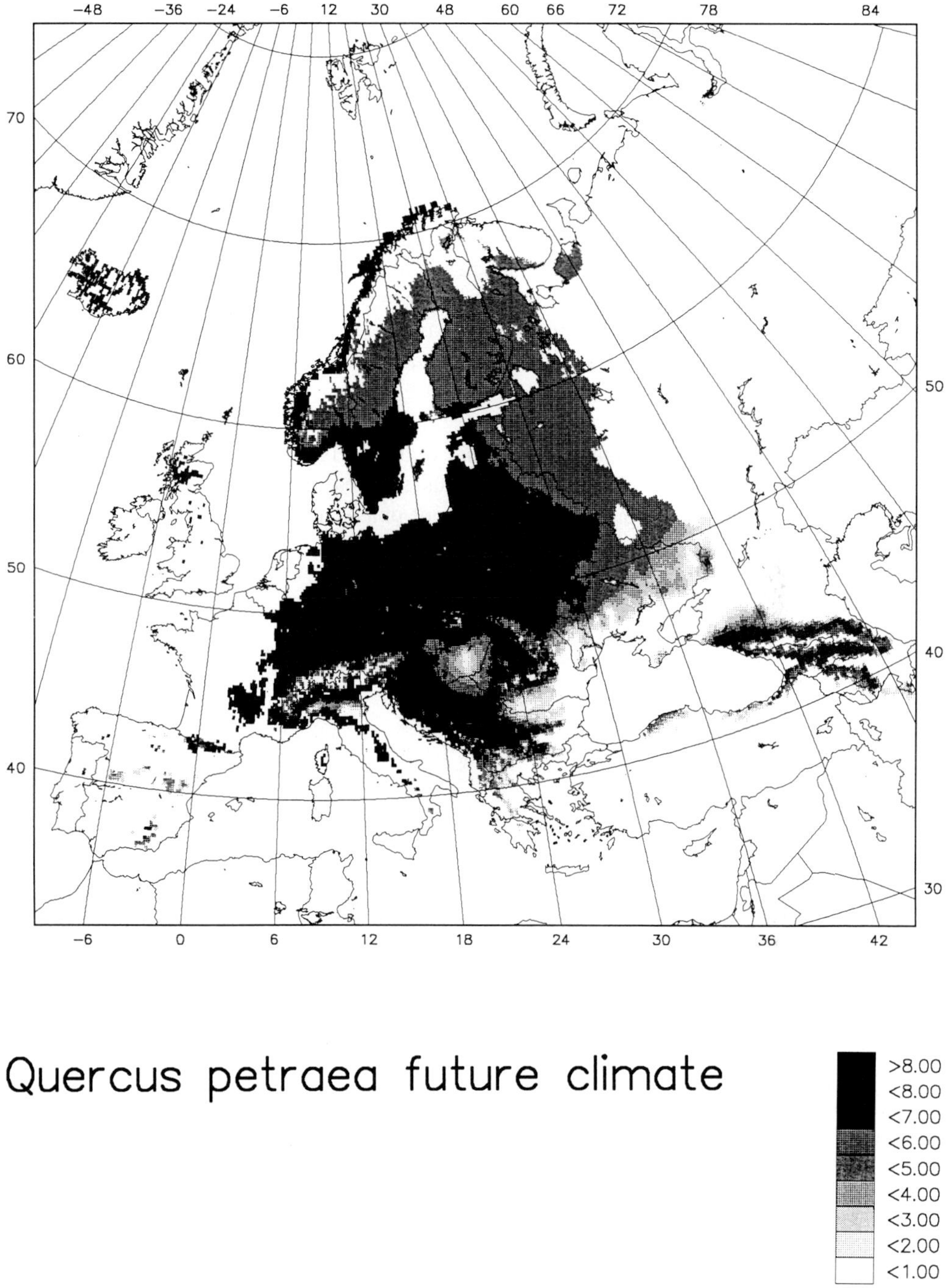

Fig. 2. (continued)

3.3.3. *Tyresta, Stockholm*

Tyresta lies in the boreo-nemoral zone of central Sweden. Today *Picea* and *Pinus* dominate (Figure 4c), but a number of deciduous species (including *Betula* and *Quercus* spp.) also contribute significantly to the total biomass. The warming phase leads to a rapid decline of the conifers, as well as a brief decline in biomass and a corresponding increase in the deciduous species already on site. The major component of the new forest however is *Fagus sylvatica*, which was not originally present. Biomass increases, a substantial part of this increase being due to *Fagus*.

However if no immigration of new species is allowed, the forest composition under the new climate becomes dominated by *Quercus* and *Betula* spp. (Figure 7). Biomass then declines rapidly during the transient phase, recovering only after ≈150 yr to a level slightly less than would be achieved in the presence of *Fagus*.

3.3.4. *Tiveden, Örebro*

Picea is dominant in the simulation of this region (Figure 4e), though *Pinus* forms quite a large part of the biomass. Deciduous species are present though in minor amounts. *Picea* declines rapidly in the transient phase, as new recruitment fails, and

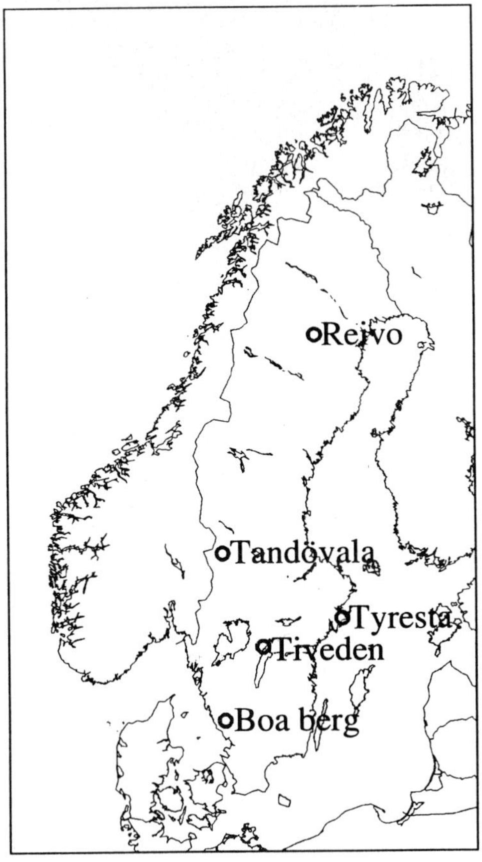

Fig. 3. Locations of the five sample sites.

Pinus is outcompeted by the increasing presence of deciduous species. *Fagus* immigrates and becomes the dominant species. Here, however, there is somewhat less biomass under the changed climate than was shown for the present climate.

3.3.5. *Boa Berg, Halland*

This site lies near the border between the boreo-nemoral and nemoral zones (Figure 3). *Picea* is still the major component but *Pinus* and some deciduous species, including *Fagus*, make up the rest of the forest (Figure 4f). Once the warming begins there is a complex transitional response in which *Picea* and *Pinus* decline rapidly but *Pinus* remains in the forest (in small amounts) even under the new warmer climate. *Fagus* increases during the transient phase, only to decline abruptly later due to failure of chilling. *Fagus* is then replaced by a *Betula-Quercus* forest of lower biomass than before the warming began.

4. Discussion and Conclusions

The models predict large changes in the boreal and boreo-nemoral forests of northern Europe. STASH, as a static model, predicts these changes by modelling present and potential equilibrium ranges of species over large regions. STASH also calculates a rough index of the potential productivity of each species. Some boreal species (e.g. *Picea abies*) are more productive in the warmer areas of their ranges than in colder. There is no decline in productivity towards the southern boundary. These observations, reproduced by STASH, run counter to the parabolic response curve to GDD as used in early forest gap models (Botkin *et al.*, 1972). There is no basis for a maximum GDD limit (Prentice *et al.*, 1991; Prentice *et al.*, 1993; Sykes *et al.*, in press). However the boreal conifers do show abrupt southern and western limits which may be explained by other mechanisms including chilling requirements and increasing competition from deciduous species. Our models embody reasonable hypotheses about these limits, but more experimental work on the mechanisms is needed in order to reduce the associated uncertainties.

Many boreal species show range shifts under the changed climate scenario. These changes are of such a magnitude as to give a completely different structural and compositional character to much of landscape now occupied by boreal or boreo-nemoral forest. The true boreal forest of Fennoscandia may after a climate change become restricted to the far north and the mountains.

The forest gap model approach using FORSKA2 indicates some of the possible routes that different regions might follow in response to a changing climate. Forest dynamics simulated at five representative sites show a variety of behaviour. The speed of the changes can be strongly influenced by the future disturbance and management regimes as well as by an interaction with species availability (Sykes and Prentice, unpubl. ms.). In the north, species already in the forest may respond favourably to a warmer climate. Further south there is a greater potential for loss of current species and the possibility of immigration for warmth-requiring deciduous species. Natural immigration may take some time. Simulations for Tyresta, with immigration suppressed, show that there is at least a possibility that species already in the region

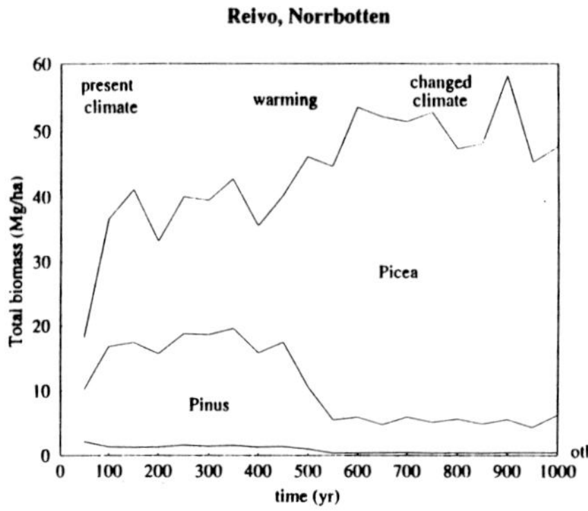

(a) Near Reivo, Norbotten, northern Sweden.

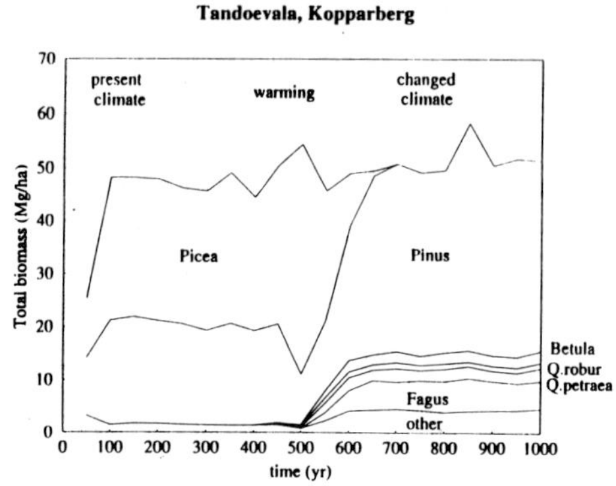

(b) Near Tandövala, Kopparberg, north-central Sweden.

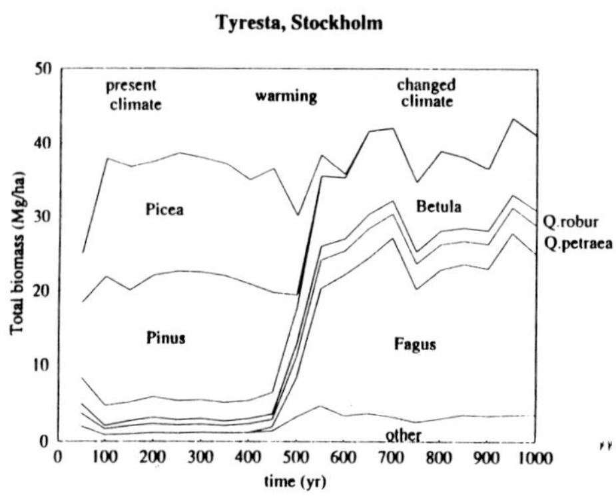

(c) Near Tyresta, Stockholm, central-south Sweden.

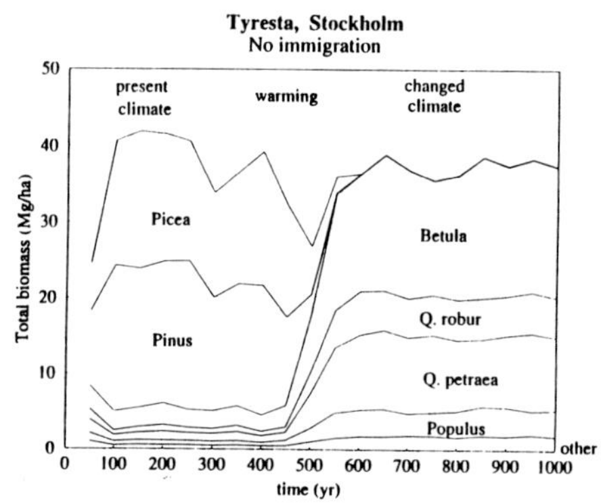

(d) as (c), but only those species present in the region under the *current* climate were available for establishment in the future.

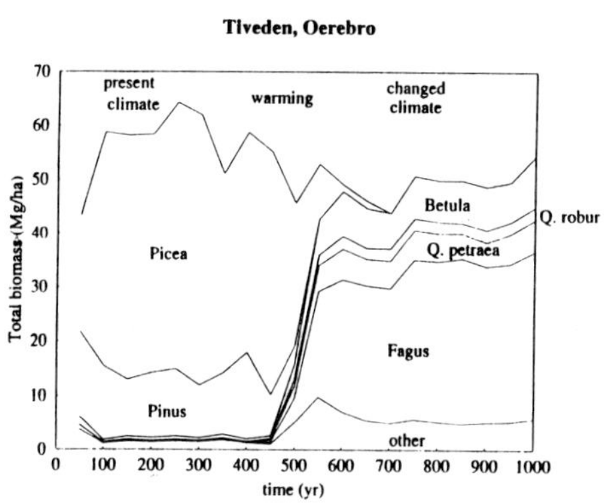

(e) Near Tiveden, Örebro, southern Sweden.

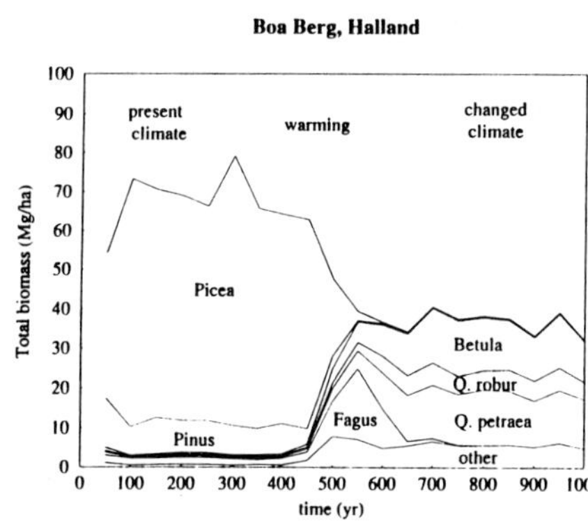

(f) Near Boa Berg, Halland, southwestern Sweden

Fig. 4. Simulation of forest landscape at the five sites in Sweden shown in Figure 3. A linear change in climate was imposed during the 100 year period marked "warming". After that, the climate remained constant at the mean 2×CO_2 climate for each location.

could maintain the previous biomass despite a major change in dominance. Further south, where we have not considered the possibility that still more warmth-demanding species might invade, the existing species will not maintain present biomass in the changed climate, so there could be substantial net losses of forest biomass. Such effects could be mitigated by importing and planting species from other regions. The example of *Fagus*' behaviour in the Boa Berg simulation illustrates the potential for surprise in this approach. Managing forests under climate change will be risky, unless climate can be reliably projected 100 yr ahead.

Both modelling approaches have been used to give some indications of the possible changes that could occur in the north European boreal forest. The two models provide us with different but complementary insights into how species respond to climate in general, and how the potential effects of climate change in these northerly regions could affect forest landscapes. The picture they give is presumably incomplete and disregards (among many other things) the direct physiological effects of changed CO_2 that would be expected to modify several aspects of the tree species' response. Nevertheless, the models indicate quite drastic changes due to climate acting through basic mechanisms that will continue to operate, regardless of further modifying factors, and which imply major consequences for the future of both natural and managed forests.

Acknowledgment

This research was supported by a grant from the Swedish Environmental Protection Board (SNV).

References

Botkin, D. B., Janak, J. F. and Wallis. J. R.: 1972, *J. Ecol.* **60**, 849-873.

Cannell, M. G. R. and Smith, R. I.: 1986, *J. Appl. Ecol.* **23**, 177-191.

Hansen, J., Fung, I., Lacis, A., Rind, D., Russell, G., Lebedeff, S., and Ruedy, R.: 1988, *J. Geophys. Res.* 93, 9341-9364.

Huntley, B. and Prentice, I. C.: 1993, *In:* Wright Jr., H. E., Kutzbach, J. E., Webb, T. III, Ruddiman, W. F., Street-Perrott, F. A. and Bartlein, P. J. (eds.), *Global Climates Since the Last Glacial Maximum*, University of Minnesota, Minneapolis, USA, pp. 136-168.

Hustich, H.: 1960, *In:* Sömme A. (ed.) *The Geography of Norden*, Heinemann, London pp 54-62.

Hutchinson, M. F.: 1989, *Need for Climatic and Hydrologic Data in Agriculture in Southeast Asia* 89/5, CSIRO, Canberra, Australia, pp. 95-104.

Leemans, R. and Cramer, W.: 1991, *The IIASA Climate Database for Mean Monthly Values of Temperature, Precipitation and Cloudiness on a Terrestrial Grid.* RR-91-18, IIASA, Laxenburg, Austria, 62 pp.

Manabe, S. and Wetherald, R. T.: 1987, *J. Atmos. Sci.* **44**, 1211-1235.

Mitchell, J. F. B.: 1983, *Quart. J. Roy. Met. Soc.* **B 109,** 113-152.

Murray, M. B., Cannell, M. G. R. and Smith, I.: 1989, *J. Appl. Ecol.* **26**, 693-700.

Prentice, I. C., Cramer, W., Harrison, S. P., Leemans, R., Monserud, R. A. and Solomon, A. M.: 1992, *J. Biogeogr.* **19**, 117-134.

Prentice, I. C. and Leemans, R.: 1990, *J. Ecol.* **78**, 340-355.

Prentice, I. C., Sykes, M. T. and Cramer, W.: 1991, *Global Ecol. Biogeogr. Lett.* **1**, 129-135.

Prentice, I. C., Sykes, M. T. and Cramer, W.: 1993, *Ecol. Modelling* **65**, 51-70.

Price, D. T., Apps, M. J., Kurz, W. A., Prentice, I. C. and Sykes, M. T.: 1993, *In:* Ung, C-H. (ed.) *Forest Growth Models and Their Uses.* Canadian Forest Service Laurentian Forestry Centre, Sainte-Foy, Canada, pp. 251-264.

Sakai, A. and Weiser, C. J.: 1973; *Ecology* **54**, 118-126.

Schlesinger, M. E. and Zhao, Z. C.: 1989, *J. Clim.* **2**, 459-495.
Sjörs, H.: 1965, *Acta Phytogeographica Suecica* **50**, 48-63.
Sykes, M. T.: 1994, *Biolog* **2-3**, 17-24, Oslo, Norway.
Sykes, M. T., Prentice, I. C. and Cramer, W.: 1995, A bioclimatic model for the potential distributions of north European tree species under present and future climates. *J. Biogeogr.* (In press).
Sykes, M. T. and Prentice, I. C.: Climate change, tree species distributions and forest dynamics: a case study in the mixed conifer/northern hardwoods zone of northern Europe. (Unpublished manuscript).
Sykes, M. T. and Prentice, I. C.: 1995, *In:* Apps, M. J. and Price, D. T. (eds.) *Role of Global Forests and Forest Management in the Global Carbon Cycle*, Springer-Verlag, Heidelberg, Germany. (In press).
Wolfe, J. A.: 1979, *Prof. Pap.U.S. Geol. Surv.* **1106**, 37 pp.
Woodward, F. I.: 1987, *Climate and Plant Distribution.* Cambridge University Press, Cambridge, UK.

DISTURBANCE IMPACTS ON FOREST TEMPORAL DYNAMICS

C. LI[1] and M. J. APPS[2]

[1]*Ontario Forest Research Institute, P.O. Box 969, 1235 Queen Street East, Sault Ste. Marie, Ontario P6A 5N5, Canada.*
[2]*Northern Forestry Centre, Canadian Forest Service, Natural Resources Canada, 5320-122 Street, Edmonton, Alberta, T6H 3S5, Canada.*

Abstract. Contagious disturbances are widespread and an important component of the life-cycle of the Canadian boreal forest. Models which simulate the dynamics of these forests must include a representation of disturbance influences. Spatial models differ from non-spatial models in the use of area-based rather than point-based sampling designs. Distance-dependent disturbance propagation accounts for spatial correlation among disturbed stands and its inclusion in spatial models may generate more realistic forest temporal dynamics than can be generated by non-spatial models. A comparison of results from simulations using age-dependent and distance-dependent disturbance propagation can provide further insight on the role contagion in structuring forest temporal dynamics.

Keywords. DISTURBANCE, CONTAGION, TEMPORAL DYNAMICS, MODELLING APPROACH, SPATIAL MODEL, FIRE BEHAVIOUR, DISTANCE-DEPENDENT FIRE PROPAGATION, SUCCESSION, SAMPLING THEORY.

1. Introduction

The importance of forest temporal dynamics, particularly forest succession, for understanding the structure and function of ecosystems and for practical forest resource management has long been recognized (Climenton, 1916). Recent concern for the relationship between possible global climate change and alterations in global carbon budgets has increased interest in the role of northern forest dynamics in the global carbon cycle (e.g., Apps and Kurz, 1991).

Clementsian forest succession theory assumes that a forest tends towards a sustained climax species assemblage in a highly ordered sequence, determined by climate and soil conditions. This succession is controlled by two functions: exploitation, the rapid colonization of recently disturbed areas, and conservation, the slow accumulation of energy and matter.

Two additional functions can be added to this scheme (Holling, 1986, 1992): release, the release of accumulated biomass and nutrients by disturbances such as forest fires and insect pests, and reorganization, the post-release redeployment of resources such as nutrients in the ecosystem. Thus a renewal cycle starts with exploitation, grades slowly into conservation, followed by very rapid release, rapid reorganization, and a return to a new cycle of exploitation. Duration of individual renewal cycles is therefore a key element in forest temporal dynamics, with disturbances functioning as a timer for successive processes. Disturbances also generate spatial patterns, controlling spatial scales.

Understanding disturbance regimes is a prerequisite to modelling Canadian boreal forest dynamics. This is because the interval between disturbances is typically short enough that the forest species assemblage rarely achieves a climax. A non-spatial model, FORSKA2 (Prentice *et al.*, 1993), in which disturbances are described by a function of stand age only, had been employed to simulate the forest temporal

Water, Air and Soil Pollution **82**: 429–436, 1995.

dynamics across a transect through Canada's boreal forest (Price and Apps, 1995, this volume). However, the simulation results (Price and Apps, 1993) did not satisfactorily describe the observed patterns of species abundance and composition. One possible reason for the discrepancy is the rather simplistic representation of disturbance in the model. The aim of the present work was to examine whether the inclusion of spatial connectivity associated with contagious disturbances would have any significant effect on the temporal dynamics as expressed by change in biomass over time.

"Disturbance" is defined in various ways (Apps, 1993; Li and Apps, 1995; White and Pickett, 1985), but usually refers to catastrophic events which trigger the renewal cycle change to a release phase. Li and Apps (1995) distinguish two types of disturbances, based on their process characteristics. In type I disturbances, the probability of disturbance of a given patch is a function only of conditions in that patch, and not of the probability of disturbance in neighboring patches. Such disturbances are often climate-related and determined by local climate and topography. In type II disturbances, the probability of disturbance of a given patch is a function of both conditions in that patch and conditions in neighboring patches, such that events in one patch may 'infect' neighboring patches. Such disturbances are contagious, may be self-propagating, and include such common disturbances as fire and insect infestations, which are known to be important in shaping the mosaic pattern of the Canadian boreal forest. The two types of disturbances require different modelling approaches. While type I disturbances can be adequately modelled in a non-spatial scheme, a spatial approach is required for modelling type II disturbances (Li and Apps, 1995).

In this paper, we present a summary of recent work in modelling the effects of type II disturbances. The results suggest that more attention must be paid to spatial interactions when simulating the temporal dynamics of boreal forests. We discuss the implications of the assumption of distance-dependent propagation of type II disturbances, and demonstrate the suitability of a spatial approach (one which includes meso-scale contagion processes; Li *et al.*, 1995) to modelling them.

2. Non-Spatial Modelling Approach

Non-spatial models assume that a large area can be adequately represented by aggregating a large number of independent small areas; this uses the central limit theorem in which the statistics of a population (or system) can be approximated by increasing the sample size of its components. Well-developed point-sampling theory (Cochran, 1963) provides a comprehensive sampling design methodology, with calculations of error of estimation. The theory has been extended from populations with normal distributions to those with Poisson or negative binomial distributions (e.g., Southwood, 1978).

Shugart *et al.* (1992) document the strategies that are most often used in this point-sampling modelling approach. They suggest a coarse classification to stratify the biome(s) to be modelled by individual stand models, with the results from the lower stratum aggregated to represent the higher strata. This approach simplifies a large land surface to a set of region types, each of which can be modelled separately to reduce the computational complexity of the problem.

Several gap models have used this strategy to simulate landscapes, including FORSKA2 (Prentice *et al.*, 1993) and its predecessor FORSKA (Leemans and Prentice, 1989), which were developed for the Scandinavian boreal forest. Landscape applications of gap models, including FORSKA, usually assume that the landscape is sufficiently homogenous to be adequately represented by the aggregation of 100 small plots. Each plot is subjected to disturbances according to a Weibull function, and thus is independent of the other plots. It is explicitly assumed that the small scale results (plot level) can be linearly "scaled up" to the landscape level (King, 1991; Shugart *et al.*, 1992). King (1991) provides an excellent, critical review of methods currently used for scaling up. The simplest method, usually employed with gap models such as FORSKA2, is the averaging of typically 100 independent small plots.

There are several difficulties with this approach. First, the independence of the individual sample plots makes the model appear weak in its representation of type II disturbances. Second, according to point-based sampling theory, the appropriate sample size is determined by the statistical distribution of the studied population. However, little attention has been given to the actual distribution of forest structures, resulting in difficulties distinguishing sampling error (stochastic events) from systematic error (flawed sampling design) in model results. Third, as noted by King (1991), scaling up by lumping assumes that modelled system properties do not change with scale—i.e., that the modelled larger-scale system behaves like the average smaller-scale system. King further points out that this assumption only holds strictly if the system is linear and if a non-linear equation is required to describe the system, the assumption can introduce a bias error.

Shugart *et al.* (1992) point out the need for the development of different strategies to extend responses of individual ecosystems to global change to the biome and global scales. One possible strategy is spatial modelling, i.e., area-based sampling designs.

3. Spatial Modelling Approach

3.1. CANADIAN FOREST FIRE BEHAVIOR

The extensive body of research on forest fire behavior makes it ideal for studying the characteristics of type II disturbance. There are three types of fires in the Canadian boreal forest: (1) ground fire, (2) surface fire, and (3) crown fire. Lightning-ignited crown fire is responsible for ca. 90 per cent of the total area burned in the boreal region (Johnson, 1992). Fuel accumulation is a major consideration in the probability of a lightening strike causing a fire. Fuel type is important in determining the rate of fire spread. The Canadian Forest Fire Behaviour Prediction System (Alexander *et al.*, 1984) recognizes 13 fuel types: seven in coniferous stands, one in deciduous stands, two in mixed wood stands, and three categories of slash (Forestry Canada Fire Danger Group, 1992). Different rates of spread result from combinations of fuel moisture codes—Fine Fuel (fuels less than two centimeters in diameter directly on the forest floor), Duff (F layer of the soil humus layer) and Drought (the deepest and most compact Duff layer and large downed logs). The Canadian Forest Service provides maps of fuel moisture codes, fire weather index, and initial spread index; these maps are used to identify high risk areas for fire in the near future (Baker *et al.*, 1991).

A fire disturbance can be divided into two stages: ignition and contagion or self-propagation (Li and Apps, 1995). Quantitative studies of fire have shown that it is a complex process: ignition is determined not only by the availability of fuel and ignition, but also by fuel moisture; fire spread is affected by these factors and also by wind and topography. Therefore two sets of probabilities are needed to simulate fire disturbances: one for fire ignition, another for fire propagation (Li *et al.*, 1995).

Although the determination of fire ignition can be similar in all cell-based models, self-propagation can be expressed in different ways and the resulting simulated fire impacts may also differ. Here we examine distance-dependent fire propagation.

3.2. DISTANCE-DEPENDENT PROPAGATION ASSUMPTION

The distance-dependent propagation assumption is derived from studies of fire shapes. While low rate of spread fires result in circular patches, high rate of spread fires result in elliptical patches (see Alexander, 1985; Anderson, 1983; and Van Wagner, 1969). The length-to-breadth ratio of elliptical fire patches is a function of wind speed (Alexander, 1985; Johnson, 1992). Final fire shape is not related to forest stand conditions, rather the probability of a given stand being affected by a fire is mainly a function of its location relative to the point of ignition. Johnson and Larsen (1991) found a negative exponential distribution of fire frequency with stand age in the Kananaskis River watershed of Western Canada, suggesting a constant hazard function and implying that fire spread is not sensitive to the age of forest stands. Once a crown fire starts, it is likely to spread to surrounding trees regardless of their ages (B. Lee, Canadian Forest Service, 1994, pers. comm.). Therefore, in such cases it is appropriate to represent fire contagion by a distance-dependent function. For improved realism, this function will depend on such spatial characteristics as topography and hydrology. The assumption does not account for detailed fire propagation processes, but is derived from observed fire scar shapes. For simulations using an annual or seasonal time-step, the distance-dependent assumption may be a useful simplification of the complex dynamic processes of fire spread.

The distance-dependent function can account for spatial correlations in fire disturbed forest stands. An immediate question is: how important is this spatial correlation for forest temporal dynamics? To answer this question we tested the hypothesis that there is no significant difference in the forest temporal dynamics resulting from spatial and non-spatial modeling approaches (see Li and Apps, 1995 for details). To do this, a simple gridded landscape model (50 × 50 cells) was run for two scenarios, with each cell representing a homogenous forest stand in terms of age. In the first scenario, representing a non-spatial approach, disturbances affect only single cells. The affected cells are randomly selected and thus independent of each other. In the second, representing a spatial approach, each disturbance attacks an initial cell (disturbance center), but may also affect surrounding cells; the probability of contagion is a function of the distance from the disturbance center.

In the simulations, the age of all cells which have not been disturbed is incremented at the end of each time step while the age of disturbed cells is reset to 1. To eliminate the influence of initial conditions, simulations were run for a 3000-year period but only the records of the last 500 years were used for analysis. To ensure comparability of the results, the long-term-average disturbance probability for each cell must be the same in both scenarios. This is achieved by manipulating the disturbance interval parameter values of a Weibull function.

The two model approaches yield significantly different mean biomass results (Li and Apps, 1995). With age-dependent disturbance initiation and distance-dependent disturbance propagation, the simulated mean biomass was significantly higher than that obtained from random disturbance simulations, though the average disturbance return interval was adjusted to the same value. Table I shows the difference in estimated biomass, represented by volume, for 3 different mean disturbance intervals. With a 100-year disturbance return interval, the differences in mean biomass was 22.8 % of the total biomass, representing a significant potential error in the estimates.

We investigated the reasons for these differences, and found that the probability of disturbance as a function of stand age was different. We performed regression analysis on the simulation results from the two models and found that with spatial correlation between disturbed cells, the probability rises less than 20% as rapidly with stand age than it does with the non-spatial model. In other words, the probability of disturbance for a given cell is not only a function of its age, but also a function of its distance from a disturbance center. Consequently, more cells survive to old age (and higher biomass) in the simulation results from a spatial model than do in the non-spatial model.

TABLE I

Errors in estimated mean biomass (represented by volume in m^3 ha^{-1}) for three mean disturbance intervals for age-dependent disturbance initiations.

Mean disturbance interval	Mean biomass from spatial model	Mean biomass from non-spatial model	Differences in estimates (%)
50	42.4	33.4	21.3
100	80.4	62.1	22.8
150	118.3	90.8	23.3

Although these numerical results apply only for our hypothetical forest they show that when contagious disturbances operate, age may not be the only state variable of importance and a spatial variable, such as distance from disturbance centre, may also be needed. The large differences in simulation results suggests that a non-spatial model, which may be regarded as a point-based modelling approach, may not be adequate for investigating impacts of contagious disturbances on forest dynamics.

The situation is somewhat different and the differences in simulated results less pronounced when the disturbance initiation is independent of age. Table II shows that these differences are an order of magnitude lower than with age-dependent initiation considered above. At least for the hypothetical forest we considered, it appears that for this type of disturbance the random simulation design (non-spatial model) works adequately.

TABLE II

Errors in estimated mean biomass (represented by volume in $m^3 ha^{-1}$) for three mean disturbance intervals for age-independent disturbance initiations.

Mean disturbance interval	Mean biomass from spatial model	Mean biomass from non-spatial model	Differences in estimates (%)
50	50.2	49.2	2.0
100	96.5	94.3	2.3
150	142.8	139.5	2.3

While the assumption of distance-dependent propagation accounts for spatial correlation among disturbed stands, age-dependent propagation considers the detailed processes of contagion (Antonovski *et al.*, 1992; Holling *et al.*, 1995; Li *et al.*, 1995; Peterson, 1994). Models employing age-dependent disturbance propagation assume that the probability of ignition of a given stand due to the burning of an adjacent stand is a function of the stand age, that is, the probability of disturbance is related to stand conditions. The shape of this function need not be positive correlated to stand age. For example, younger or recently burned stands may accumulate fine fuels fast enough to be re-ignited (G. Peterson and P. Marples, 1995, pers. comm.).

Models of forest landscape dynamics that assume age-dependent propagation can generate results different from those of traditional models (Li *et al.*, 1995). In such models, the processes for each of disturbance event are simulated in detail and follows the contagious spread from the location where the disturbance is initiated, to the immediately neighboring stands, providing that conditions are suitable for propogation, and so on, until adjacent stand conditions are unsuitable for further propogation. With such a spatially explicit approach, the sizes and shapes of disturbed regions will be determined by pre-existing forest stand conditions and mosaic patterns, rather than a built-in pre-assumption of the models. This modelling approach can be called area-based sampling designs because each simulation run is for a landscape, rather than for an isolated plot. When using area-based sampling designs, one must consider not only the key characteristics of the dynamics within a single plot, but also the interactions among those plots mediated by the type II disturbances. This approach appears to be more appropriate for simulating the temporal dynamics of the Canadian boreal forest than traditional non-spatial models.

4. Conclusions

In this paper we show that spatial correlation among disturbed stands may result in simulated forest temporal dynamics that differ significantly from those generated by traditional non-spatial models. Further detailed consideration of disturbance contagion processes (i.e., age-dependent propagation) may be needed for simulations of the long-term dynamics of forest landscapes.

Whether a model employs a distance- or age-dependent assumption is a function of research focus. While a distance-dependent assumption focuses on the impact of type II processes, an age-dependent assumption emphasizes the dynamics of the disturbance process itself. This is partly a matter of temporal scales: the simulation of the propagation of fire across the landscape may require time steps of hours to days. The impact of the disturbance on the forest life cycle is typically assessed over a period of years and decades. Thus if a realistic distance-dependent disturbance function can be established, it can provide significant simplification for the simulation of longer-term dynamics. However, the error due to this approximation should be carefully examined by comparing the results of simulations using distance and age-dependent disturbance propagation. This comparison will assist in the selection of an appropriate modelling approach. We suggest that this comparison should be a priority.

The choice of modelling approach is a function of research purpose but it is also a trade-off between computing power required and the desired degree of realism of the simulation. For example, to investigate the influence of type II disturbances on a landscape structure and the evolution of forest landscape patterns affected by type II disturbances, a detailed simulation of disturbance processes will be necessary. Consequently age-dependent disturbance propagation should be incorporated in these simulation models. However, if temporal dynamics is the major concern, our results show that spatial correlations among disturbed stands must still be considered. When there is a practical limitation on computing power, then a distance-dependent disturbance propagation assumption may serve.

Acknowledgments

We thank Celina Campbell for critical reviews of an early draft and very helpful improvements to this manuscript. The valuable comments from two anonymous reviewers are gratefully acknowledged. We also thank Buzz Holling, Garry Peterson, Paul Marples, and David Price for discussions during the workshop "Cross-scale modelling of ecosystem dynamics" in October, 1993 in Gainesville, Florida. This work was supported in part by Green Plan funding within the Department of Natural Resources Canada and the Forest Landscape Ecology Program of the Ontario Forest Research Institute.

References

Alexander, M. E.: 1985, *In: Proceedings of the Eighth Conference on Fire and Forest Meteorology*, Society of American Foresters, Bethesda, USA, SAF Publication **85**-04, pp. 287-304.

Alexander, M. E., Lawson, B. D., Stocks, B. J. and Van Wagner, C. E.: 1984, *User Guide to the Canadian Forest Fire Behavior Prediction Systems: Rate of Spread Relationships*. (Interim Edition) 76 pp.

Allen, T. F. H. and Starr, T. B.: 1982, *Hierarchy: Perspectives for Ecological Complexity*. University of Chicago Press, Chicago, USA, 310 pp.

Anderson, H. E.: 1983, US Department of Agriculture, Forest Service, Research paper INT-**305**, 26 pp.

Antonovski, M. Ya., Ter-Mikaelian, M. T. and Furyaev, V. V.: 1992, *In:* Shugart, H. H., Leemans, R. and Bonan, G. B. (eds.) *A Systems Analysis of the Global Boreal Forest*. Cambridge Univ. Press, Cambridge, UK, pp. 373-403.

Apps, M. J.: 1993, *World Resour. Rev.* **5**, 41-65.

Apps, M. J. and Kurz, W. A.: 1991, *World Resour. Rev.* **3**, 333-343.

Baker, W. L., Egbert, S. L. and Frazier, G. F.: 1991, *Ecol. Model.* **56**, 109-125.

Climenton, F. E.: 1916, *Plant Succession: An Analysis of the Development of Vegetation*. Carnegie Institution of Washington, Washington, USA, **242**, 512 pp.

Cochran, W. G. (ed.): 1963, *Sampling Techniques* (2nd edn.), Wiley, New York, USA 413 pp.

Forestry Canada Fire Danger Group: 1992, *Development and Structure of the Canadian Forest Fire Behavior Prediction System*, Information Report **ST-X-3**, Ottawa, Canada, 63 pp.

Forman, R. T. T. and Godron, M.: 1986, *Landscape Ecology*. Wiley, New York, USA, 619 pp.

Holling, C. S.: 1986, *In:* Clark, W. C. and Munn, R. E. (eds.) *Sustainable Development of the Biosphere*. Cambridge University Press, Cambridge, UK, pp. 292- 317.

Holling, C. S.: 1992, *In:* Shugart, H. H., Leemans, R. and Bonan, G. B. (eds.) *A Systems Analysis of the Global Boreal Forest,* Cambridge University Press, Cambridge, UK, pp. 170-191.

Holling, C. S., Peterson, G. D., Marples, P., Sendzimir, J. J., Redford, K., Gunderson, L. and Lambert, D.: 1995, *In:* Walker, B. H. and Steffen, W. L. (eds.) *Global Change and Terrestrial Ecosystems*. Cambridge University Press, Cambridge, UK. (In press).

Johnson, E. A.: 1992, *Fire and Vegetation Dynamics: Studies from the North American Boreal Forest*. Cambridge University Press, Cambridge, UK, 129 pp.

Johnson, E. A. and Larsen, C. P. S.: 1991, *Ecology* **72**, 194-201.

King, A. K.: 1991, *In:* Turner, M. G., and Gardner, R. H. (eds.) *Quantitative Methods in Landscape Ecology: The Analysis and Interpretation of Landscape Heterogeneiyy*. Springer-Verlag, New York USA, pp. 479-517.

Leemans, R. and Prentice, I. C.: 1989, *FORSKA, A General Forest Succession Model*. Medd. Vaxtbiol. Inst. **2**, Uppsala, Sweden, 45 pp.

Li, C. and Apps, M. J.: 1995, *Ecol. Model*. (In press).

Li, C., Ter-Mikaelian, M. T. and Perera, A.: 1995, (Unpublished manuscript).

Peterson, G. D.: 1994, Spatial modeling of fire dynamics in the Manitoban boreal forest. M.Sc. Thesis, Univ. Florida, Gainsville, USA.

Prentice, I. C., Sykes, M. T. and Cramer, W.: 1993, *Ecol. Model*. **65**, 51- 70.

Price, D. T. and Apps, M. J.: 1993, *World Resour. Rev*. **5**, 15-31.

Price, D. T., and Apps, M. J.: 1995, *Water, Air, Soil Pollut*. **82**, 203-214.

Shugart, H. H., Smith, T. M. and Post, W. M.: 1992, *Annu. Rev. Ecol. Syst*. **23**, 15-38.

Southwood, T. R. E.: 1978, *Ecological Methods: With Particular Reference to the Study of Insect Populations*. Second Edition. Chapman and Hall.

White, P. S. and Pickett, S. T. A.: 1985, *In:* Pickett, S. T. A. and White, P. S. (eds.) *The Ecology of Natural Disturbance and Patch Dynamics*. Academic Press, Inc. pp. 3-13.

Van Wagner, C. E.: 1969, *For. Chron*. **45**, 103-104.

PREDICTING THE EFFECTS OF CLIMATE CHANGE ON FIRE FREQUENCY IN THE SOUTHEASTERN CANADIAN BOREAL FOREST

Y. BERGERON[1] and M. D. FLANNIGAN[2]

[1]*Groupe de Recherche en écologie forestière, Université du Québec à Montréal, Montréal, Québec H3C 3P8, Canada*
[2]*Canadian Forest Service, Petawawa National Forestry Institute, Chalk River, Ontario K0J 1J0, Canada*

Abstract. Although an increasing frequency of forest fires has been suggested as a consequence of global warming, there are no empirical data that have shown climatically driven increases in fire frequency since the warming that has followed the end of the "Little Ice Age" (~1850). In fact, a 300-year fire history (AD 1688–1988) from the Lac Duparquet area (48°28'N, 79°17'W) shows a significant decrease both in the number and extent of fires starting 100 years ago during a period of warming. To explore this relationship between climatic change and fire frequency we used daily data from the Canadian Atmospheric Environment Service's General Circulation Model to calculate components of the Canadian Forest Fire Weather Index (FWI) System for the 1×CO_2 and 2×CO_2 scenarios. The average FWI over much of eastern Canada, including the Lac Duparquet region, decreased under the 2×CO_2 simulation, whereas FWI increased dramatically over western Canada. According to these results, fire frequency would decrease over the southeastern boreal forest which is in agreement with the empirical data from the fire history. Our results stress the importance of large regional variability and call into question previous generalisations suggesting universal increases in the rate of disturbance with climate warming.

Keywords. BOREAL FOREST, WILDFIRE, CLIMATE CHANGE, GCM.

1. Introduction

Whether or not it is related to the 'Greenhouse Effect', an increase in temperature has already been observed in the northern hemisphere. Abrupt change in tree radial growth (Payette *et al.*, 1985; Fritts, 1991; Archambault and Bergeron, 1992), glacier recession (Luckman and Osborn, 1979; Luckman, 1988) and tree-line fluctuations (Scott *et al.*, 1987) have been attributed to increasing temperatures since the end of the Little Ice Age (~1850). For the last century, many historical climate analyses (Boden *et al.*, 1990) point to an increase of about 0.6 °C in the global mean temperature. Rising temperatures were observed from the 1880s until the 1940s which was followed by a cooling to the mid-1960s and a warming since 1970 which now exceeds the 1940s level. For Canada, a significant warming has occurred which has been most pronounced over the Mackenzie district where the mean temperature has risen 1.7 °C during the last century (Environment Canada, 1992).

Climatic change and its effects on forest disturbance dynamics is presently a major concern. In the boreal forest, fire is a major disturbance (Payette, 1992) and a change in fire regime resulting from climatic change might have a greater impact on the forest that climatic change *per se.*

Although variations in fire frequency driven by climate change have been reported for the Holocene (Filion, 1984; Clark, 1990; Johnson and Larsen, 1991), the effect of recent warming on fire frequency has proven difficult to document. Decreases in fire frequency during the last century were observed in many studies (Heinselman, 1973;

Water, Air and Soil Pollution **82**: 437–444, 1995.

Cwynar, 1977; Van Wagner, 1988), however, the timing of these decreases corresponds to an increase in human activities and they were attributed to fire suppression.

Recently, Bergeron (1991) reported a decrease in fire frequency on islands on Lac Duparquet in the southern boreal forest of Quebec. Fire frequency during the period between 1870 and 1989 was 34% lower than in the preceding 74 years. The observed decrease was attributed to changes in the climate because fires are not suppressed on the islands and fires were shown to be locally ignited. More recently Bergeron and Archambault (1993) using a dendroclimatological approach attributed this decrease to a reduced frequency of drought periods conducive to fire. Since climate may control fire regimes over large areas, they suggested that the decrease in fire frequency would have affected all the Quebec southern boreal forest fringe.

These empirical results contrast with the predicted effects of climate warming. An increasing rate of forest disturbances, including fires, has been postulated (Overpeck *et al.*, 1987). Based on twentieth century climate data, Clark (1988), postulated that the warming should have brought about an increase in fire frequency for Minnesota. For Canada, Flannigan and Van Wagner (1991) predicted a possible 46% increase in seasonal fire severity rating, with a possible similar increase in area burned in a $2\times CO_2$ climate. These contradictory results between observed and predicted effects call into question the accuracy of the models to predict fire danger or may be indicative of large regional variability in response to climate change.

Using the daily data from the Canadian Atmospheric Environment Service's General Circulation Model we present here simulations predicting the components of the Canadian Forest Fire Weather Index (FWI) System for the $1\times CO_2$ and $2\times CO_2$ scenarios. Our objective is to compare the modelled results with field data on fire frequency from the southeastern boreal forest (Lac Duparquet region). Further, we will investigate the hypothesis suggested by Bergeron and Archambault (1993) that warming may lead to a decrease in drought periods conducive to fires in the southeastern boreal forest. Additionally, we briefly evaluate regional variation of the components of the FWI System across Canada under the $2\times CO_2$ scenario.

2. Methods

The Canadian General Circulation Model (GCM) was used in this study to evaluate the effect of climate change on the fire regime. This model has a transform grid spacing of 3.75° by 3.75° with upgraded physics including full diurnal and annual cycles (Boer *et al.*, 1992, McFarlane *et al.*, 1992). Daily data rather than monthly data were used in this study as the fire regime can change dramatically over time periods much shorter than a month. The Canadian GCM was selected because of the availability of daily data. Temperature, specific humidity, precipitation and wind speed were obtained every 12 h (0000 and 1200 GMT) for 9 years for the $1\times CO_2$ and $2\times CO_2$ simulations. The specific humidity was converted into relative humidity. We used the maximum daily temperature, relative humidity, 24 h precipitation and the 12 h (1200 GMT) mean wind speed to calculate components of the Canadian Fire weather Index (FWI) System (Van Wagner, 1987) during the fire season (01 April to 31 October). The Fire Weather Index System comprises three fuel moisture codes and three fire behaviour indexes.

The three moisture codes represent the Fine Fuel Moisture Content (FFMC), loosely compacted organic matter (Duff Moisture Code) and the deep layer of compact organic matter (Drought Code). The three fire behaviour indexes, which are derived from the moisture codes and the surface wind speed, indicate the rate of initial fire spread (Initial Spread Index), total available fuel (Build Up Index) and the intensity of spreading fire (Fire Weather Index). Components of the FWI System were calculated for both simulations and then compared by taking the ratio of 2×CO_2 divided by 1×CO_2. Extreme maximums in addition to the mean values for the 9 years of the FWI components were used in this analysis. Extremes were used in this study because only a few days with extreme fire weather conditions are responsible for most of the area burned by forest fires. The importance of extremes is another reason to use daily GCM data as opposed to monthly data. Comparisons of the meteorological variables used in the calculation of the FWI components were done between the 1×CO_2 and 2×CO_2 simulations. Transient conditions that would occur between the two simulations were not addressed in this study. To test the hypothesis of Bergeron and Archambault (1993) who suggest the number of drought periods are reduced in a warmer climate we calculated the number of days with precipitation greater than or equal to 1.5 mm from the two data sets. We used this threshold as values less than 1.5 mm would be intercepted by the canopy (Rutter, 1975).

3. Results and Discussion

The FWI System is calculated by combining temperature, relative humidity, 24 h precipitation and the wind speed. Figure 1 shows the increase in mean temperature for Canada between the 2 simulations (2×CO_2 – 1×CO_2). Most of Canada including the Lac Duparquet region show increases of 3–4 °C. Temperature increases considered alone will lead to higher FWI values. The ratio (2×CO_2/1×CO_2) of precipitation amounts (Figure 2) and relative humidity (not shown) show that precipitation and relative humidity increase (ratio greater than 1.00) over our study area. Increases in precipitation and relative humidity will lead to decreases in FWI. Mean wind speed is essentially unchanged between the two runs. From the response of the input weather variables for the calculation of the FWI (temperature, precipitation, relative humidity and wind speed) to the projected climatic change (2×CO_2) it is unclear which trend (increasing FWI due to increasing temperature or decreasing FWI due to increasing precipitation and relative humidity) will prevail.

The 9 year average FWIs across Canada were calculated for the 1×CO_2 and 2×CO_2 simulations (not shown). In terms of validation for the FWI data, the relative pattern for the 1×CO_2 simulation is similar to a map of FWI across Canada (Simard, 1973) which was prepared using actual observations from 364 stations for 10 years (1957–66), although our absolute values are lower than those calculated by Simard. Our results for the mean FWI calculated from the 1×CO_2 scenario were compared against the mean FWI values calculated using observations for 29 stations across Canada for the period 1953–80 (Harrington *et al.*, 1983). The mean FWI calculated from the 1×CO_2 simulation were reasonably well correlated with the observed mean FWI (correlation coefficient 0.66) and we feel that the FWIs from the simulation are a reasonable approximation of the recent past. Figure 3 shows the ratio of the 2×CO_2 values

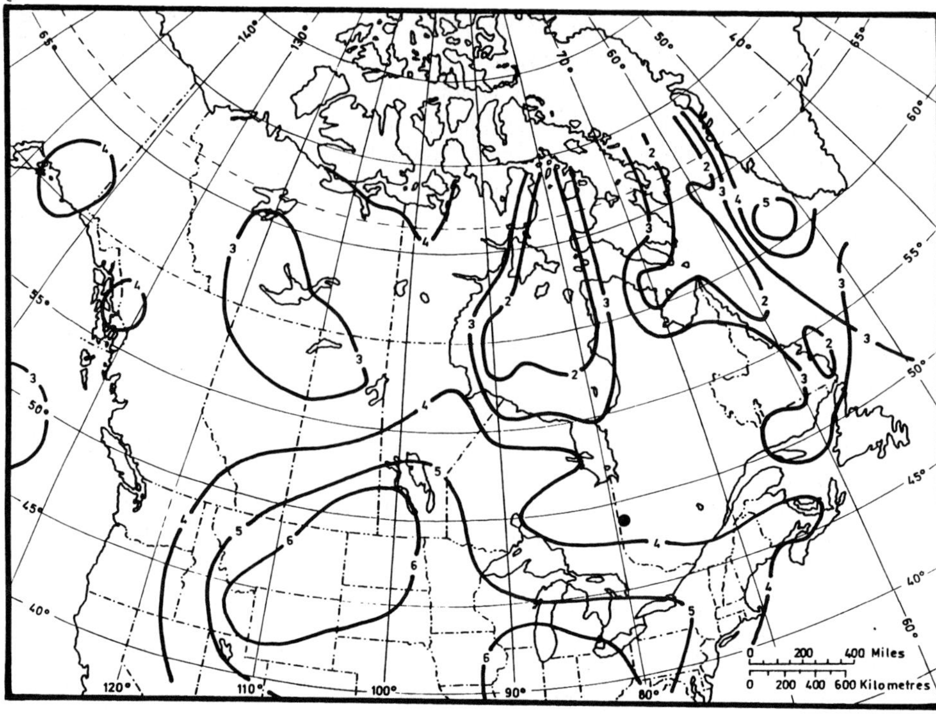

Fig. 1. Mean temperature (°C) for $2\times CO_2$ – $1\times CO_2$ simulations. The black dot (●) indicates the location of Lac Duparquet on all the figures.

divided by the 1×CO2 values. Much of eastern Canada, including the Lac Duparquet region, has a ratio below 1.00 (Figure 3) indicating that the FWI has decreased despite the warmer temperatures associated with the $2\times CO_2$ climate. The ratio of extreme maximum values of the FWI (not shown) for the 9 year period have a similar pattern to Figure 3. The other components of the FWI System show similar results to those shown in Figure 3.

The FWI represents the intensity of a spreading fire. Thus, lower intensity fires would on average yield reduced area burned given everything else remains unchanged (ignition agents, fire suppression activities etc.). Increasing fire severity would be anticipated with increasing temperatures if the other factors such as relative humidity, wind speed and precipitation amount remain unchanged. Precipitation amounts and relative humidity are somewhat higher over the Lac Duparquet region in the $2\times CO_2$ simulation as compared to the $1\times CO_2$ (Figures 1 and 2) which has more than offset any increases in the FWI due increased temperature. These results differ from those obtained from Flannigan and Van Wagner (1991) who found an increase in seasonal fire severity rating (a derivative of the FWI) at 6 stations across Canada. However, that study used monthly anomalies from three GCMs (Geophysical Fluid Dynamics Laboratory (GFDL), Goddard Institute for Space Studies (GISS), and Oregon State University (OSU)) whereas the present study used daily GCM data from the Canadian GCM.

Figure 4 shows the difference in the number of days in June with precipitation over 1.5 mm for the two model runs. The month of June was selected as it has the most fire

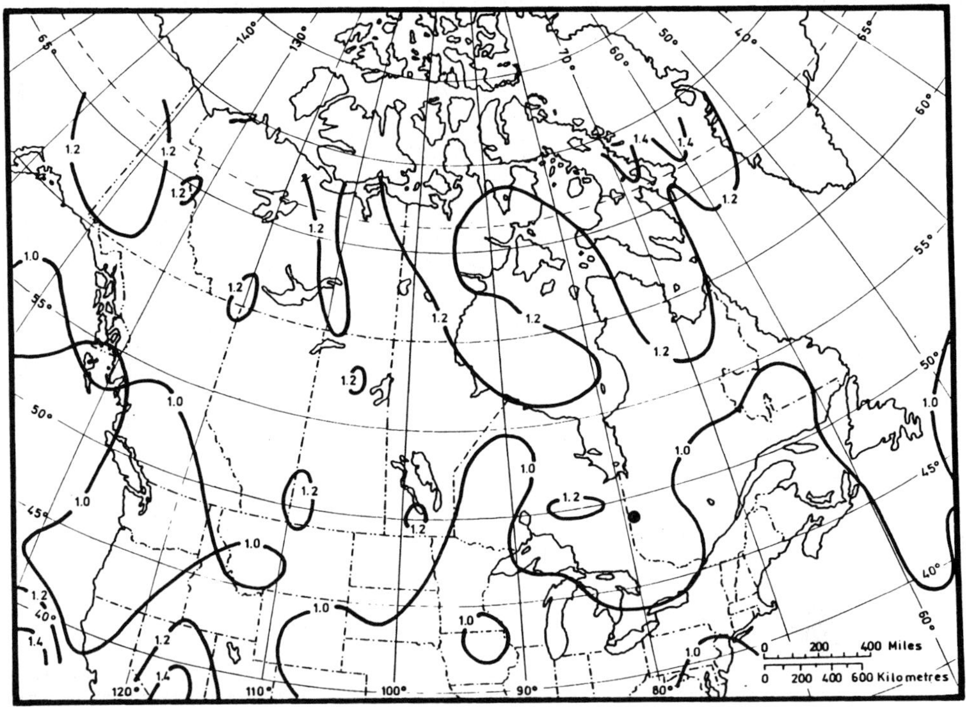

Fig. 2. Precipitation ratio ($2\times CO_2$ /$1\times CO_2$).

activity for northwestern Quebec (Harrington, 1982). There are almost 2 more days of precipitation per month for the $2\times CO_2$ simulation at this location. This result lends credence to the hypothesis that the fire frequency can be reduced even in a warmer climate if there are fewer periods of drought. Flannigan and Harrington (1988) have demonstrated a link between area burned and frequency of drought.

Fire activity is often linked to the location of the upper ridge associated with warm and dry conditions at the surface (Newark, 1975). Conversely, there is a lack of fire activity associated with upper troughs that usually result in wet and cool conditions at the surface. Given the variation in wavelengths in the upper flow at least one ridge and one trough are typically found over North America during the summer. On average the upper ridge is found over western North America typically running along the west coast and an upper trough is found over eastern North America anchored over Hudson's Bay. (This is an average position, a great deal of variability is found on daily weather charts.) In a warming climate the upper ridge would be expected to amplify while the trough would weaken. The trough would remain anchored over Hudson's Bay because of the relatively cold temperatures associated with this large water body. (This is consistent with the results shown in Figure 1 showing much lower temperature increases over Hudson's Bay.) A weakening of the upper trough over eastern Canada would probably result in the northward movement of the polar front. This would allow more frequent incursions of a warm and humid airmass. This would lead to higher relative humidity and more precipitation which is consistent with our results. Thus, a

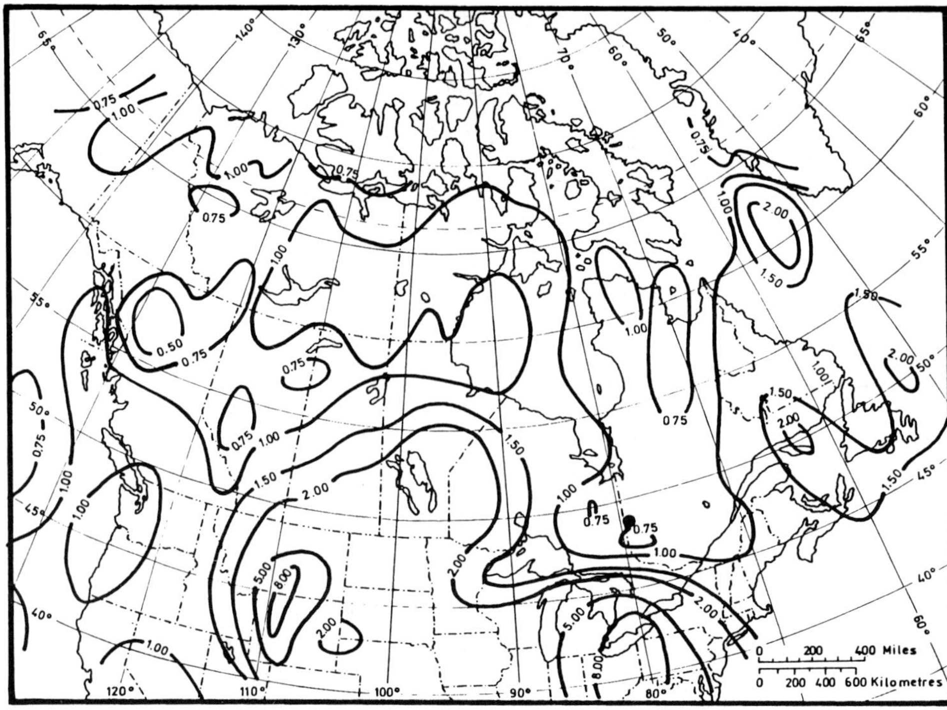

Fig. 3. FWI ratio ($2\times CO_2/1\times CO_2$).

reduced frequency of drought periods would be anticipated which would inhibit fire ignition and spread and has already been observed in the last 100 years over the southern boreal forest in the Lac Duparquet area. However, this relationship between warming and decreased frequency of drought found in the recent past and suggested by the GCM does not preclude the possibility that warmer and drier conditions can occur.

Climatic warming would allow many species to extend their range limits northward (Liu, 1990; Solomon, 1986). This would mean that the southern boreal forest might be replaced by many species that are currently found in the present day mixed-wood forest (Rizzo and Wiken, 1992). This area would be available for expansion because of the warmer climate or because of the change to a less severe fire regime. The change in vegetation may alter the fire regime as many of the broad leaved tree species of the mixed woods forest are not predisposed to fire. The possible decrease of fire frequency over the boreal forest may hasten vegetational change as the boreal forest has been maintained by semi-regular fire occurrence. Red pine (*Pinus resinosa* Ait.) and white pine (*Pinus strobus* L.) may increase in abundance as fire intensity decreases. Bergeron and Dubuc (1989) suggest that if disturbance frequency is reduced eastern white cedar (*Thuja occidentalis* L.) and balsam fir (*Abies balsamea* (L.) Mill.) would also increase in abundance.

While these results indicate a reduced FWI over the Lac Duparquet region the results also highlight a large regional variation across Canada. We will restrict our discussion to non-mountainous regions of Canada as the GCM grid is too coarse to

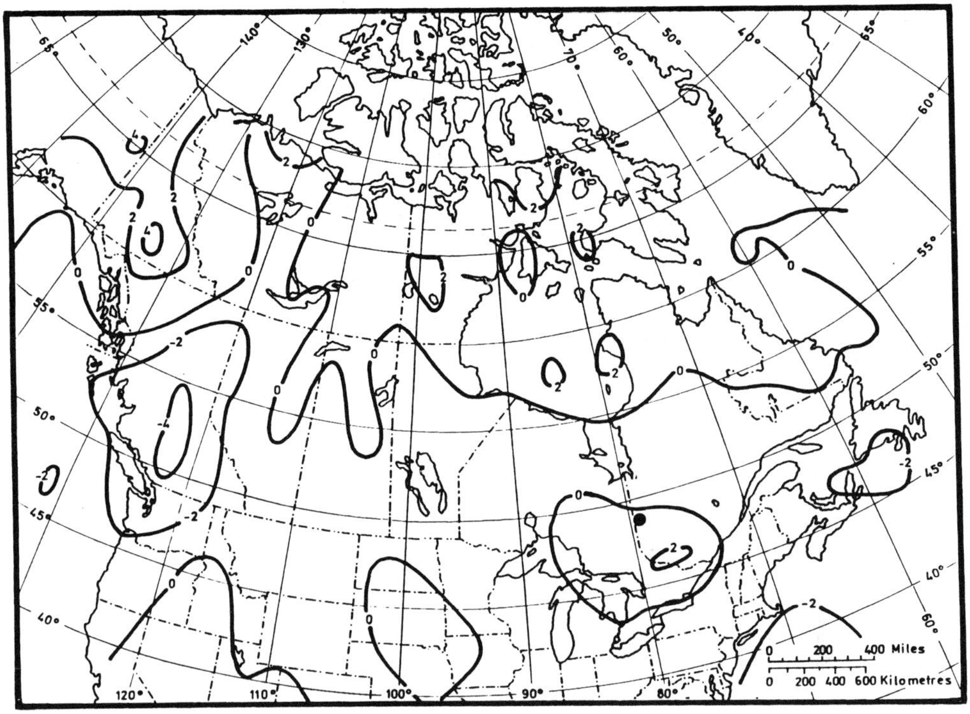

Fig. 4. Number of days in June with precipitation of 1.5 mm or greater from the simulations ($2{\times}CO_2$ - $1{\times}CO_2$).

obtain credible estimates for mountainous areas. The most striking feature of the regional variation is the dramatic increase in FWI (Figure 3) and all the other FWI components with the climatic change over central Canada. These results do corroborate some aspects of the results of other studies that suggest increased disturbance with climate warming (Flannigan and Van Wagner, 1991; Overpeck *et al.*, 1990). These increases in the FWI occur over western Canada where historically most of the fire activity occurs (Harrington, 1982; Higgins and Ramsey, 1992). Therefore, if fire intensities decrease where fire is a relatively minor problem, such as eastern Canada, and fire intensities increase where fire can be a major problem, such as western Canada, then climate change can still result in dramatic changes in the fire regime on a continental scale.

4. Summary

Results from our simulations are consistent with field data collected in the southeastern boreal forest that indicate a decrease of fire frequency with climate warming. Our results also support the hypothesis of Bergeron and Archambault (1993) that climatic change may lead to a decrease in drought periods in the southeastern boreal forest. Finally, it is clearly evident from this study that there is a large regional variation in the components of the Fire Weather Index and that some regions show a decrease in the FWI.

References

Archambault, S. and Bergeron, Y.: 1992. *Can. J. For. Res.* **22**, 674-682.

Bergeron, Y.: 1991. *Ecology* **72**, 1980-1992.

Bergeron, Y. and Archambault, S.: 1993. *The Holocene* **3**, 255-259.

Bergeron, Y. and Dubuc, M.: 1989. *Vegetatio* **79**, 51-63.

Boden, T. A., Kanciruk, P. and Farrell, M. P.: 1990. *Trends '90: A Compendium of Data on Global Change.* ORNL/CDIAC-36. Department of Energy. Washington D.C., USA.

Boer, G. J., McFarlane, N. A. and Lazare, M.: 1992. *J. Climate* **5**, 1045-1077.

Clark, J. S.: 1988. *Nature* **334**, 233-235.

Clark, J. S. 1990. *Ecol. Monogrphs* **60**, 135-159.

Cwynar, L. C.: 1977. *Can. J. Bot.* **55**, 1524-1538.

Environment Canada: 1992. *The State of Canada's climate: Temperature Change in Canada 1895-1991. State of the Environment Report No.* 92-2. Environment Canada, Ottawa, Canada. 36 pp.

Filion, L.: 1984. *Nature.* **309**, 543-546.

Flannigan, M. D. and Harrington, J. B.: 1988. *J. Appl. Meteorol.* **27**, 441-452.

Flannigan, M. D. and Van Wagner, C. E.: 1991. *Can. J. For. Res.* **21**, 66-72.

Fritts, H. C.: 1991, *Reconstructing large-scale climate patterns from tree-ring data.* University of Arizona Press, Tucson, USA.

Harrington, J. B.: 1982, *A statistical study of area burned by wildfire in Canada 1953-80.* Canadian Forestry Service Information Report PI-X-16. Petawawa National Forest Institute, Chalk River, Canada. 32 pp.

Harrington, J. B., Flannigan, M. D. and Van Wagner, C. E. 1983, *A study of the relation of components of the Fire Weather Index to monthly provincial area burned by wildfire in Canada 1953-80.* Canadian Forestry Service Information Report PI-X-25. Petawawa National Forest Institute, Chalk River, Canada. 65 pp.

Heinselman, M. L.: 1973, *Quaternary Res.* **3**, 329-382.

Higgins, D. G. and Ramsey, G. S.: 1992, *Canadian Forest Fire Statistics.* Canadian Forest Service Information Report PI-X-107E/F. Petawawa National Forest Institute, Chalk River, Canada.

Johnson, E. A. and Larsen, C. P. S.: 1991, *Ecology* **72**, 194-201.

Liu, Kam-Biu.: 1990, *Ecol. Monographs* **60**, 179-212.

Luckman, B. H.: 1988, *Arctic and Alpine Res.* **11**, 52-77.

Luckman, B. H. and Osborne, G. D.: 1979, *Quaternary Res.* **11**, 52-77.

McFarlane, N. A., Boer, G. J., Blanchet, J.-P., and Lazare, M.: 1992, *J. Climate* **5**, 1013-1044.

Newark, M. J.: 1975, *Atmosphere* **13**, 26-33.

Overpeck, J. T., Rind, D. and Goldberg, R.: 1990, **343**, 51-53.

Payette, S.: 1992, In: Shugart, H. H, Leemans, R. and Bonan, G. B. (eds.). *A Ssystems Analysis of the Boreal Forest.* Cambridge University Press. Cambridge, UK.

Payette, S., Filion, L., Gauthier, L., and Boutin, Y.: 1985, *Nature* **315**, 135-138.

Rizzo, B. and Wiken, E., 1992. *Climatic Change* **21**, 37-55.

Rutter, A. J.: 1975, In: Monteith, J. L. (ed.). *Vegetation and the Atmosphere: Vol. 1, Principles,* Academic Press, New York, USA.

Scott, P. A., Hansell, R. I. C. and Fayle, D. C. F.: 1987, *Arctic and Alpine Res.* **19**, 45-51.

Solomon, A. M.: 1986. *Oecologia* **68**, 567-579.

Simard, A. J.: 1973, *Forest fire weather zones of Canada. Environment Canada, Forestry Service. Map with text.*

Van Wagner, C. E.: 1987, *Can. For. Serv. For. Tech. Rep.* **35**.

Van Wagner, C. E.: 1988, *For. Chron.* **64**, 182-185.

EFFECTS OF CLIMATE CHANGE ON INSECT DEFOLIATOR POPULATION PROCESSES IN CANADA'S BOREAL FOREST: SOME PLAUSIBLE SCENARIOS

R. A. FLEMING[1] and W. J. A. VOLNEY[2]

[1]*Forest Pest Management Institute, Canadian Forest Service, P.O. Box 490, Sault Ste. Marie, Ontario P6A 5M7, Canada*

[2]*Northern Forestry Centre, Canadian Forest Service, 5320-122 St., Edmonton, Alberta T6H 3S5, Canada*

Abstract. Insect populations have a substantial impact on Canada's forest. They are a dominating disturbance factor and during outbreaks they can cause tree mortality over vast areas of forest. If the predicted climate changes take effect, the damage patterns caused by insects may be drastically altered, especially for the many insects whose occurrence in time and space is severely limited by climatic factors. This possibility substantially increases the uncertainties associated with the long-term planning of pest control requirements, with hazard rating models, with depletion forecasts, and with projections for the sustainability of future timber supplies. Moreover, because insect damage affects the rates of various processes in nutrient and biogeochemical cycling, potential changes in damage patterns can affect ecosystem resilience. This paper presents a number of plausible scenarios that describe how some key processes in the boreal forest's insect defoliator outbreak systems may respond to climate change. The spruce budworm, *Choristoneura fumiferana* Clem. (Lepidoptera: Tortricidae), is used as an illustrative case study throughout. The potential importance of phenological synchrony in the dynamical interactions between species is emphasised. It is argued that natural selection may be a particularly important process in the response of insects to climate change and that climate change may already be influencing some insect lifecycles. The importance of threshold effects, rare but extreme events, and transient dynamics is emphasised, and the inadequacy of 'equilibrium' models for forest:pest systems noted. We conclude by discussing approaches to developing forecasts of how one of the boreal forest's insect defoliator-based disturbance regimes, as a whole, might respond to climate change.

Keywords. *Choristoneura fumiferana* (LEPIDOPTERA: TORTRICIDAE), DISTURBANCE REGIMES, TROPHIC INTERACTIONS, CLIMATE CHANGE, BOREAL FOREST.

1. Introduction

Current projections suggest that the world's climate is changing at rates much faster than rates ever experienced before (Houghton *et al.*, 1990). Under these circumstances, many ecosystems probably will not be able to respond as integral units to the change. Rather, certain parts or species will probably respond faster than others, producing new combinations, both in the centre and at the edge of an ecosystem's present range, where it meets neighbouring ecosystems. Changes in disturbance regimes such as insect outbreak systems are seen as the driving force that will bring about many of the anticipated changes in forest ecosystem function and composition (Apps, 1995). This view recognises that insects are important contributors to nutrient and carbon cycling, to energy flow, and to biomass decomposition (Haack and Byler, 1993). For example, insect feces decompose more quickly than fallen foliage and, in attacked trees, defoliation increases the rates of litterfall and nutrient leaching from damaged foliage. As a result, elements such as calcium, nitrogen, phosphorus, and potassium cycle more quickly (Schowalter *et al.*, 1986).

Insects already exert a major influence on forest productivity. For instance, annual forest losses from tree mortality due to insect attack in Canada are estimated at 51.0

Water, Air and Soil Pollution **82**: 445–454, 1995.

million m^3 per year. These losses are 1.5 times those due to wildfire and amount to 1/3 of the annual harvest volume (Hall *et al.*, 1994). (These figures do not include insect-caused growth loss when tree mortality does not occur. Piene (1980) has shown that such losses can be substantial). The spruce budworm, *Choristoneura fumiferana* Clem. (Lepidoptera: Tortricidae), is responsible for more losses in Canada than any other insect.

2. The Spruce Budworm and Climate Change

The spruce budworm occurs throughout most of the range of white spruce, *Picea glauca* (Moench) Voss., in Canada's boreal forest and records of severe defoliation in the Northwest Territories extend to within 150 km of the arctic circle (Volney and Cerezke, 1992). This native insect attacks balsam fir, *Abies balsamea* (L.) Mill., and spruce, *Picea* spp., and can kill almost all the trees in dense, mature stands of fir during uncontrolled outbreaks. These periods of high population densities typically last 5–15 years; between outbreaks the budworm can remain rare for 20–60 years.

To understand how insect defoliators might respond to climate change, it is important to consider their life histories and trophic relationships. In its lifecycle, the spruce budworm undergoes one generation a year, overwintering as a tiny second instar larva, emerging in early May, and then passing through four feeding instars before pupating in late June. Mating, moth dispersal, and oviposition occur in July. In mid-August the non-feeding first instar larvae hatch and move to overwintering sites on the branches.

The task of forecasting the possible effects of climate change on the spruce budworm can be separated into two parts, the first dealing with the direct effects of climate on the species' per capita growth rate, and the second dealing with indirect effects mediated through interactions and feedbacks with other species. Climate directly affects budworm survival (Lucuik, 1984), and to some extent, fecundity (Harvey, 1983b). On the other hand, self-regulation (brought about by competitive interactions among individuals) limits the extent to which climate-induced gains in survival and fecundity produce gains in the per capita growth rate. There is little indication of competition among spruce budworm, however, except among feeding larvae at very high densities (e.g., Fleming and Van Frankenhuyzen, 1992; Sanders, 1991). Hence, limitations on climate-induced increases in the spruce budworm's rate of population growth must come mainly through indirect effects such as those associated with trophic interactions. The net result is that increased per capita growth rates are associated with warm, dry conditions (Wellington, 1948).

2.1. Trophic Interactions

Like many other herbivores, including almost 50% of forest insect pests (Martineau 1984), the spruce budworm has evolved a life history by which its greatest nutritional demand is synchronised with the time when developing, rather then mature, host plant foliage is most accessible. As foliage matures, leaf growth slows but photosynthesis continues, and this can result in increased leaf carbohydrate content relative to nitrogen concentration (Little, 1970). The excess carbohydrates are often used to form secondary metabolites (Bryant *et al.,* 1983) which either purposefully (e.g., Rhoades and Cates, 1976; Haukioja 1980), or incidentally (Tuomi *et al.*, 1988), impede herbivory. In addition, because animals need tissue concentrations of 7–14% nitrogen while plants typically provide only 0.5–4%, dietary nitrogen may frequently limit herbivore growth (Mattson, 1980; White, 1984). Thus, developing plant tissue offers the advantage to the herbivore of being low in secondary metabolites (defensive chemicals), high in nitrogen (nutritional value), and low in fibre (which can limit digestibility). Under climate warming, however, higher temperatures should increase the rates of foliage development and hence reduce the 'window of opportunity' when the herbivore has access to immature foliage. This may reduce breeding success in certain homeothermic herbivores (Sedinger and Flint, 1991), but the effect on poikilothermic herbivores, which also develop faster at higher temperatures, is less certain. For instance, simulation of individual-based models of the phenological development of spruce budworm and host tree foliage suggest that the insect is well synchronised with white spruce (Volney and Cerezke, 1992). By combining models of feeding and phenological development, Régnière and You (1991) concluded that the budworm was so well synchronised with balsam fir that weather made little difference to the final amount of defoliation.

Increased atmospheric CO_2 levels are another aspect of climate change, but there is no evidence of any direct effects on insect herbivores (Fajer *et al.*, 1991). Plants, however, will probably increase their photosynthetic rates and, in the absence of any corresponding increase in nitrogen concentration, the carbohydrate content of their foliage as well (Rogers *et al.*, 1983). Scriber and Slansky (1981) report that low nitrogen concentrations induce Lepidoteran larvae to either increase their feeding rates (presumably to meet their dietary nitrogen requirements) or reduce their growth rates. Generally, however, current evidence (Ayres, 1993) suggests that although changes to host plant quality due to increased atmospheric CO_2 may have some effects on various aspects of insect growth and survival, these effects are trivial compared to those due to other sources of variation in host plant quality (e.g., tree-to-tree variability, foliage maturity).

Climate change may also affect the phenological relationships between the budworm and its natural enemies. These relationships are particularly important because, according to the most popular theory of budworm population dynamics (Royama, 1992), it is a group of numerically responding natural enemies that drives the budworm's outbreak cycle. Royama (1992) specifically identifies two univoltine parasitoids as key members of this group. Simple models provide a general tactical approach to studying how climate change might influence host-parasitoid dynamics. Hassell *et al.* (1993) describe a model in which both the host insect and its parasitoid have four stages in their lifecycle: egg, larva, pupa, and adult. Only the host larva is susceptible to attack and only the adult parasitoid is capable of attack. Hence, the rate of attack depends on the synchrony between the parasitoid's adult stage and the host's larval stage. It was assumed that, because of differences in the biological constraints on host and parasitoid physiology, as the climate warmed, the host larvae would emerge progressively earlier relative to the emergence of the adult parasitoids. Simulations showed that this effectively provided a 'temporal refuge' whereby many hosts escaped any possibility of attack by passing completely through their susceptible larval stage and pupating before the first adult parasitoid emerged. The earlier the host's phenological development relative to the parasitoid's, the greater the host's density at the stable equilibrium which characterises the model's long term dynamics. Eventually, however, the host's development was so much faster than the parasitoid's that the host completely escaped from the parasitoid's regulation.

3. Extreme Weather Events

3.1. Late Spring Frosts

Cerezke and Volney (in press) note that the white spruce forests of western Canada have long supported chronic spruce budworm outbreaks. The longest outbreak observed in the transcontinental boreal forest was first detected near Namew Lake on the Manitoba/Saskatchewan border in 1951. This outbreak came to an end in 1969. Tree-ring records suggest that outbreaks may last longer. An outbreak near Wabasca in central Alberta apparently began in 1941 and was observed to end in 1969. In a southern boreal forest district, the Spruce Woods Provincial Forest of southwestern Manitoba, moderate-to-severe defoliation was reported every year from 1938 to 1979 (Hardy *et al.,* 1986) when observations were interrupted. In addition, according to Cerezke and Volney (in press), damaging population densities have been observed every year since 1986 when observations were resumed.

In their review of outbreaks in the prairie provinces and the Northwest Territories, Cerezke and Volney (in press) call attention to the repeated association of spring frosts with the end of outbreaks in that region. There is a clear record of these frosts in June affecting the Namew Lake outbreak after 18 consecutive years of moderate-to-severe defoliation. The outbreaks in central Alberta and the Mackenzie valley in the Northwest Territories also ended in 1969 when late spring frosts destroyed the current year's crop of developing foliage. By contrast, the Spruce Woods outbreak to the south remained unaffected in 1969 where there were no late frosts. There are other instances when late spring frosts brought an end to outbreaks.

This evidence suggests that weather events can play a significant role in catastrophic declines in population density on a regional scale. As the climate warms, as it appears to have in the Spruce Woods case, the frequency of late spring frosts will decline and population densities will remain high for longer periods in northern white spruce stands. Not surprisingly, the balsam fir, which can not survive chronic defoliation has either been eliminated from these stands or remains a minor component.

3.2. Drought

Climate change projections suggest that much of Canada's boreal forest, the southwestern part in particular, will experience generally drier conditions with an increased probability of droughts and heat waves (Hengeveld, 1991). Since the survival and fecundity of insect herbivores often improves on drought stressed host plants (Mattson and Haack, 1987), Canada's boreal forest may suffer increasing insect pest problems as the climate warms.

Plants stressed by drought provide an ideal thermal environment for many insect pests. This is partly because stomatal closure reduces transpirational cooling so that plants suffering drought stress are typically 2–4 °C warmer than well-watered plants (Mattson and Haack, 1987), and partly because of the higher air temperatures and lower humidities generally associated with drought. For instance, as the microenvironment of the spruce budworm warms towards its optimum, the insect usually develops more quickly (e.g., Lysyk, 1989), suffers less early instar mortality due to cool wet springs (e.g., Lucuik, 1984), and has greater fecundity (e.g., Sanders *et al.*, 1978). In addition, because the optimum temperature for budworm development of about 26.6 °C (Hudes and Shoemaker, 1988) exceeds that of many of its natural enemies [e.g., the microsporidian parasite *Nosema fumiferanae* (Wilson, 1974), the solitary endoparasitoid *Apanteles fumiferanae* (Nealis and Fraser, 1988), and the entomophthoralean fungal pathogen *Erynia (Zoophthora) radicans* (Perry and Fleming, 1988)], budworm mortality due to natural enemies may decline during drought conditions.

There is also evidence that drought stress may affect the quality of host plants to the benefit of insect herbivores. For example, late-instar spruce budworm larvae are very sensitive to sucrose concentration which acts as a feeding stimulant, and thus helps the insect attain greater body size Harvey (1974), and possibly also helps the insect find and accept suitable host material. Mattson and Haack (1987) report that balsam fir sucrose concentrations are typically near 0.004M, but are almost three times this level in moisture-stressed trees. Consequently, the level in moisture-stressed fir falls in the range of the peak feeding response of sixth-instar spruce budworm larvae, which occurs between 0.01–0.05M (Albert *et al.*, 1982). This may explain why Mattson *et al.* (1983) found increased spruce budworm growth in conditions of low to moderate moisture stress.

Thus, if the climate becomes generally warmer and drier, with more frequent droughts, some increase in spruce budworm survival and fecundity can be expected. This may have little effect on the insect's population dynamics other than to produce some increase in average densities over the long term. At the other extreme, however, synergistic interaction between drought induced changes in host plant quality and drought induced changes in the insect's thermal environment may allow the budworm to improve its performance to the point that it can escape natural enemy regulation more easily. Under these circumstances, outbreaks may become more frequent (Mattson and Haack, 1987) and more severe.

4. Natural Selection

Because of their enormous population sizes, many insects are preadapted to climate change. For instance, Crawford and Jennings (1989) estimate a 'low density' spruce budworm population at 100 000 fourth instar larvae per ha. Since outbreaks involve densities 1000 times greater (Royama, 1994) and may extend over as much as 72 Mha (Hardy *et al.*, 1986), this represents approximately 7.2×10^{15} insects! With insect mutation rates typically reported at 10^{-5} to 10^{-4} per generation (Sager and Ryan, 1961), it is clear that billions of very rare genotypes will always be present in the population. Some of these genotypes are bound to be 'preadapted' to a warmer climate.

In addition, if the insect's geographic range is large, common genotypes from the warmer areas may readily invade the cooler areas as the climate warms. For instance, Harvey (1983a,b) found a genetically-based cline in spruce budworm egg weights which appears to be an adaptation to winter conditions. Northern females tend to put their egg-laying resources into fewer (150 *versus* 250 per 100 mg moth) but larger (0.22 *versus* 0.16 mg) eggs than females from the southern part of the range. The larger egg is necessary to give the young insect enough food reserves to survive the long northern winters (Harvey, 1985). If climate warming shortens northern winters, however, one can expect selection in the north to favour the 67% fecundity advantage of the southern genotypes, all else being equal. Thus, natural selection need not operate through spectacular die-offs in responding to climate change. Rather, it may be subtle, non-lethal effects that bring about much of the anticipated changes in the abundance and distribution of organisms.

The projections for future climates suggest a long trend toward warmer temperatures (Houghton *et al.,* 1990). This represents many generations of directional selection, especially for insects, since many have at least one generation per year. Under these circumstances, genotypes better adapted to a warmer climate can be expected to represent an ever increasing proportion of their populations, and consequently, their populations can be expected to become progressively better adapted to warmer climates. This has implications for closed environment experiments in which insects (or other organisms) are exposed to conditions simulating climates projected for the distant future. Because any opportunity for evolutionary change is effectively excluded by such experiments, the results may substantially underestimate a species' ability to deal with those projected climates. Another implication is that, because the mean temperature in the northern hemisphere has risen by about 0.4 °C since 1964 (Jones and Wigley, 1990), some insects may have already responded.

Fleming and Tatchell (1994a,b) searched for evidence of accelerated phenological development as an indicator of recent insect response to climate warming. They used the unique records of daily suction trap catches from the Rothamsted Insect Survey. The records for five of the most thoroughly sampled aphid species at each of the eight longest running trap sites were examined to see if flight periods were occurring earlier in more recent years. They found that these flight periods had moved ahead by an average of about 3–7 days over the last 25 years. The consistency in the direction of these trends among all species-site combinations was highly significant ($P < 0.0001$), as might be expected with climate warming.

5. Transient Dynamics

Many of the large scale models used to project the effects of climate change on the structure of the boreal forest are 'equilibrium' models (e.g., Smith *et al.*, 1992). Such models project future equilibrium forest conditions while implying that the 'paths' by which the forest approaches those conditions (i.e., the transient dynamics) are inconsequential. This implication is questionable for systems exhibiting thresholds or multiple equilibria (Isaev and Khlebopros, 1979).

Antonovsky *et al.* (1990), for instance, explored the dynamics of a simple forest:pest model which allows both thresholds and multiple equilibria. In this model the forest has two age classes and only older trees are attacked by the pest. Figure 1 illustrates this model's long-term dynamics for parameter values estimated for the spruce budworm–balsam fir interaction. Depending on the relative rates of senescence and regeneration, the

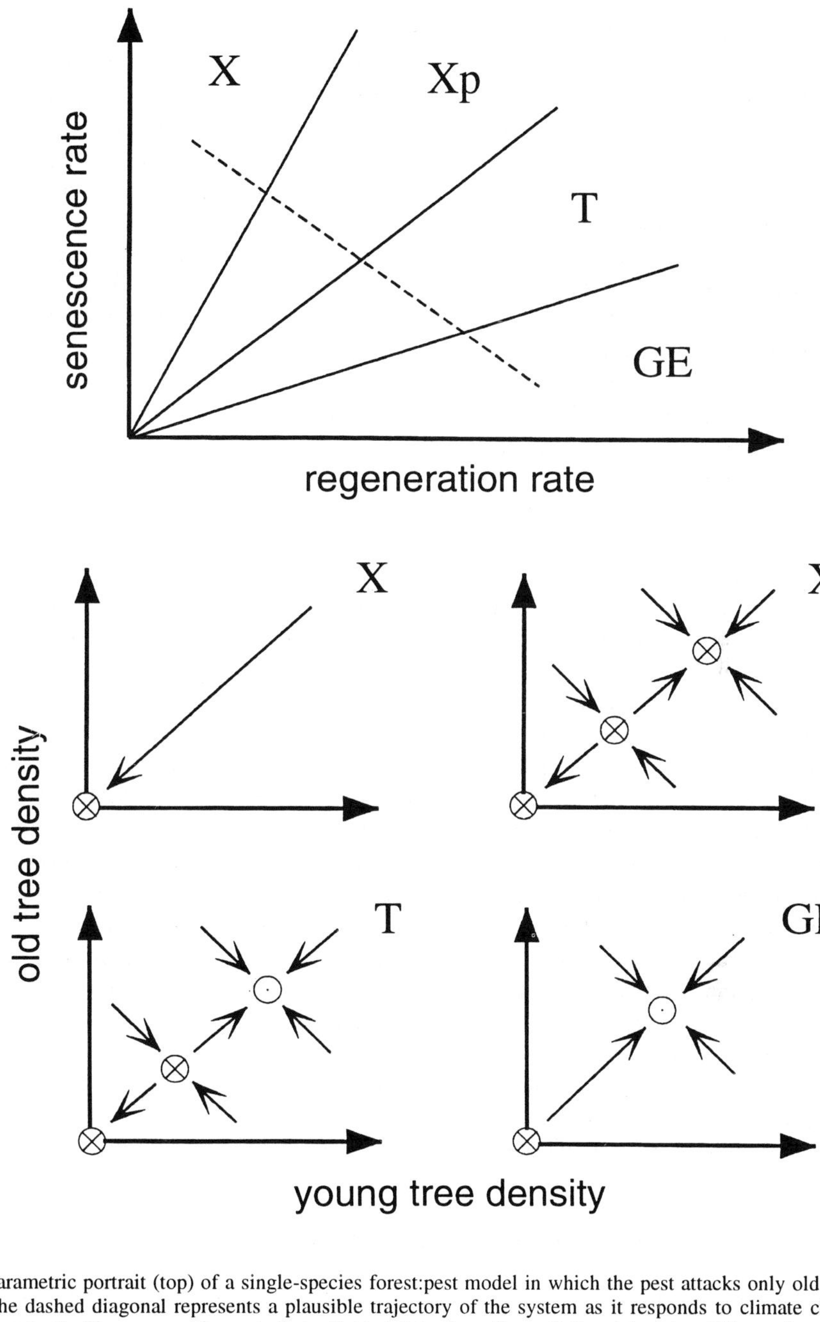

Fig. 1. Parametric portrait (top) of a single-species forest:pest model in which the pest attacks only old trees. The dashed diagonal represents a plausible trajectory of the system as it responds to climate change (see text). The parametric portrait is divided into 4 sections distinguished by different long-term dynamics of the forest-pest system: X, eventual extinction for the forest and pest; X_p, extinction for the pest, possible persistence for the forest; T, system persistence provided the forest exceeds a low density threshold; GE, system persistence at a global equilibrium. Four phase planes corresponding to the different sections of the parametric portrait are shown below. In the phase planes the pest density is represented by the z-axis coming up out of the page. Small circles enclosing an X, or dot, respectively indicate equilibria where small pest densities are drawn down toward extinction, or up towards a stable forest-pest equilibrium. (After Antonovsky *et al.*, 1990).

model permits four qualitatively different types of long-term behaviour (Figure 1, upper plot). These behaviours are (according to the lettering identifying each area on the parametric portrait): system extinction (X), pest extinction and forest survival above a threshold forest density (X_p), system persistence above a threshold forest density (T), and system persistence (GE). The four phase planes at the bottom of Figure 1 (with lettering matching that of the relevant area of the parametric portrait, above) show details of these behaviours. These phase planes have the young and old tree densities plotted on the x- and y-axes, respectively. Pest densities are plotted along the z-axis which rises up from the page. Small circles enclosing an X, or dot, indicate respectively, equilibria where small pest densities are drawn down toward extinction, or up towards a stable forest:pest equilibrium.

In Figure 1, the dashed diagonal line represents plausible trajectories of the model system as it responds to climate change. For instance, at a site above the northern boundary of balsam fir where regeneration rates are negligible and senescence rates would presumably be high if there were any trees, the system is in area X of the parametric portrait. Assuming it is the cold climate that makes the site unfavourable for balsam fir, increasing rates of regeneration and decreasing rates of senescence can be expected as the climate warms (Rizzo and Wicken, 1992). Consequently, the system at this northern site moves along the dashed diagonal in the parametric portrait from the upper left towards the lower right. When the system enters parametric area X_p and then T, however, tree densities do not exceed the threshold so balsam fir does not establish. Only when the model system enters parametric area GE does the forest and then the pest become established.

Contrast this scenario with one for a densely forested southerly site where the model system starts in parametric area GE. Then, as the climate warms, the boreal forest on the site may be replaced by temperate deciduous forest (Rizzo and Wicken, 1992; Sargent, 1988). In other words, the site is expected to become less favourable for balsam fir (and other boreal species) causing balsam fir's regeneration rates to decline and senescence rates to increase. If these rate changes occur, the model system will traverse the parametric portrait along a path from the lower right towards the upper left. (The dashed line is an example of such a path). But because the forest exceeds the low density threshold when the system enters parametric area T (and then X_p), the whole model system persists in area T (and the forest persists in area X_p). Thus, for exactly the same senescence and regeneration rates, but depending on the direction of approach, the model indicates that the forest:pest system could become either completely extinct or remain completely intact in parametric area T.

6. Summary and Conclusion

Using the spruce budworm as an illustrative case study, some of the potentially important processes underlying insect population dynamics in the context of climate change were briefly described. Plausible scenarios were developed to explain how climate change might affect these processes and what the consequences might be for boreal forest:defoliator systems. This exercise pointed to a number of characteristics of forest:pest interactions which may have important influences on how climate change affects the boreal forest's insect-based disturbance regimes. First, insects often vary substantially from species to species in their life histories, in their trophic relationships, and in the nature of their 'outbreak' patterns. Second, insect trophic interactions and population dynamics are often characterised by multiple and complex feedback loops which are sensitive to weather conditions. Third, rare and extreme weather events may be more important than long term trends in determining the consequences of climate change, at least for some insects. Fourth, natural selection implies (a) that the genotypic composition of future insect populations may differ substantially from the species' current state, and (b) that some innocuous insect species could evolve into major pests of the future. Last, forecasting the equilibrium conditions of forest:pest systems may not be enough; the probable existence of thresholds and multiple attractors means that one may also need to forecast the transient dynamics.

Developing sound forecasting procedures will depend on how well the effects of climate change on some of the processes and characteristics of boreal forest:defoliator interactions described above can be integrated into large scale models. Ideally such large scale models would then provide a basis for projecting the dynamics of the boreal forest's insect-based disturbance regimes into the future. From a practical point of view, however, it may be impossible to deal thoroughly with all of these processes and characteristics for each important boreal forest:defoliator interaction. Hence, the most promising strategy will rely on a judicious interdisciplinary mix of available research approaches (many of which have been referred to above) and results (e.g., from other insect pest:host plant interactions).

Acknowledgments

We thank the Canadian Forest Service Green Plan for financial support and Tim Burns for his help in drawing the figure.

References

Albert, P. J., Clearley, C., Hanson, F., and Parisella, S.: 1982, *J. Chem. Ecol.* **8,** 233-239.

Antonovsky, M. Y., Fleming, R. A., Kuznetsov, Y. A., and Clark, W. C.: 1990, *Theor. Popul. Biol.* **37,** 343-367.

Apps, M. J.: 1995, *Water, Air, and Soil Pollut.* (this volume).

Ayres, M. P.: 1993, *In:* Kareiva, P. M., Kingsolver, J. G., and Huey, R. B. (eds.), *Biotic Interactions and Global Change,* Sinauer Associates Inc., Sunderland, USA, pp. 75-94.

Bryant, J. P., Chapin III, F. S., and Klein, D. R.: 1983, *Oikos* **40,** 357-368.

Cerezke, H. F., and Volney, W. J. A.: 1995, *In:* Ives, W.G.H. (ed.), *Forest Insect Pests in Canada,* Canadian Forest Service, Northwest Region, Ottawa, Canada. (In press).

Crawford, H. S., and Jennings, D. T.: 1989, *Ecology* **70,** 152-163.

Fajer, E. D., Bowers, M. D., and Bazzaz, F. A.: 1991, *Ecology* **72,** 751-754.

Fleming, R. A., and Tatchell, G. M.: 1994a, *In:* Leather, S. R., Watt, A. D., Mills, N. J., and Walters, K. A. F. (eds.), *Individuals, Populations and Patterns in Ecology,* Intercept, Andover, UK, pp. 63-71.

Fleming, R. A., and Tatchell, G. M.: 1994b, *In:* Harrington, R. and Stork, N. E. (eds.), *Insects in a Changing Environment,* Academic Press, London, UK, pp. 479-482.

Fleming, R. A., and van Frankenhuyzen, K.: 1992, *Can. Ent.* **124,** 1101-1113.

Haack, R. A., and Byler, J. W.: 1993, *J. Forestry* **91,** 32-37.

Hall, J. P., and Moody, B. H.: 1994, *Forest Depletions Caused by Insects and Diseases in Canada, 1982-1987,* Information Report ST-X-8, Canadian Forest Service, Ottawa, Canada 14 pp.

Hardy, Y., Mainville, M., and Schmitt, D. M.: 1986, *An Atlas of Spruce Budworm Defoliation in Eastern North America, 1938-1980,* U.S.D.A. For. Serv., Cooperative State Research Service. Miscellaneous Publ. No. 1449. U.S.D.A., Washington, USA, 52 pp.

Harvey, G. T.: 1974, *Can. Ent.* **106,** 353-365.

Harvey, G. T.: 1983a, *Can. Ent.* **115,** 1103-1108.

Harvey, G. T.: 1983b, *Can. Ent.* **115,** 1109-1117.

Harvey, G. T.: 1985, *Can. Ent.* **117,** 1451-1461.

Hassell, M. P., Godfray, H. C. J., and Comins, H. N.: 1993, *In:* Kareiva, P. M., Kingsolver, J. G., and Huey, R. B. (eds.), *Biotic Interactions and Global Change.* Sinauer Associates Inc., Sunderland, USA, pp. 402-423.

Haukioja, E.: 1980, *Oikos* **35,** 202-213.

Hengeveld, H.: 1991, *Understanding Atmospheric Change: A Survey of the Background Science and Implications of Climate Change and Ozone Depletion,* SOE Report 91-2, Atmospheric Environment Service, Environment Canada, Ottawa, Canada, 68 pp.

Houghton, J. T., Jenkins, G., and Ephraums, J. J. (eds.): 1990, *Climate Change: The IPCC Scientific Assessment,* Cambridge University Press, Cambridge, UK, 403 pp.

Hudes, E. S., and Shoemaker, C. A.: 1988, *Environ. Entomol.* **17,** 97-108.

Isaev, A. S., and Khlebopros, R. G.: 1979, *In:* Norton, G. A., and Holling, C. S. (eds.), *Pest Management.* Pergamon, Oxford, UK, pp. 317-339

Jones, P. D., and Wigley, T. M. L.: 1990, *Sci. Amer.* **263**, 66-73.

Little, C. H. A.: 1970, *Can. J. Bot.* **48,** 2021-2028.

Lucuik, G. S.: 1984, *Can. Ent.* **116,** 1077-1084.

Lysyk, T. J.: 1989, *J. Econ. Entomol.* **82,** 1161-1168.

Martineau, R.: 1984., *Insects Harmful to Forest Trees,* Canadian Forest Service, Govt. Publ. Centre, Ottawa, Canada, 261 pp.

Mattson, W. J.: 1980, *Oecologia* **81,** 186-191.

Mattson, W. J., and Haack, R. A.: 1987, *Bioscience* **37,** 110-118.

Mattson, W. J., Slocum, S. S., and Koller, C. N.: 1983, *Spruce Budworm Performance in Relation to Foliar Chemistry of its Host Plants,* U.S.D.A. For. Serv. Gen. Tech. Rep. NE-85, pp. 55-56.

Nealis, V. G., and Fraser, S.: 1988, *Can. Ent.* **120,** 197-204.

Perry, D. F., and Fleming, R. A.: 1988, *Can. J. Bot.* **67,** 1657-1663.

Piene, H.: 1980, *Forest Sci.* **26,** 665-673.

Régnière, J., and You, M.: 1991, *Ecol. Modelling* **54,** 277-297.

Rhoades, D. F., and Cates, R. G.: 1976, *Rec. Adv. Phytochem.* **10,** 168-213.

Rizzo, B., and Wicken, E.: 1992, *Climate Change* **21,** 37-56.

Rogers, H. H., Thomas, J. F., and Bingham, G. E. 1983, *Science* **220,** 428-429.

Royama, T.: 1992, *Analytical Population Dynamics,* Routledge, Chapman and Hall, New York, USA, 387 pp.

Sager, R., and Ryan, F. J.: 1961, *Cell Heredity,* Wiley, New York, USA, 411 pp.

Sanders, C. J.: 1991, *In:* van der Geest, L. P. S., and Evenhuis, H. H. (eds.), *Tortricid Pests, Their Biology, Natural Enemies and Control,* Elsevier, Amsterdam, The Netherlands, pp. 579-620.

Sanders, C. J., Wallace, D. R., and Lucuik, G. S.: 1978, *Can. Ent.* **107,** 1289-1299.

Sargent, N. E.: 1988, *Climatological Bull.* **22(3),** 23-34.

Schowalter, T. D., Hargrove, W. W., and Crossley, D. A., Jr.: 1986, *Annu. Rev. Entomol.* **31,** 177-196.

Scriber, J. M., and Slansky, F.: 1981, *Ann. Rev. Ent.* **26,** 183-211.

Sedinger, J. S., and Flint, P. L.: 1991, *Ecology* **72,** 496-502.

Smith, T. M., Leemans, R., and Shugart, H. H.: 1992, *Climate Change* **21**, 367-384.

Tuomi, J., Niemelii, P., Chapin, F. S., III, Bryant, J. P., and Siren, S.: 1988, *In:* Mattson, W. J., Levieux, J., and Bernard-Dagan, C. (eds.), *Mechanisms of Woody Plant Defenses Against Insects: Search for Patterns,* Springer-Verlag, New York, USA, pp. 416.

Volney, W. J. A.: 1989, *Agricultural Zoology Reviews* **3,** 133-156.

Volney, W. J. A., and Cerezke, H. F.: 1992, *Can. J. For. Res.* **22,** 198-205.

Wellington, W. G.: 1948, *Can. Ent.* **80,** 56-82.

White, T. C. R.: 1984, *Oecologia* **63,** 90-105.

Wilson, G. F.: 1974, *Can. J. Zool.* **52,** 59-63.

GLOBAL CARBON DYNAMICS OF HIGHER LATITUDE FORESTS DURING AN ANTICIPATED CLIMATE CHANGE: ECOPHYSIOLOGICAL VERSUS BIOME-MIGRATION VIEW

G. H. KOHLMAIER, Ch. HÄGER, A. NADLER,
G. WÜRTH and M. K. B. LÜDEKE

Institut für Physikalische und Theoretische Chemie der Johann Wolfgang Goethe-Universität, Marie Curie-Str. 11, D-60439 Frankfurt, Germany

Abstract. The response of the vegetation and soils of the higher latitude forests and tundra ecosystems to an anticipated climate change is investigated using two alternative approaches to calculate the resulting change in the total carbon content (TCC) of the vegetation and the soils: On the one hand a BGC (bio-geochemical-cycle) model, in this case the FBM (Frankfurt Biosphere Model), where the ecosystem response is entirely due to the ecophysiological response of the vegetation and the ecological response of the soils. On the other hand a biome or "rule-based" model, in this case the BIOME model, which allows for the determination of the occurrence of a specific biome type from a given climatic situation assuming equilibrium conditions. Within the FBM prognosis net primary production and TCC are reduced both for needle leaved and broad leaved forests if the CO_2-fertilisation effect is not taken into account. When the CO_2-fertilisation effect is taken into consideration NPP, standing biomass and soil carbon are increased in a future greenhouse climate. Although there is a considerable shift of the biomes in response to the greenhouse climate within the BIOME approach, the TCC in the investigated northern biomes stayed more or less constant. This is due to a decrease in biomass in the southern regions of today's temperate forests compensating the biomass increase by the northward shift of the taiga border.

Keywords. CARBON DYNAMICS, FBM, BIOME, FUTURE GREENHOUSE CLIMATE, CO_2-FERTILISATION, BIOME MIGRATION.

1. Introduction

The chemical composition of the atmosphere is assumed to be the main cause of the variation of the global climate since the beginning of industrialisation. The concentrations of relevant greenhouse gases are predicted to increase continuously during the next decades. Climate model calculations using atmospheric and oceanic General Circulation Models (GCMs) show that an increase in surface air temperatures as well as changes in precipitation regimes and cloudiness are to be expected due to the rising greenhouse gas concentrations. The role of the terrestrial biosphere responding to these alterations is discussed intensively with respect to the changes of net primary production NPP, total biomass and soil carbon (C) content.

To assess the future global ecosystem C content two alternative modelling approaches were used. On the one hand the biome (or "rule-based") models which focus on the description of biome shifts and are based on a mapping of climate parameters on biome types (Smith *et al.*, 1992; Monserud and Leemans, 1992). On the other hand the global "Bio-Geochemical Cycle (BGC)" models which focus on the description of matter fluxes (e.g., C and water fluxes) within the ecosystems and between the ecosystems and the atmosphere (Raich *et al.*, 1991; Esser, 1991; Lüdeke *et al.*, 1994).

Both approaches represent important response patterns of the vegetation/soil systems in the presence of a global climate change. The BGC models tend to describe

Water, Air and Soil Pollution **82**: 455–464, 1995.

effects directly related to ecophysiology (e.g., changes in biomass pools or LAI caused by changes in photosynthesis and respiration) acting on the intermediate timescale, while the rule-based models describe the results of biome migration processes depending on the changed climate and therefore act on the long-term timescale. Attempts to couple these aspects are being pursued, but a consistent solution has not yet been obtained. Therefore at the present time a detailed comparison and discussion of the results obtained by both model approaches is a useful tool to assess the complex response pattern of ecosystems to global warming.

In this study we investigate the response of the northern hemispheric deciduous and evergreen forests and the tundra ecosystems to a 3×CO_2 climate as calculated by the GCM of the MPI in Hamburg.

2. Description of the Models

The selected BGC (bio-geochemical-cycle) model, in this case the FBM (Frankfurt Biosphere Model, Kindermann *et al.*, 1993; Lüdeke *et al.*, 1994), is a seasonal high resolution process model for vegetation and soils in which photosynthesis, autotrophic and heterotrophic respiration, and transpiration are driven by surface temperature, precipitation, radiation (including cloud cover effects) and atmospheric CO_2 concentration. These fluxes are balanced considering four mass compartments which describe the C content of the woody and green biomass, as well as the soil C and water content. Therefore the model is a coupled system of four nonautonomous ordinary differential equations which are parameterised for 32 different vegetation types. The area covered by these vegetation types is given by a vegetation map. With a constant climate the vegetation and the soils approach a climax state for which the climax biomass and soil C content are given. This calculation can be performed for actual as well as for future climate. Thus the ecosystem response is entirely due to the direct consequences of ecophysiological vegetation response and the ecological response of the soils (see Appendix for further information).

The most advanced biome, or rule-based, model is the BIOME model (Prentice *et al.*, 1992), which allows for the determination of the occurrence of a specific biome type from a given climatic situation assuming equilibrium conditions. Therefore a five dimensional "climate space" is defined considering growing degree days (based on two different definitions), mean temperature of the coldest and warmest months and a

TABLE I

FBM simulation runs and assumptions of vegetation's response on climate change.

Model Run	Description
I	1×CO_2 equilibrium run (present situation)
II	3×CO_2 equilibrium run ("business as usual") with climate as calculated by ECHAM2 (Perlwitz, 1992; Claussen, 1993), without CO_2-fertilisation
III	Run II with CO_2-fertilisation and assuming a rise of maximum photosynthesis rate, α

measure of water availability. The latter is calculated from the results of a simple bucket model for the soil water content, driven by seasonal rainfall and evapotranspiration. Five-dimensional subspaces, confined by planes which are parallel to planes stretched by two base vectors respectively, are mapped to the biome types.

To evaluate the resulting vegetation maps in terms of C storage, we assigned fixed values of standing biomass and soil C content to the biomes (values according to Plöchl and Cramer, 1994).

3. Simulations

As input data for both models we used the following data sets: the climate map compiled by Shea (1986) as reference for present day climatic conditions and the output of a GCM developed at the Deutsches Klimarechenzentrum (ECHAM2; Claussen, 1993; Perlwitz, 1992). The GCM was run for 100 model years under rising atmospheric CO_2 conditions, that means an increase of atmospheric CO_2 concentration from 350 ppm (1985) towards 1050 ppm (2085). All remaining greenhouse gases like CH_4 and others are described in CO_2 equivalents, the described simulation is called the "business as usual scenario" of the ECHAM2 model. From the year 2085 the CO_2 concentration was held constant at 1050 ppm for the following 30 model years. The averaged climate of these 30 years was compared with the present climate as calculated by the same GCM to obtain the climate anomalies of a future greenhouse atmosphere. We then used these anomalies to calculate a greenhouse climate on the basis of Shea's map (Lüdeke *et al.*, 1994b). Compared with the present situation the GCM predicts for the investigated regions an increase of summer temperatures by more than 3 K and an increase of winter temperatures by more than 5 K. Precipitation will increase by 10–20% in a future greenhouse climate.

The area considered for the FBM-calculations consists of $1° \times 1°$-grid elements covered by one of the eight vegetation types listed in Table II, first column. On this area we performed three FBM simulation runs (for the definition see Table I)

For the BIOME calculations we considered the area covered by one of biome types 1–8 listed in Table III, first column, *either* in the $1{\times}CO_2$ climate *or* in the $3{\times}CO_2$ climate. The biome types 9–11, (Table III, first column), migrate into this area caused by the transition from the $1{\times}CO_2$ to the $3{\times}CO_2$ conditions. In this calculation the fixed total carbon density (biomass + soil C per unit area) for every biome type is given in Table III, second column.

4. Results and Discussion

Figures 1a and 1b show the northward migration of biomes as a response to a $3{\times}CO_2$ climate as calculated by the BIOME model: tundra ecosystems and cold deciduous forests (e.g., larch forests) are mainly replaced by taiga, while cool coniferous forests and cold mixed forests will be shifted northward by about 5° and in the south some biomes will immigrate into the investigated region.

The results of the evaluation in terms of total carbon content (TCC) are listed in Table III. The TCC of Tundra will decrease by 78%, the TCC of cold deciduous forests and cool grass and shrubs will decrease by about 50%. The simulations give a slight increase of TCC in Taiga, a moderate increase in cool mixed forests, temperate mixed

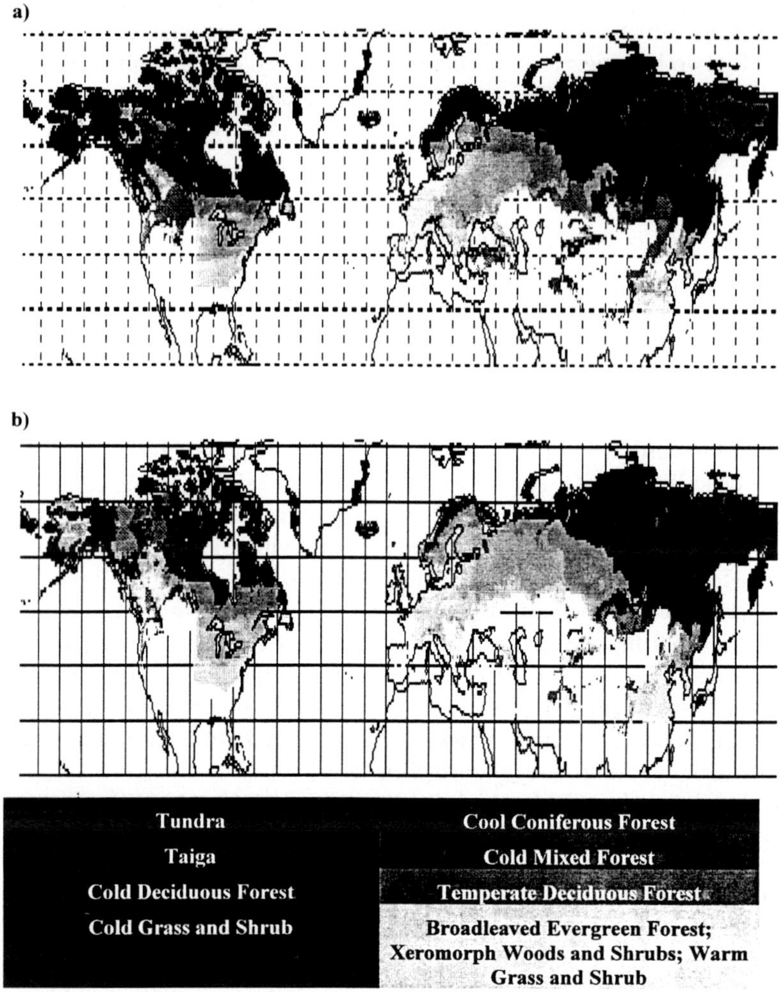

Fig. 1. Results of the rule-based BIOME model (Prentice and Cramer, 1992). (a) Biome distribution under $1\times CO_2$ conditions (present situation); (b) Biome distribution under $3\times CO_2$ conditions ("business as usual").

forests, and temperate deciduous forests (15–27%) and an remarkable increase in cool conifer forest (45%). Three biomes will immigrate from the south into the region considered: broadleaved evergreen forest, warm grass and shrub, and xeromorph woods. The total carbon content in the whole region will, however, only slightly increase (by about 3%), because the drastic shifts of the biomes partly compensate each other in terms of TCC.

TABLE II

Resulting Total Carbon Content (TCC) for the vegetation types investigated within the three different simulation runs using the FBM.

Vegetation type	TCC (Pg C) Run I	Run II	Run III
Temperate/subpolar needleleaved forest	243.49	185.50	239.16
Cold deciduous forest, with evergreens	141.12	84.120	145.39
Cold deciduous forest, without evergreens	114.54	77.553	117.65
Evergreen needleleaved woodland	15.333	13.027	21.481
Cold decidous woodland (larch forest)	23.600	21.320	31.603
Evergreen needleleaved/microphyllous shrubland/thicket	4.340	3.380	5.100
Cold decidous subalpine/subpolar shrubland, dwarf shrubland	1.915	1.886	3.119
Arctic/alpine tundra, mossy bog	68.912	56.428	100.06
Total	613.25	443.21	663.56

This result of the BIOME model neglects a possible CO_2-fertilisation effect which could affect both the rules for selecting the biome types and the TCC densities. Therefore it has to be compared with the results of the FBM model runs I and II (see Table II). Here we calculated a considerable decrease in TCC of 28% (this surprising result is discussed in detail below). With respect to the different time scale foci of the two models, we can therefore conclude that on the intermediate time scale (decades) governed by mainly ecophysiological responses, the northern boreal and temperate ecosystems will be a source of C, while on the longer time scale the TCC tends to equal the present situation, providing the CO_2-fertilisation effect would not be operative at the ecosystem level.

Another interesting result is the change in the latitudinal TCC distribution as calculated by the FBM (Figure 2). Here we find that even in the BGC model the northward shift of the biomes of high C density as predicted by the BIOME model is qualitatively reproduced by the strong decrease of the forest biomass in the south and, for instance, the increase of the biomass of northern woodlands up to values which are typical for forest ecosystems. In this way migration effects on biomass are partly simulated by the quantitative FBM response to climate which could be interpreted qualitatively as a transition in vegetation type (e.g., forest → bushland).

In general we find an increase in photosynthesis caused by temperature which is overcompensated by the increase in autotrophic respiration, leading to a net decrease of NPP caused by the pure temperature change. A further decrease in NPP is due to increased water limitation. The latter occurs because the enhanced evapotranspiration is not balanced by the slightly increased rainfall. The remaining two fertilisation effect related components, an increase of maximum photosynthesis rate and a shift in the temperature optimum of photosynthesis towards higher temperatures (Long and Drake, 1992), which are proved for the plant level, overcompensate the climate caused NPP decrease.

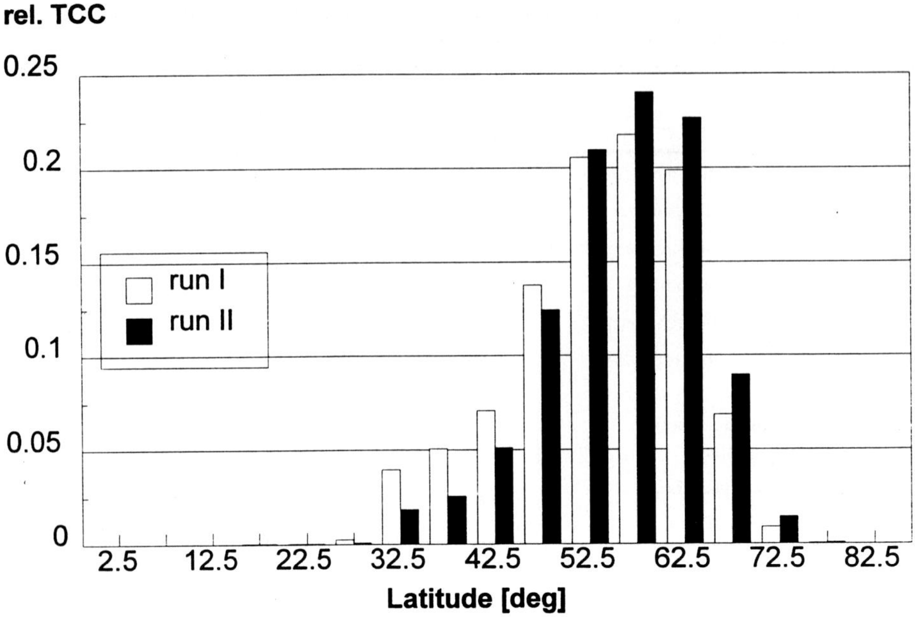

Fig. 2. Relative latitudinal NPP distributions in 5° belts as simulated in two model runs under different climate conditions. model Run I: 1×CO_2 climate (present situation); model Run II: 3×CO_2 climate ("business as usual scenario" of the ECHAM2 model (Perlwitz, 1992; Claussen, 1993)). See text and Table I for detailed description of the model runs.

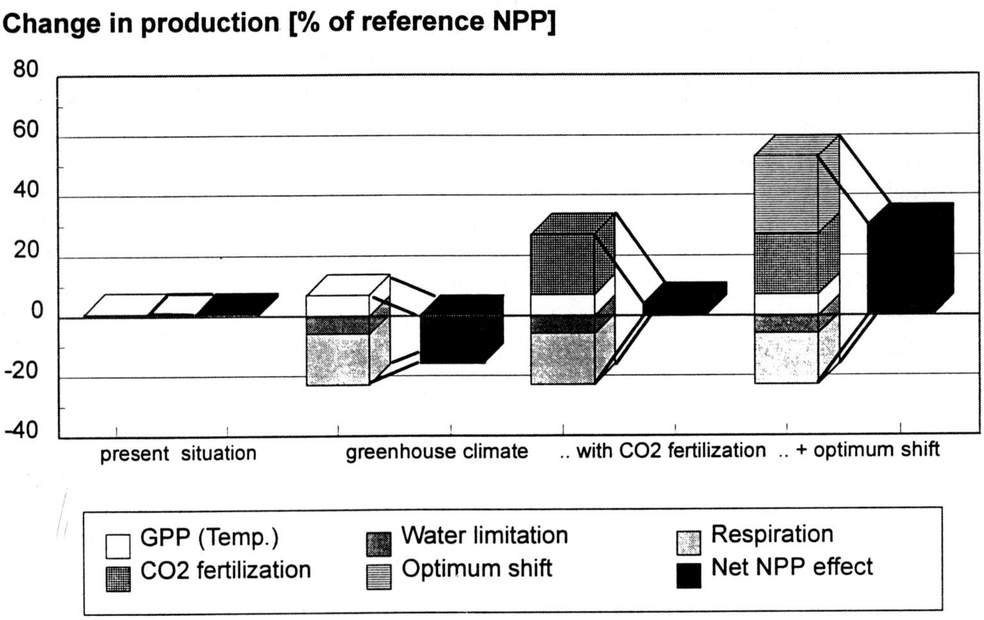

Fig. 3. Net NPP effect under different assumptions of the vegetation's reaction to climate change and gross effect of the assumptions on primary production and respiration.

TABLE III

Results of calculations performed with the BIOME model: Evaluation of the biome distribution's total carbon content (TCC) under different climate conditions with assigned carbon (C) densities to each biome.

Biome	Assigned total C density (kg C m^{-2})	TCC "1×CO_2" (Pg C)	TCC "3×CO_2" (Pg C)
Tundra	19.46	136.8	30.1
Taiga	27.5	367.9	378.1
Cold deciduous forest	13.4	48.5	22.0
Cool grass and shrub	9.16	40.5	21.8
Cool conifer forest	27.5	93.5	136.5
Cold mixed forest	24.6	12.9	16.4
Cool mixed forest	24.6	143.4	169.0
Temperate deciduous forest	38.92	193.5	222.5
Broadleaved evergreen forest	25.65	0.0	22.8
Warm grass and shrub	9.16	1.8	42.8
Xeromorph woods and shrub	18.75	0.0	6.6
Total	-	1038.6	1068.6

We conclude:

- The results of the FBM are very sensitive to whether a CO_2-fertilisation effect is included or not. Only with CO_2-fertilisation, the total carbon content (TCC) is increased (by 8%) upon a predicted change in climate. Disregarding this effect, TCC will decrease by 27%.
- As a comparison the BIOME model of Prentice *et al.* (1992) was evaluated. There is a shift in the biomes to higher latitudes, an effect which is reproduced by the FBM run in terms of TCC. If a constant C density is assigned to each biome, there will be a slight increase in TCC by 3%. This long term prediction does not incorporate CO_2-fertilisation, because this version of the BIOME model has not yet been published.

Acknowledgments

We thank the European Union and the German Ministry of Research and Technology for financial support. Furthermore we thank Franz-W. Badeck, Ralf D. Otto, Silke Dönges, Jürgen Kindermann, Torsten Lang, Ulrich Jäkel, Axel Klaudius, Peter Ramge and Stefan Habermehl for their considerable support of the FBM model development and the numerical evaluations.

References

Claussen, M.: 1993, Report No. 115. Max Planck-Institut für Meteorologie, Hamburg, Germany, 30 pp.

Esser, G.: 1991, *In:* Esser, G. and Overdieck, D. (eds.). *Modern Ecology. Basic and Applied Aspects* Elsevier, Amsterdam, The Netherlands, pp. 3-30.

Fung, I. Y, Tucker, C. J. and Prentice, K. C.: 1987, *J. Geophys. Res.* **92** (D3), 2999-3015.

Kindermann, J., Lüdeke, M. K. B., Badeck, F.-W., Otto, R. D., Klaudius, A., Häger, C., Würth, G., Lang, T., Dönges, S., Habermehl, S. and Kohlmaier, G. H.: 1993, *Water, Air, Soil Pollut.* **70**, 675-684.

Kirschbaum, M. U. F.: 1993, *Tellus* **45B**, 321-334.
Kohlmaier, G. H., Benderoth, G., Klaudius, A. and Janecek, A.: 1989, *In:* WMO (ed.), *Extended Abstracts of Papers presented at the 3rd International Conference on Analysis and Evaluation of Atmospheric* CO_2 *Data Present and Past* No. 59, WMO/TD-No. 340, Geneva, Switzerland, pp. 274-279.
Kohlmaier, G. H., Lüdeke, M., Janecek, A. and Benderoth, G.: 1991, *In* Schneider, S. H. P. and Boston, J. (eds.), *Scientists on Gaia*, MIT Press, Cambridge, MA, USA, pp. 223-239.
Long, S. P. and Drake, B. G.: 1992, *In:* Baker, N. R., Thomas, H. and Elsevier, S. P. (eds.), *Crop Photosynthesis: Spatial and Temporal Determinants*, Vol. 12, Elsevier, Amsterdam, The Netherlands, pp. 69-103.
Lüdeke, M. K. B., Badeck , F. W., Otto, R. D., Häger, Ch., Dönges, S., Kindermann, J., Würth, G., Lang, T., Jäkel, U., Klaudius, A., Ramge, P., Habermehl, S. and Kohlmaier, G. H.: 1994a, *Climate Research* **4**(2), 143-166.
Lüdeke, M. K. B., Dönges, S., Otto, R. D., Kindermann, J., Badeck, F.-W., Ramge, P., Jäkel, U., and Kohlmaier, G. H.: 1994b, *Tellus* **47B** (1/2), 191-205.
Monserud, R. A. and Leemans, R.: 1992, *Ecol. Mod.* **62**, 275-293.
Perlwitz, J.: 1992, *Annales Geophysicae Assembly of the European Geophysical Society*. Edinburgh, UK, 6-10 April 1992.
Plöchl, M. and Cramer, W.: 1994, *Tellus* **47B** (1/2), 240-250.
Prentice, I. C., Cramer, W., Harrison, S. P., Leemans, R., Monserud, R. A. and Solomon, A. M.: 1992, *Journal of Biogeography* **19**, 117-134.
Raich, J. W., Rastetter, E. B., Melillo, J., Kicklighter, D. W., Steudler, P. A., Peterson, B. J., Grace, A. L., Moore, B. and Vörösmarty, C. J.: 1991, *Ecological Applications*, **1**(4), 399-429.
Shea, D. J.: 1986, *Climatological Atlas: 1950-1979. Surface Air Temperature, Precipitation, Sea-Level Pressure, and Sea-Surface Temperature (45°S-90°N).* NCAR Technical Note 269+STR. NCAR, Boulder, USA, 143 pp.
Smith, T. M., Leemans, R. and Shugart, H. H.: 1992, *Climatic Change* **21**, 367-384.
Thornthwaite, C. W.: 1948, *Geogr. Rev.* **38**, 55-94.

Appendix: Short Description of the Frankfurt Biosphere Model

CARBON ALLOCATION AND PHENOLOGY

As shown in Figure A1, it is assumed that the assimilate production, C_{ASS}, is determined by the mass of compartment GC, reflecting the amount of leaves, the actual soil water content SW as well as by the external driving variables temperature T and irradiance I. This flux is to be partitioned according to present needs of the plant organs, namely the build-up and maintenance of the photosynthesising tissue and of the feeder roots, (represented by GC) on the one hand, and the build-up and maintenance of stems, branches and coarse roots (represented by RC) on the other. Furthermore, assimilates have to be translocated in order to fill particular storage organs, which are included in the C mass of the GC compartment as well.

The partitioning of the C assimilation flux, C_{ASS}, into the GC and RC compartments is derived in its seasonal and long term patterns from two basic assumptions:

- The vegetation tends to maximise the amount of photosynthesising tissue (contained in the GC compartment).
- It is possible to identify a function $RC = \Omega(GC) = \xi GC^{\kappa}$ determining the minimum amount of RC needed to support and maintain the given amount of GC.

We distinguish four different phenophases:

1. Shooting phase

The C gain from photosynthesis is greater than the C loss. The system allocates most of the assimilates to the GC compartment until the trajectory reaches $\Omega(GC)$, maximising its production ability.

2. Secondary growth

The system is forced to allocate simultaneously into the GC and RC compartment according to the $\Omega(GC)$ function.

3. Leaf shedding

At the end of the vegetation period, when unfavourable weather conditions do not allow biomass increase (e.g., drought, cold), a leaf abscission phase reduces the GC compartment to a remaining amount of feeder roots and assimilate store, which is proportional to the annual maximum of GC and characterised by the function $RC = \Theta(GC) = \nu GC^{\kappa}$.

4. Dormancy phase

When the trajectory reaches the $\Theta(GC)$ curve the dormancy phase starts. During this phase the biomass losses, as defined by the RC respiration and the total litter production C_{BH}, are distributed among the compartments so that the systems trajectory follows the $\Theta(GC)$ curve. This phase ceases when the weather conditions allow a net biomass increase, assuming a total conversion of stored assimilates into leaf biomass and feeder roots.

CALCULATION OF FLUXES

The net uptake of CO_2 by plants is determined by a balance of two processes: C assimilation, C_{ASS} (i.e., the gross photosynthetic C fixation) and autotrophic respiration, C_{GA} and C_{RA}. As assimilation and respiration show different seasonal courses and different temperature responses, we think it is more convenient to model these processes separately instead of simulating the NPP directly.

UPTAKE OF CO_2

The effective C assimilation rate, C_{ASS}, can be considered as a product function of a term dependent on light and canopy structure, a temperature dependent term, a soil water dependent term, and a term dependent on the atmospheric CO_2 concentration. The dependence on irradiance and leaf area index, LAI, which is correlated with the GC compartment is modelled taking into account the light attenuation in the canopy. For the temperature dependence an optimum curve is used characterised by the minimum, maximum, and optimum temperature of photosynthesis. The dependence of photosynthesis on water availability, represented by the soil water content SW, is assumed to follow a saturation curve which is zero at the permanent wilting point and approaches one for field capacity. The response to increasing atmospheric CO_2 is simulated by a Michaelis-Menten type function depending on the intra-leaf CO_2 concentration and the CO_2 compensation point.

RELEASE OF CO_2 - AUTOTROPHIC RESPIRATION

Autotrophic respiration, C_{GA} and C_{RA} respectively, is modelled similarly for both compartments, GC and RC, depending on the compartment size and an exponential function of the temperature corresponding to a constant Q_{10} value for each ecosystem type.

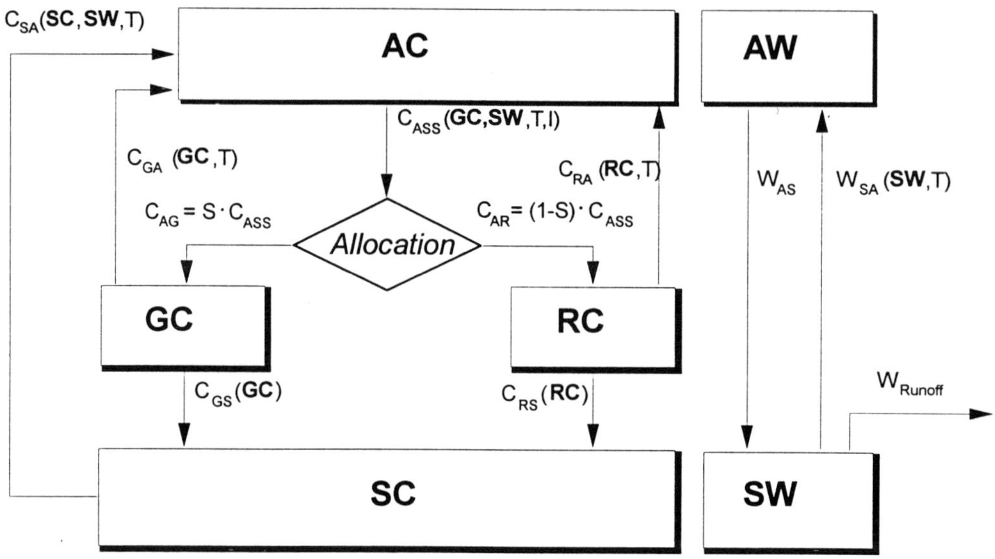

Fig. A1.Flow chart and model structure. Bold capital letters represent the reservoirs of C (second letter C) and water (second letter W) as follows: AC: atmospheric C; GC: C content of green biomass and feeder root biomass plus assimilate store; RC: C content of remaining biomass of biota (GC + RC = BC); SC: C content of litter, humus and dead biomass; AW: water in the atmosphere; SW: soil water in the rooting zone. The capital letters C and W represent C and water fluxes. The indices indicate sources and sinks of these fluxes. The functional dependence of the fluxes on the driving variables and pool sizes is given in parentheses (T: hourly air temperature; I: hourly photosynthetic active radiation (PAR)). W_{AS} is the daily precipitation, W_{SA} is daily actual evapotranspiration. The factor S represents the fraction of total assimilation C_{ASS} that is allocated to GC.

LITTER PRODUCTION

The litter production, C_{GS} and C_{RS}, is assumed to be proportional to the respective compartment size. For the GC compartment of the deciduous vegetation types an additional constant rate litter production occurs during the abscission phase.

RELEASE OF CO_2 - HETEROTROPHIC RESPIRATION

For the climate response of the decomposition of dead organic matter, C_{SA}, we extended the model of Fung *et al.* (1987) taking into consideration a linear dependence on compartment size and introducing a soil moisture factor analogous to the moisture dependence of photosynthesis.

WATER FLUXES

Due to the close relation between assimilation and transpiration, the actual evapotranspiration, W_{SA}, is calculated by the product of potential evapotranspiration after Thornthwaite (1948), W_{PET}, and the soil water dependent function as used in the calculation of assimilation.

W_{Runoff} comprises both surface runoff and drainage. It is taken as the surplus water when the soil water content reaches field capacity.

PATTERN AND CHANGE OF A BOREAL FOREST LANDSCAPE IN NORTHEASTERN CHINA

H. TIAN[1], H. XU[2] and C. A. S. HALL[1]

[1]*College of Environmental Science and Forestry, State University of New York, Syracuse, NY 13210, USA*
[2]*Forest Ecology Laboratory, Beijing Forestry University, Beijing , P. R. China*

Abstract. We have studied the spatial and temporal patterns of a boreal forest landscape in northeastern China using aerial photographs from 1955 (before logging) through 1980. The primary forest landscape is a complex mosaic of four tree species, and three age classes for each tree species, for a total of eighteen patch types. In addition, the complexity of the primary forest landscape shows great variation in the number, size and areal composition of patches. The forest landscape has been fragmented obviously by deforestation. There is an average increase in patch number from 291.6 to 438.0 patches per 10 000 ha, and an average decrease in patch size from 34.3 to 22.8 ha. Most of the logged areas are small in size, an average of 10.9 ha. The areal composition of different patch types has been changed greatly since the 1950s. The hardwood forest areas (especially) have increased significantly ($p = 0.05$); coniferous *Larix gmelinii* decreased significantly ($p = 0.05$), and *Pinus sylvestris* var. *mongolica* remained about the same. The area of old-age class in the dominant species, *Larix gmelinii*, has decreased significantly ($p = 0.05$). In addition, young and middle age classes of all tree species occupy a larger proportion of the area in 1980 than in 1955. Due to deforestation and recovery, therefore, many of these ecosystems are in early succession stages.

Keywords. BOREAL FOREST, CHINA, DEFORESTATION, FRAGMENTATION, LANDSCAPE PATTERN, PATCH.

1. Introduction

As one of the Earth's largest biomes, the boreal forests are likely to show the largest natural ecosystem responses or feedback to projected climate change (Botkin and Simpson, 1990; Kauppi *et al.*, 1992; Apps *et al.*, 1993) are especially susceptible to future land use change, and have extensive potential for biomass burning (IGBP, 1994). The boreal forests in northeastern China, a southern part of the Siberian boreal biome, are distributed along the borders between Russia and China. This region is important to our understanding of the possible shift of boreal forests due to projected climate change. Since the 1950s the landscape pattern has been changing rapidly because large scale, planned and non-planned timber cuttings were conducted. It is therefore important to understand the boreal forests in the context of both global climate and land use change. In addition, these ecosystems are also a good model for understanding the changes in other boreal regions.

The ecological phenomena at the landscape scale have interested geographers and ecologists for a long time. However, the scientific study of pattern and process at the landscape scale began just two decades ago (Forman and Godron, 1986). The interest is driven greatly by global change studies (IGBP, 1994). During the last decade, many studies have tried to understand change in landscape pattern and its impact on landscape function in both wilderness (e.g., Romme, 1982; Knight, 1987) and man-dominated landscapes (e.g., Turner, 1987; Zipperer *et al.*, 1990; Costanza *et al.*, 1990; Hall *et al.*, 1995). This research has shown that phenomena at the landscape level should be studied by theory and methodology appropriate to a landscape scale. However, this field is far from mature because of the diversity and complexity of

Water, Air and Soil Pollution **82**: 465–476, 1995.

landscapes on the Earth's surface. We believe that at this time theoretical integration is not useful because of a lack of data. A thorough case study is probably most useful and important.

The boreal forest landscape is a complex mosaic of patches of heterogeneous-aged tree species, fire-slash, marshes, rivers and bare soil. This mosaic is formed by the interaction of vegetation growth, physical environment and natural disturbance (Foster, 1985; Pickett and White, 1985; Tian and Pan, 1991). In addition, since the 1950s large areas have been affected by timber harvesting (Xu, 1990). Present forest patterns reveal the impact of both natural and anthropogenic disturbances. These factors increase the complexity of landscape patterns greatly, and increase the difficulty in identifying landscape elements.

Previous studies on the boreal forest of China focused on either forest management or only a limited area and perspective (Gu, 1986; Xu, 1990). For the purposes of forest resource management, the Chinese Department of Forestry sponsored the survey of the forest resource in both the 1950s and the 1980s. In the last decade, several studies of *Larix gmelinii* population dynamics and regeneration patterns have been done (Meng, 1988; Fan *et al.*, 1992; Xu and Du, 1993). Their results are essentially derived from observation and experiment at small sites or plots. However, there remains a lack of effective data to provide a detailed and comprehensive explanation of the impact of deforestation on forest landscape.

In 1986, we set up this study to obtain more detailed and systematic data over a large region. Our primary objective was to understand the patterns and processes of the boreal forest at the landscape scale. We used aerial photography extensively because it allowed us to get data from a large area for several points in time, and also provided an initial data base to compare human impacts. Moreover, the resolution of aerial photographs is well suited for the identification of major tree species and the age classes of forest patches in this landscape. First, we obtained a spatial data base from aerial photographs and a topological map. This spatial data base was then refined by a field survey and the use of some historic data from a forest survey. We then analysed the change of the boreal forest in terms of landscape components such as diversity, dominance and patch features.

2. The Study Area

The study area is located in the northeastern part of China, between 121°01'E and 122°49'E and between 51°13'N and 52°29'N (Figure 1). The region belongs to the Inner Mongolia Autonomous Region and is administered by the Mangui Forest Administration Bureau. To the east lies the main body of the Daxinganling Range and Heilongjiang province.

The climate of the region is characterised by long, dry and cold winters and short, rainy summers. The average temperature is -6 °C, with a maximum of 31 °C and a minimum of -48.3 °C. The average annual precipitation is 445 mm, concentrated in July and August. The average annual relative humidity is 73%. Early frost begins at the middle of September and late frost ends at the beginning of June. The frost-free period is only 80 to 100 days per year.

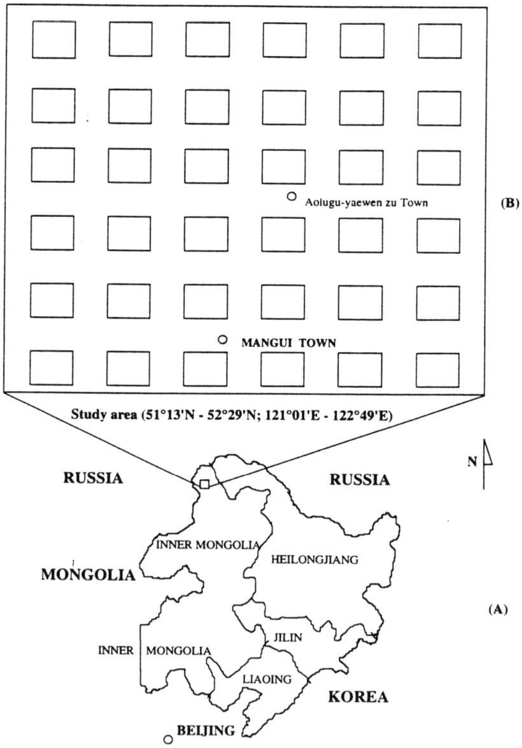

Fig. 1. (a) The location of study area in the Northeastern China, scale: 1:16 000 000; and (b) the distribution of samples in study area, scale: 1:290 000.

The soils of the region are characterised by brown coniferous forest soil developed from slope-accumulated parent material. The soil layer is only 20–30 cm deep. The surface organic matter is semi-dissolved. Its pH value is about 5.0. Under some types of the forest there exists light grey soil. In the river valleys and at the foot of mountains, there are wetland soils. Permafrost soil is found everywhere and can be as thick as 50–70 cm.

The region consists of moderate to low hills. Average elevation is less than 1000 m, with a maximum of 1297 m and a minimum of 520 m. Seventy percent of the slopes are less than 15°, while 27% are between 16° and 25°. Steep slopes (greater than 25°) constitute about 3% of the total area.

The major species are: *Larix gmelinii* (L.) Rupr., *Betula platyphylla* (L.) Suks., *Pinus sylvestris* var. *mongolica* (L.) Litv. and *Populus davidinia* (L.) Dode. The vegetation is characterised by the bright coniferous composed mainly of *Larix gmelinii*. *Larix gmelinii* is found in various topographical areas, while *Pinus sylvestris* var. *mongolica* is found mainly on dry south facing slopes. These two species are found either separately or mixed. The understory plants include *Rhododendron dauricum*, *Ledum palustre varangustrum*, and *Vaccinium vitis-idaea*. In addition, a number of lichens, mosses, and herbaceous plants are found. Pines are found at high altitude regions.

TABLE I

Classification of patch types in the boreal forest landscape.

Patch type[a]	Age	Patch type[a]	Age or description
O-LG	> 100	O-P	> 40
M-LG	50–100	M-P	20–40
Y-LG	0–50	Y-P	0–20
O-PSVM	> 100	Fire-slash	Newly burned patches
M-PSVM	50–100	Cutting-blank	Newly cut patches
Y-PSVM	0–50	Marsh	All marsh area
O-BP	> 40	River	All water area
M-BP	20–40	Soil bare	No forest area
Y-BP	0–20	Human	All constructions

[a] O, M, Y: Old, Middle, Young age classes; LG: *Larix gmelinii*; PSVM: *Pinus sylvestris* var. *mongolica*; BP: *Betula platyphylla*; P: *Populus davidinia*.

Fire, the major factor of natural disturbance, occurs quite frequently. In addition to fire disturbance, another important natural disturbance, wind fall, does not affect as large an area.

Since large scale logging was introduced to the region, human disturbance has become the major, if not most important, element of disturbance. Logging, construction of highways, railroads, and human settlements have changed the landscape of the region significantly. Changes in the patterns of the forest landscape necessarily also change the way in which the forest functions.

3. Methods

Basically, we classified types of patches, and then identified these patches through aerial photograph interpretation and ground investigation. Using the data of patch types, number, size and areal composition, we give an analysis of the pattern of patchiness or mosaic in the primary landscape and its change over time.

3.1. Patch Classification

In accordance with the structural characteristics of the boreal forest, we chose three characteristics to classify different patches, (1) composition of tree species, (2) age classes of trees, and (3) historical background of the community. The classifications of forest landscape patches are listed in Table I. The age structure is far more complicated than the species composition, and more difficult to distinguish. To make the study easier, we classified all trees into age classes, i.e., old trees, middle aged trees and young trees.

3.2. Sampling and Distribution

We used aerial photographs to represent two points in time, one in 1955 (taken by the remote sensing agency of the former Soviet Union), the other in 1980 (taken by the

Forest Survey Group of Heilongjiang Province). Both missions completely covered the area of interest. All aerial photographs were scaled to match 1:25 000 topographic maps.

We sampled 36 pairs of photographs for each time period. The area of each sample is 3.8 km x 3.8 km, with sampling spots located evenly within the study area. The pattern of sampling is shown in Figure 1. The total sampled area is about 30% of the total study area.

3.3. Aerial Photography Interpretation and Ground Investigation

We took each pair of photographs and examined them with various tools, principally a stereoscope, parallax bar, crown closure reference, and dot-grid square. We measured and identified various patches directly from the aerial photographs based on the following major indices: (1) tree species; (2) tree height (m); (3) crown closure (0–100%); (4) patch size (ha); (5) patch shape; (6) elevation (m); (7) water.

We designed a field study to couple the aerial photography interpretation with ground truthing investigation to get more detailed information on the sampled areas. Major field studies included: (1) locating the boundary of the study area and sampling spots; (2) identifying tree species and determining their features on aerial photographs, and identifying the features of age classes for each kind of tree; (3) delineating the boundary of various forest patches; (4) recording the main features of topography; (5) identifying the burned areas.

4. Results and Analysis

4.1. Pattern of the Primary Forest Landscape

4.1.1. Areal composition of patches

When we compared the relative area of the four tree species, we found that *Larix gmelinii* patch is the dominant component occupying 58.33% of the total landscape area. *Larix gmelinii* is thus the matrix of the boreal landscape. Young and old mature *Larix gmelinii*, fire-slash, and marsh areas are the four major components in the landscape (Table II). Together they cover 76.66% of the total landscape area. The species covering less area, *Betula platyphylla*, 7.27%, *Pinus sylvestris* var. *mongolica*, 4.42%, *Populus davidinia*, 1.7%, and relative area of each patch type clearly reveals an uneven distribution of patch areas in the landscape, ranging from 20 ha/10 000 ha to 3386 ha/10 000 ha (Table II).

4.1.2. Patch numbers

Table II shows a significant difference in the number (or density) and proportion of each patch type, and an uneven distribution pattern. Old *Larix gmelinii* patches are the most numerous at 77.12 patches/10 000 ha. Other numerous patches are: (1) young *Larix gmelinii* patches, 58.87 patches/10 000 ha; (2) middle age class *Larix gmelinii* patches, 24.53 patch/10 000 ha, young *Betula platyphylla*, 26.88 patches/10 000 ha and fire-slash, 37.67 patches/10 000 ha; (3) old *Pinus sylvestris* var. *mongolica* patches,

12.17 patches/10 000 ha and marsh, 16.29 patches/10 000 ha. All other patches have less than 10 patches/10 000 ha.

The number of *Larix gmelinii* patches, 160.52 patches/10 000 ha, is significantly greater than that of the other three species, *Pinus sylvestris var. mongolica* (16.49 patches/10 000 ha), *Betula platyphylla* (38.46 patches/10 000 ha), and *Populus* (12.37 patches/10 000 ha).

In addition, there is an uneven distribution of patch numbers among age classes of different tree species. The order of patch numbers for the three age classes is as follows: in *Larix gmelinii* old > young > middle; in *Pinus sylvestris* var. *mongolica*, old > middle and old > young; in *Betula platyphylla,* young > middle > old.

TABLE II

Areal composition, number and size of different patch types and their proportions, and size variation in the primary boreal forest landscape, 1955 (per unit of 10 000 ha).

Patch type	Area (ha)	Areal Proportion (%)	Number (patch)	Number proportion (%)	Size (ha/patch)	Size variation (SD)
O-LG	3386 AD	33.86	77.12 A	26.45	42.91 B	302.78
M-LG	937 D	9.37	24.53 D	8.41	38.18 B	219.28
Y-LG	1510 B	15.10	58.87 B	20.19	25.66 C	227.05
O-PSVM	311 EF	3.11	12.17 EF	4.17	25.55 C	124.07
M-PSVM	73 G	0.73	2.16 G	0.74	34.04 BC	73.67
Y-PSVM	58 G	0.58	2.16 G	0.74	26.97 C	54.60
O-BP	51 G	0.51	2.55 G	13.19	19.86 C	84.09
M-BP	198 FG	1.98	9.03 FG	3.10	21.88 C	110.47
Y-BP	478 E	4.78	26.88 D	9.22	17.78 C	268.00
O-P	106 G	1.06	4.32 G	4.24	24.50 C	202.00
M-P	24 G	0.24	1.77 G	1.48	12.92 D	69.58
Y-P	40 G	0.40	6.28 G	2.15	6.41 DE	88.14
Fire-slash	1537 B	15.37	37.67 C	12.92	48.80 B	277.67
Marsh	1233 C	12.33	16.29 E	5.59	75.68 A	119.24
River	39 G	0.39	1.96 G	0.67	20.09 C	130.81
Soil bare	20	0.20	7.85 FG	2.69	2.26 E	41.67

* A different letter means significant difference (p = 0.01). The same letter means no significant difference between different patch types. The symbols of patch types are the same as for Table I.

4.1.3. Patch sizes

The average patch size is 34.30 ha, but the variability in patch size is very large, ranging from 0.5 to 800 ha. The marsh patches have the largest average size, 75.68 ha. This is because of their distribution along riverine and lowland areas where there are very large continuous habitats suited for marsh formation. The smallest patch type is bare soil, with a mean size of 2.26 ha.

Forest fire is an important natural factor in the boreal forest landscape. It not only drives the complexity of tree age classes but also the variability of patch size among age class patches. The average size of burned patches is 40.8 ha, but its variance is very large. The smallest burned patch is 0.96 ha, the largest one 680.71 ha.

In general, the patch sizes of different tree species age classes are less than or equal to the size of burned patches. However, the size and variability also depends on the resistance of tree species to forest fire, the ability of tree species to recover from previous forest fire disturbance and the adaptation of tree species to the environment. For example, there is a significant difference in patch size for each tree species (Table II), as follows: *Larix gmelinii* > *Pinus sylvestris* var. *mongolica* > *Betula platyphylla* > *Populus davidinia*. This is a result of the different responses of various tree species to fire disturbance and the varying physical environment. *Larix gmelinii* has the largest average area not only because of its ability to spread seed and recover after fire disturbance but also because of its resistance to forest fire (especially old *Larix gmelinii*). There is a difference in the size between the various age classes of the same species. In *Larix gmelinii,* the older the age of the patch, the larger the average size of the patch. The average size of young, middle and old age *Larix gmelinii* patches is 25.66, 38.18, 43.91 ha, respectively. The percentage deviation of size in young, middle and old age class *Larix gmelinii* patches is 227.05%, 219.28%, 302.78%. The major reason that old age *Larix gmelinii* patches have the largest size and the highest percentage deviation is their resistance to forest fire which allows them to maintain a large area despite both small and large fire disturbances. This characteristic also enables them to retain small remnant patches after a huge fire disturbance. In addition, as plant community develops (succession), *Larix gmelinii* replaces other plant species, (e.g., *Betula platyphylla*), and extends into many areas where it grows into large and old trees if the frequency of forest fires is low. For these reasons there is large variability in the patch sizes of this species, from 0.5 ha to 798.20 ha. *Pinus sylvestris* var. *mongolica* shows no significant difference in size among the three age classes. As they mature the variance increases. *Betula platyphylla* also shows no difference in size among the three age classes. Their variance, however, decreases with age, which is contrary to the two tree species above. In *Populus davidinia*, the size and variance of old age classes (24.5 ha per patch; 49.49) is significantly more than middle (12.92 ha per patch; 8.98) and young age classes (6.41 ha per patch; 5.65) *Populus davidinia.*

4.2. CHANGE OF LANDSCAPE PATTERN

4.2.1. Change in area composition of patches

Table III gives the change in area composition of different patch types from 1955 to 1980. Among the 18 types, cutting-blank and human settlements patches are the only new patch types. Their areas are 739 and 11 ha/10 000 ha, respectively. The area of old

Larix gmelinii is still the highest (2048 ha/10 000 ha), but its area in 1980 is significantly less than that of 1955, decreasing by 1338 ha/10 000 ha. This decrease is clearly a result of timber harvesting. In 1955, the area of fire-slash is greater than all other types except old *Larix gmelinii.* But in 1980, the area of fire-slash is only 82 ha/10 000 ha. This is significantly less than that of 1955, a decrease of 1455 ha/10 000 ha. In contrast to old *Larix gmelinii*, the decrease in fire-slash patch area is a result of the regeneration of diverse tree species.

There are 7 types of patches, i.e., middle age class *Larix gmelinii*, old, middle and young *Betula platyphylla*, middle and young *Populus davidinia*, and bare soil patches which show 1980 areas significantly higher than those of 1955. Many young *Larix gmelinii* are translated into middle *Larix gmelinii*, and many middle and young *Betula platyphylla* are translated into high age class *Betula platyphylla.* The increase in area of young *Betula platyphylla* can be attributed to this species' ability to invade fire-slash and cutting-blank area and quickly regenerate. As a result, three age classes of *Betula platyphylla* greatly increase in the landscape. Though the area of young class *Larix gmelinii*, young, middle and old class *Pinus sylvestris* var. *mongolica*, and old class *Populus* increase, there is no significant difference in their areas between 1955 and 1980. The area of different age class patches of *Pinus sylvestris* var. *mongolica*

TABLE III

Areal composition, number and size of different patch types and their proportions, and size variation in the deforested boreal forest landscape, in 1980 (per unit of 10 000 ha).

Patch type[a]	Area	Change[b]	Number (patches)	Change[b] (patches)	Average size (ha)	Change[b] (ha per patch)
O-LG	2048	-1338*	83.39	6.27	30.56	-13.35*
M-LG	1432	495*	63.58	39.05*	33.41	-4.77
Y-LG	1573	63	68.48	9.61	15.97	-9.69*
O-PSVM	315	4	16.29	4.12	23.35	-2.20
M-PSVM	95	2	2.75	0.59	30.35	-3.69
Y-PSVM	58	27	1.37	-0.79	27.65	0.68
O-BP	386	335*	17.86	15.31**	21.64	1.78
M-BP	619	421*	27.86	18.83**	22.70	0.32
Y-BP	1011	533*	68.87	41.99**	11.78	-6.00*
O-P	157	51	10.79	6.47**	15.52	-8.98**
M-P	107	83*	3.14	1.37*	15.03	2.11
Y-P	61	21	3.92	-2.36	5.60	0.81
Fire-slash	82	-1455*	5.69	-31.98*	14.34	-26.46**
Cut-blank	739	739*	24.53	125.00	10.92	
Marsh	1163	-70	16.09	-0.20	72.50	-3.43
River	39	0	2.35	0.39	19.46	-3.63
Soil bare	77	57*	19.43	11.58**	3.96	1.70**
Human settlements	11	11	1.57	8.00	49.01	

[a] Symbols are as in Table I.
[b] Significance level: * ($p = 0.05$), ** ($p = 0.01$)

remains in relatively stable proportion because of a policy that limits the cutting of this tree species in China. Marsh and river are relatively stable patches. A few area of marsh decrease, and river area does not change. Although the area of bare soil is very low, the increase in its area is significant ($p = 0.05$), because it is a result of deforestation activities. It may greatly increase the fragmentation of the landscape.

If we consider only tree species, the area of *Larix gmelinii* patches decreased by 780 ha/10 000 ha from 1955 to 1980. *Pinus sylvestris* var. *mongolica* remains about the same area. The area of *Betula platyphylla* patches significantly increase by 1289 ha/10 000 ha, for a total area of 2016 ha/10 000 ha. The area of *Populus davidinia*also significantly increases by 155 ha/10 000 ha. In summary, since 1955, the areal components of the different patch types in the boreal forest landscape have changed greatly. There is a decrease in the area of coniferous forest, an increase in the area of hardwood forest, a decrease in old mature *Larix gmelinii* and fire-slash areas, and the creation of two new patch types.

4.2.2. Change in patch number

In contrast to patch size, the number of patches increased by 146.38 patches/10 000 ha, from 291.6 patches/10 000 ha in the primary landscape to 438.0 patches/10 000 ha. We also found by comparing deforested with primary landscapes that two patch types increase, cutting and human settlements. The two new types of patch have cut up the landscape, which results directly in the increase in the number of patches. Total patch number has increased significantly since 1955. However, different patch types show very different pattern in the change of patch number. From Table III, four patch types, young *Pinus sylvestris* var. *mongolica*, young *Populus davidinia*, fire-slash and marsh, decrease in number. Fire-slash patches decreases the most, by 31.98 patches/10 000 ha (84.89%). This is because most fire-slash patches are replaced by patches of other tree species. Except for the four above, other patches tend to increase in number after deforestation. For patch types such as middle class *Larix gmelinii*; old, middle and young *Betula platyphylla*, old and middle *Populus davidinia*, cutting-blank, bare soil and human settlements there is a significant increase in the number of patch ($p = 0.01$). Young *Betula platyphylla* increase the most, by 41.99 patches/10 000 ha. Old *Betula platyphylla* demonstrate the highest percentage increase from 1955 to 1980. The increase in the number of old Betula platyphylla is a result of the transfer of many young and middle *Betula platyphylla* patches from 1955 to 1980. Many young Betula platyphylla patches in 1980 are a result of cutting-blank patches and niche partitioning formed by competition with other species when regenerating on the fire-slash. Only middle-aged *Larix gmelinii* significantly ($p = 0.05$) increase in the number of patches, by 39.05 patches/10 000 ha (159.19%). This is because of low patch number in 1955 and the transfer of many young *Larix gmelinii* patches. Though the patch numbers of young age *Larix gmelinii* increase by only 9.61 patches/10 000 ha, most patches in 1980 are new patches derived from regeneration or succession after forest fire disturbance. Old *Larix gmelinii* increases by only 6.27 patches/10 000 ha (8.13%). However, in 1980, the number of cutting-blanks (a result of cutting old *Larix gmelinii*) is 125 patches/10 000 ha. Cutting thus greatly increases the porosity of the landscape and has little role in decreasing the connectivity of old *Larix gmelinii*. Though *Pinus sylvestris* var. *mongolica* and *Populus davidinia* are found on completely different niches, the number of their patches tend to show similar changes. Old and middle class in these two patch types increase from 1955 to 1980.

4.2.3. Change in patch size

From 1955 to 1980, the average size of patches decreased by 11.49 ha, from 34.3 ha per patch in primary landscapes to 22.83 ha per patch in deforested landscapes. Table III shows that fire-slash decrease the most in average size, by 26.46 ha per patch. In addition, patch size of old and young *Larix gmelinii*, young *Betula platyphylla* and old *Populus davidinia* decrease in size. Only bare soil, cutting-blank and human settlements increase greatly in size. Other patches do not show a significant decrease or increase in patch size from 1955 to 1980. Fire-slash decreases in size because most burned areas have been replaced over this period by the regeneration of new forests, and because there were only a few events of forest fire during the period. Old *Larix gmelinii* shows a significant decrease in size ($p = 0.05$) in direct response to cutting. The patch size in young *Larix gmelinii* and young *Betula* also decreases. This is because of the small size of clear-cuts and the niche partitioning during regeneration following fire disturbance. The decrease in size of old *Populus davidinia* patches may be caused by the building of railways, roadways and timber transport roads. The patch size of bare soil also increase due to deforestation. There are two forces that cause patch size to decrease, i.e., niche partitioning or succession, and human activities (especially forest cutting and the building of roads).

5. Discussion and Conclusions

The primary boreal forest landscape is a mosaic dominated by *Larix gmelinii* with other tree species(*Pinus sylvestris* var. *mongolica*, *Betula platyphylla*, and *Populus davidinia*), and fire-slash, marsh, river and bare soil areas. Fire disturbance and tree species regeneration (or succession) result in a complex age structure (Tian and Pan, 1991; Fan *et al.*, 1992). The landscape is thus heterogeneous in space and dynamic in time. It is formed by the interaction of the biological processes of the trees, the natural disturbance regime, (especially fire) and the physical environment (Urban *et al.*, 1987; Hall *et al.*, 1992).

The number and area of different patch types show great differences. The dominant component, old *Larix gmelinii*, occupies the largest area and is the greatest in number. *Larix gmelinii* plays an important role in maintaining the stability and sustainability of the boreal forest landscape because of its resistance to fire disturbance and facilitation of regeneration after natural disturbance. How to manage and use old *Larix gmelinii* is a key to ensure the sustainability of the boreal forest landscape.

As in other forest landscapes of the world, the landscape fragmentation is obvious after deforestation (Burgess, 1988; Skole and Tucker, 1993). This is seen in the increase in patch numbers and decrease in patch sizes. Forest cutting not only creates many small logged areas ("holes"), but also cuts other patches into several smaller patches. The regeneration strategy of tree species in the large area play an important role in the spatial partitioning of patches. It results in decreasing patch size and increasing patch number.

The fragmentation of a landscape directly influences the function of ecosystems and the distribution of species (Forman and Godron, 1986). Some recent studies have shown that edges created by forest fragmentation alter physical environments, such as microclimate conditions (Chen *et al.*, 1993; Xu and Du, 1993), and affect the rates of the growth and mortality in some tree species (Chen *et al.*, 1992; Xu and Du, 1993). Species that require a large specific patch for survival will decrease. Conversely, species that require an edge environment will increase. The loss of large animals such as the tiger, is a response to forest fragmentation. In addition, when *Betula platyphlla*, a pioneer tree species, is added to large old *Larix gmelinii* patches, the heterogeneity of the landscape increases, but *Betula platyphylla* is an easily burned species. In this case the increase in heterogeneity thus speeds the spread of forest fire. The decrease in the patch size of old *Larix gmelinii* and the interspersion of *Betula platyphylla* are thus two major reasons for the large areas of fire-slash when a forest fire follows deforestation. We can say then that the conservation of some large areas of old *Larix gmelinii* will not only preserve some large animals but also maintain the stability of the boreal forest landscape. In traditional forestry, however, the size, shape and distribution of a clear-cut is only considered as it relates to the regeneration of a tree species (Xu, 1990). Most logged area are thus small in size, in this case an average of 10.92 ha per patch. Questions about the optimal size, shape and distribution of clear-cuts have been debated for a long time. Landscape management for multiple purposes should move beyond the tradition (Franklin and Forman, 1987). However the answer to these questions greatly depends on our understanding of the structure and function of forest landscape (Burgess and Sharp, 1981; Franklin and Forman, 1987).

The areal composition of different patch types has changed greatly since 1955. The hardwood forest areas especially have increased significantly. Coniferous *Larix gmelinii* decreased significantly, while *Pinus sylvestris* var. *mongolica* remained about same. Except for old mature *Larix gmelinii,* the area of old age classes of other species increased somewhat. The most obvious change is that the area of young and middle-age classes in all tree species in 1980 occupy larger proportions than in 1955. This reflects many ecosystems in early successional stages due to deforestation.

Pioneer tree species in early successional stages, such as *Betula platyphylla*, grow very quickly. In addition, the increase in middle-age classes leads to an increase in total carbon uptake (Kurz and Apps, 1993). Plants in early succession stages have high absorption of CO_2 because of the processes of plant community succession (Bormann and Likens, 1979; Tian and Qi, 1990). We define this as the "early succession effect" This does not mean, however, that there must be a net carbon storage in early successional stages, because both ecosystem production and respiration increase during these stages (Waring and Schlesinger, 1985; Borman and Likens, 1979). We think, however, that the early succession effect is one of the factors that have enhanced biospheric metabolism since the 1970s (Hall *et al.*, 1993). This also implies that through conservation of the timber in logged forests, it is possible to manage forests in a way to increase their role as a sink for atmospheric carbon.

Acknowledgments

This study was supported by the State Education Commission of Peoples Republic of China and the United States Department of Energy, under contract No. DEFG0290 ER61080. We wish to thank Shiping Xiong of Bejing Forestry University, for his assistance in data collection and processing. We especially thank Dr. Ye Qi for in-depth discussion on these topics, Dr. Robert L. Burgess and two anonymous reviewers for their helpful comments, and Jamie Cass for his helpful editing.

References

Apps, M. J., Kurz, W. A., Luxmoore, R. J., Nilsson, L. O., Sedjo, R. A., Schmidt, R., Simpson, L. G. and Vinson, T. S.: 1993, *Water, Air, and Soil Pollut.* **70**, 39-54.

Bormann, F. H. and Likens, G. E.: 1979, *Pattern and Process in a Forested Ecosystem*, Springer-Verlag, New York, 253 pp.

Botkin, D. B. and Simpson, L. G.: 1990, *Biogeochemistry* **9**, 161-174.

Burgess, R. L.: 1988, *Can. J. Bot.* **66**, 2687-2690.

Burgess, R. L. and Sharpe, D. M. (eds): 1981, *Forest Island Dynamics in Man-Dominated Landscape*, Springer-Verlag, New York, USA, 310 pp.

Chen J., Franklin, J. F. and Spies, T. A.: 1992, *Ecol. Appl.* **2(4)**, 387-396.

Chen J., Franklin, J. F. and Spies, T. A.: 1993, *Agric. For. Meteorol.* **63**, 219-237.

Costanza, R., Sklar, F. H. and White, M. L.: 1990, *Bioscience* **40(2)**, 91-107.

Fan, Z., Xu, H. and Yu, R.: 1992, *Scientia Silvae Sinicae* **28(1)**, 2-11.

Forman, R. T. T. and Godron, M.: 1986, *Landscape Ecology*, Wiley, New York, USA, 619 pp.

Foster, D. R.: 1985, *J. Ecology*, **73**, 517-534.

Franklin J. F. and Forman, R. T. T.: 1987, *Landscape Ecology*. **1**, 5-18.

Gu, Y.: 1986, *The Forest Resources of Daxinganling Range*, Chinese Forestry Press, Beijing, P. R. China.

Hall, C. A. S., Stanford, J. A. and Hauer, F. R.: 1992, *OIKOS* **65**, 377-390.

Hall, C. A. S., Tian, H. and Qi, Y.: 1993, *World Resour. Rev.* **5(2)**, 207-213.

Hall, C. A. S., Tian, H., Qi, Y., Pontius, G., Cornel, J. and Uhlig, J.: 1995, *CDIAC Publications*, Oak Ridge Natl. Lab., Oak Ridge, USA, pp. 1-4.

IGBP: 1994, *IGBP in Action: Work Plan 1994 - 1998*, IGBP Report No. **28**, Stockholm, Sweden, 151 pp.

Kauppi, P. E., Mielikainen, K. and Kuusela, K.: 1992, *Science* **256**, 70-74.

Knight, D. A.: 1987. *In:* Turner, M. G. (ed.), *Landscape Heterogeneity and Disturbance*, Springer-Verlag, New York, USA, pp. 59-83.

Kurz, W. A. and Apps, M. J.: 1993, *Water, Air, and Soil Pollut.* **70**, 163-176.

Meng, X.: 1988, *Acta Beijing Forestry University* **10(1)**, 40-47.

Pickett, S. T. A. and White, P. S.: 1985, *The Ecology of Natural Disturbance and Patch Dynamics*. Academic Press, New York, USA, 472 pp.

Romme, W. H.: 1982, *Ecol. Monogr.* **52**, 199-221.

Skole, D. L. and Tucker, C. J.: 1993, *Science* **260**, 1905-1910.

Tian, H. and Pan, S.: 1991, *In:* Xiao, D. (ed.) *Landscape Ecology: Theory, Methods and Application*, Chinese Forestry Press, Beijing, P. R. China, pp. 47-51.

Tian, H. and Qi, Y.: 1990, *In:* Shijun Ma (ed.), *Advances in Modern Ecology*, Scientific Press, Beijing, P. R. China, pp. 90-100.

Turner, M. G.: 1987, *Landscape Ecology* **1**, 29-36.

Turner, M. G.: 1989, *Annu. Rev. Ecol. Syst.* **20**, 171-191.

Urban, D. L., O'Neill, R. V. and Shugart, H. H.: 1987, *Bioscience* **37(2)**, 119-127.

Waring, R. H. and Schlesinger, W. H.: 1985, *Forest Ecosystems: Concepts and Management*, Academic Press, Orlando, 340 pp.

Xu, H.: 1990, *Acta Beijing Forestry University* **12(supp. 3)**, 1-8.

Xu, H. and Du, Y.: 1993, *Scientia Silvae Sinicae* **29(4)**, 298-306.

Zipperer, W. C., Burgess, R. L. and Nyland, R.: 1990, *For. Ecol. Man.* **36**, 103-117.

DESCRIPTION OF THE CANADIAN REGIONAL CLIMATE MODEL

D. CAYA[1], R. LAPRISE[1], M. GIGUÈRE[1], G. BERGERON[1],
J. P. BLANCHET[1], B. J. STOCKS[2], G. J. BOER[3] and N. A. McFARLANE[3]

[1]*Cooperative Centre for Research in Mesometeorology and Université du Québec à Montréal, CP 8888, Succursale Centre-Ville, Montréal Québec, H3C 3P8, Canada*
[2]*Forestry Canada, Ontario Region, Sault-Sainte-Marie, Ontario P6A 5M7, Canada*
[3]*Canadian Centre for Climate Modelling and Analysis, University of Victoria, British Columbia, Canada*

Abstract. Simulation of regional climate by limited-area model coupled with global low-resolution model is becoming a standard approach to achieve high-resolution climate projections at a computationally affordable cost. A regional climate model (RCM) based on a state-of-the-art numerical formulation is under development at the Université du Québec à Montréal. A brief description of this Canadian RCM (CRCM) is given. The characteristics of its dynamical formulation and the processes involved in its physical parameterisation are described. Results of a short simulation show that CRCM develops fine-scale features of climatological interest.

Keywords. MODEL, SIMULATION, REGIONAL, CLIMATE, MODELLING, GCM, RCM, LAM.

1. Introduction

Global general circulation models (GCMs) are the main tools available today for the simulation of climate. The very long integrations required for climate studies, however, impose a strong constraint on the maximum spatial resolution at which GCMs can be run. Half a day of supercomputer CPU time is needed for a one-year integration of a typical GCM with spatial resolution of the order of 300 km. This limited spatial resolution is currently identified as one of the major limitations in the ability of GCMs to properly simulate the hydrological cycle (IPCC, 1992). Because computing time required to run a GCM increases approximately with the third power of the horizontal grid resolution (fourth power if the concomitant vertical resolution is included), running a GCM at a resolution of the order of 30 km would require roughly two days of supercomputer computation per simulated day (Giorgi and Mearns, 1991). This spatial resolution would be required for a good representation of hydrology related processes. Since the computing power necessary for global fine-scale climate modelling will not be available in a near future, alternative approaches have to be explored.

Two major constraints are related to climate simulation: long integrations of the order of a few decades and global coverage of the Earth. One approach to refine the coarse resolution of GCMs consists in of limiting the spatial coverage of the high-resolution domain while maintaining global low-resolution coverage. In this approach, a high-resolution limited-area model (LAM) is nested in a low-resolution global model over the region of interest. Large-scale information from the global low-resolution model is transferred to the LAM by forcing its lateral boundaries with the values of the low-resolution global model. In this one-way nesting technique, the LAM receives its lateral boundary values from the outputs of the global model. This approach offers the

Water, Air and Soil Pollution **82**: 477–482, 1995.

possibility of increasing the horizontal resolution of climate models by nearly an order of magnitude while keeping the computational cost at an affordable level because the fine grid is of limited area. Regional climate simulation for limited area is then possible. The potential of regional climate modelling by nesting a high-resolution regional model in a lower resolution general circulation model (GCM) has been demonstrated in several papers by Giorgi (e.g., Giorgi *et al.*, 1993).

This paper is intended to give a brief description of a regional climate model (called the Canadian RCM or CRCM) currently under development at the Université du Québec à Montréal (UQAM). Preliminary results of a two-month integration are also presented to illustrate the ability of the model to represent regional-scale characteristics of some climatological fields.

2. Model Description

CRCM results from coupling the complete subgrid-scale parameterisation package of the second-generation Canadian GCM (GCMII, McFarlane *et al.*, 1992) with a very efficient dynamical model. This model, based on semi-Lagrangian and semi-implicit integration scheme, is the Cooperative Centre for Research in Mesometeorology's Mesoscale Compressible Community (CCRM-MC2) model (Bergeron *et al.*, 1994; Tanguay *et al.*, 1990). A one-way nesting procedure is used to drive the CRCM with GCMII data, i.e. CRCM gets information from GCMII through its lateral boundary, but it does not influence GCMII. Using GCMII as driving model for the CRCM permits both driving and driven models to share the same subgrid-scale parameterisation package, which facilitates the transfer of information between the two models. Another important advantage of using GCMII subgrid parameterisation is that all geophysical data files (soil type, vegetation, bare soil albedo, etc.) needed for physics parameterisation are global, and configuring the model for a specific area of the globe is thus simpler.

The current configuration of the model consists of a grid of 120 by 120 points covering a $(5400\ \text{km})^2$ area centred over the state of Vermont. The various fields are projected on a polar-stereographic grid with a resolution of 45 km at 60°N. In the vertical, 10 unequally spaced terrain-following Gal-Chen coordinate (Gal-Chen and Somerville, 1975) levels, from the ground up to 30 km, are used. Simulations with a 30-level version are also in progress. The following fields at the boundaries are driven by GCMII: wind components, temperature, pressure and water vapour. They are interpolated from GCMII archived data onto the CRCM lateral boundaries. Initial conditions are also taken from GCMII archived data. A sponge zone of ten gridpoints on the boundary of the domain is used to blend CRCM fields with the information received from GCMII. The nesting method used is inspired from that of Davies (1976) and refined by Robert and Yakimiw (1986) and Yakimiw and Robert (1990). Over a distance of ten gridpoints from the lateral boundaries, the variables of the regional model are gently and gradually blended with the those of the driving (global) model. By the time the CRCM lateral boundaries are reached, the values of the driving model variables are imposed. Throughout the rest of the grid (free zone), the variables of the regional model are not affected explicitly by those of the global model.

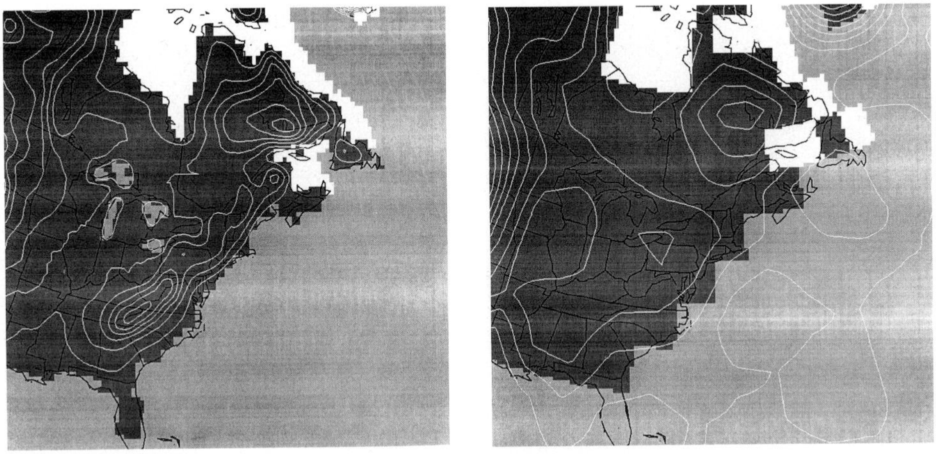

(a) (b)

Fig. 1. January land-sea-ice mask and topography of the 45 km CRCM (a) and T32 GCMII (b) over North-America. Land gridpoints appear in dark grey, open water in light grey and sea ice in white. Topographic contours are drawn at 100 m intervals. The spectral GCM is projected on a Gaussian latitude–longitude grid with a resolution of 3.75° × 3.75° and interpolated on the CRCM polar-stereographic grid for display (the distance between gridpoints on the Gaussian grid is approximately 450 km).

The dynamics of the CRCM are based on the complete non-hydrostatic Euler equations. Therefore, the dynamical formulation does not restrict spatial scales at which the model can be run. In fact, this model is used also at the millimetre scale for research on supersonic flow, at a few hundred metres for moist convection problems and at synoptic scales for weather forecasting. The integration of the complete Euler equations is made affordable by the efficiency of the three dimensional semi-Lagrangian semi-implicit integration scheme. The efficiency of the integration scheme also permits a 20-minute timestep at 45-km resolution, which is a very long timestep for non-hydrostatic numerical model at this resolution. For comparison, GCMII uses the same timestep with a transform Gaussian grid of approximately 450-km resolution and an Eulerian explicit model with 45-km resolution usually runs with a 90-second timestep.

The physical parameterisation package (imported from GCMII) takes into account the following processes: vertical turbulent fluxes of momentum, mountain-wave drag, radiation absorption and emission (solar and terrestrial) by the atmosphere and by the surface, release (absorption) of latent heat when condensation (evaporation) occurs, precipitation (solid and liquid), convection (dry and moist), variation of the surface albedo (defined for the two spectral bands and function of ground cover type, soil moisture, snow cover and snow age), cloud cover (evaluated from the prognostic water vapour and temperature fields), surface energy budget, soil moisture regime, vegetation and soil characteristics. The physical package of GCMII (and therefore CRCM) also includes a ocean mixed-layer model and a thermodynamic sea-ice model. For the experiment described in this paper, both of these models were turned off and climatological values of sea-surface temperature (SST) and ice-coverage were used.

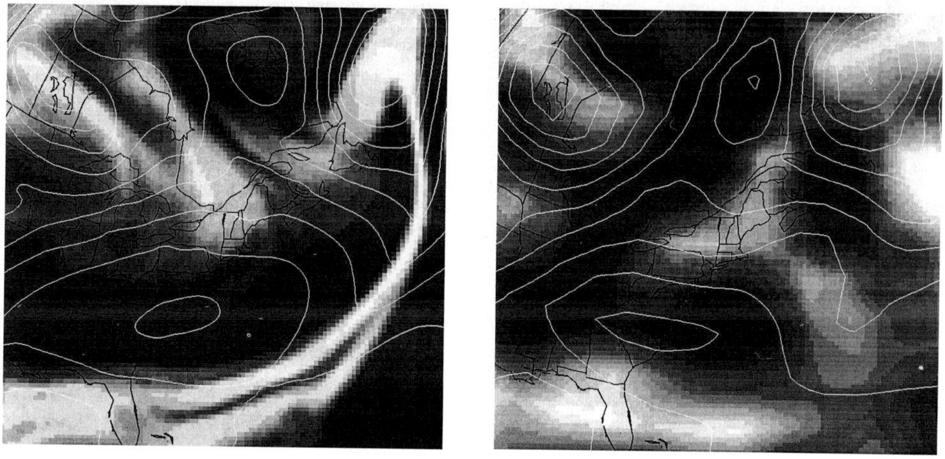

(a) (b)

Fig. 2. Relative humidity at 875 hPa and mean sea level (MSL) pressure for 14 January at 12Z for CRCM (a) and GCMII (b). Humidity field is presented in grey scale colour with dry area in black and wet area in white. MSL pressure contours are drawn at 4 hPa intervals. Minimum pressure in the cyclone over the East coast of Labrador are 985 hPa for CRCM and 987 hPa for GCMII.

3. Simulated Data and Discussion

The CRCM was integrated for a two-month period with the various model fields archived at six-hour intervals. The archived data for this experiment are time series of the following three dimensional fields: air temperature, pressure, water vapour, winds, clouds, and heating rates (solar and terrestrial). The following bi-dimensional surface fields are also archived: surface albedo, downward surface terrestrial radiation flux, solar flux absorbed at the surface, ground temperature, sensible heat flux, latent heat flux, precipitation, surface pressure, snow depth, liquid and frozen soil water contents.

Initial and CRCM time-dependent lateral boundary conditions were taken from archived files of a GCMII simulation that was initialised from January FGGE observed data. The CRCM simulation began the first day of the twelfth month (1 December of the first year) of a GCMII simulation. Sea surface temperature and ice cover are prescribed from climatological data. Average values of selected fields for the second month of the CRCM simulation (January) are here presented over a $(4500\ \text{km})^2$ region excluding the nesting zone. It is important to note that a climatological analysis is not intended here, a two-month simulation being by no way long enough to have any statistical significance. The investigation was more directed onto the detection of regional characteristics in the various output fields of the CRCM compared to that of the driving GCM.

Figure 1 is used to highlight the differences between two model representations of the same world. Land-sea-ice mask of the 45-km resolution CRCM (a) and T32 (32 waves spectral model with roughly 450-km resolution) GCMII (b) over North America are shown. The spectral GCM is projected on a Gaussian latitude–longitude grid with a resolution of $3.75° \times 3.75°$. The Florida peninsula, the Great Lakes, James Bay and

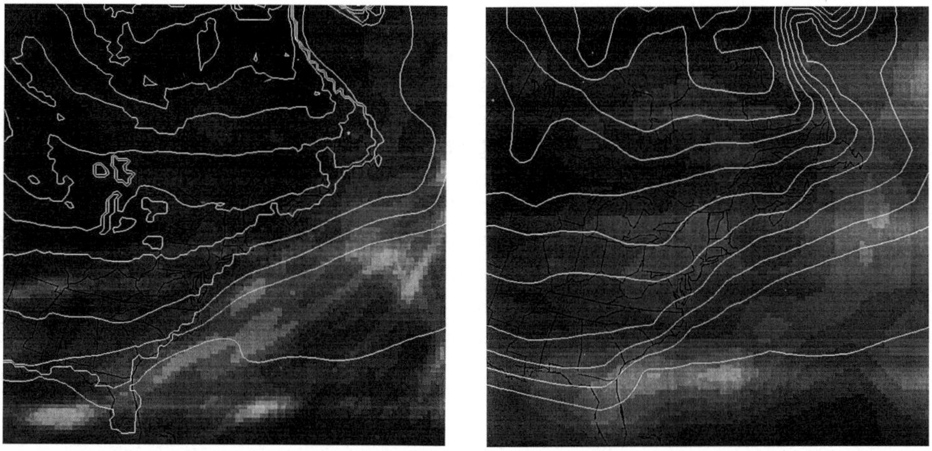

(a) (b)

Fig. 3. Monthly mean ground temperature and precipitation rate for CRCM (a) and GCMII (b). Monthly mean precipitation field is presented in grey scale colour with weak precipitation in black and strong precipitation in white. Ground temperature (and sea surface temperature over open water) contours are drawn at 5 °C intervals. Ground temperature on the Eastern side of the Florida peninsula is 23 °C in both CRCM and GCMII (SST are prescribed in both models).

other geographical characteristics of comparable (and smaller) scales cannot be reproduced at the coarse resolution of GCMII. The $1^\circ \times 1^\circ$ resolution of the reference file used to define the land-sea-ice mask of the CRCM is apparent in the finer details of the coastline. James Bay, Florida and the Great Lakes are now present in the model. Also on Figure 1, the topography of CRCM (a) and GCMII (b) are presented. Note that negative values are present over the ocean because of the spectral analysis made in GCMII. The increased resolution in CRCM allows a much more realistic representation of topography. In particular, sharper gradients are apparent in the Appalachian ridge and Labrador high lands, where maximum heights are almost twice as high as in the GCM.

It appears that even with a short simulation, many regional aspects of hydrology related fields (that are not well reproduced by a GCM) appear in the CRCM simulation. In particular, the classical comma shape in the distribution of humidity over a low-pressure system (located on the East coast of Labrador) is clearly visible in the CRCM output (Figure 2a); the coarse resolution GCMII (Figure 2b) cannot reproduce the strong gradient in the frontal zone. Mean sea level pressure is also contoured in white in Figure 2. Although the central pressure of the system is similar for both models (985 hPa for CRCM and 987 hPa for GCMII), the configuration of the cyclone is different. Again, the frontal zone in CRCM shows stronger gradients. This kind of structure in the CRCM humidity field develops rapidly after the initiation of a CRCM simulation. In order to evaluate the time needed by the CRCM to modify the humidity field from the patchy GCMII distribution to the more structured CRCM field, an experiment was conducted in which CRCM was initialised with GCMII data at a time when a deep cyclone was located over the area covered by the CRCM grid. It was observed that the frontal zone begins its development during the first 12 hours of CRCM simulation. After 48 hours, the observed CRCM distribution is attained.

The monthly mean precipitation and surface temperature for this single January CRCM simulation are presented in Figure 3a and corresponding fields for GCMII are presented in Figure 3b. The relative warm surface temperature of the Great-Lakes (with respect to surrounding temperatures) is clearly apparent in the CRCM field. One can also observe the strong temperature gradient over the ice edge near the Labrador coast. The distribution of the monthly mean precipitation rate is, again, more detailed in the CRCM than that simulated by the GCMII. The sharper definition of the various storm tracks that affected the east coast of North America within the month are apparent in the monthly mean precipitation field. The maximum mean precipitation rate observed is approximately 8 mm d^{-1} in CRCM and 7 mm d^{-1} in GCMII. Despite the similarity in the maximum values, the distributions are very different with much larger gradient in the CRCM.

4. Conclusion

These preliminary CRCM results demonstrate the ability of the nesting technique for the simulation of regional characteristics of the atmospheric flow. All the differences between simulations of the GCMII and the CRCM are directed toward what is expected for regional climate. A more detailed analysis of the performance of CRCM will be presented in Caya and Laprise (1995). Quantitative validation over much longer period of time, however, will be necessary to assess the performance of CRCM for regional climate simulation. These long integrations are now in progress and will be reported in later publications.

References

Bergeron, G., Laprise, R. and Caya, D.: 1994, *Formulation of the Mesoscale Compressible Community (MC2) Model.* Internal Report from Cooperative Centre for Research in Mesometeorology, Montréal, Canada, 165 pp.

Caya, D. and Laprise, R.: 1995, The Canadian Regional Climate Model. (Unpublished manuscript).

Davies, H. C.: 1976, *Quart. J. Roy. Met. Soc.* **102**, 405-418.

Gal-Chen, T. and Sommerville, R. C.: 1975, *J. Comput. Phys.* **17**, 209-228.

Giorgi, F., Bates, G. T. and Nieman, S. J.: 1993, *J. Climate* **6**, 75-95.

Giorgi, F. and Mearns, L. O.: 1991, *Rev. Geophys.*, **29**, 191-216.

Houghton, J. T., Jenkins, G. J. and Ephraums J. J.: 1990 (eds.), *Climate Change, the IPCC Scientific Assessment.* Intergovernmental Panel on Climate Change, Cambridge University Press, Cambridge, UK, 365 pp.

McFarlane, N. A., Boer, G. J., Blanchet J. P. and Lazarre, M.: 1992, *J. Climate*, **5**, 1013-1044.

Robert, A. and Yakimiv, E.: 1986, *Atmos.-Ocean,* **24**, 369-385.

Tanguay, M., Robert, A. and Laprise, R.: 1990, *Mon. Wea. Rev.* **118**, 1970-1980.

Yakimiv, E. and Robert, A.: 1990, *Atmos.-Ocean* **28**, 466-472.

ASPEN BARK PHOTOSYNTHESIS AND ITS SIGNIFICANCE TO REMOTE SENSING AND CARBON BUDGET ESTIMATES IN THE BOREAL ECOSYSTEM

V. I. KHAROUK[1], E. M. MIDDLETON[2], S. L. SPENCER[3], B. N. ROCK[3] and D. L. WILLIAMS[2]

[1]*Sukachev Forest Institute, Russian Academy of Sciences, Academgorodok, Krasnoyarsk 660036, Russia.*
[2]*Laboratory for Terrestrial Physics, National Aeronautics and Space Administration/Goddard Space Flight Center, Greenbelt, Maryland 20771, USA.*
[3]*Complex Systems Research Center, Institute for the Study of Earth, Oceans, and Space, University of New Hampshire, Durham, New Hampshire 03824, USA.*

Abstract. Aspen bark was investigated for photosynthetic function, pigment content, and spectral characteristics during the 1993–1994 Boreal Ecosystem-Atmosphere Study (BOREAS) summer field campaigns in the boreal zone of Saskatchewan, Canada. Parameters related to photosynthetic function were similar for bark and leaves: chlorophyll (Chl) concentration; fluorescence responses; and spectral reflectance. Similar increases along a vertical gradient from base to tree top were observed for incident photosynthetically active radiation (PAR), photosynthetic pigment content, photosynthetic capacity, and spectral reflectance variables. Since transmittance of aspen bark periderm was 20–30% in the blue, and 50–60% in the red Chl absorption bands, the PAR available to the photosynthetic cortical layer in the natural, canopy environment (<1000 μmol m^{-2} s^{-1}) was sufficient to support positive net assimilation (<8–10 μmol CO_2 m^{-2} s^{-1}) under ideal conditions (e.g., light, temperature, saturating CO_2), a rate approximately 30–50% that of leaves. However, the respiring tissues comprising the greater fraction of bark tissue bias the balance of CO_2 exchange in favour of respiration for the whole bark. Therefore, net photosynthesis under ambient conditions on the whole bark was, in general, negative. The total bark surface area was estimated to contain 17–40% of the whole tree Chl. The contribution of the bark surface area fraction of the full canopy (leaves plus bark) increased with age (<60 years), with a similar trend expected for bark in total tree (and stand) photosynthesis. A spectral reflectance variable, the red edge inflection point (REIP), was related to total bark Chl content ($r^2 = 0.74$). A better predictive relationship ($r^2 = 0.82$) for total bark Chl was observed using a spectral index calculated from the reflectance ratio of two narrow wavebands (R3/R2: R2 and R3 are between 0.715–0.726 μm and 0.734–0.747 μm, respectively), which may have greater utility in landscape remote sensing. The bark spectra for Chl-containing bark should improve understanding of carbon balance in aspen forests, based on landscape-level radiative transfer simulations.

Keywords. BARK PHOTOSYNTHESIS, ASPEN, BOREAS, BARK CHLOROPHYLL.

1. Introduction

Aspen is the dominant deciduous tree in the boreal forests of the northern hemisphere. Along with poplar and cottonwood, the several aspen species are members of the genus *Populus L.*, in the willow family. Two aspen species are dominant/co-dominant within their respective ranges, *Populus tremuloides* in North America and *P. tremula* in Russia. These competitive and fast-growing pioneer species are colonisers after boreal forests are clear-cut or burned. They are also found over a wide environmental range, from semi-desert to pretundra at 69°N (Perala, 1990). In Saskatchewan, Canada, where this study was conducted, the volume of aspen growing stock is 39.5%, almost equal to the 42.1% stock of all coniferous species (Forestry Canada, 1993).

The presence of chlorophyll in the bark tissue enables bark photosynthesis. Aspen bark has the highest chlorophyll (Chl) content of all species examined (Kharouk and

Water, Air and Soil Pollution **82**: 483–497, 1995.

Tershov, 1982), although bark Chl has been verified in more than 60 woody and shrub species (Sanio, 1860; Scott, 1907; Gundersen, 1954; Brayon, 1973; Kharouk and Tershov, 1982). Chl is concentrated in a thin, outer cortical layer of the bark (Figure 1) that is within 1 mm of the exterior surface. This layer is comparable in thickness to a leaf and comprises 1-5% by volume of the whole bark. Photosynthesis in aspen bark has been examined as well (e.g., Pearson and Lawrence, 1958a, 1958b; Shepard, 1975; and Schaedle and Brayman, 1986) in studies which typically concluded that since bark respiration usually exceeds photosynthesis, the latter is not important to either the carbon (C) balance or survival strategies of trees (Brayman and Schaedle, 1982). A satisfactory explanation for the presence and maintenance of bark photosynthetic function has not been offered.

We suggest that the bark Chl/photosynthesis fraction is important to the whole tree C budget, contributing to aspen's high competitive ability under natural conditions. The fact that net CO_2 exchange is typically negative or close to zero for the complete bark tissue may not be the most relevant factor. The unique physical placement of the Chl-containing cortex, sandwiched between the transparent periderm and the underlying, non-photosynthetic secondary phloem, is likely an adaptive trait that allows exposure to incident photosynthetically active radiation (PAR, 0.4–0.7 μm) *and* provides access to a considerable source of CO_2 that is produced by the adjacent, respiring tissues. This process probably represents an efficiency function whereby respired CO_2 is recycled for further use by the tree (Schaedle and Foote, 1971). In leaves, stomatal closure under drought conditions limits both water loss and CO_2 assimilation. However, in bark—which lacks stomata—CO_2 assimilation may continue through lenticels (Figure 1) under drought conditions >30 times more severe than those tolerated by leaves (Shepard, 1975).

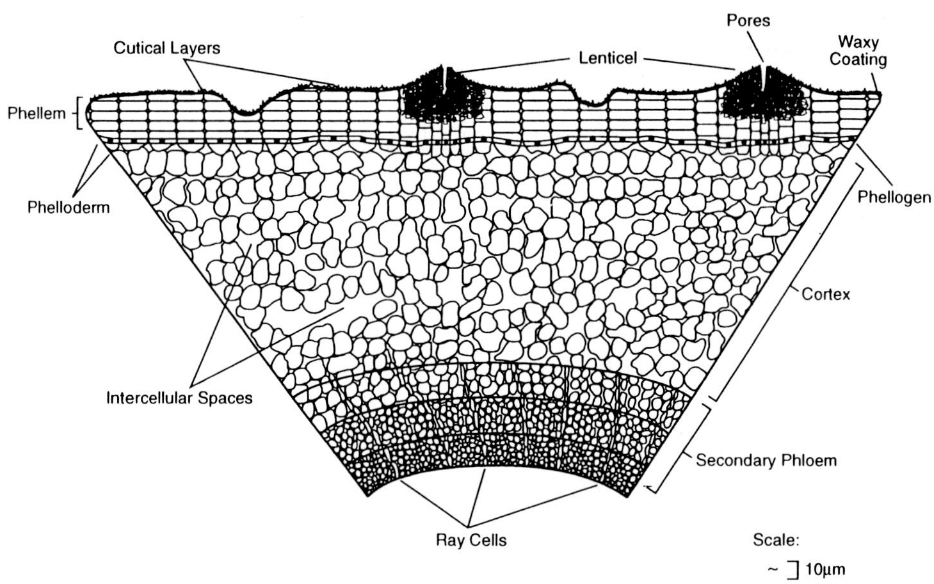

Fig. 1. The anatomy of aspen bark is presented schematically. Chlorophyll is found in cortical cell layers beneath the periderm, which includes the phellem, the phellogen, and the phelloderm.

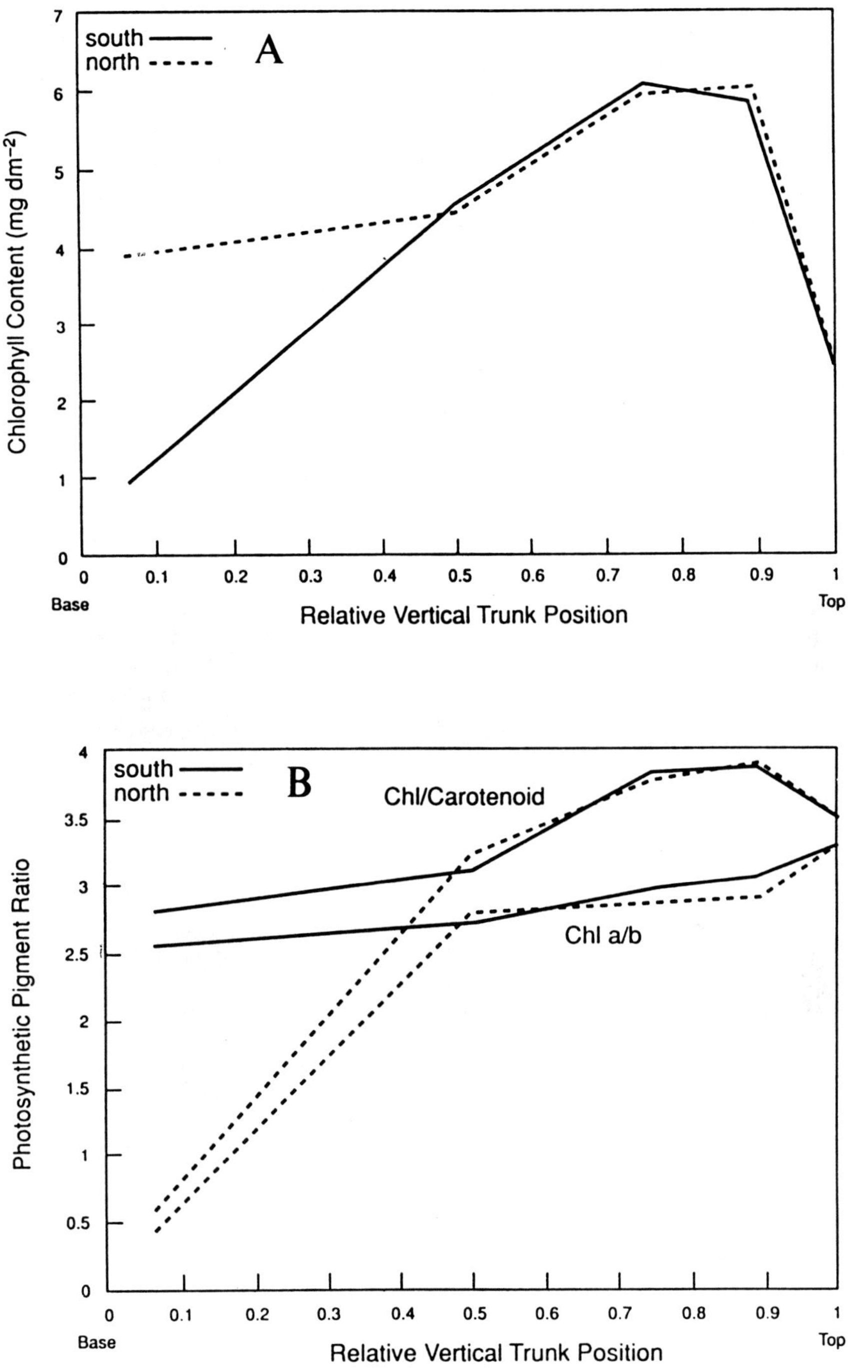

Fig. 2. Representative data for pigment gradients along a vertical height gradient at the BOREAS Old Aspen Study Site (OAS) in mid-summer (July): (A) total chlorophyll (Chl *a* + Chl *b*) content; (B) two pigment ratios (Chl *a/b*; Total Chl/carotenoids). Tree height = 22.7 m, age = 57 years. Separate curves are shown for the north-facing (dashed lined) and south-facing (solid line) trunk surfaces, along a vertical height gradient.

Moreover, the presence of Chl in the bark may significantly influence the spectral characteristics of aspen stands in remote sensing observations. One of the objectives of the Boreal Ecosystem-Atmosphere Study (BOREAS) is to utilise transfer and ecosystem process models to simulate the absorption of PAR and to estimate the C balance of forests, including those dominated by aspen (NASA, BOREAS Experimental Plan, 1994). These simulations will utilize spectra acquired via remote sensing systems (satellite, aircraft and helicopter) and spectral representations of bark as neutral "grey bodies." From our measurements obtained while participating in BOREAS, we will demonstrate the potential importance of addressing the phenomenum of active photosynthetic tissue in aspen bark to the characterization of PAR absorption and C balance of aspen-dominated landscapes.

The current study was conducted on quaking aspen (*P. tremuloides*) in the boreal forests chosen for the southern BOREAS study site in Saskatchewan, Canada. The principal goals of this study were to: (1) evaluate the potential role of aspen bark in the canopy C balance; and (2) relate the spectral characteristics of aspen bark to bark chlorophyll content.

2. Materials and Methods

The study was conducted in the boreal zone of northern Saskatchewan, Canada, during the 1993-1994 BOREAS summer field campaigns. Although measurements were made in August 1993 and throughout the period between May and September in 1994, our most intensive measurements were made between late June to early September, 1994 at four sites: (1) a "young" aspen site (designated as YAS1) with trees approximately 10 years old; (2) a second "young" aspen site (YAS2), 30 years old; (3) an "old" aspen site (OAS), 60 years old; and (4) a mixed-wood aspen site (MAS), 60 years old. The characteristics of these stands are summarised in Table I.

TABLE I

Summary of characteristics for the four boreal forest study sites[1] in Saskatchewan, Canada. Means (and 95% confidence limits) are given. DBH is diameter at breast height.

Study site/abbreviation	Age (yr)	Height (m)	DBH (cm)	Basal Area (m^2 ha^{-1})
Young Aspen #1/YAS1	10	3.3 (0.6)	2.1 (0.8)	10.7 (2.2)
Young Aspen #2/YAS2	30	12.7 (6.3)	8.2 (2.8)	35.2 (7.5)
Mature Aspen/OAS	60	17.2 (2.6)[2]	19.4 (3.9)[2]	22.0[3] (4.62)
Mixed Wood/MAS	60	19.8 (1.5)	19.2 (1.6)	11.0 (1.0) aspen
				3.0 (2.2) black spruce
				14.0 (3.2) stand

1 The "young" aspen and mixed wood sites (sites 1, 2 and 4) are clustered (±6 km) south of Candle Lake, SK; the "old" aspen site (site 3) is located in Prince Albert National Park, northwest of Prince Albert, SK. Two of these sites served as BOREAS flux tower sites: YAS1 (Site ID = D6H4T, at 53°40'N , 105°15'W) and OAS (Site ID = C2B7T, at 53°38'N, 106°12'W).

2 Stand data collected in August 1993 by Sasan Saatchi (Jet Propulsion Laboratory/California Institute of Technology, Pasadena, CA 91109.

3 Stand data reported in BOREAS Experimental Plan, Version 3.0 (May, 1994).

Laboratory measurements were made on field-acquired aspen bark samples of different ages and trunk positions, including optical properties, fluorescence, Chl content, and photosynthesis. Laboratory measurements were also made on aspen leaves. Additionally, a few trees (n = 7) were harvested to provide allometric data for quantifying branch, leaf and bark components.

2.1. FIELD MEASUREMENTS

At each site, bark and leaf samples were acquired and measurements (mean ±20% error) were made for diameter at breast height (DBH, measured at 1.3 m), tree height, and basal area along a randomly selected transect spanning several clones. For YAS2 and MAS, data were calculated from six sample points of variable radius, separated by 30 m along the transect, that were acquired using the standard prism method (basal area factor = 10) (Avery and Burkhart, 1983). In YAS1 (mean mid-summer LAI, 3.94), too dense to apply the prism method, three circular plots (2 m radius) were selected for manual census along the transect. Three trees were destructively sampled at YAS1 and YAS2, and one each at OAS and MAS. Additionally, the midday incident PAR received on a sunny mid-summer day at the canopy top and transmitted PAR received at different vertical positions along the trunk were measured at YAS1 on 5 trees, using a PAR quantum sensor (ADC LCA-4, Analytical Development Co., Ltd., Hoddesdon, England).

Total bark surface area per harvested tree was determined by summing the areas estimated for trunk sections and branch size classes. The trunk's bark surface area was determined from measurements of trunk diameters made at 1 m intervals, treating these sections as cylinders, and summing the calculated surface areas per cylinder. Branches were stratified into five or fewer size classes (dependent on tree size) based on the branch length; bark and leaf areas were determined for a subset (~33%) of each class. The average branch bark area per class was multiplied by the number of branches per class, and classes were summed to obtain the total branch bark area per tree. For specialised laboratory measurements, bark samples were taken from both south- and north-facing trunks of each harvested tree at several positions along a vertical (height) gradient from the base to the tree-top; the number of samples were dependent on tree height. For YAS1, samples were taken at the base, 1/2 Ht. (~1.3 m), and 3/4 Ht.; for YAS2, samples were taken at the base, 1.3 m, and mid-stem (1/2–3/4 Ht.); six positions were examined for the taller stands (at base, 1.3 m, 1/2 Ht., 3/4 Ht., 7/8 Ht., and 15/16 Ht.).

Leaf areas for each harvested tree were later determined: for YAS1, where the trees were relatively small, the complete one-sided leaf area per tree was measured in the laboratory; for YAS2 and MAS, leaves were collected from those branches used to calculate bark surface area only; for the OAS tree, leaf area was estimated using stand data provided by S. Saatchi (NASA, BOREAS Experiment Plan, 1994). Both the total bark surface area and total one-sided leaf area per tree were normalised to the projected area of the crown on the ground (equal to the crown diameter), to determine bark area index (BAI) and leaf area index (LAI).

Trunk sections obtained in the tree harvests, and other bark and leaf samples, were placed in ziplock bags with wet paper towels and stored in a cooler with icepacks for transport to the field laboratory located at Paddockwood School in Paddockwood, SK. There, they were refrigerated until measurements were made (≤ 2 days).

2.2. Laboratory measurements

One-sided leaf area per tree from the young and mixed (YAS1, YAS2, and MAS) sites was determined using a Leaf Area Meter (Model LI-3100; LI-COR, Lincoln, Nebraska, USA). Fresh weights were also measured on the total available leaf set per tree. For a leaf subset (n = 30), both fresh weights and one-sided leaf area were determined in order to calculate the fresh wt./leaf area ratio. This ratio was used in conjunction with the whole tree leaf fresh weights to derive the whole tree leaf areas. For YAS2 and MAS, the total tree leaf area was estimated by extrapolation from the average leaf area per branch and the number of branches.

Bark samples were examined at the positions described in "Field Measurements" above. Bark photosynthetic pigment content (Chl *a* and *b*, and total carotenoids) was determined from 1 cm^2 bark disks removed by a cork borer and placed in closed vials containing 4 mL dimethyl sulfoxide (DMSO). Pigment content was also determined on 1 cm^2 discs removed from each leaf in the leaf subset (n~30) described above. Extractions required 1–3 days at room temperature in the dark (Theisen and Rock, 1994). Vials were then frozen for 1–3 weeks, until further processing could be conducted on a Beckman spectrophotometer at the University of New Hampshire. Chl and carotenoid concentrations were calculated by standard absorption equations (Hiscox and Israelstam, 1979) and expressed per unit surface area.

Spectral measurements were made on bark sections $(\sim 10\ cm)^2$ from the trunk using two methodologies/instruments: (1) a sample subset was processed at the field lab using a spectroradiometer (0.4–1.0 μm; resolution, ~0.01 μm; model SE-590, Spectron Engineering, Denver, Colorado, USA) and an attached integrating sphere (Model LI-1800-12S; Li-Cor, Lincoln, Nebraska, USA) to determine spectral optical properties (reflectance and transmittance); and (2) spectral reflectance measurements were made on all leaf and bark samples at the University of New Hampshire using the Visible/Infrared Intelligent Spectrometer (VIRIS) (Rock *et al.*, 1994) with a spectral range of 0.4–2.5 μm (resolution: 0.002 μm between 0.4–1.1 μm; 0.004 μm for >1.1 μm). For both leaf and bark samples, a fluorescence spectrophotometer (Theisen and Rock, 1994) was used to measure several fluorescence parameters (the Kautsky effect; the time period for delayed light emission; and the emission spectrum, 0.55–0.80 μm) resulting from an excitation pulse at 0.470 μm. Photosynthetic capacity was measured in mid and late summer as a function of incident light (with 10% carbon dioxide, CO_2, in air) on bark disks (2.54 cm^2) freshly removed from the trunk sections, using an oxygen evolution system (Hansatech Corporation, Kings Lynn, England) at 25 °C. Due to variable thickness of the bark disks (1 cm), their lower portions comprising non-photosynthetic tissue were removed to standardise disk volume placed in the instrument's chamber. For comparison, measurements were made on bark disks removed from trees of two co-occurring species, fir (*Abies balsamea* (L.) Mill.) and spruce (*Picea mariana* (Mill.) BSP). The photosynthetic capacity measurements represent potential maximum rates since resistance to diffusion are overcome in the high CO_2 environment, and temperature and light levels are controlled. These measurements provided light curves and dark respiration values, and enabled comparisons with leaves (Middleton *et al.*, 1995). In addition to laboratory measurements, net bark photosynthesis was determined under ambient conditions in the field *in situ* for twigs 1–10 years of age from 5 trees. With whole twig sections placed in a small chamber (made for conifer shoots), net photosynthesis was measured as carbon dioxide exchange (flow rate, 200 mL min^{-1}; relative humidity, 50%) using a portable infra-red gas analyser (ADC LCA-4, Analytical Development Co., Ltd., Hoddesdon, England).

3. Results

3.1. Bark Pigment Content

The average Chl concentration in the aspen bark of twigs and stems was similar to that of the leaves (all sites: 4.15 ±0.43 mg dm^{-2}). Chloroplasts were present in the region between ~1.3 m and the tree top, and increased along the stem (or trunk) height gradient. For all trees examined, the total Chl content (mg Chl dm^{-2}) increased between the base and 3/4 Ht positions on both the north and south-facing trunks, as shown in Figure 2A. The Chl content was maximal in the upper trunks (~6 mg dm^{-2}), a level 1.5 times that of a typical leaf, achieving similar levels in both north and south-facing bark surfaces. Chl content declined in the bark of the lower trunk, markedly (and linearly) in the north-facing trunks, whereas much higher levels (~4 mg dm^{-2}) were maintained in south-facing trunks. Similar gradients (Figure 2B) were observed for two pigment ratios, the Chl *a/b* and Chl/carotenoid ratios. For the Chl *a/b* ratio (Chl/carotenoid ratio), values were lower in bark than in leaves: 2.75 ±0.44 (3.34 ±0.60) and 4.4 ±0.29 (4.47 ±0.29), respectively.

3.2. Spectrometry

The reflectance spectra of aspen bark were similar to those of leaves (Figures 3A, 3B), as determined with the VIRIS spectroradiometer. Some differences between north and south sides of the stem were observed (Figure 3A) that were correlated with the Chl distributions described above. Also, a blueward shift of the red edge inflection point (REIP, the wavelength of the red reflectance minimum; Rock *et al.*, 1988) revealed lower Chl content in the trunk base and in twigs. REIP and total bark Chl content were linearly related ($r^2 = 0.74$), as shown in Figure 4A. A stronger relationship ($r^2 = 0.82$) was achieved between a spectral index (R3/R2) and bark Chl content; R2 and R3 are the average reflectances for the wavebands between 0.715–0.726 μm and 0.734–0.747 μm, respectively (Figure 4B).

The transmittances of aspen bark periderm in the maximal/minimal Chl absorption bands in the blue/red spectral regions were 20–30% and 50–60%, respectively (Figure 5). At the YAS1, direct incident PAR measurements were acquired in mid-summer (mean LAI = 3.94) at the crown top and at the bark surface, indicating PAR transmittance through the leaf layer was 70 ±13% on sun-lit bark and 15 ±1% on shaded bark. Comparable values at mid-crown (~DBH or 1/2 Ht, depending on tree height) were 80 ±20% and 14 ±1%, and at the base of the crown were 73 ±23% and 22 ±0.6%, respectively. These data demonstrate a similarity between the observed gradients along the trunk surface for PAR and photosynthetic pigments (Figure 2A, 2B). They also demonstrate that PAR available to bark within the crown was 14–80% of PAR intercepted by the uppermost leaf layer; this was sufficient to support bark photosynthesis beneath the overstory, in the same light environment supporting dense understory shrub (hazelnut, *Corylus cornuta* Marsh.) or grass communities. An unpublished simulation (Huemmrich, 1994, pers. comm.) using an approach recently

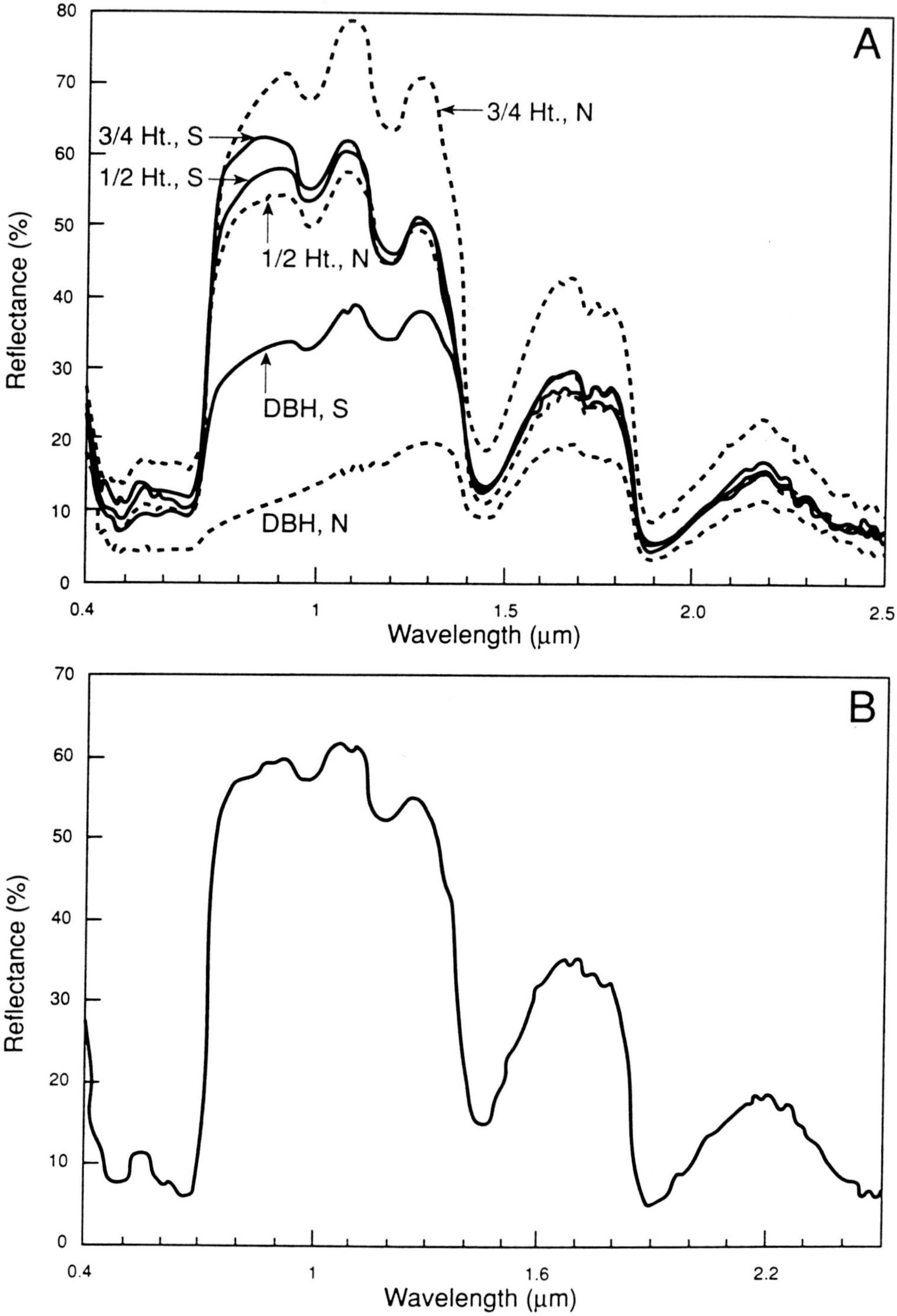

Fig. 3. Representative reflectance spectra acquired with the VIRIS spectroradiometer, for aspen bark and leaves for the tree presented in Fig. 2: (A) bark spectra acquired along the vertical height gradient at positions DBH (~1.3 m), 1/2 Ht., and 3/4 Ht. for both north-facing (dashed lined) and south-facing (solid line) trunk surfaces; and (B) Average leaf spectra from leaves of same tree (n = 15).

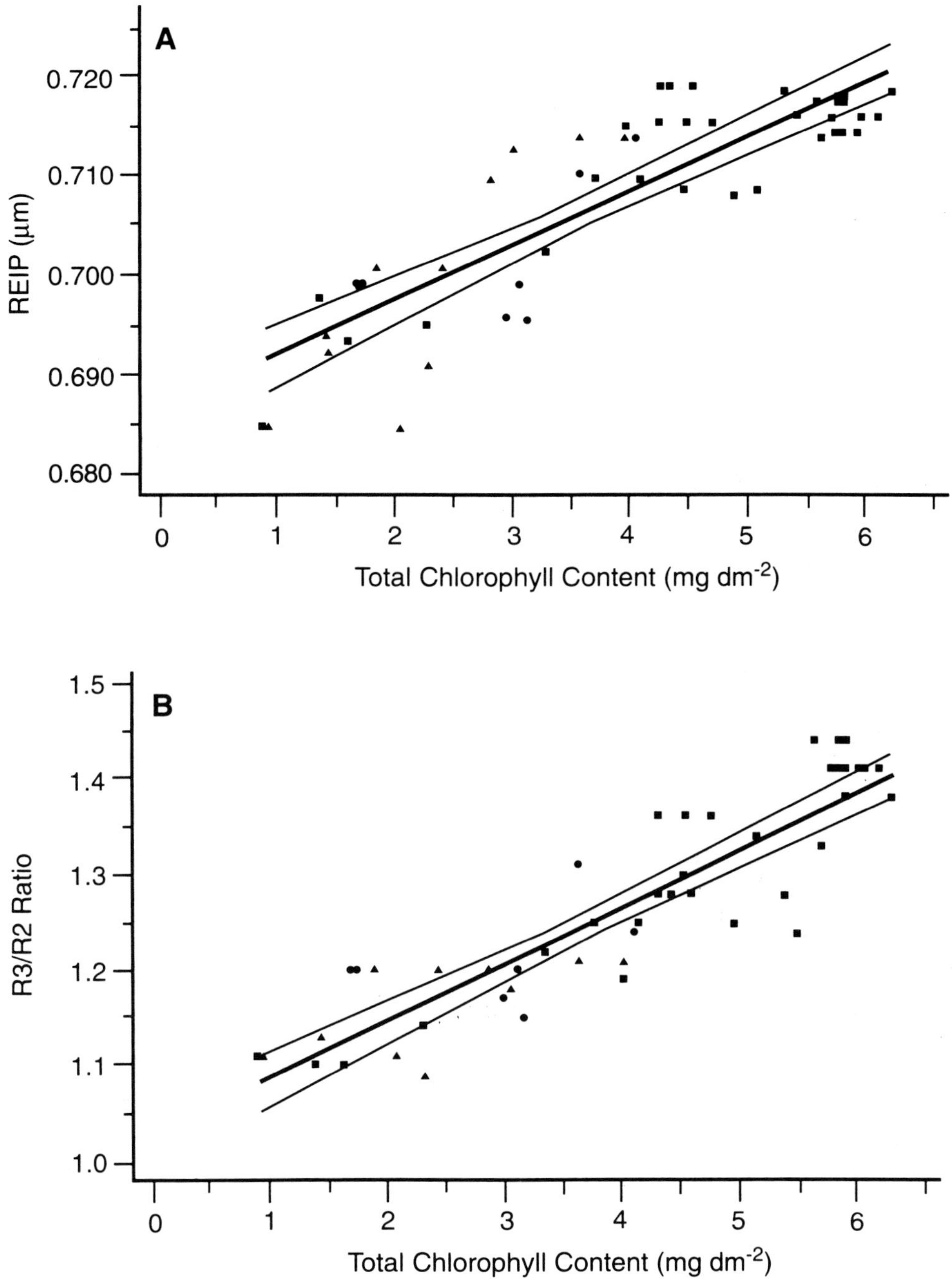

Fig. 4. Regression analyses with 95% confidence intervals for two spectral variables derived from the VIRIS reflectance spectra as a function of mid-summer (July, 1994) total bark Chl content (n = 50): (A) the "red edge inflection point" or REIP ($r^2 = 0.74$), where REIP = 0.055(Chl) + 0.686; and (B) a spectral ratio (R3/R2) for the average reflectance in wavebands between 0.715–0.726 µm (R2) and 0.734–0.747 µm (R3) ($r^2 = 0.82$), where R3/R2 = 0.0599(Chl) + 1.026.
Symbols indicate the source: YAS1 (●), YAS2 (▲), and OAS (■).

described by Goetz and Prince (1995) shows that average light interception by stem boles is approximately 60% of the above-canopy radiation, in agreement with our measured PAR levels at the bark surface.

3.3. BARK PHOTOSYNTHESIS

There were no significant differences for any measured Chl *a* fluorescence parameter (Kautsky effect, delayed light emission, or fluorescence spectra) for bark versus leaves, indicating similarity of photosynthetic function. For example, emission spectra (Figure 6) show Chl fluorescence maxima at 0.69 and 0.74 μm, with the magnitude at these peaks related to total Chl content.

Maximal photosynthetic capacity was evaluated as a function of bark position along the stem (YAS1, OAS, and MAS). The mean photosynthetic capacity increased from 8.4 μmol m^{-2} s^{-1} at DBH to 12.5 μmol m^{-2} s^{-1} at the upper trunk, an increase that mirrors that of the increasing Chl content over this gradient. Potential maximum rates of 10–12 μmol m^{-2} s^{-1} were attained by mid-trunk (1/2 Ht.) aspen bark of south-facing trunks, under light-saturated and high internal CO_2 conditions (Figure 7). Under these ideal conditions, bark photosynthetic capacity was ~30–50% of comparable values for leaves (Middleton *et al.*, 1995). Positive net photosynthetic capacity was maintained under diminishing incident PAR down to 250 μmol m^{-2} s^{-1}, the light compensation point; below this PAR intensity, gas exchange was dominated by respiration. For comparison, similar data for fir and spruce were obtained (Figure 7) indicating that while both possess normal photosynthetic function, neither coniferous species attained the bark photosynthetic capacity exhibited by aspen. However, in CO_2 exchange measurements conducted on aspen twigs from the upper canopy under ambient conditions, net photosynthesis was either negative or close to zero.

Total bark and total tree Chl content were estimated for twigs, stems, and leaves in conjunction with the biometry data. The estimated total leaf area per tree always exceeded the total bark area (LAI>BAI). The largest BAI/LAI ratio observed (0.68) was for mature (60 yr) trees. The fraction of whole tree photosynthetic surface area attributed to bark surface, or total tree Chl content associated with bark, increased with tree age (Figure 8) from ~28%, 33–37%, and 42–44% for trees 10, 30 and 60 years-old, respectively.

4. Discussion and Conclusions

Evaluating the role of aspen bark in the C balance of stands and forests requires information on bark and leaf pigment content, absorbed, transmitted, and reflected PAR, photosynthesis, and relative leaf/bark amounts per tree. Age- and phenology-dependent factors must also be considered. Then, whole tree estimates of Chl content, photosynthetic function, and C uptake can be made, followed by extrapolations to stands and forests for each phenologic period and season. In this study, we considered all of these issues for mid-summer, the season when C assimilation (g CO_2 m^{-2}) by bark is greatest (Foote and Schaedle, 1978), at the individual tree level. We recognise that our results are preliminary and that more research is needed to fully describe and quantify the role of bark in C budgets of aspen-dominated stands. Although our sample size was small, it is typical of many ecophysiological studies. This is related to the

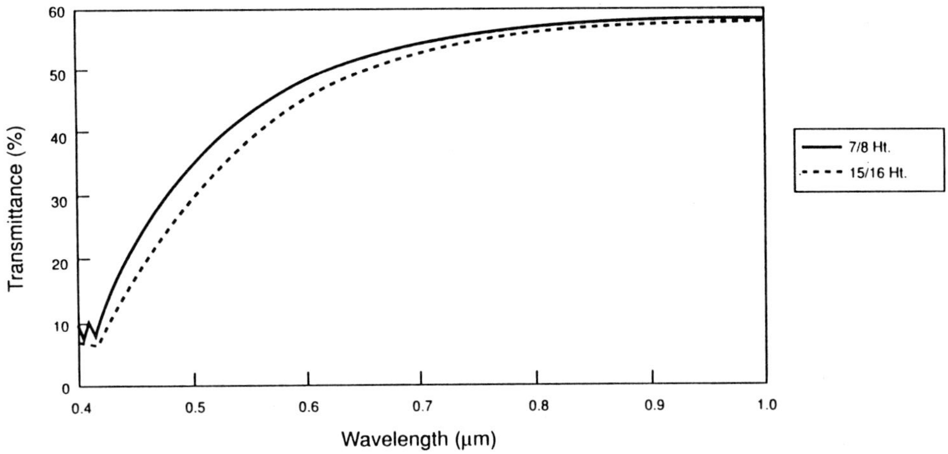

Fig. 5. Average transmittance spectra acquired with the Spectron SE-590 spectroradiometer, of aspen phellema from OAS tree at 7/8 Ht. (——) and 15/16 Ht. (------) positions at the top of the vertical height gradient.

difficulties involved in obtaining approvals for tree harvests in well-studied research areas and/or projects such as BOREAS, and in obtaining whole tree allometric and physiological data.

This study supports the view that bark photosynthesis must be included to quantify C budgets in aspen stands. The aspen bark measurements, including Chl content, Chl *a* fluorescence, photosynthetic capacity, and spectral reflectance/transmittance, are all indicative of active photosynthetic activity and/or potential. We found that the photosynthetic layer in the outer cortex of stem and twigs contains Chl at levels equal to or greater than aspen leaves on a unit area basis. The Chl-containing bark surface of aspen trees comprises 22–44% of the total tree surface, and contains up to 42% of the total tree Chl. Since Chl and photosynthetic activity were observed along the whole trunk system (trunk, branches, twigs), and since the bark Chl fraction increased with age of the bole (≤ 60 years), the role of bark in C balance should increase with stand maturation. Using the available data for our study sites (Table I; NASA, BOREAS Experiment Plan), we calculate the average bark input to whole tree C balance as 10–15% during the mid-summer vegetative period, and suggest that a larger, undetermined fraction might be attributed to bark under environmental or phenological conditions where leaf contributions are limited. Our mid-summer estimate is slightly larger than the 5–10% estimated by Schaedle and Foote (1971), based on the contribution of the stem only. Subsequently, Foote and Schaedle (1978) estimated that aspen bark tissue assimilated 59% of the annual total CO_2 during summer (June–August).

Synthesis and maintenance of photosynthetic apparatus are highly specialised and energetically expensive functions, for which it is assumed the tree derives some benefit. Substantial photosynthetic capacities for bark, especially in the trunk's middle and

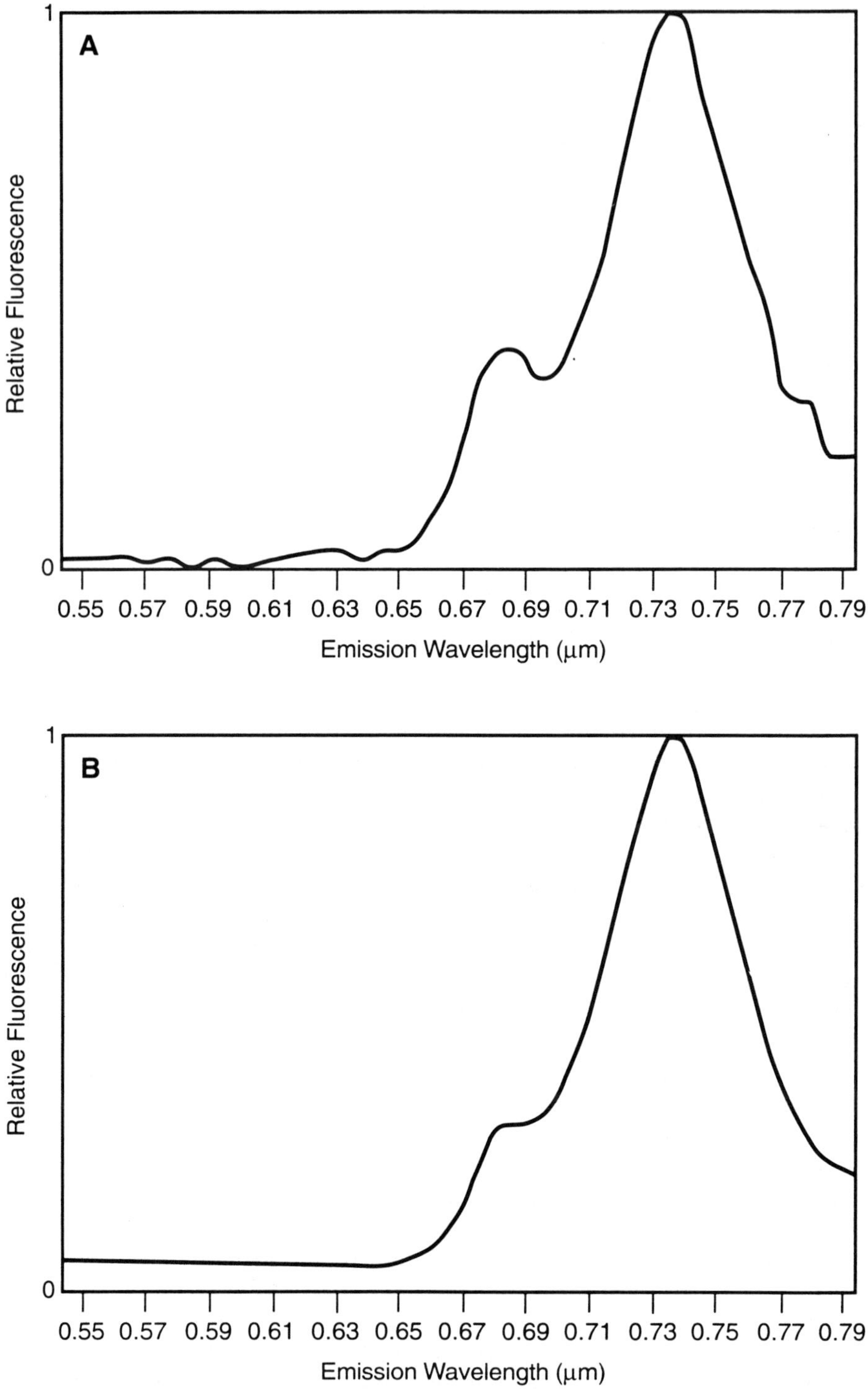

Fig. 6. Representative Chl *a* emission spectra are shown for aspen samples from OAS, for (A) leaf; and (B) bark. The excitation wavelength = 0.470 μm.

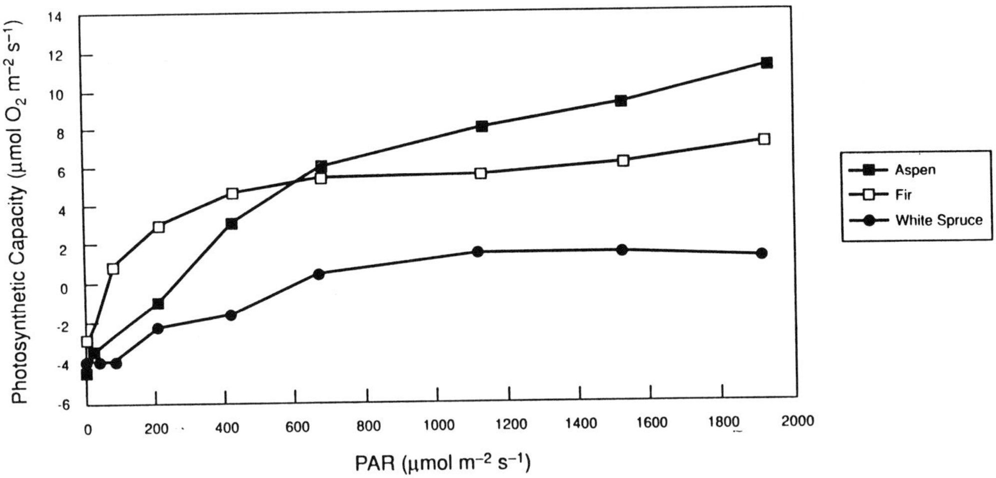

Fig. 7. Average photosynthetic capacity, determined as oxygen evolution (μmol O_2 m^{-2} s^{-1}), from the mid-section of south-facing trunks, for three boreal species in early September (9/6/94): quaking aspen (■), fir (□), and white spruce (●).

upper portions, were observed at PAR intensities found in the natural environment. (With instantaneous maximum PAR irradiance of 2300 μmol m^{-2} s^{-1} measured above the canopy, it might be possible for the upper cortex to receive PAR > 1000 μmol m^{-2} s^{-1}.) Though only ~1/3 the photosynthetic capacity of leaves, the bark fraction might provide crucial photosynthate during spring/fall leaf-off periods. Also, bark photosynthesis may be especially important for survival and recovery of trees in stands where leaves are damaged by pests and pollution (Shepard, 1975), or when leaf contributions are limited by environmental factors (e.g., drought). Furthermore, bark photosynthesis may contribute to a competitive advantage among clones within aspen forests or among individual trees growing in mixed stands, as shown in two examples from Russia (Kharouk and Terskov, 1982): (a) some aspen clones whose bark exhibits a more intensive green colour (i.e., higher bark Chl content) are more resistant to fungal diseases and are faster-growing; and (b) the bark of two boreal species which co-occur with aspen (*P. tremula*) in Siberia, *Abies sivirica* and *Betula verrucosa*, contain only 8–10% of total tree Chl, consistent with our Canadian comparisons of *P. tremuloides* to fir and spruce.

Although we measured considerably lower photosynthetic capacity in bark relative to leaves, the photosynthetic capacity of whole bark may not provide the information required since substantially lower estimates result from inclusion of the considerable (by weight or volume) non-photosynthetic, and respiring tissue. This approach is akin to expressing leaf photosynthetic capacity on a whole branch (area, volume, or dry wt.) basis, an unacceptable method. Future studies should address separate determinations of CO_2 exchange for the Chl-containing cortical layer versus the underlying non-photosynthetic tissue.

The relationship between Chl content and the R3/R2 spectral ratio, using narrow (0.011–0.015 μm) wavebands in the far-red spectral region, has advantages for remote sensing observations over the more restrictive REIP, which requires hyperspectral resolution in the red spectral region. However, either spectral parameter should provide a method for remote estimation of total Chl (leaf plus bark) in aspen stands and other

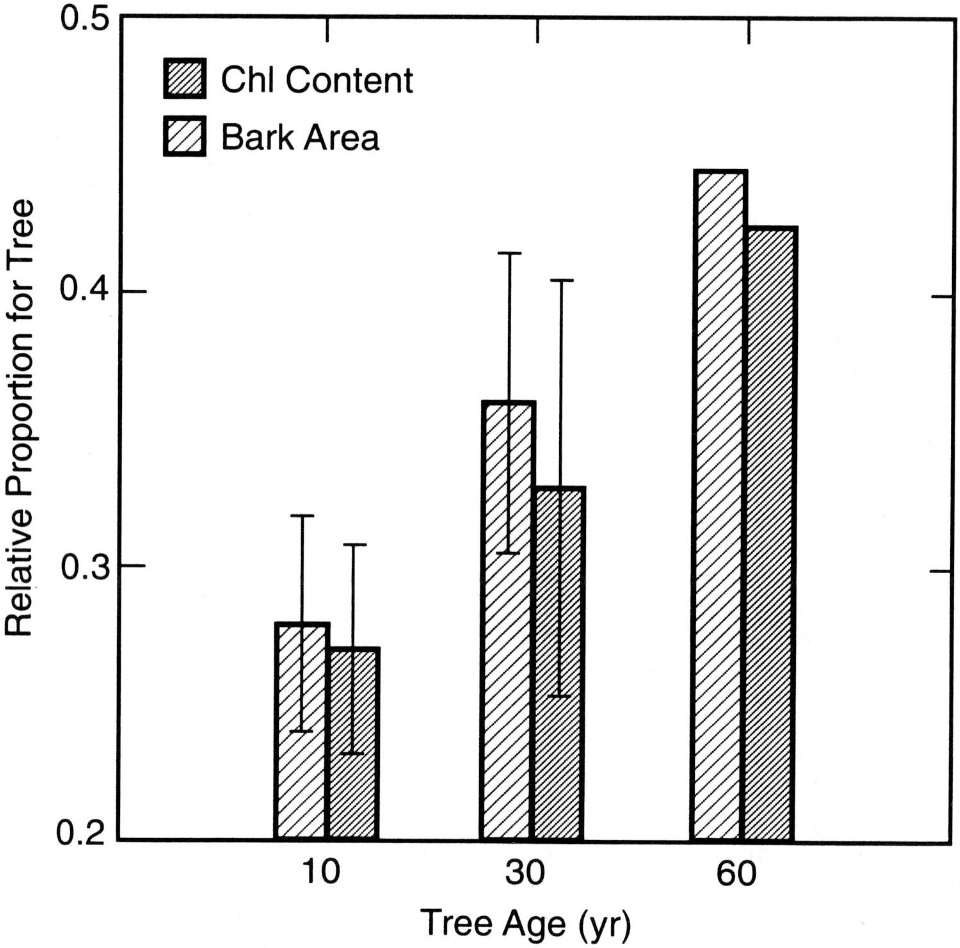

Fig. 8. The relative proportion of bark and Chl allocated to bark tissue for whole trees is estimated as a function of tree age (10 and 30 yr age classes, n = 3; for 60 yr class, n = 1): (A) bark (chlorophyll-containing) surface area : total tree surface area (one-sided leaf area plus bark surface area); and (B) Chl content of whole bark : Chl of whole tree.

forests (Rock *et al.*, 1988). Early evidence of this was given by Covington (1975), who reported a gradient in aspen bark Chl along an altitudinal transect that was also observed in the remotely measured spectral reflectance.

Our findings have shown that the spectral characteristics of aspen bark differ considerably from the "grey body" representations typically utilised in radiative transfer models. Also, since the bark and leaf canopy fractions have different C assimilation capacities, the partitioning of canopy Chl into leaf and bark strata should improve C assimilation estimates. Remote sensing technology must be relied upon, especially in vast and largely inaccessible regions such as the boreal biome, for landscape- and regional-scale studies of C budgets. In these studies, estimates of forest productivity and C exchange currently rely on spectral indices obtained from remote satellite/aircraft sensors; these spectral ratios are used to indirectly estimate C assimilation (Sellers *et al.*, 1992) through correlation with chlorophyll and photosynthetic capacity.

Acknowledgments

The authors thank Stephen S. Chan (Science Systems and Applications, Inc., Lanham, MD 20706, USA) for optical measurements on bark tissue, the Complex Systems Research Center at the University of New Hampshire for access and use of the VIRIS instrument and laboratory/analysis facilities, Joe H. Sullivan (Botany Dept., University of Maryland) for *in situ* gas exchange measurements, and the Biospheric Sciences Branch (NASA/GSFC) for use of the spectroradiometer. We especially thank BOREAS for providing the field laboratory at Paddockwood School. This research was supported under NASA BOREAS funding (RTOP 462-33), which enabled travel and sponsorship of V. I. Kharouk for six months at the University of New Hamphire.

References

Avery, T. E. and Burkhart, H. E.: 1983, *Forest Measurements*, 3rd ed., McGraw-Hill, New York, 331 pp.
Brayman, A. and Schaedle, M.: 1982, *Plant Physiol.* **69**, 911-915.
Brayon, O. V.: 1973, *Fluorescent Microscopy of Plant Cells and Tissues*. Kiev, Russia 142 pp. [In Ukrainian.]
Covington, W. W.: 1975, *Ecology* **56**, 715-720.
Forestry Canada: 1993, *National Forestry Database*. Can. Council For. Mins., Ottawa, Canada.
Foote, K. C. and Schaedle, M.: 1978, *For. Sci.* **24**, 569-573.
Goetz, S. J. and Prince, S. D.: 1995, Remote sensing of net primary production in boreal forest stands. *Agric. For. Meteorol.* (In press).
Gundersen, A.: 1954, *Torrey Bot. Club B.* **81**, 210-214.
Hiscox, J. D. and Israelstam, G. F.: 1979, *Can. J. Bot.* **57**, 1332-1334.
Johansen, D. A.: 1940, *Plant Microtechnique*, New York, USA, McGraw-Hill, 523 pp.
Kharouk, V. I. and Terskov, I. A.: 1982, *Pigments in Tree Tissues*, Nauka, Novosibirk, Russia, 88 pp. [In Russian.]
Middleton, E. M., Sullivan, J. H., Chan, S. S., Bovard, B., DeLuca, A. and Cannon, T.: **15** pp. (Unpublished manuscript.)
National Aeronautics and Space Administration (NASA): 1994, *Boreal-Ecosystem Atmosphere Study (BOREAS) Experimental Plan*, V. 3.0 (May), Goddard Space Flight Center, Greenbelt, MD, USA.
Pearson, L. C. and Lawrence, D. B.: 1958a, *Amer. J. Bot.* **45**, 383-387.
Pearson, L. C. and Lawrence, D. B.: 1958b, *Minn. Acad. Sci.* **25-26**, 101-107.
Perala, D. A.: 1990, *In: Silvics of North America, Hardwoods*. Washington D.C., USA, **2**, pp. 555-569.
Rock, B. N., Hoshizaki, T. and Miller, J. R.: 1988, *Rem. Sensing Environ.* **24**, 109-128.
Rock, B. N., Williams, D. L., Moss, D. M., Lauten, G. N. and Kim, M.: 1994, *Rem. Sensing Environ.* **47**, 176-189.
Sanio, C.: 1860, *Jahrb. Wiss. Bot.* **2**, 39-109.
Schaedle, M. and Brayman, A.: 1986, *Tree Physiol.* **1**, 53-56.
Schaedle, M. and Foote, K. C.: 1971, *For. Sci.* **17**, 308-313.
Scott, D. G.: 1907, *Amer. Bot.* **21**, 327-439.
Sellers, P. J., Berry, J. A., Collatz, G. J., Field, C. B. and Hall, F. G.: 1992, *Rem. Sensing. Environ.* **42**, 1-30.
Theisen, A. F. and Rock, B. N.: 1994, *In:* International Symposium on Spectral Sensing Research '94 (ISSSR), 10-15 July 1994, San Diego, USA, pp. 737-748.

USING AERIAL PHOTOGRAPHY AND SATELLITE IMAGERY TO MONITOR FOREST COVER IN WESTERN SIBERIA

V. N. SEDYKH

Forest Institute, Russian Academy of Sciences, Siberian Branch, Zhukovskogo st 100/1, Novosibirsk 630082, Russia

Abstract. Correlations between picture structure, geological and geomorphological terrain and forest structure of the western Siberian plain offer new complex computer-based approaches and methods for using satellite imagery for remote sensing of forest conditions, types and age. A hierarchy of geographically defined forest units has been developed from satellite imagery; monitoring of different forest parameters is recommended within these units. The current condition of the forest of western Siberia has been evaluated through remote sensing combined with ground truthing. This serves as the basis for predicting its natural course of development and for recommendations for long-term planning of forest resource usage taking into consideration current and anticipated changes in the environment

Keywords. BOREAL FOREST, REMOTE SENSING, SATELLITES, AERIAL PHOTOGRAPHY, SIBERIA, COVER CLASSIFICATION.

1. Introduction

For the last 20 years, the forest cover of western Siberia has been subject to intensive human impacts connected both with the direct exploitation of forest resources and with industrial development of oil-and-gas fields and their influences on forest habitat. In order to prevent negative changes of forest cover, to ensure against deterioration in their appearance, to eliminate serious impacts, and to define the problem it is necessary to recognise the state of the forest vegetation and the reasons for changes in it. The application of remote sensing methods and data with aerial photography and satellite imagery is a promising route to the solution of the problem.

Accumulated experience and knowledge in the use of aerial photographs designed for comprehensive forest estimation work is inadequate to perform an evaluation of the state and dynamics of forest cover over large areas (Sedykh, 1991). Hence it is necessary first to develop the scientific basis and the methods of using satellite imagery in combination with aerial photography to solve the problem of regional forest cover monitoring in western Siberia.

2. Role of Aerial Photography and Satellite Imagery in Monitoring Forest Cover

Complex transformations of the forest cover of western Siberia necessitates the establishment of a general problem of recognition of all the possible states of forest vegetation and explanations of the causes of changes in these states. Its solution should be a basic activity of forest monitoring directed at rapid evaluation of new forest cover states and at correcting human economic activity.

Monitoring is defined as an observation of two groups of parameters:

1. Indicators of the state of forest cover in a landscape

Water, Air and Soil Pollution **82**: 499–507, 1995.

   (a) landscape parameters: i.e., the areas occupied by various vegetation types and forest formations, their changes in ecology and age;
   (b) phytocoenotic parameters comprising forest cover: i.e., species composition, morphostructure, number of individuals, single tree size indices, crown closure, volume density, etc.
2. Indicators of forest vegetation conditions in parameters of landscapes and forest types. Here a two-fold hierarchy of observation systems is worthwhile based on the resolution of images:
   (a) for the detection of changes of landscape macroparameters, satellite images at 1:1 000 000–1:200 000 scale and direct observation from earth orbiting satellites are more acceptable.
   (b) for the detection of changes in the composition of forest phytocoenoses and their complexes and to reveal the mechanism of these changes, aerial photographs at 1:50 000 and larger scale are more suitable.

Changes in landscape macroparameters can be observed throughout Siberia and can be used to rate large regions. To observe the second group of parameters it is advisable to use field sites (reference polygons) within each landscape unit.

To keep a record of different forest states both in landscape parameters and in phytocoenotic parameters, regional methods of forest interpretation are necessary which do not yet exist in the western Siberia area. This is an obstacle to the wide use of aerial and satellite images for forest cover monitoring.

Methods of identification and measurement relying on reference patterns in aerial photographs are still applicable. With the advent of satellite photos this principle of interpretation becomes still more fruitful. Its proficiency has increased because the application of satellite pictures greatly increased the range of application of interpretation patterns, both direct and indirect, the combined use of which increases the reliability of the identification of the objects.

In an effort to describe the patterns in images and to use these patterns to identify forest objects, four groups of direct and three groups of indirect indicators have been determined.

Indirect indicators are an application of direct indicators in images of one object for interpretation of another on the basis of landscape interrelationships such as:

1. determining of locational relationships (interface, neighbourhood, combination, surroundings) of objects in an area and in some compound landscapes;
2. determining that an object is a sub-unit of a larger area;
3. relevance of conjugation (coincidence) of objects of different landscape compounds.

An evaluation of the potential usefulness of the combined use of selected direct and indirect interpretation signs reveals a high degree of reliability of forest object identification.

3. Forest Ecology Evaluation from Satellite Imagery

The afforestation of an area is one of the most important indices of the ecological landscape which is characterised by its water supply, drainage, erosion and forest-forming processes. Based on the observation that forests occur on stable soils in moderately hilly terrain, a correlation analysis has been performed between relief density (total relief combined with mean slope, defined mathematically below) and afforestation in order to use the latter for obtaining quantitative information not directly sensed by satellite imagery.

Statistical processing of large-scale aerial survey and topographic maps reveals a strong correlation between afforestation (*P*) and relief density. Among different signs of relief structure, as suggested by Chentsov (1948), the indices of height dispersion (σ_h), reflecting the total relief, and mid-slope tangent, tan(φ), reflecting interfluve density, proved to be the most valuable for calculation.

Correlation coefficients for mid-slope tangent and height dispersion with afforestation are 0.816 and 0.809 respectively. The strengths of the correlations are reflected by empirical formulae showing the indices as functions of relief and afforestation.

$$\sigma_h = 0.93 \exp(2.27P) \quad [1]$$

$$\tan(\varphi) = 0.003 \exp(2.60P) \quad [2]$$

For convenience they have been integrated into a single generalised index of relief density:

$$R = \sqrt{\sigma_h \tan(\varphi)} = \exp(2.45P - 3.06) \quad [3]$$

The correlation coefficient of afforestation based on this index is 0.824.

The ecological meaning of this correlation is that hillier areas have more well-drained plots and are more densely forested. Height dispersion characterises the potential of runoff of both vadose and phreatic water and the mean integrated tangent characterises the kinetic properties of this runoff. The index *R*, involving both indices, summarises the geological and geomorphological conditions which provide rapid drainage of excess moisture in the upper soil layer. Conforming to Darcy's law, it can be used as an indirect index of drainage, i.e., the drainage coefficient of the area as estimated by satellite imagery.

To facilitate this, an evaluation scale has been developed for estimating relief density and drainage, which can be used as a working tool for interpreting satellite images at scales of 1:3 500 000–1:200 000.

Since different forest species respond differently to drainage and the degree of erosion in a habitat, it may be supposed that the qualitative state of afforestation may be used to estimate forest composition and ecological status. A correlation analysis has been performed for relationships between afforestation and different ecological and formational groups of forest communities. It has been found that the closest relationship can be observed between percentage of dry pine forests (Pc) in forested areas and the percentage of cedar (*Pinus sibirica*) and deciduous formations taken together in the composition of dry terrains.

The analytical expressions for these relationships are:

percentage of dry pine forests in the area (Pc):

$$\mathrm{Pc} = 0.31 - P^{1.7} \quad [4]$$

percentage of cedar and deciduous forests in a dry area (Pck):

$$\mathrm{Pck} = 1 - 0.38P^{1.6} \quad [5]$$

Calculations carried out by means of regression models permit determination of the proportion of cedar and deciduous forests (Pk) on the basis of satellite pictures and forest inventory data.

$$\mathrm{Pk} = 0.81\mathrm{P}^{0.11} - 0.31P^{1.7} \quad [6]$$

while the percentage of hydromorphic pine forests (Pch) in the forested area is:

$$\mathrm{Pch} = 1 - 0.81P^{0.1} \quad [7]$$

These relationships imply that dry pine forests occur on more intersected surfaces (higher relief density) and thus better drained areas and the less drained areas are mainly covered with cedar forests, and their derivative forests and swampy pine forests. This points to the fundamental importance of geological factors in the formation of forest cover structure.

Such relationships are of theoretical and practical significance in geomorphology, hydrology, soil science, geobotany and specifically in the solution of the tasks of monitoring forest cover using satellite images at scales of 1:3 500 000–1:200 000, and for estimating forest vegetation conditions of taiga areas.

4. Remote Sensing of Forest Successional Stage

The change of species arising in post-fire succession is the most wide-spread kind of dynamics of the forest-forming process in western Siberia. The evaluation of this phenomenon should be a permanent activity through the use of aerial and space-based monitoring of forest cover.

Morphostructural analysis of postfire deciduous and cedar stands allow the determination of specific interrelations between the vertical and horizontal projections of complex forest communities. The relationship between morphostructure of the stands and the character of their aerial and satellite images is regular and steady. Its variability within the limits of forest vegetation conditions of the same type is associated with age changes of forest communities.

The temporal dynamics of species composition, tree number and tree size in the stands all affect the vertical and horizontal projections of the stands and thus the hue and texture of the stands are modified as well. This property of synchroneity between stand morphology dynamics and images of the stand structure was used to develop methods of remote detection of pyrogenic deciduous and cedar successions, which is useful for estimating forest dynamics by remote sensing. The use of post-fire forest dynamics as an interpretation scheme requires accompaniment of each stage by illustrative aerial images in proper scale and its description by indirect interpretation indicators.

At this stage of development, the interpretation indicators can be used with black and white aerial photographs at scales of 1:15 000–1:50 000 to study the spatial relationships of the disposition of cedar and deciduous forests, for interpretation of lithology, surficial deposits and soils, for assigning areas for thinning, for developing long-term plans for forest exploitation in connection with the areal tendency of change of species, and to predict forest cover dynamics.

5. Evaluation of Soils and Soil-Forming Rocks by Satellite Imagery

The forest cover of the taiga area is defined by a combination of forest communities of differing forest types and their derivatives, the appearance of which is related to fire, windfalls, and other exogenic factors. Based on this and principles of genetic typology developed by Kolesnikov (1956), one should suppose that forest communities of different phytocoenotic structure and morphology and having different image indicators can be associated with the same soils and soil-forming rocks. To determine the distribution of soil and lithological units, it is necessary to first ascertain all the stands which are involved in the region being mapped. Therein lies the difficulty in interpreting soils, parent materials, and therefore forest types and is determined by the state of knowledge on successional dynamics and its relation to forest vegetation and soil types.

As a check on the similarity of soils and parent materials, a study was performed on areas covered by deciduous and cedar stands, which are significantly different in phytocoenotic structure on satellite images. Studies of soil profiles under birch and cedar forests reveal similarities in granulometry and chemistry typical of loamy and light loamy podzolic surface-gleyed soils of the middle taiga of western Siberia. The differences in soils are mainly due to fire and post-fire biogenic impacts.

The similarities of the habitats of two different forest-formers permit us to consider that birch and cedar stands comprise separate stages of regenerative and age dynamics of cedar forests which develop after fire on loamy surface-gleyed podzolic soils. This gives us grounds to associate all the genetically related stands of this dynamic with podzolic surface-gleyed soils and loamy parent materials in analysing black and white aerial photographs. This also applies to lightly textured soils where multi-species forest stands develop. Thus it is necessary to know forest-forming processes when developing methods for interpreting soils and parent materials in taiga areas.

6. Mensurational Interpretation of Aerial and Satellite Pictures

Due to the pixelation of images in small-scale pictures, measurements of morphological elements of stand structure are impossible. Thus it is necessary to develop other methods and principles of mensurational interpretation.

Recognising that image indicators of stands change uniquely with age, depending on the change of stand morphostructure, mensurational interpretation becomes possible both visually and with computers, by using image indicators of reference stands as additional information with existing yield tables.

Yield tables have been compiled including descriptions of image interpretation keys for stages of development of birch and cedar stands, using common methods and the author's own experience.

The tables have been tested using 1:50 000 scale aerial photos and field studies. In more than 70% of the analysed forest units, the results differ from the field data by not more than 20% for major mensurational indices such as composition, mean height, mean diameter and basal area density.

To support automatic mensurational interpretation, relations expressing the changes of texture and average hue are derived to complement other stand parameter/yield relationships. For this purpose one needs to measure texture and hue of forest communities in definite age intervals, and the entire successional and genetic series should be included wherever possible.

The combined use of these curves allows us to determine stand age using images, beginning from the age of 30 years. It is possible not only to determine stand age automatically but also to obtain other mensurational indices by means of computerised yield tables.

As was shown by comparison with field data, the spatial borders of development stages (with the exception of initial ones) determined by computer analysis and quantitative methods of interpretation coincide at a level of 75%.

The suggested method for mensurational evaluation of stands by computer will be useful in aerial and space-based monitoring of forest cover.

7. Interpretation of Forest Type by Aerial and Space-based Monitoring

One of the important problems in remote sensing is the development of methods for interpreting forest types, i.e., basic taxonomic units of forest cover within the limits of which the nature of forests is investigated and monitoring of their state should be conducted. Geographical and genetic typology developed by Kolensnikov (1956) is useful for solving this problem; this typology forms the theoretical basis for developing methods of interpretation of forest types.

The basic unit of geographical and genetic classification by Kolesnikov (1956) and his followers is forest type, understood as a stage in the forest-forming process which consists of genetically combined stands which develop within the limits of a definite type of edaphic condition. This type is identified by geomorphological evidence. Relief and forest communities have been found to be not only reflected in the images but to determine all the morphological properties of images. Individually and in combination they can be effectively used as diagnostic signs for interpreting and evaluating the forest types.

In evaluating and classifying forest vegetation habitats seven criteria have been used combined with relief and vegetation signs which are evident in satellite images. The first classification criterion consists of evaluating the degree of moisture of the forest habitat. According to this criterion the forest plots are subdivided into auto-morphic, semi-hydromorphic and hydromorphic. In this case auto-morphic plots are subdivided into dry, fresh and moist according to the drainage network. Hydromorphic plots are subdivided into wet with stagnant or percolating moisture.

The second criterion consists of dividing habitats by trophic principles combined with relief types evident in satellite images. According to soil and bedrock texture which characterise the fertility of habitats, the forest units are subdivided into coarse and fine sandy, loamy sandy and bistratified.

The third criterion of habitat classification characterises the degree of continuity of ecological regime of habitat that is conditioned by micro-relief. According to this criterion the habitats are subdivided into *homogenous*, in which the micro-relief is weakly pronounced and plant cover is relatively uniform; *mosaic*, where developed micro-relief controls the distribution of species of the herb-shrub layer excluding the distribution of trees; and *complex* (heterogeneous) where plant cover is fragmented into different associations because of sharp differences in micro-relief.

The fourth criterion of habitat division includes the degree of drainage, subdivided into well-drained, normal, insufficient, weak, very weak and percolated. The state of drainage is completely controlled by geological and geomorphological conditions of habitats and is therefore easily determined by remote sensing.

The fifth criterion, which is well defined by stereoscopic models of the location, has been introduced for evaluation of relief of forest units by indicators of elevation, relief density and degree of slope These determine different hydrological regimes of habitat.

The sixth criterion is depth to ground water table. The depth to ground water has been estimated in metres with the following ranges: 0 m, ≤1 m, 1–5 m, 5–10 m and >10 m. Ground water tables are determined by remote sensing combined with topographic maps indicating the elevations of plots in relation to swamp surface and water level.

The seventh criterion is soil characteristics according to which genetic soil type is determined on the basis of established relations between trophic and hydrological regimes, relief of habitat and vegetation.

For subdividing forest vegetation within the limits of the determined forest vegetation conditions, the criteria of forest vegetation classification are used, i.e., the basic forest-forming species, site class, primary and secondary forests, age states and forest mensurational indices, such as mean height, and basal area density.

The above-mentioned indicators of forest vegetation are rather reliably determined by remote sensing. This together with relief evidence permits a solution to the problem of forest type interpretation. According to these criteria, 42 types have been classified in the region of the Ob River reflecting the whole diversity of ecological regimes and state of stands in seven mature forest formations.

An interpretation table has been compiled for forest types as a working tool for use with black and white aerial photographs and satellite images. This is used in sequence with images at different scales from small to large. The table conveniently shows the state of landscape relation with relief, and vegetation and image references which have been used for diagnosis of forest types. Spatial location of forest types is determined by indirect indicators, and their temporal state is identified by direct indicators. A necessity for working with the interpretation table is the use of images at scales such as 1:1 000 000, 1:200 000 and 1:50 000–1:15 000 with image references of stages, types and relief forms.

8. Areal Units of Forest Cover Observation

Along with the development of interpretation methods for forest types, regional forest cover monitoring requires delimiting areal units of forest cover from satellite images. These units consist of forest types and their complexes. Their solution requires areal subdivision of forest cover so that each taxonomic unit possesses: (1) relative uniformity of geological and geomorphological structures, (2) structural and functional unity of forest cover, and (3) the ability to recognise the areal unit as a whole and the forest cover indicators to be observed in the imagery.

According to these requirements seven hierarchically subordinated areal units of forest cover have been classified within the limits of the latitudinal region of the Ob River. Monitoring here of different forest parameters is recommended.

The largest unit of the first level is the latitudinal region of the Ob River. The territory is well recognised from satellite photos at scales of 1:10 000 000–1:3 500 000 independent of the season of survey. The main monitoring indicator is afforestation, changes of which give information about large structural transformations of forest cover.

The unit of the second level corresponds to the borders of associations of relief stages, which are isolated in satellite images at scales of 1:3 500 000–1:1 000 000. The main monitoring indicator is afforestation and percentage of deciduous and coniferous forests.

The units of the third level are distinguished in satellite images at scales of 1:1 000 000–1:500 000. Main indicators of vegetation condition are afforestation and proportional area ratio of light, dark coniferous and deciduous forests.

While classifying the units of the fourth level, the watersheds, macro-slopes, plots of the valleys, i.e., macro-forms of relief stages, serve as the criteria. Forest cover indicators are afforestation and proportional of the area occupied by forests. The scale of images is 1:500 000–1:200 000.

The units of the fifth level present the association of large relief forms. Plant criteria are represented by ecological modifications of the forest formation by moisture class such as auto-morphic, semi-hydromorphic and hydromorphic. The scales of images are 1:200 000–1:100 000. Monitoring indicators are afforestation, proportions of forest formations and their ecological subdivisions.

The units of the sixth level have been distinguished by relief forms such as bottoms, slopes and watersheds of local rivers. The scale of aerial and satellite images is in the range of 1:1 000 000–1:50 000. The forest cover indicators are afforestation, area ratios of forest formations and groups of forest types, age states of the stands in three age categories, i.e., young growth, medium age and maturing, and mature and old growth forests.

Forest types, units of the seventh level occupy the elements of relief forms and lithological varieties. The scale of images is 1:50 000–1:15 000. The forest cover state indicators are afforestation, proportions of forest formations, forest types and different mensurational indices of the stand types.

For the purpose of estimating changes in forest cover, it is proposed to compile a database, which should contain data relating to ecological indices of forest vegetation, as well as information on available and expected exogenic impacts, forest and swamp-forming processes, fire-prone vegetation conditions and endogenous forest dynamics. A spatial and temporal model built on the basis of Kolesnikov's (1956) genetic typology by established mensurational methods for compiling yield tables provides a forecasting tool.

The value of the data obtained on the natural development of forest cover is evident. The probability of appearances of these states, which are reproduced by means of spatial and temporal models, will always be relative. The reliability of the forecast will be always increased by means of continuously adjusting the model using remote sensing from subsequent surveys.

9. Conclusions

Interpretation of images from aerial and satellite surveys conducted with the methods described in combination with field studies, reveals that relief density and the geological composition of surficial deposits are the governing factors in the formation of ecological regimes of forest habitats. These determine the present structure of forest cover and the structure of images of aerial and satellite images.

Using these techniques, it has been found that auto-morphic small-leafed birch and aspen forests are more widespread in the central part of western Siberia (34.0%), where they arose after fires in cedar forests. Along with auto-morphic cedar forests (16.1%) they account for 50% of dry forests and are located on weakly and moderately hilly geomorphological areas with gently hillocky relief formed of late quaternary deposits. Dry pine forests (17.2%) presumably occur on elevated intersected plots of floodplain terraces, having gently crested relief and underlain by sands, and on strongly intersected relief forms of interfluves with bistratified deposits. Most often they are common in asymmetric interfluves. Swamp forests of all formations occupy 33% of the forested area, and are spread over all the relief stages, but mainly occupy lower flat plots with peat accumulations of variable thickness.

Such a distribution of forest formations bears witness to the leading role of geological and geomorphological factors in the forest ecology in the plains of western Siberia, and the importance of these factors in maintaining the stability of historically established forest cover structures. In this connection, controlling the geology in remote sensing monitoring of forest cover, especially in the area of oil-and-gas complexes, should be a basic activity. Taking into account the increasing severity of ecological problems associated with disturbance of the substrate during the exploitation of the natural resources of the taiga creates a demand for the development of a new type of geological knowledge, i.e., ecological geology, oriented to the tasks of forest monitoring and environmental protection in northern areas.

It has been shown from aerial and satellite images that the change of species arising as a result of fires in cedar and pine forests and subsequent postfire successions is the most common dynamic of forest-forming processes in the central part of western Siberia. One hundred- to 130-year old birch and aspen forests with understory are spread over an extensive area accounting for 95% of the dry deciduous forests or 31% of the total forested area. According to the forecast, one should expect in 40–60 years a 30–35% increase of cedar forests at the expense of deciduous stands, such that cedar will occupy 50% of the forest-covered area. In 100 years in the absence of exogenic factors, the forest cover of dry areas will mainly consist, according to the forecast, of stands of two basic forest-formers, i.e., cedar and pine.

As for the expected rise in global temperature, such a phenomenon is expected to be deleterious for the forest cover. Fuel accumulation can bring about continuous fires in coniferous forests and destruction of valuable mature stands over large areas.

The resulting prediction generates a need for a constructive review of many solutions concerning all aspects of economic activity in the forests of western Siberia.

Acknowledgments

The author wishes to acknowledge Russian, U.S. and Canadian colleagues who assisted with the preparation of this paper, specifically T. A. Stone. He also greatly appreciates financial assistance from the USDA Forest Service and from the Canadian Forest Service (Edmonton) to attend the IBFRA '94 Boreal Forests and Global Change conference.

References

Chentsov, V. N.: 1948, Morphological indicators on small-scale morphological maps. *In: Reprints Geogr. Inst. USSR Acad. Sci.*, No. **39**, 281-306.

Kolesnikov, B. P.: 1956, Far Eastern cedar forests. *In: Rep. Far Eastern Br. USSR Acad. Sci., Bot. Ser.*, Vol. **2**(4), 264 pp.

Sedykh, V. N.: 1991, Satellite monitoring of forest cover. Nauka, Novosibirsk, Russia, 237 pp. [In Russian]

MONITORING PRIMARY PRODUCTION FROM EARTH OBSERVING SATELLITES

S. D. PRINCE, S. J. GOETZ and S. N. GOWARD

Department of Geography, University of Maryland, College Park, MD 20742-8225, USA

Abstract. Satellite remotely sensed observations provide a unique capacity to monitor biospheric activity at regional and global scales. Spectral vegetation indices from these observations capture the heterogeneity of growth patterns across the Earth's surface including the significant decline in vegetation activity above 50°N latitude. Conversion of these observations to meaningful assessments of vegetation growth, particularly changes within and between years, requires more than a simple linear transform between the spectral vegetation index and forest production. Although satellite observations of vegetation indices are essential, appropriate environmental variables and models are also necessary. The approach is illustrated by a study of the southern Boreal forest in northern Minnesota. A conceptual and methodological framework for monitoring primary production from remotely sensed observations alone is proposed and results of its application in the Global Production Efficiency Model (GLO-PEM) for global boreal forests are discussed.

Keywords. PRIMARY PRODUCTION, REMOTE SENSING, BOREAL FOREST, ABSORBED PHOTOSYNTHETICALLY ACTIVE RADIATION, LIGHT USE EFFICIENCY, RESPIRATION, ASPEN, SPRUCE.

1. Introduction

Estimates of the global extent of boreal forests vary from 7.5% (Miller *et al.*, 1983) to 15.5% (Rodin *et al.*, 1975) of the continental land surface and, according to some estimates, carry out about 8% of the world's terrestrial primary production (Whittaker, 1975; Woodwell *et al.*, 1978). In order to understand the response of the vegetation in boreal forests to future climate change we need a better understanding of the current primary production, including its spatial distribution, interannual variation and overall amount.

Boreal forests as a biome are extensive and yet virtually unmanaged, except for forestry mainly along the southern fringe. There is little comprehensive information on the vegetation and environment throughout the region. The only means by which this information may be acquired is by some form of remote sensing. Aerial surveys have been used for many years to map forests and measure tree stocks. More recently the availability of multi-temporal and multi-spectral radiometer measurements from spacecraft has enabled measurement and monitoring of additional, important variables such as leaf area index (LAI) and phenology (Hall *et al.*, 1995; Badhwar *et al.*, 1986) These studies are at an early stage in the boreal biome but they already promise significant additional information with which to measure and monitor boreal forests; for example recent studies using these techniques have suggested that the forests are subject to significant interannual variations in spectral vegetation indices (SVI) and net primary production (NPP) (Cihlar *et al.*, 1991; Goward and Prince, 1995; Hall *et al.*, 1991), possibly caused by variations in cloud cover and evaporative demand (Elliot-Fisk, 1988).

There are some special challenges to the successful application of remote sensing in the boreal biome. These include remote sensing science issues such as low sun angles, extreme atmospheric effects, and the presence of large numbers of small water bodies

Water, Air and Soil Pollution **82**: 509–522, 1995.

and snow backgrounds that can be expected to influence measurements of vegetation reflectances because of the spatial resolution of current sensors (fields of view of 1 km^2 or more). There are also production modelling issues such as the presence of evergreens that are green yet non-photosynthetic throughout the winter, the presence of green and photosynthetic stems, the high biomass to NPP ratio, lichen and bryophyte backgrounds, and the common occurrence of epiphytic lichens. In addition, the algorithms that are used to infer most of the variables from remotely sensed data have been developed in environments that are somewhat different from boreal forest.

In the present paper we summarise the current remote sensing approaches to the measurement of primary production and illustrate one of these using a study carried out in the boreal forest of northern Minnesota. Next we analyse the variability in the conversion efficiency of absorbed photosynthetically active radiation into biomass, a key variable in the production efficiency approach. Finally we report the first results of a new primary production model that incorporates respiration and the effects of environmental variables on efficiency. The model uses remotely sensed data exclusively and is therefore well suited to regional and global studies, especially where few field measurements are available.

2. Monitoring Primary Production - The Remote Sensing Contribution

2.1. Remote sensing and traditional NPP models

In studies using Advanced Very High Resolution Radiometer (AVHRR) data from the NOAA series of meteorological satellites, strong regional patterns of SVIs have been noted, especially across the world's major ecotones such as the Sahel (Tucker *et al.*, 1985; Prince, 1991b), the boreal forest (Justice *et al.*, 1985), and the Pacific Northwest (Runyon *et al.*, 1993). The seasonal sums of SVIs thus seem able to detect at least some features of continental scale variation in NPP (Goward *et al.*, 1985) and also to highlight spatial and interannual variations in NPP at the regional scale (Justice, 1986). More recently, remotely sensed SVIs have been used in a new class of NPP model called "Production Efficiency Models" (PEM) (Prince *et al.*, 1994). The distinctive features of the PEM approach are the use of spatially comprehensive and high temporal frequency observations of the actual vegetation, thus avoiding the problems of interpolation between often widely-separated field observations. The PEM approach starts with a functional characteristic of vegetation, namely the amount of absorbed photosynthetically active radiation (APAR) but may also include less aggregated variables such as soil water status and temperature.

2.2. Production efficiency models

The PEM approach builds on crop-scale studies in which it has been shown that NPP can be modelled using remotely sensed measurements of APAR and a quasi-constant conversion "efficiency" (ε) that relates biomass production to APAR (Prince, 1991a; Landsberg *et al.*, 1995) Because the range of values of ε appeared to be small (Monteith, 1977) it was suggested that it might be treated as a conservative quantity and hence, at least under stated conditions, remote sensing measurements of APAR may allow a direct monitoring of NPP. The basis of the approach to measuring APAR using remote sensing is the relation between APAR and certain SVIs such as the normalised difference vegetation index [NDVI = (IR – R)/(IR + R)], where IR is the near infrared and R the red radiance measured by a remote radiometer, for example mounted on a satellite) (Kumar and Monteith, 1982; Sellers, 1985; Asrar *et al.*, 1986; Sellers, 1987; Goward and Huemmrich, 1992).

Although early work suggested that ε is relatively conservative across vegetation types, recent studies indicate that this quotient may vary in the presence of variable environmental conditions and for plants having different biosynthetic pathways and respiration rates (Jarvis and Leverenz, 1983; Russell *et al.*, 1989; Prince, 1991b; Runyon *et al.*, 1994). To address these and other problems Prince (1991a) has developed an extended form of the Production Efficiency Model that explicitly considers the plant respiratory component and the effect of multiple physiological stresses on the efficiency.

$$P_{(n)} = \Sigma_t \sigma_t \, \varepsilon^*_{(g),t} \left(N_t S_t\right) Y_{g,t} Y_{m,t} d_t \qquad [1]$$

where:

$P_{(n)}$ = NPP (g m^{-2})

σ_t = a functional relationship accounting for the effect of environmental variables on the unstressed value of $\varepsilon^*_{(g),t}$, owing to stomatal closure, temperature etc. (dimensionless)

$\varepsilon^*_{(g),t}$ = potential, gross energy fixation per unit APAR (g MJ^{-1})

$N_{(t)}$ = proportion of incident PAR absorbed by canopy (dimensionless).

$S_{(t)}$ = incident PAR (MJ m^{-2})

Y_m = proportion of assimilate not used in maintenance respiration (R_m).

Y_m = $(1 - R_m/P_g)$ (dimensionless), where P_g = gross production (photosynthesis – photorespiration).

Y_g = efficiency of conversion of assimilate into biomass, taking account of growth respiration R_g. Hence $Y_g = (1 - R_g/Y_m P_g)$ (dimensionless).

d = proportion of biomass lost in death, decay and grazing (dimensionless).

Note that the original use of ε (Monteith, 1972) is equivalent to $\sigma_t \varepsilon^*_{(g),t} Y_{g,t} Y_m d_t$ and is hereafter referred to as $\varepsilon_{(n)}$.

The model represented by Eq. [1] consists of an energy harvesting component (N_t, S_t) and an unstressed (potential) value of ε expressed in terms of gross production $\varepsilon^*_{(g),t}$. Respiratory losses of carbon (C) are represented by separate growth and maintenance respiration terms (Y_g, Y_m). Losses of biomass caused by death, grazing and decay, if NPP is not corrected for these, are accounted for by *d*.

Environmental factors that reduce the conversion of APAR into plant material (represented by σ) whatever their origin and mechanism, reduce the value of $\varepsilon^*_{(g),t}$. This functional relationship may be complex; in fact it subsumes much of the substance of leaf-level models. At present we have only some indication of the appropriate form of this relationship from a rule-based model developed for OTTER (Runyon *et al.*, 1994). The full details of the factors that influence ε are not yet clear. The innovation represented in Eq. [1] is to define ε in the form of a potential, or maximum gross rate of energy fixation.

3. Application to Superior National Forest, Minnesota

The stability of ε and possible causes of variability in ε both within and between species were explored in a study of a 2280 km^2 portion of the North American boreal forest at its southern terminus in the Superior National Forest of north-east Minnesota (Figure 1) (Goetz and Prince, 1995).

3.1. METHODS

The relationship of annual incident photosynthetically active radiation (PAR) interception by the canopy to annual above-ground NPP (AANPP) was measured in 30 quaking aspen (*Populus tremuloides*) and 29 lowland black spruce (*Picea mariana*) stands representing a range of density and age classes (Woods *et al.*, 1991; Hall *et al.*, 1992). LAI, AANPP and measurements of other site characteristics were available, as were measurements of incident PAR and a series of LANDSAT Multi-spectral Scanner (MSS) images. The SVI used was the greenness vegetation index (G_i) (Kauth *et al.*, 1979), extracted from the satellite images for each of the sites. G_i was calculated with MSS channels 1–4 weighted with the following coefficients: -0.284, -0.660, 0.577, 0.388. Temporal profiles of G_i were fitted to the observations using the model of Badhwar (1984) with growing degree-day instead of time as the abscissa (i.e., thermal time). A relatively simple canopy radiative transfer model (Hall *et al.*, 1995) was used to estimate the relationship between G_i and APAR. Three components were modelled:

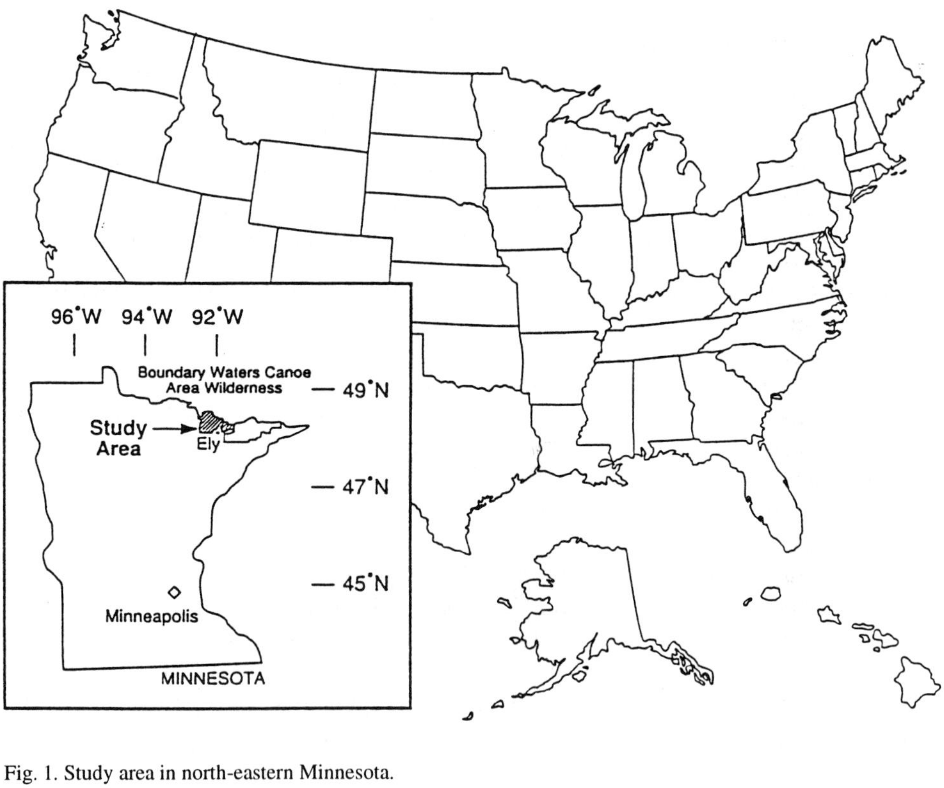

Fig. 1. Study area in north-eastern Minnesota.

proportional reflectance from the illuminated background, f_{bg_i}, illuminated canopy, f_{c_i}, and shadowed background, f_{bg_s}. Canopy reflectance and transmittance calculations were parameterized with measured values of LAI and leaf-level optical properties, then fractional PAR interception (f_{ipar}) by the canopy was estimated as:

$$f_{ipar} = (1 - f_{bg_i})(1 - \tau_{c_i}) \quad [2]$$

where τ_{c_i} is transmittance through the illuminated canopy in the PAR wavelengths, as calculated with the Kubelka-Munk (1931) equation for transmission. The canopy reflectance component of the model was modified in the aspen stands to incorporate an additional layer representing the understory vegetation within each stand. Simulated values of G_i were then calculated for the range of stand densities encountered by adjusting variables such as LAI and canopy closure.

The modelled relationships between simulated G_i and canopy f_{ipar} were used to obtain canopy f_{ipar} from stand-level G_i measured by the MSS in each of the stands. Annual PAR interception was estimated for only the tree component of the canopy, that is the part for which NPP estimates were available. Canopy PAR interception (IPAR) was calculated as the product of f_{ipar} and incident PAR. IPAR is effectively equal to APAR at LAI > 1 (Prince, 1991a). The annual sum of IPAR (ΣIPAR) was calculated by summing the daily values estimated from the fitted relationship with degree days.

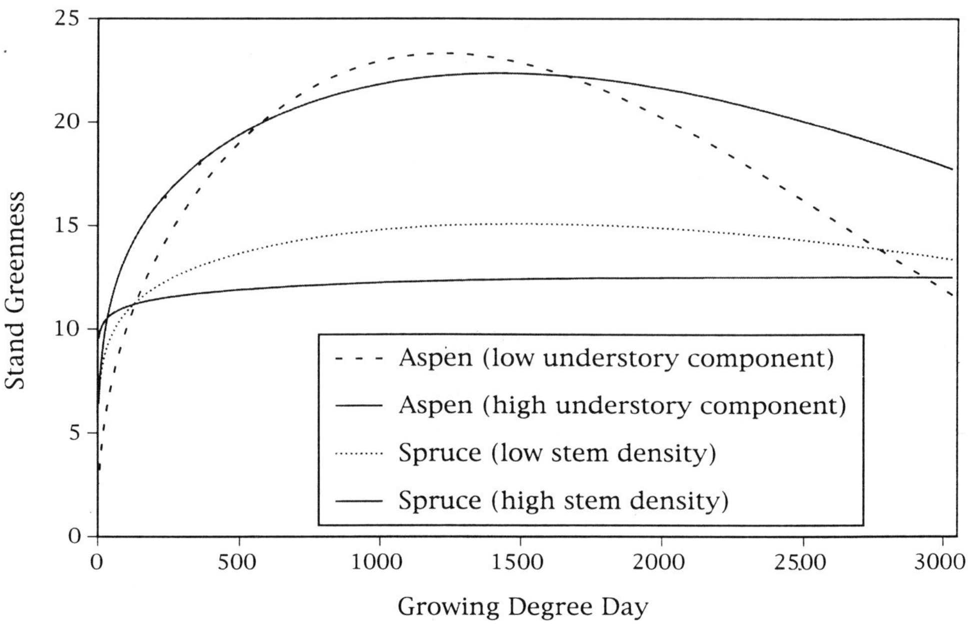

Fig. 2. Temporal profiles of the greenness index (G_i) plotted in thermal time (growing degree-days) for aspen (*Populus tremuloides*) with low and high understory components and for black spruce (*Picea mariana*) with low and high stem densities. Smooth curves were fitted to the MSS observations using the Badhwar model (see text).

3.2. RELATIONSHIP OF ΣIPAR TO NPP

Phenological variation in spruce stands, as depicted by variations in G_i, was negligible in closed-canopy stands and slight in discontinuous stands (Figure 2), mainly caused by variations in solar zenith angle and hence changes in illumination of the moss background vegetation. In contrast, aspen stands exhibited large phenological variation in G_i, particularly during leaf expansion and senescence. The results of canopy model simulations of G_i and f_{ipar} are shown in Figure 3. Two aspen stands are shown (the same stands as depicted in Figure 2) to indicate the range of f_{ipar} relative to G_i. The understory in aspen stands strongly influenced the relationship between canopy f_{ipar} and G_i. Spruce stands were found to exhibit an inverse relationship between G_i and f_{ipar}. In spruce, as stand density decreases from a closed to a discontinuous canopy, the amount of illuminated *Sphagnum* moss increased, which increased G_i. As a result, an increase in G_i reflected a *decrease* in canopy f_{ipar}.

The relationship between NPP and ΣIPAR is shown in Figure 4. ΣIPAR accounted for only 59% of the variability in NPP for all stands, and the residuals were large. In spruce stands, ΣIPAR was a better predictor of AANPP ($r^2 = 0.69$) with an RMS error less than 100 g m^{-2} yr^{-1}. In aspen stands ΣIPAR was not a strong predictor of AANPP ($r^2 = 0.49$); stratification into young and mature stands improved the correlation in young stands ($r^2 = 0.77$) but reduced it in mature stands ($r^2 = 0.18$).

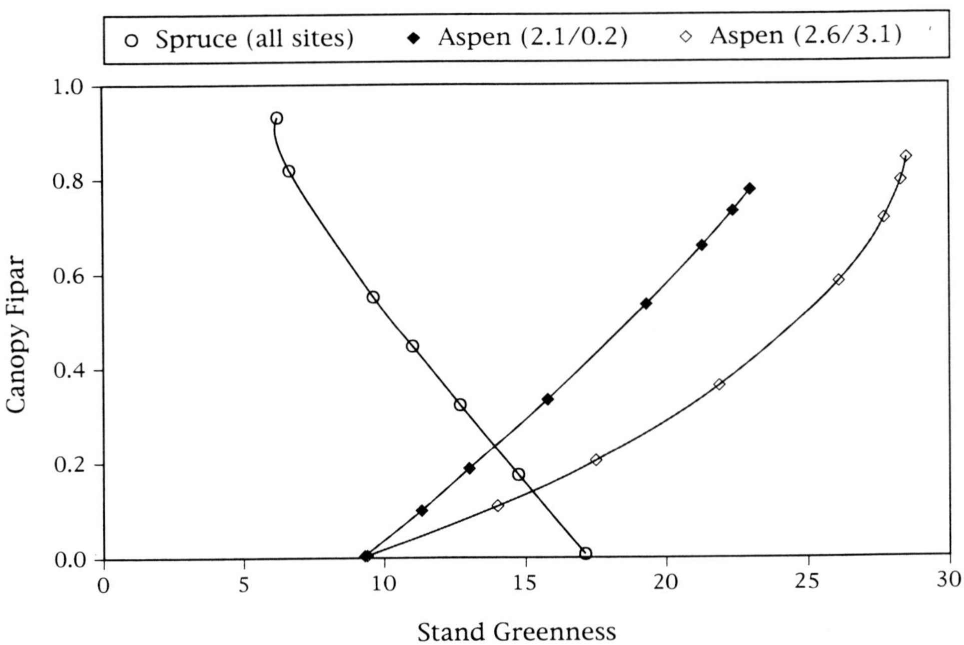

Fig. 3. Overstory f_{ipar} as a function of greenness for a range of aspen (*Populus tremuloides*) and black spruce (*Picea mariana*) stands predicted by a canopy radiative transfer model (see text). The LAI values used for the canopy components are given for the aspen stands (canopy/understory) the spruce line applies to all spruce sites but there are individual lines for each aspen site. The two aspen sites shown are same as in Figure 3 and cover the range of understory LAI values observed.

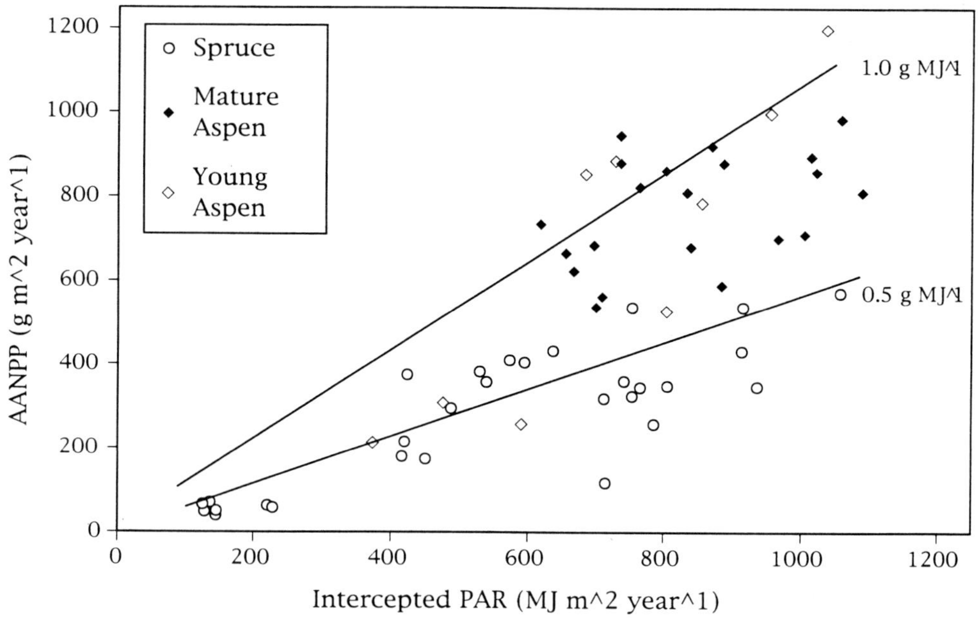

Fig. 4. Annual interception of PAR versus above-ground average annual NPP. Lines indicate constant values of dry matter yield of intercepted PAR ($\varepsilon_{(n)}$) of 0.5 and 1.0 g MJ^{-1}.

$\varepsilon_{(n)}$ was found to vary in the range 0.17–0.89 g MJ^{-1} in spruce stands, and 0.44–1.29 g MJ^{-1} in aspen stands. Hunt and Running (1992) reported simulated values of $\varepsilon_{(n)}$ for spruce of 1.03–1.59 g MJ^{-1} and aspen of 0.68–1.54 g MJ^{-1} at two sites in central Canada. Assuming 40% below-ground allocation in both species, the values for Superior National Forest become 0.73–2.15 g MJ^{-1} for aspen and 0.28–1.48 g MJ^{-1} for spruce, which are similar to Hunt and Running's simulated values.

Using the median values of $\varepsilon_{(n)}$ observed for above-ground production in the aspen and spruce stands, an estimate of the areal extent of the different cover types from Hall *et al.* (1991), and assuming 50% C content in dry matter, the net annual C uptake was 256 g C m^{-2} for the 2280 km^2 study area. Uptake by broad-leaf deciduous areas was estimated at 363 g C m^{-2} and uptake by needle-leaf evergreen areas was 125 g C m^{-2}. The net contribution of ecosystem components other than these forest classes (including peatlands and mixed cover types) are not included in these estimates.

3.3. Causes of Variation in ε

The reliability of these estimates of production are subject not only to the accurate estimation of APAR, but also to variability in estimates of $\varepsilon_{(n)}$. The within species variations in $\varepsilon_{(n)}$ observed here cover over half the range reported for all forest ecosystems. The causes of the observed variability between stands, other than climatic variability and differences in C pathway (which did not differ between the stands studied in Superior National Forest) include five likely possibilities: (1) competitive effects of other components of the stand, such as understory and ground cover vegetation; (2) variation in C losses to death, decay, and herbivory; (3) variation in below-ground allocation; (4) variation in stresses associated with limiting resources; (5) variation in respiration.

Cause (1) is unlikely since there was no correlation of $\varepsilon_{(n)}$ with the presence of understory and field layers. (2) can be dismissed because omission of parts of the C fixed is unlikely to vary between sites. (3) cannot be eliminated, but there is conflicting information on below-ground allocation (Sanantonio *et al.*, 1977; Alban *et al.*, 1978; Raich and Nadelhoffer, 1989). Most existing NPP models assume a constant ratio of above- to below- ground allocation, but at least one study of aspen suggests an age-dependence (Ruark and Bockheim, 1987).

In relation to cause (4), much of the variability in aspen NPP has been shown to be closely associated with site quality variables such as soil depth, texture, nutrients and moisture holding capacity (Lieffers and Campbell, 1984; Verma *et al.*, 1986) In comparison, black spruce is a stress-adapted species usually associated with low nutrient availability, resulting in a close correlation of NPP and LAI (Van Cleve *et al.*, 1983; Viereck *et al.*, 1983; Woods *et al.*, 1991).

These considerations suggest that the differences in NPP should be evident in foliage display. A positive correlation was observed between ΣIPAR and NPP in spruce and young aspen stands in the present study which supports the concept of a relationship between stress and foliage display. If this were a complete explanation, we would expect resource availability to have little effect on ε; however, periodic moisture stress may be regulated predominantly through stomatal control rather than through diminished LAI (Verma *et al.*, 1986; Sucoff, 1982). Such an effect would decouple PAR interception and NPP in the short-term, in contrast to longer-term regulation through foliage display.

Finally cause (5): variation in the respiration demands among stands may arise from increases in the costs associated with maintenance of non-photosynthetic material (Biscoe *et al.*, 1975; McCree, 1975). The proportion of assimilate lost through respiration processes may increase with stand age as the proportion of woody to foliar biomass increases (Waring, 1983). Although, in spruce, there is no indication of variation in ε related to stand age, model simulations suggest that age-dependent variations in respiration are the primary determinant of variability in ε in aspen stands (Hunt and Running, 1992).

It is likely that a combination of drought stress and respiration demands caused the observed variations in $\varepsilon_{(n)}$. The variability of $\varepsilon_{(n)}$ suggests that resource-use optimisation via functional convergence (Field, 1991) may not result in a narrow range of $\varepsilon_{(n)}$ at least in these species. Goetz and Prince (1995) discuss these results and their implications in more detail.

4. Incorporation of Environmental Physiology into Production Efficiency Models: the GLO-PEM Model

Although $\varepsilon_{(n)}$ may vary, as shown by the results from Superior National Forest, the concept of a constant ε may still be useful for modelling NPP. If respiration and environmental stress functions can be specified using, for example, Eq. [1] and if the driving variables of each of the new functions can be measured, then ε may be made constant. Similarly, if respiration can be modelled, $\varepsilon_{(n)}$ can be redefined as $\varepsilon_{(g)}$. One such approach uses biomass to model maintenance respiration (Hunt, 1994; Prince and Goward, 1995). Models of environmental stress and ε have been described by Landsberg *et al.* (1995). In one study, stress functions were used to model NPP in six field sites along the OTTER transect across Oregon (Runyon *et al.*, 1994). The result was a reasonable fit to field measurements of NPP using what was, in effect, an unstressed, net "efficiency" ($\varepsilon^*_{(n)}$).

The variation in $\varepsilon_{(n)}$ found in Superior National Forest and in other studies shows that large errors will result from assumptions of a single, global value (Heimann *et al.*, 1989). Attempts to control variation in $\varepsilon_{(n)}$ by land cover and climatic stratification suffer from a serious shortage of reliable estimates of $\varepsilon_{(n)}$ at an appropriate spatial scale for the Earth's major land-cover types (Ruimy *et al.*, 1994). Also, as both the Superior National Forest study and Prince and Goward (1995) have shown, there can be as much variation in $\varepsilon_{(n)}$ within one biome as there is among biomes. A variant on the stratification approach has been used in the CASA model (Potter *et al.*, 1993) in which $\varepsilon_{(n)}$ values for each biome are obtained by inversion of the model using a set of field test sites for which NPP and climate variables are available. Application of the CASA values of $\varepsilon_{(n)}$ to the Superior National Forest data, however, resulted in a serious underestimation of NPP (Goetz and Prince, 1995) which suggests that stand-level variation in $\varepsilon_{(n)}$ is at least as important as biome-level variation.

Recent progress in bioclimatology and remote sensing has demonstrated that most of the variables needed to evaluate $P_{(n)}$ using Eq. [1], may be extracted from multi-spectral observations of radiance from the Earth's surface (Goward *et al.*, 1994; Goward and Dye, 1995; Prince and Goward, 1995). Canopy light absorption may be derived from combinations of visible and near infrared spectral reflectance measurements as already discussed. Incident solar radiation and, more specifically, incident PAR may be estimated from measurements in the ultraviolet or visible wavelengths (Eck and Dye, 1990; Pinker, 1990; Dye, 1992). Methods have recently been proposed to infer, from remotely sensed data, the environmental variables, including soil moisture, atmospheric water vapour pressure deficit and air temperature, that reduce $\varepsilon^*_{(g)}$, (Goward *et al.*, 1994; Goward and Dye, 1995; Prince and Goward, 1995). These methods utilise thermal as well as optical radiances measured by the AVHRR sensor.

The GLO-PEM GLObal Production Efficiency Model incorporates both respiration and environmental stress models to adjust to the actual value $\varepsilon_{(n)}$ (Prince and Goward, 1995). Furthermore the potential value ($\varepsilon^*_{(g)}$) is calculated from well-established physiological measurements of non-saturated, leaf-level quantum yields for C3 and C4 species, taking the temperature dependence of the quantum yield into account. Thus GLO-PEM does not depend on any calibration steps nor inversions from field observations, rather it specifies the NPP entirely from physiological principles.

5. Primary Production of Global Boreal Forests

The GLO-PEM estimates of boreal forest NPP, $\varepsilon_{(n)}$, and respiration are shown in Figures 5–7 respectively. The results indicate a northward decline in NPP from the boreal forest to the tundra and a southward decline to agricultural lands in the mid continents. Higher values of NPP continue southwards on the western margins of the continents. Values range from over 2000 g m^{-2} yr^{-1} to under 750 g m^{-2} yr^{-1}. The lower end of the range is significantly higher than was observed in Superior National Forest, however the values are for 64 km^2 areas which are unlikely to be entirely occupied with unproductive stands of species such as *Picea mariana*. The spatial heterogeneity shown in the GLO-PEM images is partly a result of failure of the current algorithms to produce a value (black pixels) but also it seems likely to be a result of natural heterogeneity of the region. Observations in Superior National Forest confirm the extreme range of production in the boreal forests.

Using a global land-cover map (DeFries and Townshend, 1994) the mean and total NPP for those classes that include boreal forest are shown in Table I together with all other global land-cover classes. Tundra has a mean NPP of 130 g m^{-2} yr^{-1}, high latitude deciduous evergreen forest and woodland (a class composed mainly of *Larix* that is found in Eastern Siberia and a small part of Alaska) 781 g m^{-2} yr^{-1}, and coniferous evergreen forest and woodland 833 g m^{-2} yr^{-1}. The global NPP totals for these three classes are 1.01, 4.57, and 11.08 Pg C yr^{-1} respectively. Respiration is estimated to be about 30% of NPP in all three classes. The mean dry matter yield of APAR ($\varepsilon_{(n)}$) for each class is 0.11, 0.67, and 0.66 g C MJ^{-1} for tundra, high latitude deciduous forest and woodland, and coniferous evergreen forest and woodland, respectively.

It is important to recognise that GLO-PEM is in an early stage of development and several significant problems remain to be solved (Prince and Goward, 1995). Nevertheless the model does illustrate the rich source of information that can be derived from a relatively simple sensor, such as the AVHRR. Although the results reported here have a spatial resolution of 64 km^2 the success of the model encourages evaluation of the 1 km^2 AVHRR data sets. In this way a spatial resolution closer to the scale of individual stands and landscape processes can be obtained. Furthermore the results suggest that variation in $\varepsilon_{(n)}$ can be explained in terms of respiration and the effects of environmental variables on the potential efficiency, and that these effects can be effectively modelled using satellite data. Thus the contributions of the extensive, but inaccessible boreal forests to global C assimilation may become amenable to routine monitoring.

6. Summary and Conclusions

Planetary-scale assessments of terrestrial biospheric activity may be significantly improved with the use of remotely sensed observations. Of the variety of approaches which have been proposed, the concept of production efficiency has strong appeal. This is because the production efficiency model requires only modest information inputs of a form reasonably derived from remote sensing measurements.

The PEM concept, as originally proposed, suggested a simple conservative, empirical relation between sunlight absorbed and plant production. Further examination of this concept has suggested that to define a conservative efficiency rate, the effects of environmental stressors and respiration much be taken into account. The work in northern Minnesota summarised here shows that, at this boreal forest site, accounting of drought stress and respiration is required to use the PEM concept.

TABLE I

GLO-PEM estimates of global net primary production, respiration and actual carbon (C) yield of APAR by biome class for 1987. Boreal forest occurs mainly in classes 2 and 3, but transitional woodlands also occur in class 4.

Biome class[a]	Class description	Number of 8 × 8 km pixels in class	Number of NPP values in class	% Pixels in class missing	Class mean efficiency[b] (std. dev.) g C MJ^{-1}	Class mean respiration (std. dev.) g C m^{-2} yr^{-1}	Class mean NPP (std dev.) g C m^{-2} yr^{-1}	Class total NPP Pg C yr^{-1}
0	Water	8 761 628						
1	Broadleaf evergreen forest	206 516	194 383	5.88	0.71 (0.35)	409 (443)	1206 (507)	15.95
2	Coniferous evergreen forest and woodland	207 822	154 195	25.80	0.66 (0.75)	283 (376)	833 (729)	11.08
3	High latitude deciduous forest and woodland	91 396	67 992	26.61	0.67 (0.86)	266 (346)	781 (653)	4.57
4	Tundra	121 411	20 532	66.18	0.11 (0.46)	45 (140)	130 (387)	1.01
5	Mixed coniferous forest and woodland	105 728	87 323	17.41	0.64 (0.61)	313 (388)	920 (675)	6.23
6	Wooded grassland	329 278	265 376	19.41	0.43 (0.46)	220 (293)	645 (571)	13.61
7	Grassland	309 906	197 521	36.26	0.40 (0.71)	97 (173)	280 (419)	5.56
8	Bare ground	256 619	32 780	87.23	0.19 (0.80)	69 (33)	19 (93)	0.32
9	Cultivated	206 354	139 908	32.20	0.41 (0.53)	160 (242)	468 (533)	6.18
10	Broadleaved deciduous forest and woodland	69 617	48 817	29.88	0.41 (0.52)	187 (275)	549 (594)	2.45
11	Data unavailable	8 997	4 101	54.42	0.55 (0.84)	195 (316)	573 (733)	0.33
12	Shrubs and bare ground	173 400	118 095	31.89	0.50 (0.93)	53 (102)	151 (250)	1.68
Total land pixels		2 087 044	1 331 023	36.22				68.97

a Source is DeFries and Townshend (1994).
b Class Mean Efficiency is denoted ε_g in text.

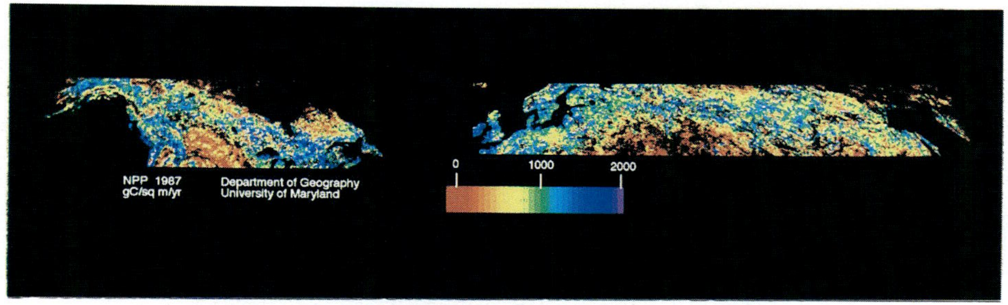

Fig. 5. GLO-PEM modelled NPP for global boreal regions.

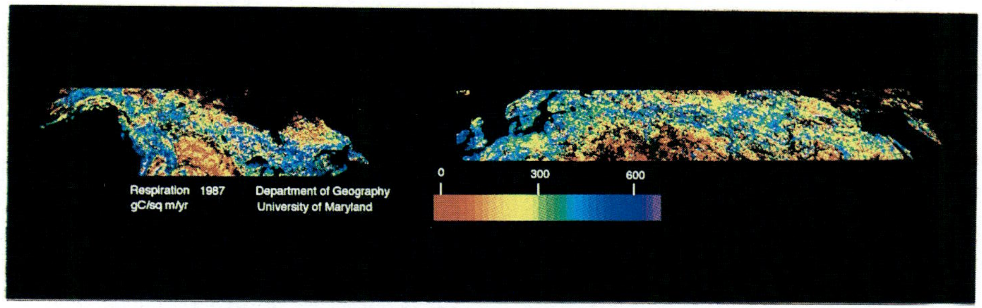

Fig. 6. GLO-PEM modelled dry matter yields of APAR ($\varepsilon_{(n)}$) for global boreal regions

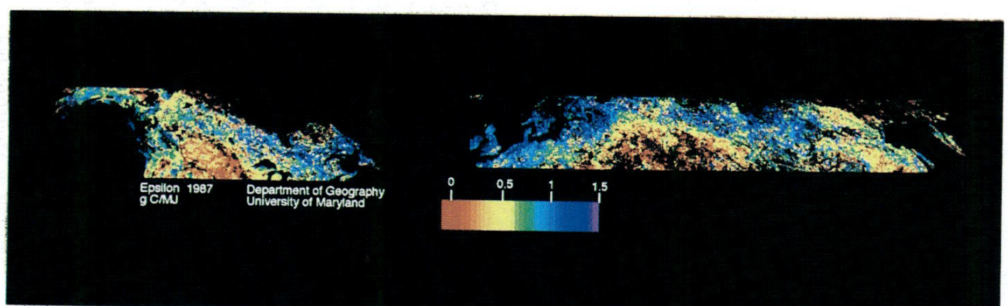

Fig. 7. GLO-PEM modelled respiration for global boreal regions.

A first attempt to employ this "constrained" PEM approach, strictly using satellite remote sensing observations (GLO-PEM) produces circumpolar estimates of boreal forest net primary production which compare favourably with other studies of this region, although further work is needed to refine and validate this approach. The major incentive to do so is that GLO-PEM provides an effective means to examine inter-annual variations in biospheric activity in this climatically-sensitive region.

The simplicity of the PEM approach to biospheric modelling also makes it highly suitable for inclusion in predictive models of future global climate changes. The capability to model the future state of the biosphere is of particular importance in the boreal forest region as current climate models forecast large environmental changes. The observed sensitivity of primary production processes to environmental stress in this region indicates that future changes may affect global scale processes.

Acknowledgments

GLO-PEM was programmed by Shelly Thawley. This work was partly supported by NASA grants NAGW-1967 (SDP) and NAG-52293 (SNG). The GLO-PEM model used the NOAA/NASA Pathfinder AVHRR Land Data Set (James and Kalluri, 1994).

References

Alban, D. H., Perala, D. A. and Schlaegel, B. E.: 1978, *Can. J. For. Res.* **8**, 290-299.

Asrar, G., Kanemasu, E. T., Miller, G. P. and Weiser, R. L.: 1986, *IEEE Trans. Geosci. Remote Sens.* **GE-24**, 76-82.

Badhwar, G. D.: 1984, *Remote Sens. Environ.* **16**, 175-182.

Badhwar, G. D., Macdonald R. B., Hall F. G. and Carnes, J. G.: 1986, *IEEE Trans. Geosci. Remote Sens.* **GE-24** (3) 322-326.

Biscoe, P. V., Scott, R. K. and Monteith, J. L.: 1975, *J. Appl. Ecol.* **12**, 269-293.

Cihlar, J., St.-Laurent, L. and Dyer, J. A.: 1991, *Remote Sens. Environ.* **35**, 279-298.

DeFries, R. S. and Townshend, J. R. G.: 1994, *Int. J. Remote Sens.* **15**(17) 3567-3586.

Dye, D.: 1992, Satellite estimation of the global distribution and interannual variability of photosynthetically active radiation, Ph.D. Dissertation, University of Maryland, College Park, MD, USA, 180 pp.

Eck, T. and Dye, D.: 1990, *Remote Sens. Environ.* **38**, 135-146.

Elliot-Fisk, D. L.: 1988, *In:* Barbour, M. G. and Billings, W. D., (eds.) *North American Terrestrial Vegetation,* Cambridge University Press, Cambridge, pp. 33-62.

Field, C.: 1991, *In:* Mooney, H. A., Winner, W. E. and Pell, E. J. (eds.) *Responses of Plants to Multiple Stresses,* Academic Press, New York, pp. 35-65.

Goetz, S. J. and Prince, S. D.: 1995, *Agr. Forest Meteorol.* (In press).

Goward, S. N. and Dye, D.: 1995, In: Gholtz, H. L., Nakane, K. and Shimoda, H. (eds.) The Use of Remote Sensing in Modeling Forest Productivity at Scales from the Stand to the Globe. Kluwer, New York. USA. (In press).

Goward, S. N. and Huemmrich, K. F.: 1992, *Remote Sens. Environ.* **39**, 119-140.

Goward, S. N. and Prince, S. D.: 1995, *Glob. Ecol. Biogeog. Let.* (In press).

Goward, S. N., Tucker, C. J. and Dye, D. G.: 1985, *Vegetatio* **64**, 3-14.

Goward, S. N., Waring, R. H., Dye, D. G. and Yang, J.: 1994, *Ecol. Appl.* **42**, 322-343.

Hall, F. G., Botkin, D. B., Strebel, D. E., Woods, K. D. and Goetz, S. J.: 1991, *Ecology* **72**, 628-640.

Hall, F. G., Huemmrich, K. F., Strebel, D. E., Goetz, S. J., Nickeson, J. E. and Woods, K. D.: 1992, NASA Technical Memorandum **104568**.

Hall, F. G., Shimabukuro Y. and Huemmrich, K. F.: 1995, *Ecol. Appl.* (In press).

Heimann, M. C., Keeling, D. and Tucker, C. J.: 1989, *In:* Peterson, D. H. (eds.) *Aspects of climate variability in the Pacific and Western Americas.* American Geophysical Union, *Geophys. Monogr.* **55**, 277-303.
Hunt, E. R.: 1994, *Int. J. Remote Sens.* **15**, 1725-1730.
Hunt, E. R. and Running, S. W.: 1992, *Can. J. Remote Sens.* **18**, 126-133.
James, M. E. and Kalluri, S. N. V.: 1994, *Int. J. Remote Sens.* **15**(17) 3347-3364.
Jarvis, P. G. and Leverenz, J. W.: 1983, *In:* Lange, O. L., Nobel, P. S., Osmond, C. B. and Ziegler, H. (eds.) *Ecosystems Processes: Mineral Cycling, Productivity and Man's Influence.* Springer-Verlag, New York, USA, pp. 234-280.
Justice, C. O.: 1986, *Int. J. Remote Sens.* **7**, 1385-1390.
Justice, C. O., Townshend, J. R. G., Holben, B. N. and Tucker, C. J.: 1985, *Int. J. Remote Sens.* **6**, 1271-1318.
Kauth, R. J., Lambeck, P. F. , Richarson, W., Thomas, G. S. and Pentland, A. P.: 1979, *LACIE Symposium*, Johnson Space Center, Houston, USA, pp. 705-721.
Kubelka, P. and Munk, F.: 1931, *Zeit. Tech. Phys.* **12**, 593-605.
Kumar, M. and Monteith, J. L.: 1982, *In:* Smith, H. (ed.) *Plants and the Daylight Spectrum.* Academic Press, London, pp. 133-144.
Landsberg, J. J., Prince, S. D., Jarvis, P. G., McMurtrie, R. E., Luxmore, R. and Medlyn, B. E.: 1995, *In:* Gholz, H. L., Nakane, K. and Shimoda, H. (eds.), *The Use of Remote Sensing in the Modeling of Forest Productivity at Scales from the Stand to the Globe.* Kluwer, New York, (In press).
Lieffers, V. J. and Campbell, J. S.: 1984, *Can. J. For. Res.* **14**, 610-616.
McCree, K. J.: 1975, *In:* Black, R. H. and Black, C.C. (eds.) *CO_2 Metabolism and Plant Productivity* University Park Press, Madison, USA, pp. 177-184.
Miller, P. C., Kendall, R. and Qechel, W. C.: 1983, *Simulation* **40**, 119-130.
Monteith, J. L.: 1972, *J. Appl. Ecol.* **9**, 747-766.
Monteith, J. L.: 1977, *Phil. Trans. Roy. Soc. Lon.* **281**, 277-294.
Pinker, R. T.: 1990, *Paleogeography, Paleoclimatology, Paleoecology (Global and Planetary Change Section)* **82**, 321-342.
Potter, C. S., Randerson, J. T., Field, C. B., Matson, P. A., Vitousek, P. M., Mooney, H. A. and Klooster, S. A. 1993, *Global Biogeochem. Cycl.* **7**, 811-841.
Prince, S. D.: 1991a, *Int. J. Remote Sens.* **12**, 1313-1330.
Prince, S. D.: 1991b, *Int. J. Remote Sens.* **12**, 1301-1311.
Prince, S. D. and Goward, S. N.: 1995, *Glob. Ecol. Biogeog. Let.* (In press).
Prince, S. D., Justice, C. O. and Moore, B.: 1994, *Monitoring and Modelling of Terrestrial Net and Gross Production.* IGBP DIS/GAIM Report **1**, 56 pp.
Raich, J. W. and Nadelhoffer, K. J.: 1989, *Ecol. Appl.* **20**, 1346-1354.
Rodin, L. E., Bazilevich, N. I. and Rozov, N. N. (eds.) 1975, *Productivity of the World's Main Ecosystems.* General Assembly of the Special Committee for the International Biological Program, National Academy of Sciences, Washington D.C., pp. 13-26.
Ruark, G. A. and Bockheim, J. G.: 1987, *Pedobiologia* **30**, 207-217.
Ruimy, A., Saugier, B. and Dedieu, G.: 1994, *J. Geophys. Res. Atmos.* **99**, 5263-5283.
Runyon, J., Waring, R. H., Goward, S. N. and Welles, J. M.: 1993, *Ecol. Appl.* **4**, 226-237.
Russell, G., Jarvis, P. G. and Monteith, J. L.: 1989, *Plant Canopies: Their Growth, Form and Function.* Cambridge University Press, Cambridge, UK, 178 pp.
Sanantonio, D., Herman, R. K. and Overton, W. S.: 1977, *Pedobiologia* **17**, 1-31.
Sellers, P. J.: 1985, *Int. J. Remote Sens.* **6**, 1335-1371.
Sellers, P. J.: 1987, *Remote Sens. Environ.* **21**, 143-183.
Tucker, C. J., Vanpraet, C. L., Sharman, M. J. and Van Ittersum, G.: 1985, *Remote Sens. Environ.* **17**, 233-249.
Van Cleve, K., Oliver, L., Schlentner, R., Viereck, L. and Dyrness, C. T.: 1983, *Can. J. For. Res.* **13**, 747-766.
Viereck, L. A., Dyrness, C. T., van Cleve, K. and Foote, M. J.: 1983, *Can. J. For. Res.* **13**, 703-720.
Waring, R. H.: 1983, *Adv. Ecol. Res.* **13**, 327-354.
Whittaker, R. H.: 1975, *Communities and Ecosystems.* Macmillan, New York, USA, 385 pp.
Woods, K. D., Fieveson, A. H. and Botkin, D. B.: 1991, *Can. J. For. Res.* **21**, 974-989.
Woodwell, G. M., Whittaker, R. H., Reiners, W. A., Likens, G. E., Delwiche, C. C. and Botkin, D. B.: 1978, *Science* **199**, 141-146.

EPILOGUE

FIRST NATION'S PERSPECTIVE ON THE BOREAL FOREST

D'ARCY LINKLATER

Nelson House First Nation, Nelson House, Manitoba R0B 1A0, Canada

It is indeed an honour for me to be included in this very important publication on the Boreal Forest and Global Change. I welcome this opportunity not only because of my aboriginal responsibility as keeper of this sacred land, but also to acknowledge the numerous scientists and researchers, who authored and co-authored the many fine papers in this book.

I think that it is about time that we explore that common goal of solidarity for our sacred Mother Earth. Our understanding as First Nation people is this: if we take care of Mother Earth then she will take care of us. Western scientists must learn our one-with-the-Earth philosophy. *"We are part of Mother Earth, Mother Earth is part of us. We are one with Mother Earth."* When all the four colours of the human family come together, then we will begin to heal.

Our people have lived with the land and have watched grandfather sun rise from the same place since time immemorial. We have been existing and growing naturally in this land for many, many years, that is why we are called indigenous.

Our people also have a way of acquiring knowledge and we want to share our Cree philosophy, our circle of life. When we talk about our circle of life, we speak of our four true colours. Each of these colours represents a race of people.

In the East is the colour Red which represents the Red Man, all aboriginal people in North America. Their gift from the Creator is Vision and Land.
In the South is the colour Yellow which represents the Oriental race. Their gift is Knowledge and Air.
In the West is the colour Black which represents the Black race. Their gift is Water.
In the North is the colour White. This represents the Caucasian race. Their gift is Movement and Fire.

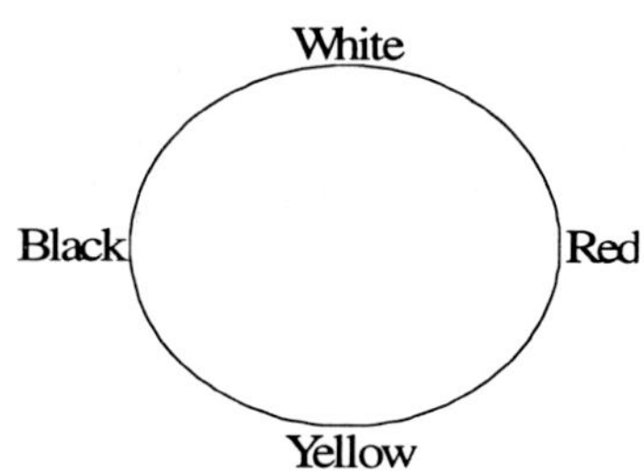

It is prophesied that one day all races of mankind will come together in the centre of the SACRED CIRCLE to live in peace, harmony and co-exist as united peoples.

These colours give us direction in our every day lives. We are fascinated with these colours because they stand for the unity of mankind.

The legends of our culture are rich with images of people, plants and animals working together, shaping the natural environment and sustaining life.

To understand the environmental aspect of our life, one must understand that our relationships to the land are holistic and spiritual. Our spirituality plays an important role in our lives. It is because of this relationship and the importance of our environment that we continue to seek to live in harmony with our surroundings.

We have maintained our traditional life styles and values for centuries in the face of tremendous adversity. Our people have been mandated with foreign cultures, technologies, and development which the governments call progress. These developments have destroyed large potions of our traditional lands, destroyed the forest, polluted the air, and displaced our lives.

All this has been done at the cost of our plants and animals lives. Our people have also shared in the human cost in the same way our wildlife and environment have paid the price.

In spite of all these changes and paternalism, we continue to live with the land. We will continue monitoring a balance between the developed society and subsistence lifestyle of our forefathers before us, we maintain their vision of land management.

We were managing our land long before the Europeans came and today, its continuance is a matter of our survival. Our vision is certainly not the kind of conservation and management we have now, one in which our land continues to be raped, our land and wildlife flooded, and our lives disrupted: one in which our people with their traditional ecological knowledge is ignored. We can only envision a most depressing future for our children.

However, we believe that our knowledge and western science can be integrated and the present situation can be turned around.

We also believe that we have a significant role to play, with a body of knowledge to impart. We know that western science cannot succeed without our input and co-operation.

It is with these truths that we call upon you to respect our science as keepers of this land.

As we proceed and continue in this period of global warming, let us ask the Creator for understanding and sympathy, for a clear vision and sensitivity, for wisdom, strength and courage. For we, as a human family are the only creatures that have strayed from creation.

We must respect and dignify each other, to give ourselves the freedom, the right and the opportunity to enrich our material, cultural and spiritual mode of life...and to create unity and harmony in this Sacred Land.

Thank you, Ekosani, and we leave you with this song.

FOUR DIRECTIONS CREE SONG

This is the translation of our Four Directions Song, as it pertains to Native spirituality, culture, values and its principles. Also to acknowledge and give thanks to the creator for his guidance.

Grandfather,

who sits in the east, thank you for a new day, a new beginning and for the healing.

Grandfather,

who sits in the south, thank you for the growth, physically, mentally, spiritually, and emotionally.

Grandfather,

who sits in the west, thank you for the day you have given us, a time to reflect the things we have done and a time to rest.

Grandfather,

who sits in the north, thank you for granting us your infinite wisdom, that we may continue to survive as a Nation.

Written by: Councillor Andrew Wood
Nelson House First Nation

Nelson House First Nation

LIST OF CONFERENCE PARTICIPANTS

D. AIKMAN
Repap Manitoba
Bay 3-20 Berens Road
Thompson, MN
R8N 1X2

E. ALCORN
Acadia University
Dept. of Biology
Wolfville, NS
Canada
B0P 1X0

V. ALEXEYEV
V.N. Sukachev Institute
Russian Academy of Sciences
Siberian Branch
Academgorodok
Krasnoyarsk 660036
Russia

K. ANDREASSEN
Norwegian Forest Research Institute
Høgskoleveien 12
N-1432 Ås
Norway

S. M. ANDRUS
USDA Forest Service
14th & Independence SW
P.O. Box 90690
Washington, DC 20090-6090
USA

M.J. APPS
Canadian Forest Service
5320 - 122nd Street
Edmonton, AB
Canada
T6H 3S5

A.G. BARR
Environment Canada
National Hydrology Research Centre
11 Innovation Blvd.
Saskatoon, SK
Canada
S7N 3H5

J. BENSON
Saskatchewan Environment &
Resource Management
Box 3003
McIntosh Mall
Prince Albert, SK
Canada
S6V 6G1

Y. BERGERON
Université du Québec à Montréal
CP 8888, Succ. A.
Montréal, QC
Canada
H3C 3P8

E. BEUKER
The Finnish Forest Research Institute
Punkaharju Research Station
Finlandiantie 18
Fin-58450 Punkaharju
Finland

S. BILL
Weyerhaeuser Company
WWC-2H6
Tacoma, WA 98477
USA

S. BONDRUP-NIELSEN
Centre for Wildlife &
Conservation Biology
Acadia University
Wolfville, NS
Canada
B0P 1X0

O.T. BOUMAN
Prince Albert Model Forest Assoc. Inc.
Box 2406
Prince Albert, SK
Canada
S6V 7G3

P. BRADLEY
Oregon State University
Dept. of Civil Engineering
202 Apperson Hall
Corvallis, OR 97331
USA

G. BRASSARD
Canadian Forest Service
351 St. Joseph Blvd. - 20th Floor
Hull, QC
Canada
K1A 1G5

J.C. BRISSETTE
USDA Forest Service
NE Forest Experiment Stn.
5 Godfrey Drive
Orono, ME 04473
USA

P.J. BURTON
University of British Columbia
Dept. of Forest Science
270-2357 Main Mall
Vancouver, BC
Canada
V6T 1Z4

G.A. CHIBISOV
Arkhangelsk Inst. of Forestry
& Wood Chemistry
Ul Nikitova 13
Arkhangelsk 163062
Russia

R. COX
Canadian Forest Service
P.O. Box 4000 College Hill
Fredericton, NB
Canada
E3B 5P7

S. CRITES
University of Alberta
Dept. of Biological Sciences
Room CW405
Edmonton, AB
Canada
T6G 2E1

E. CUMMING
University of Saskatchewan
Dept. of Biology
Saskatoon, SK
Canada
S7N 0W0

S.G. CUMMING
University of British Columbia
Dept. of Forest Sciences
270-2357 Main Mall
Vancouver, BC
Canada
V6T 1Z4

M.A. DALMAN
Prince Albert National Park
P.O. Box 100
Waskesiu Lake, SK
Canada
S0J 2Y0

I. DANILIN
V.N. Sukachev Institute
Russian Academy of Sciences
Siberian Branch
Akademgorodok
Krasnoyarsk 660036
Russia

K. DION
Environment Canada
11 Innovation Blvd.
Saskatoon, SK
Canada
S7N 3H5

M. DREWS
Environment Canada
200 Twin Atria Bldg.
4999-98 Avenue
Edmonton, AB
Canada
T6B 2X3

J. DURNIN
Prince Albert National Park
Box 100
Waskesiu Lake, SK
Canada
S0J 2Y0

L. EDENIUS
Swedish University of
Agricultural Sciences
Dept. of Animal Ecology
S-901 83 Umeå
Sweden

J.W. EHNES
University of Manitoba
116 Ethelbert Street
Winnipeg, MB
Canada
R3G 1V6

H.T. EPP
Saskatchewan Environmental
Resource Management
3085 Albert Street
Regina, SK
Canada
S4S 0B1

G. FENTON
Prince Albert National Park
Box 100
Waskesiu Lake, SK
Canada
S0J 2Y0

M. FITZSIMMONS
Prince Albert National Park
Box 100
Waskesiu Lake, SK
Canada
S0J 2Y0

M. FLANNIGAN
Canadian Forest Service
Petawawa National Forestry Institute
P.O. Box 2000
Chalk River, ON
Canada
K0J 1J0

R.A. FLEMING
Canadian Forest Service
P.O. Box 490, 1219 Queen Street E
Sault Ste. Marie, ON
Canada
P6A 5M7

H. FOELLMI
The World Bank
1818 H Street N.W.
Washington, DC 20433
USA

P. FONTAINE
Assembly of Manitoba Chiefs,
500-286 Smith Street
Winnipeg, MB
Canada
R3C 1K4

C. FRIES
Swedish University of
Agricultural Science
Dept. of Silviculture
Umeå S-90183
Sweden

A.M.F. GJEDTJERNET
Norwegian Forest Research Institute
Høgskoleveien 12
Ås N-1432
Norway

S.B. GONZALES
University of Alberta
Dept. of Renewable Resources
Earth Sciences Building
Edmonton, AB
Canada
T6G 2H1

P. GOODE
Sentar Consultants Ltd.
300-333 25th |Sreet E
Sakatoon, SK
Canada
S7K 0L4

R. GRANGER
Environment Canada
11 Innovation Blvd.
Saskatoon, SK
Canada
S7N 3H5

K.J. GREENWAY
Alberta Environment Centre
Bag 4000
Vegreville, AB
Canada
T9C 1T4

N. GRILLER
University of Saskatchewan
Crop Science & Plant Ecology
Saskatoon, SK
Canada
S7N 0W0

S.D. GRINDAL
University of Regina
Dept. of Biology
Regina, SK
Canada
S4S 0A2

A. GROOT
Canadian Forest Service
P.O. Box 490, 1219 Queen St. E
Sault Ste. Marie, ON
Canada
P6A 5M7

R. GUAY
Regent Instruments Inc.
165 Fatima Avenue
Québec City, QC
Canada
G1P 2C7

CH. HÄGER
Johann Wolfgang Goethe-Universität
Institut für Physikalische
und Theoretische Chemie
Marie-Curie-Straße 11
D-60439 Frankfurt am Main
Germany

J. HÄGGMAN
Finnish Forest Research Institute
Punkaharju Research Station
Finlandiantie 18
Fin-58450 Punkaharju
Finland

J.P. HALL
Canadian Forest Service
351 St. Joseph's Blvd. - 20th Floor
Hull, QC
Canada
K1A 1G5

D.H. HALLIWELL
Canadian Forest Service
5320 - 122nd Street
Edmonton, AB
Canada
T6H 3S5

I.B. HEDIN
FERIC
2601 East Mall
Vancouver, BC
Canada
V6T 1Z4

N. HEDSTROM
Environment Canada
11 Innovation Blvd.
Saskatoon, SK
Canada
S7N 3H5

F. HEINZELMANN
University of British Columbia
Dept. of Forest Science
270-2357 Main Mall
Vancouver, BC
Canada
V6T 1Z4

D.G. HERR
Canadian Forest Service
Petawawa National Forestry Institute
P.O. Box 2000
Chalk River, ON
Canada
K0J 1J0

K. HOBSON
Environment Canada
115 Perimeter Road
Saskatoon, SK
Canada
S7N 0X4

J.C. HOGENBIRK
Laurentian University
Dept. of Biology
Ramsey Lake Road
Sudbury, ON
Canada
P3E 2C6

J.L. HOM
USDA Forest Service
P.O. Box 6775
Radnor, PA 19087-8775
USA

D. HOWES
Repap Manitoba
Box 1590
The Pas, MB
Canada
R9A 1L4

P.A. HURDLE
Canadian Forest Service
5320 122nd Street
Edmonton, AB
Canada
T6H 3S5

L. HUTCHINSON
SASK Power
Interagency Regulations & Affairs
Box 1560
Saskatoon, SK
Canada
S7K 3R3

A. ISAEV
Centre for the problems of ecology
& forest productivity of the
Russian Academy of Sciences
Novocheriomushkinskaia Str. 69
Moscow 117418
Russia

D. JACKSON
University of Saskatchewan
Dept. of Soil Science
5E74 Agriculture Bldg.
Saskatoon, SK
Canada
S7N 0W0

K. JARDINE
Greenpeace Atmospheric Campaign
185 Spadina Avenue
Toronto, ON
M5T 2C6

T. KARJALAINEN
University of Joensuu
Faculty of Forestry
P.O. Box 111
Fin-80101 Joensuu
Finland

V.I. KHAROUK
V.N. Sukachev Forest Institute
Russian Academy of Sciences
Siberian Branch
Akademgorodk
Krasnoyarsk 660036
Russia

T.P. KOLCHUGINA
Oregon State University
Dept. of Civil Engineering
202 Apperson Hall
Corvallis, OR 97331
USA

B. KOROL
University of Saskatchewan
Plant Science & Crop Ecology
Saskatoon, SK
Canada
S7N 0W0

G. KOROVIN
International Forest Institute
Novocheriomushkinskaia Str. 69
Moscow 117418
Russia

O.N. KRANKINA
Oregon State University
Dept. of Forest Science
Corvallis, OR 97331
USA

E. KULCSAR
Alberta Forest Products Assn.
200-11738 Kingsway Ave.
Edmonton, AB
Canada
T5G 0X5

S.N. KULSHRESHTHA
University of Saskatchewan
Dept. of Economics
Saskatoon, SK
Canada
S7N 0W0

W.A. KURZ
ESSA Technologies Ltd.
1765 W 8th Ave.
Vancouver, BC
Canada
V6J 5C6

J. LANE
Weyerhaueser Canada
P.O. Box 1720
Prince Albert, SK
Canada
S6V 5R2

G. LANGEN
Prince Albert Model Forest Assoc. Inc.
Box 2406 77 - 11th Street W.
Prince Albert, SK
Canada
S6V 7G3

N. LARIONOVA
Caterpillar Overseas SA
Pokrovsky Blvd.. 4/17 ap 13
Moscow, 101000
Russia

C.P.S. LARSEN
SUNY at Buffalo
Dept. of Geography
Wilkeson Quadrangle
Buffalo, NY 14261-0023
USA

M.B. LAVIGNE
Canadian Forest Service
P.O. Box 4000 College Hill
Fredericton, NB
Canada
E3B 5P7

L. LAVOIE
Universite du Quebec à Rimouski
300 Allée des Ursulines
Rimouski, QC
Canada
G5L 3A1

P. LEE
Alberta Environment Centre
Bag 400
Vegreville, AB
Canada
T9C 1T4

T. LETCHFORD
BC Ministry of Forests
Red Rock Research Station
RR#7, RMD #6
Prince George, BC
Canada
V2N 2J5

C. LI
Ontario Forest Research Institute
P.O. Box 969 - 1235 Queen St. E
Sault Ste. Marie, ON
Canada
P6A 5N5

D. LINKLATER
Nelson House First Nation
General Delivery
Nelson House, MB
Canada
R0B 1A0

D. LOUSIER
BC Forest Service
Prince George Forest Region
1011 4th Avenue
Prince George, BC
V2L 3H9

R.F. LOWERY
Weyerhaueser Company
WWC-2F2
Tacoma, WA 98477
USA

J.C. LUO
Forest Resource & Environment Centre
Beijing Forestry University
Beijing, 100083
China

D. MACISAAC
Canadian Forest Service
5320 - 122nd Street
Edmonton, AB
Canada
T6H 3S5

E. MÄLKÖNEN
Finnish Forest Research Institute
P.O. Box 18
Vantaa 01301
Finland

P. MARSH
Environment Canada
11 Innovation Blvd.
Saskatoon, SK
Canada
S7N 3H5

J.-P. MARTEL
Canadian Pulp & Paper Association
#1900 - 1155 Metcalfe Street
Montréal, QC
Canada
H3B 4T6

E. MATTHEWS
NASA GISS NY
2880 Broadway
New York, NY 10025
USA

F.-R. MENG
University of New Brunswick
Faculty of Forestry
Fredericton Campus
Fredericton, NB
Canada
E3B 6C2

B. MILLER
Environment Canada
11 Innovation Blvd.
Sakatoon, SK
Canada
S7N 3H5

I. MONTEITH
Prince Albert Model Forest Assoc. Inc.
Box 2406 77-11th Street W
Prince Albert, SK
Canada
S6V 7G3

D. MOUNT
Repap Manitoba
Box 1590
The Pas, MB
Canada
R9A 1L4

B.J. MUIR
Prince Albert National Park
Waskesui Lake, SK
Canada
S0J 2Y0

I.A. NALDER
University of Alberta
Dept. of Renewable Resources
751 General Services Building
Edmonton, AB
Canada
T6G 2H1

T.A. NERBAS
University of Saskatchewan
Dept. of Soil Science
5E74 Agriculture Bldg.
Saskatoon, SK
Canada
S7N 0W0

R.G. NESDOLY
University of Saskatchewan
Dept. of Soil Science
5E74Agriculture Bldg.
Saskatoon, SK
Canada
S7N 0W0

J.W. NOLAN
Alberta Environmental Centre
Bag 4000
Vegreville, AB
Canada
T9C 1T4

F. OBERLE
Daishow-Marubeni
International Ltd.
Postal Bag 2200
Peace River, AB
Canada
T8S 1Y4

R.J. ORYNIK
Weyerhaeuser Canada
P.O. Box 1720
Prince Albert, SK
Canada
S6V 5T3

K.H. ÖSTERBERG
Swedish Environmental
Protectional Agency
Solna S-171 85
Sweden

N. PARKER
Environment Canada
Twin Atria Building
#200, 4999-98 Ave.
Edmonton, AB
Canada
T8A 1W3

L. PELESHOK
Alberta Environmental Centre
Bag 4000
Vegreville, AB
Canada
T9C 1T4

A.-M. PELLETIER
Model Forest Program
Science & Sustainable Dev.
Natural Resources Canada
351 St. Joseph Blvd. - 20th Floor
Hull, QC
Canada
K1A 1G5

C.H. PENG
Canadian Forest Service
5320 - 122nd Street
Edmonton, AB
Canada
T6H 3S5

M.E. PENNER
Canadian Forest Service
Petawawa National Forestry Institute
Box 2000
Chalk River, ON
Canada
K0J 1J0

D.J. PENNOCK
University of Saskatchewan
Dept. of Soil Science
5E74 Agriculture Bldg.
Saskatoon, SK
Canada
S7N 0W0

M.A. PHARE
Centre for Indigenous
Environmental Resources
74 Sherburn St.
Winnipeg, MB
Canada
R3G 2K2

A. PIDWERBESKI
Prince Albert National Park
Box 100
Waskesiu Lake, SK
Canada
S0J 2Y0

E. POHTILA
The Finnish Forest Research Institute
Unioninkatu 40 A
Fin-00170 Helsinki,
Finland

J.W. POMEROY
Environment Canada
11 Innovation Blvd.
Saskatoon, SK
Canada
S7N 3H5

D.T. PRICE
Canadian Forest Service
5320 - 122nd Street
Edmonton, AB
Canada
T6H 3S5

S.D. PRINCE
University of Maryland
Geography Dept.
College Park, MD 20742-8225
USA

R. QUINN
Sask. Env. Resource Management
3085 Albert Street
Regina, SK
Canada
S4S 0B1

G.J. RADSTAAK
Alberta Environmental Centre
Bag 4000
Vegreville, AB
Canada
T9C 1T4

H.M. RASK
University of Saskatchewan
Dept. of Soil Science
5E74 Agriculture Bldg.
Saskatoon, SK
Canada
S7N 0W0

B. REDMANN
University of Saskatchewan
Dept. Crop Science & Plant Ecology
Saskatoon, SK
Canada
S7N 0W0

L.F. RILEY
Canadian Forest Service
Criteria & Indicators Secretariat
351 St. Joseph's Boulevard
Hull, QC
Canada
K1A 1G5

G. ROBITAILLE
Service canadien des forêts
C.P. 3800, 1055 rue du P.E.P.S.
Ste-Foy, QC
Canada
G1V 4C7

V.A. ROJKOV
Dokuchaev Soil Institute
Pyzhevsky Line 7
Moscow 109017
Russia

B. ROLLET
Museum National d'Histoire Naturelle
CTFT
1, rue René Panhard
75013 Paris Cédex
France

L.D. ROY
Alberta Environmental Centre
Bag 4000
Vegreville, AB
Canada
T9C 1T4

D. RYERSON
Alberta Environmental Centre
Bag 4000
Vegreville, AB
Canada
T9C 1T4

D.L. SACHS
Oregon State University
Dept. of Forest Science
Corvallis, OR 97331
USA

S. SANDBERG
USDA Forest Service
3200 SW Jefferson
Corvallis, OR 97331
USA

J. SCHIECK
Alberta Environmental Centre
Bag 4000
Vegreville, AB
Canada
T9C 1T4

F.K.A. SCHMIEGLELOW
University of British Columbia
Dept. of Zoology
6720 University Bldg.
Vancouver, BC
Canada
V6T 1Z4

S.H. SCHNEIDER
Stanford University
Dept. of Biological Sciences
Gilbert Building 109
Stanford, CA 94305-5020
USA

L. SCHWENDENMANN
Im Frondel 24
Muenster Sarmsheim,
55424
Germany

V.N. SEDYKH
Forest Institute
Russian Academy of Sciences
Siberian Branch
Zhukovskogo St. 100/1
Novosibirsk 630082
Russia

S.R. SHEWCHUK
Saskatchewan Research Council
15 Innovation Boulevard
Saskatoon, SK
Canada
S7N 2X8

A.Z. SHVIDENKO
Int. Inst. for Applied Systems Analysis
(IIASA)
Schloss Laxenburg
Schlossplatz-1
A-2361 Laxenburg
Austria

C.W. SLAUGHTER
Northwest Watershed Research Center
USDA-Agricultural Research Service
Boise, ID 83712
USA

T.M. SMITH
University of Virginia
Clark Hall
Charlottesville, VA 22903
USA

M.C.J. ST. ARNAUD
University of Saskatchewan
Dept. of Soil Science
5C08 Agriculture Bldg.
Saskatoon, SK
Canada
S7N 0W0

B. STELFOX
Alberta Environmental Centre
Bag 4000
Vegreville, AB
Canada
T9C 1T4

H. STELFOX
Alberta Fish & Wildlife
9945-108th Street
Edmonton, AB
Canada
T5K 2G6

J. STEWART
USDA Forest Service
14th & Independence Ave. SW
201 14th Street S.W.
Washington, DC 20250
USA

R.B. STEWART
Canadian Forest Service
351 St. Joseph Blvd. - 20th Floor
Hull, QC
Canada
K1A 1G5

R.R. STEWART
Terrestrial & Aquatic
Environmental Managers Ltd.
P.O. Box 2647
Melville, SK
S0A 2P0

B.J. STOCKS
Canadian Forest Service
P.O. Box 490, 1219 Queen Street E
Sault Ste. Marie, ON
Canada
P6A 5M7

T.A. STONE
The Woods Hole Research Centre
P.O. Box 296
Woods Hole, MA 02543
USA

R. STOTTLEMYER
USDA Forest Service
240 Prospect Road W
Fort Collins, CO 80526
USA

K. STUART-SMITH
Alberta-Pacific Forest Industries
Box 8000
Boyle, AB
Canada
T0A 0M0

K. STURGESS
Alberta Environmental Centre
Bag 4000
Vegreville, AB
Canada
T9C 1T4

M.T. SYKES
Lund University
Global Systems Group
Dept. of Ecology
Ekologihuset
Lund S-223 62
Sweden

P.M. TARLETON
Prince Albert National Forest
Box 100
Waskesiu Lake, SK
Canada
S0J 2Y0

S. THORESEN
The Norwegian
Forest Owners' Federation
Stortingsgt. 30
PO Box 1438 Vika
N-0115 Oslo
Norway

J. THORPE
Saskatchewan Research Council
15 Innovation Blvd.
Saskatoon, SK
Canada
S7N 2X8

H. TIAN
SUNY ESF
301 Illick Hall
Syracuse, NY 13210
USA

A. TODD
Alberta Fish & Wildlife Service
6909 116th Street - 7th Floor
Edmonton, AB
Canada
T6H 4P2

J.A. TROFYMOW
Canadian Forest Service
506 Burnside Road W
Victoria, BC
Canada
V8Z 1M5

C. VAN KESSEL
University of Saskatchewan
Dept. of Soil Science
5E74 Agriculture Bldg.
Saskatoon, SK
Canada
S7N 0W0

K.C.J. VAN REES
University of Saskatchewan
Dept. of Soil Science
5E74 Agriculture Bldg.
Saskatoon, SK
Canada
S7N 0W0

W. VANDERSCHUIT
Riding Mountain National Park
Wasagaming, MB
Canada
R0J 2H0

T.M. VAREM-SANDERS
Canadian Forest Service
5320 - 122nd Street
Edmonton, AB
Canada
T6H 3S5

E.F. VEDROVA
V.N. Sukachev Institute
Russian Academy of Sciences
Siberian Branch
Academgorodok
Krasnoyarsk 660036
Russia

L.A. VIERECK
USDA Forest Service
308 Tanana Drive
Fairbanks, AK 99775-5500
USA

F.L. WALLEY
University of Saskatchewan
Dept. of Soil Science
5E74 Agriculture Bldg.
Saskatoon, SK
Canada
S7N 0W0

G.G. WANG
University of Winnipeg
Dept. of Biology
515 Portage
Winnipeg, MB
Canada
R3B 2E9

J.R. WANG
University of British Columbia
Faculty of Forestry
270-2357 Main Mall
Vancouver, BC
Canada
V6T 1Z4

S. WASEL
Alberta-Pacific Forest Industries
P.O. Box 8000
Boyle, AB
Canada
T0A 0M0

J. WATSON
Royal Society of Canada
Canadian Global Change Program
P.O. Box 9734
Ottawa, ON
Canada
K1G 5J4

J. WEIR
Prince Albert National Park
Box 100
Waskesiu Lake, SK
Canada
S0J 2Y0

D.E. WICKLAND
NASA Headquarters
NASA HQ, Code SEP 03
300 East Street SW
Washington, DC 20546
USA

J. WISNIEWSKI
Wisniewski & Associates
P.O. Box 1291
McLean, VA 22101
USA

A. WOOD
Nelson House First Nation
General Delivery
Nelson House, MB
Canada
R0B 1A0

D. WULFSOHN
University of Saskatchewan
Agricultural & Bioresource
Engineering Department
Saskatoon, SK
Canada
S7N 0W0

C.H. YANSA
University of Saskatchewan
Dept. of Geological Sciences
Saskatoon, SK
Canada
S7N 0W0

C.W. YEATMAN
Consultant, Forest Genetics
Box 721
98 Glendale Ave
Deep River, ON
Canada
K0J 1P0

LIST OF REVIEWERS

Andreassen, K., Norway
Benson, J., Canada
Birdsey, R.A., USA
Bradley, P., USA
Brissette, J.C., USA
Bugmann, H., Germany
Burton, P.J., Canada
Campbell, I.D., Canada
Cannell, M.G.R., UK
Chen, J., USA
Chen, J.M., Canada
Coxworth, A., Canada
Cramer, W., Germany
Cumming, S.G., Canada
Davis, E., Canada
Day, R., Canada
Eckerberg, K., Sweden
Epp, H.T., Canada
Flannigan, M., Canada
Friend, A.D., UK
Gal, J., Hungary
Golding, D.L., Canada
Harding, R.J., UK
Hawkins, C.D.B., Canada
Hedin, I.B., Canada
Hogg, E.M., Canada
Hunt, E.R., USA
Isaev, A., Russia
Johnson, E.A., Canada
Johnson, T.G., USA
Karjalainen, T., Finland
Klein, K.K., Canada
Kohlmaier, G.H., Germany
Kolchugina, T.P., USA
Krankina, O.N., USA
Krasowski, M.J., Canada
Kulshreshtha, S.N., Canada
Kurz, W.A., Canada
LaBau, V.J., USA
Landhäusser, S., Canada
Larsen, C., USA
Lavigne, M.B., Canada
Lawson, B.D., Canada
Leckie, D.G., Canada
LeNet, E., France
Li, C., Canada
Linder, S., Sweden
Livingston, N.J., Canada
Lowe, J.J., Canada
MacDonald, S.E., Canada
MacDonald, G.M., Canada
MacIsaac, D.A., Canada
Maini, J.S., Canada
Malcolm, D.C., UK
Mälkönen, E., Finland
Mattson, W.J., USA
Maynard, D., Canada
Meng, F.-R., Canada
Nalder, I.A., Canada
Nilsson, L.O., Sweden
Oberle, F., Canada
Peterson, E.B., Canada
Plamondon, A., Canada
Pluth, D.J., Canada
Pohtila, E., Finland
Poso, S.J., Finland
Prince, S.D., USA
Redman, R., Canada
Rizzo, B., U.S.A
Rojkov, V.A., Russia
Roulet, N., Canada
Rowe, S., Canada
Running, S., USA
Ryan, K., USA
Sachs, D.L., USA
Sampson, R.N., USA
Schmiegelow, F.K.A., Canada
Sedykh, V.N., Russia
Sirois, L., Canada
St. Arnaud, M.C.J., Canada
Stathers, R.J., Canada
Stocks, B.J., Canada
Stone, T.A., USA
Sykes, M.T., Sweden
Swallow, S., USA
Tait, D.E., Canada
Trofymow, J.A., Canada
Turner, D., USA
Van Cleve, K., USA
Van der Kamp, B., Canada
Verma, S.B., USA
Viereck, L.A., USA
Vinson, T.S., USA
Wakimoto, R.H., USA
Weetman, G.F., Canada
White, W., Canada
Williams, D.L., USA
Winjum, J.K., USA
Wisniewski, J., USA
Woodward, F.I., UK
Zoltai, S.C., Canada

AUTHOR INDEX

SUBJECT INDEX